STALLCUP'S ELECTRICAL DESIGN BOOK

2005

National Fire Protection Association
Quincy, Massachusetts

Written by James Stallcup, Sr.
Edited by James Stallcup, Jr.
Design, graphics and layout by Billy G. Stallcup

Copyright ©2005 by Grayboy, Inc.

Published by the National Fire Protection Association, Inc.
One Batterymarch Park
Quincy, Massachusetts 02169

All rights reserved. No part of the material protected by this copyright notice may be reproduced or utilized in any form without acknowledgment of the copyright owner nor may it be used in any form for resale without written permission from the copyright owner and publisher.

Notice Concerning Liability: Publication of this work is for the purpose of circulating information and opinion among those concerned for fire and electrical safety and related subjects. While every effort has been made to achieve a work of high quality, neither the NFPA nor the authors and contributors to this work guarantee the accuracy or completeness of or assume any liability in connection with the information and opinions contained in this work. The NFPA and the authors and contributors shall in no event be liable for any personal injury, property, or other damages of any nature whatsoever, whether special, indirect, consequential, or compensatory, directly or indirectly resulting from the publication, use of or reliance upon this work.

This work is published with the understanding that the NFPA and the authors and contributors to this work are supplying information and opinion but are not attempting to render engineering or other professional services. If such services are required, the assistance of an appropriate professional should be sought.

National Electrical Code® and *NEC®* are registered trademarks of the National Fire Protection Association, Inc.

NFPA No.: SED05
ISBN: 0-87765-667-3
Library of Congress Card Catalog No.: 2004109210

Printed in the United States of America
05   06   07   08   09   5   4   3   2   1

Introduction

James Stallcup, Sr. has reinvented the electrical design book that catapulted him to the forefront of the electrical industry as a **prominent author**, **effective instructor** and **respected authority** on the application of the NEC.

Everyone from Engineers, Electrical Contractors, Inspectors, Electricians and Instructors of the Code have anticipated the arrival of this book. The large workbook format allows a masterful blending of valuable Design Tips, NEC Loops, Examples, Quick Calcs and effective illustrations with authoritative Code references. Because of the abundant amount of detailed information included, it is the most comprehensive design book of its kind.

Stallcup's Electrical Design Book explains the purpose of the National Electric Code (NEC) and more particularly, its use as it applies to the design and installation of electrical wiring systems and equipment.

While the substance of design is found in the NEC, the art of the design is found in the applicability of that same NEC. With the advancement of today's techology and ever increasing liabilities, effective electrical design must now, more than ever, consider the use of certified products, energy conservation, economy vs. quality, anticipated load growth, local codes, special applications of electrical equipment and the use and interpretation of NFPA and IEEE standards that relate to special areas, etc. For better understanding and interpretation of these advancements, considerable effort has been made by the author to condense the more complicated rules pertaining to the design, installation and selection of wiring methods and equipment.

For the convenience of the reader, the Design Book not only contains discussions and explanations of code rules, but also includes detailed illustrations and sample calculations that will help tremendously in understanding and becoming proficient in the application of the NEC. The Design Book also points out common industry problems and shows in detail the proper procedures and techniques to use in order to ensure proper code compliance. Design Tips, Calculation Tips and guidelines for "rule of thumb" methods for instances where a fast and approximate design answer is needed are also provided.

In order for the reader to measure his or her design skills or to prepare for a maintenance, journeyman or master electrician examination or for inspector certification, an ample number of quizzes, tests and final examinations have been included.

With the wealth of information found in this book, it will make a valuable addition to the library of any consultant, serve as an excellent reference to the seasoned professional and will also provide valuable introduction material to the beginner.

Table of Contents

Chapter One
Electrical Codes and Standards 1-1

Chapter Two
Electrical Safety and First Aid 2-1

Chapter Three
Dress Standards 3-1

Chapter Four
Electrical Systems 4-1

Chapter Five
Working Clearances 5-1

Chapter Six
Services 6-1

Chapter Seven
Switchboards and Panelboards 7-1

Chapter Eight
Conductors 8-1

Chapter Nine
Overcurrent Protection Devices 9-1

Chapter Ten
Equipment Over 600 Volts 10-1

Chapter Eleven
Grounding 11-1

Chapter Twelve
Designing Wiring Methods 12-1

Chapter Thirteen
Installing Wiring Methods 13-1

Chapter Fourteen
Branch-circuits 14-1

Chapter Fifteen
Feeder-circuits 15-1

Chapter Sixteen
Receptacle Outlets 16-1

Chapter Seventeen
Lighting and Switching Outlets 17-1

Chapter Eighteen
Motors 18-1

Chapter Nineteen
Compressor Motors 19-1

Chapter Twenty
Transformers 20-1

Chapter Twenty-one
Hazardous (Classified) Locations 21-1

Chapter Twenty-two
Residential Calculations 22-1

Chapter Twenty-three
Commercial Calculations 23-1

Chapter Twenty-four
Industrial Calculations 24-1

Appendix A-1

Abbreviations A-7

Glossary of Terms A-11

Index I-1

1

Electrical Codes and Standards

The *National Electrical Code* (NEC) shall be used in conjunction with other Codes and Standards when designing and installing electrical systems if they are to be considered safe, dependable and reliable. This chapter discusses such Codes and Standards and outlines how they are utilized with the rules and regulations of the NEC.

HISTORY OF THE NATIONAL ELECTRICAL CODE

The *National Electrical Code* (NEC) was first published in 1897 under the sponsorship of the National Conference of Electrical Rules, the membership of which consisted of delegates from interested national associations. The 1897 NEC was the forerunner of a Standard that has been responsible in saving many lives and many dollars in property. This organization was disbanded in 1911, at which time the National Fire Protection Association (NFPA) took over the great responsibility of writing and publishing the NEC, which has been continued since that time. The NEC has been adopted by the American National Standards Institute (ANSI) as a National Consensus Document.

NFPA set up a National Electrical Code committee, which is headed by the technical correlating committee and its members work with the various code making panels. Each code making panel has the responsibility of certain Articles during the proposal and comment stage of upgrading the NEC. The Technical correlating committee and Code making panels include dedicated men from all phases of the electrical industry. Members are assigned to the particular code making panel where it is felt they have the most expertise. The NEC is a minimum standard used for electrical installations and for the protection of life and property.

PURPOSE OF THE NEC

Electricity can be very hazardous, if proper precautions are not adhered to. The NEC is intended to give guidelines for proper installations, in order to safeguard personnel and property from the dangers of electricity.

NEC requirements are essential for safety and compliance. However, such requirements may not necessarily result in efficient, convenient or adequate service if not used as designed.

For example, provisions for future expansion of the electrical system may not be provided. However, the electrical system is essentially free from hazards that may be encountered during the use of its elements. If future needs are taken into consideration at the time of the original design and installation and adequate measures taken to provide for the increased usage of electricity, such hazards and overloading should be greatly eliminated.

The NEC is not intended to be used for design specifications nor as an instruction manual for untrained persons. The requirements of the NEC will, however, serve as rules that may be used to help design electrical systems properly.

The NEC is usually adopted by government agencies and becomes effective January 1 of the year that it is published. Its rules and regulations govern wiring methods and equipment used in residential, commercial and industrial facilities.

Design Tip: There may be additional requirements by local agencies and such amendments must be checked for they are usually more restrictive and require, in some cases, larger components to be used in completing electrical systems.

APPLYING THE NEC

The NEC is intended to be used as a mandatory standard by governmental bodies and inspection authorities. All facets of the electrical industry apply the NEC as a minimum standard for the installation of electrical systems.

LAYOUT OF THE NEC

The rules and regulations of the NEC are laid out by Chapters, Articles, Parts and Sections which contain the requirements for design, installation, use and maintenance of electrical systems. The arrangement of the Chapters and what they pertain to are as follows:

- Introduction
- Chapters 1 through 4 - General Requirements
  Chapter 1 - General
  Chapter 2 - Wiring and Protection
  Chapter 3 - Wiring Methods and Materials
  Chapter 4 - Equipment for General Use
- Chapter 5 - Special Occupancies
- Chapter 6 - Special Equipment
- Chapter 7 - Special Conditions
- Chapter 8 - Communications Systems
- Chapter 9 - Tables
  Chapters 1 through 4 apply generally
  Chapters 5 through 7 supplement or modify the general rules in Chapters 1 through 4
  Chapter 8 is independent of other chapters except as referenced
- Annex G - Administration and Enforcement

See Figure 1-1 for a detailed illustration of how the NEC is laid out.

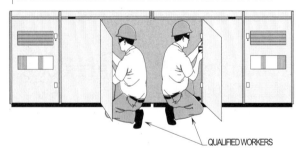

Figure 1-1. Electrical workers and maintenance personnel shall be trained on applying the rules of the NEC as a means of providing safety for electrical or nonelectrical workers that work around or near electrical circuits and components.

ADMINISTRATION AND ENFORCEMENT ANNEX G

Annex G covers code administration and enforcement rules. Included in this Annex, located at the back of the NEC, are provisions for existing nonconforming installations and appointment requirements for the authority of electrical inspectors and inspection procedures. Another provision is the creation of an electrical board having the power to appoint, regulate and discipline inspectors. This Annex also has requirements for plan review, permits and record keeping. For this Annex to be enforceable, its provisions shall be adpoted by a jurisdiction. Until it is adopted, its rules may only function in an advisory capacity. **(See Figure 1-2)**

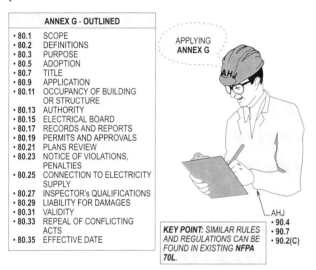

Figure 1-2. Annex G contains a standarized set of administrative requirements which may be utilized by adopting jurisdictions.

INTRODUCTION AND CHAPTER ONE ARTICLES 90; 100; 110

For many years in the electrical industry, **Chapter 1** has been called the "get acquainted" chapter. Everyone in the electrical industry who has the responsibility of applying the rules of the NEC must know the information and requirements of **Chapter 1**. When using the definitions and rules of **Chapter 1**, the "get acquainted chapter," the user of the NEC must use either **Article 90** or one of the Articles of the 100 series.

ARTICLE 90

Article 90 contains requirements which gives the purpose of the NEC and other pertinent information that shall be utilized with such purpose.

For example, what is "covered" and what is "not covered" by the NEC is found in **Article 90**. The word "purpose" simply means the intent, aim or goal of the requirements of the NEC. **(See Figure 1-3)**

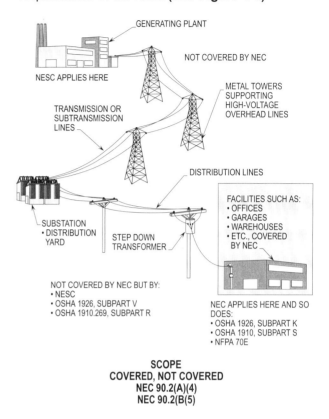

Figure 1-3. Article 90 contains requirements which gives the purpose of the NEC and other pertinent information that shall be utilized with such purpose.

CHAPTER 1 ARTICLE 100

Article 100 breaks down the language barrier for the user of the NEC. When users of the NEC fully understand the meanings of the words, terms or phrases that are used, they have a much better chance to comprehend and properly interpret the rules of the NEC.

For example, if the designer wanted to run two services to a building and he or she was to obtain the definition of a building from Webster's Dictionary instead of **Article 100**, the designer would be very confused as to whether an individual service could be provided for such a building due to different definitions. The definition in **Article 100** is defined specifically for the installation of wiring methods and equipment in a building.

Commonly defined terms are not included in **Article 100**. Only the terms essential to the proper application or enforcement of the NEC rules are listed. **(See Figure 1-4)**

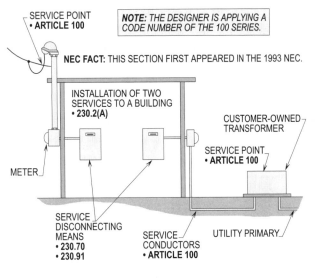

DEFINITIONS
NEC ARTICLE 100

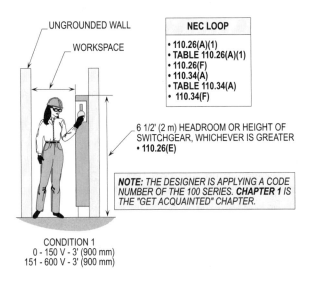

DEPTH OF WORKING SPACE
TABLE 110.26(A)(1), CONDITION 1

Figure 1-4. Article 100 contains definitions that are essential to the proper application and enforcement of the NEC.

The definitions for systems of 600 volts or less are listed in **Part I** and for over 600 volts are listed in **Part II of Article 100**.

Design Tip: For a definition to be eligible to be included in **Article 100**, such definition shall appear in at least two or more Articles of the NEC. Otherwise, definitions appear in front of an Article, such as 240.2 for definitions used in **Article 240**.

CHAPTER 1
ARTICLE 110

Article 110 of the NEC covers the requirements for electrical installations. The rules of **Article 110** apply to all types of electrical installations that come after **Chapter 1** unless there is an exception granted or it is supplemented or modified by a rule in **Chapters 5, 6** or **7** in the NEC.

For example, the clearance in front of a 277/480 volt panelboard is 3 ft. (900 mm) per **Table 110.26(A)(1)**, if the electrical installation falls under the provisions of **Condition 1** to such Table. **(See Figure 1-5)**

Design Tip: When determining clearances in front of electrical installations for 600 volts or less or over 600 volts, **Article 110** of the 100 series in **Chapter 1** shall be used.

Figure 1-5. Article 110 contains requirements for electrical installations except those granted, supplemented or modified by a rule in **Chapters 5, 6** or **7** in the NEC.

CHAPTER 2
ARTICLES 200 - 285

Chapter 2 with its accompanying Articles is used by designers to design the elements and components of branch-circuits, feeder-circuits and service equipment that make up the premises electrical system. The title of **Chapter 2** is "Wiring and Protection" which pertains to designing loads for selecting wiring methods and equipment and sizing overcurrent protection devices to protect such elements as conductors, parts and equipment.

When designers are in the planning stage and calculations have to be made by which the elements are determined and equipment is selected, the designer shall apply one of the Articles of **Chapter 2**. In other words, one of the Articles in the 200 series shall be chosen.

For example, if the designer wants to know how many disconnects are allowed for each service, the subject is services and disconnects, the Article is **230, 230.71** and **230.72** are applied.

When designing the electrical system, the designer must think **Chapter 2** and use one of the 200 series Articles which deals with his or her particular design.

If one of the Articles in **Chapter 2** can't be used to design a certain installation, the designer will be referred to another Chapter that covers such installation.

Electrical Codes and Standards

For example, 220.50 in the NEC refers the designer to Article **430 (430.24, 430.25** and **430.26)** for calculating motor loads. **(See Figure 1-6)**

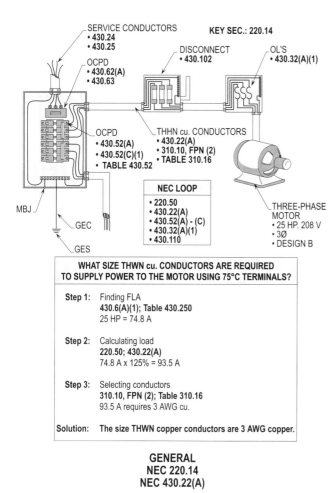

Figure 1-6. Chapter 2 is used by designers to design the elements and components of branch-circuits, feeder-circuits and service equipment. (See **220.50**)

Design Tip: The first six Articles of **Chapter 2** are used for designing the electrical system and the last three are used for protecting people, wiring and equipment.

CHAPTER 3
ARTICLES 300 - 398

Chapter 3 with its accompanying Articles is used by installers to install the electrical wiring methods and equipment after they have been designed by the designer per **Chapter 2** in the NEC. Electricians cannot install electrical wiring methods and equipment to comply with NEC rules without reading, studying and applying the requirements of **Chapter 3**.

The rules electricians must use for routing raceways and cable systems, mounting and securing boxes, outlet boxes, junction boxes, panelboards, switchgear, etc. are outlined.

For example, how often does a run of EMT between a process machine and panelboard have to be supported?

Section 358.30(A) requires supports to be provided within 3 ft. (900 mm) of the machine and panelboard and at intervals of 10 ft. (3 m) in the run of EMT between such equipment. **(See Figure 1-7)**

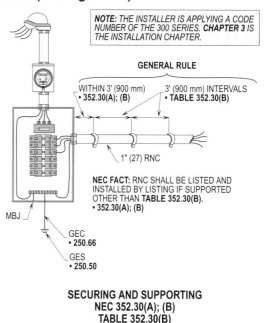

Figure 1-7. Chapter 3 is used by installers to install the electrical wiring methods and equipment after they have been designed by **Chapter 2**.

CHAPTER 4
ARTICLES 400 - 490

Chapter 4 with its accompanying Articles is used by users and maintainers. The electrical equipment in **Chapter 4** is current consuming and requires electricity to operate. Therefore this Chapter is usually applied after the design and installation of such systems and equipment have been complete.

For example, what Article and Section in the 400 series is used to replace a 120 volt ballast (1000 volts or less output) in an electrical discharge luminaire (lighting fixture)?

Section 410.73(E) of **Article 410** in **Chapter 4** is used to replace the ballast of 1000 volts or less. The maintainer may use the rules of **410.73** when dealing with replacement parts in electric-discharge luminaires (lighting fixtures). **(See Figure 1-8)**

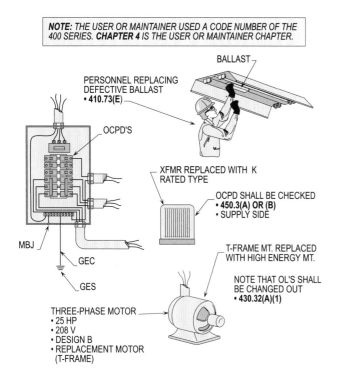

Figure 1-8. Chapter 4 with its accompanying Articles is used by users and maintainers.

CHAPTER 5
ARTICLES 500 - 555

Chapter 5 deals with special occupancies and the first four Articles deal with identifying specific classifications and general provisions. Rules and regulations pertaining to classifying these areas are discussed. The remaining Articles cover specific provisions for specific locations such as commercial garages, aircraft hangers, service stations, bulk storage plants, spray applications, etc., as well as health care facilities.

Design Tip: When special occupancies are involved, one of the Articles of the 500 series shall be used based upon the type of occupancy listed in **Chapter 5**. **(See Figure 1-9)**

CHAPTER 6
ARTICLES 600 - 695

Chapter 6 is completely devoted to special equipment that requires special requirements for design and installation. In other words, the needed rules cannot be found in **Chapter 4** which covers "equipment for general use."

If an engineer is designing the electrical system for an elevator, he or she shall apply the rules and regulations of **Article 620** in **Chapter 6**.

For example, to design the disconnecting means to disconnect power to the machine room for elevator equipment, the designer shall consult with **620.51** in the NEC for requirements which are related to sizing, selecting and locating this disconnect.

Design Tip: When the requirements for special equipment are not found in **Chapter 4**, the designer shall use one of the Articles found in the 600 series of **Chapter 6**. **(See Figure 1-10)**

CHAPTER 7
ARTICLES 700 - 780

Chapter 7 contains requirements for special conditions that arise from the particular type, use and application of the power source used to supply circuits and equipment. The number used to identify Articles and Sections of **Chapter 7** is 700. When special conditions pertaining to the power source, circuits and equipment are addressed, one of the Articles of the 700 series shall be used.

For example, the designer wants to know if the wiring of an emergency system may be routed in the same raceway as the wiring of the normal power source.

Special conditions are needed to verify this requirement. Article 700 of **Chapter 7** is used and **700.9(B)** in the NEC shall not permit the emergency wiring and normal wiring to be mixed in a raceway system. **(See Figure 1-11)**

Design Tip: Designers must take notice and carefully review how predominantly the **source**, the **supply** and the **circuit(s)** are applied in each Article of **Chapter 7** to ensure the reliability of such wiring methods and equipment. Remember these "three words" are main "key words" in applying the rules and regulations of **Chapter 7**.

CHAPTER 8
ARTICLES 800 - 830

Chapter 8 covers communications circuits and equipment and is independent of the other Chapters except where they are specifically referenced.

Communications systems regulated by the Articles in **Chapter 8** are telephone, telegraph, radio and alarm systems. The requirements basically cover systems that connect to a central alarm station to sound some kind of alarm.

Electrical Codes and Standards

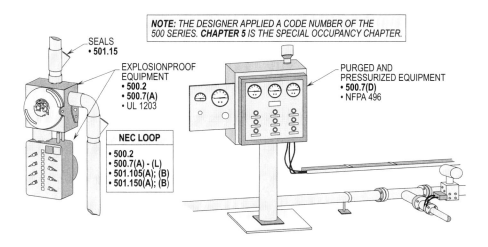

PROTECTION TECHNIQUES
NEC 500.7

Figure 1-9. **Chapter 5** deals with special occupancies and the first four Articles deal with identifying specific classifications and general provisions.

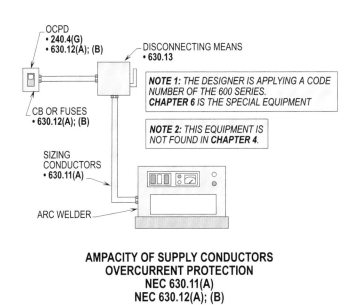

AMPACITY OF SUPPLY CONDUCTORS
OVERCURRENT PROTECTION
NEC 630.11(A)
NEC 630.12(A); (B)

Figure 1-10. **Chapter 6** deals with special equipment that requires special requirements for design and installation.

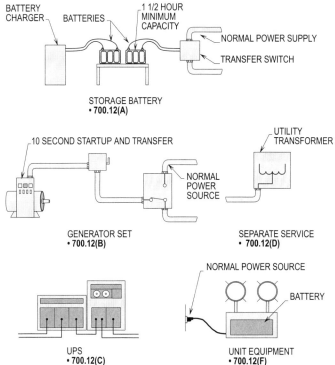

GENERAL REQUIREMENTS
NEC 700.12(A) THRU (F)

For example, an alarm could communicate to a central control room that the ventilation system has failed in a hazardous location. **(See Figure 1-12)**

Design Tip: When requirements for communications systems are needed, one of the Articles in the 800 series in **Chapter 8** shall be utilized by the designer.

Figure 1-11. **Chapter 7** contains requirements for special conditions that arise from the particular type, use and application of the power source used to supply circuits and equipment.

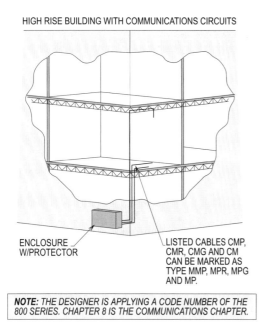

**CABLE USES AND PERMITTED SUBSTITUTIONS
TABLE 800.154**

Figure 1-12. Chapter 8 covers communications circuits and equipment.

CHAPTER 9
TABLES 1 - 12(B)

Chapter 9 contains Tables which are mainly used to size raceways, based upon the conductor arrangement. **Annex D** contains examples to illustrate how calculations are made for various occupancies, motor circuits, elevator systems, etc. Designers may use these examples to give them basic procedures for calculating the load for a simple occupancy, branch-circuit or feeder-circuit.

> **For example:** What size EMT is required for 3 - 4 AWG THHN and 1 - 6 AWG THHN copper conductors?
>
> **Step 1:** Finding sq. in. area
> **Table 5, Ch. 9**
> 4 AWG = .0824 sq. in.
> 6 AWG = .0507 sq. in.
>
> **Step 2:** Calculating sq. in. area
> **Table 5, Ch. 9**
> 3 x .0824 sq. in. = .2472 sq. in.
> 1 x .0507 sq. in. = .0507 sq. in.
> Total = .2979 sq. in.
>
> **Step 3:** Selecting EMT
> **Table 4, Ch. 9**
> .2979 sq. in. requires 1"(27)
>
> **Solution: A 1 in. (27) EMT is required.**

Annex C shall be permitted to be used per AHJ to size the raceway where the conductors are the same size. There are approximately 26 numbered Tables in **Annex C** that shall be permitted to be used for the purpose of selecting raceways based upon the same size conductors. See **Note (1) to the Tables in Chapter 9**.

> **For example:** What size EMT is required for 4 - 4 AWG THHN copper conductors?
>
> **Step 1:** Finding number
> **Table C1 in Annex C**
> 4 - 4 AWG THHN requires 1" (27)
>
> **Solution: A 1 in. (27) EMT is required.**

See **Figure 1-13** for a step-by-step procedure for sizing a raceway system.

> **Design Tip:** All the Tables and Examples shall be reviewed carefully by designers, installers, maintainers and inspectors in order to be familiar with such information that may be pertinent with the duty and performance of one's job. The "new" Tables may require different size conduits based on the type of conductors and conduit used.

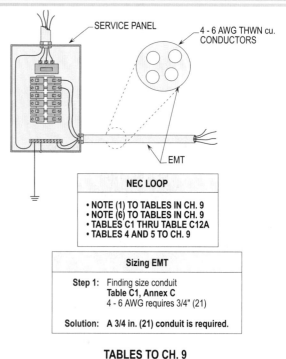

**TABLES TO CH. 9
TABLES C1, ANNEX C**

Figure 1-13. Chapter 9 contains Tables that shall be permitted to be used to size raceways based upon the conductor arrangement and in **Annex D** examples are available to illustrate how calculations are made.

PURPOSE OF OSHA

The OSHA Safety Standards, **29 CFR 1910, Subpart S** and **1926, Subpart K (Electrical)** is designed to help protect employees from electrical hazards during the construction, repair and maintenance of such systems. These Standards are the industrial standards utilized by employers to ensure that their employees have a safe workplace.

APPLYING OSHA 1910, SUBPART S

OSHA defines a workplace as a location where one or more persons performing a work task are employed. OSHA puts the sole responsibility on the employer to see that employees are trained to use safety related work practices. Employees shall adhere to these procedures and implement the practices in their respective workplace. **OSHA 1910, Subpart S** requires these safety related work practices to be applied so that employees are safeguarded from injury due to electrical hazards while they are working around or near electrical circuits and components. This required training shall be permitted to be classroom training, on the job training or both. OSHA requires electrical circuits and components to be maintained so that they continue to comply with the provisions listed in **Subpart S, Electrical**. These rules, if applied, prevent employees from being exposed to electrical shock and other dangerous hazards due to energized electrical circuits and components. **(See Figure 1-14)**

> **Design Tip: OSHA 1910, Subpart S** covers four main categories with requirements designed to protect employees in his or her workplace and they are as follows:
>
> • Design and installation safety,
> • Safety-related work practices,
> • Safety-related maintenance requirements and
> • Special equipment.

APPLYING OSHA 1926, SUBPART K

OSHA 1926, Subpart K contains rules and regulations that are necessary to protect employees from electrical hazards posing significant risks in construction site workplaces.

Subpart K has incorporated, by reference, all relevant requirements pertaining to the NEC which directly affects employee safety in construction workplaces. By placing such NEC requirements in this OSHA Standard, there is no need to reference the NEC to obtain such requirements.

The rules and regulations in **Subpart K of OSHA 1926** have been designed to protect electrical personnel performing work in workplaces of construction sites while the requirements in the NEC are used to design, install and inspect the wiring methods and equipment used to build the electrical system. **(See Figure 1-15)**

> **Design Tip: OSHA 1926, Subpart K** covers four main categories with requirements designed to protect construction workers in his or her workplace and they are as follows:
>
> • Installation safety requirements,
> • Safety-related work practices,
> • Safety-related maintenance and environmental consideration (construction) and
> • Safety requirements for special equipment (construction).

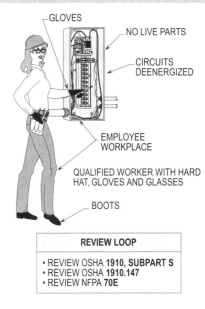

Figure 1-14. OSHA 1910, Subpart S deals with design and installation safety, safety-related work practices, safety-related maintenance requirements and special equipment.

APPLYING OSHA 1926, SUBPART V

Subpart V of OSHA 1926 deals with work practices to be used during the construction, operation and maintenance of electric power generation, transmission and distribution facilities. Work clearances, deenergizing methods and emergency procedures are covered and related in such a manner as to provide safe and reliable working conditions for personnel serving high-voltage systems.

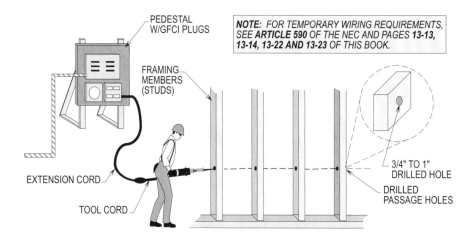

OSHA 1926, SUBPART K

Figure 1-15. OSHA 1926, Subpart K deals with installation safety requirements, safety-related work practices, safety-related maintenance and environmental considerations (construction) and safety requirements for special equipment (construction).

As used in **Subpart V**, the term "construction" includes the erection of new electric transmission and distribution lines and equipment, and the alteration, conversion and improvement of existing electric transmission and distribution lines and equipment.

Existing electric transmission and distribution lines and electrical equipment need not be modified to conform to the requirements of applicable standards until such work as described above is performed on such lines or equipment.

The standards set forth are minimum requirements for safety and health. Employers may require adherence to additional standards which are not in conflict with the requirements contained in **OSHA Subpart V. (See Figure 1-16)**

Safety Tip: Working clearances, deenergizing methods and emergency procedures are covered in **Subpart V**, including recommended practices. Special tools that are used for climbing are discussed and special in-service caring techniques are outlined. Safe operating rules to be followed before operating mechanically related equipment are highlighted. Proper ways to tie down loads in transit to prevent shifting are covered in detail and unloading of such materials with safety tips are presented. The correct procedures of grounding and bonding overhead lines and equipment to protect workers are shown. Overhead and underground wiring techniques are outlined. Safety related work practices involved with such systems are discussed.

APPLYING OSHA 1910, SUBPART R

OSHA 1910, Subpart R covers the work practices to be used during the operation and maintenance of electric power generation, transmission and distribution facilities. Employees engaged in the construction of electric power transmission or distribution systems are protected by the provisions of **OSHA 1926, Subpart V**. Employee training techniques on existing facilities are covered and related to in such a manner as to provide safe and reliable working conditions for personnel maintaining high-voltage electrical systems.

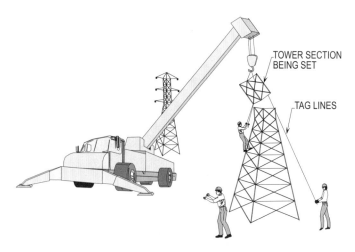

OSHA 1926, SUBPART V

Figure 1-16. OSHA 1926, Subpart V deals with work practices to be used during the construction, operation and maintenance of electric power generation, transmission and distribution facilities.

Electrical Codes and Standards

OSHA 1910, Subpart R covers five main areas of operation and maintenance which are electric power generation, control, transformation, transmission and distribution lines and equipment. These provisions apply to:

- Power generation,
- Transmission and distribution installations,
- Including related equipment for the purpose of communications or metering and
- Which are accessible only to qualified employees.

> **Design Tip:** The types of installations covered by this paragraph include the generation, transmission and distribution installations of electric utilities, as well as equivalent installations of industrial establishments. Supplementary electric generating equipment that is used to supply a workplace for emergency, standby or similar purposes only is covered under **Subpart S** of this Part. (See **paragraph (a)(1)(ii)(B) in OSHA 1910.269**)

OSHA 1910.303 through 1910.308 in Subpart S has never actually covered the operation and maintenance of electric power generation, transmission and distribution installations under the exclusive control of electric utilities or industrial facilities. **Subpart S** specifically exempts work performed by qualified persons on or directly associated with such systems and lines regardless who owns or controls them per **OSHA 1910.331(c)(1) in Subpart S**.

In contrast, telecommunications workers facing similar hazards have always been covered under the rules and regulations of the telecommunications standard **OSHA 1910.268**. This regulation protects employees performing communications work from two major hazards of falling and electric shock. (**See Figure 1-17**)

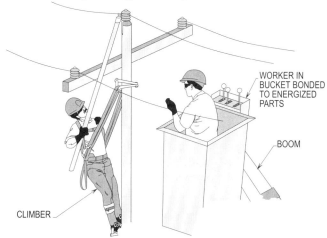

OSHA 1910, SUBPART R

Figure 1-17. OSHA 1910, Subpart R deals with work practices to be used during the operation and maintenance of electric power generation, transmission and distribution facilities.

PURPOSE OF NFPA 70E

NFPA 70E (Electrical Safety Requirements for Employees Workplaces) has been developed by NFPA by the request of OSHA.

NFPA 70E covers rules and regulation procedures on installing equipment, providing working clearances, guarding live parts, wiring design and protection, grounding connections, wiring methods, damp and wet locations, specific-purpose equipment, electric signs, cranes and hoists, electric welders, portable electric equipment, hazardous locations and special systems. **NFPA 70E** seems to apply directly to electrical personnel.

APPLYING NFPA 70E

NFPA 70E is a companion document to the NEC. It basically covers the same industries as the NEC. **NFPA 70E** addresses electrical safety rules and regulations that are needed for safeguarding employees in the workplace. Three major installations are covered as follows:

- Electrical utilization installations including carnival and parking lots, mobile homes and recreational vehicles, and industrial substations.
- Conductors and wiring methods which connect installations to a supply of electricity.
- Other outside premises wiring.

These three major installations are in addition to **OSHA 1910, Subpart S** and are intended to be utilized to provide a safe work area for employees in his or her workplace. They are not to be used as requirements for design or installation. (**See Figure 1-18**)

> **Safety Tip:** OSHA covers similar requirements in **OSHA 1910.302 through 1910.308 in Subpart S**. OSHA uses **NFPA 70E** to help keep **Subpart S** rules more current.

OSHA's main thrust in using **NFPA 70E** is to provide directions to employees on how to perform their work so that electrical hazards may be avoided. Information on working on or near energized parts, working on or near deenergized parts, the safe deenergizing of systems, lockout-tagout procedures, the safe use of tools and equipment, personal and other protective equipment, safe approach distances for overhead lines, switching of protective devices after operation, safety training and safe voltage-measurement practices are covered in detail.

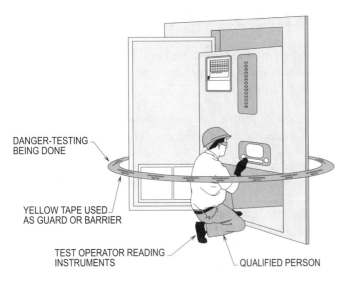

NFPA 70E, CH. 4

Figure 1-18. NFPA 70E deals with electrical safety rules and regulations that are needed for safeguarding employees in the workplace.

PURPOSE OF NFPA 70B

As in **NFPA 70E**, **70B (The Recommended Practice for Electrical Equipment Maintenance)** is a companion standard to the NEC. It contains recommended practice information for the proper maintenance of electric equipment.

NFPA 70B covers maintenance of industrial-type electric systems and equipment. Such systems and equipment are typical of those installed in commercial buildings, industrial plants and multifamily dwellings. The standard makes it clear that it does not cover consumer appliances or home-type equipment and its rules are not intended to supersede manufacturer recommended maintenance procedures.

APPLYING NFPA 70B

NFPA 70B includes twenty-two chapters and ten appendices which cover proper preventive maintenance procedures. The standard covers important items such as detailed maintenance techniques for switchgear, circuit breakers, cables, motor control centers, rotating equipment, wiring devices and other related electric equipment.

Specific details of electrical testing in areas such as insulation resistance, protective device testing, fault-gas analysis and many other such modern testing procedures are discussed. The maintenance of equipment which is subject to long intervals between shutdowns and the methods that are to be used for deenergizing equipment are highlighted and discussed.

Maintenance Tip: The appendices of **NFPA 70B** are equipped with information on walk-through inspections, instruction techniques, symbols, diagrams and recommended test sheets and forms that are designed to be used for maintenance programs. **(See Figure 1-19)**

NFPA 70B, CH. 18

Figure 1-19. NFPA 70B covers important items such as detailed maintenance techniques for switchgear, circuit breakers, cables, motor control centers, rotating equipment, wiring devices and other related equipment.

PURPOSE OF NFPA 79

The purpose of **NFPA 79** is to provide detailed information for the application of electrical/electronic equipment, apparatus or systems supplied as part of industrial machinery that will promote safety to life and property.

Information in this standard is not intended to limit or inhibit the advancement of the state of the art designing techniques of such equipment. Each type of machine has unique requirements that shall be accommodated to provide adequate safety for the operator.

APPLYING NFPA 79

The rules of **NFPA 79** apply to the electrical/electronic equipment, apparatus or systems of industrial machinery operating from a nominal voltage of 600 volts or less and commencing at the place of connection of the supply to the electrical equipment of the machines.

Electrical Codes and Standards

Modern machine tool electrical equipment may vary from that of single-motor machines, such as drill presses, that perform simple, repetitive operations, and that of very large, multimotored automatic machines that involve highly complex electrical control systems, including electronic and solid-state devices and equipment. Generally, these machines are especially designed, factory-wired and tested by the builder and then erected in the plant in which they will be used. Because of their importance to plant production and their usually high cost, they are customarily provided with safeguards and other devices not often incorporated in the usual motor and control application as contemplated by the NEC.

Although these machines may be completely automatic, they are constantly attended, when operating, by highly skilled operators. The machine usually incorporates many special devices to protect the operator, protect the machine and building against fires of electrical origin, protect the machine and work in process against damage due to electrical failures and protect against loss of production due to failure of a machine component. To provide these safeguards, it may be preferable to deliberately sacrifice a motor or some other component, rather than to chance injury to the operator, the work or the machine. **(See Figure 1-20)** It is due to proper supervision and qualified personnel, that **NFPA 79** allows such procedures when operating plant machine tool electrical equipment.

Design Tip: It is because of the above consideration that **NFPA 79** varies from the basic concepts of motor protection and other protection techniques as required by the provisions in the NEC.

PURPOSE OF THE NESC

The **NESC (National Electrical Safety Code)** was developed in 1913 by the National Bureau of Standards, became a consensus standard and is used in the industry as such today. The **NESC** is an American National Standard published by IEEE. It provides practical rules for safeguarding personnel during the installation, operation or maintenance of electric supply and communications lines and associated apparatus and equipment. The NESC contains three main general rules and they are as follows:

- All electric supply and communications lines and equipment shall be designed, constructed, operated and maintained to comply with its requirements.

- The utilities, authorized contractors or other entities performing design, construction, operation or maintenance procedures for electric supply or communication lines or equipment by the **NESC** are responsible for complying with specific requirements.

- For all conditions not specified in the **NESC**, construction and maintenance procedures should be performed and accepted with good practices based upon given local conditions. **(See Figure 1-21)**

Design Tip: The NESC covers supply and communications lines, equipment and associated work practices used by both public and private electric supply, communications, railway or similar utilities. Similar systems which are under the control of qualified persons are also covered.

PURPOSE OF THIRD PARTY CERTIFICATION

Underwriters laboratories (UL) and other such laboratories are nonprofit organizations supported by the manufacturers who submit merchandise, that is, for the testing of merchandise for safety when being used.

UL or any other approved testing laboratory label assures the user that the manufacturer of the item has submitted samples of the item to the laboratory for testing and was found to meet the required minimum safety standards.

Manufacturers submit samples of their products to a laboratory such as **UL** for testing before they are manufactured on a large scale. It is then presumed that the manufacturer will maintain the same quality in the future production of the same item. Such testing facilities also perform on-site inspections to ensure that the manufacturer continues to manufacture the product by their standards.

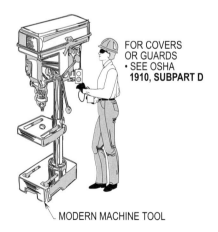

NFPA 79, CH. 4

Figure 1-20. NFPA 79 covers electrical/electronic equipment, apparatus or systems of industrial machinery operating from a nominal voltage of 600 volts or less.

1-13

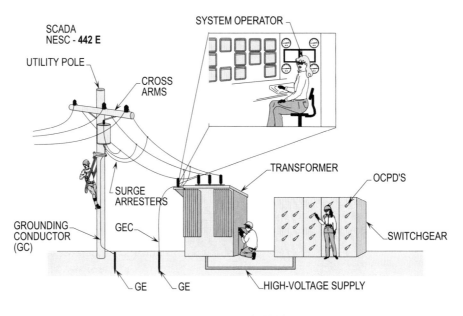

Figure 1-21. The **NESC** provides practical rules for safeguarding personnel during the installation, operation or maintenance of electric supply and communications lines and associated apparatus equipment.

For example, if the product passes the exhaustive tests of **UL** in accordance with established standards, it is added to the official **UL** published list and the product is then known as "**Listed by Underwriter's Laboratories, Inc.**"

Such **UL** items usually have a **UL** label attached directly to the product. In some cases, this **UL** "label" is molded or stamped into the merchandise.

Note, this same procedure is used by other testing laboratories that are approved by OSHA.

An item approved by Testing Laboratories, however, does not mean that the item is approved for all uses. Rather, a testing label means that the item or device, as labeled, is safe only for the purpose for which it was intended. **(See Figure 1-22)**

Design Tip: Once equipment has been approved, it is not to be modified in any way, but used as instructed by the manufacturer.

For example, the wiring or any of the other elements are not to be changed out or rearranged in a manner that does not comply with the listing and labeling of the nameplate on the equipment or accompanying literature.

THIRD PARTY CERTIFICATION
- UL
- FM
- ETL
- CSA
- MET
- D,S,G
- NOTE THAT THERE ARE ABOUT 15 TESTING LABS NOW APPROVED BY OSHA.

Figure 1-22. UL or any other approved testing laboratory label assures the user that the product being installed is listed and labeled to perform the function intended.

UL WHITE BOOK

The **UL white book**, known in the electrical industry as the **"General Information Electrical Construction, Hazardous Locations and Electrical Heating and Air Conditioning Equipment Directory"**, provides a service for the classification of products that have been determined to meet the appropriate requirements of the applicable international publication(s). For those products which comply with the requirements of an international publication(s), the Classification Marking may appear in various forms as authorized by **UL**.

For example, a form may include and illustrate the word "Classified" with a control number assigned by **UL**. The product name as indicated in this Directory under each of the product categories is generally included as part of the Classification Marking text but may be omitted when, in **UL's** opinion, the use of the name is superfluous and the Classification Marking is directly and permanently applied to the product by stamping, molding, ink stamping, silk screening or similar processes.

Separable Classification Markings (not part of a nameplate and in the form of decals, stickers or labels) will always include four elements; **UL's** name and/or symbol, the word "Classified", the product category name and a control number.

The complete Classification Marking will appear on the smallest unit container in which the product is packaged when the product is of such a size that the complete Classification Marking cannot be applied to the product or when the product size, shape, material or surface texture makes it impossible to apply any legible marking to the product. When the complete Classification Marking cannot be applied to the product, no reference to **UL** on the product is permitted. **(See Figure 1-23)**

> **Design Tip:** Designers, installers and inspectors use the **UL white book** to determine if certain wiring methods and equipment are indeed listed and labeled and are apt for a particular installation according to their condition of use.

APPROVAL

OSHA requires conductors and equipment to be accepted only when approved. The definition of approved is to be acceptable to the authority having jurisdiction (AHJ) per **Article 100** in the NEC. **OSHA 1910.303(a)** and NEC **110.2** now have the same requirements considering the acceptability of components, wiring methods and equipment. The AHJ will usually require third party certification of all equipment. Third party certification is where a qualified testing laboratory performs tests to verify if the equipment will do what it is designed to do and still be safe to the user. The only problem with accepting equipment that is not certified by a third party is that the authority having jurisdiction takes the sole responsibility that the equipment is safe.

> **Design Tip:** OSHA demands third party certification by an independent testing laboratory for any and all electrical equipment designed and installed in industrial locations. Custom-made equipment is acceptable where the equipment needed is not available to be purchased and is not listed and labeled by an NRTL. **(See Figure 1-24)**

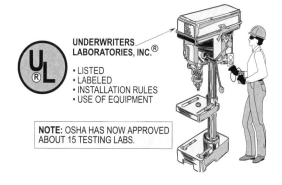

NEC 110.2
NEC 110.3(B)

Figure 1-24. OSHA requires the conductors and equipment to be accepted only when approved.

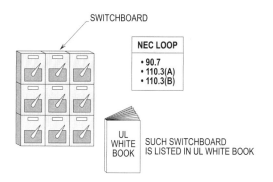

Figure 1-23. The **UL white book** provides a service for the classification of products that have been determined to meet the appropriate requirements of the applicable international publication(s).

Codes and Standards
• NFPA 70E - 400.1
• NEC 90.7
• NEC 110.2 and NEC Art. 100
• NEC 110.3(B)
• OSHA 1910.303(a) and 1910.399
• OSHA 1926.403(a)

EXAMINATION, INSTALLATION AND USE OF EQUIPMENT

All electrical equipment and work must be related to the Occupational Safety and Health Administration (OSHA) regulations and comply to the provisions of standards and installation instructions of qualified testing laboratories that have tested and certified the equipment. Such Nationally Recognized Testing Laboratories (NRTL) and certifying organizations are:

- Underwriters Laboratories (UL)
- Factory Mutual Engineering Corp. (FM)
* Electrical Testing Laboratories (ETL)
* MET Testing Company, Inc.
* Dash, Straus, and Goodhue, Inc.
* Canadian Standards Association (CSA)
* (Limited Certification) Note, there are others

EXAMINATION

The NEC, OSHA and **NFPA 70E** does not include detailed requirements for wiring inside of electrical equipment. Information pertaining to the internal wiring of equipment is usually found in individual standards that contain special requirements for installing the equipment concerned. It is not the intent of the NEC and OSHA to take away the authority of inspectors to examine and approve equipment. However, the NEC, OSHA and **NFPA 70E** indicate that the rules pertaining to the internal construction of equipment are usually investigated and approved by a qualified testing laboratory. The NEC does not make a flat rule to the effect that all electrical equipment shall have third party certification of equipment by independent testing laboratories per **90.7, 110.2 and 110.3**. OSHA, on the other hand, does require electrical equipment to have third party certification for the safety of personnel using such equipment per **Subpart S to 1910.303(a)**. NFPA 70E recognizes the same rule in **400.3(A)** and OSHA recommends the requirements in **NFPA 70E** to be enforced to provide safety for employees in employee workplaces.

In judging and evaluating electrical equipment that is installed in the employee workplace, the following steps shall be followed:

- Judge the wiring installation to ensure that it complies with the provisions of **NFPA 70E**. Evaluate the equipment for suitability for installation and use according to nameplate ratings and instructions. List the appropriate standard for such equipment.

For example, every product listed in the **UL Electrical Construction Materials Directory (Green Book)** is required to be installed as described in the application data given with the listing in the book. It becomes the responsibility of the AHJ to decide and determine the suitability of electrical equipment that has not been tested and certified by a qualified testing laboratory.

- Mechanical strength and durability of enclosures, raceways, etc. enclosing electrical parts and wiring shall be adequate to provide proper protection of such elements.

- Electrical insulation shall be installed by **Table 310.13** and loaded according to the ampacities of **Table 310.16** for systems rated at 2000 volts or less. Insulation of conductors may be checked with any good insulation resistance tester.

- Equipment and insulation shall be sized and selected to operate safely under normal conditions of use and also under abnormal conditions without adverse heating effects. (See **310.10** in the NEC).

- Electrical equipment shall be equipped with shields, etc. to disguise the arcing effect and limit the burning effect of the arc. (See **110.18** in the NEC).

- Electrical equipment with components shall be classified by its type, size, voltage, current capacity and specific use. An evaluation shall be made to verify that the type and size of equipment is being supplied with the proper voltage and is loaded to a value no greater than it is designed to carry. (See **90.7** and **110.3(B)** in the NEC and **OSHA 1910.399**).

Codes and Standards
• NFPA 70E - 400.3(A)
• NEC 110.3(A)
• OSHA 1910.399
• OSHA 1910.303(b)(1)
• OSHA 1926.403(b)(1)

INSTALLATION AND USE

OSHA'S relationship to electrical equipment is very clear in **1910, Subpart S, Electrical**. Electrical equipment shall have third party certification for the essential safety of the equipment and components used to assemble electrical installations. For the safety of personnel utilizing such equipment, OSHA requires the following:

Utilization equipment. Utilization equipment is equipment that utilizes electric energy for electronic, electromechanical, chemical, heating, lighting or similar purposes.

Approval. The conductors and equipment required or permitted by **Subpart S** shall be acceptable only if approved.

Electrical Codes and Standards

Approved. Acceptable to the authority enforcing **Subpart S**. The authority mentioned is the Assistant Secretary of Labor for Occupational Safety and Health. The definition of "Acceptable" indicates what is acceptable to the Assistant Secretary of Labor, and therefore approved within the meaning of **Subpart S**.

Acceptable. An installation or equipment is acceptable to the Assistant Secretary of Labor, and approved within the meaning of **Subpart S**: where it is accepted, certified, listed, labeled or otherwise determined to be safe by a qualified testing laboratory, such as, but not limited to, Underwriters Laboratories Inc. and Factory Mutual Engineering Corp. or with an installation or equipment of a kind which no qualified testing laboratory accepts, certifies, lists, labels or determines to be safe, where it is inspected or tested by another federal agency, or by a state, municipality or local authority responsible for enforcing the safety provisions of **NFPA 70E** and found in compliance with the provisions of the NEC as applied in **Subpart S**, or custom-made equipment or related installations which are designed, fabricated for, and intended for use by, a customer, where it is determined to be safe for its intended use by its manufacturer on the basis of test data which the employer keeps and makes available for inspection to the Assistant Secretary and his or her authorized representatives in the field (OSHA inspectors). Custom-made equipment should be designed and built by the provisions listed in **UL 508** or one of the C series published by ANSI, based on the type equipment involved. The customer specifies to the manufacturer the function that their equipment must perform. **(See Figure 1-25)**

Labeled. Equipment or materials to which has been attached a label, symbol or other identifying mark of an organization that is acceptable to the AHJ and concerned with product evaluation, that maintains periodic inspection of production of labeled equipment or materials, and by whose labeling the manufacturer indicates compliance with appropriate standards or performance in a specified manner.

Listed. Equipment, materials or services included in a list published by an organization that is acceptable to the AHJ and concerned with evaluation of products or services, that maintains periodic inspection of production of listed equipment or materials or periodic evaluation of services, and whose listing states that the equipment, material or services either meets appropriate designated standards or has been tested and found suitable for a specified purpose.

Accepted. An installation is "accepted" where it has been inspected and found by a qualified testing laboratory to conform to specified plans or to procedures of applicable codes.

Certified. Equipment is "certified" where it:
- Has been tested and found by a qualified testing laboratory to meet national standards or be safe for use in a specified manner, or
- Is of a kind whose production is periodically inspected by a qualified testing laboratory, and
- It bears a label, tag or other record of certification.

> **Design Tip:** OSHA requires all electrical equipment, where available, to be listed and labeled by a Nationally Recognized Testing Laboratory. UL and FM are laboratories without limited certification which are accepted for industrial locations. However, if such needed electrical equipment is not listed, labeled and available, then custom-made equipment that complies with the requirements of the Assistant Secretary is acceptable. OSHA recognizes a NRTL while the NEC recognizes a qualified testing laboratory (QTL).

There is a conflict in the standards due to the change in the 1981 NEC which changed the wording from NRTL to QTL.

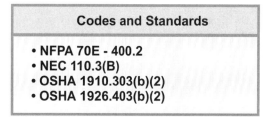

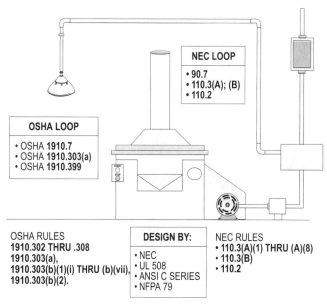

LISTED OR CUSTOM-MADE EQUIPMENT
OSHA 1910.399   NEC 110.2

Figure 1-25. An installation or equipment is acceptable to the Assistant Secretary of Labor, and approved within the meaning of **Subpart S**; where it is accepted, certified, listed, labeled or otherwise determined to be safe by a nationally recognized testing laboratory.

PURPOSE OF CUSTOM-MADE ELECTRICAL EQUIPMENT

OSHA requires all electrical equipment, where available, to be listed and labeled by a NRTL. However, if such electrical equipment is not listed, labeled and available, then custom-made equipment that complies with the requirements of the Assistant Secretary of Labor is acceptable. (See **OSHA 1910.399**)

Custom-made equipment or related installations is equipment that is designed, fabricated for and intended for use by a customer, where it is determined to be safe for its intended use by its manufacturer on the basis of test data which the employer keeps and makes available for inspection to the Assistant Secretary and his or her authorized representatives in the field (OSHA inspectors). Custom-made equipment should be designed and built by the provisions listed in the **NEC, UL 508, NFPA 79** or one of the C series published by ANSI, based on the type or equipment involved. The customer specifies to the manufacturer the function that their equipment shall perform. **(See Figure 1-25)**

INSTALLING AND MAINTAINING ELECTRICAL EQUIPMENT

It is the sole responsibility of personnel who have been designated to install and maintain electrical equipment, including all electrical apparatus, to do so in a safe and reliable manner. Such personnel do not have the authority to wire in and modify the electrical element, where they do not comply with the **NEC**, OSHA and various testing laboratory's related standards. This rule also applies to custom-made equipment.

OSHA'S RELATIONSHIP TO ELECTRICAL EQUIPMENT OSHA 1910.7, 1910.303(a), 1910.399(a), NEC 110.2, 110.3(A); (B) AND ARTICLE 100 (DEFINITIONS OF APPROVAL)

The requirements set forth in **OSHA 29 CFR, 1910.302 through 1910.308** pertains to electrical equipment installed before and after certain mandatory dates listed in **Subpart S**. All electrical utilization systems and pieces of equipment that were designed, installed and inspected after March 15, 1972 shall comply with all the provisions according to **OSHA 1910.302 through 1910.308**. All major replacement, modification, repair or rehabilitation performed after March 15, 1972 on such systems and equipment shall comply with these requirements. These requirements also include equipment installed before March 15, 1972. The above rules are retroactive to such electrical systems and equipment.

There are certain provisions in **OSHA 1910.302 through 1910.308** that are intended to apply only to electrical systems and equipment installed after April 16, 1981 which are not meant to be retroactive. These nonretroactive requirements and the electrical installations and equipment they apply to are listed in **OSHA 1910.302(b)**.

> **Design Tip:** Major replacements, modifications or rehabilitation as outlined in OSHA regulations include work similar to that involved where a new building or facility is built, a new wing is added or an entire floor is renovated.

PURPOSE OF UL 508

UL 508 contains basic requirements for products covered by Underwriters laboratories, Inc. under its Follow-Up Service for this category within the rules and regulations of the Standard. These requirements are based upon sound engineering principles, research, records of tests, field experience and an appreciation of the problems of manufacture, installation and use derived from consultation with and information obtained from manufacturers, users, inspection authorities and others having specialized experience. They are subject to revision as further experience and investigation may show is necessary or desirable.

A product which complies with the text of **UL 508** will not necessarily be judged to comply with the Standard if, when examined and tested, it is found to have other features which impair the level of safety contemplated by these requirements.

A product employing materials or having forms of construction differing from those detailed in the requirements of this Standard may be examined and tested according to the intent of the requirements and, if found to be substantially equivalent, may be judged to comply with the Standard.

APPLYING UL 508

The requirements of **UL 508** cover industrial control devices, and device accessories thereto, for starting, stopping, regulating, controlling or protecting electric motors. Such requirements also cover industrial control devices or systems that store or process information and are provided with an output motor control function(s). This equipment is for use in ordinary locations in accordance with the NEC.

Requirements also cover industrial control panels which are

assemblies of industrial control devices and other devices associated with the control of motor operated and related industrial equipment. Examples of devices in an industrial control panel are disconnecting means, motor branch-circuit protective devices, temperature control devices, electrical instruments, etc.. **(See Figure 1-26)**

For example, an industrial control panel for the control of metalworking machine tools or plastics machinery is judged on the basis of compliance with the applicable requirements of this Standard as well as the requirements in the **Standard for Industrial Machinery, NFPA 79**.

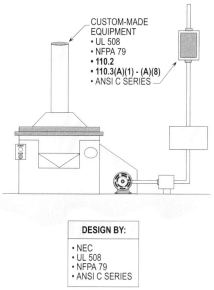

Figure 1-26. UL 508 covers industrial control devices, and device accessories thereto, for starting, stopping, regulating, controlling or protecting electric motors.

> **Design Tip:** Electrical personnel, due to the type of apparatus or equipment involved, may need to use other Codes and Standards in addition to **UL 508**.

UL 508 covers five main categories with requirements designed to judge and evaluate industrial control devices and accessories and they are as follows:
- Construction,
- Performance,
- Rating,
- Marking and
- Instruction and use.

> **Design Tip:** The rules and regulations in **UL 508** cover devices rated at 1500 volts or less. Such rules and regulations must be reviewed very carefully before attempting to apply them to the design of such apparatus.

AUTHORITY HAVING JURISDICTION

The "AHJ" is the organization, office or individual responsible for "approving" equipment, installations or procedures.

> **Design Tip:** The phrase "AHJ" is used in NFPA documents in a broad manner since jurisdictions and "approval" agencies vary, as do their responsibilities. Where public safety is primary, the AHJ may be a federal, state, local or other regional department or individual such as a fire chief, fire marshal, chief of a fire prevention bureau, labor department, health department, building official, electrical inspector or others having such authority. For insurance purposes, an insurance inspection department, rating bureau or other insurance company representative may be the "AHJ". In many circumstances, the property owner or his designated agent assumes the role of the "AHJ" at government installations or county areas not requiring a permit and inspection. (See definition of AHJ in **Article 100**)

The AHJ has the option of permitting alternative methods where specific rules are not established in the NEC. This allows the local authority to waive specific requirements in industrial occupancies, research and testing laboratories, and other occupancies where the specific type of installation is not contemplated in the NEC requirements.

The AHJ may waive a new NEC requirement during the interim period between the acceptance of a new edition of the NEC and the availability of a new product, construction or material redesigned to comply with the increased safety required by the new NEC edition. It is difficult to establish a viable future effective date in the NEC because the time needed to change existing products and standards and to develop new materials and test methods is not usually known at the time of adoption of the new requirement in the NEC.

It is also the responsibility of the local authority (AHJ), enforcing the NEC, to interpret the specific rules and regulations of the NEC. **(See Figure 1-27)**

By special permission, the AHJ has the responsibility of giving the engineer, contractor, electrician, etc. written consent of his or her decision to waive a requirement in the NEC pertaining to a specific installation.

Figure 1-27. The "AHJ" is the organization, office or individual responsible for "approving" equipment, installations or procedures.

Name

Date

Chapter 1: Electrical Codes and Standards

Section   Answer

_____   T   F   1. The NEC is a minimum standard used for electrical installations and for the protection of life and property.

_____   T   F   2. The NEC is usually adopted by government agencies and becomes effective September 1 of the year that it is published.

_____   T   F   3. The NEC does not contain provisions necessary for safety.

_____   T   F   4. The layout of the NEC contains an arrangement of 9 Chapters with Articles and Sections which contains the requirements for design, installation, use and maintaining of electrical systems.

_____   T   F   5. **Chapter 2** of the NEC pertains to designing loads for selecting wiring methods and equipment and sizing overcurrent protection devices to protect such elements as conductors, parts and equipment.

_____   T   F   6. **Chapter 7** covers communication circuits and equipment.

_____   T   F   7. OSHA requires electrical circuits and components to be maintained so that they continue to comply with the provisions listed in **Subpart S, Electrical**.

_____   T   F   8. **Subpart V of OSHA 1926** deals with work practices to be used during the construction, operation and maintenance of electric power generation, transmission and distribution facilities.

_____   T   F   9. **NFPA 70B** is a companion document to be used for electrical safety rules and regulations that are needed for safeguarding employees in the workplace.

_____   T   F   10. The UL white book is known in the electrical industry as the "General Information Electrical Construction, Hazardous Locations and Electrical Heating and Air-conditioning Equipment Directory."

_____ _____   11. **Chapter 5** in the NEC deals with _____ occupancies.

_____ _____   12. **Chapter 7** in the NEC deals with requirements for special _____ that arise from the particular type, use, and application of the power source used to supply circuits and equipment.

_____ _____   13. **Chapter 8** in the NEC deals with requirements for _____ circuits and equipment.

_____ _____   14. The rules and regulations in **Subpart K of OSHA 1926** have been designed to protect electrical personnel performing work in workplaces of _____ sites.

_____ _____   15. **NFPA 70E** has been developed by NFPA by the request of _____.

Section	Answer

16. The purpose of **NFPA 79** is to provide detailed information for the application of _____ and electronic equipment, apparatus or systems supplied as part of industrial machinery that will promote safety to life and property.

17. The _____ of UL or any other testing laboratory assures the user that the manufacturer of the item has submitted a sample of the item to the laboratory for testing and was found to meet the required minimum safety standards.

18. OSHA requires the conductors and equipment to be accepted only when _____.

19. The requirements of **UL 508** covers _____ control devices, and devices accessory thereto, for starting, stopping, regulating, controlling or protecting electric motors.

20. The _____ has the option of permitting alternative methods where specific rules are not established in the NEC.

21. **Chapter 1** for many years in the electrical industry has been called the _____ Chapter.
    - (a) Design
    - (b) Installation
    - (c) Get acquainted
    - (d) Special conditions

22. **Chapter 5** in the electrical industry has been called the _____ Chapter.
    - (a) Design
    - (b) Special occupancies
    - (c) User and maintainer
    - (d) Special conditions

23. OSHA defines a workplace as a location where _____ or more persons performing a work task are employed.
    - (a) One
    - (b) Two
    - (c) Three
    - (d) Four

24. **Subpart V of OSHA 1926** deals with work practices to be used during the construction, operation and maintenance of electric power _____ facilities.
    - (a) Generation
    - (b) Transmission
    - (c) Distribution
    - (d) All of the above

25. OSHA demands third party certification by an _____ testing laboratory for any and all electrical equipment designed and installed in industrial locations.
    - (a) Independent
    - (b) Acceptable
    - (c) None of the above
    - (d) All of the above

2

Electrical Safety and First Aid

Because of the hazard of electric shock or burns from an arc or blast when employees are working near or performing work on energized lines and equipment, workers may suffer electrocution or severe injury on the job.

The employer shall provide training or require that his or her employees are knowledgeable and proficient in the procedures involving emergency situations and first-aid fundamentals. Medical services, first-aid and job briefing procedures are covered and related in such a manner that, if followed, will provide a basic concept of first-aid for personnel servicing electrical wiring and equipment. Note that only trained qualified personnel shall be permitted to administer first-aid to hurt employees.

THE FLOW OF ELECTRICITY

When an electrical current passes through the body, there is a path drawn from the two points of contact where it enters and exits. Along this path appears the greatest magnitude of electrical current. Damage to tissues or organs along this current pathway dictates the severity of the injury.

> **For example,** a common path is current that enters on one of the arms and exits through the other arm or a leg. This type of shock is very dangerous for it can cause severe burning, respiratory arrest or ventricular fibrillation when it crosses the chest area.

ELECTRIC SHOCK

Electric shock is usually associated with alternating current (AC). However, there is still the danger of shock with direct current (DC). Other factors that enhance electric shock potential are as follows:

- Wet and/or damp locations,
- Ground/grounded objects,
- Current loop from source back to source,
- Path of current through body/duration of contact,
- Area of body contact and pressure of contact,
- Physical size/condition/age of person,
- Type and/or amount of voltage,
- Personal protective equipment/gloves/shoes,
- Metal objects such as watches, necklaces, rings, etc. and
- Miscellaneous

   (a) Poor workplace illumination,
   (b) Color blindness,
   (c) Lack of training, knowledge and
   (d) No safe work procedures.

Safety Tip: The above conditions may have a direct effect on how much current the voltage can push through a body from phase-to-phase conditions or phase-to-ground conditions.

DANGER OF ELECTRIC SHOCK

Mild exposures to electric shock can be very painful and may cause muscular contractions, respiratory arrest, severe tissue burning or ventricular fibrillation, all of which can be fatal.

MUSCULAR CONTRACTIONS

The electrical stimulation of nerves along the current path in the body can cause the muscles in that area to involuntarily contract.

For example, if a person held on to a deenergized conductor and had someone gradually increase the current flow, a sensation would be detected when the current reaches 1 mA. As the current is increased, pain and heat will increase, until at about 16 mA, the person would not be able to let go of the conductor because the forearm muscles would involuntarily contract due to current flow. If the current path went across the chest, the diaphragm muscles would contract at about 20-40 mA, preventing the person from breathing. This magnitude of current is fatal, unless it is interrupted. It is often referred to as "let-go" current.

Safety Tip: If current is interrupted in time, artificial respiration is usually suggested as the correct treatment and may restore breathing and prevent death or brain damage.

RESPIRATORY CONTROLS

Respiratory arrest may also be caused by a shock passing through the spinal cord to the lower portion of the brain, causing the respiratory controls of the central nervous system to shut down. In some cases, respiratory arrest may continue even after interrupting the current flow, which is called persistent respiratory arrest. Unless ventricular fibrillation has taken place (there is no pulse), artificial respiration is the correct treatment until medical help arrives.

TISSUE BURNING

As indicated above, the areas of major damage from an electrical shock in the body lie along the current path. The current does not follow the blood vessels or other parts of the circulatory system, it strikes a path through the body and burns any tissue or organ along that path. This can lead to internal bleeding or severe organ damage and requires prompt medical attention as fast as possible.

VENTRICULAR FIBRILLATION

Another result of electric current passing through the chest is a disruption of the natural rhythm of the heart, or ventricular fibrillation. When this happens regular heart action ceases, there is no pulse and blood stops circulating to the organs. This condition is nearly always fatal unless it is treated promptly.

Normal heart function can sometimes be restored through a defibrillator, an apparatus which applies electric impulses to the heart muscles in an attempt to restore regular pumping action of the heart.

Safety Tip: While waiting for medical assistance, it is appropriate to administer cardiopulmonary resuscitation (artificial respiration and chest compression). In some cases, the fibrillation may reverse itself spontaneously and it is important that oxygen is supplied to vital organs during this time.

FREEING VICTIM

An electric shock generally occurs when a person touches a live wire with his hand, a noninsulated foot or any other part of his body. When freeing the victim, the rescuer should separate the victim from the contact by utilizing a long, dry stick, a dry rope or length of dry cloth. The rescuer needs to be especially careful and not attempt to free the victim with his bare hands alone; the rescuer might become another victim if proper protection is not used.

ARTIFICIAL RESPIRATION

When a victim's natural breathing is inadequate or ceases from electric shock, artificial respiration should be applied immediately to maintain an open airway through the mouth and nose (or through the stoma) and restore breathing by maintaining an alternating increase and decrease in the expansion of the chest. Artificial respiration should always be continued by the following recommended practices until:

- The victim begins to breathe for him/herself,
- Pronounced dead by a doctor and
- Dead beyond any doubt.

MOUTH-TO-MOUTH RESUSCITATION

Mouth-to-mouth resuscitation is a technique used to ensure the victim of an air passage and increase the volume of air exchange in his or her lungs. There are nine techniques suggested when applying mouth-to-mouth resuscitation to a victim and the following should be used:

- Determine consciousness of victim.
- Tilt the victim's head back so that his/her chin is pointing upward.
- For about 5 seconds, listen and feel for air to be exhaulted and look at the victim's chest to see if it rises and falls.
- If victim is not breathing, use the thumb and index finger to pinch the victim's nostrils shut.
- Use the following techniques to blow air into the victim's mouth:

  (a) Open your mouth wide,
  (b) Take a deep breath,
  (c) Blow air into the victim's mouth with your mouth sealed tightly around the victim's mouth,
  (d) Initially, give four quick, full breaths without allowing the lungs to fully deflate (empty) between each breath,
  (e) For about 5 to 10 seconds, listen and feel for air to be exhaulted and look at the victim's chest to see if it rises and falls,
  (f) Provide at least on breath every 5 seconds, if there is a pulse and no breathing and
  (g) Moderate resistance to blowing will be felt if the airway is clear.

- Watch victim's chest to see when it rises.
- Stop blowing when the victim's chest has expanded, look and listen for air to be exhaulted.
- Watch victim's chest to see when it falls.
- Repeat recommended technique.

HEART COMPRESSION

The best technique of performing emergency first-aid is a combination of artificial respiration and artificial circulation by external heart or cardiac compression. Only properly trained personnel should perform this technique.

Inflate the victim's lungs about 5 times by artificial respiration then check the pulse of the victim in the neck or on the side of the Adam's apple. If the victim's pulse did not start, the rescuer must start cardiac compression in addition to artificial respiration in order to restore the circulation. Two rescuer's should perform this technique. One rescuer should perform the artificial respiration and the other rescuer should rhythmically compress the victim's heart. **(See Figure 2-1)**

INHALATION TREATMENT

Inhalation treatment should be used on victim's to make breathing failures less likely. The victim should breath pure oxygen from a tank when assistance is obtained from police or fire department or the emergency crew of the local electric or gas company. The artificial breathing method and inhalation treatment should be used together for best results.

KEEPING VICTIM WARM

Victims of electric shock should be kept warm by using a blanket or some loose clothing to prevent a sudden chill or pneumonia.

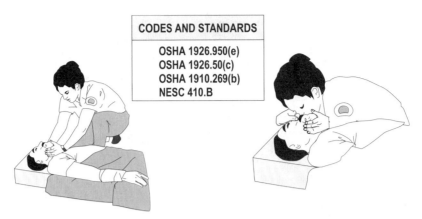

To open the airway:
The tongue is attached to the lower jaw and floor of the mouth. In an unconscious victim, the head flexes, the jaw relaxes and the tongue drops back to obstruct the air passage. To restore breathing, the first thing to do is open the airway. The simplest and quickest method is:

- Lift the neck with one hand.
- Tilt the head backward into maximum extension with the other hand this draws the tongue forward and opens the airway. This should always be the first maneuver performed. Sometimes this alone will allow the victim to resume spontaneous breathing.

Note 1: Only qualified trained personnel should perform artificial respiration on a hurt victim.

To restore breathing:
If the victim does not resume spontaneous breathing, start mouth-to-mouth respiration.

- First pinch the nose shut with your thumb and index finger to prevent the escape of air. Your cheek can also be used to seal the victim's nose.
- Then open your mouth widely, take a deep breath, place your mouth over the victim's mouth to get a good seal and inflate their lungs with about twice the amount of a normal breath.
- Remove your mouth and allow them to exhale.

Note 2: Employers must have their people, who have the responsibility of performing first-aid to hurt employees, trained by Red Cross instructors or other qualified agencies.

Figure 2-1. The best technique of performing emergency first-aid is a combination of artificial respiration and artificial circulation by external heart or cardiac compression. Only properly trained personnel should perform this technique.

Name Date

Chapter 2. Electrical Safety and First Aid

Section	Answer		
_____	T F	1.	When an electrical current passes through the body, there is a path drawn from the two points of contact where it enters and exits.
_____	T F	2.	Electric shock is usually associated with direct current (DC).
_____	T F	3.	The electrical stimulation of nerves along the current path in the body can cause the muscles in that area to involuntarily contract.
_____	T F	4.	Artificial respiration can be applied when there is not a pulse.
_____	T F	5.	When freeing the victim from electrical shock, the rescuer should separate the victim from the contact by utilizing a long, wet stick, a wet rope or length of wet cloth.
_____	T F	6.	Mouth-to-mouth resuscitation is a technique used to ensure the victim of an air passage and increase the volume of air exchange in his lungs.
_____	T F	7.	A combination of artificial respiration and artificial circulation by external heart or cardiac compression shall be performed by only properly trained personnel.
_____	T F	8.	If a victim's pulse does not start, only one rescuer must start cardiac compression in addition to the artificial respiration in order to restore circulation.
_____	T F	9.	Inhalation treatment should be used on victim's to make breathing failures less likely.
_____	T F	10.	Victim's from electric shock should be kept warm using a blanket or some loose clothing to keep him or her warm to prevent a sudden chill or pneumonia.

3

Dress Standards

Employees working in locations where electrical hazards exist shall be provided with and use electrical protective equipment that is suitable for parts of the body to be protected and for the specific work to be performed. All protective equipment shall be maintained in a safe reliable condition and periodic checks shall be made to ensure such equipment remains in a safe and usable condition. Note that applying the requirements of **NFPA 70E** to specific job functions, use the latest edition published by NFPA.

HEAD PROTECTION

Nonconductive head protection shall be worn if there is danger of head injury from electric shock, burns or flying or falling objects resulting from electrical explosions. OSHA requires all employees to wear hard hats or other suitable means to protect the head which is always vulnerable to injury. Helmets used for the protection of employees against impact and penetration of falling or flying objects shall comply with **ANSI Z89.1, Safety Requirements for Industrial Protective Helmets, 1986. (See Figure 3-1)**

NFPA 70E - 130.7(C)(1)
OSHA 1910.135

Figure 3-1. Nonconductive head protection shall be worn if there is danger of head injury from electric shock, burns or flying or falling objects resulting from electrical explosion.

HARD HATS

Hard hats shall be approved for the purpose of electrical work. Metal hard hats are not acceptable. Hard hats are required in the following cases:

- Work is being performed above the head,

- Elevated work is being performed,

- Underground work or work in confined spaces is being performed,

- Work is being performed in electrical substations or switchgear,

- Work is being performed on or near energized electrical distribution equipment and

- In all construction areas. **(See Figure 3-2)**

EAR PROTECTION

The appropriate type of ear protection shall be worn when working in areas where such protection is needed.

EYES AND FACE PROTECTION

All employees shall wear protective equipment for the eyes or face anytime there is danger of injury due to electrical arcs or flashes, or from flying or falling objects resulting from electrical explosion. Eye and face protection equipment shall comply with **ANSI Standard Z87.1, Practice for Occupational and Educational Eye and Face Protection, 1989**. (See Figure 3-3)

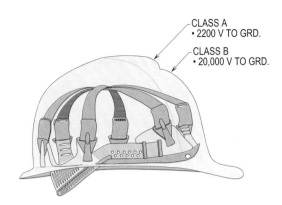

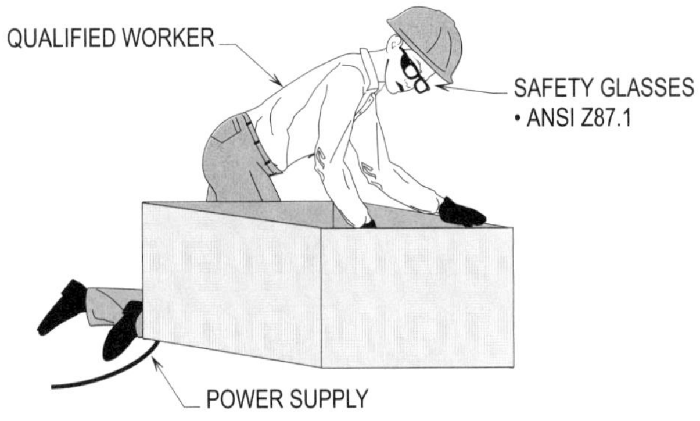

HARD HAT AND SUSPENSION SYSTEM
ANSI Z89.1

Figure 3-2. Hard hats shall be approved for the purpose of electrical work being performed.

NFPA 70E - 130.7(C)(2)
OSHA 1910.133

Figure 3-3. All employees shall wear protective equipment for the eyes or face anytime there is danger of injury due to electrical arcs or flashes, or flying or falling objects resulting from electrical explosion.

FACTOR IN PROTECTION

To protect the eyes from potentially flying or falling objects, the following types of personal protective equipment shall be used:

- Face shields,
- Windows and
- Safety goggles.

Where there is potential exposure to arc flash conditions, they shall protect the eyes from the resulting thermal and luminous energy.

PROTECTIVE GOGGLES AND GLASSES

There are two main types of protective glasses and goggles that are available in a variety of styles to protect against a wide range of hazards. Spectacle type safety glasses are designed for protection from the front only with some models having side shields for the protection of objects that fly or bounce. Eye-cup goggle safety glasses are designed to provide protection from all directions. **(See Figure 3-4)**

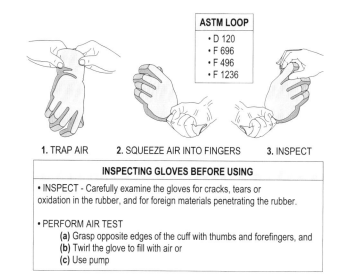

Figure 3-5. Employees shall test gloves before they are used.

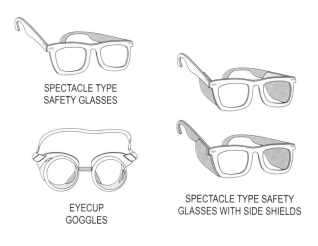

Figure 3-4. Types of safety glasses and goggles used to protect the employee against a wide range of hazards.

HAND AND ARM PROTECTION

Rubber gloves shall be tested each time they are used and shall be worn under leather gloves to protect them against mechanical damage, oil and grease.

Employees shall wear outer clothing with full length sleeves, with the sleeves rolled down for better protection. The main function of rubber sleeves is to protect employees from electrical hazards when they are working on or in proximity to live circuits and parts in or near electrical equipment. **(See Figure 3-5)**

FOOT AND LEG PROTECTION

Dielectric overshoes shall be required where insulated footwear must be depended on as the primary personal protection against step and touch potentials. Appropriate foot and leg protection shall be worn. Rubber insulated mats shall not be permitted to be used as an alternate to insulated shoes. **(See Figure 3-6)**

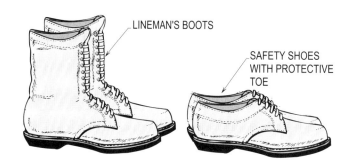

Figure 3-6. Appropriate foot and leg protection shall be worn to protect personnel from electrical hazards.

OVERALL PROTECTION

To perform a job safely, the proper protective clothing and tools shall be used in a safe manner. Safety should always be practiced by electricians when working with electrical circuits.

Electricians should be familiar with the standards pertaining to the type of protective equipment that they shall wear and how such equipment shall be cared for. **(See Figures 3-7(a) and (b))**

STANDARDS ON PROTECTIVE EQUIPMENT

Subject	Number and Title
Head Protection	ANSI Z89.2 - 1971, Safety Requirements for Industrial Protective Helmets for Electrical Workers, Class B
Eye and Face Protection	ANSI Z87.1 - 1998, Practice for Occupational and Educational Eye and Face Protection
Blankets	ANSI/ASTM D1048 - 1999, Standard Specification for Rubber Insulating Blankets
Hoods	ANSI/ASTM D1049 - 1983, Specifications for Rubber Insulating Covers
Line Hoses	ANSI/ASTM D1050 - 1990, Standard Specification for Rubber Insulating Line Hoses
Sleeves	ANSI/ASTM D1051 - 2002, Standard Specification for Rubber Insulating Sleeves
Gloves	ANSI/ASTM D120 - 2002, Standard Specification for Rubber Insulating Gloves
Mats	ANSI/ASTM D178 - 1988, Specifications for Rubber Insulating Matting

(a)
SPECIFICATIONS FOR DETERMINING IF PROTECTIVE EQUIPMENT IS SAFE TO USE FOR A PARTICULAR JOB FUNCTION.

STANDARDS ON PROTECTIVE EQUIPMENT

Subject	Number and Title
Head Protection	ANSI Z89.1 - 1997, Requirements for Protective Headwear for Industrial Workers
Eye and Face Protection	ANSI Z87.1 - 1998, Practice for Occupational and Educational Eye and Face Protection
Rubber Insulating Gloves and Sleeves	ASTM F 496-02, Standard Specfication for In-Service Care of Insulating Gloves and Sleeves
Line Hose and Covers	ASTM F 478-99, Standard Specification for In-Service Care of Insulating Line Hose and Covers
Blankets	ASTM F 479-95, Standard Specification for In-Service Care of Insulating Blankets
Insulating Matting	ASTM D 178-88, Specifications for Rubber Insulating Matting
Leather Protectors Gloves and Mittens	ASTM F 696-02, Standard Specification for Leather Protectors for Rubber Insulating Gloves and Mittens

(b)
SPECIFICATIONS TO BE UTILIZED TO PROVIDE IN-HOUSE SERVICE CARE OF PROTECTIVE EQUIPMENT USED BY PERSONNEL.

Figures 3-7(a) and (b). If a job is to be performed safely, the proper protective clothing and tools shall be used in a safe manner and cared for, so they will be safe to use again.

Name Date

Chapter 3. Dress Standards

Section	Answer			
_____	T	F	1.	Nonconductive head protection shall be worn if there is danger of head injury from electric shock, burns or flying or falling objects resulting from electrical explosion.
_____	T	F	2.	Metal hard hats are acceptable when performing electrical work.
_____	T	F	3.	The appropriate type of ear protection shall be worn when working in areas where such protection is needed.
_____	T	F	4.	All employees shall wear protective equipment for the eyes or face anytime there is danger of injury due to electrical arcs or flashes.
_____	T	F	5.	Face shields shall not be required to be worn to protect the eyes from potentially flying or falling objects.
_____	T	F	6.	Spectacle type safety glasses shall be designed to provide protection from all directions.
_____	T	F	7.	Eye-cup goggles shall be designed to provide protection only from the front.
_____	T	F	8.	Rubber gloves shall be tested each time they are used.
_____	T	F	9.	Rubber insulated mats shall be permitted to be used as an alternate to insulated shoes.
_____	T	F	10.	To perform a job safely, the proper protective clothing and tools shall be worn and used in a safe manner.

4

Electrical Systems

The distribution of electrical power for electrical systems to a premises consists of three-phase transmission lines with a typical voltage of 2400 or 13,800 volts across each phase.

For example, the voltage from phase-to-phase is 4160 volts for a three-phase wye configuration transformed at the substation output. This distribution power of 2400/4160 volts is stepped down by transformers to voltages suitable for each premises served. The electrical wiring system loads consist of three parts known as: the service, the feeders (subfeeders) and the branch-circuits. The computed load of the service conductors is found from the total volt-amperage of all the branch-circuits and feeders. Contained in this Chapter are the types of electrical systems and voltages used to supply premises wiring.

Electric power is produced and distributed to the premises by the electric public utility companies almost exclusively as alternating current. Private generating systems in industries which require close control of motor speeds continue the use of direct current even thought it has largely been discontinued by supplying utilities. In recent years, the use of three-phase (alternating current) services has increased rapidly. Single-phase service is used mainly for power systems supplying facilities requiring smaller loads.

SYSTEMS OF 600 VOLTS OR LESS

Systems of 600 volts or less are usually called utilization systems. Such systems are used within a facility to supply power to lighting, general purpose outlets and utilization equipment.

SINGLE-PHASE SYSTEMS

Single-phase systems are available in two and three-wire services. Dwelling units, small business establishments and stores use three-wire services for most single-phase applicants.

120 VOLT, SINGLE-PHASE, TWO-WIRE SYSTEMS

Two-wire circuits consist of 120 volts between the ungrounded (phase) conductor and a grounded (neutral) conductor. 120 volt, two-wire, single-phase systems are used to supply equipment such as signboards and television boosters and other related loads. **(See Figure 4-1)**

208/120 V, SINGLE-PHASE, THREE-WIRE SYSTEMS

Electrical supply systems of 208/120 volts are usually used for individual dwelling units in apartment complexes which are supplied by 208/120 volt, four-wire, three-phase systems. At a tap can, individual 208/120 volt, three-wire, single-phase systems are routed to the service equipment in each dwelling unit. This same arrangement of utility and service supply may be used for small commercial buildings. **(See Figure 4-2)**

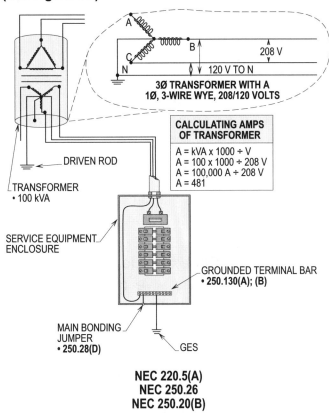

Figure 4-2. 208/120 volt, single-phase, three-wire systems are usually used for individual dwellings units in apartment complexes or small commercial buildings which are supplied by 208/120 volt, four-wire, three-phase systems.

240/120 V, SINGLE-PHASE, THREE-WIRE SYSTEMS

Electrical supply systems of 240/120 volt, single-phase, three-wire are also used in residential and small commercial occupancies to provide 120 volts for lighting and small appliance loads, and 240 volts, single-phase for heavy appliances and small motor loads. The third wire gives additional capacity over a 120 volt, two-wire system which has been largely superseded in new installations. Systems of 240/120 volts will have two 120 volt conductors to the grounded (neutral) conductor and 240 volts between the two ungrounded (phase) conductors. **(See Figure 4-3)**

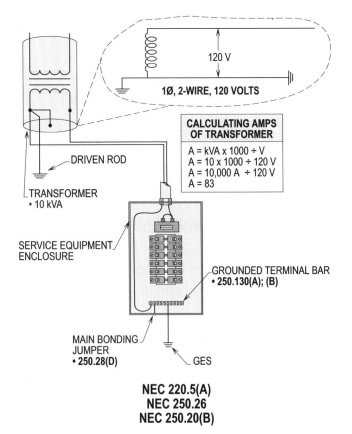

Figure 4-1. 120 volt, single-phase, two-wire systems consist of 120 volts between the ungrounded (phase) conductors and a grounded (neutral) conductor.

Electrical Systems

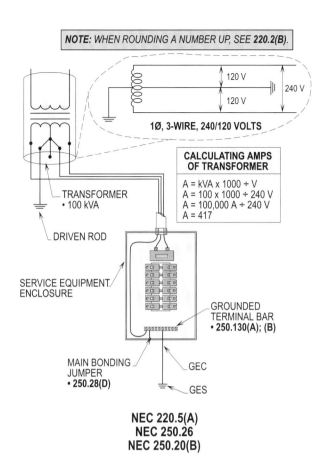

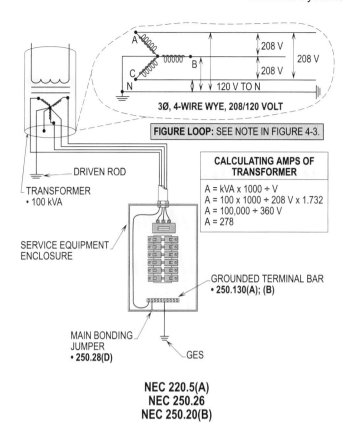

Figure 4-3. 240/120 volt, single-phase, three-wire systems consist of two 120 volt conductors to the grounded (neutral) conductor and 240 volts between the two ungrounded (phase) conductors.

Figure 4-4. 208/120 volt, three-phase, four-wire systems consist of three 120 volt ungrounded (phase) conductors to the grounded (neutral) conductor and 208 volts between any two of the ungrounded (phase) conductors.

208/120 V, THREE-PHASE, FOUR WIRE SYSTEMS

Electrical supply systems of 208/120 volt, three-phase, four-wire are generally used for small industrial plants, office buildings, stores and schools. They provide 120 volts, single-phase, for lighting and appliances, and 208 volts, three-phase, for various types of power loads. These systems are known in the electrical industrial as wye-connected systems. Systems of 208/120 volt will have three 120 volt ungrounded (phase) conductors to the grounded (neutral) conductor and 208 volts between any two of the ungrounded (phase) conductors. **(See Figure 4-4)**

Design Tip: The fourth conductor of this system provides almost 50 percent greater capacity than a single-phase, three-wire electrical system.

240/120 V, THREE-PHASE, FOUR-WIRE SYSTEMS

Electrical supply systems of 240/120 volt, three-phase, four-wire are available using three transformers (closed delta) or two transformers (open delta). Closed delta connected transformers offer a greater range of capacity than open delta connected units and provide additional advantages in servicing 120/240 volt single-phase and 240 volt three-phase loads.

CLOSED DELTA SYSTEMS

Systems of 240/120 volt, three-phase, four-wire delta provides single-phase or three-phase, 240 volt supply for power between ungrounded (phase) conductors, and 120 volt, single-phase between two of the ungrounded (phases) and grounded (neutral) conductors. Voltage between the middle phase and the grounded (neutral) is 208 volts (high Leg). The total load is not equally balanced over the three phases. Such systems are usually used where motor loads are large compared to lighting, receptacle and appliance loads. **(See Figure 4-5)**

4-3

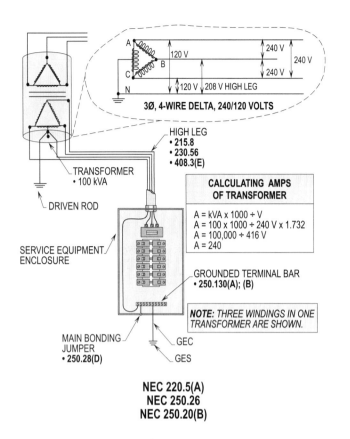

Figure 4-5. Systems of 240/120 volt, three-phase, four-wire delta provides single-phase or three-phase, 240 volt supply for power between ungrounded (phase) conductors, and 120 volt, single-phase between the ungrounded (phases) and grounded (neutral) conductors.

> **Design Tip:** Where there is three individual transformers and there is a failure of one of the single-phase supply transformers, the system may operate open-delta at approximately 58 percent capacity.

OPEN DELTA SYSTEMS

The 240/120 volt, three-phase, four-wire, open-delta system is the same type of system as a closed-delta system.

The main difference is that only two single-phase transformers are used instead of three, and the total capacity is approximately 58 percent of a three-transformer unit system.

Open-delta systems are usually used where the initial load may be expected to be increased some date after the original installation has been completed. **(See Figure 4-6)**

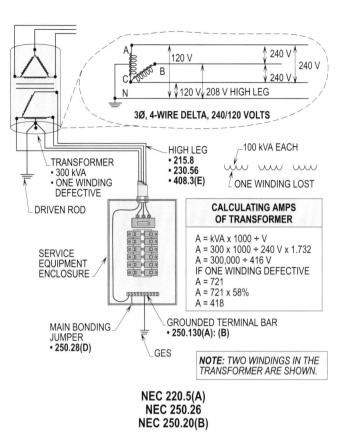

Figure 4-6. Systems that are 240/120 volt, three-phase, four-wire, open-delta are the same type of system as a closed-delta system. The main difference is that only two single-phase transformers are used instead of three. The total capacity is approximately 58 percent of a three-transformer unit system.

> **Design Tip:** The system capacity can be nearly doubled just by adding one single-phase transformer to the bank of two single-phase transformers. However, there must be three individual transformers for this rule to apply.

480/277 V, THREE-PHASE, FOUR-WIRE SYSTEMS

Electrical supply systems of 480/277 volt, three-phase, four-wire are called wye connected systems. Such systems are used for circuits in large industrial plants and large commercial facilities. It provides 480 volts, three-phase, for motor and machinery loads and 277 volts, single-phase for fluorescent lighting systems and other 277 volt related loads. Equipment and conductor sizing is one big advantage that this system provides. **(See Figure 4-7)**

Electrical Systems

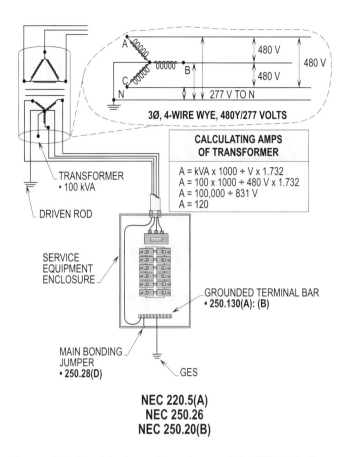

Figure 4-7. Electrical supply systems of 480/277 volt, three-phase, four-wire are called wye connected systems.

For example, electric power can be delivered at 480/277 volts with less than half the conductor size (cu. or alu.) that is needed at 208/120 volts.

SYSTEMS OF 600 VOLTS

Voltages of over 600 volts are usually referred to as distribution systems. They are used in large industrial facilities to supply large motors and other related type equipment and apparatus. Depending on the voltage, larger pieces of equipment may be supplied by the higher voltage, while lighting, general purpose outlets and appliances may be supplied by lower voltage systems that are stepped down by transformers to the desired load.

SYSTEMS OF 2.4 TO 15 kV

Electrical supplying systems of 2.4 to 15 kV are usually used for primary power distribution in industrial plants. All plants use a primary voltage of this rating except some of the very large chemical plants, steel mills, etc.

When the utility voltage is below 15,000 volts, selecting the primary voltage is no problem because the *National Electrical Code* and *National Electrical Safety Code*, allow 15,000 volts in buildings. The higher voltage 2400 to 15,000 volts may be routed in cables and conduits or in interlocked-armored cables to the load-center substations and there transformed down to utilization voltage.

SYSTEMS OF 4160 VOLTS

A problem that often arises is why the selection of 4160 volts over a lower voltage. There are two major reasons. The first is that 4160 volt systems have lower cost, and second, provides greater allowances for expansion and growth. In the concernment of cost, 4160 volt switchgear for a given interrupting rating costs less than lower voltage switchgear such as 2400 volt systems.

Systems of 4160 volts are three-phase and are equipped with three phase conductors. **(See Figure 4-8)**

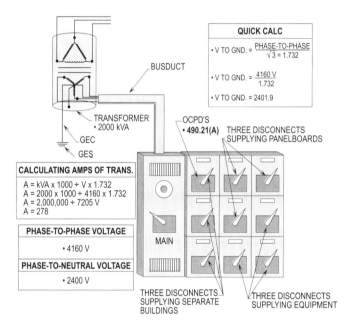

Figure 4-8. Systems of 4160 volts are three-phase and are equipped with three phase conductors.

Design Tip: Generally, more kVA per circuit may be carried at 4160 volts than at 2400 volts, which results in fewer circuits. Therefore, reducing the cost of the switchgear. Conductor costs are usually less at 4160 volts than at 2400 volts because less copper is required due to the smaller load amps supplied.

SYSTEMS OF 13,800 VOLTS

In large industrial plants, much of the power is usually used by large motors. Therefore, the motor cost as well as the system cost must be considered to really get a true overall system cost when choosing the primary voltage. There are certain levels of voltage which are suitable for large industrial plant motors. Such voltage levels are 69,000 volts through 12,470 volts respectively. The voltage level selected depends upon the size and ratings of the plant's service equipment, motors and the possibility of future expansion. The electrical supply of these voltages are usually three-phase, four-wire systems.

The larger the electrical system becomes, the more factors there are that cause the cost of a 69,000 volt system to increase faster than that for a 13,800 system. This consideration must be evaluated before such system voltage is selected.

Designers usually select 12,470, 13,200 or 13,800 volt systems and route this voltage to transformers that step the voltage down to desired levels. **(See Figure 4-9)**

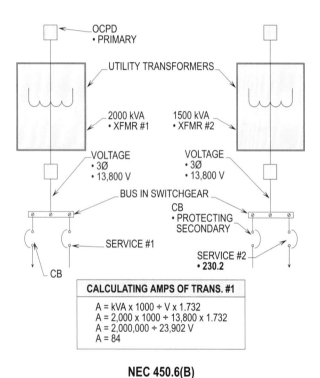

NEC 450.6(B)

Figure 4-9. Designers usually select 12,470, 13,200 or 13,800 volt systems and route this voltage to transformers that step the voltage down to desired levels.

SYSTEMS OF LESS THAN 50 VOLTS

For a designer or installer to fully understand the requirements pertaining to designing and installing low-voltage electrical systems, he or she must review different Articles and Sections in the NEC. **Article 720** covers Circuits and Equipment operating at less than 50 volts. **Article 725** deals with Class 1, Class 2 and Class 3 circuits which may fall under the rules of low-voltage systems of less than 50 volts. **Article 411** contains requirements for low-voltage lighting systems of 30 volts or less which regulates such systems whether installed indoors or outdoors.

SYSTEMS OF LESS THAN 50 VOLTS
ARTICLE 720

The rules and regulations of low-voltage electrical systems rated at less than 50 volts are stricter than other low-volt systems. Conductors, lampholders, receptacles and protection devices shall be a certain size and rating.

SYSTEMS INSTALLED IN HAZARDOUS LOCATIONS
720.3

Low-voltage systems of less than 50 volts shall comply with all the requirements of **Articles 500 through 517** if they are installed in such locations. **Section 500.1** makes it very clear that all voltage levels utilized in hazardous (classified) locations shall be considered dangerous and shall comply with certain installation rules to ensure that such systems are not capable of igniting an explosive mixture of gases and air.

> **Design Tip:** It must be understood that low-voltage systems alone do not render a circuit incapable of igniting flammable atmospheres. Under some conditions, an ordinary flashlight using two 1 1/2 volt "D" cells can be a source of ignition in hazardous (classified) locations. Even nonincendive circuits can produce a dangerous ignitable source of energy if components are damaged.

CONDUCTORS
720.4

Conductors used for low-voltage systems of less than 50 volts shall not be smaller than 12 AWG copper or equivalent. Conductors for appliance branch-circuits supplying more than one appliance or appliance receptacle shall not be smaller than 10 AWG copper or equivalent. Due to the energy output of such systems, conductors shall be large enough to carry normal currents and fault-grounds or short-circuit currents if any should occur. **(See Figure 4-10)**

LAMPHOLDERS
720.5

Standard lampholders used in systems of less than 50 volts shall have a rating of not less than 660 watts. Note that a similar requirement appears in **210.21(A)** for heavy duty lampholders. In both cases, there are stricter rules for such lampholders based upon voltage and ratings. **(See Figure 4-11)**

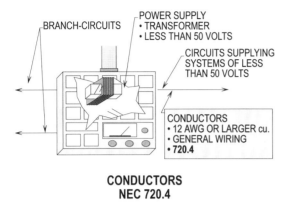

Figure 4-10. Conductors used for low-voltage systems of less than 50 volts shall not be smaller than 12 AWG copper or equivalent.

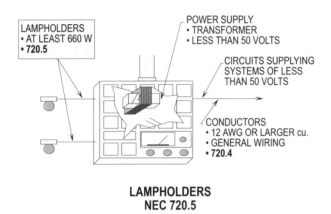

Figure 4-11. Standard lampholders used in systems of less than 50 volts shall have a rating of not less than 660 watts.

RECEPTACLES
720.6 AND 720.7

Receptacles shall have a rating of not less than 15 amperes when used for general-purpose. Receptacles of not less than a 20 ampere rating shall be utilized in kitchens, laundries and other locations where portable appliances are likely to be cord-and-plug connected. **(See Figure 4-12)**

> **Design Tip:** Receptacles shall be sized with capacity ratings which are capable of carrying normal currents and abnormal currents due to short-circuit conditions.

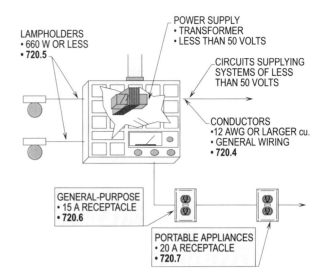

Figure 4-12. Receptacles shall have a rating of not less than 15 amperes when used for general-purpose.

OVERCURRENT PROTECTION
720.8

Overcurrent protection devices (OCPD's) protecting conductors shall comply with **240.4**. OCPD's sized and selected to protect equipment and the elements of equipment shall meet the provisions of **240.3**. OCPD's and conductors shall be computed and sized per **210.19(A)(1)** and **210.20(A) through (D)** in the NEC. **(See Figure 4-13)**

BATTERIES
720.9

When storage batteries are used as the power source for systems operating at less than 50 volts, they shall be designed and installed in accordance with **Article 480** in the NEC.

For example, ventilation systems shall meet all the rules of **480.9(A)** and **NFPA 30** to determine if the location is hazardous or nonhazardous, **480.9** must be carefully reviewed and adhered to. **(See Figure 4-14)**

GROUNDING
720.10

The grounding of low-voltage systems rated less than 50 volts shall comply with the grounding rules of **250.20(A)** and **250.114**. **Section 250.114** covers the requirements for cord-and-plug connected equipment. Review these rules very carefully before cord-and-plug connections of equipment rated less than 50 volts are used. **(See Figure 4-15)**

> **Design Tip:** Indoor AC circuits of less than 50 volts shall be grounded if the circuit is taken from a transformer having an ungrounded primary circuit or a primary circuit of over 150 volts-to-ground. Otherwise, they shall be permitted to be operated ungrounded.

For example, a low-voltage bell-ringing circuit in a building with a 240/120 volt service supply does not have to be grounded, since the circuit is taken from a transformer with a grounded primary circuit of only 120 volts-to-ground. However, low-voltage AC circuits installed outdoors and run overhead shall be grounded under any conditions of use.

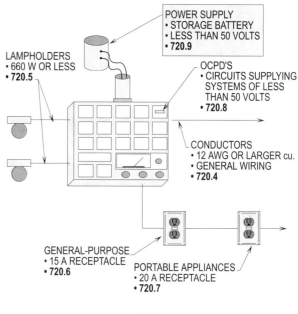

Figure 4-14. When storage batteries are used as the power source for systems operating at less than 50 volts, they shall be designed and installed in accordance with **Article 480**.

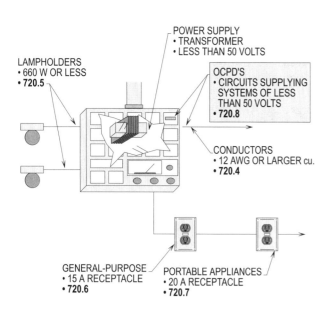

Figure 4-13. Overcurrent protection devices (OCPD's) protecting conductors shall comply with **240.4**.

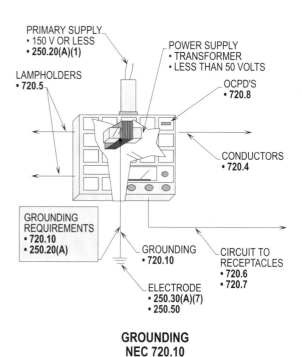

Figure 4-15. The grounding of low-voltage systems rated less than 50 volts shall comply with the grounding rules of **250.20(A)** and **250.114**.

CLASS 1, 2, AND 3 CIRCUITS
725.2

Class 1, 2 and 3 circuits per the NEC are classified as Remote-Control, Signaling and Power-Limited Circuits. The NEC defines such circuits as that portion of the wiring system between the load side of the OCPD or the power-limited supply and all connected equipment. Class 1, Class 2 or Class 3 remote control, signaling or power-limited circuits are characterized by their usage and electrical power limitation which differentiates them from light and power circuits. These circuits are classified in accordance with their respective voltage and power limitations.

CLASS 1 CIRCUIT CLASSIFICATIONS AND POWER SOURCE REQUIREMENTS
725.21

Class 1 circuits are divided into two types, power-limited and remote-control and signaling circuits. Power-limited Class 1 circuits are limited to 30 volts and 1000 volt-amperes. Class 1 remote-control and signaling circuits are limited to 600 volts, but there aren't any limitations on the power output of the source. Note that the rules pertaining to circuits of less than 50 volts only are reviewed in this chapter.

Class 1 power-limited circuits are supplied from a power source that has a rated output of not more than 30 volts and a power limitation of 1000 volt-amps. Class 1 power-limited circuits have a current limiter on the power source that supplies them. This limiter is an overcurrent protection device that restricts the amount of supply current to the circuit in the event of an overload, short-circuit or ground-fault. These Class 1 circuits shall be permitted to be supplied from a transformer or other type of power supply such as generators or batteries.

Class 1 remote control or signaling circuits shall be permitted to operate at up to 600 volts and have no limitation on the power rating of the source. Class 1 systems, generally, shall meet most wiring requirements for power and light circuits. Class 1 remote control circuits are commonly used in motor controllers that operate mechanical processes, elevators, conveyors and equipment that is controlled from one or more remote locations. Class 1 signaling circuits are used in nurses' call systems in hospitals, electric clocks, bank alarm systems and factory call systems. **(See Figure 4-16)**

Power limitations of power sources used for Class 1 circuits can be reviewed in **Tables 11(A) and Table 11(B) in Chapter 9** of Tables.

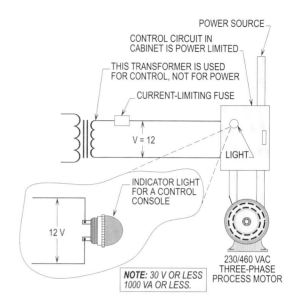

CLASS 1 CIRCUIT CLASSIFICATIONS AND POWER SOURCE REQUIREMENTS NEC 725.21(A); (B)

Figure 4-16. Class 1 remote control or signaling circuits shall be permitted to operate up to 600 volts with or without power limitations.

CONDUCTORS OF DIFFERENT CIRCUITS IN THE SAME CABLE, CABLE TRAY, ENCLOSURE OR RACEWAY
725.26(A); (B); 300.3(C)(1)

Class 1 circuits shall be permitted to occupy the same cable, enclosure or raceway without regard to whether the individual circuits are AC or DC current, provided all conductors are insulated for the maximum voltage of any conductor in the cable, enclosure or raceway. Class 1 circuits and power supply circuits shall be permitted to occupy the same cable, enclosure or raceway only in situations where the equipment power system is functionally associated. **(See Figure 4-17)**

Section 725.26(B)(2) clarifies that they shall be permitted to be mixed where installed in factory- or field-assembled control centers. **Section 725.26(B)(3)** allows mixing for underground conductors in a manhole if all of the following conditions are complied with:

- The power-supply or Class 1 circuit conductors are in a metal-enclosed cable or Type UF cable,

- The conductors are permanently separated from the power-supply conductors by a continuous firmly fixed nonconductor, such as flexible tubing, in addition to the installation on the wire and

- The conductors are permanently and effectively separated from the power supply conductors and securely fastened to racks, insulators or other approved supporting means.

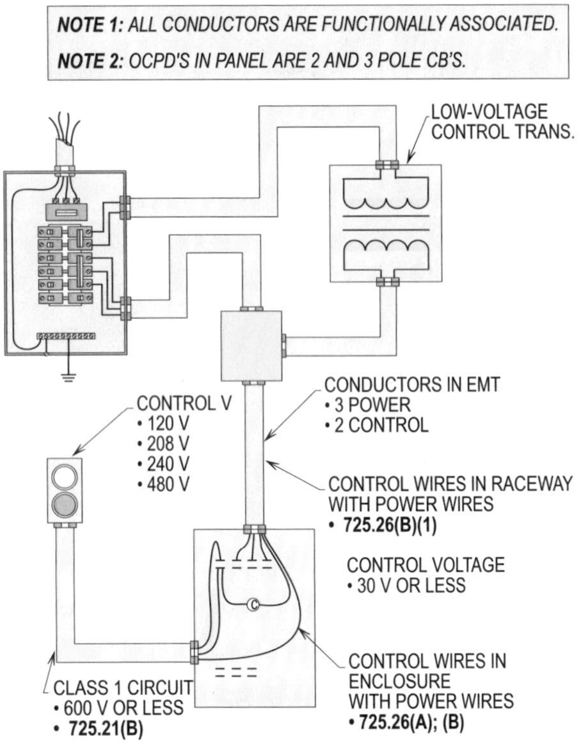

CONDUCTORS OF DIFFERENT CIRCUITS IN THE SAME CABLE, CABLE TRAY, ENCLOSURE OR RACEWAY NEC 725.26(A); (B)

Figure 4-17. Class 1 circuits shall be permitted to occupy the same cable, enclosure or raceway without regard to whether the individual circuits are AC or DC current, provided all conductors are insulated for the maximum voltage of any conductor in the cable, enclosure or raceway.

NUMBER OF CONDUCTORS IN CABLE TRAYS AND RACEWAY AND DERATING
725.28(A); (B); (C)

Where only Class 1 circuit conductors are in a raceway, the number of conductors shall be permitted to be determined by the provisions of **300.17**. The derating factors given in **Article 310, Table 310.15(B)(2)(a)** to Ampacity Tables of 0 to 2000 volts, apply only if such conductors carry continuous loads in excess of 10 percent of the ampacity of each control conductor routed through the raceway system.

The number of power-supply conductors and Class 1 circuit conductors pulled through a raceway based upon the rules of **725.26** shall be determined per **300.17** in the NEC. The derating factors given in **Article 310, Table 310.15(B)(2)(a)** to Ampacity Tables of 0 to 2000 Volts apply to the following conditions:

- To all conductors where the Class 1 circuit conductors carry continuous loads in excess of 10 percent of the ampacity of each conductor and where the total number of conductors is four or more.
- To the power-supply conductors only, where the Class 1 circuit conductors do not carry continuous loads in excess of 10 percent of the ampacity of each conductor and where the number of power-supply conductors is four or more. **(See Figure 4-18)**

Design Tip: Class 1 circuit conductors installed in cable tray systems shall comply with the rules and regulations of **392.9 through 392.11 of Article 392** in the NEC.

CLASS 1 CIRCUIT CONDUCTORS
725.27(A); (B)

Conductors of 18 AWG and 16 AWG shall be permitted to be used, if they supply loads that do not exceed the ampacities given in **402.5** and are installed in a raceway, an approved enclosure or a listed cable. Conductors larger than 16 AWG shall not be permitted to supply loads greater than the ampacities given in **310.15**. Flexible cords shall comply with the design and installation requirements of **Article 400**.

Insulation on conductors shall be suitable for 600 volts. Conductors larger than 16 AWG shall comply with the requirements of **Article 310**.

Conductors in sizes 18 AWG and 16 AWG shall be Type FFH-2, KF-2, KFF-2, PAF, PAFF, PF, PFF, PGF, PGFF, PTF, PTFF, RFH-2, RFHH-2, RFHH-3, SF-2, SFF-2, TF, TFF, TFFN, TFN, ZF or ZFF. However, conductors with other types and thicknesses of insulation shall be permitted to be used if listed for Class 1 circuit use.

POWER SOURCES FOR CLASS 2 AND 3 CIRCUITS
725.41(A); (B)

Class 2 and Class 3 circuits are defined by two Tables, one for AC current and one for DC current. In general, a Class 2 circuit operating at 24 volts with a power supply durably marked "Class 2" and not exceeding 100 volt-amperes is the type most commonly used.

A Class 2 circuit is defined as that portion of the wiring system between the load side of a Class 2 power source

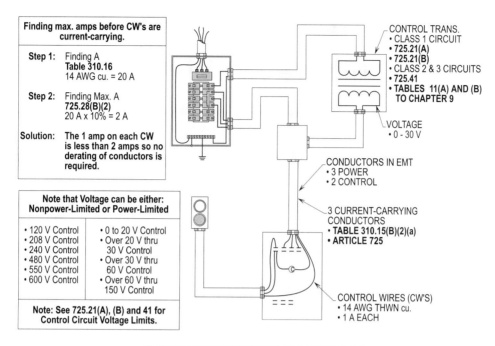

Figure 4-18. The derating factors given in **Table 310.15(B)(2)(a)** to Ampacity Tables of 0 to 2000 volts, apply only if such control conductors carry continuous loads in excess of 10 percent of the ampacity of each conductor routed through the raceway system. For determining if grounded (neutral) conductors are current-carrying, **See Figure 8-8**.

and the connected equipment. Due to its power limitations, a Class 2 circuit is considered safe from a fire initiation standpoint and provides acceptable protection from electric shock.

A Class 3 circuit is defined as that portion of the wiring system between the load side of a Class 3 power source and the connected equipment. Due to its power limitations, a Class 3 circuit is not considered safe from a fire initiation standpoint. Since higher levels of voltage and current than for Class 2 circuits are permitted, additional safeguards are specified to provide protection from an electric shock hazard that might be encountered.

Power for Class 2 and Class 3 circuits is limited either inherently (in which no overcurrent protection is required) or by a combination of a power source and overcurrent protection.

The maximum circuit voltage is 150 volts AC or DC for a Class 2 inherently limited power source and 100 volts AC or DC for a Class 3 inherently limited power source. The maximum circuit voltage is 30 volts AC and 60 volts DC for a Class 2 power source limited by overcurrent protection, and 150 volts AC or DC for a Class 3 power source limited by overcurrent protection. **(See Figure 4-19)**

For example, heating system thermostats are commonly Class 2 systems and the majority of small bell, buzzer and annunciator systems are Class 2 circuits. Class 2 also includes small intercommunicating telephone systems in which the voice circuit is supplied by a battery and the ringing circuit by a transformer.

Class 2 and 3 systems do not require the same wiring methods as power, light and Class 1 systems. There are cases when a 2 in. (50 mm) separation is required between these systems.

See Figure 725.41 in the NEC for a comparison of Class 2 and 3 remote control, signaling and power-limited circuits.

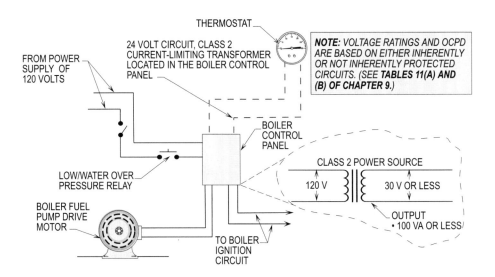

POWER SOURCES FOR CLASS 2 AND 3 CIRCUITS
NEC 725.41(A); (B)

Figure 4-19. The maximum voltage is usually 100 volts AC or DC for a Class 2 inherently limited power source, and 150 volts AC or DC for a Class 3 inherently limited power source.

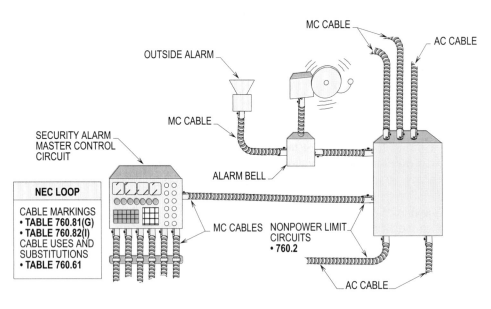

NONPOWER-LIMITED FIRE ALARM CIRCUITS
NEC 760.2

Figure 4-20. Nonpower-limited fire protective signaling circuits may include circuits that are part of a central station signaling system, a sprinkler water flow alarm or a local fire alarm in a building.

FIRE PROTECTIVE SIGNALING CIRCUITS
760.2

A fire protective signaling circuit is defined as that portion of the wiring system between the load side of the overcurrent device or the power-limited supply and the connected equipment of all circuits powered and controlled by the fire alarm system. Fire alarm circuits are classified as either nonpower-limited or power-limited.

NONPOWER-LIMITED FIRE ALARM CIRCUIT
760.2

Nonpower-limited fire protective signaling circuits may include circuits that are part of a central station signaling system, a sprinkler water flow alarm or a local fire alarm in a building. Voltages for these circuits range up to 600 volts. These circuits shall be permitted to be located in the same enclosure, cable or raceway as Class 1 circuits if the insulation on all of the wires within that enclosure are rated for the highest voltage of any conductor therein.

Power supply conductors are not usually permitted in the same enclosure, cable or raceway as fire protective conductors because a fault or overcurrent condition in the power supply conductor could damage the fire protective circuits. This would cause the fire protective signal circuit to malfunction and perhaps not transmit a needed alarm or fire signal. However, power supply conductors and fire protective signaling circuits shall be permitted to occupy the same enclosure if they are connected to the same equipment. **(See Figure 4-20)**

POWER-LIMITED FIRE ALARM CIRCUIT
760.2

Since power-limited conductors are usually light gauge wire with low-voltage rating and operate at lower voltages and power ratings than Class 1 circuits, power circuits and nonpower-limited circuits, special measures shall be taken to keep these conductors physically separate. Generally, the power-limited circuit conductors shall be separated from these other circuits by at least 2 in. (50 mm). However, the different circuits shall be permitted to be closer:

- If the light, power, Class 1 or nonpower-limited circuit is in a raceway or in a sheathed, metal-clad, or Type UF cable or

- If the power-limited circuit conductors are separated from the other circuits by a nonconductor, such as porcelain tubes or flexible tubing, in addition to the conductor insulation.

Because of the differing operating voltages and insulation levels of power-limited signaling circuits, the conductors shall only be permitted to be located where not subject to damage or interference from other types of circuits. Class 3 circuits and power-limited fire protective signal circuits operate at similar voltages and power levels. Therefore, the conductors and cables of two or more power-limited fire protective signaling circuits or Class 3 circuits shall be permitted to occupy the same enclosure. However, power-limited fire protective signaling circuits and Class 2 circuits differ in operating voltages and insulation and shall be permitted in the same enclosure only when the insulation of the Class 2 conductors is comparable to that of the fire protective circuit. **(See Figure 4-21)**

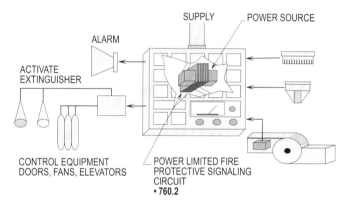

**POWER-LIMITED FIRE ALARM CIRCUIT
NEC 760.2**

Figure 4-21. Because of the differing operating voltage and insulation levels of power-limited signaling circuits, their conductors shall only be permitted to be located where not subject to damage or interference from other types of circuits. (For installation rules, see **760.54**)

FIRE ALARM CIRCUIT IDENTIFICATION
760.10

Identification of fire protective signaling circuits is required so that these systems will not be interfered with during maintenance operations. Because these are essential systems, this requirement is intended to protect signaling circuits while work is being performed on other systems or while the signaling circuit itself is being serviced, avoiding damage to the circuit and creating false alarms. Therefore, fire alarm systems shall be properly identified.

For power-limited fire protective signaling circuits, the marking shall indicate that the circuit is a power-limited fire protective signaling circuit. This rule is intended to ensure that power-limited fire circuits, which operate at lower power and voltage levels, are not confused with other circuits operating at higher voltages.

INTRINSICALLY SAFE SYSTEMS
504.2

An intrinsically safe circuit (ITSC) is defined as an assembly of interconnected intrinsically safe apparatus, associated apparatus and interconnecting cables in that those parts of the system that may be used in hazardous (classified) locations. A circuit in which any spark or thermal effect is incapable of causing ignition of a mixture of flammable or combustible material in air under prescribed test conditions is considered an ITSC. **(See Figure 4-22)**

Associated apparatus is an apparatus in which the circuits are not necessarily intrinsically safe themselves but affect the energy in the ITSC's and are relied upon to maintain intrinsic safety. Associated apparatus may be either:

- Electrical apparatus that has an alternative-type protection for use in the appropriate hazardous (classified) location or

- Electrical apparatus not so protected that shall not be used within a hazardous (classified) location.

SIMPLE APPARATUS

An electrical component or combination of components of simple construction with well defined electrical parameters that does not neither generate more than 1.5 volts, 100 milliamps, 25 milliwatts or a passive component that does not dissipate more than 1.3 watts and is compatible with the intrinsic safety of the circuit in which it is used. Examples of such apparatus are switches, thermocouples, light-emitting diodes (LED's), connectors and resistance temperature devices (RTD's).

EQUIPMENT APPROVAL
504.4

All intrinsically safe apparatus and associated apparatus shall be listed except simple apparatus, as described on control drawings. **(See Figure 4-23)**

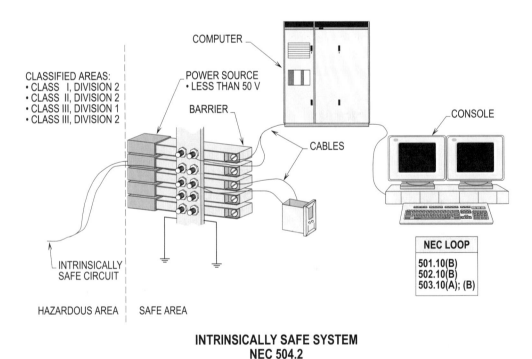

INTRINSICALLY SAFE SYSTEM
NEC 504.2

Figure 4-22. An intrinsically safe circuit is defined as an assembly of interconnected intrinsically safe apparatus, associated apparatus and interconnecting cables in that those parts of the system that may be used in hazardous (classified) locations.

Electrical Systems

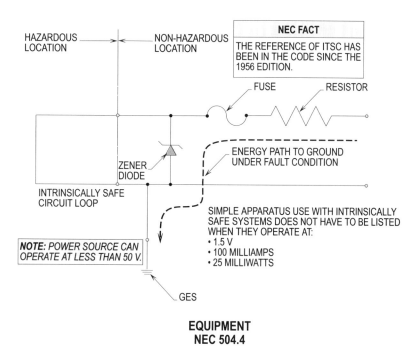

Figure 4-23. All intrinsically safe apparatus and associated apparatus shall be listed except simple apparatus, as described on control drawings.

CONTROL DRAWING(S)
504.10(A); 500.2

Intrinsically safe apparatus, associated apparatus and other equipment shall be installed in accordance with the control drawing(s).

EQUIPMENT INSTALLATION
504.10(B)

Intrinsically safe apparatus, associated apparatus and other equipment shall be permitted to be installed if they are in accordance with the control drawing(s) except a simple apparatus that does not interconnect intrinsically safe circuits.

Intrinsically safe and associated apparatus shall be permitted to be installed in any hazardous (classified) location for which it has been identified. **(See Figure 4-24)**

Design Tip: In most installations, general-purpose enclosures shall be permitted for intrinsically safe apparatus. However, they shall be installed in either Class I, Division 2 or nonhazardous locations.

WIRING METHODS
504.20

Intrinsically safe apparatus and wiring can be installed using any of the wiring methods suitable for unclassified locations, including **Chapters 7** and **8**. Sealing shall comply with the provisions of **504.70**, and separation shall meet the requirements of **504.30**.

OTHER (NOT IN RACEWAY OR CABLE TRAY SYSTEM)
504.30(A)(3)

Conductors and cables of intrinsically safe circuits that are not in raceways or cable trays shall be separated at least 2 in. (50 mm) and secured from conductors and cables of any nonintrinsically safe circuits.

The **Ex. 2 to 504.30(A)(3)** does not require such separation where either (1) all of the intrinsically safe circuit conductors are in Type MI or MC cables or (2) all of the nonintrinsically safe circuit conductors are in raceways or Type MI or MC cables where the sheathing or cladding is capable of carrying fault-current to ground. **(See Figure 4-25)**

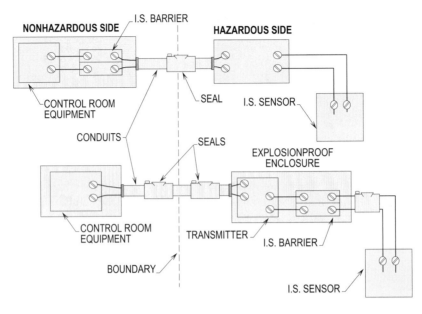

Figure 4-24. Intrinsically safe and associated apparatus shall be permitted in any hazardous (classified) location for which it has been approved.

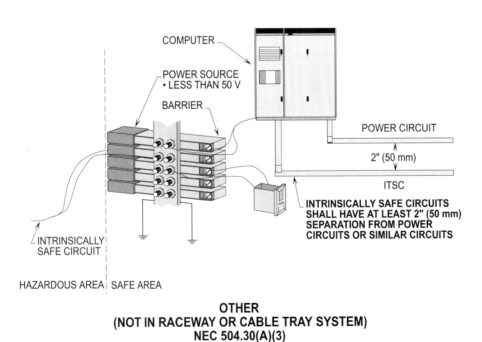

Figure 4-25. Conductors and cables of intrinsically safe circuits that are not in raceways or cable trays shall be separated at least 2 in. (50 mm) and secured from conductors and cables of any nonintrinsically safe circuits.

IN RACEWAYS, CABLE TRAYS AND CABLES
504.30(A)(1)

Conductors of intrinsically safe circuits shall not be permitted to placed in any raceway, cable tray or cable with conductors of any nonintrinsically safe circuit.

Ex. 1 to 504.30(A)(1) recognizes that where conductors of intrinsically safe circuits are separated from conductors of nonintrinsically safe circuits by a distance of at least 2 in. (50 mm) and secured, or by a grounded metal partition or an approved insulating partition.

Ex. 2 to 504.30(A)(1) also recognizes that where either (1) all of the intrinsically safe circuit conductors or (2) all of the nonintrinsically safe circuit conductors are in grounded metal-sheathed or metal-clad cables where the sheathing or cladding is capable of carrying fault-current to ground. **(See Figure 4-26)**

Design Tip: The use of separate wiring compartments for the intrinsically safe and nonintrinsically safe terminals is the preferred method of complying with this requirement. Physical barriers such as grounded metal partitions, approved insulating partitions or approved restricted access wiring ducts separated from other such ducts by at least 3/4 in. (19 mm) shall also be permitted to be used to help ensure the required separation of the wiring.

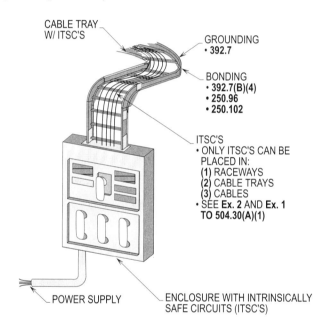

IN RACEWAYS, CABLE TRAYS AND CABLES
NEC 504.30(A)(1)

Figure 4-26. Conductors of intrinsically safe circuits shall not be permitted to be placed in any raceway, cable tray or cable with conductors of nonintrinsically safe circuits.

WITHIN ENCLOSURES
504.30(A)(2)

Conductors and cables of intrinsically safe circuits shall be separated at least 2 in. (50 mm) from conductors of any nonintrinsically safe circuits or as required per **504.30(A)(1)**. Conductors shall be secured so that any conductor that might come loose from a terminal cannot come in contact with another terminal. **(See Figure 4-27)**

FROM DIFFERENT INTRINSICALLY SAFE CIRCUIT CONDUCTORS
504.30(B)(1); (B)(2)

Different intrinsically safe circuits shall be in separate cables or be separated from each other by one of the following means:

- The conductors of each circuit are within a grounded metal shield, or

- The conductors of each circuit have an insulation with a minimum thickness of 0.01 in. (0.25 mm)

- The clearance between two terminals for connection of different intrinsically safe circuits shall be at least 0.25 in. (6 mm), unless this clearance is permitted to be reduced by the control drawing.

IDENTIFICATION OF TERMINALS AND WIRING
504.80(A); (B)

Labels shall be suitable for the environment where they are installed and consideration given where they are exposed to chemicals and sunlight.

Intrinsically safe circuits shall be identified at terminal and junction locations in such a manner that will prevent unintentional interference with the circuits during testing and servicing.

Raceways, cable trays and open wiring for intrinsically safe system wiring shall be identified with permanently affixed labels with the wording "Intrinsic Safety Wiring" or equivalent. The labels shall be so located as to be visible after installation and placed so that they are readily traced through the entire length of the installation. Spacing between labels shall be provided at 25 ft. (7.5 m) intervals. **(See Figure 4-28)**

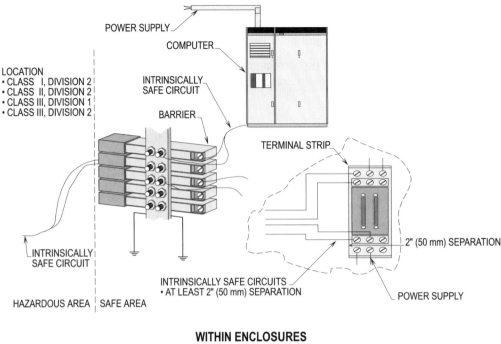

WITHIN ENCLOSURES
NEC 504.30(A)(2)(a); (b)

Figure 4-27. Conductors of intrinsically safe circuits shall be separated at least 2 in. (50 mm) from conductors of any nonintrinsically safe circuits when installed within enclosures.

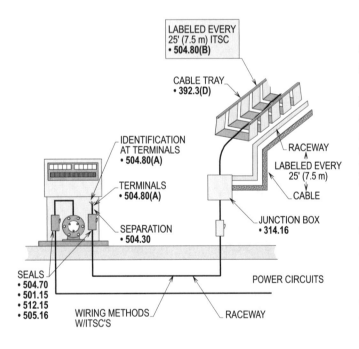

IDENTIFICATION OF TERMINALS AND WIRING
NEC 504.80(A); (B)

Figure 4-28. Spacing between labels on raceways and cables shall be provided at 25 ft. (7.5 m) intervals

The **Ex. to 504.80(B)** permits circuits run underground to be identified where they become accessible after emergence from the ground.

COLOR CODING
504.80(C)

Color coding shall be provided to identify intrinsically safe conductors, where they are colored light blue and no other conductors are colored light blue. Likewise, color coding shall be permitted in identifying raceways, cable trays and junction boxes where they are colored light blue and contain only intrinsically safe wiring systems. **(See Figure 4-29)**

GROUNDING
504.50

Intrinsically safe apparatus, associated apparatus, cable shields, enclosures and raceways, if of metal, shall be grounded.

If a connection to a grounding electrode is required, the grounding electrode shall be as specified in **250.50(A)(1) through (A)(4)** and shall comply with **250.30(A)(7)**. **Sections 250.52(A)(5), (A)(6) and (A)(7)** shall not be permitted if electrodes specified in **250.50(A)(1) through (a)(4)** are available. **(See Figure 4-30)**

Shielded conductors or cables, if used, shall be grounded, except where a shield is part of an intrinsically safe circuit.

Electrical Systems

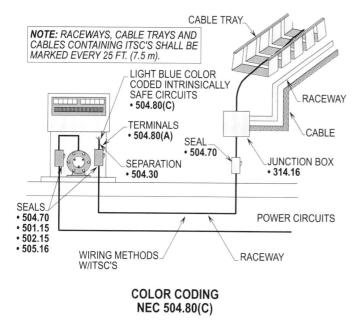

Figure 4-29. Light blue color coding shall be provided to identify intrinsically safe conductors.

BONDING
504.60

In hazardous (classified) locations, intrinsically safe apparatus shall be bonded in the hazardous (classified) location according to the provisions of **250.100**.

In nonhazardous locations, metal raceways used for intrinsically safe system wiring in hazardous locations shall have associated apparatus bonded in accordance with **501.30(A), 502.30(A) or 503.30(A) or 505.25**, where applicable. **(See Figure 4-31)**

NONINCENDIVE CIRCUITS

Nonincendive circuits (NICC's) are defined as a circuit in which any arc or thermal effect produced, under intended operating conditions of the equipment or due to opening, shorting or grounding of field wiring, is not capable, under specified test conditions, of igniting the flammable gas, vapor or dust-air mixture.

Simply put, NICC's are not capable of causing ignition under normal conditions of operation. The concept is the same as with other equipment operating in Class I, Division 2 locations.

For example, a fault in the equipment that makes it capable of causing ignition is very unlikely to occur at the same time as a flammable atmosphere in a Division 2 location is present. (See **500.2**)

FOUR TYPES OF NONINCENDIVE CIRCUITS

There are four types of systems that may be used for designing and installing a nonincendive system. Such systems are as follows:

- Nonincendive circuits
- Nonincendive contacts
- Nonincendive components
- Nonincendive equipment

NONINCENDIVE CIRCUITS

A nonincendive circuit is a low-energy circuit and the opening, shorting or grounding of one or more of the circuit conductors will not result in release of energy that is capable of causing ignition.

> **Design Tip:** Such circuits allow wiring methods to be used, which under normal conditions does not release sufficient energy to ignite a specific ignitable atmospheric mixture by opening, shorting or grounding. Any wiring method permitted in Class I, Division 2 locations shall be permitted to be utilized for these circuits. For such a wiring method, see **Article 727** in the NEC and review the requirements very carefully.

NONINCENDIVE CONTACTS

Nonincendive contacts are not always utilized in a nonincendive circuit. However, they are used in a circuit that is considered a low-energy circuit.

An Ex. to the NEC includes a definition of nonincendive contacts that pertains to contacts used in meters, instruments and relays, with a separate set of rules for signaling, alarm, remote-control and communications systems. This requirement permits general-purpose enclosures to be used, if current interrupting contacts are in circuits that under normal conditions do not release sufficient energy to ignite a specific ignitable atmospheric mixture. See **501.105(B)(1), Ex.** in the NEC for more information concerning this rule.

Stallcup's Electrical Design Book

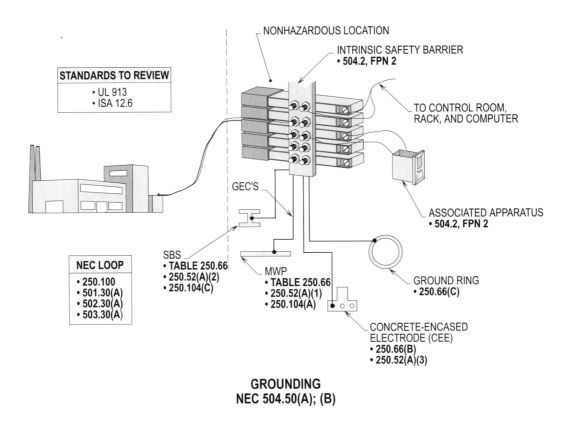

Figure 4-30. Intrinsically safe apparatus, associated apparatus, cable shields, enclosures and raceways, if of metal, shall be grounded.

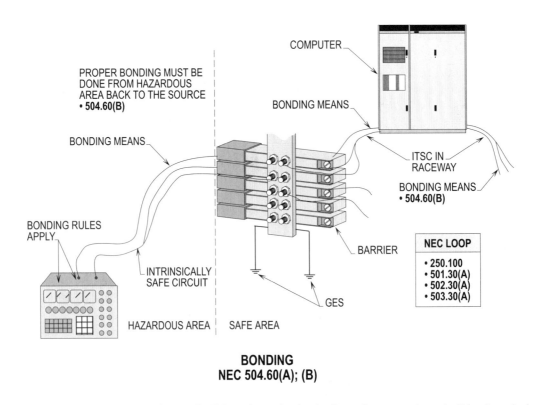

Figure 4-31. In hazardous (classified) locations, intrinsically safe apparatus shall be bonded.

NONINCENDIVE COMPONENTS

Nonincendive components are contacts which are installed in such a manner that ignition is precluded, even though in the open, the contacts may be incendive.

Nonincendive components are designed in such a manner to absorb energy and prevent its release that could cause arcing and heating.

Nonincendive components are components having contacts for making and breaking an incendive circuit. They are installed within an enclosure that is so constructed that the contacts are not capable of causing ignition. Danger only occurs, if these components are hit and damaged by a forklift, etc. Then enough energy could be released to ignite an atmospheric mixture.

NONINCENDIVE EQUIPMENT

Nonincendive equipment usually will not comply with all the requirements needed to be classified as intrinsic safety. This is due to the absence of protective barriers. Care must be exercised when designing and using nonincendive equipment in Class I, Division 2 locations. Such locations are usually adjacent to or above Division 1 locations.

An accident might occur which could damage the enclosure housing nonincendive contacts and such accident may result in a fault releasing (ignition-capable) energy at the same time an explosive mixture of gas and air is present around the equipment.

Design Tip: It is due to the far out fact that an explosive mixture of gas and air being present at the time a fault capable of releasing ignitable energy occurs that nonincendive equipment is allowed to be used in Class I, Division 2 locations.

WIRING METHODS AND ENCLOSURES (RECOMMENDED REQUIREMENTS)

Nonincendive circuits and wiring shall be permitted to be installed using any of the wiring methods suitable for unclassified locations, including **Chapters 7 and 8**. Sealing shall comply with the provisions of **501.15(A)** and separation shall comply with **UL 1604, ANSI S12.12** or manufacturer recommendations.

Design Tip: General-purpose enclosures are usually used for such installations per **501.105(B)(1), Ex.'s (3) and (4)**.

See **Figure 4-32** for a detailed illustration on requirements pertaining to nonincendive circuits and enclosures.

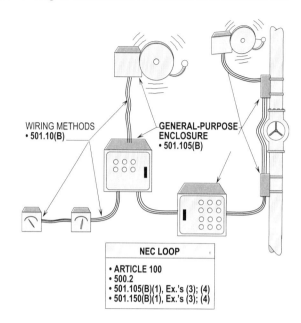

WIRING METHODS AND ENCLOSURES (RECOMMENDED REQUIREMENTS)

Figure 4-32. General-purpose enclosures are usually used when installing nonincendive circuits and wiring.

IDENTIFICATION OF TERMINALS AND WIRING (RECOMMENDED REQUIREMENTS)

It is recommended that labels be suitable for the environment in which they are installed and consideration must be given when they are exposed to chemicals and sunlight.

Nonincendive circuits shall be identified at terminal and junction locations in such a manner that will prevent unintentional interference with the circuits during testing and servicing.

Raceways, cable trays and open wiring for nonincendive system wiring shall be identified with permanently affixed labels with the wording "Nonincendive Wiring" or equivalent. The labels shall be so located as to be visible after installation and placed so that they are readily traced through the entire length of the installation. It is recommended that spacing between labels be provided at 25 ft. (7.5 m) intervals. **(See Figure 4-33)**

Nonincendive circuits run underground should be identified where they become accessible after emergence from the ground.

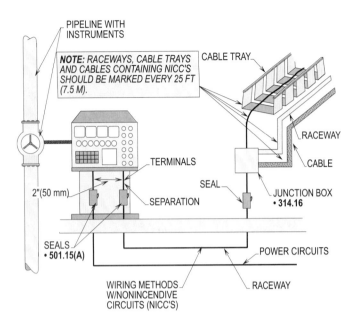

IDENTIFICATION OF TERMINALS AND WIRING (RECOMMENDED REQUIREMENTS)

Figure 4-33. Raceways, cable trays and open wiring for nonincendive system wiring should be identified with permanently affixed labels with the wording "Nonincendive Wiring" or equivalent at 25 ft. (7.5 m) intervals.

COLOR CODING (RECOMMENDED REQUIREMENTS)

It is recommended that color coding (yellow) should be provided to identify nonincendive conductors and where they are colored yellow no other conductors are allowed to be colored yellow. Likewise, color coding is recommended to be used to identify raceways, cable trays and junction boxes where they are colored yellow and contain only nonincendive wiring systems, this color code scheme distinguishes them from other wiring methods. **(See Figure 4-34)**

LIGHTING SYSTEMS OPERATING AT 30 VOLTS OR LESS
ARTICLE 411

The rules and regulations of this Article covers lighting systems operating at 30 volts or less and their associated components. Designers and installers using these requirements will design and install safe, dependable and reliable lighting systems operating at 30 volts or less.

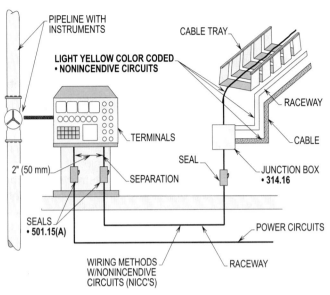

COLOR CODING (RECOMMENDED REQUIREMENTS)

Figure 4-34. It is recommended that color coding (yellow) should be provided to identify nonincendive conductors.

LIGHTING SYSTEMS OPERATING AT 30 VOLTS OR LESS
411.2

Such a lighting system is a lighting system consisting of an isolating power supply operating at 30 volts (42.2 Vpk) or less. Under any load condition, one or more of the secondary circuits, are limited to 25 amperes (each) maximum while supplying luminaires (lighting fixtures) and associated equipment identified for such use. **(See Figure 4-35)**

LISTING REQUIRED
411.3

Lighting systems operating at 30 volts or less shall be listed for the purpose. In other words, they shall be examined for safety by the rules in **90.7**, and installed by the regulations of **110.3(A)**. Such systems shall be installed and used by the requirements concerning listing and labeling of **110.3(B)**. Basically, such lighting systems shall be installed, used or both, in accordance with any instructions included in the listing or labeling of the manufacturer. For approval of such lighting systems, see **110.2** and the definitions of approved, identified, labeled and listed per **Article 100** in the NEC. **(See Figure 4-36)**

Electrical Systems

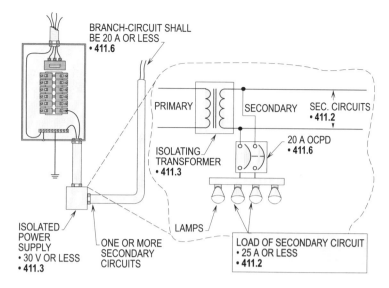

Figure 4-35. Under any load condition, one or more of the secondary circuits, are limited to 25 amperes (each) maximum while supplying luminaires (lighting fixtures) and associated equipment identified for such use.

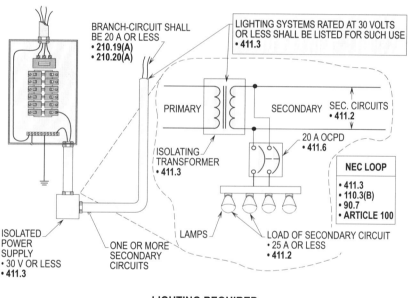

Figure 4-36. Lighting systems operating at 30 volts or less shall be listed for the purpose.

LOCATIONS NOT PERMITTED
411.4

Lighting systems operating at 30 volts or less shall not be installed if they are concealed or extended through a building wall, unless a wiring method per **Chapter 3** is utilized and installed using wiring supplied by a Class 2 power source and installed in accordance with **725.52**. Such lighting systems shall not be permitted within 10 ft. (3 m) of pools, spas, fountains or similar locations, except as allowed by **Article 680**. **(See Figure 4-37)**

SECONDARY CIRCUITS
411.5

Secondary circuits of lighting systems rated at 30 volts or less shall not be grounded per **250.20(A)** and **411.5** in the NEC.

All secondary circuits shall be insulated from the branch-circuit by the installation of an isolation transformer.

Exposed bare conductors and current-carrying parts of such systems shall be permitted for indoor installations only. Bare conductors shall not be installed less than 7 ft. (2.1 m) above the finished floor, unless specifically listed for a lower installation height and equipped with proper safeguards. **(See Figure 4-38)**

BRANCH-CIRCUIT
411.6

Lighting systems operating at 30 volts or less shall be supplied from a maximum 20 ampere branch-circuit. This requirement limits the number of outlets and prevents greater sized branch-circuits with a greater number of outlets installed throughout a building. **(See Figure 4-39)**

> **Design Tip:** See **210.3** for branch-circuit rating rules and **210.19(A)(1)** for ampacity requirements for conductors based upon load. Review **210.20(A)** for regulations concerning the selection of the overcurrent protection devices.

HAZARDOUS (CLASSIFIED) LOCATIONS
411.7

Lighting systems of 30 volts or less installed in hazardous (classified) locations shall comply with the provisions of **Articles 500 through 517** which are in addition to this Article.

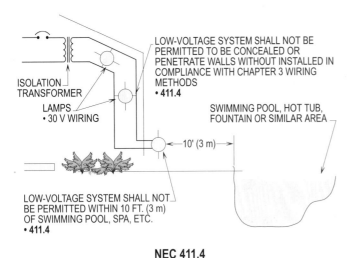

Figure 4-37. Lighting systems shall not be permitted within 10 ft. (3 m) of pools, spas, fountains or similar locations, except as allowed by **Article 680**.

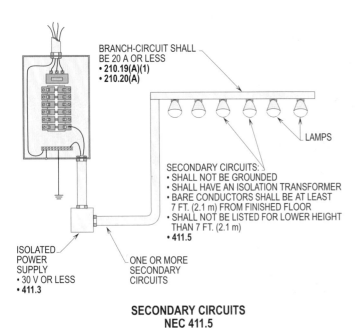

Figure 4-38. Secondary circuits of lighting systems rated at 30 volts or less shall not be permitted to be grounded.

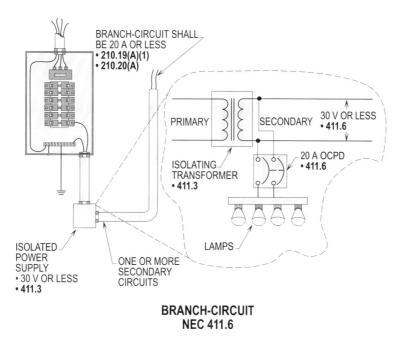

Figure 4-39. Lighting systems operating at 30 volts or less shall be supplied from a maximum 20 ampere branch-circuit.

Name   Date

Chapter 4. Electrical Systems

Section	Answer		
_____	T F	1.	Single-phase systems are only available in two-wire services.
_____	T F	2.	208/120 volts, three-phase, four-wire systems provide 120 volts for lighting and appliances and 208 volts for various types of power loads.
_____	T F	3.	Electrical supply systems of 480/277 volt, three-phase, four-wire are called delta connected systems.
_____	T F	4.	Electrical supply systems of 2.4 to 15 kV are usually used for primary power distribution in industrial plants.
_____	T F	5.	Conductors used for low-voltage systems of less than 50 volts shall not be permitted to be smaller than 10 AWG copper or equivalent.
_____	T F	6.	Class 1, 2 and 3 circuits shall be classified as remote-control, signaling and power-limited circuits.
_____	T F	7.	Fire alarm circuits shall be classified as either nonpower-limited or power-limited.
_____	T F	8.	Intrinsically safe apparatus and associated apparatus shall be permitted to be installed in any hazardous (classified) location for which it has been identified.
_____	T F	9.	Conductors of instrinsically safe circuits shall be permitted to be placed in any raceway, cable tray or cable with conductors of any nonintrinsically safe circuit.
_____	T F	10.	Yellow color coding shall be used to identify intrinsically safe conductors.
_____ _____		11.	Nonincendive contacts are always utilized in a nonincendive circuit.
_____ _____		12.	A yellow color coding shall be permitted to identify nonincendive circuits installed in a Class I, Division 2 location.
_____ _____		13.	Lighting systems operating at 30 volts or less (**Article 411**) shall not be listed for the purpose.
_____ _____		14.	Secondary circuits of lighting systems rated at 30 volts or less shall be grounded for safety reasons.
_____ _____		15.	Lighting systems operating at 30 volts or less shall be supplied from a maximum 20 ampere branch-circuit.
_____ _____		16.	Electrical supply systems of 208/120 volts single-phase may be used for _____ dwelling units in apartment complexes or other small facilities.

Section	Answer

17. Electrical supply systems of 240/120 volt, three-phase, four-wire are available using _____ transformers (closed delta) or _____ transformers (open delta).

18. Voltages of over _____ volts are usually referred to as distribution systems.

19. Conductors used for low-voltage systems of less than 50 volts (for appliance branch-circuits) supplying more than one appliance or appliance receptacle shall not be permitted to be smaller than _____ AWG copper or equivalent.

20. Conductors used for low-voltage systems of less than 50 volts for receptacles shall have a rating of not less than _____ amperes when used for general-purpose.

21. Class 1 power-limited circuits are limited to 30 volts and _____ volt-amperes.

22. The maximum circuit voltage is _____ volts AC or DC for a Class 2 inherently limited power source.

23. Conductors of intrinsically safe circuits shall be separated at least _____ from conductors of any nonintrinsically safe circuits.

24. Raceways, cable trays and open wiring for nonincendive system wiring shall be identified with permanently affixed labels with the wording _____ wiring or equivalent.

25. Lighting systems operating at 30 volts or less shall not be permitted within _____ ft. of pools, spas, fountains or similar locations.

26. Two-wire circuits consist of _____ volts between the ungrounded (phase) conductor and a grounded (neutral) conductor.

    (a) 120
    (b) 240
    (c) 208
    (d) 300

27. Standard lampholders for system of less than 50 volts shall have a rating of not less than _____ watts.

    (a) 550
    (b) 660
    (c) 750
    (d) 1000

	Section	Answer

28. Receptacles of not less than _____ ampere rating shall be utilized in kitchens, laundries and other locations where portable appliances are likely to be cord-and-plug connected for systems of less than 50 volts.

    (a) 15
    (b) 20
    (c) 30
    (d) 40

29. Class 1 remote control or signaling circuits shall be permitted to operate up to _____ volts and with no limitation on the power rating of the source.

    (a) 120
    (b) 240
    (c) 480
    (d) 600

30. The maximum circuit voltage is _____ volts AC or DC for a Class 3 inherently limited power source.

    (a) 30
    (b) 50
    (c) 100
    (d) 150

31. Intrinsically safe circuits shall be separated from conductors of nonintrinsically safe circuits by a distance of at least _____ in. and secured.

    (a) 1.97
    (b) 2
    (c) 12
    (d) 24

32. Intrinsically safe circuits shall be identified by color coding the conductors _____.

    (a) Yellow
    (b) Light Blue
    (c) Red
    (d) Purple

33. Lighting systems operating at 30 volts or less shall be permitted to be installed with bare conductors that are less than _____ ft. above the finished floor.

    (a) 6
    (b) 6 1/2
    (c) 7
    (d) 10

Section	Answer

34. The maximum circuit voltage for a Class 3 power source limited by overcurrent protection is _____ volts AC or DC.

    (a) 100
    (b) 120
    (c) 150
    (d) 600

35. It is recommended that spacing between labels for nonincendive circuits be provided at _____ ft. intervals.

    (a) 10
    (b) 15
    (c) 20
    (d) 25

36. What is the amperage for a 120 volt, single-phase, two-wire, 20 kVA transformer?

37. What is the amperage for a 208/120 volt, single-phase, three-wire, 150 kVA transformer?

38. What is the amperage for a 240/120 volt, single-phase, three-wire, 150 kVA transformer?

39. What is the amperage for a 208/120 volt, three-phase, four-wire, 150 kVA transformer?

40. What is the amperage for a 240/120 volt, three-phase, four-wire, 150 kVA closed delta transformer?

41. What is the amperage for three 240/120 volt, three-phase, four-wire, 150 kVA closed delta transformers? (one XFMR was lost on closed delta and the system recommended open delta)

42. What is the amperage for a 480/277 volt, three-phase, four-wire, 150 kVA transformer?

43. What is the amperage for a 4160 volt, three-phase, four-wire, 1500 kVA transformer?

44. What is the amperage for a 13,800 volt, three-phase, four-wire, 1500 kVA transformer?

5

Working Clearances

All electrical equipment shall be provided with sufficient working space to allow safe access for servicing the equipment without exposing electrical workers to shock hazards. Workspace shall be maintained around and about all electrical equipment where parts may be serviced while energized.

There are three main clearances required for electrical equipment to ensure protection for personnel from electric shocks and burns. The first clearance is to maintain a measurement of 30 in. (750 mm) wide in front of the equipment and secondly a clearance of at least 3 ft. (900 mm) is required outwardly in front of all electrical equipment. Thirdly, a minimum headroom clearance of 6 ft. 6 in. (2 m) or the equipment height, whichever is greater, is required.

Design Tip: For electrical equipment installed before the 1978 NEC, the minimum clearance is only 2 ft. 6 in. in front of such electrical equipment. The 30 in. (750 mm) wide rule has been in the NEC since the 1971 edition. Headroom clearance has been required since the 1965 NEC.

WORKING SPACE (600 VOLTS OR LESS)
110.26

All electrical equipment shall be provided with sufficient working space to allow safe access for servicing the equipment without exposing electrical workers to shock hazards. Workspace shall be maintained around and about all electrical equipment where parts may be serviced while energized.

> **Design Tip:** Equipment not extending more than 6 in. (150 mm) shall be permitted to be mounted below or above another piece of equipment, if they comply with the height requirements per **110.26(A)(3)**. **(See Figure 5-3(a))**

SPACES ABOUT ELECTRICAL EQUIPMENT
110.26(A)(1) THRU (A)(3)

A minimum working space of 30 in. (750 mm) width shall be required in front of electrical equipment operating at 600 volts or less. This space permits sufficient room to avoid contact of elbows from contacting live parts and metal parts at the same time while working on the equipment. Equipment doors and hinged panels shall have at least a 90° opening provided in the workspace. This opening allows electrical workers to have adequate room to repair, adjust or reset overcurrent protection devices without placing his or her body between the panel door and panelboard. **(See Figure 5-1)**

> **Design Tip:** The 90° opening of the door has been required since the 1987 NEC.

Working space shall not be required in the back of electrical equipment where there are not any removable or adjustable parts such as circuit breakers, fuses or switches mounted on the back of the equipment. All connections and service areas for maintenance shall be accessible from other locations other than the back of the equipment. If rear access is necessary to work on nonelectrical parts on the back of enclosed equipment, a minimum working space of 30 in. (762 mm) horizontally shall be provided.

The inspection authority has the power to make exceptions for smaller workspaces, where in his or her judgment there is no hazard involved. For less working space to be allowed, the installation of the equipment shall provide adequate accessibility to electrical workers. Such space shall be approved by the AHJ per **110.26(A)(1)(b)** and **90.4**.

> **Design Tip:** Section **110.26(A)(1)(b)** in the NEC only applies where there are uninsulated parts operating at a voltage no greater than 30 volts RMS, 42 volts peak or 60 volts DC.

Note: If working on live parts, **110.16** requires a marking to be placed on equipment that warns personnel of possible arc blast.

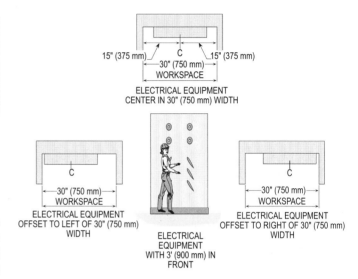

**SPACES ABOUT ELECTRICAL EQUIPMENT
NEC 110.26(A)(2)**

Figure 5-1. A minimum workspace (width) of 30 in. (750 mm) shall required in front of electrical equipment. The workspace may be centered or offset, if water pipe or other types of piping, etc. are present.

DEPTH OR WORKING SPACE CONDITIONS 1, 2 AND 3
TABLE 110.26(A)(1)

There are different clearances for workspace required in front of electrical equipment based on the type of material that the wall directly in front of the equipment is made of or if there are any live parts opposite the equipment. The clearances are based on two voltage levels to ground which are 150 volts or less to ground or over 150 volts to 600 volts-to-ground.

For example, the clearances of a 120/208 volt, three-phase, four-wire system falls under the 150 volts or less to ground while a 277/480 volt, three-phase, four-wire system is guarded by the 151 to 600 volts-to-ground.

DEPTH OR WORKING SPACE CONDITION 1
TABLE 110.26(A)(1)

Condition 1 is where the electrical equipment is installed in or on one wall with the wall on the opposite side being an insulated wall. An insulated wall is constructed of wood or metal studs with the wallboard consisting of sheetrock, wood panels, etc. An electrical worker making contact with the insulated wall while touching live parts is isolated from the grounded slab or earth, therefore, **Condition 1** allows less working space. **(See Figure 5-2)**

makes contact with the wall while touching a live part or conductor, a circuit is made, he or she could easily be electrocuted. Because this danger is present, a greater workspace is required for the voltages ranging from 151 to 600 volts. **(See Figure 5-3)**

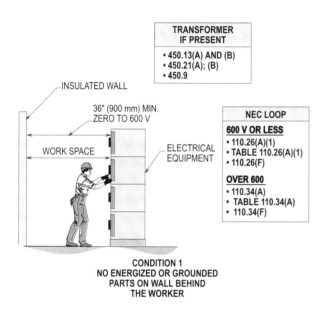

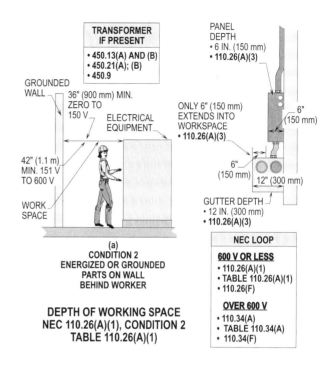

Figure 5-2. Clearances in front of electrical equipment with an insulated wall opposite equipment.

Design Tip: Until the 1965 NEC was published, Condition 1 only required a 2 1/2 ft. workspace in front of electrical equipment.

DEPTH OF WORKING SPACE CONDITION 2
TABLE 110.26(A)(1)

Condition 2 is where the electrical equipment is installed on one wall with the wall on the opposite side being a conductive wall. A conductive wall (grounded) is constructed of concrete, brick or tile where touched may connect the body to earth ground. If an electrical worker accidentally

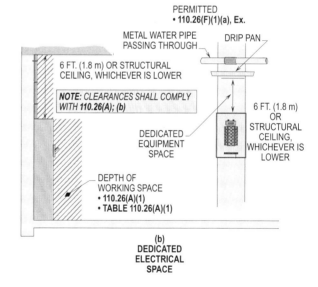

Figures 5-3. Figure 5-3(a) outlines clearances in front of electrical equipment with a grounded wall opposite equipment. Figure 5-3(b) describes clearance rules above electrical equipment.

DEPTH OR WORKING SPACE
CONDITION 3
TABLE 110.26(A)(1)

Condition 3 is where the electrical equipment is installed in or on one wall with the wall on the opposite side having electrical equipment mounted or set on it. With the electrical equipment installed in this manner, there are live parts on both sides of the room. Electricians and maintenance workers are subjected to phase-to-phase voltage or phase-to-ground voltage when servicing the equipment. Where electrical equipment is mounted or set on opposite walls directly across and in front of each other, electrical workers may be exposed to live parts with panelboard covers removed. Because electrical workers could be exposed to a fatal shock from live parts on both sides of the working space, a greater clearance shall be required for their safety. **(See Figure 5-4(a))**

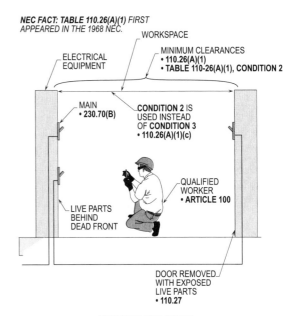

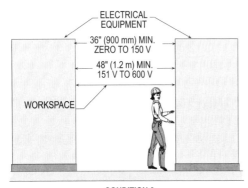

Figure 5-4(a). Clearances in front of electrical equipment opposite other electrical equipment.

EXISTING BUILDINGS
110.26(A)(1)(c)

In existing buildings where electrical equipment is being replaced, **Condition 2** working clearances shall be applied between dead-front switchboards, panelboards or motor control centers, which are located across the aisle from each other. However, for this rule to apply, conditions of maintenance and supervision shall be ensured by written procedures which have been adopted to prohibit equipment on both sides of the aisle from being open at the same time. In addition, only authorized, qualified and trained personnel shall be permitted to service such installations. **(See Figure 5-4(b))**

Figure 5-4(b). Section **110.26(A)(1)(c)** allows the use of **Condition 2** instead of **Condition 3** when working between existing pieces of electrical equipment with dead fronts that have been removed.

CLEAR SPACES
110.26(B)

Electrical equipment shall be located where adequate working space is accessible to take voltage measurements, check continuity of circuits, adjust or replace defective overcurrent protection devices and tighten loose connections. Working space in front of electrical equipment shall be free from storage of materials, etc. Mains and overcurrent protection devices shall be accessible to the users in case of emergencies as well as disconnecting means of equipment. Any and all exposed live parts located in a passageway or open space shall be suitably guarded. **(See Figure 5-5)**

ENTRANCE TO WORKING SPACE
110.26(C)

Electrical equipment shall be designed and installed to have at least one entrance to the workspace. Easy and fast access to electrical devices is essential. Special considerations shall be given to electrical equipment that has 1200 amps or more of bus containing overcurrent devices, switching devices or control devices. Such equipment shall have a clearance of 24 in. (610 mm) wide and 6 1/2 ft. (2 m) high at each end for safe exit in case of a ground-fault. Where the entrance to the working space has personnel door(s), the door(s) shall open in the direction of egress and be equipped with panic bars, pressure plates or other devices that are normally latched but open under simple pressure. If there are not two ways to enter and leave the workspace of such equipment, a worker could be trapped at one end of the workspace by a ground-fault with fire between him and the exit. **(See Figure 5-6)**

Design Tip: This requirement has been in the NEC since the 1978 edition.

However, the workspace shall be permitted to be doubled per **110.26(C)(2)(b)** and **Table 110.26(A)(1)** based on **Conditions 1, 2 or 3** and only one entrance is required. With the workspace doubled in front of the equipment, a worker may move back out of the endangered workspace and exit along the length of the equipment.

The deeper workspace in front of the equipment provides a safe route to exit (unobstructed per **110.26(C)(2)(a)**, without providing two entrances. **(See Figure 5-7)**

The ability to exit these areas using swing out doors with panic bars, pressure plates, etc. is very beneficial to an electrical worker, if he or she is hurt from an arc blast.

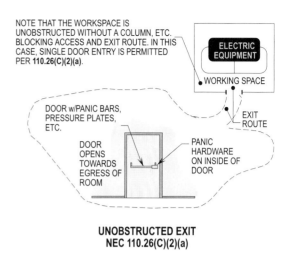

Figure 5-6. This illustration shows easy exit for personnel shall be provided for large equipment by installing panic hardware on door(s).

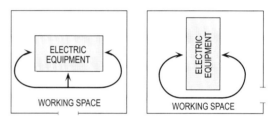

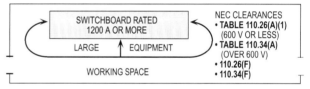

Figure 5-7. Access and entrance to working space for electrical equipment shall be provided.

ILLUMINATION
110.26(D)

Service equipment, switchboards, panelboards and motor control centers installed indoors shall be provided with adequate lighting for the safety of electrical workers servicing such equipment from the front or rear when live parts are accessible. Luminaires (lighting fixtures) shall be permitted to be incandescent or fluorescent as long as they provide the proper lighting for parts to be serviced.

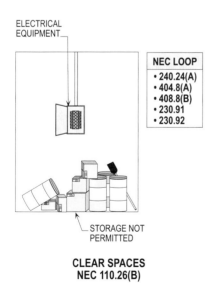

Figure 5-5. Storage shall not be permitted in front or below electrical equipment.

Luminaires (lighting fixtures) shall have a headroom clearance of at least 6 1/2 ft. (2 m) to give personnel sufficient room to stand in front of electrical equipment without the threat of their head or head gear contacting metal, etc. **(See Figure 5-8)**

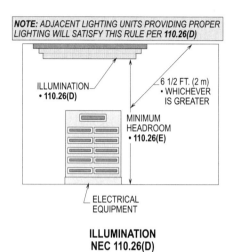

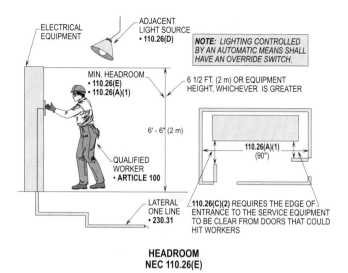

Figure 5-8. Proper lighting shall be provided for the servicing of electrical equipment.

Figure 5-9. Adequate headroom space shall be provided based upon the electrical equipment height or the 6 1/2 ft. (2 m) rule, whichever is greater. Also, 90° rule shall be applied properly.

Design Tip: Additional luminaires (lighting fixtures) shall not be required where the workspace is illuminated by an adjacent light source. An automatic means, by itself, shall not be permitted to be used to control the illumination in electrical equipment rooms or can be as permitted by **210.70(A)(1), Ex. 1**. Illumination was not required in the workspace until the 1965 NEC.

Design Tip: The minimum headroom or working space shall not be permitted to be less than 6 1/2 ft. (2 m) or the height of the equipment, whichever is greater. Headroom in the workspace was not required until the 1965 NEC.

DEDICATED EQUIPMENT SPACE
110.26(F)

In the following locations, switchboards, panelboards and distribution boards which are installed for the control of light and power circuits, battery-charging supplied from light and power circuits and motor control centers shall be located in dedicated spaces and protected from damage:

- Indoor
- Outdoor

INDOOR
110.26(F)(1)

The following conditions of use shall be considered when equipment is installed indoors in a dedicated equipment space:

- Dedicated electrical space
- Foreign systems
- Sprinkler protection
- Suspended ceilings

HEADROOM
110.26(E)

A minimum headroom clearance of 6 1/2 ft. (2 m) shall be maintained from the floor or platform up to the luminaire (lighting fixture) or any overhead obstruction. This overhead workspace is mandatory and applies especially to service equipment, switchboards, panelboards and motor control centers. The purpose of the overhead workspace is to protect employees from accidentally contacting grounded objects with their bodies (head, hands, etc.) while touching live parts and completing a circuit to ground which could cause a fatal electric shock. Electricians or maintenance workers should never have to stoop or bend down to gain access to service, repair, replace or modify components inside electrical equipment as previously mentioned. See **110.26(A)(2)** and **90.4** for application of 90° rule. **(See Figure 5-9)**

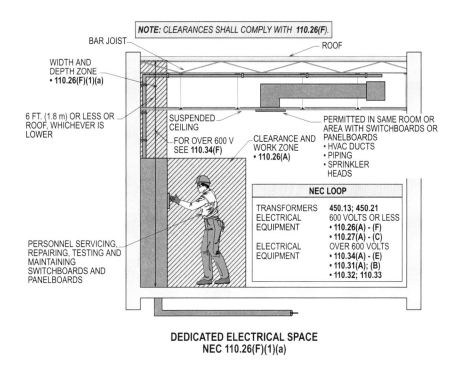

Figure 5-10. Switchboards shall be installed in a dedicated space extending 6 ft. (1.8 m) above the equipment from the floor or to the structural ceiling (roof), whichever is lower.

DEDICATED ELECTRICAL SPACE
110.26(F)(1)(a)

Switchboards shall be installed in a dedicated space equal to the width and depth of the equipment and extending from the floor to a height of 6 ft. (1.8 m) or to the structural ceiling (roof), whichever is lower. No piping, ducts or equipment foreign to the electrical installation shall be permitted to be installed in this dedicated space since leakage of water or condensation could damage the equipment enclosure. Any space extending 6 ft. (1.8 m) above the dedicated space shall not be considered dedicated space. **(See Figure 5-10)**

Suspended ceilings with removable panels shall be permitted within the 6 ft. (1.8 m) zone per **110.26(F)(1)(a), Ex.** This dedicated space is considered by electricians to be an area were they may run and connect cables and raceways without foreign items (other than electrical) blocking their access directly above equipment. **(See Figure 5-11)**

FOREIGN SYSTEMS
110.26(F)(1)(b)

The area above the dedicated space shall be permitted to contain foreign systems, provided protection is installed to avoid damage to the electrical equipment from condensation, leaks or breaks in such foreign systems. Otherwise, equipment not associated with the electrical equipment shall not be permitted to be installed in the 6 ft. (1.8 m) area above the electrical equipment. **(See Figure 5-12)**

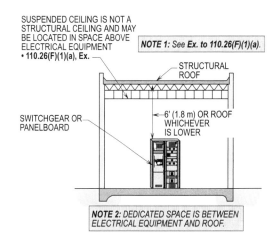

Figure 5-11. A suspended ceiling shall not be considered a structural ceiling and shall be permitted to be located in the dedicated space.

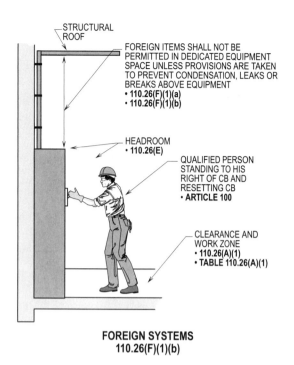

FOREIGN SYSTEMS
110.26(F)(1)(b)

Figure 5-12. This illustration shows the foreign items permitted in the dedicated electrical space.

SPRINKLER PROTECTION
110.26(F)(1)(c)

Sprinkler protection shall be permitted to be installed in this dedicated space to protect the equipment from fire hazards.

> **Design Tip:** For further information on sprinkler systems, see **NFPA 13, 4-4.14.**

SUSPENDED CEILINGS
110.26(F)(1)(d)

A suspended ceiling such as a dropped, suspended or similar ceiling shall be permitted to be installed in this dedicated space. The suspended ceiling does not add strength to the building structure, therefore, shall not be considered a structural ceiling. **(See Figure 5-11)**

OUTDOOR
110.26(F)(2)

Electrical equipment that is installed outdoors shall be in adequate enclosures or covers that will afford adequate mechanical protection from vehicular traffic, accidental contact by unauthorized personnel or accidental spillage or leakage from piping systems. The working space clearance around such equipment shall comply **110.26(A)**. No architectural appurtenance or other equipment shall be permitted to be located in this working space clearance around equipment.

# GUARDING OF LIVE PARTS
110.27(A)

The general rule of protecting live parts (energized) from accidental contact is by installing them in a complete enclosure which provides a dead front. Sometimes it is not practical to construct enclosures to house large control panels, etc. and in such cases if the apparatus is rated at 50 volts or more, then suitable guards or isolation shall be provided by one of the following rules:

- **Section 110.27(A)(1)** permits the live parts in electrical equipment that are not mounted in a completely enclosed enclosure to be installed in a room, vault or similar enclosure that is accessible only to qualified personnel. **(See Figure 5-13)**

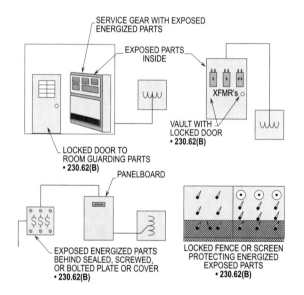

LIVE PARTS GUARDED AGAINST ACCIDENTAL CONTACT
NEC 110.27(A)(1)

Figure 5-13. Live parts of electrical equipment shall be protected and guarded from unqualified persons.

- **Section 110.27(A)(2)** permits live parts to be separated by permanent partitions or screens so located that only qualified persons have access and reach of the live parts. Openings in partitions or screens shall be designed to prevent accidental contact with live parts or bring conductive objects in contact with live parts.

Working Clearances

Design Tip: The safety thought is making the live parts in the equipment accessible without obstruction and giving special attention to the prevention of contacting conductive materials such as metal conduit, pipes, etc. carried by a passing worker. **(See Figure 5-14)**

- **Section 110.27(A)(4)** permits the live electrical parts to be elevated at least 8 ft. (2.5 m) above the floor.

For example, the live parts may be mounted or set on a platform which is attached to a utility pole or on the side of a building, etc. and this type of installation shall be considered safely guarding live parts. **(See Figure 5-15)**

PREVENT PHYSICAL DAMAGE 110.27(B)

Section 110.27(B) recognizes that many times electrical equipment is located in work areas where the work activity around it might damage the equipment. In such cases, the equipment shall be properly protected with enclosures or guards that provide the necessary strength to prevent any damage to the electrical equipment. **(See Figure 5-16)**

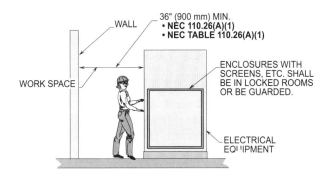

Figure 5-14. Live parts in electrical equipment shall be guarded or enclosed.

- **Section 110.27(A)(3)** permits exposed live parts in equipment to be located on a suitable balcony, gallery or platform that is high enough or designed in such a manner to keep unqualified personnel out.

For example, a well designed location of a balcony which is accessible only to qualified personnel who have been given the responsibility to maintain such equipment fully complies as a suitable means of guarding live parts. **(See Figure 5-15)**

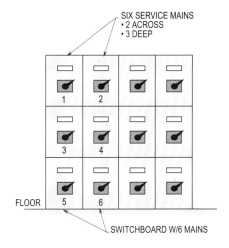

Figure 5-16. Totally enclosed gear protects live parts and equipment from physical damage.

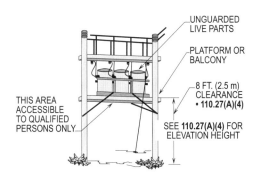

Figure 5-15. Live parts shall be considered protected where they are elevated above grade and are out of reach from the general public.

WARNING SIGNS 110.27(C)

Section 110.27(C) requires guarded locations to be posted with warning signs, giving warning that only qualified personnel are allowed to enter and service the electrical components. Any dangers that might exist shall be posted to warn employees to be careful. **(See Figure 5-17)**

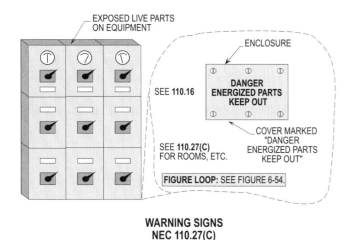

Figure 5-17. Electrical rooms with exposed live parts supplied by 600 volts or less systems shall be posted with warning signs that only qualified persons are allowed to enter.

ENTRANCE AND ACCESS TO WORKSPACE (OVER 600 V) 110.33

There shall be at least one entrance that provides a minimum of 24 in. (610 mm) by 6 1/2 ft. (2 m) to gain access to the working space of high-voltage equipment (over 600 volts). This is a minimum and more should be provided when possible. Where the equipment is over 6 ft. (1.8 m) in width, there shall be one entrance at each end. **(See Figure 5-18)**

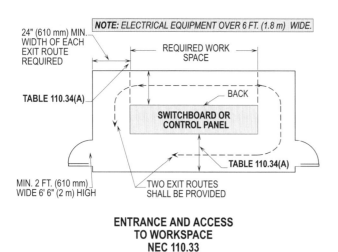

Figure 5-18. Electrical equipment shall be provided with proper exits for personnel servicing such equipment.

Design Tip: The 1984 NEC required at least one entrance of 24 in. by 6 1/2 ft. wide to be provided for electrical equipment that was over 4 1/2 ft. in width.

If bare or insulated parts of more than 600 volts, nominal, are located adjacent to such entrances, there shall be suitable means designed to guard them.

When electric equipment is installed on platforms, balconies, mezzanine floors or in attic or roof rooms or spaces, there shall be permanent ladders or stairways installed for access. There is an OSHA regulation that requires ladders to extend 3 ft. (900 mm) above the platform, etc. to which they give access to service equipment. **(See Figure 5-19)**

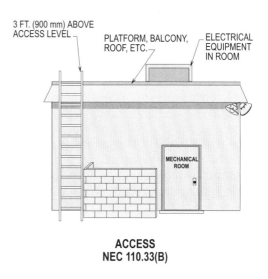

Figure 5-19. Where electrical equipment shall have a ladder to gain access to a platform, balcony, etc., the side rails of the ladder shall extend 3 ft. (900 mm) above the access area.

OVER 600 VOLTS NOMINAL 110.34(A)

The installation requirements of conductors, overcurrent protection devices and equipment are usually more stringent for systems rated over 600 volts. Conductors are normally shielded and special terminations are required where they terminate to the equipment, etc. Greater clearances around electrical equipment shall be required due to the threat of flash over voltage from live parts to grounded metal. High-voltage switches and overcurrent protection devices shall be marked to indicate the circuit or equipment that they supply and control. The rules in the NEC makes it very clear that **110.22** applies to systems rated over 600 volts as well as systems rated at 600 volts or less. **Article 490** in the NEC is used to design, install and inspect systems over 600 volts.

Working Clearances

ENCLOSURE FOR ELECTRICAL INSTALLATIONS
110.31

In locations where entrance and accessibility is controlled by lock and key or other acceptable and approved means, these locations shall be considered as accessible to only qualified personnel. Door and covers of enclosures used solely for pull boxes, splice boxes or junction boxes shall be locked, bolted or screwed on. Underground box covers that weigh over 100 lbs. (45.4 kg) shall be considered as complying with this requirement. Such locations that are involved are as follows:

- Electrical installations in vaults.
- Electrical installations in rooms.
- Electrical installations in closets.
- Locations surrounded by a:
  (a) Wall,
  (b) Screen or
  (c) Fence.

The design and construction of enclosures shall be suitable to the nature and degree of the hazards involved.

For example, an enclosure installed in an area exposed to gas is normally required to be classified for Class I, Division 1 locations with a Group D rating.

Design Tip: A wall, screen or fence that is less than 6 ft. (1.8 m) tall shall not be considered as preventing access to equipment unless other features are provided with a degree of isolation equivalent to a 7 ft. (2.1 m) fence. Consider a fence of 6 ft. (1.8 m) in height and equipped with 1 ft. (300 mm) of barbed wire at the top of the fence. This addition of the barbed wire is the extra feature needed to prevent easy access of unqualified personnel. **(See Figure 5-20)**

FIRE RESISTIVITY OF ELECTRICAL VAULTS
110.31(A)

The walls, roof, floors and doorways of vaults containing conductors and equipment over 600 volts, nominal, shall be constructed of materials that have adequate structural strength for the conditions, with a minimum fire rating of 3 hours. The floors of vaults in contact with the earth shall be of concrete that is not less than 4 in. (102 mm) thick, but where the vault is constructed with a vacant space or other stories below it, the floor shall have adequate structural strength for the load imposed on it and a minimum fire resistance of 3 hours. For the purpose of the section, studs and wallboards shall not be considered acceptable. **(See Figure 5-21)**

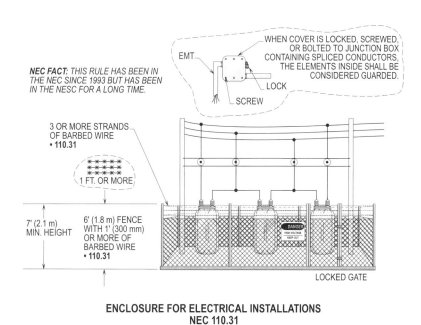

ENCLOSURE FOR ELECTRICAL INSTALLATIONS
NEC 110.31

Figure 5-20. A wall, screen or fence installed properly prevents easy access of unqualified personnel.

INDOOR INSTALLATIONS
110.31(B)

Electrical equipment installed indoors shall meet certain requirements where places are accessible to unqualified persons and places are accessible to qualified persons only.

Indoor electrical installations shall be made with metal-enclosed equipment that are accessible to unqualified persons. Metal-enclosed switchgear, unit substations, transformers, pull boxes, connection boxes and other similar associated equipment shall be marked with appropriate caution signs where unqualified persons have access. Ventilating or similar openings shall be so designed that foreign objects are deflected from energized parts. **(See Figure 5-22)**

Electrical installations installed indoors containing exposed live parts shall be accessible to qualified persons only and comply with **110.34, 110.36** and **490.24**.

OUTDOOR INSTALLATIONS
110.31(C)

Outdoor electrical installations shall comply with **Parts I, II and III of Article 225** where accessible to unqualified persons. Electrical installations installed outdoors, containing exposed live parts, shall be accessible to qualified personnel only, and shall comply with the provisions of **110.34, 110.36 and 490.24**. It is essential that the live parts of high-voltage systems be totally enclosed in the equipment or be isolated to prevent access by the general public or unqualified employees. High-voltage traveling through the body to ground can do great damage. If it doesn't kill immediately, the internal damage usually causes death later. Only qualified personnel shall be permitted to service high-voltage systems. **(See Figure 5-23)**

ENCLOSED EQUIPMENT ACCESSIBLE TO UNQUALIFIED PERSONS
110.31(D)

Ventilating or similar openings in equipment shall be so designed that foreign objects inserted through these openings are deflected from energized parts. When exposed to physical damage from vehicular traffic, suitable guards and caution signs shall be provided. Metal-enclosed equipment located outdoors that is accessible to the general public, shall be designed so that exposed nuts or bolts are not readily removed, permitting access to live parts. Where metal enclosure equipment is accessible to the general public and the bottom of the enclosure is less than 8 ft. (2.5 m) above the finished floor or grade level, the enclosure door or hinged cover shall be kept locked. Doors and covers for electrical enclosures such as pull boxes, splice boxes or junction boxes shall be locked, bolted or screwed on. Underground box covers that weigh over 100 lbs. (45.4 kg) shall be considered as complying with this requirement. **(See Figure 5-24)**

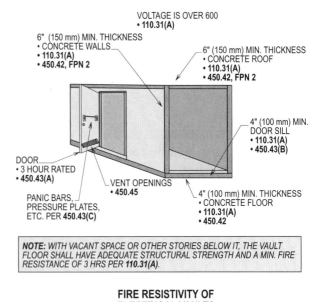

FIRE RESISTIVITY OF ELECTRICAL VAULTS NEC 110.31(A)

Figure 5-21. The walls, roof, floors and doorways of vaults containing conductors and equipment over 600 volts, nominal, shall be constructed of materials that have adequate structural strength for the conditions, with a minimum fire rating of 3 hours.

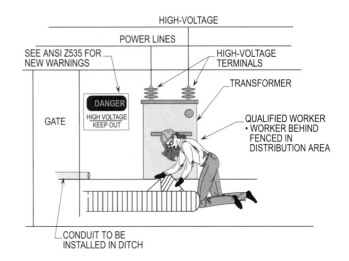

INDOOR INSTALLATIONS NEC 110.31(B)

Figure 5-22. Only qualified personnel shall be permitted to service high-voltage systems.

5-12

Working Clearances

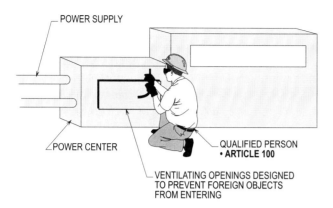

Figure 5-23. Openings in electrical equipment shall be designed to deflect foreign objects from live parts.

WORKSPACE ABOUT EQUIPMENT
110.32

There shall be sufficient clear working space about high-voltage equipment to permit ready and safe operation of such equipment. There have been minimum clearances set for workspace in and around electrical equipment and they are as follows:

- Where energized parts are exposed, the minimum clear working space shall not be permitted to be less than 6 1/2 ft. (2 m) measured vertically from the floor or platform or less than 3 ft. (900 mm) wide measured parallel to the equipment.

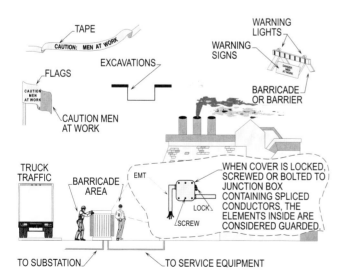

Figure 5-24. Covers for totally enclosed enclosures, with or without ventilating openings, shall be secured on or locked when accessible to the general public. Caution signs, etc. shall be used when necessary.

See Figure 5-25 for a detailed procedure on clearances about electrical equipment supplied with voltage over 600 volts.

- Space shall be adequate to permit doors and hinged panels to open a minimum of 90° so personnel may stand in front of the equipment without making contact with their body to the doors or hinged panels while touching live parts.

- See **Table 110.34(A)** for the minimum working clearances (depth) of equipment based upon on supply voltage.

MINIMUM DEPTH OF CLEAR WORKING SPACE AT ELECTRICAL EQUIPMENT

NOMINAL VOLTAGE-TO-GROUND	MINIMUM CLEAR DISTANCES		
	CONDITION 1	CONDITION 2	CONDITION 3
	(FEET)	(FEET)	(FEET)
601 - 2500 V	3 ft. (900 mm)	4 ft. (1.2 m)	5 ft. (1.5 m)
2501 - 9000 V	4 ft. (1.2 m)	5 ft. (1.5 m)	6 ft. (1.8 m)
9001 - 25,000 V	5 ft. (1.5 m)	6 ft. (1.8 m)	9 ft. (2.8 m)
25,001 - 75 kV	6 ft. (1.8 m)	8 ft. (2.5 m)	10 ft. (3.0 m)
ABOVE 75 kV	8 ft. (2.5 m)	10 ft. (3.0 m)	12 ft. (3.7 m)

**WORKSPACE ABOUT EQUIPMENT
NEC 110.34(A)
NEC TABLE 110.34(A))**

Figure 5-25. Minimum clearances (depth) of equipment based upon the supply voltage.

ENTRANCE AND ACCESS TO WORKSPACE
110.33

The entrance and access to workspace shall be considered for high-voltage electrical equipment. The entrance and access to workspace may require at least one entrance at each end or allow one entrance to the workspace.

ENTRANCE
110.33(A)

There shall be at least one entrance a minimum of 24 in. (610 mm) x 6 1/2 ft. (2 m) to provide access to the working space of high-voltage equipment (over 600 volts). This is a minimum and more should be provided when possible. Personnel door(s) at the entrance shall open in the direction of egress and be equipped with panic bars, pressure plates or other devices that are normally latched but open under simple pressure. **(See Figure 5-26)**

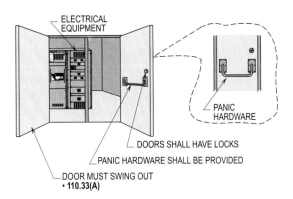

Figure 5-26. This illustration shows that personnel door(s) at the entrance shall open in the direction of egress and be equipped with panic bars, pressure plates or other devices on door(s).

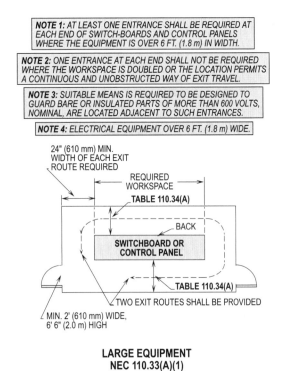

Figure 5-27. This illustration shows that electrical equipment shall be provided with proper exits for personnel servicing such equipment.

LARGE EQUIPMENT
110.33(A)(1)

There shall be one entrance at each end of switchboards and control panels where the equipment is over 6 ft. (1.8 m) in width. A single entrance to the required working space shall be permitted if either one of the following conditions is applied:

- The workspace permits a continuous and obstructed way of exit travel.
- The workspace is doubled per **110.34(A)**. This entrance shall be located so that the edge of the entrance nearest the electrical equipment is the minimum clear distance given in **Table 110.34(A)** away from such equipment. **(See Figure 5-27)**

GUARDING
110.33(A)(2)

Where bare or insulated parts or more than 600 volts, nominal, are located adjacent to such entrances, they shall be suitably guarded.

ACCESS
110.33(B)

When electrical equipment is installed on platforms, balconies, mezzanine floors or in attic roor room or spaces, permanent ladders or stairways shall be installed for access. **(See Figure 5-28)**

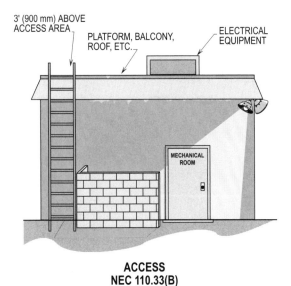

Figure 5-28. This illustration shows that electrical equipment shall have a ladder to gain access to a platform, balcony, etc. The side rails of the ladder shall extend 3 ft. (900 mm) above the access area.

WORKING SPACE AND GUARDING 110.34(A)

Table 110.34(A) in the NEC lists the minimum clear working space in front of electrical equipment. This can be such equipment as switchboards, control panels, switches, circuit breakers, motor control centers, relays, etc. If other distances are spelled out elsewhere in **NFPA 70E**, those specified distances are to be applied.

In measuring the distances given in **Table 110.34(A)**, they shall be measured from exposed live parts or if in enclosures, the distance is measured from the enclosures front or opening.

The **Ex. to 110.34(A)** does not require working space to be provided in back of electrical equipment where there are not any renewable or adjustable parts such as fuses and circuit breakers on the back and all connections are accessible from locations other than the back. If rear access is required to work on deenergized parts on the back of enclosed equipment, a minimum distance of 30 in. (750 mm) horizontally shall be required for the working space. **(See Figure 5-25)**

WORKING SPACE CONDITION 1 TABLE 110.34(A)

Condition 1 is where there are exposed live parts on one side of the equipment and there are no live or grounded parts on the opposite side of the working space. Such condition is where there is exposed parts on one side that is effectively guarded by suitable wood or other insulating materials on the opposite side.

> **For example:** What is the minimum clearance required for a switchgear supplied with 4,160 volts and installed opposite an insulated wall?
>
> **Step 1:** Finding clearance
> Table 110.34(A), Condition 1
> 4160 volts requires 4 ft. (1.2 m)
>
> **Solution:** The minimum clearance is 4 ft. (1.2 m).

WORKING SPACE CONDITION 2 TABLE 110.34(A)

Condition 2 has exposed live parts on the equipment with the other side (directly in front) having grounded parts such as concrete, brick or tile walls.

> **For example:** What is the minimum clearance for a motor control center supplied with 12,470 volts and installed opposite a concrete wall?
>
> **Step 1:** Finding clearance
> Table 110.34(A), Condition 2
> 12,470 volts requires 6 ft. (1.8 m)
>
> **Solution:** The minimum clearance is 6 ft. (1.8 m).

WORKING SPACE CONDITION 3 TABLE 110.34(A)

Condition 3 consists of exposed live parts on both sides (not guarded as in Condition 1) with the worker or operator in-between the equipment and live parts. Such a condition would be where two covers are removed from the equipment opposite each other, exposing live parts to trouble shoot an electrical problem. The electrician or maintenance worker could accidentally make contact with live parts and be in series with phase-to-phase voltage or be in series with a phase and the grounded metal of the equipment. Either condition would most likely be fatal.

> **For example:** What is the minimum clearance for equipment enclosures supplied with 13,800 volts and are mounted opposite one to another?
>
> **Step 1:** Finding clearance
> Table 110.34(A), Condition 3
> 13,800 volts requires 9 ft. (2.8 m)
>
> **Solution:** The minimum clearance is 9 ft. (2.8 m).

SEPARATION FROM LOW-VOLTAGE EQUIPMENT
110.34(B)

When low-voltage (600 volts or less) equipment is installed in a room or enclosure where there are exposed high-voltage live parts, the high-voltage equipment shall be separated from the low-voltage equipment by a suitable partition, fence or screen.

If the room or enclosure is accessible to qualified persons (only), a partition, etc. shall not be required, provided the low-voltage equipment serves only equipment within the room or enclosure per **110.34(B), Ex.**

> **Design Tip:** Many companies will not allow low-voltage in transformer vaults with high-voltage, with the exception of low-voltage buses. However, this does not include lighting and other low-voltage systems that might be required in the operation of the high-voltage equipment.

LOCKED ROOMS OR ENCLOSURES
110.34(C); Figures 5-22 and 6-53

The entrances to all buildings, rooms or enclosures containing high-voltage exposed live parts shall be kept locked, except where such entrances are under the constant observation of a qualified person. For voltages over 600 volts, warning signs shall be posted at the entrance which reads as follows:

DANGER - HIGH-VOLTAGE - KEEP OUT!

ILLUMINATION
110.34(D)

Illumination shall be provided to properly illuminate the high-voltage area for safe working conditions and the luminaires (lighting fixtures) shall be installed so that there is no danger to anyone changing bulbs, ballast's, etc. or working on the illumination system. The switching point for this illumination shall be readily accessible and in such a place that in operating the controls for the illuminations, there is no danger of coming in contact with any live parts.

> **Design Tip:** There is not a footcandle level given. **Section 110.34(D)** requires sufficient lighting to be provided when needed. **(See Figure 5-29)**

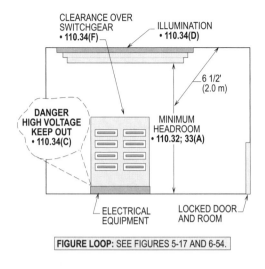

**ILLUMINATION
NEC 110.34(D)**

Figure 5-29. Electrical equipment shall have the proper lighting around the workspace for safe servicing.

ELEVATION OF UNGUARDED LIVE PARTS
110.34(E); TABLE 110.34(E)

It is permissible to install unguarded live parts at elevated heights of at least 9 ft. (2.8 m) from any and all finished grade areas. For installations prior to April 16, 1981, the minimum clearance height was 8 ft. (2.5 m), where the voltage between phases range from 601 to 6,600 volts respectively. The elevated height is based and selected from the supply voltage supplying the live parts.

> **For example:** What is the height required for unguarded live parts operating at 35,000 volts?
>
> **Step 1:** Finding height
> **Table 110.34(E)**
> 35,000 volts require
> 9 1/2 ft. (2.9 m)
>
> **Solution: The minimum clearance is 9 1/2 ft. (2.9 m).**

See **Figure 5-30** for a detailed procedure for installing live parts at elevated heights.

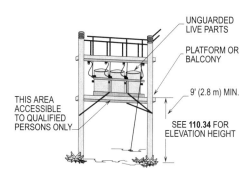

ELEVATION OF UNGUARDED LIVE PARTS
NEC 110.34(E)
NEC TABLE 110.34(E)

Figure 5-30. Electrical equipment elevated at least 9 ft. (2.8 m) above grade shall be considered guarded by height.

> **Design Tip:** For equipment installed before April 16, 1981 which operates above 25 kV, the required working clearance shall not be required to be increased beyond the value listed for 25 kV systems.

IDENTIFICATION OF DISCONNECTING MEANS
110.22

Panelboard circuit directories shall be fully and clearly filled out to indicate the load served and its location. Marking on equipment shall be painted lettering or another suitable identification. The circuit breakers in panelboards and fuses in fusible disconnects shall be marked to match the markings on equipment they serve. This is a mandatory requirement and it is to be applied to both new and old electrical systems. Adequate circuit mapping and proper identification shall be provided for expanding or alteration of all electrical systems. It is extremely important that the right disconnecting means be opened to deenergize a circuit quickly and safely when there is a threat of injury to personnel working, repairing or pulling maintenance on the equipment supplied.

Painted labeling or embossed plates fixed to the equipment complies with the requirements that disconnects be "legibly marked" and the "marking shall be durable". Glue-on paper labels or marking with crayon, ink or chalk most likely will be rejected as not being suitable permanent markings.

For example, a circuit breaker in a panelboard is marked No. 1 and supplies power to a disconnecting means in sight of a motor controller and motor.

The disconnecting means, motor controller and motor shall be legibly marked No. 1 to correspond to the marking on the circuit directory on the panelboard enclosure. Effective identification of disconnecting means and protective devices is a must by the NEC, NFPA 70E and OSHA to ensure the necessary safety for employees working in the employee workplace. **(See Figure 5-31)**

FLASH PROTECTION
110.16

Switchboards, panelboards, industrial control panels, meter socket enclosures and motor control centers in other than dwelling occupancies, that are likely to require examination, adjustment, servicing or maintenance while energized, shall be field marked to warn qualified persons of potential electric arc flash hazards. This marking shall be located so as to be clearly visible to qualified persons before examination, adjustment, servicing or maintenance of such equipment. **(See Figure 1-32)**

> **Design Tip: FPN 1 to 110.16** refers users to **NFPA 70E** for assistance in determining exposure, planning safe work practices and selecting personal protective equipment. **FPN 2 to 110.16** refers users to **ANSI Z535.4** for guidelines pertaining to designing safety signs and labels to provide warnings, etc.

Stallcup's Electrical Design Book

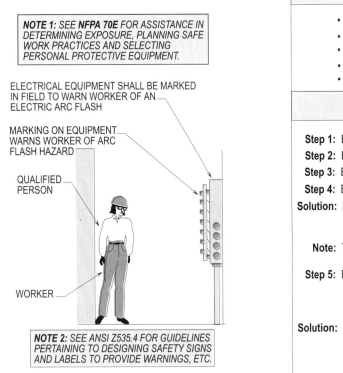

Figure 5-31. All circuits and disconnects shall be identified. OSHA regulations make **110.22** mandatory and retroactive for existing installations and for all new expanded or modernized systems - applying to switches as well as circuit breakers.

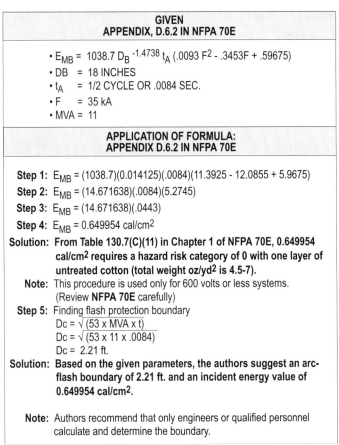

GIVEN
APPENDIX, D.6.2 IN NFPA 70E

- E_{MB} = 1038.7 $D_B^{-1.4738}$ t_A (.0093 F^2 - .3453F + .59675)
- DB = 18 INCHES
- t_A = 1/2 CYCLE OR .0084 SEC.
- F = 35 kA
- MVA = 11

APPLICATION OF FORMULA:
APPENDIX D.6.2 IN NFPA 70E

Step 1: E_{MB} = (1038.7)(0.014125)(.0084)(11.3925 - 12.0855 + 5.9675)
Step 2: E_{MB} = (14.671638)(.0084)(5.2745)
Step 3: E_{MB} = (14.671638)(.0443)
Step 4: E_{MB} = 0.649954 cal/cm^2
Solution: From Table 130.7(C)(11) in Chapter 1 of NFPA 70E, 0.649954 cal/cm^2 requires a hazard risk category of 0 with one layer of untreated cotton (total weight oz/yd^2 is 4.5-7).
Note: This procedure is used only for 600 volts or less systems. (Review **NFPA 70E** carefully)
Step 5: Finding flash protection boundary
Dc = √ (53 x MVA x t)
Dc = √ (53 x 11 x .0084)
Dc = 2.21 ft.
Solution: Based on the given parameters, the authors suggest an arc-flash boundary of 2.21 ft. and an incident energy value of 0.649954 cal/cm^2.

Note: Authors recommend that only engineers or qualified personnel calculate and determine the boundary.

Figure 5-32. This illustration shows that markings on electrical equipment shall be provided to warn employees of hazardous energy from an arcing fault.

Name _____   Date _____

Chapter 5. Working Clearances

Section	Answer			
_____	T	F	1.	A minimum working space of 36 in. width shall required in front of electrical equipment operating at 600 volts or less.
_____	T	F	2.	If rear access is necessary to work on deenergized parts on the back of enclosed equipment, a minimum working space of 30 in. horizontally shall be required.
_____	T	F	3.	**Condition 2 to Table 110.26(A)(1)** is where the electrical equipment is installed in or on one wall with the wall on the other side having electrical equipment mounted or set on it.
_____	T	F	4.	Working space in front of electrical equipment shall be free from storage of materials, etc.
_____	T	F	5.	Luminaires (lighting fixtures) shall have a headroom clearance of at least 7 1/2 ft. to give personnel sufficient room to stand in front of electrical equipment.
_____	T	F	6.	At least one entrance with a minimum of 24 in. by 6 1/2 ft. shall be provided for access to the working space of high-voltage equipment.
_____	T	F	7.	Electrical installations installed outdoors containing exposed live parts shall be required to be accessible to qualified personnel only.
_____	T	F	8.	**Condition 1 to Table 110.34(A)** is where the exposed live parts on the equipment with the other side (directly in front) having grounded parts such as concrete, brick or tile walls.
_____	T	F	9.	Illumination shall not be required to be provided to illuminate the high-voltage equipment area for safe working.
_____	T	F	10.	Panelboard circuit directories shall be fully and clearly filled out to indicate the load served and its location.
_____	_____		11.	Equipment doors and hinged panels shall have at least a _____ opening provided in the workspace. (600 volts or less)
_____	_____		12.	A 600 volt or less switchgear with a bus rated _____ amps shall have an exit at each end. (General Rule)
_____	_____		13.	Live electrical parts shall be considered protected where elevated at least _____ ft. above grade. (600 volts or less)
_____	_____		14.	The minimum clearance for a motor control center supplied with 12,470 volts and installed opposite a concrete wall is _____ ft..
_____	_____		15.	The height required for unguarded live parts operating at 35,000 volts shall be _____ ft..

Section	Answer
_____	_____

16. The minimum clearance in front of electrical equipment with an ungrounded wall opposite the equipment shall be _____ in..

   (a) 30
   (b) 36
   (c) 48
   (d) None of the above

17. The minimum clearance in front of electrical equipment opposite other electrical equipment with a voltage of 151 volts to 600 volts shall be _____ in..

   (a) 30
   (b) 36
   (c) 48
   (d) None of the above

18. Covers for an underground junction box that weigh over _____ lbs. shall be considered as accessible to only qualified personnel.

   (a) 20
   (b) 50
   (c) 100
   (d) 150

19. The minimum clearance required for a switchgear supplied with 4,160 volts and installed opposite an insulated wall shall be _____ ft.. (corner grounded)

   (a) 4
   (b) 6
   (c) 9
   (d) 10

20. The minimum clearance required for equipment enclosures supplied with 13,800 volts and are mounted opposite one to another shall be _____ ft.. (system is corner grounded)

   (a) 4
   (b) 6
   (c) 9
   (d) 10

6

Services

The service equipment consists of two main parts: the service conduit and service conductors, the panelboard and the overcurrent protection devices. The wiring methods for the service consists of four main parts: the service, the feeders, subfeeders and the branch-circuits. The size of the service-entrance conductors and equipment can be calculated by adding the total volt-amps of all the branch-circuits loads and dividing by the configuration of voltage utilized.

The utility company supplies power to a building utilizing two methods: an overhead service drop or an underground lateral. The service conductors usually enter a building between the meter base, CT can or splice can and service equipment. Conductors shall be permitted to be installed in conduit or cables designed for such use.

NUMBER OF SERVICES PERMITTED TO A BUILDING
ARTICLE 230, PART I

Only one service drop or lateral shall be permitted to be installed to a building. However, a second service drop or lateral shall be permitted to supply large loads and larger buildings. High-rise buildings permit additional service(s) to be installed to supply large loads that are present in multiple-occupancy buildings. Any additional service drops or laterals for the building shall have a reciprocal plaque denoting locations of each service.

NUMBER OF SERVICE DROPS OR LATERALS
230.2; FIGURES 6-1 THRU 6-7

To accommodate the load requirements for a facility, the following seven conditions shall permit more than one service to be installed:

FIRE PUMPS
230.2(A)(1)

To ensure against the interruption of power to fire pumps, a additional service shall be permitted to be installed. Fire pump overcurrent protection shall be set above the pump motor's locked rotor current rating including other accessory loads per **230.90(A), Ex. 4** with running overload protection being eliminated per **240.4(A)**. Disconnecting means rated 1000 amps or more supplied by a 277/480 volt service does not require ground-fault protection per **230.95, Ex. 2** and **240.13(3)**. **(See Figure 6-1)**

EMERGENCY LIGHTING OR POWER SYSTEMS
230.2(A)(2)

Emergency lighting or power systems shall be permitted to be supplied by a additional service. Emergency lighting or power systems will provide the needed power for lighting and equipment for the safety of occupants to exit if the main service is interrupted for any reason such as loss of the normal power. For further information, see **700.12(D)**. **(See Figure 6-2)**

SPECIAL OCCUPANCIES
230.2(B)(1)

By special permission, more than one service drop or underground lateral shall be permitted to be installed, provided there is no space available for service equipment to be mounted and OCPD's are readily accessible in each occupancy. For further information, see **230.40, Ex. 1** and **230.71(A)**.

When multiple tenants occupy individual units in an apartment complex or other types of buildings, the service drop or underground lateral shall be permitted to be classified as a multiple occupancy and served by more than one service. **(See Figure 6-3)**

BUILDING COVERING A LARGE AREA
230.2(B)(2)

Two or more services shall be permitted for buildings that cover a large area. Buildings that cover large areas usually have loads located a great distance from the service equipment. Service drops or underground laterals shall be permitted to supply buildings with panelboards and switchboards located in different areas because it is more practical than trying to use feeder-circuits which will present voltage drop problems due to their long length. **(See Figure 6-4)**

Design Tip: High-rise buildings are not included in this exception.

CAPACITY REQUIREMENTS
230.2(C)(1); (C)(2)

Load requirements in excess of 2000 amps shall be permitted to be supplied with more than one service to accommodate large loads that have been added. An additional service shall be permitted if the utility company requires more than one due to the size of the loads being added and the way they shall be served. **(See Figure 6-5)**

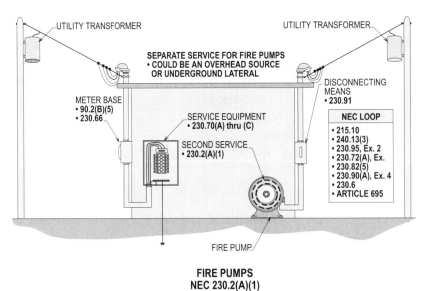

FIRE PUMPS
NEC 230.2(A)(1)

Figure 6-1. To ensure against the interruption of power to fire pumps, a additional service shall be permitted to be installed.

Services

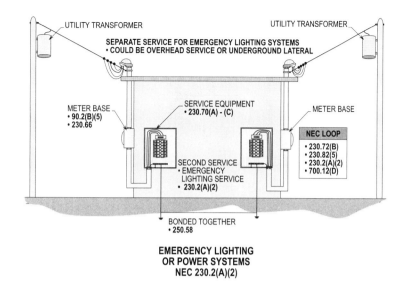

Figure 6-2. Emergency lighting or power systems shall be permitted to be supplied by a additional service.

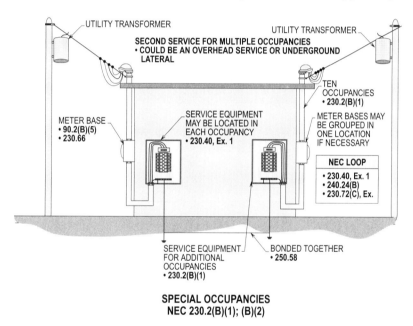

Figure 6-3. By special permission, more than one service drop or ungrounded lateral shall be permitted to be installed, provided there is space available for service equipment to be mounted and OCPD's are readily accessible to each occupancy.

DIFFERENT CHARACTERISTICS
230.2(D)

By special permission, a building of different voltages shall be permitted to have two or more service drops or underground laterals. A building operating on a special rate schedule permits an additional service to supply specific loads or equipment such as computers, etc. **(See Figure 6-6)**

For example, a service drop or lateral may consist of a 277/480 volt service and a 120/208 volt service supplying a building. The 120/208 volt service handles general-purpose outlets and equipment while the 277/480 volt service takes care of special equipment.

UNDERGROUND SETS OF CONDUCTORS
230.2

A building shall be permitted to be supplied with more than one set of underground laterals. Sections **230.40, 230.71** and **230.72** require all laterals to be grouped and located adjacent to one another. When 1/0 AWG and larger conductors are used, this type of installation shall be permitted. At the supply end, the 1/0 AWG conductor shall be connected in parallel and shall be permitted to be run and terminated to six or less main circuit breakers or disconnecting switches used exclusively for supplying an individual service. **(See Figure 6-7)**

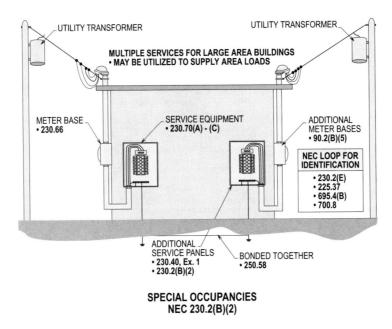

Figure 6-4. Two or more services shall be permitted for buildings that cover a large area.

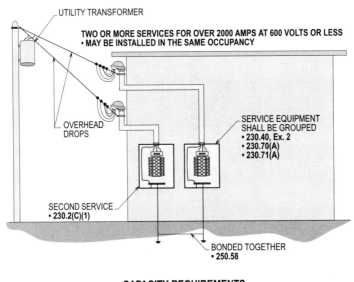

Figure 6-5. Load requirements in excess of 2000 amps shall be permitted to be supplied with more than one service.

ONE BUILDING OR OTHER STRUCTURE NOT TO BE SUPPLIED THROUGH ANOTHER
230.3

Service conductors supplying a building or structure shall not be permitted to pass through the interior of another building or structure.

Service conductors installed in a duct or conduit and under at least 2 in. (50 mm) of concrete or encased with at least 2 in. (50 mm) of concrete or brick covering shall be considered outside of the building and shall be permitted to pass under, in or on a building or structure to supply another building or structure. See **230.6** for the rules and regulations pertaining to this requirement.

Design Tip: A feeder-circuit with an OCPD ahead of it does not fall under this rule. Therefore, feeder-circuits shall be permitted to be routed through the building to another building without the need to apply **230.6**.

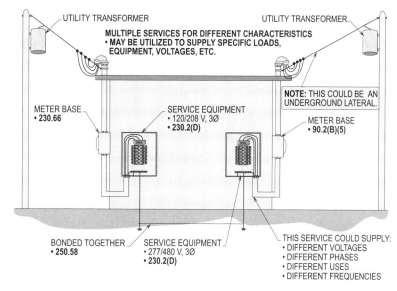

Figure 6-6. By special permission, a building of different voltages shall be permitted to have two or more service drops or underground laterals.

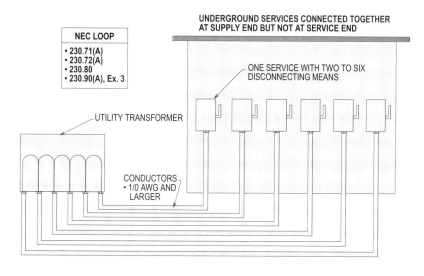

Figure 6-7. A building shall be permitted to be supplied with more than one set of underground laterals.

SELECTING INSULATION AND AMPACITY OF SERVICE CONDUCTORS
ARTICLE 230, PART I

Table 310.16 lists the different ampacities of conductors based on 60°C, 75°C or 90°C terminals per **110.14(C)**. These allowable ampacity ratings are applied and used based on conductors consisting of not more than three current-carrying conductors in a raceway and ambient temperatures of 30°C and 86°F. Ampacity ratings vary with the type of insulation used to insulate the conductor. The type of insulation selected and the terminal rating of the OCPD used determines the ampacities of the conductors. The designer shall consider and allow for adjustment factors due to four or more current-carrying conductors and correction factors if the surrounding ambient exceeds 86°F. All these factors shall be evaluated when selecting conductors to calculate allowable ampacities to supply service loads.

CONDUCTORS CONSIDERED OUTSIDE THE BUILDING
230.6

Service conductor installed in a duct or conduit and under 2 in. (50 mm) of concrete or encased with at least 2 in. (50 mm) of concrete or brick covering shall be considered outside of the building. In other words, these conductors have never entered the building and an OCPD does not have to be installed ahead of these conductors. Service conductors installed in a conduit and buried at least 18 in. (450 mm) in the earth beneath a building or other structure shall be considered outside the building. **(See Figure 6-8)**

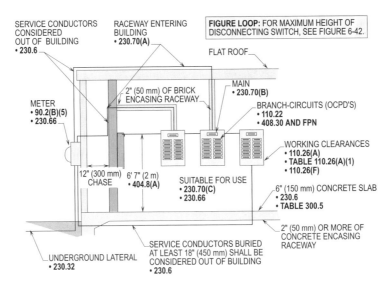

CONDUCTORS CONSIDERED OUTSIDE THE BUILDING
NEC 230.6

Figure 6-8. Service conductors installed in a duct or conduit and under 2 in. (50 mm) of concrete or encased with at least 2 in. (50 mm) of concrete or brick covering shall be considered outside of the building.

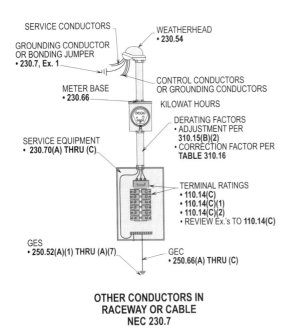

OTHER CONDUCTORS IN RACEWAY OR CABLE
NEC 230.7

OTHER CONDUCTORS IN RACEWAY OR CABLE
230.7

When installing service-entrance conductors in a service raceway or service-entrance cable, no other conductors shall be permitted. The following are two exceptions to the rule:

- When installing service-entrance conductors, the grounding conductors and bonding jumpers shall be permitted to occupy the same service raceway.

- Control conductors for load management with overcurrent protection (OCP) shall be permitted to occupy the same service raceway. **(See Figure 6-9)**

WIRING METHODS
230.43

Figure 6-9. When installing service-entrance conductors in a service raceway or service-entrance cable, no other conductors shall be permitted without an **Ex.** is applied.

The following wiring methods shall be permitted to be used when designing and installing service-entrance conductors:

- Open wiring on insulators
- Type IGS cable
- Rigid metal conduit
- IMC
- ENT
- RNC
- Service-entrance cables
- Wireways
- Busways
- Auxiliary gutters
- RNC
- Cable bus
- Type MC cable
- Mineral-insulated and metal-sheathed cables
- Flexible metal conduit and LFMC
   (a) In lengths not exceeding 6 ft. (1.8 m) per **250.102(A); (B); (C); (E)**
- Liquidtight flexible nonmetallic conduit
   (a) In lengths not exceeding 6 ft. (1.8 m) per **250.102(A); (B); (C); (E)**

CLEARANCE FROM BUILDING OPENINGS
230.9

A clearance of 3 ft. (900 mm) shall be required from windows designed to be opened, including porches, platforms, etc. A clearance of 3 ft. (900 mm) shall not be required for service conductors attached above windows for they shall be considered out of reach. Such clearance shall not be required for windows that do not open. **(See Figure 6-10)**

VEGETATION AS SUPPORT
230.10

Vegetation such as trees shall not permitted to be used as a means of support for overhead service conductors. **(See Figure 6-11)**

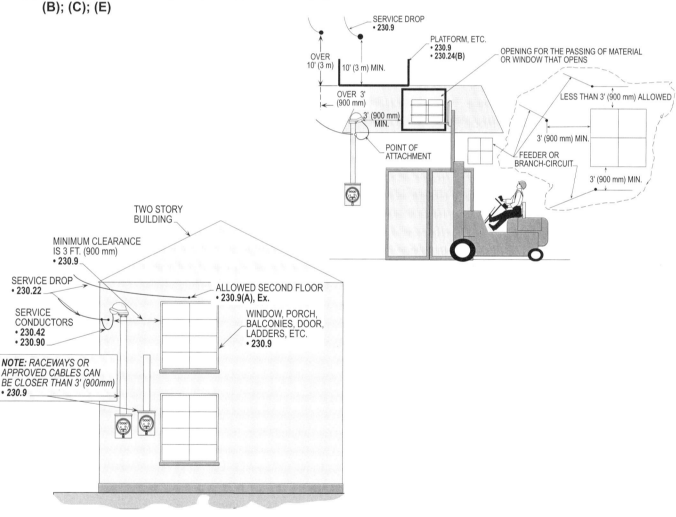

Figure 6-10. A clearance of 3 ft. (900 mm) shall be required from windows designed to be opened, including porches, platforms, etc.

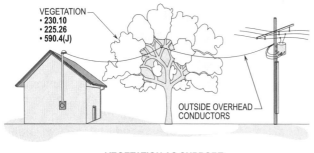

Figure 6-11. This illustration shows a tree being used as a support for overhead conductors. 'NOT ALLOWED'

OVERHEAD SERVICES
ARTICLE 230, PART II

Service drops to a building shall be permitted to be supplied from an overhead supply from the utility pole to the attachment to the building. Such shall be considered an overhead service drop. A minimum clearance from finished grade and the size and rating of the service drop conductors shall be observed when installing these conductors. The designer and installer shall consider who and what has access under these conductors. Such access determines the height of service drops from finished grade.

INSULATION OR COVERING
230.22

Service-entrance conductors shall be insulated to protect against short-circuit conditions (phase-to-phase) or ground-faults (phase-to-ground) when the service drop consists of individual conductors that supply the load of the building. The conductor insulation shall be approved for the type of service drop to be installed. Where multiconductors are installed for the service drop, a grounded (neutral) conductor shall be permitted to be uninsulated per 230.22, Ex. Thermoplastic or thermosetting insulation shall be installed where applicable, review **Table 310.13** for the different types of insulation and their specific condition of use. **(See Figures 6-12(a) and (b))**

SIZE AND RATING
230.23

Service-drop conductors shall not be smaller than 8 AWG copper or 6 AWG aluminum. Service-drop conductors shall be permitted to be 12 AWG hard-drawn copper where installed to serve small loads of only one branch-circuit. Such installations of small loads are: phone booths, small polyphase motors, etc. that do not draw large currents.

Design Tip: The service-drop grounded (neutral) conductor shall not be smaller than the minimum size calculated per **220.22** for normal current flow and for ground-fault conditions per **250.24(C)(1) and (C)(2)**.

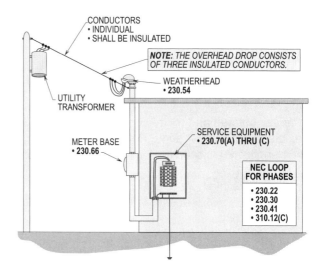

Figure 6-12(a). The conductor insulation shall be approved for the type of service-drop to be installed.

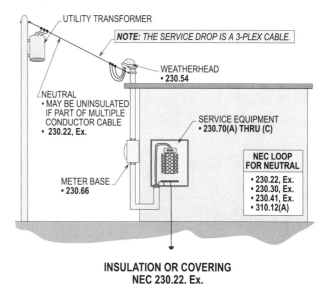

Figure 6-12(b). Where multiconductors are installed for the service-drop, a grounded (neutral) conductor shall be permitted to be uninsulated.

CLEARANCES
230.24

Service-drop conductors shall not be readily accessible for voltages of 600 volts or less. To ensure the protection of conductors and the safety of the general public, proper clearances shall be provided underneath such conductors based on people travel or vehicle travel.

ABOVE ROOFS
230.24(A)

A clearance of 8 ft. (2.5 m) shall be provided for service-drop conductors passing over roofs. At least a 3 ft. (900 mm) clearance shall be maintained in all directions for vertical clearances. These clearances shall be permitted to be reduced by applying any one of the following three exceptions: **(See Figures 6-13(a) thru (c))**

OVERHANGING PORTION OF THE ROOF
230.24(A), Ex. 3

Service-drop conductors shall be permitted to be installed for overhanging roofs provided no more than 6 ft. (1.8 m) of such conductors does not pass over more than 4 ft. (1.2 m) of the roof. The attachment of the service-drop conductors shall have a clearance of at least 18 in. (450 mm) from the conduit and roof line. **Note,** for size of conduit containing service-entrance conductors, check with local electrical ordinances and utility specifications. **(See Figure 6-13(c))**

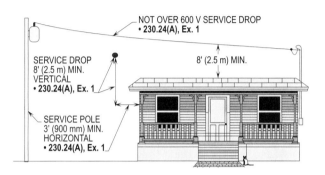

ROOFS SUBJECT TO PEDESTRIAN OR VEHICULAR TRAFFIC
NEC 230.24(A), Ex. 1

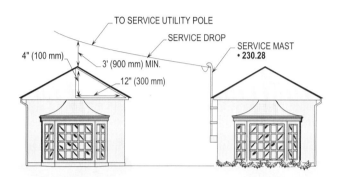

VOLTAGE NOT EXCEEDING 300 VOLTS
NEC 230.24(a), Ex. 2

Figure 6-13(a). Service-drop conductors shall have minimum clearances, complying with **230.24(B)** where they are installed above roofs subject to accessibility.

Figure 6-13(b). Service-drop conductors shall be permitted to pass above roofs where voltages between conductors does not exceed 300 volts and are installed at a height of 3 ft. (900 mm), where the roof has a slope of at least 4 in. (100 mm) by 12 in. (300 mm).

ROOFS SUBJECT TO PEDESTRIAN OR VEHICULAR TRAVEL
230.24(A), Ex. 1

Service-drop conductors shall have minimum clearances complying with **230.24(B)** where they are installed above roofs subject to pedestrian or vehicular traffic. Conductors shall have a vertical clearance of at least 10 ft. (3 m) above flat roofs that are accessible to pedestrians. **(See Figure 6-13(a))**

VOLTAGE NOT EXCEEDING 300 VOLTS
230.24(A), Ex. 2

Service-drop conductors shall be permitted to pass above roofs where voltages between conductors does not exceed 300 volts and are installed at a height of 3 ft. (900 mm), where the roof has a slope of at least 4 in. (100 mm) by 12 in. (300 mm). Service-drop conductors shall be installed at a clearance of 8 ft. (2.5 m) when the voltage between conductors exceed 300 volts, regardless of the slope of the roof. **(See Figure 6-13(b))**

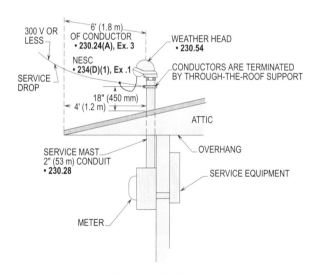

OVERHANG PORTION OF THE ROOF
NEC 230.24(A), Ex. 3

Figure 6-13(c). Service-drop conductors shall be permitted to be installed for roof overhangs provided no more than 6 ft. (1.8 m) of such conductors does not pass over more than 4 ft. (1.2 m) of the roof.

Stallcup's Electrical Design Book

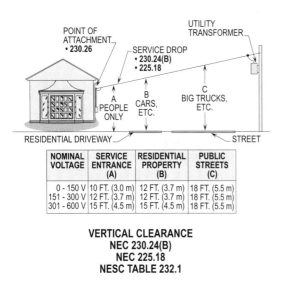

Figure 6-14. The vertical clearances from the finished grade to the service-drop conductors shall be determined by the voltage-to-ground.

VERTICAL CLEARANCE FROM GROUND 230.24(B)

The vertical clearances from the finished grade to the service-drop conductors shall be determined by the voltage-to-ground. The following are the clearances required for each service voltage for service-drop conductors crossing or penetrating property or roof. Note that there are three levels of voltage (due to ground) by which these clearances shall be based. **(See Figure 6-14)**

150 VOLTS OR LESS TO GROUND

Service-drop conductors crossing property where people have access shall have a clearance of at least 10 ft. (3 m) from finished grade. This vertical height shall be considered safe where only people have access under such conductors.

OVER 150 VOLTS TO 300 VOLTS-TO-GROUND

Service-drop conductors crossing residential property or driveways shall have a clearance of 12 ft. (3.7 m) from finished grade. Only a car or pickup truck, etc. should have access under these conductors.

OVER 300 VOLTS-TO-GROUND

Service-drop conductors crossing residential property or driveways shall have a clearance of 15 ft. (4.5 m) from finished grade where the voltage exceeds 300 volts-to-ground. The higher voltage-to-ground dictates a greater height to protect people and equipment from such voltage.

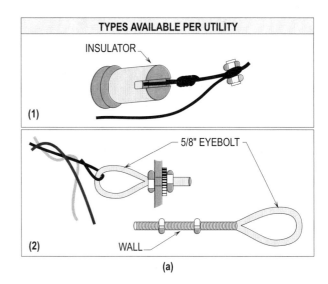

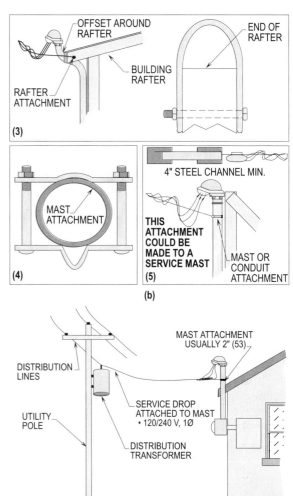

Figures 6-15(a), (b), and (c). Service-drop cables shall be permitted to be installed by attachment to the service mast pipe or to the building.

SUBJECT TO TRUCK TRAFFIC

Service-drop conductors crossing public streets, alleys, roads, parking areas subject to traffic, driveways on other than residential property and other land such as cultivated, grazing, forest and orchard shall have a clearance of 18 ft. (5.5 m) from finished grade. This higher height shall be required to ensure safety for vehicles with greater heights and loads passing under such conductors.

POINT OF ATTACHMENT
230.26

The minimum point of attachment for service-drop conductors shall be 10 ft. (3 m) from finished grade. The point of attachment shall be installed to provide the minimum clearances per **230.24**. The minimum point of attachment for service-drop conductors shall be installed based on the voltage level to ground.

> **Design Tip:** The point of attachment shall be made to the attachment fitting itself or the drip loop, whichever is greater.

MEANS OF ATTACHMENT
230.27

Service-drop cables shall be permitted to be installed by attachment to the service mast pipe or to the building. The point of attachment for service-drop conductors shall be permitted to be on a wall or rafter with fittings approved for the attachment usually dictated by utilities. **(See Figures 6-15(a) through (c))**

SERVICE MASTS AS SUPPORTS
230.28

Service-drop cables shall be supported with service masts of adequate strength to prevent damage to building structures. Service masts of adequate strength will help prevent the strain on pipe where service-drop conductors are covered with ice or snow or subject to high winds. At least a 2 in. (53) metal rigid or IMC conduit is required by most utility companies.

SUPPORTS OVER BUILDINGS
230.29

Poles shall be installed on each side of the building for service-drop conductors passing over buildings. The clearance height of these poles shall be installed by the rules and regulations of **230.24**. Service-drop conductors shall be permitted to be supported by an A-frame, cradle or cross-arm which will provide a sufficient means to attach conductors to a building. **(See Figure 6-16)**

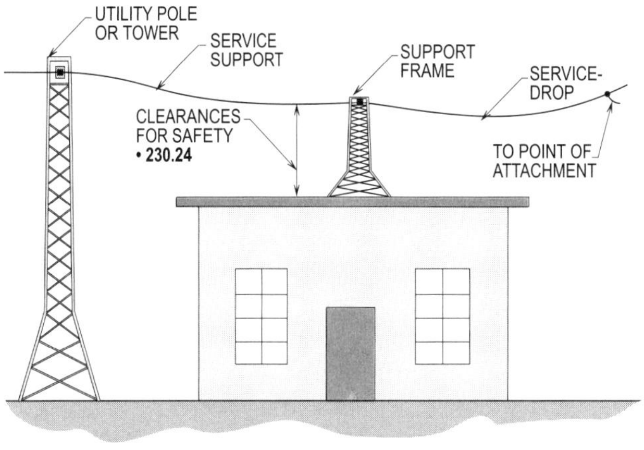

SUPPORTS OVER BUILDING
NEC 230.29

Figure 6-16. Poles shall be permitted to be installed on each side of the building for service-drop conductors passing over buildings.

UNDERGROUND SERVICE-LATERAL CONDUCTORS
ARTICLE 230, PART III

A lateral is an underground service supplying electrical power to a building. Laterals shall be permitted to be terminated to a pad-mounted transformer or run underground and up a pole and terminated to an overhead transformer. An underground lateral shall be permitted to be installed inside or outside of a building with terminations in a terminal box or meter can. The conductors run between the terminal box or meter can and service equipment shall be considered service-entrance conductors and shall be installed as such. **(See Figure 6-17)**

INSULATION
230.30

The conductor insulation shall be approved for the purpose if they are to be used as an underground service lateral. The rule calls for ungrounded (phase) conductors to be insulated so they will be protected from short-circuits and ground-faults.

UNINSULATED GROUNDED CONDUCTOR
230.30, Ex.

A copper grounded (neutral) conductor shall be permitted to be bare when used for an underground service lateral utilizing a cable assembly or installed in conduit. If soil conditions will not deteriorate the copper, a bare copper conductor shall be permitted to be installed directly in the ground. Aluminum or copper clad bare conductors shall be permitted to be installed only when used in a cable assembly routed in conduit or approved for direct burial. **(See Figure 6-18)**

SIZE AND RATING
230.31

Service lateral conductors shall not be permitted to be smaller than 8 AWG copper or 6 AWG aluminum. However, service-drop conductors shall be permitted to be 12 AWG hard-drawn copper where installed to serve small loads of only one branch-circuit. Such installation of small loads are: phone booths, small polyphase motors, etc. that require very small currents. **(See Figure 6-19)**

> **Design Tip:** Service-drop grounded (neutral) conductor shall not be permitted to be smaller than the minimum size calculated by **220.22**, and sized for fault-currents per **250.24(C)(1)** and **250.24(C)(2)** as required by **230.31(C)**.

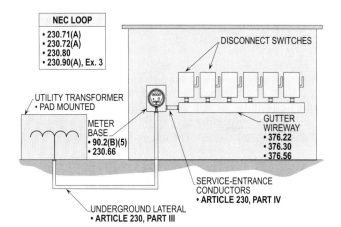

Figure 6-17. A lateral is an underground service supplying electrical power to a building.

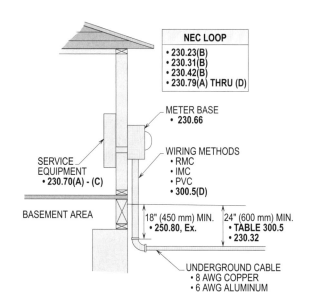

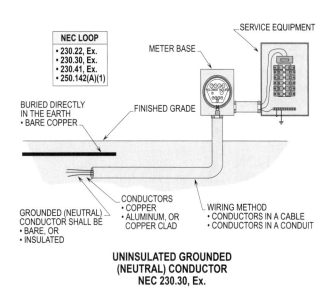

Figure 6-18. A copper grounded (neutral) conductor shall be permitted to be bare when used for an underground service lateral utilizing a cable assembly or installed in conduit.

Figure 6-19. Service lateral conductors shall not be permitted to be smaller than 8 AWG copper or 6 AWG aluminum.

SERVICE-ENTRANCE CONDUCTORS
ARTICLE 230, PART IV

Service-entrance conductors shall be sized and installed large enough to supply the calculated loads of the premises. Service-entrance conductors installed in a raceway or cable shall be insulated to protect from short-circuits and ground-faults, except the grounded (neutral) conductor shall be permitted to be bare under certain conditions of use.

NUMBER OF SERVICE-ENTRANCE CONDUCTOR SETS
230.40

Only one set of service-entrance conductors shall be supplied by a service-drop or service lateral. The following three exceptions allow service-entrance conductors to be installed by more than one set:

MORE THAN ONE OCCUPANCY
230.40, Ex. 1

Each occupancy or group of occupancies in a multiple occupancy building shall be permitted to have their service equipment supplied with one set of service-entrance conductors. Note that one to six service mains shall be permitted to be installed at such locations per **230.71(A)** and **230.72(A)**.

TWO TO SIX DISCONNECTS IN SEPARATE ENCLOSURES
230.40, Ex. 2

Two to six service disconnecting means in separate enclosures shall be permitted to be installed with one set of service-entrance conductors from a single service-drop or underground lateral.

SINGLE-FAMILY DWELLING UNITS
230.40, Ex. 3

Single-family dwelling units and separate structures shall be permitted to have one set of service-entrance conductors routed to their service equipment from a single service-drop or underground lateral.

TWO-FAMILY DWELLING OR MULTIFAMILY DWELLING
230.40, Ex. 4

A set of service-entrance conductors shall be permitted to be run to a two-family or multifamily dwelling to supply "common area" circuits such as lighting, central alarm and signal circuits.

SUPPLY SIDE OF THE NORMAL SERVICE DISCONNECTING MEANS
230.40, Ex. 5

A set of service-entrance conductors shall be permitted to be connected to the supply side of the normal service disconnecting means to supply each or several systems covered per **230.82(4) or (5)**.

INSULATION OF SERVICE-ENTRANCE CONDUCTORS
240.41

The conductor insulation shall be approved for the purpose when utilized in an underground service lateral to supply power to a building or structure.

SIZE AND RATING
230.42(B)

Service-entrance conductors shall be installed to carry the total calculated load of the premises per **Article 220**. Service-entrance conductors shall be sized and selected from the ampacities per **Tables 310.16 through 310.19**, including all applicable Notes.

The ampacity for ungrounded (phase) conductors used for service-entrance shall be sized and selected by the minimum requirements as follows:

- For limited loads of not less than 15 amps, a 15 amp service-entrance disconnecting means shall be permitted to be used per **230.79(A)**.

> **Design Tip:** Limited loads such as a motor on a single branch-circuit shall be permitted to be supplied with a 15 amp service-entrance disconnecting means.

- Not more than two, two-wire branch-circuits shall be permitted to be supplied with a 30 amp disconnecting means and a 10 AWG or 8 AWG copper service-entrance conductors per **230.79(B)**.

- A minimum 100 amp, three-wire service disconnect means shall be installed for each one-family dwelling per **230.79(C)**. The minimum size service-entrance conductors that shall be permitted to be installed are 4 AWG THWN copper and 2 AWG THWN aluminum if **Table 310.15(B)(2)(a)** is applied.

> **Design Tip:** For other types of insulation permitted, see **Table 310.15(B)(6)**.

- A minimum 60 amp service disconnect means shall be installed for a single-family dwelling with a total calculated load which allows such size per **230.79(D)**. The minimum size service-entrance conductors that are usually installed are 6 AWG THW or THWN copper or 4 AWG THW or THWN aluminum.

The ampacity for grounded (neutral) conductors for service-entrance shall be permitted to be sized and selected by the requirements listed in **220.22** and **250.24(C)(1)**. **(See Figure 6-20)**

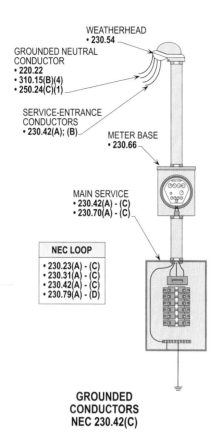

Figure 6-20. Service-entrance conductors shall be installed to carry the total calculated load of the premises per **Article 220**. Service-entrance conductors shall be sized and selected from the ampacities per **Tables 310.16 through 310.19** and **Table 310.13**.

SPLICED CONDUCTORS
230.46

Service-entrance conductors shall be permitted to be spliced or tapped in the following locations in accordance with **110.14, 300.5(E), 300.13** and **300.15**:

- Metering equipment enclosures shall be permitted to be installed with spliced service-entrance conductors per **230.46**. **(See Figure 6-21)**
- Service-entrance conductors shall be permitted to be tapped for two to six disconnecting means if grouped in a common location per **230.46**. **(See Figure 6-22)**
- Service-entrance conductors shall be permitted to be spliced where installing one wiring method to another per **230.46**. **(See Figure 6-23)**
- Service-entrance conductors shall be permitted to be spliced where an existing meter for a building has been relocated from inside to the outside of the building per **230.46**. **(See Figure 6-24)**
- Various sections and fittings shall be permitted to be installed for the connection and assembly of busways per **230.46**. **(See Figure 6-25)**
- Listed underground splice kits shall be permitted to be installed for existing service-entrance conductors per **230.46**. **(See Figure 6-26)**

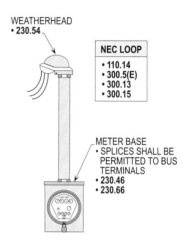

Figure 6-21. Metering equipment enclosures shall be permitted to be installed with spliced service-entrance conductors.

PROTECTION OF OPEN CONDUCTORS AND CABLES AGAINST DAMAGE - ABOVE GROUND
230.50

Service-entrance cable shall be installed and protected by rigid metal conduit, IMC, EMT, RNMC (schedule 80) or by other approved methods if subject to physical damage. Service-entrance cable shall be provided with physical protection if installed near awnings, shutters, driveways or coal chutes.

SERVICE CABLES
230.50(A)

Service-entrance cable shall be installed and protected by rigid metal conduit, IMC, EMT, RNMC (schedule 80) or by other approved methods if subject to physical damage. Such wiring methods are found in **230.43(1) through (16)** in the NEC.

Services

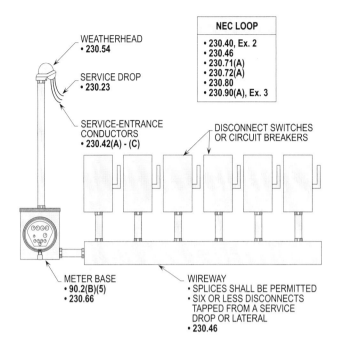

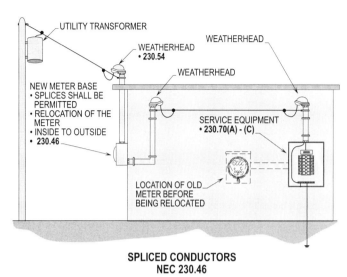

Figure 6-22. Service-entrance conductors shall be permitted to be tapped for two to six disconnecting means if grouped in a common location.

Figure 6-24. Service-entrance conductors shall be permitted to be spliced where an existing meter for a building is relocated from inside to the outside of the building.

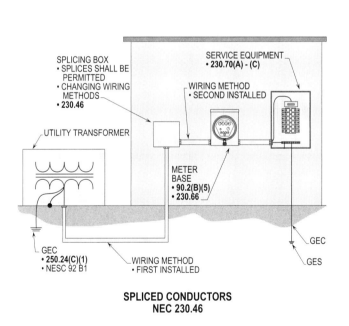

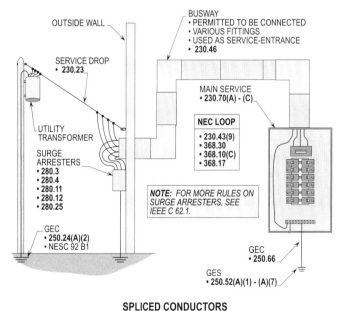

Figure 6-23. Service-entrance conductors shall be permitted to be spliced where installing one wiring method to another.

Figure 6-25. Various sections and fittings shall be permitted to be installed for the connection and assembly of busways.

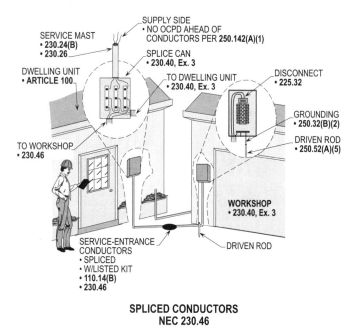

Figure 6-26. Listed underground splice kits shall be permitted to be installed for existing service-entrance conductors.

OTHER THAN SERVICE CABLE
230.50(B)

Open wiring on insulators shall be installed 10 ft. (3 m) above finished grade or be protected where exposed to physical damage.

SERVICE CABLES
230.51(A)

Service-entrance cable shall be supported at 30 in. (750 mm) intervals and within 12 in. (300 mm) of the service weatherhead when supporting cable directly to the wall. **(See Figure 6-27)**

OTHER CABLES
230.51(B)

Other cables shall be supported at 15 ft. (4.5 m) intervals on insulators if not approved to be supported directly to the wall. Other cables shall be installed with a 2 in. (50 mm) set off clearance from the wall. **(See Figure 6-27)**

INDIVIDUAL OPEN CONDUCTORS
230.51(C)

Open wiring on insulators shall be supported at specific intervals when installed inside of a building. Open wiring on insulators shall be spaced from the wall surface with a clearance to prevent the insulation from being damaged. Individually supported conductors shall have minimum clearance and support scheme complying with **Table 230.51(C)**. **(See Figure 6-28)**

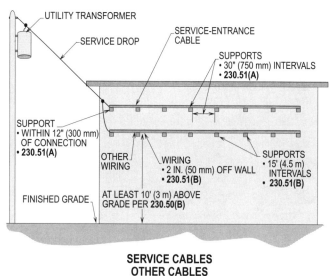

Figure 6-27. Service-entrance cables shall be supported within 30 in. (750 mm) intervals and within 12 in. (300 mm) of the service weatherhead when supporting cable directly to the wall. Other cables shall be supported at 15 ft. (4.5 m) intervals on insulators if not approved to be supported directly to the wall. Other cables shall be installed with a 2 in. (50 mm) set off clearance from the wall.

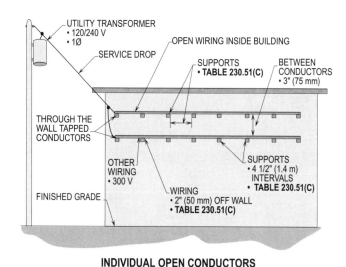

Figure 6-28. Open wiring on insulators shall be supported at specific intervals when installed inside of a building.

Services

INDIVIDUAL CONDUCTORS ENTERING BUILDINGS OR OTHER STRUCTURES
230.52

Roof bushings shall be installed where individual conductors enter a building through a roof. Nonabsorbent insulating tubes shall be installed where individual conductors enter a building through a wall. To prevent rain from entering the tubes, bushings shall be installed in a slanted direction and service-entrance conductors shall be provided with drip loops. **(See Figure 6-29)**

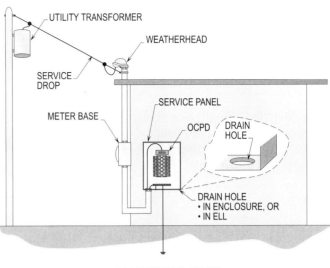

**RACEWAYS TO DRAIN
NEC 230.53**

Figure 6-30. When installing rigid metal conduit, IMC, EMT, PVC, etc. all fittings shall be raintight with a drain hole provided in the service ELL at the bottom of the run to protect service-entrance conductors.

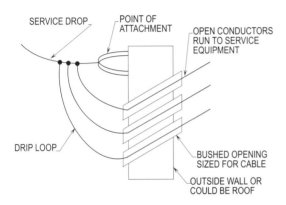

**INDIVIDUAL CONDUCTORS ENTERING
BUILDINGS OR OTHER STRUCTURES
NEC 230.52**

Figure 6-29. Roof bushings shall be installed where individual conductors enter a building through a wall.

RAINTIGHT SERVICE HEAD
230.54(A)

If used, a raintight (weatherproof) service head shall be installed at the top of the service-entrance raceway. Individual holes through the service head allows for the termination and protection of the service-drop and service-entrance conductors after they are spliced together.

SERVICE CABLE EQUIPPED WITH RAINTIGHT SERVICE HEAD OR GOOSENECK
230.54(B)

RACEWAYS TO DRAIN
230.53

When installing rigid metal conduit, IMC, EMT, PVC, etc. all fittings shall be raintight with a drain hole provided in the service ELL at the bottom of the run to protect service-entrance conductors. A means to drain condensation shall always be provided to prevent corrosion of metal. **(See Figure 6-30)**

Service-entrance cable shall be installed in the form of a gooseneck and then taped and painted or plastic coated for protection from the weather. The entrance of rain through a service head shall be kept to a minimum. When installing weatherproof fittings, care shall be taken to make good tight connections and use sealing material properly. **(See Figure 6-31)**

OVERHEAD SERVICE LOCATIONS
230.54

When installing service cable or individual conductors in a raceway, the service-entrance shall be installed above the service-drop conductors.

SERVICE HEADS AND GOOSENECKS ABOVE SERVICE-DROP ATTACHMENT
230.54(C)

The service-drop shall not be permitted to be installed above the service-entrance conductors where the overhang of the roof is too low and prevents proper installation of the service-drop. The **Ex. to 230.54(C)** requires the service head to be connected within 24 in. (600 mm) from the point of attachment when it is impracticable to locate above the point of attachment.

> **Design Tip:** When connecting the service-drop to the point of attachment, a straight-line distance is used to determine the 24 in. (600 mm) distance. **(See Figures 6-32(a) and (b))**.

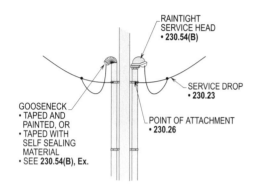

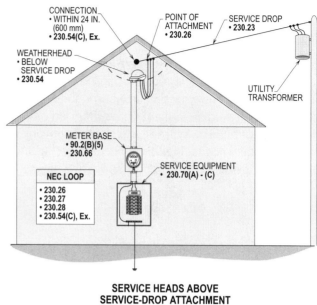

Figure 6-31. Service-entrance cable shall be installed in the form of a gooseneck and then taped and painted or plastic coated for protection from the weather.

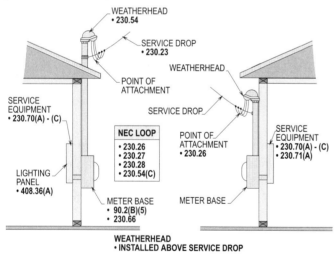

Figure 6-32(a). The service-drop shall not be permitted to be installed above the service-entrance conductors where the overhang of the roof is too low and prevents proper installation of the service-drop.

Figure 6-32(b). When connecting the service-drop to the point of attachment, a straight-line distance is used to determine the 24 in. (600 mm) distance. **Note:** Get the approval of the AHJ for the above installation.

SERVICE EQUIPMENT TERMINATION (RECOMMENDED PROCEDURE)

Terminal boxes, cabinets, disconnecting switches, panelboards, switchgear or similar enclosures shall be permitted to be installed for the termination of any service raceway or cable located in the building. Live parts of the service equipment shall be enclosed to protect personnel from injury. When using open-type switchboards, a terminating box shall not be required for the termination of service-entrance conductors. Conductors shall be protected with bushings where leaving the conduit.

SERVICE CONDUCTOR WITH THE HIGHER VOLTAGE-TO-GROUND 230.56

When installing a three-phase, four-wire, delta-connected service, where the midpoint of one phase winding is grounded, one phase leg will have a higher voltage-to-ground. An outer finish marked orange or identified by other effective means shall be used for the conductor at the higher voltage-to-ground. This identification will help prevent an electrician from connecting 120 volt loads to the 208 volt-to-ground (high-leg), for great damage may occur to 120 volt equipment if supplied by the high-leg. **(See Figure 6-33)**

6-18

Services

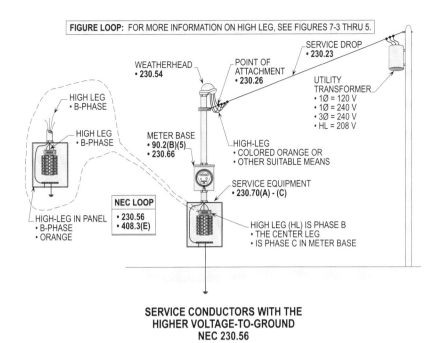

Figure 6-33. When installing a three-phase, four-wire, delta-connected service, where the midpoint of one phase winding is grounded, one phase leg will have a higher voltage-to-ground. An outer finish marked orange or identified by other effective means shall be used for the conductor with a higher voltage-to-ground.

SERVICE EQUIPMENT - GENERAL ARTICLE 230, PART V

Energized parts of service equipment shall be enclosed so that personnel are not exposed to accidental contact with live parts. Energized parts shall be guarded for the protection of personnel where not enclosed, an electrical worker could make contact with live parts or create a short-circuit or ground-fault condition. Service equipment shall be grounded and bonded to ensure the safe and effective operation of the electrical system and clear any circuit subjected to a short-circuit.

SERVICE EQUIPMENT - ENCLOSED OR GUARDED
230.62

Energized parts of service equipment shall be enclosed so that personnel are not exposed to accidental contact with live parts while near or working on such equipment. **Sections 110.27(A), 110.33** and **110.34(A)** requires energized parts to be guarded to protect electrical personnel from dangerous arcs due to short-circuits and to the threat of electrical shock hazards. (See **pages 6 through 9 of chapter 5**)

GROUNDING AND BONDING (RECOMMENDED ITEMS TO BE GROUNDED AND BONDED)

Service equipment and elements shall be grounded per **250.80**. The following elements of the service equipment shall be grounded in such a manner as to provide an effective grounding path:

- Service equipment
- Meter base housing
- Service disconnecting means and enclosure
- Metal sheath of cable
- Metal armor of cable
- Grounded (neutral) conductor
- Grounded (phase) conductor

(**See Chapter 11** for grounding rules)

AVAILABLE SHORT-CIRCUIT CURRENT 110.9; 110.10

Service equipment shall be designed and installed to withstand the available fault-current that could be delivered to service equipment terminals from the utilities or customers transformer. Available fault-currents are caused by phase-to-phase short-circuits or phase-to-ground faults. Circuit breakers, fuses and other current-carrying elements shall be designed and installed to have an interrupting capacity equal to the short-circuit current that the equipment may be called upon to carry. Equipment shall be capable of handling and carrying high currents until the circuit is cleared. Current-limiting fuses or service rated circuit breakers shall be permitted to be installed to help alleviate this problem. The let-through current of these fuses shall be limited to a value that will not damage the service equipment and its components.

Overcurrent protection devices shall be installed and selected to ensure proper interrupting rating and provide protection of components from the power company's transformer in case of a short-circuit condition. The overcurrent protection shall be designed and installed so that the interrupting rating has the ability to interrupt or open the circuit under short-circuit or ground-fault conditions.

The rating of the supply voltage and the resistance of the conductors, due to length and size, may limit the available fault-current to a safe level, if designed properly. A greater short-circuit or ground-fault will occur at the service equipment than a short-circuit down stream on a feeder or branch-circuit. Fault-current is reduced by the accumulated resistance of the conductors and circuit elements located between the service and equipment. The point where the short-circuit and ground-fault occurs in the electrical system determines the amount of fault-current that will flow.

Higher values of fault-current will be produced for short runs of conductors to the service equipment. This would be the case if, the transformer is located close to the building. Lower amounts of fault-current will be produced, in most cases, to a safe value for long runs of conductors from the transformer to the service equipment. In this case, the transformer is located some distance away from the building and the service equipment. **(See Figure 6-34)**

MARKING
230.66

Service equipment rated at 600 volts or less shall be marked to identify as being suitable for use as service equipment. Individual meter socket enclosures shall not be considered as service equipment.

SERVICE EQUIPMENT - DISCONNECTING MEANS
ARTICLE 230, PART VI

A means to disconnect the service shall be provided at each service. All ungrounded (phase) conductors shall be disconnected from the power supply. The service disconnecting means (one or more) shall be permanently marked and each grouped in a common location when more than one service supplies a building.

LOCATION
230.70(A)

The service disconnecting means shall be located in a readily accessible location for all premises. The service disconnecting means shall be located as close as possible to where the service-entrance conductors enter the premises. Service disconnecting means shall not be permitted to be located in bathrooms. The service disconnecting means shall be located in accordance with **230.70(A)(1)** where a remote-control device(s) is used to activate the service disconnecting means. **(See Figure 6-35)**

> **Design Tip:** The service disconnecting means shall be permitted to be located on the outside or inside wall, such location complies with **230.70(A)**.

MAXIMUM NUMBER OF DISCONNECTS
230.71

No more than six fusible switches or six circuit breakers mounted in a single enclosure, a group of separate enclosures, or in or on a panelboard or switchboard shall be installed for each set of service-entrance conductors. A power panel shall be permitted to have up to six circuit devices in a suitable enclosure. Six circuit breakers in a single enclosure shall be permitted to be used to supply air-conditioners, water heaters, heating units, ranges, dryers and general-purpose circuits for lighting, receptacles and appliances. **(See Figure 6-36)**

A main circuit breaker shall be permitted to installed ahead of other circuit breakers in a panel and one throw of the hand shuts the power OFF to all other circuit breakers and loads served. The disconnecting means for the service shall also be permitted to be six separate disconnecting switches with fuses or six separate circuit breakers, each in single enclosures.

> **Design Tip:** The entire service shall be designed so the service devices are shut OFF with six throws of the hand. **(See Figure 6-36)**

Services

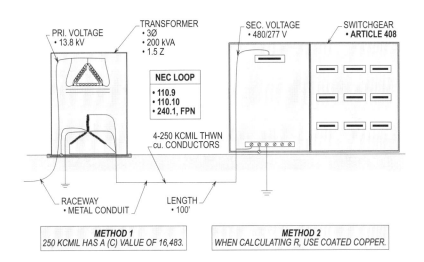

AVAILABLE SHORT-CIRCUIT CURRENT (ASCC) AT THE TERMINALS OF THE SERVICE EQUIPMENT

Calculating ASCC at service equipment using METHOD 1

Step 1: Calculating FLC of transformer
FLC = kVA x 1000 ÷ V x √3
FLC = 200 kVA x 1000 ÷ 480 V x 1.732
FLC = 241 A

Step 2: Calculating ASCC at transformer
ASCC = TLFC ÷ Z
ASCC = 241 A ÷ .015 (1.5%)
ASCC = 16,067 A

Step 3: Calculating F value using manufacturer Table of C values Table B of Appendix
F = 1.732 x L x I(ASCC) ÷ C x V(L TO L)
F = 1.732 x 100' x 16,067 ÷ 16,483 x 480
F = 2,782,804 ÷ 7,911,840
F = .352

Step 4: Calculating M value
M = 1 ÷ 1 + F
M = 1 ÷ 1 + .352
M = .740

Step 5: Calculating line to line ASCC at service equipment
ASCC = ASCC x M
ASCC = 16,067 x .740
ASCC = 11,890 A

Solution: The ASCC at the terminals of the service equipment is 11,890 amps.

AVAILABLE SHORT-CIRCUIT CURRENT (ASCC) AT THE TERMINALS OF THE SERVICE EQUIPMENT

Calculating ASCC at service equipment using METHOD 2

Step 1: Calculating ASCC of transformer
ACSS = TFLC ÷ Z
ASCC = 241 A ÷ .015 (1.5%)
ASCC = 16,067 A

Step 2: Calculating R of 100' of 250 KCMIL Table A of Appendix
R = .054 x 100 ÷ 1,000
R = .0054

Step 3: Calculating R of transformer
R = V to GRD. ÷ ASCC
R = 277 V ÷ 16,067 A
R - .01724

Step 4: Calculating Total R
R of 250 KCMIL   = .0054
R of transformer   = .01724
Total R   = .02264

Step 5: Calculating ASCC of service equipment
ASCC = V to GRD. ÷ total R
ASCC = 277 V ÷ .02264
ASCC = 12,235 A

Solution: The ASCC at the terminals of the service equipment is 12,235 amps.

AVAILABLE SHORT-CIRCUIT CURRENT
NEC 110.9; 110.10

Figure 6-34. Service equipment shall be designed and installed to withstand the available fault-current that could be delivered to service equipment terminals from the utilities or customers transformer.

Stallcup's Electrical Design Book

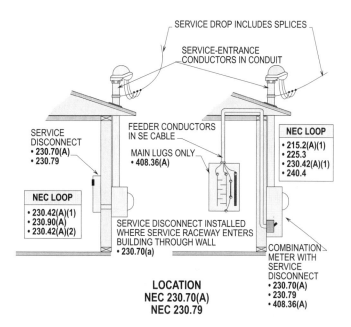

Figure 6-35. The service disconnecting means shall be located in a readily accessible location for all premises.

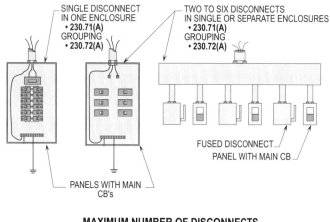

Figure 6-36. No more than six fusible switches or circuit breakers mounted in a single enclosure, a group of separate enclosures or in or on a panelboard or switchboard shall be installed for each set of service-entrance conductors.

GROUPING OF DISCONNECTS
230.72

The service disconnecting means shall be grouped in a common location and be readily accessible so the entire service may be disconnected with six throws of the hand. Dwelling units, apartment complexes, commercial facilities or other types of premises shall be permitted to have their service designed with two to six disconnecting means. The disconnecting means shall be marked to identify the load they serve. Disconnects shall be grouped in a common location where six or less are installed. A main disconnect shall be installed ahead of the disconnects if more than six disconnects are utilized. The following are two exceptions that allow switches to be ungrouped: **(See Figure 6-37)**

- Services installed per **230.2**
- Remote water pumps

SIMULTANEOUS OPENING OF POLES
230.74

The disconnecting means shall disconnect all service-entrance conductors from the power supply simultaneously. Ungrounded (phase) conductors shall be disconnected by a disconnecting switch or circuit breaker that is capable of disconnecting such conductors without rupturing. Grounded (neutral) conductors shall be permitted to be disconnected by the service-entrance disconnecting means. **(See Figure 6-38)**

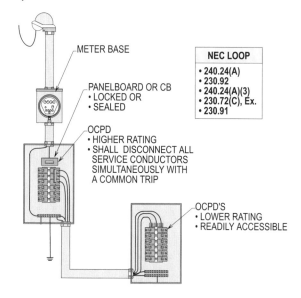

Figure 6-38. The disconnecting means shall disconnect all service-entrance conductors from the power supply simultaneously.

DISCONNECTION OF GROUNDED CONDUCTOR
230.75

If the grounded (neutral) conductor is removed for any reason, a means shall be provided to disconnect the grounded (neutral) conductor safely. The disconnection of

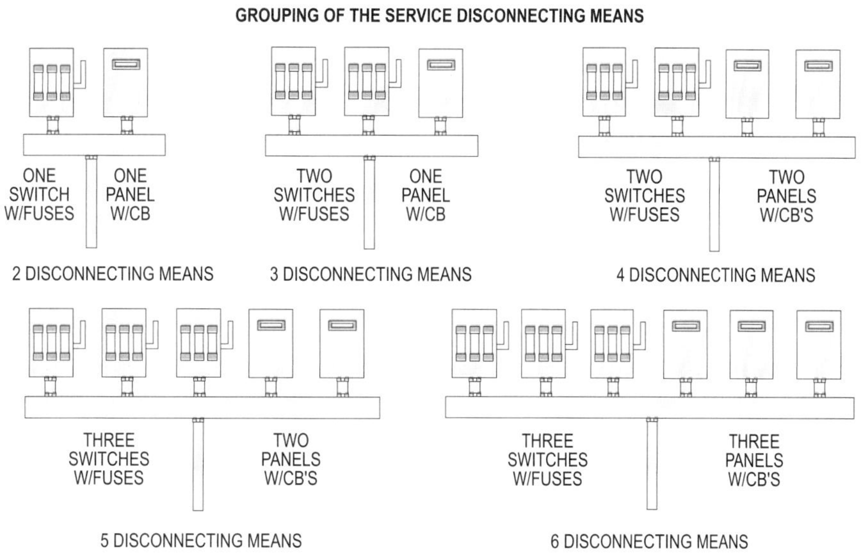

Services

GROUPING OF THE SERVICE DISCONNECTING MEANS

[Diagrams showing 2, 3, 4, 5, and 6 disconnecting means configurations]

GROUNDING OF DISCONNECTS
NEC 230.72

Figure 6-37. The service disconnecting means shall be grouped in a common location and readily accessible so the entire service may be disconnected with six throws of the hand.

the grounded (neutral) conductor is usually removed from the terminal bar or bus manually by using proper tools designed for such purpose. Note, the grounded (neutral) conductor shall be permitted to be terminated in a different vertical section than the ungrounded (phase) conductors if the switchboard is of the multisection type. **(See Figure 6-39)**

RATING OF SERVICE DISCONNECTING MEANS
230.79

The main disconnecting means shall have a rating large enough to carry the calculated load per **Article 220**. The rules are as follows per **Subdivisions (A), (B)** and **(C)** to **230.79**.

ONE CIRCUIT INSTALLATION
230.79(A)

The service disconnecting means shall be installed at a rating of 15 amps for single branch-circuit loads supplying limited loads only.

TWO-CIRCUIT INSTALLATION
230.79(B)

The service disconnecting means shall be installed at a rating of 30 amps for one or two, two-wire branch-circuit loads. Note, these loads are small loads which require small amounts of current (amps) to operate.

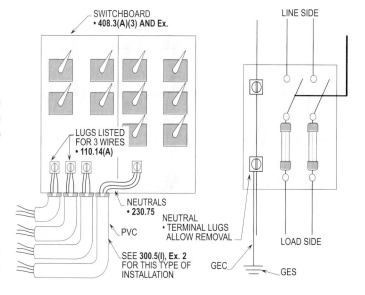

DISCONNECTION OF GROUNDED CONDUCTOR
NEC 230.75

Figure 6-39. If the grounded (neutral) conductor is removed for any reason, a means shall be provided to disconnect the grounded (neutral) conductor safely.

ONE-FAMILY DWELLING
230.79(C)

A three-wire service disconnecting means shall be installed at a rating of 100 amps for each dwelling unit. The service disconnecting switch shall be installed at a rating of at least 60 amps for all other installations below 100 amps. For applications of this rule, check with AHJ.

COMBINED RATING OF DISCONNECTS
230.80

The combined rating for one to six switches or circuit breakers shall not be permitted to be less than the required rating for a single switch or circuit breaker. For further information, see **230.90(A), Ex. 3**.

For example, a number of conductors servicing six service disconnects has an ampacity of 1200 amps. This service shall be permitted to have six switches protected with 200 amp fuses respectfully. Now, consider a number of conductors supplying six service disconnects that have a total ampacity of 765 amps. The total number of OCPD's (6) shall not be permitted to exceed 700 or 800 amps. (General Rule)

EQUIPMENT CONNECTED TO SUPPLY SIDE OF SERVICE DISCONNECT
230.82

The following are certain types of equipment that shall be permitted to be connected to the supply (line) side of the service disconnecting means:

- Cable limiters and other current-limiting devices
- Meters and meter sockets not rated over 600 volts with grounded enclosures
- Meter disconnect switches not rated over 600 volts with grounded enclosures
- Instrument transformers (current and voltage), high-impedance shunts, load management devices and surge arresters
- Taps for specific use
- Solar photovoltaic systems, fuel cell systems or interconnected electric power production sources
- Control circuits
- Ground-fault protection systems or transient voltage surge suppressors

Design Tip: There are not any OCPD's between the transformer, conductors and loads served. Designers must not confuse this tap rule with **240.21** and **240.90** which deals with conductors having an upstream OCPD.

TRANSFER EQUIPMENT RATED AS SERVICE EQUIPMENT

A transfer switch shall be used to disconnect the main service before transferring and connecting the alternate source of power. For further information, see **Article 100**. Transfer switches shall be permitted to be of the three-pole or four-pole type. If a three-pole transfer switch is used, the alternate source shall not be considered a separately derived system. When a four-pole transfer switch is used, because the grounded (neutral) conductor can be switched, the alternate source of power shall be considered a separately derived system. **(See Figure 6-40)**

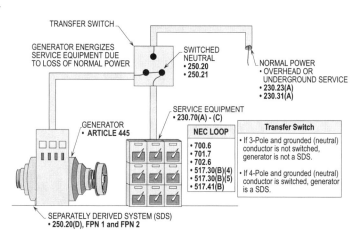

TRANSFER EQUIPMENT RATED
AS SERVICE EQUIPMENT
ARTICLE 100

Figure 6-40. A transfer switch shall used to disconnect the main service before transferring and connecting the alternate source of power.

NUMBER OF SUPPLIES
225.30

A separate disconnecting means shall be installed if more than one building is on the same property and under the same management. In other words, each separate building shall have a disconnecting means to completely disconnect all ungrounded (phase) conductors supplying or passing through such building or structure per **225.31**. **(See Figure 6-41)**

LOCATION
225.32

Each building shall have the disconnecting means installed in a location, either outside or inside, in or on each building in a readily accessible location and as close as possible where the conductors enter in the building. This disconnect shall be permitted to be provided with OCP or be of the nonfused or nonautomatic type.

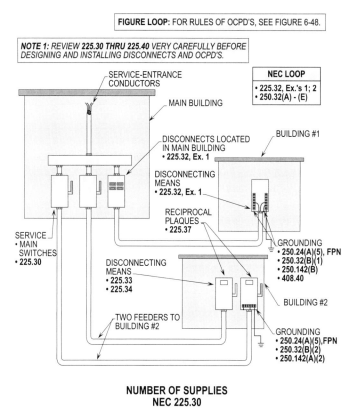

Figure 6-41. A separate disconnecting means shall be installed if more than one building is on the same property and under the same management. However, disconnects shall be permitted to be located in the main building per **Ex. 1 to 225.32**.

SUITABLE FOR SERVICE EQUIPMENT
225.36

The disconnecting means shall be rated as a type suitable for use as service equipment. A set of three-way or four-way snap switches shall be permitted to be installed to disconnect garages or outbuildings on residential property. For further information regarding this rule, see **225.36, Ex.** and **250.32(A), Ex.**

SERVICE EQUIPMENT - OVERCURRENT PROTECTION ARTICLE 230, PART VII

The general rule requires each ungrounded (phase) conductor to be provided with overcurrent protection devices to protect the service equipment. The rating of service-entrance conductors shall not be permitted to be exceeded when sizing overcurrent protection devices to protect the allowable ampacities of such conductors. However, there are exceptions which permit the OCPD to exceed the ampacity of the conductors.

UNGROUNDED CONDUCTOR
230.90(A)

The ampacity rating of service-entrance conductors shall not be permitted to be exceeded when installing overcurrent protection devices. A set of fuses shall be permitted where all the fuses required to protect all ungrounded (phase) conductors of a circuit. Single-pole circuit breakers, grouped in accordance with **230.71(B)**, shall be considered one protective device. **(See Figure 6-42)**

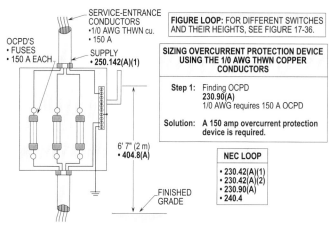

Figure 6-42. The ampacity rating of service-entrance conductors shall not be permitted to be exceeded when installing overcurrent protection devices.

MOTORS
230.90(A), Ex. 1

Motor conductors shall be permitted to be overfused for the start-up current per **430.52, 430.62** or **430.63**. Overcurrent protection devices shall be sized to carry the high inrush current until the motor accelerates up to its running speed. **(See Figure 6-43)**

FUSES AND CIRCUIT BREAKERS
230.90(A), Ex. 2

When the setting or rating does not correspond to the allowable capacity of service conductors, the next higher standard fuse or circuit breaker shall be permitted to be installed. This type of installation shall be permitted to be installed per **240.4(B)** or **(C)**. The standard ratings of fuses or circuit breakers are listed in **240.6(A)**. **(See Figure 6-44)**

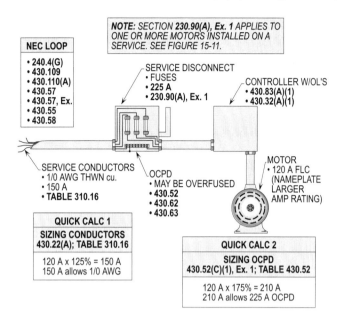

Figure 6-43. Overcurrent protection devices shall be sized to carry the high inrush current until the motor accelerates up to its running speed.

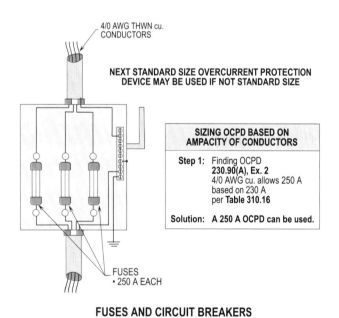

Figure 6-44. When the setting or rating does not correspond to the allowable capacity of service conductors, the next higher standard fuse or circuit breaker shall be permitted to be installed.

SIX CIRCUIT BREAKERS OR SIX SETS OF FUSES
230.90(A), Ex. 3

The number of overcurrent protection devices permitted for a service shall be one to six circuit breakers or six sets of fuses installed in a service-entrance panel. **Note,** that there shall be permitted to be up to six separate enclosures with a single CB or single set of fuses in each. **(See Figure 6-45)**

For example, a 400 amp service-entrance could be installed with six 70 amp circuit breakers. The total calculated rating would be 420 amps, which exceeds the 400 amp rating of the service-entrance. The service shall be permitted to be protected at 420 amps per **240.4(B)**, but the ampacity of the service conductors is limited to the load which shall not be permitted to exceed 400 amps per **230.42(A)** and **(B)**.

FIRE PUMPS
230.90(A), Ex. 4

The overcurrent protection device shall be sized high enough to carry the locked-rotor current for a service supplying a fire pump. The overcurrent protection device shall not clear until a short-circuit or ground-fault develops. This designing technique allows the motor to pump water as long as possible to fight the fire. **(See Figure 6-46)**

TABLE 310.15(B)(6) TO AMPACITY TABLES OF 0 - 2,000 VOLTS
230.90(A), Ex. 5

This exception permits smaller size service-entrance conductors to be installed than the calculated load of the dwelling unit. However, to obtain this smaller rating than calculated by the standard or optional method, **310.15(B)(6)** to ampacity Tables 0-2000 volts shall be applied. **(See Figure 6-47)**

NOT IN GROUNDED CONDUCTOR
230.90(B)

Grounded (neutral) conductor or ungrounded (phase) conductors shall be permitted to be fused unless it is installed in a common trip circuit breaker. An individual fuse shall not be permitted to be installed for the grounded (neutral) conductor or ungrounded (phase) conductor, for all circuit conductors would not open at the same time if an individual fuse was to blow.

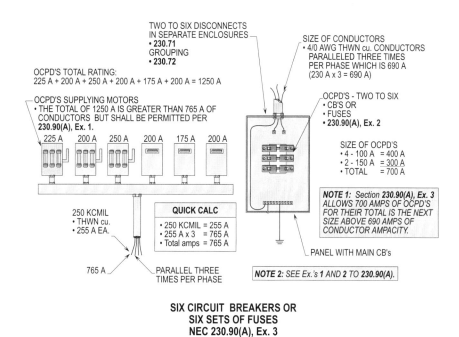

Figure 6-45. The number of overcurrent protection devices permitted for a service shall be one to six circuit breakers or six sets of fuses installed in a service-entrance panel or separate panelboards and switches.

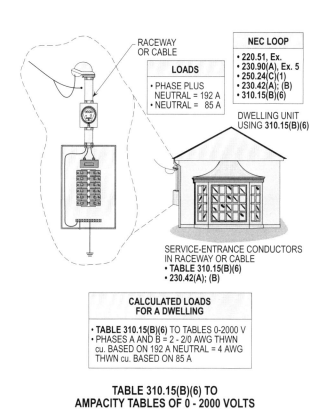

Figure 6-47. This exception permits smaller size service-entrance conductors to be installed than the calculated load of the dwelling unit.

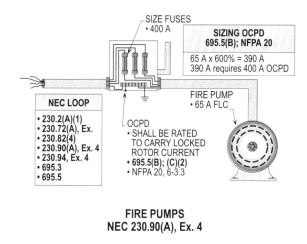

Figure 6-46. The overcurrent protection device shall be sized high enough to carry the locked-rotor current for a service supplying a fire pump.

LOCATION
230.91

The service overcurrent protection device shall be an integral part of the service disconnecting means and located immediately adjacent to the service. A fused disconnecting switch or individual CB in a panel would comply with this requirement.

UNDER SINGLE MANAGEMENT
225.32, Ex. 1

Overcurrent protection devices shall be permitted to be grouped in one building where a property consists of more than one building or structure under single management per **240.24(B), Ex. 1**. Feeder-circuits shall be permitted to be installed to supply power to each of the other buildings or structures, provided all occupants of that building have access to the disconnecting means to each building or structure. Overcurrent protection devices shall be permitted to be installed at the first building, with the disconnecting means installed at each of the other buildings or structures. **(See Figure 6-48)**

OCCUPANCY
240.24(B), Ex. 1

The main service-entrance conductors shall be permitted to be tapped without a main disconnecting means installed ahead of such disconnects. The disconnecting means and overcurrent protection devices for the service shall be provided with access for each occupant or user. However, the designer shall refer to **240.24(B), Ex. 1**, which allows the OCPD's to be grouped in a single location under certain conditions of use. **(See Figure 6-48)**

LOCKED SERVICE OVERCURRENT DEVICES
230.92

Fuses and circuit breakers supplying an individual feeder or branch-circuit and enclosures that are locked or sealed, additional fuses or circuit breakers shall be permitted to be installed but they shall have a lower rating and shall be accessible to the user. The enclosure for fuses and circuit breakers shall be permitted to be locked in the closed position or individually locked, if the downstream OCPD's are readily accessible. The question is "readily accessible" to whom, for such interpretations, see **240.24(A) and (B), 230.72(C), Ex.** and **Article 100**. **(See Figure 6-49)**

PROTECTION OF SPECIFIC CIRCUITS
230.93

A connection ahead of the meter shall be permitted to be installed ahead of the main to supply a flat rate circuit for a water heater, heating equipment or other such appliances. Overcurrent protection devices protecting the circuit shall be permitted to be locked to prevent access to unauthorized personnel. **(See Figure 6-50)**

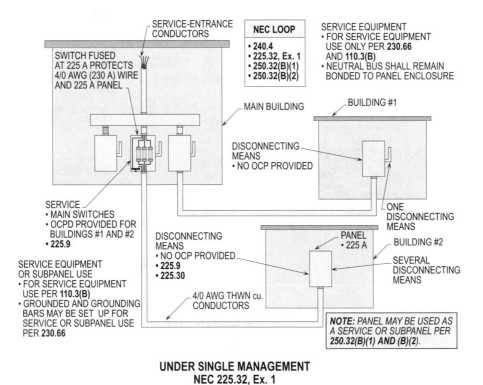

UNDER SINGLE MANAGEMENT
NEC 225.32, Ex. 1

Figure 6-48. Overcurrent protection devices shall be permitted to be grouped in one building where a property consists of more than one building or structure under single management. (For disconnecting rules, See **Figure 6-41**)

Services

GROUND-FAULT PROTECTION OF EQUIPMENT
230.95

Solidly grounded wye electrical services of more than 150 volts-to-ground, having a service disconnecting means rated 1000 amps or more, must be protected by ground-fault protection. The service disconnecting means shall be rated for the largest fuse that may be installed or the highest continuous current trip setting for which the actual overcurrent device rating (circuit breaker) may be adjusted. **(See Figure 6-51)**

SETTING
230.95(A)

The maximum setting permitted for ground-fault protection shall not be permitted to exceed 1200 amps, when the service disconnecting means exceeds 1000 amps or more. The ground-fault protection shall permit all ungrounded (phase) conductors of the faulted circuit to open, until such circuit(s) is clear of the low-magnitude fault.

> **Design Tip:** Ground-fault protection shall not be required for six OCPD's that are not equal to or greater than 1000 amps.

To prevent damage to overcurrent protection devices, busbars and service equipment, ground-fault systems shall clear a fault of 3000 amps or more within one second.

The system will be cleared once the low-magnitude fault-currents are detected and the overcurrent protection device is tripped open. Such trip will prevent low-magnitude-faults from burning and damaging the electrical system beyond use.

FUSES
230.95(B)

When installing a switch and fuse combination, the fuses shall be capable of interrupting any current higher than the interrupting capacity of the switch during a time until the ground-fault protective system causes the switch to open and clear the faulted circuit.

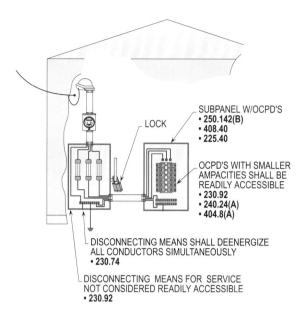

Figure 6-49. Fuses and circuit breakers supplying an individual feeder or branch-circuit located in enclosures that are locked or sealed, additional fuses or circuit breakers shall be permitted to be installed but they shall have a lower rating and shall be accessible to the user.

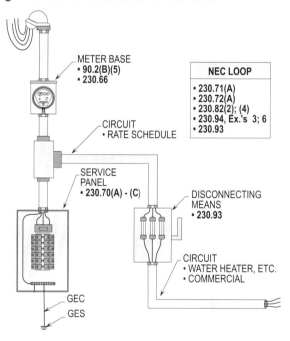

Figure 6-50. A connection ahead of the meter shall be permitted to be installed ahead of the main to supply a flat rate circuit for a water heater, heating equipment or other such appliances.

6-29

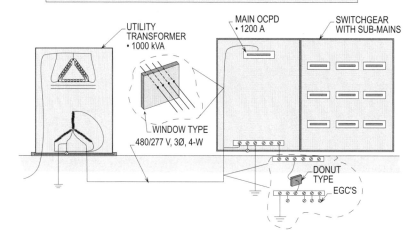

NOTE 1: TO CALCULATE THE MAXIMUM ALLOWABLE ENERGY LEVEL, ALWAYS USE 200,000 A CYCLES AS A CONSTANT.

NOTE 2: FLA OF TRANSFORMER TIMES 6 PRODUCES MAXIMUM GROUND-FAULT AMPS (MGFA). SIX (6) IS ALWAYS USED AS A CONSTANT. (RULE OF THUMB)

SIZING MAXIMUM SIZE TIME DELAY AND TRIP SETTING OF THE GROUND-FAULT SENSOR AT THE MAIN DISCONNECT

Calculating Maxium Time Delay

Step 1: Selecting cycles of max. allowable energy level (MEL)
MEL = 200,000 A cycles

Step 2: Calculating FLA of transformer
FLA = kVA x 1000 ÷ V x $\sqrt{3}$
FLA = 1000 x 1000 ÷ 480 V x 1.732
FLA = 1203 A

Step 3: Calculating max. ground-faults A (MFGA)
MFGA = Transformer FLA x 6
MFGA = 1203 A x 6
MFGA = 7218

Step 4: Calculating time delay cycles (TDC)
TDC = Step 1 ÷ Step 3
TDC = 200,000 A cycles ÷ 7218 A
TDC = 27.7 cycles

Solution: The maximum time delay cycles is 27.7 cycles.

Calculating the Trip Setting (TS)

Step 1: Calculating the setting
TS = OCPD x 20%
TS = 1200 A x 20%
TS = 240 A

Solution: The trip setting of the ground-fault sensor is 240 amp.

Note 1: The trip setting may be calculated from 10% to 50% of OCPD as long as 400 amp setting is not exceeded.

For Example: Maximum setting is 399.6 (1200A x 33.3% = 399.6 A) However, most designers start with 20%. The NEC allows a maximum setting of 1200 amps under certain conditions of use

Note 2: Manufacturers usually recommend no higher setting than 20 percent of OCPD rating.

GROUND-FAULT PROTECTION OF EQUIPMENT
NEC 230.95

Figure 6-51. Solidly grounded wye electrical services of more than 150 volts-to-ground having a service disconnecting means rated 1000 amps or more shall be protected by ground-fault protection.

PERFORMANCE TESTING
230.95(C)

The following two types of ground-fault detection are used to sense low-magnitude ground-faults:

- Donut method
- Window method

When applying the donut method to sense any current leaking to ground, a current relay shall be placed over the main bonding jumper. When applying the window method to sense any current leaking to ground, the ungrounded (phase) conductors and grounded (neutral) conductors shall be placed in the window (ground-fault sensor). The window type is the most accurate method for it reads the most leakage current available.

Design Tip: The system will be cleared once the low-magnitude fault-currents are detected and the overcurrent protection device is tripped open. It is most important that the sensor of the relay be set correctly.

SERVICE EQUIPMENT RATED OVER 600 VOLTS, NOMINAL
ARTICLE 230, PART VIII

High-voltage supply systems are installed to reduce the cost of conductors, raceways, elements and equipment. Low-voltage supply systems require a greater cost of installation. The current values will decrease, if the voltage applied is raised. The current values will increase, if the voltage applied is lowered. A high-voltage supply system installed for a large building will decrease the cost of installation by using small conductors, raceways, elements and equipment. A low-voltage supply system installed for a small facility will increase the cost of installation for conductors, raceways, elements and equipment. Load requirements determine the voltage level selected to supply power to the service equipment and downstream panelboards.

SERVICE POINT
ARTICLE 100

The service point is the point of connection between the facilities of the serving utility and the premises wiring. The service point is located ahead of the service equipment and is installed on the line side of the meter housing and service equipment. This part of the service equipment is located on the service side of the service point. **(See Figure 6-52)**

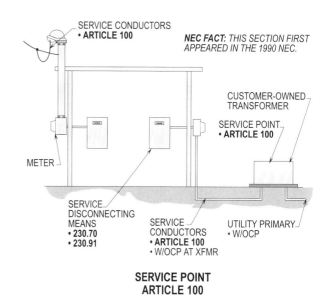

Figure 6-52. The service point is the point of connection between the facilities of the serving utility and the premises wiring.

CLASSIFICATION
ARTICLE 100, PART II

High-voltage primaries for buildings requiring large services are generally used with step-down transformers to supply loads located at various places in the building. Lighting, receptacles and other types of loads are supplied by step-down transformers. If the transformer belongs to the owner, the point of service is located at the transformer if installed outside or within a building per **Article 100**. In this case, the supply conductors to the primary side of the transformer belongs to the utility company. The service conductors are the conductors installed between the service point and the first disconnecting means for the building or structure. **(See Figure 6-53)**

WIRING METHODS
230.202(B)

The following wiring methods shall be permitted to be installed for high-voltage service-entrance conductors to protect them from physical damage per **300.37** and **300.50**:

- Rigid metal conduit (RMC)
- Intermediate metal conduit (IMC)
- Electrical metallic tubing (EMT)
- Rigid nonmetallic conduit (RNC)
- Cable trays
- Busways
- Cablebus

- Other identified raceways
- Type MV cables

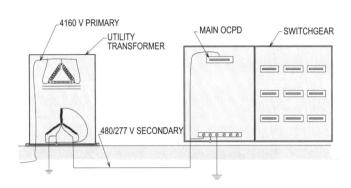

Figure 6-53. High-voltage primaries for buildings requiring large services are generally used with step-down transformers to supply loads located at various places in the building.

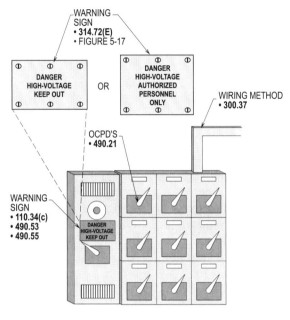

Figure 6-54. A permanent and conspicuous warning sign shall be posted where the voltage exceeds 600 volts.

LOCKED ROOMS OR ENCLOSURES
110.34(C)

A permanent and conspicuous warning sign shall be posted where the voltage exceeds 600 volts. The signs shall be posted to prevent unauthorized personnel from coming in contact with energized parts. The sign shall read as follows:

DANGER - HIGH-VOLTAGE - KEEP-OUT

Design Tip: The sign shall have a warning of "High-Voltage" and give a command "Keep Out" or "Authorized Personnel Only" etc. **(See Figure 6-54)**

ISOLATING SWITCHES
230.204

Isolating switches shall be installed for certain types of disconnecting means. Isolating switches shall be installed to open the circuit when there is no load. Equipment on the load side shall be permitted to be serviced by personnel without danger of the circuit being energized. **(See Figure 6-55)**

Design Tip: Isolating switches shall not be required to be installed for rollout or pullout equipment from switchgears or panel enclosures.

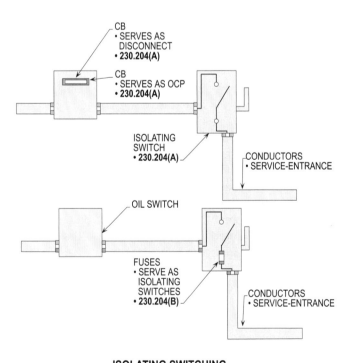

Figure 6-55. Isolating switches shall not be required to be installed for certain types of disconnecting means.

DISCONNECTING MEANS
230.205

All ungrounded (phase) conductors shall be disconnected simultaneously (common trip) by the service disconnecting means. The service disconnecting means shall be capable of being closed on a fault not less than the maximum available short-circuit current imposed on the terminals of the service equipment.

The characteristics of a fuse may contribute to the fault closing rating of the disconnecting means and serve as an aid for such closing. The disconnecting means shall be permitted to be installed ahead of or near the defined service point per **Article 100** in the NEC.

PROTECTION REQUIREMENTS
230.208

Service conductors operating at over 600 volts shall be provided with an overcurrent protection device in each ungrounded (phase) conductor. Short-circuit conditions shall be protected by properly sized overcurrent protection devices. Overload conditions shall be protected against by adding a second stage of overcurrent protection.

Overcurrent protection devices shall be permitted to be installed on the line side of the service disconnecting means or as an integral part of the service disconnecting means.

A fuse shall not exceed three (300 percent) times the ampacity of the continuous rating of the conductor to be installed. A circuit breaker shall have trip settings not to exceed six (600 percent) times the ampacity of the continuous rating of the conductor to be installed. (Also see **490.21(A) through (E)**)

High-voltage supply systems have different characteristics than low-voltage supply systems. A short-circuit will clear much faster for a high-voltage system than a low-voltage system. Percentages shall be permitted to be selected as listed above, higher percentages shall be permitted to be used to size overcurrent protection devices to protect service conductors from short-circuit conditions.

The higher voltage will clear a phase-to-phase or phase-to-ground short-circuit very rapidly. This is not always the case, when using lower voltage systems rated at 600 volts or less. **(See Figure 6-56)**

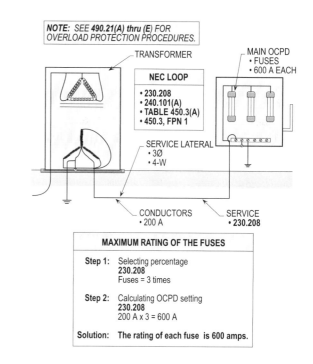

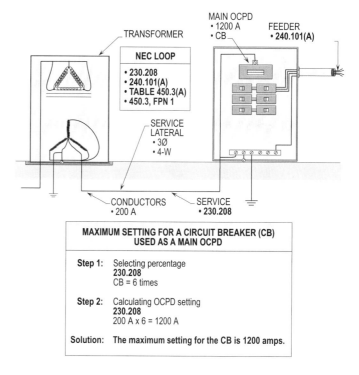

Figure 6-56. Service conductors operating at over 600 volts shall be provided with an overcurrent protection device in each ungrounded (phase) conductor.

OVERCURRENT PROTECTION AND REQUIREMENTS FOR FEEDER-CIRCUIT CONDUCTORS
240.100; 240.101

Feeder-circuits shall have a short-circuit overcurrent protection device in each ungrounded (phase) conductor. The protection device(s) shall be capable of detecting and interrupting all values of current which may occur at their location in excess of their trip setting or melting point. A fuse rated in continuous amperes not exceeding three times the ampacity of the conductor or a breaker having a trip setting of not more than six times the ampacity of the conductor shall be considered as providing the required short-circuit protection. For further information, see **490.21(A)** and **(B)** and **Article 490** which deals with high-voltage electrical systems. **(See Figure 6-57)**

> **Design Tip:** The operating time of the protective device, the available short-circuit current, and the conductor used will need to be coordinated to prevent damaging or dangerous temperatures in conductor insulation under short-circuit or ground-fault conditions.

Branch-circuits shall have a short-circuit protective device in each ungrounded (phase) conductor or comply with **490.21(A)** and **(B)**. The protective device(s) shall be capable of detecting and interrupting all values of current which may occur at their location in excess of their trip setting or melting point per **240.100(A)**. In other words, the OCPD's shall be sized to protect the conductors at their ampacities if not oversized to allow equipment with high-inrush current to start, run and operate.

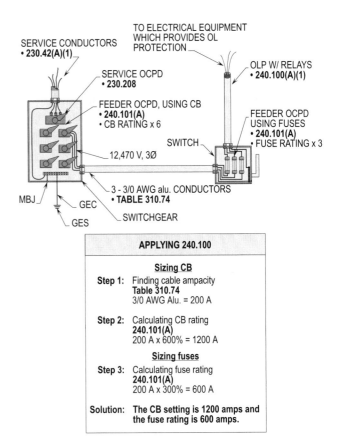

Figure 6-57. The above illustration shows the correct procedure for sizing CB's and fuses to be used to protect high-voltage electrical conductors.

Name   Date

Chapter 6. Services

Section	Answer		
_____	T	F	1. Only one service drop or lateral shall be permitted to be installed to a building.
_____	T	F	2. Emergency lighting or power systems shall not be permitted to be installed with a separate service.
_____	T	F	3. A building of different voltages shall not be permitted to be supplied with two or more service drops or underground laterals.
_____	T	F	4. Service conductors supplying a building or structure shall be permitted to pass through the interior of another building or structure.
_____	T	F	5. When installing service-entrance conductors in a service raceway or service-entrance cable, no other conductors shall be permitted. (General Rule)
_____	T	F	6. Service-drop conductors shall be permitted to be 12 AWG hard-drawn copper where installed to serve small loads of only one branch-circuit, such as a phone booth.
_____	T	F	7. Service-drop conductors crossing property where people have access shall have a clearance of at least 12 ft. from finished grade.
_____	T	F	8. The minimum point of attachment for service-drop conductors shall be 10 ft. from finished grade.
_____	T	F	9. Only one set of service-entrance conductors shall be permitted to be supplied by a service-drop or service lateral. (General Rule)
_____	T	F	10. At least a 60 amp service disconnecting means shall be installed for a single-family dwelling.
_____	_____		11. Service-entrance conductors shall be permitted to be spliced in metering equipment enclosures.
_____	_____		12. Open wiring on insulators shall be installed 10 ft. above finished grade or be protected where exposed to physical damage.
_____	_____		13. When installing a three-phase, four-wire, delta-connected service, where the midpoint of one phase winding is grounded, an outer finish marked yellow or identified by other effective means shall be used for the conductor with a higher voltage-to-ground.
_____	_____		14. Service equipment rated at 600 volts or less shall be marked to identify as being suitable for use as service equipment.
_____	_____		15. The service disconnecting means shall not be required to be located in a readily accessible location for all premises.

Section	Answer

16. The disconnecting means shall disconnect all service-entrance conductors from the power supply simultaneously.

17. A separate disconnecting means shall not be required if more than one building is on the same property and under single management.

18. The overcurrent protection device shall be sized high enough to carry the locked-rotor current for a service supplying a fire pump.

19. Overcurrent protection devices shall be permitted to be grouped in one building where a property consists of more than one building or structure under single management.

20. A connection ahead of the meter shall be permitted to be installed ahead of the main to supply a flat rate circuit for a water heater.

21. The service point is the point of connection between the facilities of the serving utility and the premises wiring.

22. A conspicuous warning sign shall not be required to be posted where the voltage exceeds 600 volts.

23. Isolating switches shall be installed for roll-out or pull-out equipment from switchgears or panel enclosures.

24. Service conductors operating over 600 volts shall not be required to be provided with an overcurrent protection device in each ungrounded (phase) conductor.

25. A high-voltage fuse shall not be permitted to exceed three times the ampacity of the continuous rating of the conductor to be installed.

26. To ensure against the interruption of power to fire pumps, a _____ service shall be permitted to be installed.

27. Service conductors shall be considered outside the building when installed in a conduit and encased with at least _____ in. of concrete or brick.

28. A clearance of _____ ft. shall be required from windows designed to be opened, including porches, platforms, etc.

29. Service-drop conductors shall not be permitted to be smaller than _____ copper or _____ aluminum.

30. Service-drop conductors shall be permitted to pass above roofs where voltages between conductors does not exceed 300 volts and are installed at a height of 3 ft., where the roof has a slope of at least _____ in. by _____ in.

31. Service-drop conductors shall be permitted to be installed for overhanging roofs provided no more than _____ ft. of such conductors does not pass over more than _____ ft. of the roof.

32. A copper neutral conductor shall be permitted to be _____ when used for an underground service lateral utilizing a cable assembly or installed in conduit.

	Section	Answer

33. Service-entrance conductors entering buildings or other structures shall be _____.

34. A minimum _____ amp, three-wire service disconnecting means shall be installed for a one-family dwelling. (General Rule)

35. Service-entrance cable shall be supported within _____ in. intervals and within _____ in. of the service weatherhead when supporting cable directly to the wall.

36. Other cables shall be supported at _____ ft. intervals on insulators if not approved to be supported directly to the wall.

37. The service disconnecting means shall be grouped in a common location and be readily accessible so the entire service can be disconnected with _____ throws of the hand.

38. The service disconnecting means shall be installed at a rating of _____ amps for single branch-circuit loads supplying limited loads.

39. The service disconnecting means shall be installed at a rating of _____ amps for one or two two-wire branch-circuit loads.

40. The disconnecting means in a separate building shall be rated as a type _____ for use as service equipment.

41. Solidly grounded wye electrical services of more than _____ volts-to-ground having a service disconnecting means rated _____ amps or more shall be protected by ground-fault protection.

42. The maximum setting for ground-fault protection shall not be permitted to exceed _____ amps, when the service disconnecting means exceeds 1000 amps or more.

43. A _____ and conspicuous warning sign shall be posted where the voltage exceeds 600 volts.

44. All ungrounded (phase) conductors rated over 600 volts shall be disconnected _____ by the service disconnecting means.

45. A CB shall have a trip setting not exceeding _____ times the ampacity of the continuous rating of the conductor to be installed. (Over 600 Volts)

46. Load requirements in a facility that are in excess of _____ amps shall be permitted to be supplied with more than one service.

    (a)   500
    (b)   1000
    (c)   1500
    (d)   2000

Section	Answer

47. A clearance of _____ ft. shall be required for service-drop conductors passing over roofs. At least a _____ ft. clearance shall be maintained in all directions for vertical clearances.

    (a)   8 and 3
    (b)   8 and 6
    (c)   10 and 3
    (d)   10 and 6

48. Service-drop conductors shall have a vertical clearance of at least _____ ft. above flat roofs that are accessible to pedestrians.

    (a)   6
    (b)   7
    (c)   8
    (d)   10

49. Service-drop conductors shall be permitted to be installed for overhanging roofs provided the attachment of the service-drop has a clearance of at least _____ in. from the conduit and roof line.

    (a)   6
    (b)   12
    (c)   18
    (d)   24

50. Service-drop conductors crossing residential property or driveways shall have a clearance of _____ ft. from finished grade.

    (a)   10
    (b)   12
    (c)   15
    (d)   18

51. Service-drop conductors crossing residential property or driveways shall have a clearance of _____ ft. from finished grade where the voltage exceeds 300 volts-to-ground.

    (a)   10
    (b)   12
    (c)   15
    (d)   18

52. Service-drop conductors crossing public streets, alleys, roads and parking areas subject to traffic shall have a clearance of _____ ft. from finished grade.

    (a)   10
    (b)   12
    (c)   15
    (d)   18

Section   Answer

53. A minimum _____ amp, three-wire service disconnecting means shall be installed for a single family dwelling.

    (a)   60
    (b)   100
    (c)   150
    (d)   200

54. Other cables used for servce conductors shall be installed with a _____ in. set off clearance from the wall.

    (a)   2
    (b)   6
    (c)   10
    (d)   12

55. No more than _____ fusible switches or circuit breakers mounted in a single enclosure, a group of separate enclosures or in or on a panelboard or switchboard shall be permitted to be installed for each set of service-entrance conductors.

    (a)   2
    (b)   4
    (c)   6
    (d)   10

7

Switchboards and Panelboards

Panelboards are intended to be installed and mounted in cabinets or cutout boxes placed "in or against a wall or partition." Panelboards are used for the control of small capacity circuits.

Switchboards are not intended to be installed and mounted in cabinets or cutout boxes. Switchboards usually are installed to stand on the floor. Switchboards are used for the control of high capacity circuits.

The space inside enclosures shall have sufficient space to terminate conductors to lugs of overcurrent protection devices or busbars. This space shall protect the conductors and equipment from physical damage when they are being terminated.

Table 312.6(A) shall be utilized to provide the proper space to terminate conductors formed into a L-bend configuration.

Table 312.6(B) shall be selected for conductors that are formed into a S or Z-bend configuration.

The minimum wire-bending space at terminals and correct width of wiring gutters in inches shall be determined by the number of conductors connected to each terminal per **408.55, Ex.'s 1 through 4**.

SWITCHBOARD AND PANELBOARDS
ARTICLE 408, PARTS II AND III

Switchboards are designed with busbars for fuses or circuit breakers which are used for protecting and supplying feeders and branch-circuits. Panelboards are designed with a permanently installed main circuit breaker or single set of fuses for disconnecting the number of circuit breakers or fuseholders installed to protect and supply feeders or branch-circuits.

SUPPORT AND ARRANGEMENT OF BUSBARS AND CONDUCTORS
408.3

Busbars shall be held firmly in place in the enclosure and shall be arranged in such a way to allow adequate space for conductors to be terminated. Busbars shall be sized with enough capacity to handle the normal currents in amps of the electrical system. In addition, they shall carry safely any type of short-circuit or ground-fault condition.

CONDUCTORS AND BUSBARS ON A SWITCHBOARD OR PANELBOARD
408.3(A)

Conductors and busbars on a switchboard, panelboard or control board shall be so located as to be free from physical damage. Service sections of such boards shall be separated from the other electrical components. These components shall be properly secured in an approved manner. Note, that conductors entering a certain section shall terminate in such section, except where it is not practical for control, interconnecting conductors or other conditions of use.

OVERHEATING AND INDUCTIVE HEATING
408.3(B)

The arrangement of busbars and conductors shall be such as to avoid overheating due to inductive effects. The inductive effect of each conductor is neutralized when grouping all conductors together. To accomplish the minimizing of heating effects, the ungrounded (phase) conductors, grounded conductors, EGC's, where used, shall be in the same enclosure, raceway, cable, cord, etc. **(See Figure 7-1)**

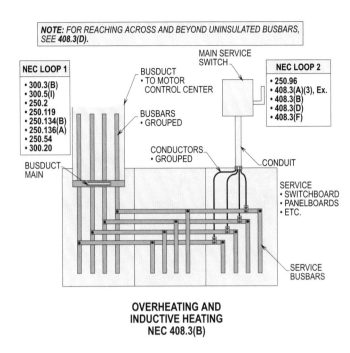

Figure 7-1. The arrangement of busbars and conductors shall be such as to avoid overheating due to inductive effects.

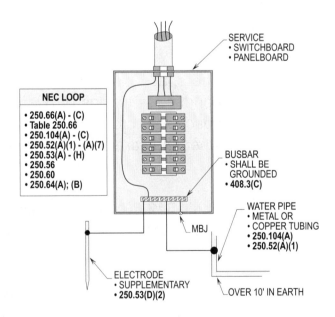

Figure 7-2. Each switchboard, panelboard or control board used as service equipment shall be provided with a grounding bar bonded to the switchboard frame or the casing of such boards.

Switchboards and Panelboards

USED AS SERVICE EQUIPMENT
408.3(C)

Each switchboard, panelboard or control board used as service equipment shall be provided with a grounding bar bonded to the switchboard frame or the casing of a such boards. For further information, see **250.28(C)**.

The main bonding jumper shall be sized per **250.28(D)** and installed within the panelboard or any section of a switchboard to which the grounded service conductor on the supply side of the switchboard or panelboard frame is connected. All sections of a switchboard shall be bonded together with an equipment bonding conductor that is sized per **Table 250.122** or **Table 250.66**. **Note**, that a screw or strap shall be permitted to be used instead of a bonding jumper per **250.28(A)**. **(See Figure 7-2)**

PHASE ARRANGEMENT
408.3(E)

Switchboards or panelboards supplied from a three-phase, four-wire, delta connected system, where the midpoint of one phase winding (neutral tap) is grounded, the voltage to the grounded neutral will have a higher voltage-to-ground than the other two phases-to-ground.

For example, the higher voltage-to-ground shall be identified by orange color or tape or tagged according to **110.15** and **230.56** and shall be connected to Phase B on a 240 volt, three-phase, four-wire delta connected system. The higher voltage-to-ground identification (high-leg, wild-leg, stinger-leg or red-leg) shall be designed to prevent 120 volt loads from being connected to 208 volts obtained from the high-leg. The 240 volt loads shall be permitted to be derived between the phases while the 120 volt loads shall be derived from two of the phases-to-ground as follows: **(See Figure 7-3)**

Phase-to-Phase V
- A to B is 240 V
- A to C is 240 V
- B to C is 240 V

Phase-to-Ground V
- A to N is 120 V
- B to N is 208 V
- C to N is 120 V

Design Tip: The center lug of utility meters shall not be connected to the B phase of a three-phase, 240 volt, four-wire, delta system. For metering purposes only, this three-phase high-leg shall be connected to Phase C. For a CT can with current transformers equipped with a remote meter shall be connected to B phase (high-leg) in the CT can for meter conductors. However, it shall be terminated to Phase C in the remote meter base. Check with local utilities for verification of this rule. **(See Figure 7-4)**

The phase arrangement on three-phase buses shall be A, B and C from the front to back, top to bottom or left to right as viewed from the front of the panelboard, switchboard or control board. The arrangement shall be viewed as Phases C, B and A from the back of such boards. With this arrangement of the Phases, the high-leg will always be placed as B and in the center. **(See Figure 7-5)**

Design Tip: The following methods of color coding shall be permitted to be used for phase arrangement but is no longer required in the NEC, except for the high-leg:

- Phase A - black, Phase B - red, Phase C - blue
Used for 120/208 volt or less systems
- Phase A - brown, Phase B - orange, Phase C - yellow
Used for over 277/480 volt systems
- Phase A - black, Phase B - orange, Phase C - Blue
Used for 120/240 volt, three-phase systems
- Phase A - black, Phase B - red or blue
Used for 120/240 volt, single-phase systems

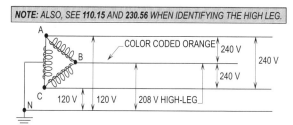

THREE-PHASE, 4-WIRE DELTA, OPEN OR CLOSED DELTA, 240/120 VOLTS

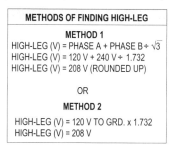

PHASE ARRANGEMENT
NEC 408.3(E)

Figure 7-3. Switchboards or panelboards supplied from a three-phase, four-wire, delta connected system, where the midpoint of one phase winding (neutral tap) is grounded, the voltage to the grounded neutral will have a higher voltage-to-ground than the other two phases-to-ground.

MINIMUM WIRE-BENDING SPACE 408.3(F)

Panelboards shall comply with the provision of **408.3(F)** which requires the minimum gutter space in the board to meet the clearance rules listed in **Tables 312.6(A) and (B)** for L, S or Z-bends.

If the lugs in the panelboard are removable, the clearance for S or Z-bends shall be permitted to be reduced from the dimensions in **Table 312.6(B)** based on the number and size of conductors connected to each lug. **(See Figure 7-6)**

To prevent damaging conductors because of over bending, panelboards shall comply with the provision of **408.3(F)** which requires the minimum gutter space in the board to meet the clearance rules in **Tables 312.6(A) and (B)** for L, S or Z-bends.

If the lugs in the panelboard are removable, the clearance shall be permitted to be reduced from the dimensions in **Table 312.6(B)** based on the number and size of conductors connected to each lug. **(See Figure 7-7)**

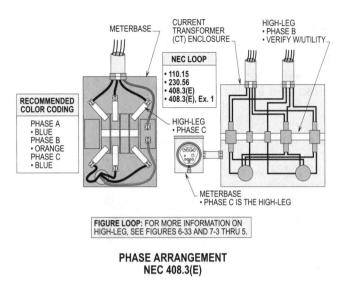

PHASE ARRANGEMENT NEC 408.3(E)

Figure 7-4. The center lug of utility meters shall not be connected to the B phase of a three-phase, 240 volt, four-wire, delta system. In a self-contained meter base, the C phase is the high-leg.

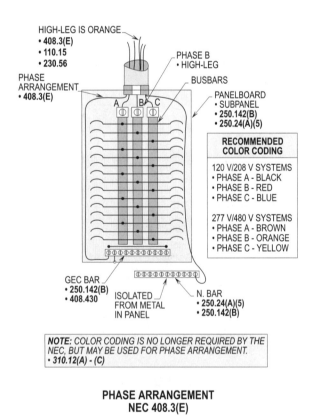

PHASE ARRANGEMENT NEC 408.3(E)

Figure 7-5. The phase arrangement on three-phase buses shall be A, B and C from front to back, top to bottom or left to right as viewed from the front of the panelboard, switchboard or control board.

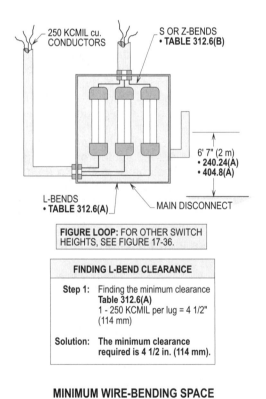

MINIMUM WIRE-BENDING SPACE NEC 408.3(F)

Figure 7-6. Panelboards shall comply with the provision of **408.3(F)** which requires the minimum gutter space in the board to meet the clearance rules listed in **Tables 312.6(A)** and **(B)** for L, S or Z-bends.

Switchboards and Panelboards

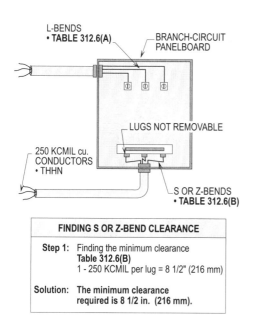

MINIMUM WIRE-BENDING SPACE
NEC 408.3(F)

Figure 7-7. To prevent damaging conductors because of over bending, panelboards shall comply with the provision of **408.3(F)** which requires the minimum gutter space in the board to meet the clearance rules in **Tables 312.6(A)** or **(B)** for L, S or Z-bends.

CLEARANCE FOR CONDUCTOR ENTERING BUS ENCLOSURES 408.5

Where conduits or other raceways enter the switchboard, sufficient space shall be provided to permit installation of conductors in the enclosure. The entering conduit or raceways, including their end fittings, shall not rise more than 3 in. (75 mm) above the bottom of the enclosure. The minimum clearance required between the bottom of the switchboard enclosure and busbars, with their supports, or other obstruction shall be 8 in. (200 mm) for insulated busbars, and 10 in. (250 mm) for noninsulated busbars. This clearance rule, if complied with, will prevent the overbending of conductors between entry and termination points. **(See Figure 7-8)**

DEDICATED ELECTRICAL SPACE 110.26(F)(1)(a)

Switchboards shall be installed in a dedicated space equal to the width and depth of the equipment and extending from the floor to a height of 6 ft. (1.8 m) or to the structural ceiling (roof), whichever is lower. No piping, ducts or equipment foreign to the electrical installation shall be installed in this dedicated space since leakage of water or condensation could damage the equipment enclosure. Any space extending 6 ft. (1.8 m) above the dedicated space shall not be considered dedicated space. Sprinkler protection shall be permitted to be installed in this dedicated space to protect the equipment from fire hazards. For further information, see **NFPA 13, 4-4.14**.

Suspended ceilings with removable panels shall be permitted within the 6 ft. (1.8 m) zone per **110.26(F)(1)(a), Ex.** This dedicated space shall be considered by electricians to be an area were they may run and connect cables and raceways without foreign items (other than electrical) blocking their access directly above equipment. **(See Figure 7-9)**

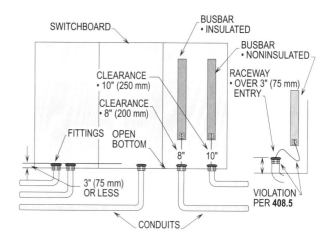

CLEARANCES FOR CONDUCTORS ENTERING BUS ENCLOSURES
NEC 408.5

Figure 7-8. Where conduits or other raceways enter a switchboard, sufficient space shall be provided to permit installation of conductors in the enclosure without damaging their insulation.

SWITCHBOARDS ARTICLE 408, PART II

Switchboards that have exposed live parts shall be installed in dry locations accessible only to qualified personnel. Totally-enclosure type switchboards shall be permitted to be installed anywhere acceptable to the AHJ. Switchboards may be used for service equipment with feeder-circuits supplying power to subpanels.

Subpanels are supplied by feeder-circuits from the main service equipment. Subpanels are installed where the wiring of branch-circuits extend lengths of great distances from the service equipment. Branch-circuit wiring is connected to the centralized subpanel which reduces the amount of wiring and copper cost required to wire electrical apparatus located in the area.

A subpanel serving lighting and receptacle outlets plus special appliance loads shall be permitted to be supplied from one of the single-pole OCPD's in the power panel. Double-pole OCPD's are usually utilized to supply larger fixed electrical appliances such as a heating unit, an A/C unit, a piece of processing equipment, etc. **(See Figure 7-10)**

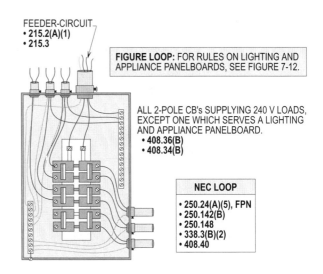

**SWITCHBOARDS
ARTICLE 408, PART II**

Figure 7-10. A power subpanel serving lighting and receptacle outlets plus special appliance loads shall be permitted to be supplied from one of the single-pole OCPD's in the power panel. Double-pole OCPD's are usually utilized to supply larger fixed electrical appliances such as a heating unit, an A/C unit, a large computer, a piece of processing equipment, a water heater, etc..

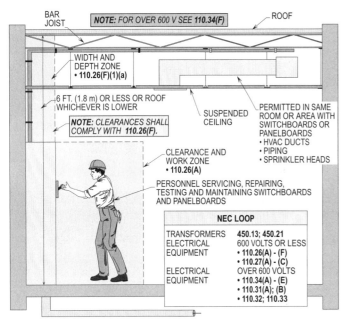

**DEDICATED ELECTRICAL SPACE
NEC 110.26(F)(1)(a)**

Figure 7-9. Switchboards shall be installed in a dedicated space extending 6 ft. (1.8 m) above the equipment from the floor or to the structural ceiling (roof), whichever is lower. No piping, ducts or equipment foreign to the electrical installation shall be installed in this dedicated space.

CLEARANCES
408.18

Switchboard that are not of the totally-enclosed type shall have a clearance of 3 ft. (900 mm) from the top of the switchboard to a ceiling constructed of wood, ceiling paper and any other type of material that will burn. This clearance shall not be required for totally-enclosed type switchboards or for a fireproof shield provided above the switchboard. Clearance around and above switchboards shall comply with the requirements of **110.26**. **(See Figure 7-11)**

LOCATION OF SWITCHBOARDS
408.20; 17

Switchboards shall be installed to reduce the possibility of communicating fire to adjacent combustible material. Switchboards installed over a combustible floor shall have suitable protection provided. In addition, only qualified persons under the proper supervision of competent personnel shall have access to service, repair or pull maintenance on the parts of such equipment.

GROUNDING OF INSTRUMENTS, RELAYS, METERS AND INSTRUMENT TRANSFORMERS ON SWITCHBOARDS
408.22

Switchboard frames and structures supporting switching equipment shall be grounded. Frames of two-wire, DC single-polarity switchboards that are insulated shall not be permitted to be grounded. Proper grounding provides equipotential planes of electrical raceways, cables and enclosures.

Switchboards and Panelboards

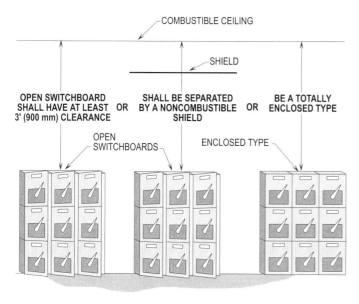

Figure 7-11. Switchboards that are not of the totally-enclosed type shall have a clearance of 3 ft. (900 mm) from the top of the switchboard to a ceiling constructed of wood, ceiling paper and any other type of material that will burn.

PANELBOARDS
ARTICLE 408, PART III

Panelboards shall have a rating equal to or higher than the minimum capacity required for the load. When designing a panelboard with a computed load of 212 amps, the panelboard shall be sized with a rating of at least 212 amps. A panelboard of 225 amps is the next standard size above 212 amps and complies with such rules.

CLASSIFICATION OF PANELBOARDS
408.34

There shall be permitted to be one to six mains in a service panel used as service equipment to supply and disconnect power to the various electrical loads. There are three types of panelboards used for such service equipment and they are as follows:

- Power panels
- Lighting and appliance panels
- Split-bus panels

Power panels installed as service equipment usually consists of one to six mains, depending on the loads. A lighting and appliance branch-circuit has a connection to the neutral of the panelboard and that has overcurrent protection of 30 amps or less in one or more conductors.

Split-bus panelboards have one to six mains in the top section with one of the mains used to supply the bottom section which is usually used for the lighting and receptacle outlet loads. Split-bus panelboards shall have a main ahead of them after the 1981 NEC was published. This in-line main shall be sized to protect the bus and loads of such panels.

POWER PANELBOARD
408.34(B); 408.34(B)

A power panel per **408.34(B)** shall be permitted to have one to six mains installed to supply electrical appliances and other electrical loads in residential, commercial and industrial facilities. A power panel has 10 percent or less of its branch-circuit overcurrent protection devices rated at 30 amps or less with neutral connected loads. Power panels are used with a great number of electrical appliances in addition to their lighting and receptacle loads.

For example, a panelboard with twenty-four single-pole branch-circuits for installing OCPD's shall be permitted to have two 30 amp or less (24 OCPD's x 10% = 2.4) overcurrent protection devices installed with neutral connections and still be classified as a power panel. Note, that a single-pole circuit breaker counts as one circuit breaker. A double-pole circuit breaker counts as two circuit breakers. A three-pole circuit breaker counts as three circuit breakers in counting the number of single-pole slots in a panelboard per **408.35** to determine its classification. **(See Figure 7-12)**

LIGHTING AND APPLIANCE BRANCH-CIRCUIT PANELBOARD
408.34(A); 408.36(A)

A lighting and appliance panelboard shall be equipped with one or two mains installed to supply power to a lighting section and power section per **408.36(A)**. A lighting and appliance panelboard has over 10 percent of its branch-circuit overcurrent protection devices rated at 30 amps or less with neutral connections per **408.34(A)**.

For example, a panel with forty-two, single-pole slots have five 20 amp overcurrent protection devices with neutral loads. This is a lighting and appliance panelboard because it has five 20 amp OCPD's with neutral loads (42 x 10% = 4.2). If only four OCPD's are present, it is classified by **408.34(B)** and **408.35** as a power panel. Lighting and appliance panelboards with one in-line main shall not be permitted to be overloaded. Therefore, they are safer to use and normally are installed because of this protection scheme. Note, see **Ex.'s to 408.36(A)** for variations of this requirement. **(See Figure 7-12)**

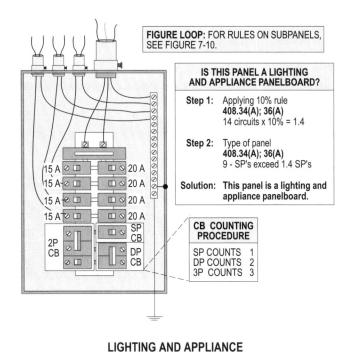

LIGHTING AND APPLIANCE
BRANCH-CIRCUIT PANELBOARD
NEC 408.34(A)
NEC 408.36(A)

Figure 7-12. A lighting and appliance panelboard has over 10 percent of its branch-circuit overcurrent protection devices rated at 30 amps or less with neutral connections.

SPLIT-BUS PANELBOARD
408.36(A), Ex. 2

The lower section served by CB 6 is used to supply power to the lighting and receptacle loads plus the small appliance loads. The five remaining mains in the top section are used to supply loads such as a heating unit, an A/C unit, a dryer, a cooktop, an oven, a range, a water heater, etc. **(See Figure 7-13)**

SNAP SWITCHES RATED AT 30 AMPS OR LESS
408.36(C)

Panelboards equipped with snap switches rated 30 amps or less shall be protected by overcurrent protection devices not exceeding the panelboards rating. Note that a 225 amp panelboard with a 225 amp in-line main shall be permitted to be used for this type of installation **(See Figure 7-14)**

SUPPLIED THROUGH A TRANSFORMER
408.36(D)

Where a lighting and appliance panelboard is supplied from the secondary of a transformer, the panelboard shall be provided with protection from the secondary side of the transformer. However, under certain design techniques and conditions of use, the overcurrent protection device shall be permitted to be installed on the primary side for a two-wire primary to a two-wire secondary per **408.36(D), Ex. (See Figure 7-15)**

For example, if the load is supplied by a single-phase, two-wire transformer with a single voltage, the secondary shall be permitted to be protected by the overcurrent devices in the primary or supply side of the transformer, provided the primary is protected as covered in **450.3(B)** and at a value that does not exceed the value determined by multiplying the transformer voltage ratio of the secondary-to-primary by the panelboard rating per **240.4(F)** and **240.21(C)(1)**.

DELTA BREAKERS
408.36(E)

Lighting and appliance panelboards shall not be permitted to be installed with delta breakers. In other words, a three-phase disconnect shall never be terminated to the buses of panelboards having less than three buses. The reasoning behind this requirement was, homeowners in the past, were overloading the rating of the panel, due to the absence of an in-line main.

BACK-FED DEVICES
408.36(F)

Plug-in type overcurrent protection devices or plug-in type main lug assemblies that are back-fed shall be secured in place by an additional fastener other than the pull-to-release type. The past history of such installed devices has proved, plugging them in only was not enough support to secure them properly.

ENCLOSURE
408.38

Panelboards shall be mounted in cabinets, cutout boxes or enclosures designed for the purpose. Dead-front panelboards shall be installed unless accessible only to qualified persons.

Switchboards and Panelboards

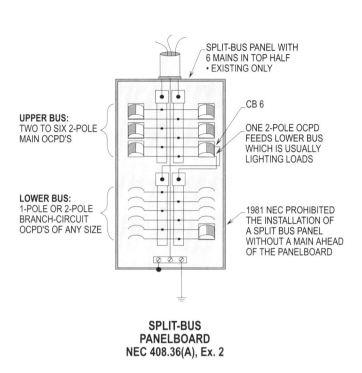

Figure 7-13. A split-bus panelboard has six mains in the upper section with one main feeding the lower section.

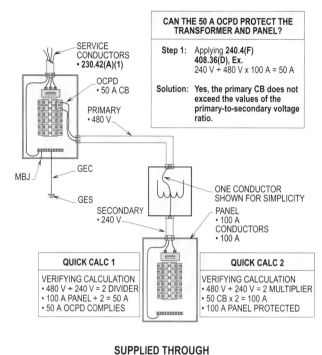

Figure 7-15. Where a lighting and appliance panelboard is supplied from the secondary of a transformer, the OCPD in the primary side shall be permitted to protect the panelboard and secondary side from overloads. (General rule)

GROUNDING OF PANELBOARDS 408.40

An equipment grounding terminal bar shall be secured inside the cabinet for the termination of all grounding conductors used to ground the metal of equipment, junction boxes, pull boxes, etc. The equipment grounding terminal bar ensures that the equipment grounding conductors are bonded to the cabinet and panelboard frame and if done correctly, provides an effective path to ground.

Panelboard enclosures or cases shall be grounded. Where installed with a metal conduit such as rigid metal conduit, electrical metallic tubing, intermediate metal conduit (any metal raceway), metal-clad cables, etc., the metal of these wiring methods shall be permitted to be used for grounding if they comply with **250.118**. An equipment grounding conductor shall be installed with a nonmetallic wiring method such as nonmetallic sheathed cable (romex or rope) or PVC. **(See Figure 7-16)**

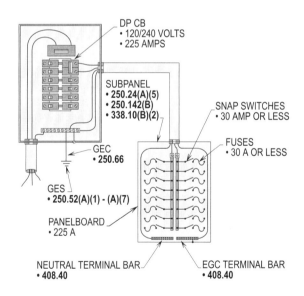

Figure 7-14. Panelboards equipped with snap switches rated 30 amps or less shall be protected by overcurrent protection devices not exceeding the panelboards rating.

The equipment grounding terminal bar and the neutral bar in service equipment shall be bonded to the panelboard case per **250.130** and **250.66** and connected to the grounding electrode system per **250.52(A)(1)** through **(A)(6)** by the grounding electrode conductor. The equipment grounding terminal bar and the neutral bar in other than service equipment shall not be permitted to be connected and bonded together in the panelboard enclosure per **250.142(B)**. Also see **250.24(A)(5)**. The neutral installed in the sub-panelboard enclosure shall be isolated back to the service equipment. Isolated equipment grounding conductors shall be permitted to pass through the panelboard enclosure per **250.146(D)**. All grounding conductors and neutrals in a subpanel shall be terminated to separate terminal bars per **408.40**. **(See Figure 7-17)**

> **Design Tip:** Equipment grounding conductors and grounded (neutral) conductors shall only terminate to a common bar at the service equipment, separately derived systems and separately fed buildings from a main building supply.

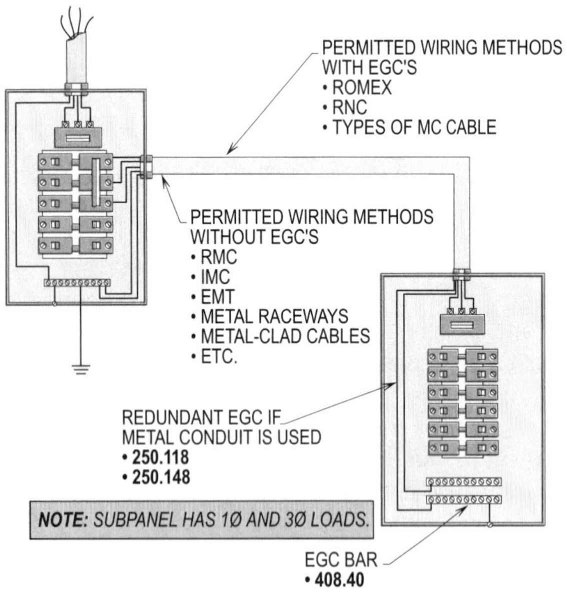

GROUNDING OF PANELBOARDS
NEC 408.40

Figure 7-16. An equipment grounding terminal bar shall be secured inside the cabinet for the termination of all grounding conductors used to ground the metal of equipment, junction boxes, pull boxes, etc.

ISOLATED EQUIPMENT GROUNDING CONDUCTOR
408.40, Ex.

Isolated equipment grounding conductors shall be permitted to pass through the panelboard with the power circuit conductors to reduce electromagnetic interference (electrical noise) if present. An isolated equipment grounding conductor shall be installed and connected directly to the equipment grounding terminal bar and neutral bar in the main service equipment or power source (separately derived system) to reduce the problem of electrical noise. Such noise, if not corrected, may disturb the operation of sensitive electronic equipment. **(See Figure 7-18)**

> **Design Tip:** This **Ex.** is necessary for isolating computer grounding systems that are cord-and-plug connected per **250.146(D)**. Also see **250.96(B)** for a similar rule when isolating the grounding system for hard-wired sensitive electronic equipment.

ISOLATED RECEPTACLES
250.146(D)

When electrical noise becomes a problem in the grounding circuit, a receptacle on the grounding circuit in which the grounding terminal or receptacle has been insulated from the mounting means or yoke shall be allowed. In this case, an insulated equipment grounding conductor shall be routed to the isolated terminal with the circuit conductors. This insulated grounding (IG) conductor shall be permitted pass through one or more panelboards in the same premises without being connected thereto as covered in **408.40, Ex.** Such IG shall be permitted to be terminated at the equipment grounding terminal at a separately derived system or the service equipment. Note that a grounding jumper (bonding jumper) shall be permitted to be used from grounded boxes to the grounding terminals of the receptacle, including boxes on conduit circuits, EMT circuits and some metal clad circuits unless the box is surface mounted so that the mounting screws may be tightened to make a secure grounding connection between the device yoke and the box. See **250.146(A)** and **(B)** for this rule.

> **Design Tip:** The grounding scheme described above is mostly used for grounding computers or solid-state counters on pieces of sensitive electronic equipment. If metal conduits or cables are used to contain the circuit conductors, the metal parts should be insulated from the equipment and then the insulated grounding conductor run back from the equipment to the source as permitted in **250.146(D)**.

Switchboards and Panelboards

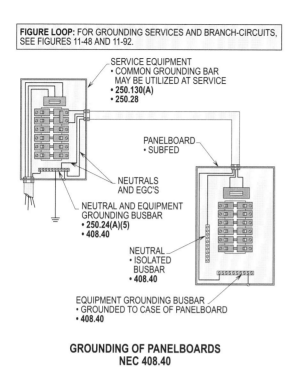

GROUNDING OF PANELBOARDS
NEC 408.40

Figure 7-17. The equipment grounding terminal bar and the neutral bar in other than service equipment shall not be permitted to be connected and bonded together in the panelboard enclosure.

RUNNING CIRCUITS TO COMPENSATE FOR HARMONIC CURRENTS

Another concern that has recently become a problem is overloaded grounded (neutral) conductors on multiwire branch-circuits. This overload is the result of harmonic currents drawn by todays typical business equipment and electronic discharge lighting systems.

To address the inherent overload produced by many of todays loads, a special AC or MC cable is available with either an oversized grounded (neutral) conductor for each three-phase, four-wire circuit or a full sized grounded (neutral) conductor (same size as the ungrounded (phase) conductors) for each ungrounded (phase) conductor in the cable. It is important to note that use of such cables does not eliminate the harmonic currents, but simply accommodates the additional neutral heating caused by the harmonic currents drawn by such loads. **(See Figures 7-19(a) and (b))**

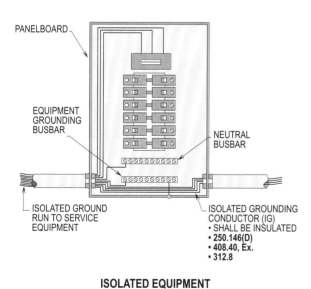

ISOLATED EQUIPMENT GROUNDING CONDUCTOR
NEC 408.40, Ex.

Figure 7-18. Isolated equipment grounding conductors shall be permitted to pass through the panelboard with the power circuit conductors to reduce electromagnetic interference (electrical noise) if present.

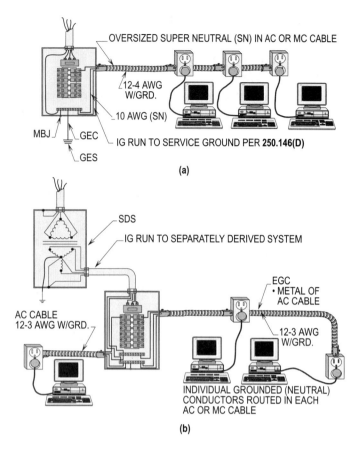

RUNNING CIRCUITS TO COMPENSATE FOR HARMONIC CURRENTS

Figures 7-19(a) and **(b).** AC and MC cables are available with an extra or oversized grounded (neutral) conductor plus an equipment grounding conductor, to supply (nearly as possible) clean and noise free power to sensitive electronic equipment. Note, circuits can also be run in metal or nonmetallic raceways.

For hard-wire circuits, see **Figures 11-112** and **11-113** of Chapter 11.

Name _____   Date _____

Chapter 7. Switchboards and Panelboards

Section	Answer		
_____	T F	1.	The arrangement of busbars and conductors shall be such as to avoid overheating due to inductive effects.
_____	T F	2.	Piping, ducts or equipment foreign to the electrical installation shall be permitted to be installed in the dedicated space extending 6 ft. from the floor above the equipment to the structural ceiling (roof), whichever is lower. (General Rule)
_____	T F	3.	Switchboards installed over a combustible floor shall not be required to be provided with suitable protection.
_____	T F	4.	The minimum clearance required between the bottom of the switchboard enclosure and busbars, with their supports or other obstruction shall be 10 in. for insulated busbars.
_____	T F	5.	Switchboard frames and structures supporting switching equipment shall be grounded.
_____	T F	6.	Not more than 42 overcurrent devices (other than those provided for in the mains) of a lighting and appliance branch-circuit panelboard shall be installed in any one cabinet or cutout box.
_____	T F	7.	Lighting and appliance panelboards shall be permitted to be installed with delta breakers.
_____	T F	8.	The equipment grounding terminal bar and the neutral bar shall be bonded to the panelboard case and connected to the grounding electrode system by the grounding electrode conductor.
_____	T F	9.	All grounding conductors and neutrals in a subpanel shall be terminated to separate terminal bars.
_____	T F	10.	Isolated equipment grounding conductors shall not be permitted to pass through the panelboard with power circuit conductors. (General Rule)
_____	_____	11.	Each lighting and appliance branch-circuit panelboard shall be individually protected on the supply side by not more than _____ main circuit breakers having a combined rating not greater than that of the panelboard.
_____	_____	12.	Each switchboard, panelboard or control board used as service equipment shall be provided with a grounding bar _____ to the switchboard frame or the casing.
_____	_____	13.	Switchboards or panelboards supplied from a three-phase, four-wire, delta connected system, where the _____ of one phase winding (neutral tap) if grounded, the voltage to the grounded neutral will have a higher voltage-to-ground than the other two phases-to-ground.

Section   Answer

14. Any space above switchboards extending above 6 ft. from the floor and above the equipment shall not be considered _____ space.

15. Switchboards shall be installed to reduce the possibility of communicating fire to adjacent _____ material.

16. Switchboards that are of the _____ enclosed type shall not be required to have a clearance above the top of the switchboard.

17. A split-bus panelboard has _____ mains in the top section with _____ main feeding the bottom section.

18. Panelboards equipped with snap switches rated _____ amps or less shall be protected by OCPD's in excess of 200 amps.

19. Where a lighting and appliance panelboard is supplied from the secondary of a transformer, the panelboard shall be provided with protection from the _____ side of the transformer. (General Rule)

20. Panelboards shall be mounted in enclosures designed for the purpose and shall be _____ front.

21. Switchboards shall be installed in a dedicated space extending _____ ft. from the floor and above the equipment to the structural ceiling (roof), whichever is lower.

    (a)   6
    (b)   20
    (c)   25
    (d)   50

22. The minimum clearance required between the bottom of the switchboard enclosure and busbars, with their supports, or other obstruction shall be _____ in. for noninsulated busbars.

    (a)   8
    (b)   10
    (c)   12
    (d)   18

23. A lighting and appliance panelboard has over _____ percent of its overcurrent protection devices rated at _____ amps or less with neutral connections.

    (a)   10, 20
    (b)   10, 30
    (c)   20, 20
    (d)   20, 30

	Section	Answer

24. Where conduits enter a switchboard bottom, the conduits, including their end fittings, shall not rise more than _____ in. above the bottom of the enclosure.

    (a) 1
    (b) 2
    (c) 3
    (d) 6

25. Switchboards that are not of the totally-enclosed type shall have a clearance of _____ ft. from the top of the switchboard to a ceiling constructed of wood, ceiling paper and any other type material that will burn.

    (a) 3
    (b) 6
    (c) 10
    (d) 25

8

Conductors

In selecting the proper size conductors and overcurrent protection devices for supplying power in a circuit from the source to the load, it is most important that the designer apply the appropriate rules of the NEC. The overcurrent protection device (OCPD) for conductors and equipment shall be sized in such a manner to open the circuit if currents reach a value that causes excessive or dangerous temperature in conductors or insulation. Such OCPD shall be selected, sized and wired, to protect both conductors and equipment or a second stage of protection shall be provided. The allowable ampacities of conductors may vary depending upon their conditions of use which are mainly based upon the number in a raceway or cable and exposure to surrounding ambient temperatures.

TEMPERATURE LIMITATION OF CONDUCTORS
310.10, FPN's (1) THRU (4)

The conductor's allowable ampacity listed in **Tables 310.16 through 310.19** in the NEC are based on four determining factors (**310.10, FPN's (1) through (4)**) and they are as follows:

- Ambient temperature
- Heat generated internally
- Dissipated into ambient medium
- Adjacent load-carrying conductors

AMBIENT TEMPERATURE
310.10, FPN (1)

The ambient temperature, may vary along the conductor length as well as from time to time and correction factors to **Table 310.16** shall be used.

As stated above, the temperature surrounding the conductor may vary from 86°F (30°C) along the conductor length. However, if the ambient temperature is different than 86°F, the allowable ampacity of the conductor as listed in **Table 310.16** shall be corrected according to the correction factors below the Table.

For example, three current-carrying copper conductors in a raceway exposed to a 40°C (104°F) ambient temperature shall have their allowable ampacities derated by 91 percent per correction factors to **Table 310.16**. **(See Figure 8-1)**

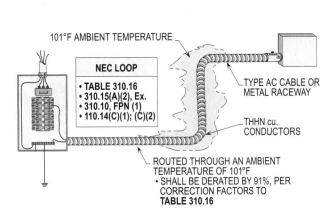

Figure 8-1. Conductors routed through an ambient temperature greater than 86°F shall be derated per correction factors to **Table 310.16** based upon surrounding temperatures. (This rule applies for conductors in cables and conduits)

WHEN TO APPLY CORRECTION FACTORS
310.15(A)(2), Ex.

Ambient temperature will not affect the allowable ampacity of conductors per **Table 310.16** if that portion of the conductors routed through the higher ambient is 10 ft. (3 m) or less and does not exceed 10 percent of the total length of such conductors.

Design Tip: The higher allowable ampacities shall be permitted to be utilized for the lower ampacity if the length is no greater than 10 ft. (3 m) or no greater than 10 percent of the circuit conductors. The 10 ft. (3 m) or 10 percent of conductor length is exposure to ambient temperatures above 86°F. **(See Figure 8-2)**

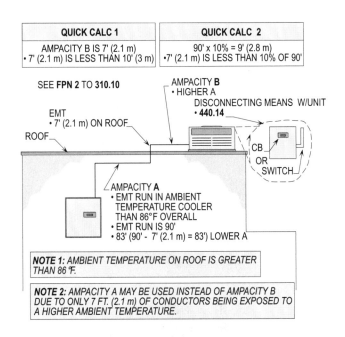

Figure 8-2. Where the total length of conductors in the EMT run (ampacity A) are 83 ft. or less, the derating factors do not have to be applied to the above circuit conductors.

HEAT GENERATED INTERNALLY
310.10, FPN (2)

Heat generated internally in the conductor as the result of load current flow, including fundamental and harmonic currents.

To comply with the above, the allowable ampacity of a conductor shall be sized to carry the load without overheating its insulation beyond the listed temperature of such insulation per **Tables 310.13** and **310.16**. A properly sized conductor serves as a heat sink to remove heat away from the terminals of devices and equipment. Conductors not sized properly may cause terminals to develop greater heat than they are designed for and improper operation may occur. To ensure conductors are adequate to supply the load without overheating, the loads used to size such conductors shall be computed per **210.19(A)(1)** and **215.2(A)(1)**.

HOW TO COMPUTE LOADS
210.19(A)(1); 215.2(A)(1)

Loads that are continuous shall have their full-load current, in amps, increased by 125 percent and this value added to noncontinuous loads at 100 percent, if any are present. **(See Figure 8-3)**

Conductors

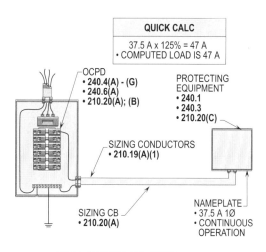

HEAT GENERATED INTERNALLY
NEC 310.10, FPN (2)

Figure 8-3. Conductors shall be sized large enough to carry the computed load supplied (47 amps) without overloading its insulation and termination points.

DISSIPATION INTO AMBIENT MEDIUM
310.10, FPN (3)

The rate of heat dissipation into the ambient medium surrounding the conductor per **110.14(C)(1)** and **(C)(2)** depends upon the surrounding temperature.

As stated above, it is clear, if the allowable ampacities of conductors are not selected according to the lowest temperature rating of devices and equipment, the higher temperature of insulation will hold heat and deteriorate internal working parts.

For example, the allowable ampacity of a 3 AWG THWN copper conductor terminated to a 75°C terminal is 100 amps based upon the 75°C, Column of **Table 310.16**. Note that it shall not be permitted to be terminated and used at 110 amps which would be the allowable ampacity of the 90°C, Column per **Table 310.16**. (Note, 85 amps at 60°C terminals)

MATCHING TEMPERATURE MARKINGS
110.14(C)(1) AND (C)(2)

The reason for the above rule is as follows:

If a conductor with 90°C insulation is connected to a terminal of 60°C, the 90°C insulation must be heated above 90°C before heat may be dissipated into the surrounding ambient medium. This heat which is held in the conductor by the higher insulation must be dissipated. As current heats up the material of the conductor, it is always seeking a cooler spot and this cooler area is usually the inside elements of the equipment and the circuit's overcurrent protection device. By matching the 90°C allowable ampacity of the conductor to the 60°C ampacities of the overcurrent protection device and equipment, the unwanted heating effects of internal components due to excessive current flow is limited. **(See Figure 8-4)**

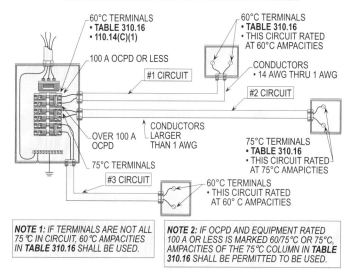

DISSIPATION INTO AMBIENT MEDIUM
NEC 310.10, FPN (3)

Figure 8-4. The above illustration shows the correct procedure for selecting allowable ampacities of conductors based upon terminal ratings. **(See Figures 8-11 and 14-16)**

ADJACENT LOAD-CARRYING
CONDUCTORS
310.10, FPN (4)

The heating effect of adjacent load current-carrying conductors, known as the proximity effect per **310.15(B)(2)(a)** to ampacity Tables 0 - 2000 volts must be considered.

As stated above, the proximity heating effect of conductors occur where conductors are bundled or pulled together in raceways and cables. When conductors are installed in this manner, their ability to dissipate heat is greatly reduced and adjustment factors per **310.15(B)(2)(a)** shall be applied to adjust for such heating and prevent damaging insulation and components of devices and equipment.

WHEN TO APPLY ADJUSTMENT FACTORS 310.15(B)(2)(a) TO TABLES 0 - 2000 VOLTS

If four or more current-carrying conductors are pulled through a raceway or bundled in a cable for a distance greater than 24 in. (600 mm), the allowable ampacity of such conductors shall be reduced (derated) by the factors (percentages) listed in **Table 310.15(B)(2)(a)**.

For example, nine current-carrying conductors in a raceway shall have their ampacities listed in **Table 310.16** reduced by 70 percent so they will not overheat their insulation ratings if they should become fully loaded. **(See Figure 8-5 and Figures 8-13 and 8-16)**

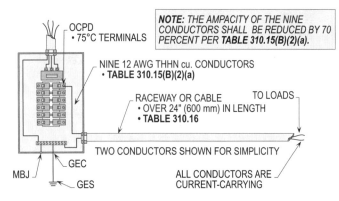

ADJACENT LOAD-CARRYING CONDUCTORS
NEC 310.10, FPN (4)

Figure 8-5. Adjustment factors per **310.15(B)(2)(a)** shall be applied where there are four or more current-carrying conductors in a raceway or cable.

DEFINITIONS
ARTICLE 100

There are certain definitions that shall be understood when calculating the allowable ampacities of conductors based upon their conditions of use.

AMPACITY
ARTICLE 100

The ampacity of a conductor is the current rating in amps that a conductor may carry continuously without exceeding its temperature rating which is determined by its condition of use. Such condition of use is too many current-carrying conductors in a raceway or cable or surrounding ambient temperature which exceeds 86°F. Note that both of these conditions may exist in a circuit.

CONTINUOUS LOAD
ARTICLE 100

A continuous load is where the current in amps may operate for a period of three hours or more. Such continuous loads are as follows:

- Indoor lighting
- Outdoor lighting
- Advertising signs
- Special outlets
- Certain types of equipment, etc.

UNGROUNDED CONDUCTORS
310.12(C)

An ungrounded (phase) conductor is a circuit conductor which is not grounded but carries current to the load. Ungrounded (phase) conductors in a service, feeder or branch-circuit are known as the ungrounded (phase) conductors.

GROUNDED CONDUCTORS
ARTICLE 100; 310.12(A)

A grounded (neutral) conductor is a system or circuit conductor that is intentionally grounded. Such grounded conductor shall be permitted to be a neutral per **220.61(A) through (C)** or a corner grounded delta conductor per **250.26** and **240.22**.

> **Design Tip:** A corner grounded delta system conductor shall always be considered current-carrying due to it being an ungrounded (phase) conductor. The neutral may or may not be current-carrying per **310.15(B)(4)(c)** to ampacity Tables 0-2000 volts.

EQUIPMENT GROUNDING CONDUCTORS
ARTICLE 100; 310.12(B)

The equipment grounding conductor (EGC) connects all the metal parts of enclosures and equipment together. It bonds such enclosures and equipment by terminating at the grounding bus at the service equipment or separately derived system. Equipment grounding conductors shall not be considered a current-carrying conductor except during a ground-fault per **310.15(B)(3)** and **(B)(5)**.

See Figure 8-6 for a detailed illustration of recommended color coding techniques.

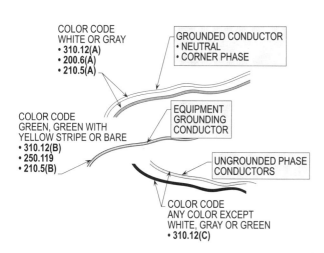

TYPES OF CONDUCTORS
NEC 310.12(A); (B); (C)

Figure 8-6. The recommendation to color code ungrounded (phase) conductors was deleted in the 1975 NEC.

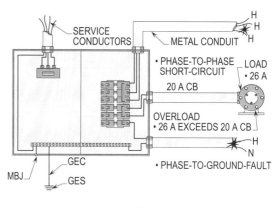

OVERCURRENT
NEC ARTICLE 100

Figure 8-7. The above illustration shows the difference between short-circuits, ground-faults and overloads in an electrical circuit.

OVERCURRENT
ARTICLE 100; 240.3; 240.4

Overcurrent is the current in excess of equipment or conductor ratings. Such overcurrents are caused by any one of the following conditions:

- Short-circuits
  (phase-to-phase)
- Ground-faults
  (phase-to-ground)
- Overloads
  (slow heat build-up)

See Figure 8-7 for a detailed illustration pertaining to short-circuits and overloads in an electrical system.

CURRENT-CARRYING CONDUCTORS
310.15(B)(4)(c) TO TABLE 310.16; 725.28(A); (B)

All ungrounded (phase) conductors shall be considered current-carrying. Neutral conductors and control circuits are current-carrying under certain conditions of use. A neutral of a four-wire, three-phase wye system shall be considered current-carrying when it carries the major portion of the load involving nonlinear related loads.

WHEN IS A NEUTRAL
CURRENT-CARRYING
310.15(B)(4)(c)

If the neutral load on Phases A, B and C have a neutral load of 100 amps respectively and 51 amps (100 A x 51% = 51 A) is harmonic related, the neutral shall be considered a current-carrying conductor. **(See Figure 8-8)**

> **Design Tip:** Any load over 50 percent is considered a major portion. However, 51 percent is normally used as the multiplier to determine the major portion in amps. (Rule of thumb method)

WHEN ARE CONTROL CIRCUITS
CURRENT-CARRYING
725.28(A) AND (B)

Control circuits shall be considered current-carrying when they carry continuously more than 10 percent of their allowable ampacities.

Twenty control circuit conductors carrying continuous current of 2 amps (20 A x 10% = 2 A) shall not be considered current-carrying conductors. Therefore, there could be 15 current-carrying conductors in a raceway and due to three being classified control circuits and three neutrals, all of which fall within the above rules, only 9 conductors shall be considered current-carrying. **(See Figure 4-18)**

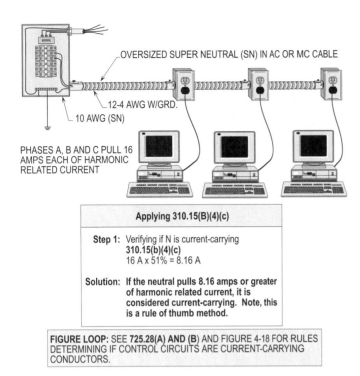

Figure 8-8. If the neutral in the above illustration is sharing harmonic loads that are considered the major portion of the circuit currents of Phases A, B and C, the neutral shall be current-carrying. (Use larger amps of A, B or C)

CONDUCTOR IDENTIFICATION 310.12

A color code scheme was recommended for ungrounded (phase) conductors prior to the 1975 NEC. Previous editions (after the 1975 NEC) only required color coding for an ungrounded (phase) conductor or grounded (neutral) conductor and the equipment grounding conductor. The high-leg of a delta system had to be colored orange or marked properly.

Conductors used in electrical wiring systems shall be identified properly to protect personnel working on such systems. It is essential to know which conductor by color represents the ungrounded (phase) conductor, the grounded (neutral) conductor and equipment grounding conductor in an electrical circuit. Because it is by color coding that conductors are connected to color coded terminals of equipment.

GROUNDED CONDUCTORS 310.12(A)

Insulated conductors of 6 AWG and smaller wire, when used as grounded (neutral) conductors, shall have a white or gray colored insulation. Where the insulated conductors are larger than 6 AWG, they shall be identified by a white or gray colored insulation, or by a distinctive marking (white) at the terminals at the time they are being installed.

The grounded (neutral) conductor used in service-entrance raceways and cables shall be permitted to be uninsulated per **230.41, Ex.**

See **240.22** for further information concerning grounded (neutral) conductors and OCPD's.

EQUIPMENT GROUNDING CONDUCTORS 310.12(B)

An equipment grounding conductor shall be permitted to be in the same raceway, cord or cable or may otherwise be run with circuit conductors. It shall be a part of the cable or cord, or run in the same raceway with the circuit conductors to keep the impedance to the lowest value. When using NM, NMC or UF cables, these cables shall contain the equipment grounding conductor as part of the cable to ground noncurrent-carrying metal parts.

The equipment grounding conductor shall be permitted to be bare, covered or insulated. Where individual covered or insulated equipment grounding conductors are run, it is required that they shall have a continuous green outer finish or a continuous green outer finish with one or more yellow stripes. Bare conductors shall be permitted to always be used for such grounding.

When the equipment grounding conductor is larger than 6 AWG copper or aluminum, this conductor shall be permanently marked at the time of installation, at each end or any other accessible point. It shall be identified by any one of the following means:

• Stripping the insulation from the entire exposed length.

• Coloring the exposed insulation green.

• Marking the exposed insulation with green colored tape or green colored adhesive labels.

See **250.119** for a detailed description of identifying the equipment grounding conductor in a circuit.

UNGROUNDED CONDUCTORS 310.12(C)

The ungrounded (phase) conductors shall be permitted to be identified with any color of insulation or tagging except white, gray or green.

Conductors

Design Tip: Until the 1975 NEC, the color coding scheme for ungrounded (phase) conductors was usually as follows:

- 120/240 volt, single-phase system
  - Phase A - Black
  - Phase B - Blue or red

- 120/240 volt, three-phase delta system
  - Phase A - Black
  - Phase B - Orange
  - Phase C - Blue

- 120/208 volt, three-phase wye system
  - Phase A - Black
  - Phase B - Red
  - Phase C - Blue

- 277/480 volt, three-phase wye system
  - Phase A - Brown
  - Phase B - Orange
  - Phase C - Yellow
  - Known as boy system

See **Figure 8-9** for a detailed illustration of recommended color code procedures for Phases A, B and C.

Section 210.5(C) requires the ungrounded (phase) conductors of multiwire circuits where more than one nominal voltage system exists in a building's wiring system to be identified. The identification of each conductor, based on the voltage, shall be marked with tape, tagging, spray paint where permitted or other effective means accepted by the AHJ.

Design Tip: This is not a color code requirement, it is a means of identification so each system's voltage may easily be determined and identified.

The means of identification shall be permanently posted at each panelboard housing such branch-circuits. **(See Figure 8-10)**

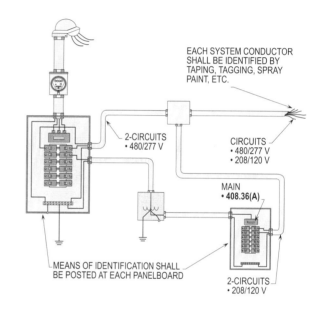

IDENTIFICATION OF
UNGROUNDED CONDUCTORS
NEC 210.5(C)

Figure 8-10. Where there is more than one system voltage supplying a premises, proper identification of circuits and panelboards shall be provided.

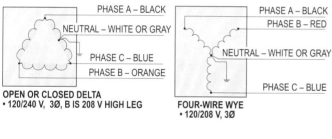
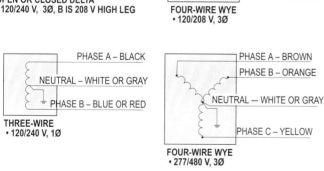

CONDUCTOR IDENTIFICATION
NEC 310.12(A); (B); (C)

Figure 8-9. Until the 1975 NEC was published, ungrounded (phase) conductors were recommended to be color coded.

CONDUCTORS FOR GENERAL WIRING
TABLE 310.13

Conductors used for general wiring shall be insulated unless they are permitted to be otherwise due to the type of installation. The conductor insulation shall be of a type which is approved for the voltage, operating temperature and location of use per **Table 310.13** in the NEC. Insulated conductors shall be rated for terminating to terminals or lugs rated at 60°C or 75°C or any combination of such per **110.14(C)** and **Table 310.16**. (For further information, **See Figure 8-20**)

TERMINAL RATINGS
110.14(C)(1)(a) AND (C)(1)(b)

The procedure used in verifying a terminal's overcurrent protection device rating is to check the rating listed on the overcurrent protection device to see if it is 60°C, 60°C/75°C or 75°C. Where the rating of the overcurrent protection device is 100 amps or less, the terminal rating is 60°C if it is not marked as mentioned above. OCPD's rated over 100 amps are rated at 75°C and may be loaded to the 75°C ampacities of conductors which are found in **Tables 310.16 through 310.19** in the NEC. Note that motors shall be permitted to be cabled with 75°C terminals and ampacities.

For example: What is the allowable ampacity of a 4 AWG THHN copper conductor connected to an overcurrent protection device supplying power to equipment where all terminals are rated at 75°C? **(See Figure 8-11 and 110.14(C) for further details.)**

Step 1: Finding ampacity
Text in book 75°C terminals -
Table 310.16
4 AWG THHN cu. has an
allowable ampacity of 85 A

Solution: **A 4 AWG THHN cu. conductor may be loaded to 85 amp with 75°C terminals. Note: See 110.14(C)(1).**

Design Tip: Conductors shall be selected according to the lowest temperature rating of the equipment and the allowable ampacities used accordingly.

For example, a piece of equipment with 60°C terminals shall be permitted to be supplied by 90°C THHN conductors. However, the allowable ampacity of the 90°C conductor shall be used at 60°C ampacities per **Table 310.16**.

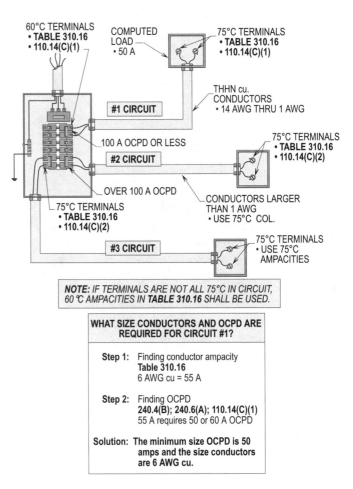

TERMINAL RATINGS
NEC 110.14(C)(1)(a); (C)(1)(b)

Figure 8-11. The allowable ampacity of conductors shall be determined by the markings on the OCPD's and electrical equipment supplied. **(See Figures 8-4 and 14-16)**

LONG AND SHORT TIME RATINGS OF CONDUCTORS 310.10

There are two current ratings of conductors in an electrical system and they are as follows:

• Long time current rating
• Short time current rating

The long time current rating of conductors shall be determined by selecting the allowable ampacity of the conductor based upon its size, material and insulation.

For example: What is the allowable ampacity of one of three 8 AWG THHN copper conductors in a metal raceway system used at 60°C?

Step 1: Finding the rating
Table 310.16
8 AWG THHN cu. = 40 A

Solution: The 8 AWG THHN copper conductors have an allowable ampacity of 40 amps at 60°C terminals.

The short time current of copper conductors shall be determined by dividing the CM rating of the conductor by 42.3 where the conductors are in the same raceway or cable assembly. If a single conductor is installed in a raceway or routed by itself, the CM rating shall be divided by 30. This method of calculating short time current ratings is a rule of thumb method and should be used only as such.

For example: What is the short time current rating of a 8 AWG THHN copper conductor in a conduit system with other conductors?

Step 1: Finding CM
Table 8, Ch. 9
8 AWG THHN cu. = 16,510 CM

Step 2: Finding the rating
I = CM ÷ 42.3
I = 16,510 CM ÷ 42.3
I = 390 A

Solution: The short time current rating is 390 amps for each 8 AWG in the conduit at 60°C terminals.

For example: What is the short time current rating of a 6 AWG THHN copper grounding electrode conductor that is run in a conduit to a water pipe?

Step 1: Finding the CM
Table 8, Ch. 9
6 AWG THHN cu. = 26,240 CM

Step 2: Finding the rating
I = CM ÷ 30
I = 26,240 CM ÷ 30
I = 875 A

Solution: The short time current rating of the 6 AWG is 875 amps.

Design Tip: The short time current rating in amps is based on a duration of 5 seconds or less.

See **Figure 8-12** for a detailed illustration of calculating and determining long and short time current ratings of conductors using the rule of thumb method.

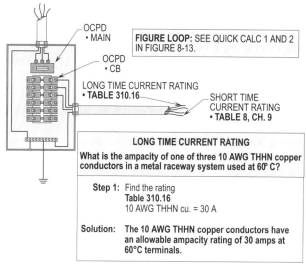

LONG AND SHORT TIME RATINGS OF CONDUCTORS NEC 310.10; TABLE 310.16

Figure 8-12. Calculating the long time current ratings and short time current ratings of conductors.

DERATING CIRCUIT CONDUCTORS TABLES 310.16 AND 310.19

Table 310.16 in the NEC shall not permit more than three current-carrying conductors to be enclosed in a raceway, cable or buried side by side in the earth. The Table states that the allowable ampacities of conductors listed are limited to a surrounding temperature of 86°F (30°C) or less. Cables or raceways enclosing more than three current-carrying conductors and routed through an ambient temperature of over 86°F are subject to derating per **310.15(B)(2)(a)** to **Table 310.16** and the correction factors beneath the Table.

THREE OR MORE CURRENT-CARRYING CONDUCTORS 310.15(B)(2)(a) TO AMPACITY TABLES 0 - 2000 VOLTS

Cables or raceways enclosing four or more current-carrying conductors shall be derated per **310.15(B)(2)(a)** to ampacity Tables 0-2000 volts. Conductors considered current-carrying by the NEC are all ungrounded (phase) conductors.

Neutrals for wye systems shall be considered current-carrying where the major portion of the load (51 percent or greater) consists of harmonic currents such as electric discharge lighting, PC's, etc. per **310.15(B)(4)(c)** to ampacity Tables 0-2000 volts. **(See Figure 8-8)**

Control circuits that are loaded to more than 10 percent of the conductors allowable ampacity ratings shall be considered current-carrying per **725.28(A), (B)** and **(C)**.

For example: What is the current-carrying ampacity of 9 - 12 AWG THHN copper conductors pulled through 3/4 in. (21) EMT? (All current-carrying)

Step 1: Finding amperage of conductor
Table 310.16
12 AWG THHN cu. = 30 A

Step 2: Finding derating factors
Table 310.15(B)(2)(a)
9 conductors requires 70%

Step 3: Finding allowable ampacity
Table 310.15(B)(2)(a)
30 A x 70% = 21 A

Solution: The allowable ampacity is limited to 21 amps for each conductor.

See **Figure 8-13** for a detailed illustration of how to calculate and determine the allowable ampacities when applying the adjustment factors of **Table 310.15(B)(2)(a)**.

DERATING FOR AMBIENT TEMPERATURE CORRECTION FACTORS TABLE 310.16

Conductors routed through ambient temperatures exceeding 86°F shall be derated according to the correction factors of **Table 310.16**. The derating factors are listed in the ampacity correction factor chart below **Table 310.16** and they shall be selected based on the ambient temperature that the conductors are exposed to. The ampacity correction factors shall be based on the material, insulation and the size of the conductors utilized.

For example: What is the allowable ampacity of 4 - 10 AWG THHN copper conductors routed through an attic in an ambient temperature of 120°F?

Step 1: Finding amperage of conductors
Table 310.16
10 AWG THHN cu. = 40 amps

Step 2: Finding correction factors
Table 310.16
120°F requires 82%

Step 3: Finding allowable ampacity
Ampacity correction factors -
Table 310.16
40 A x 82% = 32.8 A

Solution: The allowable ampacity is limited to 32.8 amps for each conductor.

See **Figure 8-14** for a detailed illustration for calculating the allowable ampacities of conductors when correction factors due to ambient temperature are applied.

APPLYING BOTH DERATING FACTORS TABLE 310.15(B)(2)(a) AND AMPACITY CORRECTION FACTORS TABLE 310.16

Four or more current-carrying conductors enclosed in a cable or raceway and routed through an ambient temperature above 86°F shall have their allowable ampacities derated twice.

Too many current-carrying conductors in a cable or raceway have problems dissipating heat into the surrounding ambient medium. The load on the conductors must heat the cable or raceway above the surrounding ambient temperature before harmful heat can be dissipated. **(See 310.10, FPN's (1) and (4))**

Derating the current-carrying ampacity of the conductors prevents overloading and deterioration of insulation.

Conductors

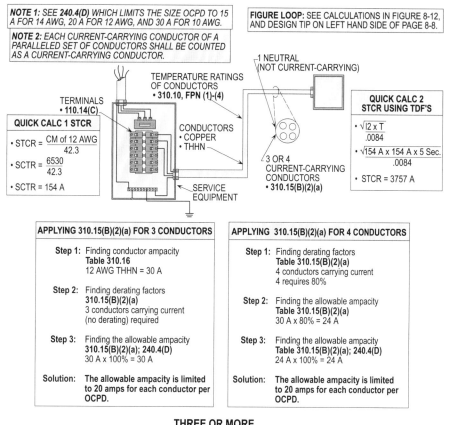

Figure 8-13. Determining the allowable ampacities of conductors based upon three or four or more current-carrying conductors pulled through a raceway or installed in a cable

For example: What is the allowable ampacity of 6 - current-carrying 14 AWG THHN copper conductors routed through an ambient temperature of 105°F?

Step 1: Finding amperage of conductors
Table 310.16
14 AWG THHN cu. = 25 A

Step 2: Finding derating factor
Table 310.15(B)(2)(a)
to Table 310.16
6 conductors requires 80%

Step 3: Finding derating factor
Correction factors to Table 310.16
105°F requires 87%

Step 4: Finding allowable ampacity
Table 310.15(B)(2)(a)
and correction factors
25 A x 80% x 87% = 17.4 A

Solution: **The allowable ampacity is limited to 17.4 amps for each conductor.**

See **Figure 8-15** for a detailed illustration of calculating and determining ampacities for four or more current-carrying conductors based upon an ambient temperature above 86°F.

APPLYING 50 PERCENT LOAD DIVERSITY FACTOR
TABLE B.310.11

A 50 percent load diversity shall be permitted to be applied to the conductor ampacity when all the conductors are not loaded at the same time, if permitted by the AHJ.

See **Figure 8-16** for a detailed illustration of applying derating factors with or without load diversity.

WHEN TO APPLY CORRECTION FACTORS
310.15(A)(2), Ex.

Ambient temperature will not effect the allowable ampacity of conductors per **Table 310.16** if that portion of the

8-11

conductors routed through the higher ambient is 10 ft. (3 m) or less and does not exceed 10 percent of the total length of such conductors.

> **Design Tip:** The higher (allowable) ampacitiy shall be permitted to be utilized for the lower (allowable) ampacity if the length is no greater than 10 ft. (3 m) or no greater than 10 percent of the circuit conductors or branch-circuit elements, whichever is less.

For example: What size overcurrent protection device is required for 5 current-carrying 12 AWG THHN copper conductors routed through an ambient temperature of 102°F?

Step 1: Finding amperage of conductors
Table 310.16
12 AWG THHN cu. = 30 A

Step 2: Finding derating factors
Table 310.15(B)(2)(a)
5 conductors require 80%
102°F requires 91%

Step 3: Finding allowable amperage
Table 310.15(B)(2)(a)
30 A x 80% x 91% = 21.8 A

Step 4: Finding OCPD
Table 310.16; 240.4(D)
12 AWG conductor with
21.8 A requires 20 A OCPD

Note: This load of 20 amps is calculated per **210.19(A)(1)** and **210.20(A)**.

Solution: Section 240.4(D) and Table 310.16 requires a 20 amp overcurrent protection device on 12 AWG conductors.

As stated above, the temperature surrounding the conductor may vary from 86°F along the conductor length. However, if the ambient temperature is different than 86°F, the allowable ampacity of the conductor as listed in **Table 310.16** shall be corrected according to the correction factors below the Table. A derating factor (adjustment) shall be applied for four or more current-carrying conductors.

For example, three current-carrying copper conductors in a raceway exposed to a 40°C ambient temperature shall have their allowable ampacities derated by 91 percent per correction factors to **Table 310.16**.

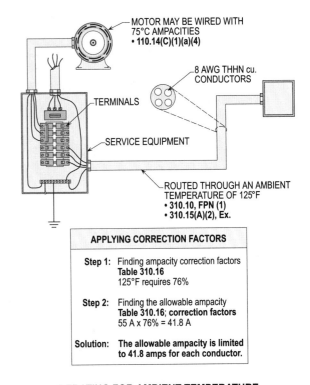

Figure 8-14. Calculating the allowable ampacity of conductors where they are routed through ambient temperatures above 86°F.

BRANCH-CIRCUIT ELEMENTS
422.10(A); (B) AND 422.11(A) - (G)

The branch-circuit conductors and overcurrent protection devices are found after the derating factors have been applied. The branch-circuit elements shall be determined based upon calculated load and this load shall be equal to or less than the allowable ampacity of the conductors after derating factors have been applied.

Conductors

For example: What size overcurrent protection device is required for 4 - 14 AWG THWN copper current-carrying conductors in a raceway or cable?

Step 1: Finding amperage of conductors
Table 310.16
14 AWG THWN cu. = 25 A

Step 2: Finding derating factors
Table 310.15(B)(2)(a)
4 conductors requires 80%

Step 3: Finding allowable amperage
Table 310.15(B)(2)(a)
25 A x 80% = 20 A

Solution: Section 240.4(D) and Table 310.16 requires a 15 amp overcurrent protection device on 14 AWG conductors.

For example: What size overcurrent protection device is required for 4 - 10 AWG THHN copper conductors that are run through an ambient temperature of 125°F? (Only three are current-carrying)

Step 1: Finding amperage of conductors
Table 310.16
10 AWG THHN = 40 A

Step 2: Finding derating factors
Correction factors to Table 310.16
125°F requires 76%

Step 3: Finding allowable amperage
Correction factors to Table 310.16
40 A x 76% = 30.4 A

Solution: Section 240.4(D) and Table 310.16 requires a 30 amp overcurrent protection device on 10 AWG conductors.

See **Figure 8-17** for a detailed illustration of sizing and selecting elements of a branch-circuit based upon conditions of use.

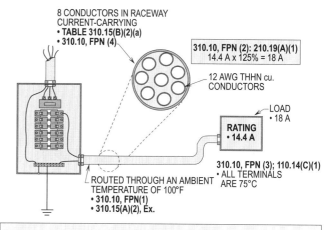

APPLYING BOTH DERATING FACTORS
TABLE 310.15(B)(2)(a)
CORRECTION FACTORS TO TABLE 310.16

Figure 8-15. Determining the allowable ampacity and size OCPD for conductors routed through an ambient temperature of more than 86°F with more than three current-carrying conductors (total of eight (8)) in a cable or raceway

INDIVIDUAL CIRCUITS 210.19(A)(1); 210.20(A)

Individual branch-circuits supply power to special appliance loads. Special appliance loads are loads such as heating units, water heaters, cooking equipment, etc. These loads require larger circuits to supply the greater amperage or VA ratings. Loads for special appliances served by individual circuits shall be determined by multiplying the amperage or VA by 100 percent for noncontinuous loads and 125 percent for continuous loads.

The following Sections in the NEC outline requirements that shall be used for sizing the branch-circuit elements of individual circuits.

- Disposals   430.22(A); 430.52; 422.10; 422.11
- Compactor   430.22(A); 430.52; 422.10; 422.11
- Dishwasher   430.22(A); 430.52; 422.10; 422.11
- Cooking Equip.   **220.55; Table 220.55; 422.33(B)**
  - Ranges
  - Cooktops
  - Ovens
- Water heater   **422.13**
- Heating units   **424.3(B)**
- A/C units   **440.22(A); 440.32; 440.62(B)**
  - Central
  - Window
- Motors   **430.6(A)(1); 430.22(A); 430.52; Table 430.52**
- Welders   **630.11; 630.12**
- X-ray   **660.6(A); 517.73**

For example: What size THWN copper conductors and overcurrent protection device is required for an individual branch-circuit to a small processing machine with a nameplate current rating of 42 amps (continuous load)?

Step 1: Finding amperage
210.19(A)(1); 210.20(A)
42 A is the circuit current rating

Step 2: Calculating load
210.19(A)(1); 210.20(A)
42 A x 125% = 52.5 A

Step 3: Selecting conductors and OCPD
Table 310.16; 240.6(A)
52.5 A requires 60 A OCPD
65 A conductor allows 70 A OCPD
per **240.4(B)**

Solution: A 60 amp overcurrent protection device and 6 AWG THWN copper conductors shall be permitted based on load or 70 amp based on conductor ampacity.

See **Figure 8-18** for a detailed illustration of how to calculate and size the elements of an individual circuit.

APPLYING 50% LOAD DIVERSITY
What is the allowable ampacity of 16 - 12 AWG THHN cu. conductors, with diversity, located in the same raceway?

Step 1: Table 310.16
12 AWG THHN cu = 30 A

Step 2: Table B.310.11
16 conductors = 70%
30 A x 70% = 21 A

Solution: The allowable ampacity of each conductor is 21 amps, however, only 8 can have this ampacity rating.

APPLYING 50% LOAD DIVERSITY
What is the allowable ampacity of each conductor when applying 50% load diversity?

Step 1: FPN to Table B.310.11
$W = I^2 \times R \times \text{\# of conductors} \times 50\%$
$W = 30^2 \times 1 \times 16 \times 50\% = 7200\ W$

Step 2: $A = W \div \text{\# of conductors} = \sqrt{W} = A$
$A = 7200\ W \div 16 = \sqrt{450} = 21.2\ A$

Solution: The allowable ampacity for 8 conductors is 21.2 amps with 50 percent diversity. See QUICK CALC for ampacity, using formula.

APPLYING DERATING FACTORS
What is the allowable ampacity of 16 - 12 AWG THHN cu. conductors, without diversity, located in the same raceway?

Step 1: Table 310.16
12 AWG THHN cu = 30 A

Step 2: Table 310.15(B)(2)(a)
16 conductors = 50%
30 A x 50% = 15 A

Solution: The allowable ampacity of 16 conductors is 15 amps.

QUICK CALC 1
APPLYING TABLE B.310.11, FPN (LOADING 8 WIRES)
- $A_2 = \sqrt{.5 \times N \div E \times A_1}$
- $A_2 = \sqrt{.5 \times 16 \div 16}$
- $A_2 = \sqrt{707 \times 30\ A}$
- $A_2 = 21.2\ A$

QUICK CALC 2
LOADING 12 WIRES
- $A_2 = \sqrt{.5 \times N \div E \times A_1}$
- $A_2 = \sqrt{.5 \times 16 \div 12 \times A_1}$
- $A_2 = 81\% \times 30\ A \times 70\%$
- $A_2 = 17\ A$

APPLYING 50 PERCENT LOAD DIVERSITY FACTOR TABLE B.310.11

Figure 8-16. Applying derating factors to determine allowable ampacities of conductors with load diversity.

Conductors

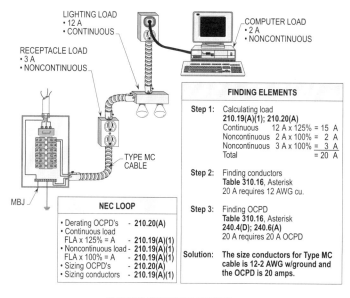

Figure 8-17. The size conductors and OCPD's shall be sized at 125 percent of the continuous loads plus 100 percent of the noncontinuous loads and derating factors, iff necessary.

CONDUCTORS IN PARALLEL 310.4

These requirements apply to copper and aluminum conductors connected in parallel so as to form a common conductor. Size 1/0 AWG or larger, comprising each phase or neutral, shall be permitted to be paralleled, that is, both ends of the paralleled conductors are connected together so as to form a single conductor. In paralleling conductors, the following conditions shall be complied with:

- They shall be the same length.
- Of the same conductor material.
- Same circular-mil area.
- Same type of insulation.
- Terminated in the same manner.
- Where run in separate raceways or cables, the raceways or cables shall have the same physical characteristics.

See Figure 8-19 for a detailed illustration pertaining to the rules of paralleling conductors 1/0 AWG and larger.

Exceptions that permit conductors of any size to be run in parallel are as follows:

- For elevator lighting (**620.12(A)(1)**)
- Conductors supplying control power to indicating instruments, contactors, relays and the like may be run in parallel in sizes smaller than 1/0 AWG provided

that the following conditions are complied with:
(a) They are in the same raceway or cable.
(b) The ampacity of each conductor is sufficient to carry the entire load shared by the parallel conductors.
(c) If one parallel conductor should happen to become disconnected, the ampacity of each conductor is not exceeded.

- Conductors supplying power for frequencies of 360 Hz and higher shall be permitted to be connected in parallel if they comply with the three rules listed above in the second bullet.

Design Tip: When necessary to run equipment grounding conductors with paralleled conductors, these equipment grounding conductors shall be treated the same as the other conductors and sized by the provisions of **Table 250.122**.

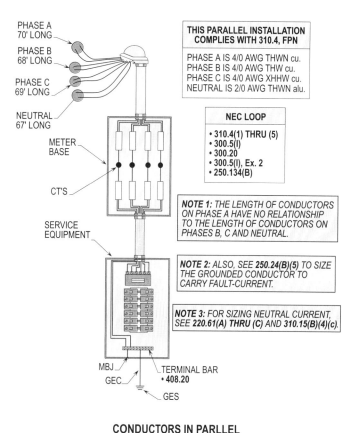

Figure 8-19. Conductors of one phase does not have to be the same as another phase, grounded (phase) conductor or grounded (neutral) conductor.

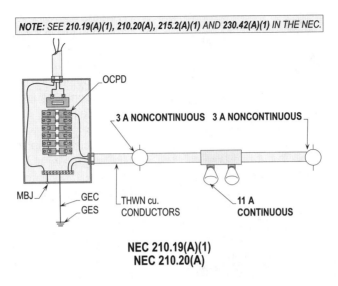

Figure 8-18. Calculating the load and sizing the elements of an individual circuit.

LOCATIONS
310.8

There are locations besides those exposed directly to the weather that conductors shall be considered to be installed in wet locations. In such locations, conductor insulation shall be either suitable for wet locations or lead covered. However, if lead covered, the insulation need not be suitable for wet locations. Such locations are as follows:

- Underground.
- In concrete slabs or masonry in direct contact with the earth, or masonry subject to moisture.
- In wet locations.
- Where condensation or accumulation of moisture within the raceway is likely to occur.

The following types of insulation are moisture-resistant:

- RHW; RHW-2; MTW
- THHW
- TW
- THW; THW-2
- THWN; THWN-2; ZW
- XHHW; XHHW-2
- Lead covered or other
- Types listed for use in wet locations

CONDUCTOR CONSTRUCTION AND APPLICATION 310.13

Table 310.13 in the NEC lists the different types of insulated conductors used for general wiring. The conductors listed shall be permitted to be used for any voltage up to and including 600 volts.

Conductor types shall be permitted to be suitable for both wet and dry locations. Some types are suitable only for dry locations. Conductors listed as suitable for dry locations shall not be permitted to be used in wet locations. Note that these are insulations for conductors that are listed for dry and damp locations. Conductors in **Table 310.13** shall be permitted to be used in any of the wiring methods recognized in **Chapter 3** if utilized as specified in their respective Tables. **(See Figure 8-20)**

Insulation resistance may vary a great deal when exposed to different temperatures. The resistance is higher at lower temperatures and steadily decreases as the temperature rises. In other words, conductors exposed to high temperature may cause insulation to have a much lower insulation reading when tested.

Design Tip: Thermoplastic insulation may stiffen at temperatures colder than minus 10°C (plus 14°F), therefore, care must be exercised during installation. Thermoplastic insulation may also be deformed at normal temperatures where subjected to pressure, requiring care to be exercised during installation and at points of support.

TYPICAL AMBIENT DESIGN TEMPERATURE 1971 NEC

The 1971 NEC included Table 310-20(c), which contained typical ambient temperatures which could be used for correcting ampacities of conductors utilizing the correction factors below Table 310-16. This Table was deleted in the 1975 NEC. However, the information concerning typical ambient temperatures in this Table may be used as a guideline when actual ambient temperatures are unknown. **(See Figure 8-21)**

TABLE 310.13 - CONDITIONS OF USE

	COLUMN 2	COLUMN 3	COLUMN 4	COLUMN 5	COLUMN 6
TYPE LETTER	INSULATION	MAXIMUM OPERATING TEMPERATURE	APPLICATIONS PROVISIONS	SIZES AVAILABLE	OUTER COVERING THHN 600 V (USED IN HOT AREAS)
THHN	FLAME-RETARDANT, HEAT RESISTANT THERMOPLASTIC	90°C	DRY AND DAMP LOCATIONS	14-1000	NYLON JACKET OR EQUIVALENT
THWN	FLAME-RETARDANT, HEAT RESISTANT THERMOPLASTIC	75°C	DRY AND WET LOCATIONS	14-1000	NYLON JACKET OR EQUIVALENT
XHHW	FLAME-RETARDANT, CROSS-LINKED SYNETHIC POLYMER	90°C 75°C	DRY AND DAMP LOCATIONS WET LOCATIONS	14-2000	NONE XHHW 600 V (USED IN COLD AREAS)
THW	FLAME-RETARDANT MOISTURE AND HEAT RESISTANT THERMOPLASTIC	75°C	DRY, DAMP, AND WET LOCATIONS ELECTRIC DISCHARGE LIGHTING EQUIPMENT • 410.31	14 - 2000	NONE THW 600 V (WET LOCATIONS, H = 75°C, THERMOPLASTIC)
TW	FLAME-RETARDANT MOISTURE AND THERMOPLASTIC	60°C	DRY AND WET LOCATIONS	14 - 2000	NONE

CONDUCTOR CONSTRUCTIONS AND APPLICATIONS TABLE 310.13

Figure 8-20. The above Table lists the most used insulations for conductors utilized in the general wiring of electrical systems. **Note:** THHN is used for hot temperature and XHHW is used for cold temperatures.

For Conductors Routed Outside:	TABLE 310.20(C) TO 1971 NEC		
	LOCATION	TEMPERATURE	MINIMUM RATING OF REQUIRED CONDUCTOR INSULATION
• In conduit, use 50°C operating temperature	WELL VENTILATED NORMALLY HEATED BUILDINGS	30°C (86°F)	* SEE NOTE BELOW
• In cable trays, use 45°C operating temperatures	BUILDINGS WITH MAJOR HEAT SOURCES AS POWER STATIONS OR INDUSTRIAL PROCESSES	50°C (113°F)	75°C (167°F)
• Note that the above operating temperatures are recommended practices.	POORLY VENTILATED SPACES SUCH AS ATTICS	45°C (113°F)	75°C (167°F)
	FURNACES AND BOILER ROOMS (MIN.) (MAX.)	40°C (104°F) 60°C (140°F)	75°C (167°F) 90°C (194°F)
	OUTDOORS IN SHADE IN AIR	40°C (104°F)	75°C (167°F)
	IN THERMAL INSULATION	45°C (113°F)	75°C (167°F)
	DIRECT SOLAR EXPOSURE	45°C (113°F)	75°C (167°F)
	PLACES ABOVE 60°C (140°F)		110°C (230°F)

NOTE: 60°C FOR UP TO AND INCLUDING 8 AWG COPPER AND UP TO AND INCLUDING 6 AWG ALUMINUM AND COPPER-CLAD ALUMINUM. 75°C FOR OVER 8 AWG COPPER AND 6 AWG ALUMINUM OR COPPER-CLAD ALUMINUM.

Figure 8-21. The above Table can be used with its ambient temperatures based upon specific locations when applying the correction factors to **Table 310.16**. **Note:** This Table is a suggested Table and not a mandatory Table when designing and selecting ampacities of conductors based upon conditions of use.

DETERMINING AMPACITIES OF OVER 600 VOLTS ELECTRICAL SYSTEMS

The ampacities for solid dielectric insulated conductors rated over 2000 through 35,000 volts shall be permitted to be utilized in **Tables 310.69 through 310.86** and **310.60**. The ampacities of conductors in these Tables are based on the Neher-McGrath method of calculating current-carrying capacity. However, the formula in **310.60(B)(1)** shall be permitted to be used to calculate the ampacities of conductors. This method can be very complex and time consuming and such calculations shall be supervised by proper engineering supervision. In some installations, the Neher-McGrath method of calculating ampacities of conductors will produce lower ampacity ratings. As a result, a lower installation cost for the consumer is possible.

DETERMINING CONDUCTOR AMPACITIES IN CABLES TABLES 310.69 THROUGH 310.86

Figure 310.60 to Tables 310.69 through 310.86 contains cables and duct banks of different numbers with detail identification numbers which can be correlated to specific ampacities Tables and their correct ampacities selected based upon voltage and size.

For example, **Table 310.85** covers the ampacities of three triplexed single insulated copper conductors and **Table 310.86** cover the ampacities of three triplexed single insulated aluminum conductors. For both Tables, the ampacity is given in two columns, one for 2001 - 5000 volts and a second for 5001 to 35000 volts. **(See Figure 8-22)**

DETERMINING CONDUCTOR AMPACITIES IN DUCT BANKS TABLES 310.69 THRU 310.86

Table 310.77 shall be permitted to be used to determine the ampacities of conductors routed through duct banks that are buried in the ground. Such ampacities are based upon the voltage, number, configuration and installation arrangement of duct banks. **(See Figure 8-23)**

DERATING DUE TO OBSTRUCTIONS 310.60(C)(2)(a) TO TABLES 310.69 THRU 310.86

Section 310.60(C)(2)(a) to Tables 310.69 through 310.86 contains information concerning derating ampacities of conductors, due to burial depth. The rule does not require derating of conductor ampacities where the duct bank burial depth to miss the obstruction is less than 25 percent of the total run located below the 30 in. (750 mm) restriction. **(See Figure 8-24)**

Conductors

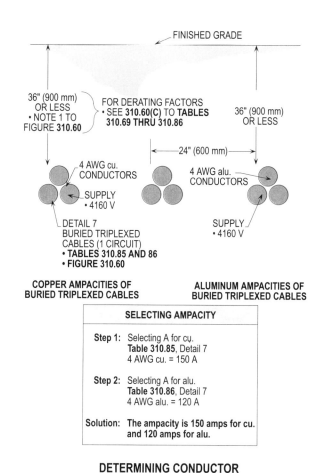

Figure 8-22. Determining ampacities of conductors utilizing the specific ampacities of **Tables 310.69 through 310.86**.

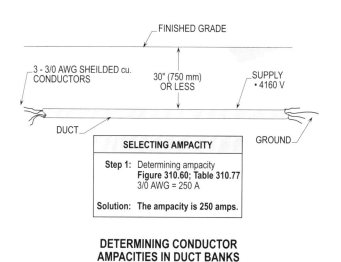

Figure 8-23. Determining ampacities of conductors utilizing the specific ampacities of **Tables 310.69 through 310.86**.

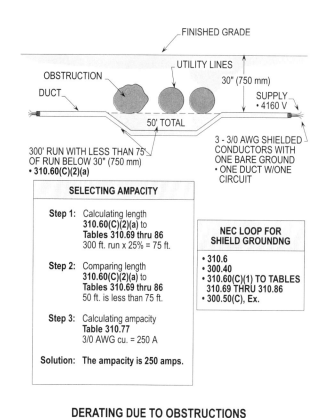

Figure 8-24. The above installation does not require the conductors ampacities to be derated due to greater burial depths to miss the obstruction.

DERATING DUE TO OBSTRUCTIONS 310.60(C)(2)(b) TO TABLES 310.69 THRU 310.86

Section 310.60(C)(2)(b) states that where the burial depth is deeper than 30 in. (750 mm) and 25 percent of the total run in a specific underground duct is exceeded, an ampacity derating factor of 6 percent increased foot of depth for all values of RHO shall be utilized. See **Figure 310.60 to Tables 310.69 through 310.86** for rules pertaining to the installation of cables. **(See Figure 8-25)**

SPACING OF CONDUITS 310.60(C)(3); Fig. 310.60

For the spacing of conduits entering enclosures, see the requirements of **310.60(C)(3)**. Note that the spacing of such conduits per **Figure 310.60** shall be permitted to be reduced without derating the ampacity of the conductors.

8-19

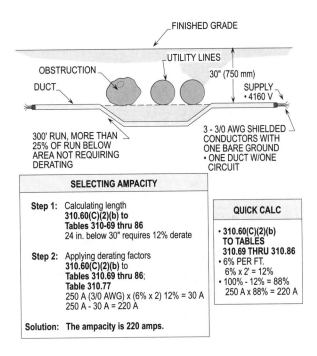

DERATING DUE TO OBSTRUCTIONS
310.60(C)(2)(b) TO TABLES 310.69 THRU 310.86

Figure 8-25. The above installation requires the conductors ampacities to be derated as shown.

Name

Date

Chapter 8. Conductors

Section   Answer

_____   T   F   1. Conductors routed through an ambient temperature greater than 87°F shall be derated per correction factors based upon surrounding temperatures.

_____   T   F   2. If a conductor with 90°C insulation is connected to a terminal of 60°C, the 90°C insulation must be heated above 90°C before heat can be dissipated into the surrounding ambient medium. (terminals rated 100 amps or less)

_____   T   F   3. A neutral of a four-wire, three-phase wye system shall be considered current-carrying when it carries the major portion of the load involving nonlinear related loads.

_____   T   F   4. Three current-carrying conductors enclosed in a cable or raceway routed through an ambient temperature above 86°F shall have their allowable ampacities derated twice.

_____   T   F   5. Conductors listed as suitable for dry locations shall be permitted to be used in damp or wet locations.

_____
_____   T   F   6. Ambient temperature will not effect the ampacity of conductors if that portion of the conductors routed through the higher ambient is _____ ft. or less and does not exceed _____ percent of the total length.

_____
_____   T   F   7. Loads that are continuous shall have their full-load currents increased by _____ percent and this value added to noncontinuous loads at _____ percent. (branch-circuit conductors)

_____
_____   T   F   8. If _____ or more current-carrying conductors are pulled through a raceway or bundled in a cable for a distance greater than _____ in., the allowable ampacity of such conductors shall be reduced (derated).

_____   T   F   9. A continuous load is where the current in amps can operate for a period of time of _____ hours or more.

_____   T   F   10. An _____ (phase) conductor is a circuit conductor which is not grounded but carries current to the load.

_____ _____   11. A grounded conductor is a system or circuit conductor that is _____ grounded.

_____ _____   12. The _____ connects all the metal parts of enclosures and equipment together.

_____ _____
_____   13. The ungrounded (phase) conductors shall be permitted to be identified with any color of insulation or tagging except gray, _____ or _____.

_____ _____   14. Where installed in raceways, conductors of size _____ AWG and larger shall be stranded.

8-21

Section	Answer

15. Copper and aluminum conductors shall be permitted to be connected in parallel in sizes _____ AWG or larger.

16. Control circuits shall be current-carrying when they carry continuously more than _____ percent of their ampacity.

   (a) 10
   (b) 20
   (c) 30
   (d) 50

17. Insulated conductors of _____ AWG and smaller, when used as grounded conductors, shall have a continuous white or natural gray insulation.

   (a) 10
   (b) 8
   (c) 6
   (d) 4

18. Where the rating of the overcurrent protection device is _____ amps or less, the terminal rating shall be 60°C if it is not otherwise marked.

   (a) 60
   (b) 100
   (c) 150
   (d) 200

19. Conductor ampacities shall be derated where the duct bank burial depth to miss the obstruction is less than _____ percent of the total run located below the _____ in. restriction.

   (a) 20, 24
   (b) 20, 30
   (c) 25, 24
   (d) 25, 30

20. An ampacity derating factor of _____ percent increased foot of depth for all values of RHO shall be utilized if burial depth is greater than **310.60(C)(2)(a) to Tables 310-69 through 86**.

   (a) 3
   (b) 6
   (c) 12
   (d) 18

21. What size THHN copper conductors and OCPD are required for a circuit with a computed load of 80 amp load connected to 60°C terminals?

22. What is the long time and short time current rating of a 6 AWG THHN copper conductor in a conduit system with other conductors used at 60°C?

23. What is the ampacity rating for 4 - 10 AWG THHN copper conductors that are current-carrying?

		Section	Answer

24. What is the ampacity rating for 4 - 10 AWG THHN copper conductors (three current-carrying) routed through an ambient temperature of 125°F? _____ _____

25. What is the ampacity rating for 8 - 10 AWG THHN copper conductors (all current-carrying) routed through an ambient temperature of 120°F? _____ _____

26. What size overcurrent protection device is required for 4 current-carrying 12 AWG THHN copper conductors in a cable? _____ _____

27. What size THWN copper conductors and OCPD are required for an individual branch-circuit to a small processing machine with a nameplate current rating of 38 amps (continuous load)? _____ _____

28. What size THHN copper conductors (2) and OCPD are required for an branch-circuit with an 6 amp noncontinuous load and 10 amp continuous load, based upon an ambient temperature of 125°F? _____ _____

29. What is the ampacity for a 6 AWG copper buried triplexed cables with a voltage supply of 4160. (terminals are 90°C per **110.40**) _____ _____

30. What is the ampacity for 3 - 4/0 AWG shielded cu. conductors with a voltage of 4160 with a 300 ft. run of conduit with 50 ft. of run below 30 in. and 25 percent? _____ _____

8-23

9

Overcurrent Protection Devices

The purpose of overcurrent protection devices is to monitor the current in a circuit and keep it at a level that will prevent overheating of conductors, elements and equipment. Excessive current flowing in an electrical circuit generates heat, which raises the circuit's temperature. Such temperature depends entirely upon the amount of current flowing through the electrical circuit.

For example, if the temperature of the conductor is high due to excessive current flow, insulation may melt and cause short-circuits or ground-faults to occur. Current flowing through a conductor generates heat which is proportional to the square of the current. In other words, if the current is doubled, the amount of heat is increased to four times the original amount. Therefore, overcurrent protection devices shall be sized and selected to match the current-carrying capacity (ampacity) of the conductor and elements of the equipment or stages of circuit protection shall be provided.

PROTECTION OF EQUIPMENT
240.3

For overcurrent protection of appliances, motors, generators, etc., it is necessary to refer to the different Articles listed in this Section. In all installations, there are three main parts to be protected and they are as follows:

- Circuit conductors,
- Circuit elements, and
- Equipment.

The fuse or circuit breaker protecting the installation shall be of a rating small enough to protect both, according to the rules and regulations in the NEC. **(See Figure 9-1)**

Design Tip: If the overcurrent protection devices are sized greater than the ampacity of the conductors to allow a motor or A/C unit to start and run (or any load per **240.4(A) through (G)**), a second stage of protection shall be required to protect such equipment.

Note that the second stage of protection shall be permitted to be overloads in the motor controller or thermal protectors in the compressor used in an A/C unit. Time delay fuses that are properly sized shall also be permitted to be utilized to provide overload protection.

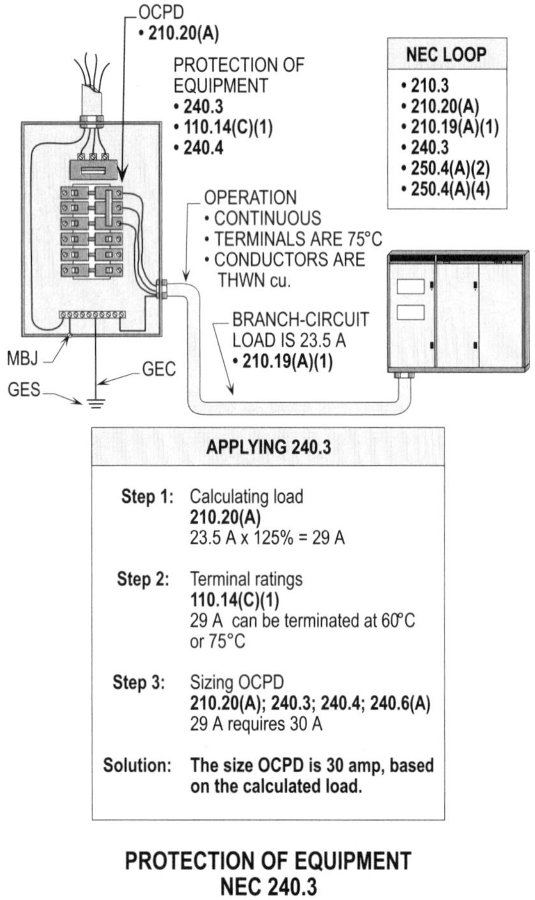

conductors, for different insulations and for different ambient-temperature conditions. The overcurrent protection device shall be sized to protect the insulation of the conductors from damage caused by the current reaching an excessive value.

For example, a 2 AWG copper conductor with THWN insulation has an ampacity of 115 amps. A 100 or 110 amp fuse or circuit breaker setting would be the largest size permitted for protection of this conductor per **240.4**. **(See Figure 9-2)**

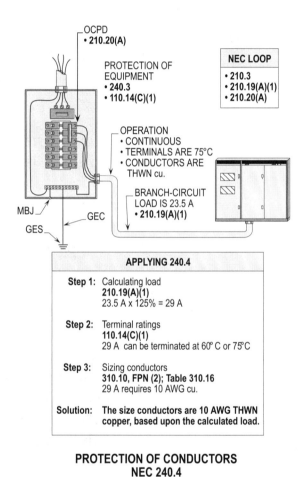

Figure 9-1. The OCPD shall be sized and selected to properly protect the electrical equipment served.

Figure 9-2. The conductors shall be sized based upon calculated load and selected to protect conductors and equipment.

For further information, see **110.14(C)(1)** and **(C)(2)** to determine if terminals and equipment ratings are mated or unmated.

However, there are exceptions (**Subdivisions (A) through (G)**) to the general rule, and in some cases the NEC allows the fuse size or circuit breaker setting to be above the ampacity of the conductor. Such exceptions listed in Parts are as follows:

PROTECTION OF CONDUCTORS
240.4

The general rule is that conductors shall be protected against overcurrent by a fuse or circuit breaker setting rated no higher than the ampacity of the conductor. Conductors have specific current-carrying ampacities for different sizes of

Overcurrent Protection Devices

- Power loss hazard
- Devices rated 800 amps or less
- Devices rated over 800 amps
- Tap conductors
- Motor-operated appliance circuit conductors
- Motor and motor control circuit conductors
- Phase converter supply conductors
- AC and refrigeration equipment circuit conductors
- Transformer secondary conductors
- Capacitor circuit conductors
- Electric welder circuit conductors
- Remote-control, signaling and power-limited circuit conductors
- Fire alarm circuit conductors

POWER LOSS HAZARD
240.4(A)

Conductor overload protection shall not be required where the interruption of the circuit would create a hazard, such as in a material handling magnet circuit. However, short-circuit protection shall be provided to clear phase-to-phase shorts or ground-faults should they occur.

DEVICES RATED 800 AMPS OR LESS
240.4(B)

If the standard current ratings of fuses or nonadjustable circuit breakers do not conform to the ampacity of the conductors being used, it shall be permitted to use the next larger standard rating when it's below 800 amps.

For example, the ampacity rating of a 2 AWG copper conductor with THWN insulation is 115 amps. There is no standard 115 amp fuse or circuit breaker setting. Nearest standard size above 115 amp is 125 amp. The 125 amp fuse or circuit breaker shall be permitted to be used in this case for protection of the conductor from short-circuits. **(See Figure 9-3)**

DEVICES RATED OVER 800 AMPS
240.4(C)

If the overcurrent protection device is greater than 800 amps, the ampacity of the conductors shall be equal to or greater than the ampacities of the overcurrent protection device per **240.6(A)**. Note that **240.6(A)** allows an OCPD, which is not listed in **240.6(A),** to be used if listed for such use. **(See Figure 9-4)**

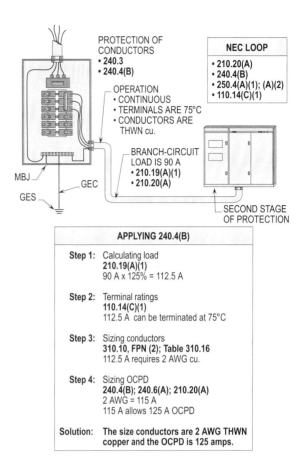

Figure 9-3. Where the ampacity of conductor(s) do not correspond with a standard device per **240.6(A)**, the next higher rating shall be permitted to be used if 800 amps or less in rating.

TAP CONDUCTORS
240.4(E)

Smaller conductors shall be permitted to be tapped from larger conductors under certain conditions. Sections **240.5, 240.21(A)** and **210.19(A)(4)** in the NEC have tap rules which permits a 18 in., 25 ft. and 100 tap respectively. Review these requirements carefully before attempting to make such tap.

MOTOR-OPERATED APPLIANCE CIRCUIT CONDUCTORS
240.4(G)

Motor-operated appliance circuit conductors are normally protected against overcurrent by the provisions listed in **Part II of Article 422**. **(See Figure 9-5)**

Stallcup's Electrical Design Book

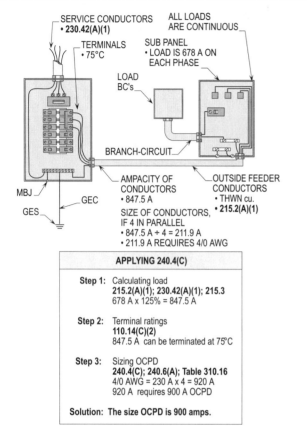

Figure 9-4. Where the ampacity of conductor(s) do not correspond with a standard device per **240.6(A)**, the next lower size shall be used if rated over 800 amps.

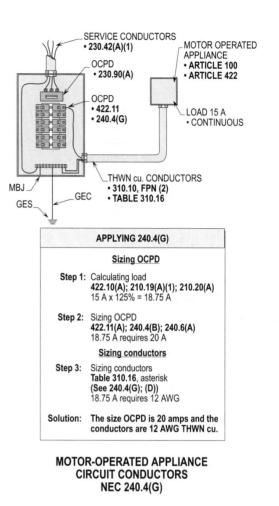

Figure 9-5. The above illustration shows the procedure for calculating the load to be used for sizing the OCPD and conductors supplying an appliance load.

MOTOR AND MOTOR CONTROL CIRCUIT CONDUCTORS
240.4(G)

Motor circuits are another exception listed in **Subdivision (G)** to the general rule where the OCPD shall be permitted to be sized above the conductors ampacity. Motor circuits shall be sized according to the requirements of **Article 430**. A study of these requirements will reveal that for motor circuits, a fuse size or circuit breaker setting in excess of the ampacity of the conductor shall be permitted by the NEC. This exception is intended to provide fuse or circuit breaker protection large enough to hold the high momentary inrush current required for starting and running the driven load. **(See Figure 9-6)**

PHASE CONVERTER SUPPLY CONDUCTORS
240.4(G)

Phase converter supply conductors for motor related loads and nonmotor loads shall be permitted to be protected against overcurrent by the rules and regulations of **455.7**. Before sizing the OCPD for such loads, review **Article 455** which contains rules much different than the requirements of **Article 430** for sizing elements of motor circuits. **(See Figure 9-7)**

AC AND REFRIGERATION EQUIPMENT CIRCUIT CONDUCTORS
240.4(G)

Circuit conductors supplying air conditioning and refrigeration equipment shall be protected against

Overcurrent Protection Devices

overcurrent by the provisions of **Parts III and VI of Article 440**. Note that these rules shall be used only for hermetically sealed motors and not individual motors per **Article 430**. **(See Figure 9-8)**

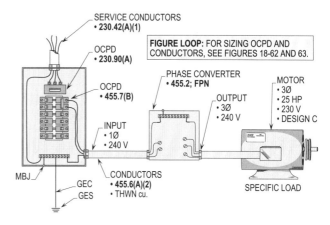

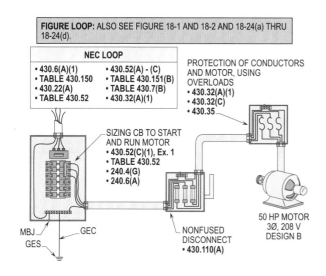

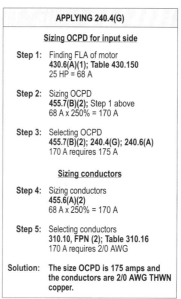

Figure 9-7. The OCPD to start and run a motor, using a phase converter, shall be permitted to have OCPD's sized with a greater rating than the circuit conductors.

TRANSFORMER SECONDARY CONDUCTORS 240.4(F)

Conductors supplied by the secondary side of a single-phase transformer having a two-wire (single voltage) secondary or a three-wire delta secondary shall be considered as protected by overcurrent protection provided on the primary (supply) side of the transformer, providing this protection is in accordance with **450.3** and does not exceed the value determined by multiplying the secondary conductor ampacity by the secondary-to-primary transformer voltage ratio. (Also, see **240.21(C)(1)** and **408.36(D), Ex.**)

Figure 9-6. Section 430.52(C)(1), Ex. 1 allows the next size OCPD to be used if the percentages of **Table 430.52** times the motor's FLA does not correspond to a standard device per **240.6(A)**.

Note that the primary side shall be a two-wire for a two-wire single-phase secondary and a three-wire for a three-wire delta secondary. **(See Figures 9-9(a) and (b))**

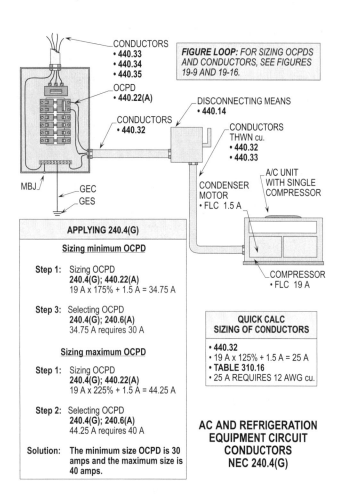

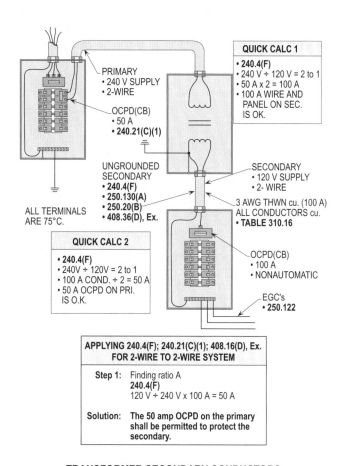

AC AND REFRIGERATION EQUIPMENT CIRCUIT CONDUCTORS
NEC 240.4(G)

TRANSFORMER SECONDARY CONDUCTORS
NEC 240.4(F)

Figure 9-8. The OCPD for the A/C units shall be permitted to be sized greater than the ampacity of the supply conductors.

Figure 9-9(a). As shown above, under certain design conditions, the primary OCPD shall be permitted to be used to protect primary and secondary sides of two-wire to two-wire transformers.

CAPACITOR CIRCUIT CONDUCTORS 240.4(G)

Capacitors draw a high inrush current when connected to the line. Circuit protection shall be permitted to be sized high enough to hold the inrush current, which normally results in overfusing the conductors. **Section 460.8(B)** allows OCPD's to be sized above the ampacities of these conductors when they supply capacitors. **(See Figure 9-10)**

ELECTRIC WELDER CIRCUIT CONDUCTORS 240.4(G)

Circuit conductors supplying welders shall be protected against overcurrent by the provisions of **Parts II and IV of**

Article 630. AC transformer and DC rectifier arc welders and conductors shall be protected per **630.12(A)** and **(B)**. Motor-generator arc welders and conductors shall also be protected per **630.12(A)** and **(B)**. Resistance welders and conductors shall be protected per **630.32(A)** and **(B)**. **(See Figure 9-11)**

REMOTE-CONTROL, SIGNALING AND POWER-LIMITED CIRCUIT CONDUCTORS 240.4(G)

Remote-control circuits have overcurrent protection sized up to three times the ampacity of the conductors per **725.23** and **725.24(C)**. Motor-control circuits shall be permitted to fused above their ampacity. A good example of a motor-control circuit is a circuit to a push-button station (start-stop) derived from a magnetic motor controller.

For motor-control circuits, the NEC allows the protection for the control circuit to be sized up to 300 or 400 percent of the ampacity of the control circuit conductors. Such protection is based upon control circuits remaining in the control enclosure or leaving the enclosure to supply a remote feed stop and start station, etc.

For further information on sizing OCPD's for motor-control circuits, see **430.72(B)(1)** and **(B)(2), Table 430.72(B)** and **725.24(C)**. **(See Figure 9-12)**

FIRE ALARM CIRCUIT CONDUCTORS 240.4(G)

Circuit conductors used in fire alarm systems shall be protected against overcurrent conditions by the provisions of **Parts II and III of Article 760**. For further information on sizing and selecting OCPD's for fire alarm circuits, see **760.23, 760.24** and **760.41** in the NEC. These circuits are special and such rules and regulations must be studied and well understood. **(See Figure 9-13)**

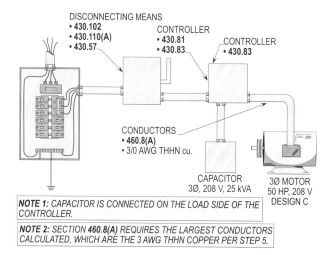

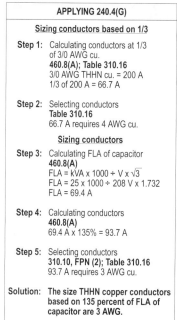

Figure 9-10. As shown above, the size conductors supplying power to capacitors are sized and selected, based either upon the FLA of the capacitor times 135 percent or 1/3 of the conductors ampacity serving the motor, whichever is greater.

PROTECTION OF FIXTURE WIRES AND CORDS 240.5

A 20 amp overcurrent protection device shall be permitted as being adequate protection for fixture wires or flexible cords or for tinsel cord, in sizes 16 AWG or 18 AWG respectively. Fixture wire taps that comply with **210.19(A)(3)** and **(A)(4)** shall be permitted to be protected when tapped from 30, 40 and 50 amp branch-circuits of **Article 210**.

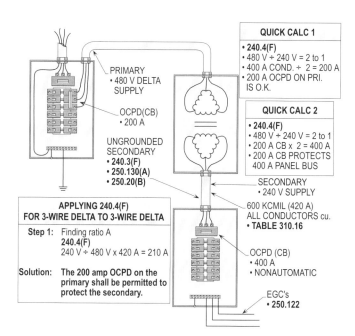

Figure 9-9(b). As shown above, under certain design conditions, the primary OCPD shall be permitted to be used to protect the primary and secondary sides of a three-wire to three-wire delta connected transformer. **(See Figure 20-8)**

ELECTRIC WELDER CIRCUIT CONDUCTORS NEC 240.4(G)

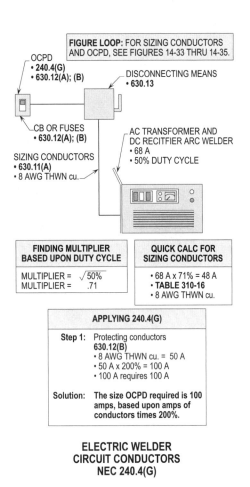

Figure 9-11. As shown above, to allow welders to operate, the OCPD shall be permitted to be sized greater than the ampacity of the conductors supplying the welder.

REMOTE-CONTROL SIGNALING AND POWER-LIMITED CIRCUIT CONDUCTORS NEC 240.4(G)

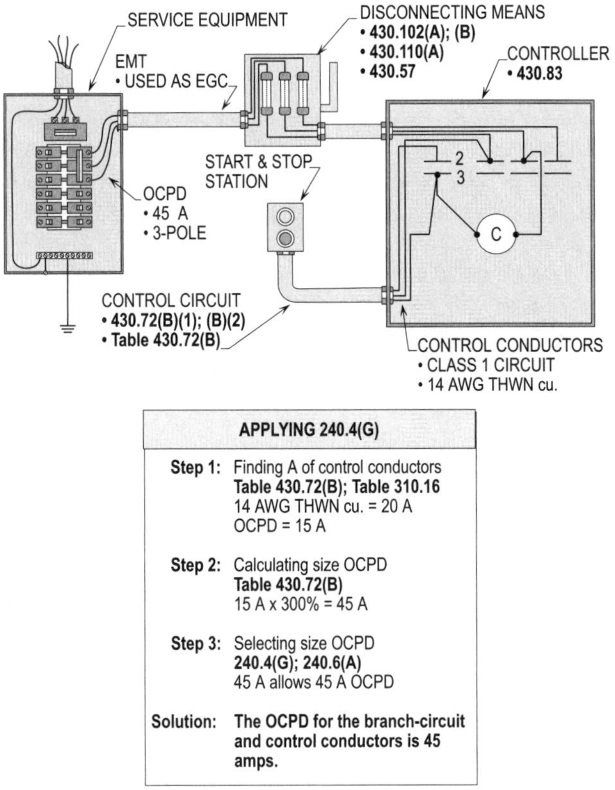

Figure 9-12. As shown above, control circuit conductors shall be permitted to be protected by the branch-circuit OCPD which is sized greater than the conductor ampacity rating.

FLEXIBLE CORD 240.5(B)(1)

Flexible cords for specific appliances shall be permitted to be protected from overcurrent conditions as follows:

- 20 amp circuits, 18 AWG cord or larger
- 30 amp circuits, cords of 10 amps capacity or greater
- 40 amp circuits, cords of 20 amps capacity or greater
- 50 amp circuits, cords of 20 amps capacity or greater

FIXTURE WIRE 240.5(B)(2)

Fixture wire shall be permitted to be protected by the overcurrent device of the branch-circuit if sized and selected per **Article 210** and complies as follows:

- 20 amp circuits, 14 AWG and greater
- 30 amp circuits, 14 AWG and greater
- 40 amp circuits, 12 AWG and greater
- 50 amp circuits, 12 AWG and greater

Design Tip: For 20 amp circuits, 18 AWG, up to a tapped length of 50 ft. and 16 AWG, up to a tapped length of 100 ft. shall be permitted to be used as fixture whips per **410.67(C)** and outside lighting standards per **240.5(B)(2)** and **410.15(B)**.

See Figure 9-14 for a detailed illustration of protecting fixture wire from overcurrent conditions.

Overcurrent Protection Devices

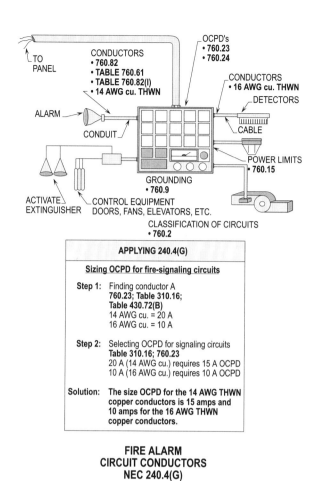

Figure 9-13. Fire-signaling circuits shall be protected from short-circuits and ground-faults.

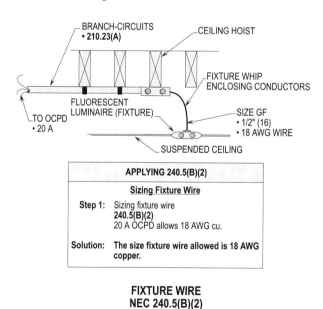

Figure 9-14. As shown above, fixture wire in whips shall be permitted to be smaller in rating than the branch-circuit OCPD and conductors.

EXTENSION CORD SETS
240.5(B)(3)

16 AWG and larger flexible cord conductors used in listed extension cord sets which are connected to a 20 amp branch-circuit per **Article 210** shall be permitted to be protected from overcurrent conditions. In other words, a second stage of protection shall not be required to be provided in addition to the branch-circuit OCPD. **(See Figure 9-15)**

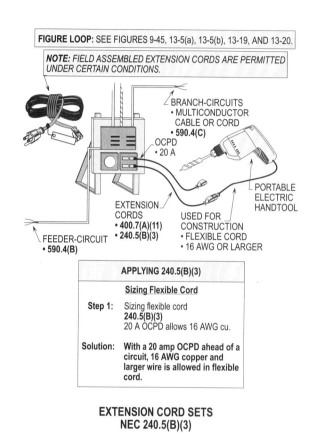

Figure 9-15. As shown above, 16 AWG and larger wire shall be permitted to be used in extension cord sets. **(Also, see Figure 13-19)**

STANDARD AMPERE RATINGS
240.6(A)

Standard ratings for fuses and circuit breakers are available as follows: 15, 20, 25, 30, 35, 40, 45, 50, 60, 70, 80, 90, 100, 110, 125, 150, 175, 200, 225, 250, 300, 350, 400, 450, 500, 600, 700, 800, 1000, 1200, 1600, 2000, 2500, 3000, 4000, 5000 and 6000 amps. For fuses only, 1, 3, 6 and 10 amps are also available as standard sizes to be used for the protection of smaller circuits such as motors rated at 8 amps or less. As for circuit breakers, the ratings above are not circuit breaker sizes but settings.

For example, a circuit breaker (CB) manufactured with a 100 amp frame shall be permitted to have settings of 15 through 100 amps respectively. A frame size of 225 amps shall be permitted have settings of 110 through 225 amps, when installed in a 225 amp panelboard.

> **Design Tip:** The setting of a circuit breaker is actually the tripping point.

For example, a circuit breaker with a 20 amp setting, will trip open when the current in amps goes above 20 amps. A 30 amp fuse will open the circuit when the circuit current in amps exceeds 30 amps.

See Figure 9-16 for a detailed illustration of selecting an OCPD based upon load.

It must be understood that the NEC does not prohibit a nonstandard rated OCPD to be used in an electrical system.

For example, a 1500 amp circuit breaker or fuse shall be permitted to be used for a circuit with a calculated load of 1575 amps when listed for such use per **240.6(A)**. This prevents the designer from having to round down to 1200 amps per **240.4(C)** and losing 375 amps respectively.

ADJUSTABLE TRIP CIRCUIT BREAKERS
240.6(B); (C)

The rating of an adjustable trip circuit breaker having external means for adjusting the longtime pickup setting for amp ratings is the maximum setting allowed by the device. Circuit breakers that have removable and sealable covers over the adjusting means, or are located behind bolted equipment enclosure doors or are located behind locked doors accessible only to qualified personnel, shall be permitted to have ampere ratings equal to the adjusted (set) longtime pickup settings.

For example, if one of these rules are not applied and an feeder-circuit supplies a lighting and appliance panelboard, the conductors for the circuit shall be capable of handling the manufacturer setting of such CB.

> **Design Tip:** It is not the intent of this section to prohibit the use of nonstandard ampere ratings for fuse and inverse time circuit breakers that have been approved by the manufacturer for such use per **240.6(A)**. (See Figure 9-17)

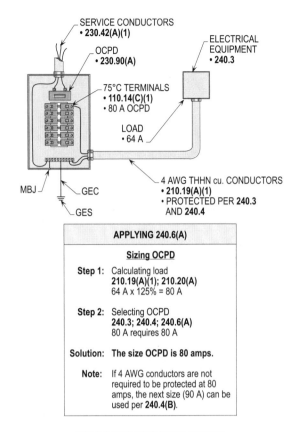

STANDARD AMPERE RATINGS NEC 240.6(A)

Figure 9-16. As shown above, a 80 amp OCPD will protect the 85 amp conductors (4 AWG THWN cu.) and 80 amp calculated load from short-circuits, ground-faults and overloads.

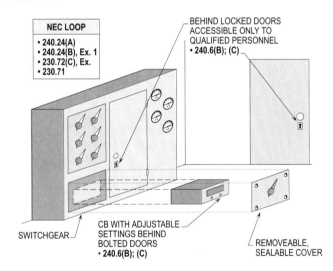

ADJUSTABLE TRIP CIRCUIT BREAKERS NEC 240.6(B); (C)

Figure 9-17. As illustrated above, the use of adjustable trip CB's is allowed only, if access for adjustment is limited to qualified personnel.

SUPPLEMENTARY OVERCURRENT PROTECTION
240.10

Electrical equipment which comes with overcurrent protection within the equipment that is provided by the manufacturer is allowed. Just because there is overcurrent protection in the equipment does not mean these devices have to be readily accessible per **240.24(A)(2)**. Such supplementary devices are designed for the purpose of detecting overcurrents and protection of individual parts or circuits of the appliance or utilization equipment and does not serve as a substitute for the branch-circuit protection. **(See Figure 9-18)**

- Coordinated short-circuit and ground-fault protection, and
- Overload indication shall be based on monitoring systems or devices.

For example, a wye system supplying power to a number of loads can have a resistance of 12 amps installed in the ground circuit, which will limit the ground-fault current to a predetermined value. A relay shall be permitted to be connected across the ground resistor, or in series with it, which actuates an alarm, if a low amperage ground-fault should occur in one phase. Such fault would not cause a shutdown, but lets it be known to personnel that a ground-fault has developed, and must be located. If necessary, arrangements shall be made to orderly shutdown the loads served and the faulted circuit found and cleared. **(See Figure 9-19)**

> **Design Tip:** Coordination is defined as properly localizing a fault condition to restrict outages to the equipment affected, accomplished by choice of selective fault-protection devices. The monitoring system may cause the condition to go to alarm allowing corrective action or an orderly shutdown, thereby, minimizing personnel hazard and equipment damage.

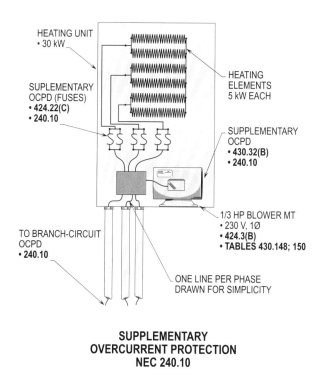

Figure 9-18. Supplementary OCPD's shall not be required to be readily accessible. However, the upstream branch-circuit OCPD shall be readily accessible.

ELECTRICAL SYSTEM COORDINATION
240.12

In industrial applications where an orderly shutdown is necessary, the coordination of overcurrent devices has always been a problem to electrical personnel. Often an orderly shutdown is needed to protect personnel from hazard(s) as well as preventing damage to equipment. To provide such coordination, the following shall be followed:

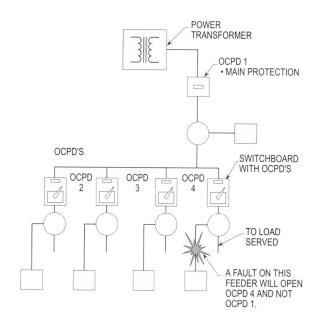

Figure 9-19. Electrical system coordination shall be provided in electrical circuits so a downstream OCPD during a fault condition will not open an upstream OCPD.

USING HANDLE TIES
240.20(B)

Circuit breakers shall open all ungrounded (phase) conductors of the circuit.

For example, when one of the trip elements operates, it trips the circuit breaker, which opens all poles of the circuit. A circuit breaker used for short-circuit protection shall have a trip unit for each ungrounded (phase) conductor, and a pole for each ungrounded (phase) conductor that opens the circuit under short-circuit and ground-fault conditions.

Design Tip: For ungrounded two-wire circuits, two single-pole circuit breakers with identified handle ties shall be permitted to be used in place of a two-pole circuit breaker.

Note that circuit breakers for three-phase circuits shall be three-pole circuit breakers with a simultaneously common trip. Circuit breakers with identified handle ties shall be permitted to be used as follows:

- Except where limited by the provisions of **210.4(B)**, individual single-pole circuit breakers, with or without identified handle ties, shall be permitted to be utilized as the protection for each ungrounded (phase) conductor of multiwire branch-circuits that serve only single-phase, line-to-neutral loads,

- In grounded systems, individual single-pole circuit breakers with identified handle ties shall be permitted to be utilized as the protection for each ungrounded (phase) conductor for line-to-line connected loads for single-phase circuits or three-wire DC circuits and

- For line-to-line loads in four-wire, three-phase systems or five-wire, two-phase systems having a grounded (neutral) conductor and no conductor operating at a voltage greater than allowed in **210.6**, individual single-pole circuit breakers with identified handle ties shall be utilized as the protection for each ungrounded (phase) conductor.

See **Figure 9-20** for a detailed illustration of using handle ties with single-pole circuit breakers.

LOCATION IN CIRCUIT
240.21

The general rule for locating fuses or circuit breakers in a circuit is that they shall be installed at the source of the circuit.

For example, where a branch-circuit taps to a feeder, the fuses or circuit breaker protecting the branch-circuit shall be provided at the point where the tap is made to the feeder-circuit.

The following requirements shall be used when making taps and locating overcurrent protection devices in circuits. The following are exceptions to the rule.

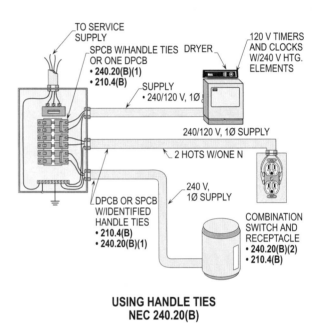

USING HANDLE TIES
NEC 240.20(B)

Figure 9-20. In certain installations and types of equipment, single-pole CB's shall be permitted to be used with identified handle ties.

BRANCH-CIRCUIT CONDUCTORS
240.21(A)

Taps to individual outlets and circuit conductors shall be permitted to be protected by the branch-circuit overcurrent protection devices when complying with the requirements of **210.19** and **210.20**.

Section 210.19 allows taps of smaller conductors to be made to larger branch-circuit conductors for certain purposes, including taps for a small range.

For example, a 14 AWG conductor shall be permitted to tap to a 20, 25 or 30 amp branch-circuit. A 12 AWG shall be permitted to tap to a 40 or 50 amp branch-circuit, and fusing shall not be required at the point of supply. **(See Figure 9-21)**

Overcurrent Protection Devices

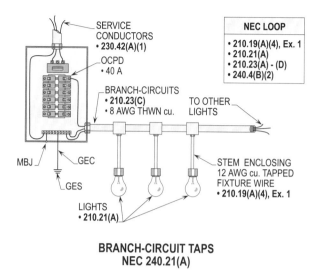

Figure 9-21. The above illustration shows 12 AWG THWN copper fixture wire being tapped from a branch-circuit having a 8 AWG THWN copper supply conductor.

TAPS NOT OVER 10 FT. (3 m) LONG 240.21(B)(1)

An example of applying this rule is a tap feeding a lighting and appliance branch-circuit panelboard. The tap shall not be permitted to be over 10 ft. (3 m) long and shall terminate at such panel. Mechanical protection by conduit, tubing or metal gutter shall be required and conductors shall be sized to carry the total load. If these requirements are followed, no overcurrent protection device shall be required at the point of such tap. **(See Figure 9-22)**

TAPS NOT OVER 25 FT. (7.5 m) LONG 240.21(B)(2)

A tap from a larger conductor to a smaller conductor shall be permitted to extend over 20 ft. (6 m) up to a distance of 25 ft. (7.5 m), provided the current-carrying capacity of the tap is at least 1/3 that of the larger conductor. Overcurrent protection shall not be required at the point of the tap, but the tap shall have overcurrent protection at the end of the run. Such tap shall be properly sized and protected from physical damage and enclosed in a raceway. **(See Figure 9-23)**

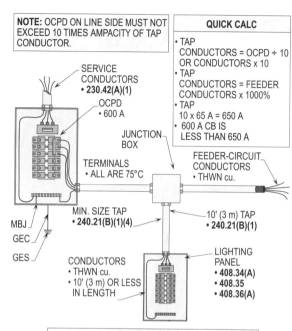

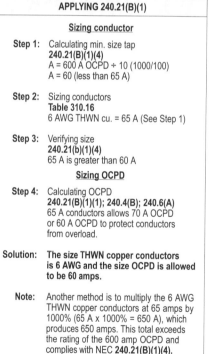

Figure 9-22. The above illustration shows the proper procedure for making a tap using the 10 ft. (3 m) tap rule.

Stallcup's Electrical Design Book

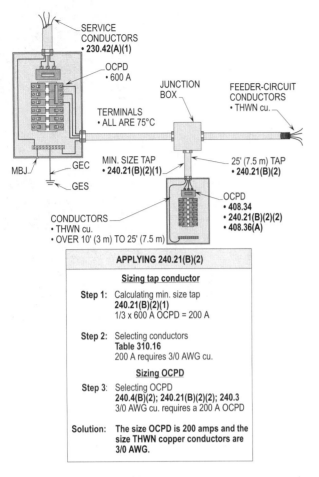

Figure 9-23. The above illustration shows the proper procedure for making a tap using the 25 ft. (7.5 m) tap rule.

TAPS SUPPLYING A TRANSFORMER (PRIMARY PLUS SECONDARY NOT OVER 25 FT. (7.5 m) LONG) 240.21(B)(3)

Transformer taps with primary plus secondary conductors which are not over 25 ft. (7.5 m) in length shall be permitted to be made without overcurrent protection at the point of such taps. The following requirements shall be applied when applying this tap rule:

- Tap conductor ampacity is at least 1/3 that of the feeder,
- Secondary conductor ampacity is at least 1/3 that of the feeder, based on the primary-to-secondary transformer ratio,
- Total length of tap is not over 25 ft. (7.5 m), primary plus secondary,
- All conductors are protected from physical damage and

- Secondary conductors terminate at a fuse or circuit breaker sized to protect the secondaries.

See Figure 9-24 for a detailed illustration of applying the requirements of this tap rule.

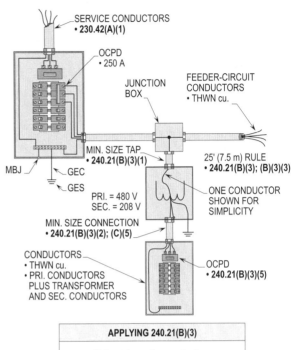

Figure 9-24. The above illustration shows the correct sizing of a primary tap plus transformer and the sizing of a secondary connection supplying a panelboard.

Overcurrent Protection Devices

TAPS OVER 25 FT. (7.5 m) LONG
240.21(B)(4)

For taps in high bay manufacturing buildings, the protection for the tap shall be permitted to be at the end of the tapped conductors, if the tap is not over 25 ft. (7.5 m) long horizontally, and not over 75 ft. (22.5 m) long vertically. Note that the total run, both horizontally and vertically, is limited to 100 ft. (30 m) or less in length. The following requirements shall be complied with when applying this tap rule:

- The ampacity of the tap conductors is at least 1/3 of the rating of the overcurrent device protecting the feeder conductors,
- The tap conductors terminate at a single circuit breaker or a single set of fuses that will limit the load to the ampacity of the tap conductors. This single overcurrent device shall be permitted to supply any number of additional overcurrent devices on its load side,
- The tap conductors are protected from physical damage by being enclosed in a raceway or other approved means,
- The tap conductors are continuous from end-to-end without splices,
- The tap conductors are sized 6 AWG copper or 4 AWG aluminum or larger,
- The tap conductors do not penetrate walls, floors or ceilings and
- The tap is at least 30 ft. (9 m) from the floor.

See **Figure 9-25** for a detailed illustration of applying the requirements of this tap rule.

TRANSFORMER SECONDARY CONDUCTORS
240.21(C)

Overcurrent protection devices of circuits shall be located at the point where the service to those circuits originates. However, it shall be permitted to connect conductors to the secondary side of transformers. Such conductors shall be designed and installed by the rules and regulations of **240.21(B)** and **(C)** in the NEC. Overcurrent protection shall be provided by **Tables 450.3(A)** or **(B)**.

PROTECTION BY PRIMARY OVERCURRENT DEVICE
240.21(C)(1)

For a detailed illustration of procedures for applying this rule, see **240.4(F)** and **Figures 9-9(a) and 9-9(b) on pages 9-6 and 7**.

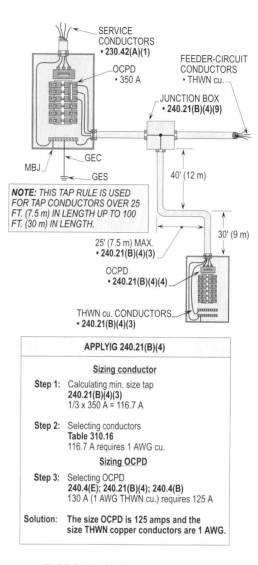

Figure 9-25. The above illustration shows the correct procedure for making a tap using the requirements of the 100 ft. (30 m) tap rule.

TRANSFORMER SECONDARY CONDUCTORS NOT OVER 10 FT. (3 m) LONG
240.21(C)(2)

Conductors shall be permitted without overcurrent protection at the transformer secondary where all the following conditions are complied with:

9-15

- Conductors do not exceed 10 ft. (3 m) in length.

- Conductors shall have a current rating not less than the combined computed loads of the circuits supplied by the conductors. Their ampacity shall not be permitted to be less than the rating of the overcurrent protection device at the termination of the secondary conductors.

- The conductors shall not be permitted to extend beyond the switchboard, panelboard, disconnecting means or control devices they supply.

- Conductors shall be enclosed in a raceway which will extend from the connection to the enclosure of an enclosed switchboard, panelboard, or control devices or to the back of an open switchboard.

See **Figure 9-26** for the proper procedure for connecting conductors to the secondary of transformers.

INDUSTRIAL INSTALLATION SECONDARY CONDUCTORS NOT OVER 25 FT. (7.5 m) LONG 240.21(C)(3)

Conductors shall be permitted to be connected to the transformer secondary of a separately derived system for industrial installations, without overcurrent protection at the connection, where all the following conditions are complied with:

- The length of the secondary conductors do not exceed 25 ft. (7.5 m) in length.

- The ampacity of the secondary conductors is not less than the secondary current rating of the transformer, and the sum of the ratings of the overcurrent devices do not exceed the ampacity of the secondary conductors.

- All overcurrent protection devices are grouped.

- The secondary conductors are protected from physical damage by being enclosed in an approved raceway or by other approved means.

> **Design Tip:** If more than 10 percent of the single-pole slots of a panelboard rated at 30 amps or less is used for branch-circuits to lighting and appliance loads with neutral connections, the panelboard shall be classified as a lighting and appliance panelboard per **408.34(A)** and **(B)** and **408.36(A)** and **(B)**. When the panelboard is classified as a lighting and appliance panelboard, OCPD is restricted to two means or less.

See **Figure 9-27** for applying the requirements of this tap rule.

OUTSIDE SECONDARY OF BUILDING OR STRUCTURE CONDUCTORS 240.21(C)(4)

Outside conductors shall be permitted to be connected at the transformer secondary, without overcurrent protection at the connection. However, the following conditions shall be complied with:

- The conductors are protected from physical damage in an approved manner.

- The conductors terminate at a single circuit breaker or a single set of fuses that limits the load to the ampacity of the conductors. This single overcurrent

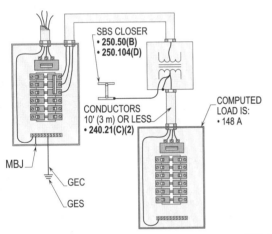

Figure 9-26. The above illustration shows the procedure for sizing a 10 ft. (3 m) connection from the secondary of a transformer.

Overcurrent Protection Devices

protection device shall be permitted to supply any number of overcurrent protection devices on its load side.

- The conductors are installed outdoors, except at the point of termination.

- The overcurrent device for the conductors is an integral part of a disconnecting means or its located immediately adjacent thereto.

- The disconnecting means for such conductors are installed in a readily accessible location either outside of a building or structure, or inside nearest the point of entrance of the conductors.

See Figure 9-28 for applying the requirements for this type of installation.

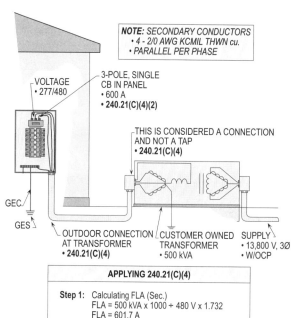

Figure 9-28. The above illustration shows the procedure to be applied when making an outside connection from the secondary side of a customer owned transformer.

SERVICE CONDUCTORS 240.21(D)

Service-entrance conductors shall be permitted to be protected by overcurrent protection devices by the provisions of **230.91**. Generally, the service overcurrent device is an integral part of the disconnecting means or it shall be installed adjacent.

For example, switches and fuses shall be either "integral" or "adjacent." A set of fuses used as service overcurrent protection shall be permitted to be separate from the disconnect, but if this is the case, the fuses shall be located "immediately adjacent to" such disconnect.

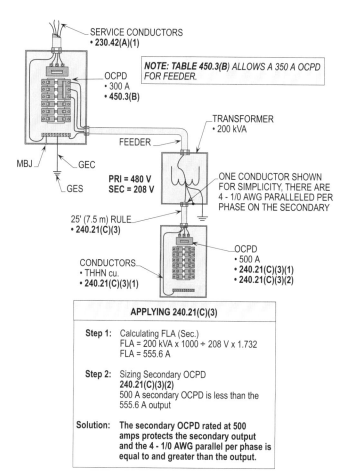

Figure 9-27. The above illustration shows the procedures to be applied when a 25 ft. (7.5 m) installation is made from the secondary side of a transformer.

Note that circuit breakers are combined with such disconnecting means and shall provide overcurrent protection for the service-entrance conductors. **(See Figure 9-29)**

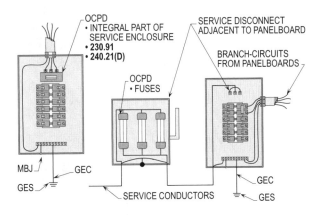

SERVICE CONDUCTORS
NEC 240.21(D)

Figure 9-29. The service-entrance conductors shall be protected by OCPD's which are either integral or adjacent to disconnect switch.

BUSWAY TAPS
240.21(E)

Section 368.17 permits a reduction in size of a busway. In such cases, an additional overcurrent protection device shall not be required for smaller busways at the point of the tap. When making a tap from a larger busway, the smaller busway shall comply with the following:

- Have a current-carrying capacity of at least 1/3 of the overcurrent setting,
- Not be over 50 ft. (15 m) long and
- Not be in contact with combustible material.

See Figure 9-30 for a detailed illustration of applying the requirements of this tap rule.

MOTOR CIRCUIT TAPS
240.21(F)

Where more than one motor is on a feeder circuit, each motor circuit is tapped from the feeder conductors. The tap shall be protected by the motor branch-circuit protection. The motor branch-circuit protection shall be located where the tap conductors terminate and the following requirements shall be complied with:

- If the length of the tap is 25 ft. (7.5 m) or less, conductor ampacity shall be at least 1/3 that of the feeder, provided that the conductors are protected from physical damage. Conduit, EMT, flexible metal conduit, or AC or MC cable, etc. provide suitable protection for the conductors.

- If the length of the tap is 10 ft. (3 m) or less, conductors with ampacity less than 1/3 that of the feeder shall be permitted to be used, provided that the conductors are in a raceway or entirely within a controller. If field installed, the OCPD on the line side of the tap conductors shall not be permitted to exceed 10 times of the tapped conductors' ampacity.

- In high bay manufacturing facilities, a tap to a feeder shall be permitted to be up to 100 ft. (30 m) in length, for a horizontal and vertical run combined, but the horizontal parts of a run shall be limited to 25 ft. (7.5 m) or less in length. The tap conductors shall comply with the following:

(a) Be unspliced,
(b) Be installed in raceways,
(c) Have an ampacity 1/3 that of the feeder,
(d) Be at least 6 AWG cu. or 4 AWG alu. and
(e) Be unspliced along their entire length.

See Figure 9-31 for a detailed illustration of applying the requirements of this tap rule.

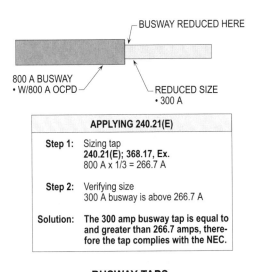

BUSWAY TAPS
NEC 240.21(E)

Figure 9-30. The above is a detailed illustration of applying the requirements for a busway tap.

NOTE 1: *OCPD ON LINE SIDE SHALL EXCEED 10 TIMES THE AMPACITY OF THE CONDUCTOR.*

NOTE 2: *USE **430.28(2)** AND Ex. WHEN USING THE 25 FT. (7.5 m) OR 100 (30 m) FT. TAP RULE FOR MOTOR CIRCUITS.*

APPLYING 240.21(F)

Sizing conductor

Step 1: Calculating min. size tap
240.21(F); 430.28(1)
A = 600 A OCPD ÷ 10 (1000/100)
A = 60

Step 2: Sizing conductors
Table 310.16
6 AWG THWN cu. = 65 A

Step 3: Verifying size
240.21(F); 430.28(1)
65 A is greater than 60 A

Sizing OCPD

Step 4: Selecting OCPD
240.4(B); 240.6(A)
65 A conductors allows a 70 A or 60 A OCPD to protect conductors from overload.

Solution: **The size THWN copper conductors is 6 AWG and the size OCPD is allowed to be 60 amps.**

Note: Another method is to multiply the 6 AWG THWN copper conductors at 65 amps by 1000% (65 A x 1000% = 650 A). This total exceeds the rating of the 600 amp OCPD and complies with **430.28(1)**.

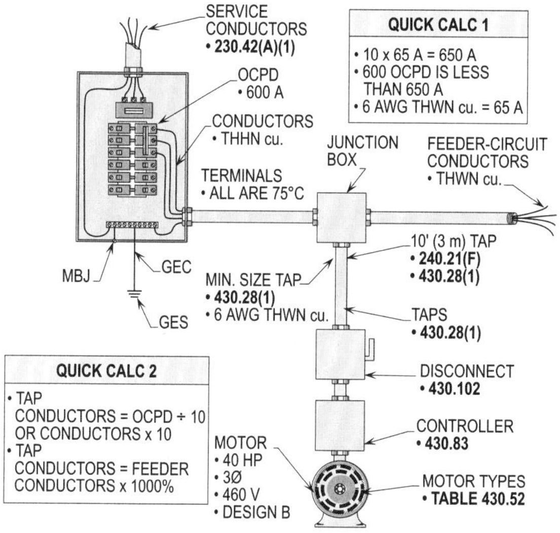

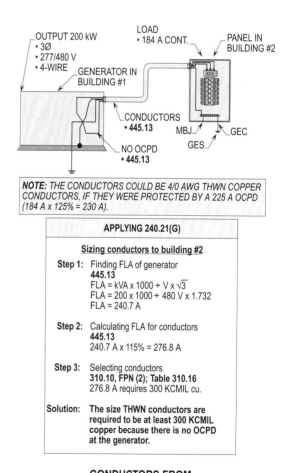

MOTOR CIRCUIT TAPS
NEC 240.21(F)

Figure 9-31. The above illustration shows the proper procedure for making a 10 ft. (3 m), 25 ft. (7.5 m) or 100 ft. (30 m) tap for motor circuits. Note that the tap illustrated is the 10 ft. (3 m) tap rule.

CONDUCTORS FROM GENERATOR TERMINALS 240.21(G)

Generator conductors shall have an ampacity equal to at least 115 percent of the nameplate current rating of the generator. The following are exceptions to the general rule:

- If the design or operation of the generator is such as to prevent overloading, an ampacity of 100 percent shall be permitted to be utilized.
- Where an integral overcurrent protection device is provided by the manufacturer, with conductors terminated to the device.

See Figure 9-32 for a detailed illustration of applying the requirements of this tap rule. Note that if the conductors are not sized based on 115 percent of the generators output in amps, an OCPD shall be provided at the generator.

NOTE: *THE CONDUCTORS COULD BE 4/0 AWG THWN COPPER CONDUCTORS, IF THEY WERE PROTECTED BY A 225 A OCPD (184 A x 125% = 230 A).*

APPLYING 240.21(G)

Sizing conductors to building #2

Step 1: Finding FLA of generator
445.13
FLA = kVA x 1000 ÷ V x $\sqrt{3}$
FLA = 200 x 1000 ÷ 480 V x 1.732
FLA = 240.7 A

Step 2: Calculating FLA for conductors
445.13
240.7 A x 115% = 276.8 A

Step 3: Selecting conductors
310.10, FPN (2); Table 310.16
276.8 A requires 300 KCMIL cu.

Solution: **The size THWN conductors are required to be at least 300 KCMIL copper because there is no OCPD at the generator.**

CONDUCTORS FROM GENERATOR TERMINALS
NEC 240.21(G)

Figure 9-32. The above illustration shows the proper procedure to apply when tapping from a generator.

SUPERVISED INDUSTRIAL INSTALLATIONS
240.90

Many of these rules apply directly to large transformer's secondary conductors. The existing NEC treated conductors connected to a transformer secondary the same as tap conductors with no upstream protection. This is technically incorrect and creates particular difficulties for large industrial installations. These difficulties do not come up in most commercial and smaller industrial installations, because transformers of the size where a problem begins to be seen either do not exist, or are installed indoors near their associated switchgear or are associated with services, which are not subject to the same rules.

In large installations, with outdoor transformers in the 1000 kVA class and larger and supplying indoor secondary switchgear, it is extremely difficult and undesirable, if not impossible, to hold the secondary conductors at 25 ft. (7.5 m) or less in length.

CONNECTIONS UP TO 100 FT. (30 m)
240.92(B)(1)(1)

Unprotected lengths of secondary conductors shall be permitted to be up to 100 ft. (30 m), if the transformer primary overcurrent device is sized at a value (reflected to the secondary by the transformer phase voltage ratio) of not more than 150 percent of the secondary conductor ampacity. **(See Figure 9-33)**

PROTECTED BY A DIFFERENTIAL RELAY
240.92(B)(1)(2)

Secondary conductors shall be permitted to be protected by a differential relay with a trip setting equal to or less than the conductor ampacity. Note that a differential relay (per **240.100(A)(1)**) provides superior short-circuit protection at a trip open value that is almost always well below the conductor ampacity. **(See Figure 9-33)**

ENGINEERING SUPERVISION
240.92(D)

Conductors shall be permitted to be protected with greater lengths, if calculations are made under engineering supervision and it is determined that the secondary conductors will be protected within recognized time versus current limits for all short-circuit and ground-fault conditions that could occur. **(See Figure 9-33)**

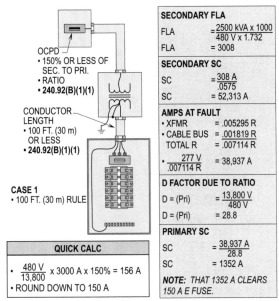

DESIGN TIP: Consider a 2500 kVA, 5.75% z transformer with a 13.8 kV to 480/277 V ratio and 500 mVA available short-circuit current on the primary for 100 circuit feet of 3000 amp cable bus, and a three-phase bolted fault at the end of the bus (worst case). About 38,937 amps will flow from the system or 1352 amps on the primary, which will clear a typical 150 E fuse and meet the maximum 150 percent requirement within .42 seconds. This time vs. current value is well within the rating of the secondary conductors. (Cable bus has a resistance of .001819 and the xfmr has a resistance of .005295.)

SHORT-CIRCUIT AND GROUND-FAULT PROTECTION
240.92(B)(1) - (B)(3)

Figure 9-33. The above illustration shows conductors being protected from short-circuit and ground-fault conditions.

GROUNDED CONDUCTORS
240.22

No overcurrent protection device shall be permitted to be connected in series with any conductor that is intentionally grounded, unless the overcurrent device operates to open the ungrounded (phase) conductors simultaneously (common trip) with such grounded conductor. This rule prohibits a fuse from being installed in the grounded conductor, because the fuse could blow and open the grounded conductor and the ungrounded (phase) conductors would still be energized. This can cause hazardous voltage levels to appear on the electrical elements of the circuit. Note that grounded conductors shall be permitted to be grounded (phase) conductors or grounded (neutral) conductors.

COMMON TRIP
240.22(1)

Where the overcurrent device opens all conductors of the circuit, including the grounded conductor, and is so designed that no pole can operate independently, the grounded conductor shall be permitted to be protected by an OCPD.

OVERLOAD PROTECTION
240.22(2)

Where required by **430.36** and **430.37** for motor running (overload) protection.

For example, running overload protection of a three-phase motor fed by a three-wire, three-phase circuit with one leg grounded may have such protection provided by installing a properly sized fuse. These fuses are usually of the time delay type, so they shall be permitted to be sized at 125 percent or less of the motors FLC in amps per **430.32(A)(1)**.

See **Figure 9-34** for a detailed illustration of providing OCP in the grounded conductor.

CHANGE IN SIZE OF GROUNDED CONDUCTOR
240.23

If a change should occur in the size of the ungrounded (phase) conductor, a similar change shall be permitted to be made in the size of the grounded (neutral) conductor, if necessary.

For changing the size of the ungrounded (phase) conductor due to voltage drop (VD), see **210.19(A)(1), FPN 4** and **215.2(A)(3), FPN 2**. For changing the size of the equipment grounding conductor, see **250.122(B)**. **(See Figures 9-35 15-7(a))**

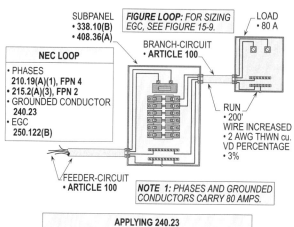

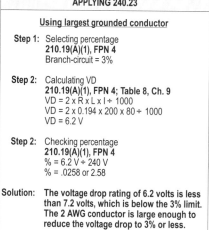

CHANGE IN SIZE OF GROUNDED CONDUCTORS NEC 240.23

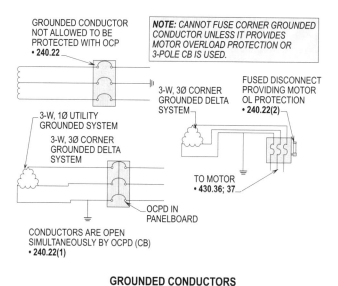

GROUNDED CONDUCTORS NEC 240.22

Figure 9-34. The grounded conductor shall not be permitted to be protected by a CB unless it opens simultaneously (common trip) with all circuit conductors. To be fused, it shall provide overload protection to a motor.

Figure 9-35. It is permissible to increase the size of the grounded (neutral) conductor, if there is a similar change in the ungrounded (phase) conductors.

READY ACCESSIBILITY OF OCPD'S 240.24(A)

For safety, the requirement that overcurrent protection devices be readily accessible is mandatory. They shall be located so that they may be readily reached in cases of emergencies or for servicing, without reaching over objects, climbing on chairs, ladders, etc.

See Figure 9-36 for a detailed illustration of OCPD's being readily accessible. However, there are four exceptions to the accessibility rule and they are as follows:

- For busways, **368.12** permits overcurrent protection for a plug-in to be out of reach from the finished grade.

- For luminaires (lighting fixtures) and appliances that are manufactured with built-in fuses, the fuses shall not be required to be readily accessible per **240.10**.

- For services, the overcurrent protection device shall be permitted to be located at the beginning of the run, at the point where the service-entrance tap is made to the service drop or lateral. In such cases, the main overcurrent protection device shall not be required to be readily accessible inside the facility. See **225.40** and **230.92** for further details concerning this type of installation.

- For overcurrent protection devices installed adjacent to motors, appliances or other types of equipment which are out of reach, such OCPD's shall not be required to be readily accessible. However, these devices shall be capable of being reached by a portable means. For further information, refer to **404.8(A), Ex. 2**.

See Figure 9-37 for a detailed illustration for applying the requirements of these rules.

OCCUPANCY 240.24(B)

The general rule does not allow overcurrent protection devices to be located where they are exposed to physical damage. Further more, shall not be installed where ignitable materials are near enough to catch fire in case of accidental sparking or overheating of such devices.

Section 240.24(B) requires each occupant of a building or apartment to have ready access to the overcurrent devices for his or her occupancy. Exceptions to this rule are made in the case of apartment buildings and guest rooms in hotels or motels where a building employee is constantly in attendance. Remember, such OCPD's shall be readily accessible to whom. Basically, they shall be readily accessible to occupants or maintenance personnel. **(See Figure 9-38)**

Sections **240.24(B), Ex. 1** and **Ex. 2** reads as follows:

- In a multiple-occupancy building where electric service and electrical maintenance are provided by the building management and where under continuous building management supervision, the service overcurrent device and feeder overcurrent devices supplying more than one occupancy shall be permitted to be accessible to authorized management personnel only.

- **Section 240.24(B), Ex. 2** allows for guest room or guest suites of hotels and motels, that are intended for transient occupancy, the overcurrent protection devices for such occupancies shall be permitted to be accessible to authorized management personnel only.

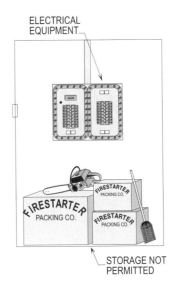

ACCESSIBILITY
NEC 240.24(A)

Figure 9-36. OCPD's shall be readily accessible without having to remove or climb over objects such as chairs, desks, etc.

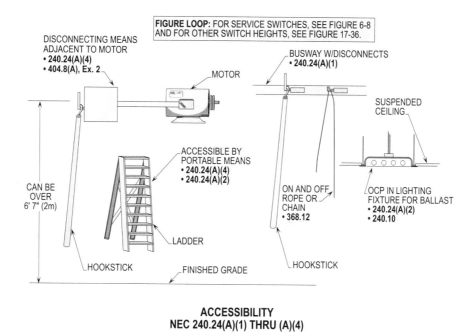

Figure 9-37. The above illustration shows installations where the OCPD's shall not be required to be readily accessible.

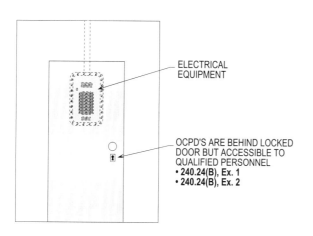

Figure 9-38. OCPD's shall be readily accessible to each occupant in a premises. However, they shall be permitted to be behind locked doors if qualified personnel are present.

FUSES
ARTICLE 240, PARTS V AND VI

Fuses for conductors with specific types of insulations are available to supply power to equipment. Conductors are sized to carry the load of the equipment in amps without deteriorating the insulation per **310.10**. The overcurrent protection devices shall be sized to protect the conductors and equipment from short-circuit, ground-faults and overloads.

The calculated load for fuses shall be determined by calculating the loads at noncontinuous duty (100 percent) or continuous duty (125 percent) and applying demand factors where applicable. Depending on the type of fuse used, a fuse will hold five times its rating for different periods of time.

SIZING
210.19(A)(1); 210.20(A); 230.42(A)(1)

Fuses shall be sized based upon the operation characteristics of the load served. The NEC defines a continuous load as a load that operates continuous for three hours or more.

The load current in amps shall be calculated by multiplying the 160 amp continuous load current by 125 percent (160 A x 125% = 200 A) to obtain the size which is 200 amps. **(See Figure 9-39)**

> **Design Tip:** An exception to the general rule allows the fuse to carry 100 percent of the continuous load current if the fuse and enclosure has been rated by specific design for 100 percent rated current operation. See the exceptions to **210.19(A)(1), 210.20(A)** and **230.42(A)(1)**.

Stallcup's Electrical Design Book

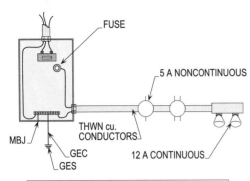

SIZING
NEC 210.19(A)(1); 210.20(A); 230.42(A)(1)

Figure 9-39. The above illustration shows the procedure for calculating the size fuse and conductors based upon continuous operation.

Fuses are equipped with a time-delay feature which is designed to allow loads requiring high inrush currents to start and run. Such loads are motors, compressors, welders, etc. For further information pertaining to these loads, see **240.4(A) through (G)**.

For example, a 200 amp time-delay fuse will hold about 5 times its rating (200 A x 5 = 1000 A) which is 1000 amps respectively. Such fuse will hold this value of 1000 amps for about 10 seconds based upon its time characteristics. A 200 amp nontime-delay fuse will hold 1000 amps for 2 seconds or less without blowing and opening the circuit due to inrush current. Therefore, a piece of equipment with an inrush current that is less than 1000 amps will start and run without blowing the fuse under normal starting and running conditions. **(See Figures 9-40(a) and (b))**

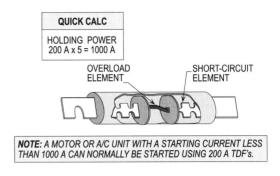

OPERATION CHARACTERISTICS
NEC 240.60(C)

Figure 9-40(a). The above illustration shows that a fuse will hold 5 times its rating for 2 to 10 seconds based upon the type used.

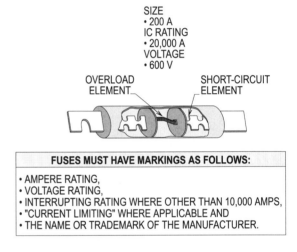

MARKING
NEC 240.60(C)

Figure 9-40(b). The above is a detailed illustration of how fuses shall be marked.

TIME AND TEMPERATURE RELATIONSHIPS

A fuse temperature has the same heating effect as for a conductor. The ampacity is rated at a standard ambient temperature for fuses and conductors. A fuse is usually 25°C with the ampacity derated for temperatures above the standard ambient temperature.

Depending on the type and class being installed, fuses consist of various current ratings that are usually based on specific time periods without opening.

For example, time-delay fuses shall withstand current higher than the rated current (overload currents) for minimum periods of time.

In a very short increment of time, a circuit shall be opened when short-circuits or ground-faults occur within a circuit without damaging the electrical elements of the circuit.

MARKING
240.60(C)

Fuses are required to be plainly marked, either by printing on fuse barrel or by a label attached to the barrel. So fuses can be sized, selected and installed correctly, the following information shall be provided:

- Ampere rating
- Voltage rating
- Interrupting rating where other than 10,000 amps
- "Current limiting" where applicable
- The name or trademark of the manufacture

DISCONNECTING MEANS FOR FUSES
240.40

Except as noted below, each set of cartridge fuses shall have an individual disconnect on the supply side, so that any one set of fuses can be individually deenergized.

For other types of fuses and for thermal cutouts, the rule applies where voltages are over 150 volts-to-ground.

If the fuses or cutouts are accessible only to qualified persons, such as maintenance electricians, these requirements shall not be permitted to be applied.

For example, one disconnect shall be permitted to serve a number of fused circuits and it shall not be necessary to provide a disconnect for each circuit.

In cases of services, where a sequence of meters, fuses and switches are utilized for services, the fuses shall be permitted to be installed ahead of the switch per **240.40** and **230.82**.

For special cases, one disconnect shall be permitted for several motors on one machine, as with metal or wood working machines per **240.40** and **430.112, Ex.**

Note that each motor shall be provided with an OCPD that allows it to start and run per **430.52**. In addition, overload protection shall be required for each motor per **430.32(A)(1)**.

Section 240.40 also permits a single disconnect where several elements of a fixed space heater are protected by individual fuses within the heating unit enclosure per **424.22(C)**. (See Figure 9-41)

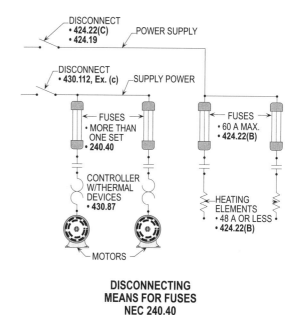

Figure 9-41. The above illustration shows different cases where one disconnect shall be permitted to be used to disconnect more than one set of fuses.

CIRCUIT BREAKERS
ARTICLE 240, PART VII

Circuit breakers and conductors with insulations are available to supply power to equipment. Conductors shall be sized to carry the load of the equipment in amps without deteriorating the insulation per **310.10**. The overcurrent protection devices shall be sized to protect the conductors and equipment from short-circuits, ground-faults and overloads.

All circuit breakers and conductors shall be designed and installed according to the latest provisions of the NEC to ensure protection and proper installation.

The calculated load for circuit breakers shall be determined by calculating the loads at noncontinuous operation (100 percent) and continuous operation (125 percent) and applying demand factors where applicable. Depending on the frame size of the unit, a circuit breaker can hold approximately three times their rating for different periods of time.

OPERATION CHARACTERISTICS

Circuit breakers operate either by thermal or magnetic

principles or any combination of such. It is a device which protects conductors from excessive current. Basically, a circuit breaker is an automatic switch that tips open and clears the circuit when the current passing through it exceeds its rating.

For example, if a current of 32 amps flows through the elements of a 30 amp circuit breaker, the element in the circuit breaker, due to excessive heat, will bend and trip open the circuit.

A large current (amps) will generate more heat than a small amount of current, so naturally a circuit breaker will trip faster on a large overload condition than a smaller one.

> **Design Tip:** A circuit breaker equipped with a magnetic plate operates both thermally and magnetically. The thermal protection clears overloads which are currents that are developed slowly on the circuit conductor. The magnetic operation clears short-circuits and ground-faults which are large currents that occur suddenly.

What makes circuit breakers so convenient to use is their ability to be used as switches and because spare ones are not necessary should a short-circuit, ground-fault or overload occur on a circuit protected by such.

SIZING
210.20(A); 215.3; 230.90(A)

Circuit breakers shall be sized based upon the operation characteristics of the load served. The NEC defines a continuous load as a load that operates continuous for three hours or more.

The load current in amps shall be calculated by multiplying the 160 amp continuous load current by 125 percent (160 A x 125% = 200 A) to obtain the size circuit breaker which is 200 amps.

> **Design Tip:** An exception to the general rule allows the circuit breaker to carry 100 percent of the continuous load current if the circuit breaker and enclosure has been rated by specific design for 100 percent rated current operation. See the exceptions to **210.20(A), 215.3** and **230.90(A).**

Circuit breakers are equipped with a time-delay feature which is designed to allow loads requiring high inrush currents to start and run. Such loads are motors, compressors, welders, etc. For further detailed information pertaining to these loads, see **240.4(A) through (G)**.

For example, a 200 amp circuit breaker will hold about 3 times its rating (200 A x 3 = 600 A) which is 600 amps respectively. Such circuit breaker will hold this value (600 amp) for a number of seconds based upon its frame size. A 200 amp circuit breaker will hold 600 amps for about 35 seconds without tripping open the circuit. Therefore, a piece of equipment with an inrush current that is less than 600 amps will start and run under normal starting conditions without tripping the circuit breaker. **(See Figure 9-42)**

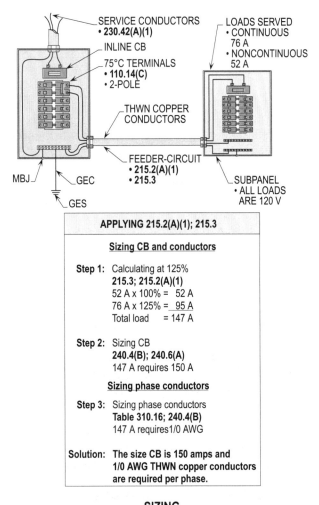

Figure 9-42. The above illustration shows the procedure for sizing the CB and conductors based upon continuous operation.

METHOD OF OPERATION
240.80

Operation by hand simply means that the circuit breaker has a handle that the operator can grasp and open or close the circuit. Circuit breakers that are pneumatic or electrically controlled may become inoperative or the contact might stick in the make position and without a handle to deenergize the circuit breaker, there would be no way to open the circuit. Should the electrical or pneumatic control fail to operate the circuit, it can be opened manually by hand using the handle on the circuit breaker.

INDICATING
240.81

When the circuit breaker is turned ON the handle shall be installed so that it is in the UP position. Circuit breakers shall be clearly marked in the open "OFF" or closed "ON" position. Circuit breaker handles that operate vertically instead of rotationally or horizontally shall be installed with the ON position of the handle in the UP position.

MARKING
240.83

The following markings shall be applied before installing circuit breakers:

- **Durable and visible.** The ampere rating shall be durable and visible after installation.

- **Location.** The ampere rating shall be molded, stamped, etched or similarly marked into the handles of circuit breakers rated 100 amps or less.

- **Interrupting rating.** Interrupting current (IC) rating other than 5000 amps shall be marked with IC rating.

- **Used as switches.** Circuit breakers used to switch 120 or 277 volt fluorescent lighting circuits shall be listed and shall be marked SWD or HID. Circuit breakers used to switch high-intensity discharge lighting circuits shall be listed and shall be marked HID.
- **Voltage marking.** Voltage rating not less than the nominal system voltage shall be marked on circuit breakers for the capability to interrupt fault currents between phases or phase-to-ground.

Design Tip: Circuit breakers used for supplementary protection shall not require an interrupting rating marked on the device.

See Figure 9-43 for a detailed illustration pertaining to the rules for marking circuit breakers.

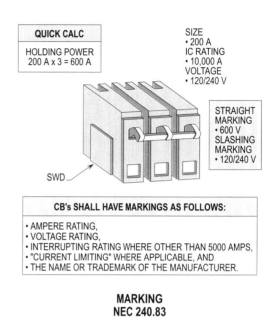

Figure 9-43. The above illustration shows the markings that shall be placed on CB's. Note that a CB will hold 3 times its rating for a period based upon its frame size and rating. A 200 amp circuit breaker will start a load with a starting current of 600 amps or less. (200 A x 3 = 600 A)

APPLICATIONS
240.85

Circuit breakers shall be permitted to be used with grounded, ungrounded or grounded neutral systems when identified with a straight voltage marking.

For example, a circuit breaker shall be permitted to be used with grounded, ungrounded or grounded neutral systems when identified with a straight voltage marking of 480 volts, etc.

Circuit breakers shall be permitted to be used with grounded neutral systems when identified with a slash voltage marking.

For example, a circuit breaker shall be permitted to be used with grounded neutral systems when identified with a slash voltage marking of 480/277 volts.

GROUND-FAULT PROTECTION OF EQUIPMENT
230.95; 240.13; 215.10

Where fault currents are available and due to the damaging effects of ground-faults, it has become necessary to prevent the injury to personnel and damage to equipment by providing ground-fault protection for equipment. The following rules shall be followed.

SIZE
230.95; 240.13; 215.10

Ground-fault protection of equipment shall be required for grounded wye connected services of more than 150 volts-to-ground and OCPD's rated 1000 amps or greater.

> **Design Tip:** This rule applies mainly for 480/277 volt wye services. However, such protection is not prohibited for 208/120 volt wye systems.

SETTING
230.95(A); 240.13; 215.10

The maximum setting permitted for ground-fault protection shall not be permitted to exceed 1200 amps, with the service disconnecting means exceeding 1000 amps or more. Ground-fault protection shall allow all ungrounded (phase) conductors of the faulted circuit to open simultaneously if a low magnitude fault develops in an ungrounded (phase) conductor.

> **Design Tip:** Ground-fault protection shall not be required for 6 - 700 amp disconnects, but shall be required for 5 - 700 amp disconnects and one 1000 amp disconnect.

For example, ground-fault protection shall be required per **230.95** if a 900 amp fuse is installed in a 1200 amp disconnect and an adjustable CB is set below 1000 amps but could later be set at 1000 amps or greater.

CLEARING TIME
230.95(A); 240.13; 215.10

To prevent damage to overcurrent protection devices, busbars and service equipment, the ground-fault system shall be required to clear a fault of 3000 amps or more within one second. The following two types of ground-fault detection are used to sense low-magnitude ground-faults:

- Donut method
- Window method

When applying the donut method to sense any current leaking to ground, a current relay shall be placed over the main bonding jumper. When applying the window method to sense any current leaking to ground, the ungrounded (phase) conductors and grounded (neutral) conductors shall be placed in the window (ground-fault sensor). The objective is to sense current leaking to ground from a ungrounded (phase) conductor.

> **Design Tip:** The system will be cleared once the low-magnitude fault-currents are detected and the overcurrent protection device is tripped open. This clearing procedure prevents damage to equipment due to low magnitude arcs produced by such leakage current.

See **Figure 9-44** for a detailed illustration for applying such rules and regulations.

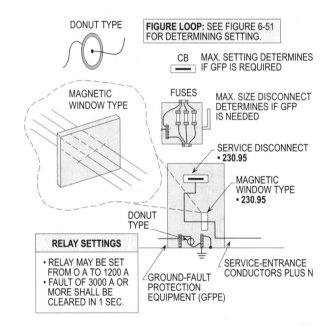

Figure 9-44. The above illustration shows the relay settings and relay requirements when installing ground-fault protection for equipment.

PERFORMANCE TESTING
230.95(C); 240.13; 215.10

Performance testing shall be performed at time of installation for the ground-fault protection system. The test shall be run according to approved test instructions that are furnished with the equipment.

A written record shall be required of the test to be available to the AHJ, before final approval of the installation.

GROUND-FAULT PROTECTION FOR PERSONNEL
590.6

Ground-fault circuit interrupters shall required to protect personnel where a leakage current develops due to a faulty electrical hand tool. Leakage currents can be caused by frayed insulation, damaged equipment, moisture, etc.

FUNCTION

A leakage current is the result of a current-carrying path developing between a current-carrying wire, equipment and ground. The function of a GFCI is to detect small currents where overcurrent protection devices such as fuses and circuit breakers cannot open or deenergize the circuit due to their being of a higher rating. These currents of a few thousandths of an ampere can be detected by a GFCI-protective device which allows the circuit to open or deenergize in a fraction of a second. Class A GFCI's will allow a circuit to open or deenergize when a ground-fault current of about 5 mA (milliamperes) is detected.

TYPES

The following are four basic categories of GFCI-protected devices:

- GFCI receptacles
  (a) Flush-mounted receptacles
  (b) Surface-mounted receptacles
  (c) Plug-in receptacles

- GFCI circuit breakers
- Portable GFCI's
- GFCI protected extension cord sets

Flush-mounted and surface-mounted type receptacles are designed and available in terminated or feed-through models. The design of terminated models will protect only their own receptacles. The design of feed-through models will protect their own receptacles and all other receptacles installed on the load side of the GFCI. Plug-in receptacles are only available to be installed in a terminated model.

Circuit breakers and GFCI's are combined devices for providing overcurrent protection and fault-current protection. The circuit breaker may be installed as an ON and OFF switch for the branch-circuit when circuits or components need servicing.

Portable equipment contains both receptacles and circuit breakers that contain GFCI's. The circuit breaker assembly supplying such equipment may or may not be a GFCI protective device.

Extension cord sets are usually protected by a GFCI in the attachment cord body.

See Figure 9-45 for the requirements of applying the rules for GFCI-protection.

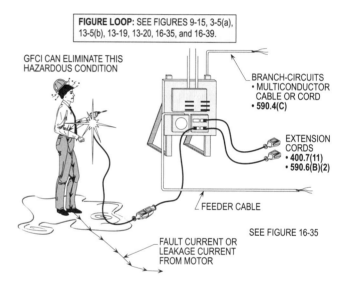

Figure 9-45. GFCI devices can detect ground-faults of 5 mA and clear the circuit and prevent electrocution of workers using electric hand tools.

OCPD'S FOR SYSTEMS OVER 600 VOLTS

OCPD's shall be sized properly to protect the elements of electrical systems rated over 600 volts. High-voltage electrical systems behave differently than systems of 600 volts or less. When short-circuits or ground-faults develop on high-voltage circuits, great damage to conductors, elements, and equipment can occur. Therefore, it is imperative to protect such systems from these hazards.

FEEDERS AND BRANCH-CIRCUITS
240.100(A); (B); (C)

Branch-circuits shall have a short-circuit protective device in each ungrounded (phase) conductor or comply with **490.21(A)** and **(B)**. The protective device(s) shall be capable of detecting and interrupting all values of current which can occur at their location in excess of their trip setting or melting point per **240.100(A)(1)**. In other words, the OCPD's shall be sized to protect the conductors at their ampacities if not oversized to allow equipment with high-inrush current to start, run and operate.

Note that the protective device(s) shall be capable of detecting and interrupting all values of current that can occur at their location in excess of their trip setting or melting point per **240.100(B)**.

Feeder and branch-circuit conductors shall have overcurrent protection in each ungrounded (phase) conductor located at the point where the conductor receives its supply or at a location in the circuit determined under engineering supervision that includes but is not limited to considering the appropriate fault studies and time-current coordination analysis of the protective devices and the conductor damage curves.

OVERCURRENT RELAYS AND CURRENT TRANSFORMERS
240.101(A)(1)

Circuit breakers used for overcurrent protection of 3-phase circuits shall have a minimum of three overcurrent relays operated from three current transformers. The separate overcurrent relay elements (or protective functions) shall be permitted to be part of a single electronic protective relay unit. On 3-phase, 3-wire circuits, an overcurrent relay in the residual circuit of the current transformers shall be permitted to replace one of the phase relays.

An overcurrent relay, operated from a current transformer that links all phases of a 3-phase, 3-wire circuit, shall be permitted to replace the residual relay and one of the phase-conductor current transformers. Where the neutral is not regrounded on the load side of the circuit as permitted in **250.184(B)**, the current transformer shall be permitted to link all 3-phase conductors and the grounded circuit conductor (neutral).

FUSES
240.100(A)(2)

When fuses are used, they shall be connected in series with each ungrounded (phase) conductor.

CONDUCTOR PROTECTION
240.100(C)

The operating time of the protective device, the available short-circuit current and the conductor used shall be coordinated to prevent damaging or dangerous temperatures in conductors or conductor insulation under short-circuit conditions.

ADDITIONAL REQUIREMENTS FOR FEEDERS
240.101(A); (B)

Feeder-circuits shall have a short-circuit overcurrent protection device in each ungrounded (phase) conductor. The protective device(s) shall be capable of detecting and interrupting all values of current which can occur at their location in excess of their trip setting or melting point. A fuse rated in continuous amperes not exceeding three times the ampacity of the conductor or a breaker having a trip setting of not more than six times the ampacity of the conductor shall be considered as providing the required short-circuit protection. For further information, see **490.21(A)** and **(B)** and **Article 490** which deals with high-voltage electrical systems. **(See Figures 9-46(a) through (d))**

> **Design Tip:** The operating time of the protective device, the available short-circuit current and the conductor used will need to be coordinated to prevent damaging or dangerous temperatures in conductors or conductor insulation under short-circuit or ground-fault conditions.

OTHER ACCESSORIES

When selecting the arrangement of the OCPD's for systems over 600 volts, see **490.21(A)** and **(B)**. Switchboard requirements are found in **Part III of Article 490**. Wiring methods and techniques are outlined in **300.37** and **300.50**, based on above ground or underground installations. Cable preparation, shielding and grounding requirements are listed in **300.40, 310.6** and **310.60(C)(1)** in the NEC.

Overcurrent Protection Devices

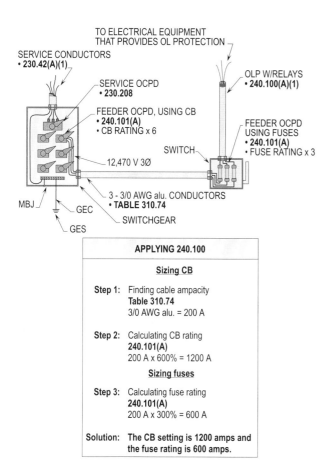

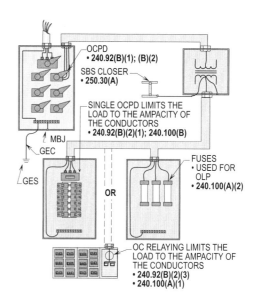

Figure 9-46(a). The above illustration shows methods of providing overload protection.

Figure 9-46(c). The overload protection (OLP) for the protection of the feeder conductors shall be permitted to be provided by installing overcurrent relays which will provide the ultimate protection scheme.

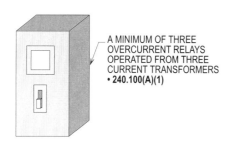

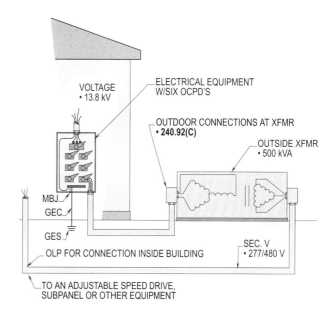

Figure 9-46(b). The above illustration shows the correct procedure for sizing CB's and fuses to be used to protect high-voltage electrical systems for feeder-circuits.

Figure 9-46(d). The above illustration shows alternate means allowed for protecting conductors connected to a transformer located outside.

Name   Date

Chapter 9. Overcurrent Protection Devices

Section	Answer		
_____	T	F	1. Conductor overload protection shall be required where the interruption of the circuit would create a hazard.
_____	T	F	2. If the overcurrent protection device is greater than 800 amps, the ampacity of the conductors shall be equal to or greater than the ampacities of the overcurrent protection device.
_____	T	F	3. The OCPD for motor circuits shall not be permitted to be sized above the conductor ampacity.
_____	T	F	4. Circuit protection for capacitors shall be permitted to be sized high enough to hold the inrush current.
_____	T	F	5. 18 AWG and larger flexible cord conductors used in field assembled extension cord sets which are connected to a 20 amp branch-circuit shall be considered protected from overcurrent conditions.
_____	T	F	6. Electrical equipment which comes with overcurrent protection within the equipment that is provided by the manufacturer is not permitted.
_____	T	F	7. In grounded systems, individual single-pole circuit breakers with identified handle ties shall be permitted to be utilized as the protection for each ungrounded (phase) conductor for line-to-line connected loads for single-phase circuits or three-wire DC circuits.
_____	T	F	8. Generator conductors shall have an ampacity equal to at least 125 percent of the nameplate current rating of the generator. (General Rule)
_____	T	F	9. No overcurrent protection device shall be connected in series with any conductor that is intentionally grounded. (General Rule)
_____	T	F	10. Overcurrent protection devices shall not be required to be readily accessible.
_____	_____		11. Fuses shall be plainly marked, either by printing on the fuse barrel or by a label attached to the barrel.
_____	_____		12. For a branch-circuit, a continuous load is a load that operates for three hours or more.
_____	_____		13. When the circuit breaker is turned ON, the handle shall be installed so that it is in the DOWN position.
_____	_____		14. Circuit breakers shall be permitted to be used with grounded, ungrounded or grounded neutral systems when identified with a slash voltage marking.
_____	_____		15. Performance testing shall be performed at the time of installation for the ground-fault protection system.

9-33

Section	Answer

_____ _____ 16. The general rule is that conductors shall be protected against _____ by a fuse or circuit breaker setting rated no higher than the ampacity of the conductor.

_____ _____
_____ _____ 17. Conductors supplied by the secondary side of a single-phase transformer having _____-wire secondary or a _____-wire delta secondary shall be considered as protected by overcurrent protection provided on the primary (supply) side of the transformer.

_____ _____ 18. Flexible cord used in extension cords made with separately _____ and installed components shall be permitted.

_____ _____ 19. Circuit breakers shall open all _____ conductors of the circuit.

_____ _____ 20. A tap feeding a lighting and appliance panelboard shall not be permitted to be over _____ ft. long. (maximum tap)

_____ _____ 21. Transformer (primary) tap conductor ampacity shall be at least _____ of the feeder's OCPD. (Primary plus Transformer and Secondary)

_____ _____ 22. No overcurrent protection device shall be permitted to be connected in _____ with any conductor that is intentionally grounded.

_____ _____ 23. Overcurrent devices shall not be located in the vicinity of easily _____ material.

_____ _____
_____ _____ 24. Ground-fault protection of equipment shall be required for grounded wye-connected services of more than _____ volts-to-ground and OCPD's rated _____ amps or greater.

_____ _____ 25. To prevent damage to overcurrent protection devices, busbars and service equipment, the ground-fault system shall be required to clear a fault of _____ amps or more within one second.

_____ _____ 26. If the standard current ratings of fuses or nonadjustable circuit breakers do not conform to the ampacity of the conductors being used, it shall be permitted to use the next larger standard rating when below _____ amps.

      (a)   100
      (b)   400
      (c)   800
      (d)   1000

_____ _____ 27. A _____ amp overcurrent protection device shall be considered as being adequate protection for fixture wire, in size 16 AWG or 18 AWG.

      (a)   15
      (b)   20
      (c)   30
      (d)   40

	Section	Answer

28. Standard ampere ratings for fuses are available in sizes of _____ amp.

    (a)   3
    (b)   6
    (c)   10
    (d)   All of the above

29. A tap from a larger conductor to a smaller conductor shall be permitted to extend up to a distance of _____ ft., provided the current-carrying capacity of the tap is at least 1/3 that of the larger conductor.

    (a)   10
    (b)   20
    (c)   25
    (d)   50

30. For taps in high bay manufacturing buildings, the protection for the tap shall be permitted to be at the end of the tapped conductors, if the tap is not over _____ ft. long horizontally, and not over _____ ft. total length.

    (a)   25; 75
    (b)   25; 100
    (c)   50; 75
    (d)   50; 100

31. Fuses shall be plainly marked with _____.

    (a)   Ampere rating
    (b)   Voltage rating
    (c)   The name of the manufacturer
    (d)   All of the above

32. Before circuit breakers are installed they shall be plainly marked with _____.

    (a)   Location
    (b)   Used as switches
    (c)   None of the above
    (d)   All of the above

33. Ground-fault detection such as the _____ method is used to sense low-magnitude ground-faults.

    (a)   Donut
    (b)   Window
    (c)   All of the above
    (d)   None of the above

34. What size OCPD is required to protect a continuous branch-circuit load of 22.5 amps?

35. What size THWN cu. conductors are required to protect a continuous branch-circuit load of 22.5 amps?

Section	Answer

36. What size OCPD is required to protect a subpanel with a load of 649 amps on each phase paralleled four times? (All load are continuous)

37. What size OCPD (CB) and THHN cu. conductors are required to protect a motor operated appliance with a 13 amp continuous load?

38. What is the minimum, next and maximum size circuit breaker for a 208 volt, 40 HP, Design B, three-phase motor?

39. What size OCPD (CB) and THWN copper conductors are required for a phase converter with a 230 volt, 20 HP, Design C, three-phase motor?

40. What is the minimum and maximum size circuit breaker allowed to start and run an A/C unit with a compressor having a full-load current rating of 18 amps?

41. What size OCPD is required to protect a two-wire to two-wire transformer with a 240 volt primary and a 120 volt secondary with 4 AWG THWN copper conductors?

42. What size OCPD is required to protect a three-wire to three-wire transformer with a 480 volt primary and a 240 volt secondary with 500 KCMIL copper conductors?

43. What size THWN copper conductors are required for a 208 volt, 20 kVA, three-phase capacitor based on the FLA of the capacitor times 135 percent or 1/3 of the conductors ampacity (3/0 AWG THWN)?

44. What size OCPD is required for an AC transformer and DC rectifier arc welder rated at 56 amps with an 80 percent duty cycle supplied with 8 AWG THWN copper conductors?

45. What size OCPD for the branch-circuit and control conductors is required for a Class 1 control circuit supplied by 14 AWG THWN copper conductors?

46. What size OCPD for fire-signaling circuits is required for an alarm supplied by 16 AWG THWN copper conductors and a detector supplied by 14 AWG copper conductors?

47. What size fixture wire is required when supplied by a 30 amp branch-circuit?

48. What size THWN copper conductors and OCPD are required for a tap using the 10 ft. tap rule with the overcurrent protection device for the feeder being tapped rated at 400 amps?

49. What size THWN copper conductors and OCPD are required for a tap using the 25 ft. tap rule with the overcurrent protection device for the feeder being tapped rated at 400 amps?

50. What size OCPD is required for the secondary and what size THWN copper conductors is needed for the primary (480 V) and secondary (208 V) taps and connection that are over 25 ft. long? The primary overcurrent protection device is rated at 200 amps.

	Section	Answer

51. What size OCPD and THWN copper conductors are required for a tap using the 100 ft. tap rule with an overcurrent protection device for the feeder being tapped rated at 300 amps?

52. What size OCPD and THWN copper conductors are required for a tap using the 10 ft. tap rule for a motor with an overcurrent protection device for the feeder being tapped rated at 400 amps? (Note that terminals are 60°C)

53. What size THWN copper conductors are required when tapping a 277/480 volt, three-phase, four-wire generator with an 250 kVA rating?

54. What size fuse and THWN copper conductors are required for a 6 amp noncontinuous load and 11 amp continuous loads? (branch-circuit)

55. What size circuit breaker and THWN copper conductors are required for a 70 amp continuous load and 56 amp noncontinuous load? (feeder-circuit)

10

Equipment Over 600 Volts

Through the years, high-voltage electrical systems have increased at an accelerating rate. The task of knowing all there is to know about high-voltage systems is particularly difficult for designers, installers, maintainers and inspectors. The term "high-voltage", as used in this chapter, refers to systems operating above 600 volts nominal, which is measured phase-to-phase. However, the electrical industry usually refers to sytems of 601 volts to 15 kV as "medium-voltage" circuits and voltage systems above 69 kV are known as "high-voltage" circuits.

Because of the potential hazards involved with high-voltage systems, extreme care and detailed planning should go into the design, installion and maintenance of such systems, specifically the application of Article 490 of the NEC as well as other pertinent and applicable Articles. The information found in this chapter will cover most of these requirements.

SERVICE POINT
ARTICLE 100

"Service point" is the "point of connection between the facilities of the serving utility and premises wiring." All equipment on the load side of that point is subject to the rules of the NEC. Equipment on the supply side is the responsibility of the power company and is not regulated by the NEC. This definition or "service point" must be used to establish where "service conductors" originating point is located. (Also, see **230.200** of the NEC.)

The definition of "service point" does identify where the NEC becomes applicable, and does pinpoint the origin of service conductors.

Conductors between the "service point" of a particular installation and the service disconnect are identified as service conductors and subject to NEC rules on service conductors per **Article 100**.

For example: This rule implies that conductors ahead of the service-entrance equipment or between the metering enclosure is considered service conductors. **(See Figures 10-1(a); (b); (c))**

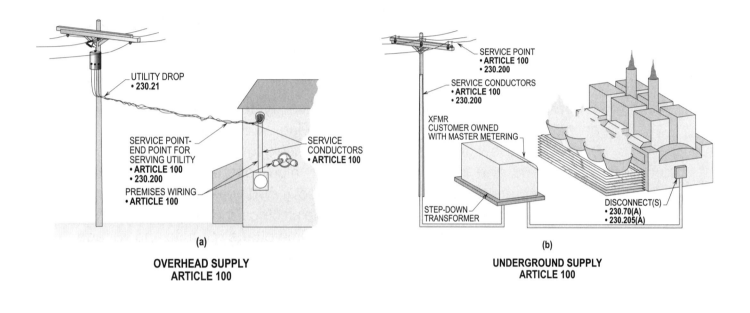

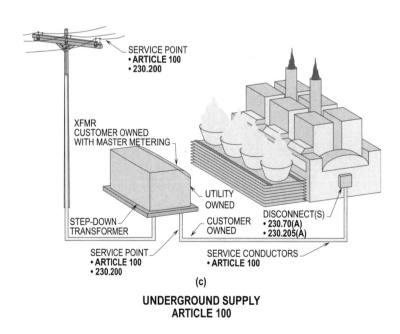

Figure 10-1. Service point is the point of connection between the facilities of the servicing utility and the premises wiring.

SERVICE CONDUCTORS

For very large buildings, such as large hotels, office buildings, and industrial sites, high-voltage primaries may be carried to step-down transformers located within the facility. Here, the primaries are the service conductors, except in the case where they enter a vault, or where they enter metal-enclosed switchgear, or where they enter a locked room accessible only to qualified personnel, in which case the secondaries within the building are the service conductors, not the primaries entering the building. If the transformers are located outside the building, the secondaries entering the building are, of course considered the service conductors.

These rules must be applied when the secondary voltage is 600 volts or less. For secondary voltages over 600 volts either the primaries or the secondaries can be classified as the service conductors, if serving an industrial complex or large commercial site. Apply **Article 100** of the NEC to classify such conductors.

SERVICE-ENTRANCE CONDUCTORS 230.202

This rule specifies the wiring methods which are allowed for use as service-entrance conductors where it has been established that primary conductors are the service conductors or where the secondary conductors are the service conductors and operate more than 600 volts.

The basic types of wiring methods permitted for high-voltage service entrances are:

(1) Rigid metal conduit
(2) Rigid nonmetallic conduit
(3) Approved cable
(4) Open wiring on insulators
(5) Cablebus
(6) Intermediate metal conduit
(7) Busways
(8) Wireways
(9) Cable trays
(10) Auxiliary gutters
(11) RNC
(12) Cablebus
(13) MC
(14) MI
(15) FMC & LFMC
(16) LFNC

See Figure 10-2 for a detailed illustration.

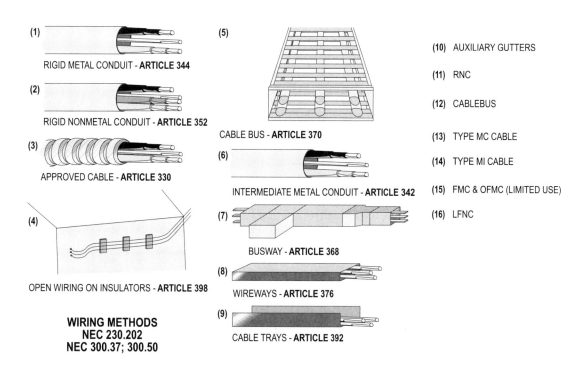

Figure 10-2. The above wiring methods can be used to enclose high-voltge conductors that supply high-voltage electrical systems. Note: Always check with AHJ for permitted wiring methods.

The following wiring methods may also be used as service-entrance conductors to wire-in high-voltage electrical systems.

(1) Cable tray systems may be used to support service cables if cables are identified for such use.

(2) Open wiring on insulators is permitted only if guarded against contact by persons, or if accessible to "qualified personnel" only.

(2)(a) Open wire services and service wires in conduit must be at least 6 AWG. If cable is used, 8 AWG is also permitted.

(2)(b) When open wiring is used, the conductors shall be solidly fastened to insulators having sufficient strength to withstand the stresses imposed on the wire and insulators in the event of a short circuit. Such stress could in some cases amount to over 1000 lbs. per ft. of conductor.

(2)(c) When open wiring is used for a high-voltage service, it shall be guarded by a fence or other suitable barrier, so that unauthorized personnel cannot contact the wire.

(3) For high-voltage services, a pothead is required, not only at the outer end but also at the inner end of a service-entrance conduit or cable.

(4) Unless conductors suitable for wet locations are used, conduit exposed to the weather or embedded in masonry shall be arranged to drain.

(5) See **Article 398** of the NEC for supporting rules pertaining to open conductors on insulators.

(6) For a complete list of wiring methods used to enclose over 600 volt conductors, see **300.37** for over installations and **300.50** for underground installations.

Note: Primaries exceeding 15,000 volts shall be brought into a vault or to metal-enclosed switchgear.

WARNING SIGNS
110.34(C); 314.72(E)

A warning sign used to warn unauthorized persons who might make contact with live parts shall always include the phrase "KEEP OUT" immediately after the phrase "DANGER HIGH VOLTAGE."

This is an mandatory factor in utilizing such a sign to act as an effective deterrent. Such wording shall always be used on signs to warn personnel that high-voltage (over 600 V) electrical systems and equipment are present and be extremely careful. **(See Figure 10-3)**

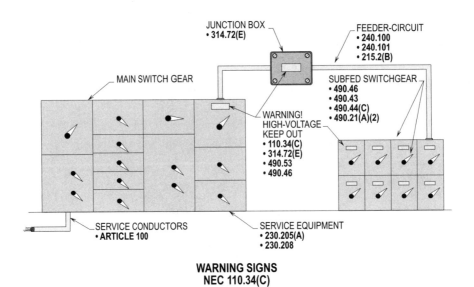

Figure 10-3. The covers on enclosures and junction boxes enclosing high-voltage elements must be permanently marked "Warning—High-Voltage—Keep Out" and such markings shall be readily visible and legible.

WARNING SIGNS

The main reason live parts of electrical systems rated over 600 volts shall be posted with warning signs is to prevent unauthorized persons from coming in contact with live parts (hots). High-voltage systems are very dangerous due to the threat of arcing or flashing across and making contact with grounded objects such as a persons body. When an individual is shocked with high-voltage, there is usually an explosion at the point where the current leaves the body.

This is due to the high pressure created by the higher voltage forcing the current to flow through the lower resistance of the body. Electrical current flowing through the resistance of one's body behaves like the current flowing through a resistive heater element. Heat is produced in the body just as in the heater element and this production of heat can severely damage the elements inside the body. Therefore, only qualified persons are allowed to be near or work on high-voltage systems.

This warning sign rule is to be applied to all pull and junction boxes housing conductors over 600 volts, which include the following wiring methods:

(1) All boxes are required to provide a complete enclosure for contained conductors or cables.

(2) All covers on boxes shall be securely fastened in place to prevent easy access of unauthorized personnel. Covers on underground boxes that weight over 100 lbs. are considered to be complying with this requirement. Boxes shall have covers mounted properly and all openings to have seals to prevent the entering of dirt, lint, etc. per **314.72**. Covers for boxes are required to be permanently marked, **"DANGER! HIGH VOLTAGE - KEEP OUT."** The marking shall be placed on the outside of the box cover and must be readily visible and legible.

Note: A warning of the high-voltage being present is required along with a command. Examples of the command are:

DANGER! HIGH VOLTAGE, KEEP OUT
DANGER! HIGH VOLTAGE
AUTHORIZED PERSONNEL ONLY

Codes and Standards
NEC 110.34(C)
OSHA 1910.269(u)(4)(iii)
OSHA 1910.269(w)(6)(ii)
NESC 441.D

ISOLATING SWITCHES
230.204; 490.21(E); 490.22

For certain types of disconnects, an isolating switch is required. Sections **230.204** and **490.21** specify the different types of disconnects approved for high-voltage services.

An air-break isolating switch shall be used between an oil switch or an air, oil, vacuum, or sulfur hexachloride circuit breaker and the supply conductors. However, removable truck panels or metal-enclosed units can be used providing to disconnect of all live parts in when the removed position.

The purpose of this supply side disconnect is to ensure safety for personnel while repairing or pulling maintenance on such equipment.

The following parts to **230.204** may be utilized to determine if an isolation switch is required ahead of the overcurrent protection device protecting the service conductors and equipment. (**See Figure 10-4**)

Note: Review **490.22** for more information on these requirements.

WHERE REQUIRED
230.204(A); 490.21(E); 490.22

When oil switches or air or oil circuit breakers are used as the disconnecting means, an additional "isolating" switch shall be installed ahead of the oil switch or the circuit breaker. "Isolating switch" is the term used for a switch that is to be opened only when there is no load on the circuit. An isolating switch does not have to have a rating high enough to break the load current. Such a switch shall open after the oil switch or the circuit breaker is opened. Opening the isolating switch deenergizes the oil switch or circuit breaker so it can be safely handled or worked on when servicing or repairs are required.

High-voltage service equipment of a type that can be rolled or pulled out from its switchgear or panel enclosure is available. An interlock permits the equipment to be removed only if the switch is in the open position. When this type of equipment is used, an isolating switch is not required. (**See Figure 10-5(a)**)

For the racking in or out of a circuit breaker, see **Figure 10-5(b)**.

Stallcup's Electrical Design Book

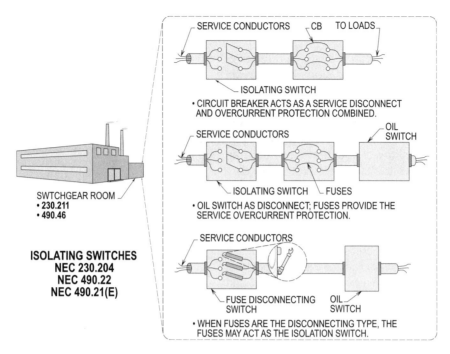

Figure 10-4. Except where roll-out equipment is used, an isolating switch is required when an oil switch or a circuit breaker is used as a service disconnect.

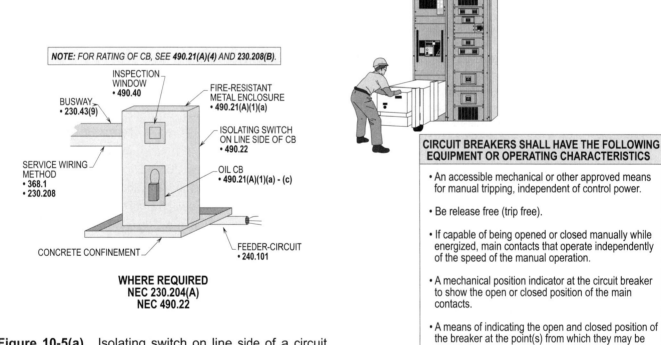

Figure 10-5(a). Isolating switch on line side of a circuit breaker used to disconnect the service conductors.

Figure 10-5(b). The above illustration shows an electrician racking out a circuit breaker.

10-6

FUSE AS ISOLATING SWITCH
230.204(B); 490.22

If a nonautomatic oil switch and fuses are used as the disconnecting means, and the fuses are of a type that open and close like a switch, no isolating switch is required. However, the disconnecting fuses shall be mounted in the line ahead of the oil switch. (**See Figure 10-6**)

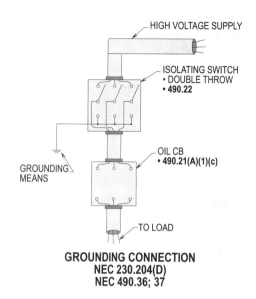

Figure 10-7. The above is one acceptable method for grounding the load side of an open isolating switch used to disconnect an oil circuit breaker.

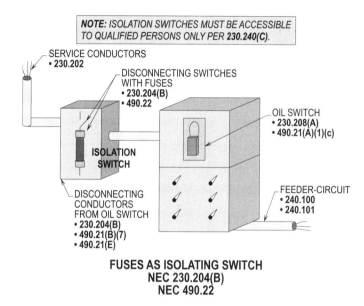

Figure 10-6. Disconnect with fuses used to disconnect (isolating switch) and protect the service conductors supplying the service equipment.

ACCESSIBLE TO QUALIFIED PERSONNEL ONLY
230.204(C); 490.32

Isolating switches shall be accessible to qualified personnel only. The main reason for this requirement is isolating switches are dangerous when opened under load. Therefore, they must never be accessible to unqualified personnel or the general public.

GROUNDING CONNECTION
NEC 230.204(D); 490.37

Isolating switches must provide means for "readily grounding" wiring on the premises. One way of accomplishing this to use a double throw switch for the isolating switch, so arranged that in one position the load connects the line, and in the other position the load connects grounding connection. (**See Figure 10-7**)

Grounding

The purpose of the grounding mentioned above serves as a safety factor in case a switch accidentally becomes energized.

When shielded-type high-voltage cables have great lengths, the cable and the shield become a fairly good sized capacitor, and such grounding diverts the charge away, thus preventing what might have been a painful shock to personnel serving such cables and equipment. See **310.6**, **310.6(C)(1)**, and **300.40**.

DISCONNECTING MEANS
230.205

A high-voltage disconnecting means and overcurrent protection device shall comply with certain rules and regulations to qualify for service equipment use.

LOCATION
230.205(A)

The basic rule requires a high-voltage service disconnect means to be located "nearest the point of entrance of the service conductors" into the building or structure being supplied. This rule is described for 600 volts or less in **230.70(A)** and **230.208** for over 600 volts is referenced per **230.205(A)**.

Section **230.205(C)** allows a feeder to an separate building on a multibuilding property under single management to have its disconnect in another building on the property. However, such a remote disconnect must be capable of being electrically opened by a control device in the building feeder circuit. In addition, the control device must be marked to show its function and provide visual indication of the "on" or "off" status of the remote service disconnect. (**See Figure 10-8**)

TYPE 230.205(B)

The service disconnecting means shall simultaneously disconnect all ungrounded conductors and must be capable of being closed on a fault equal to at least the maximum available short-circuit current in the circuit. See **230.208** and **230.205** for types permitted to be used for this purpose.

OVERCURRENT DEVICES AS DISCONNECTING MEANS 230.206

The different types of disconnects approved for high-voltage services are specified in **230.205**, and protection techniques are found in **230.208**, Overcurrent Protection (OCP) Requirements. The service disconnecting means providing the OCP must also comply with these requirements.

OCPD's

It should be understood that OCPD's can be used as a disconnecting means only where the circuit breaker or alternative for it specified in **230.208** for service overcurrent devices meets the requirements specified in **230.205**. For feeder-circuit requirements, see **240.100** and **240.101**.

Codes and Standards
NEC 230.206
NESC 171. (where applicable)
NESC 161.A (where applicable)

EQUIPMENT IN SECONDARIES 230.207 PER 1993 NEC

In existing installations, this rule allows the disconnecting means with overcurrent protection in the secondaries to be omitted if there is a circuit breaker or load interrupter in the primary side of the line, and the secondaries are connected to a common bus. If there is a vault, this primary circuit breaker or load interrupter must be operable from outside the vault and in addition it must be of proper rating to protect the secondaries.

It must be understood that only one set of secondary conductors may be installed. If two or more sets of secondary conductors are utilized, the rule must not be applied. (**See Figure 3-9**)

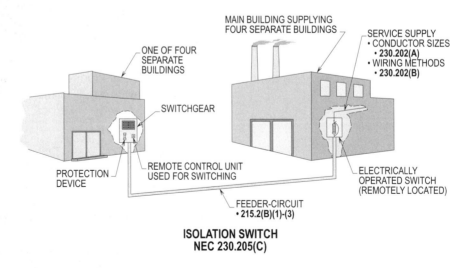

Figure 10-8. The above shows a remote control unit that is used to deenergize the electrically operated switch in the main building.

Equiment Over 600 Volts

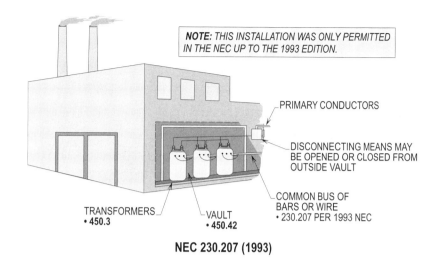

Figure 10-9. Disconnecting means with overcurrent protection can be omitted in the secondary side of the transformer, if a disconnecting means is located on the outside of the vault with only one set of secondary conductors connected to a common bus.

OCPD's

Section **230.207** and **240.4(F)** are the only NEC requirements that permit the primary overcurrent protection device to provide protection for the secondary conductors. If one of these requirements cannot be applied, the secondary conductors must have overcurrent protection provided in the secondary side.

See **450.3(A)** for protecting primary and secondary of high-voltage transformers rated at over 600 volts and **450.3(B)** for 600 volts or less.

Note: Because this type of installation does exist in older installations, it must have been installed before the 1996 NEC to comply. For this reason it is mentioned so that the user is aware that this rule did once exist.

OVERCURRENT PROTECTION REQUIREMENTS
230.208; 240.100; 240.101(A)

High-voltage systems have a different characteristic than low-voltage systems. High-voltage systems clears a short circuit faster than low-voltage systems. If a short circuit occurs on a high-voltage system, as soon as the insulation of a conductor fails, the circuit device trips open. Therefore, the elements and wiring methods of high-voltage systems (over 600 V) are treated differently than systems of 600 volts or less.

An overcurrent protection device must be provided in each ungrounded conductor for service conductors operating at over 600 volts. This overcurrent protection device protects against short-circuit currents, however it does not protect against overload currents. The overcurrent protection device must be installed on the load side of the service disconnecting means or as an integral part of the service disconnecting means. A fuse rated in continuous amps not to exceed three times the ampacity of the conductor may be installed. A circuit breaker trip setting of not more than six times the ampacity of the conductors may be installed to protect against short circuit currents. **(See Figure 10-10)**

Note: The rules for overcurrent protection and for disconnects are contained in this Section. There are two cases covered and they are as follows:

(1) When the service equipment is in a vault or consists of metal-enclosed switchgear.

(2) When the service equipment is not in a vault or does not consist of metal-enclosed switchgear.

A vault is a specially constructed room, with roof, walls, and floor of concrete or brick, and having fire-resistant doors. It must be constructed according to the specifications of **450.41** thru **450.48**. Otherwise, it is not classified as a vault. **(See Figure 10-11)**

Metal-enclosed switchgear must be in an enclosure constructed of metal of "substantial thickness."

Codes and Standards
NEC 230.208
NESC 171. (where applicable)
NESC 161.A (where applicable)

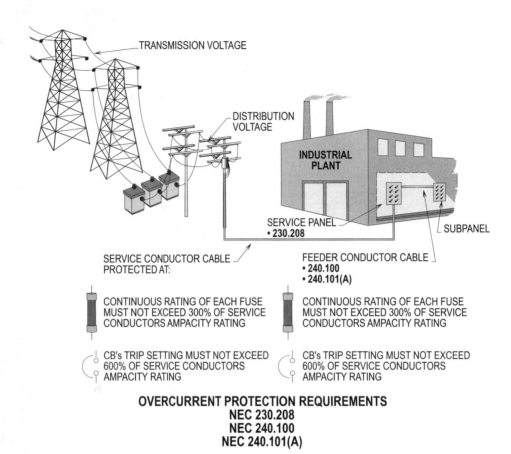

Figure 10-10. Maximum ratings or settings of the overcurrent protection device permitted for high-voltage service and feeder conductors.

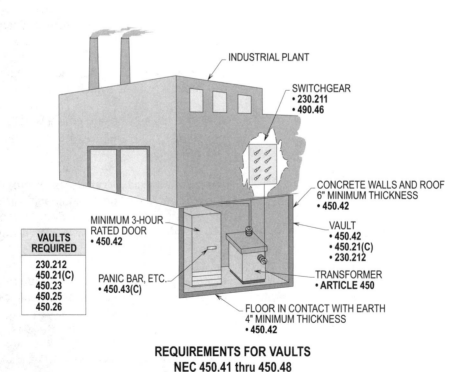

Figure 10-11. Some transformers are required to be installed in vaults because of their characteristics of use.

Equiment Over 600 Volts

IN VAULT OR CONSISTING OF METAL-ENCLOSED SWITCHGEAR
230.208(A); 490.21(E)

When the service equipment is installed in a vault or consists of metal-enclosed switchgear, the overcurrent protection and disconnecting means may be any one of the following:

(1) Oil switch and fuses,

(2) Load interrupter and fuses,

(3) Oil fuse cutout,

(4) Automatic-trip circuit breaker,

(5) Isolating switch and fuses.

An oil fuse cutout is used, it could be a disconnect type, a high-voltage fuse mounted with a hinge on one end so that it can be opened and closed, thus serves as a disconnect, as well as providing overcurrent protection.

A load interrupter is a switch or circuit breaker that is not intended to provide overload protection; therefore, fuses are required when these are used. In this case the interrupter is the disconnect, and the fuses are sized to provide the overload protection.

An "isolating" switch is not rated to interrupt load current and must not be opened under load. When an isolating switch is used as primary disconnect, it must be interlocked with a disconnect in the secondary side in such a way that it would not be possible to open the primary isolating switch, without the secondary switch is in the open position. (**See Figure 10-12**)

NOT IN VAULT OR NOT CONSISTING OF METAL-ENCLOSED SWITCHGEAR ARTICLE 490, PART II

When the service equipment is not in a vault or does not consist of metal-enclosed switchgear, the overcurrent protection and disconnecting means may be:

(1) Load interrupter and fuses located outside the building, or

(2) Automatic-trip circuit breaker located outside the building.

FUSES
490.21(B); 490.22

Interrupting rating of fuses must be at least equal to the maximum possible short circuit current available for the circuit.

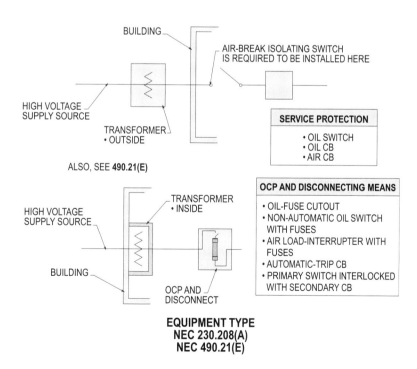

Figure 10-12. Overcurrent protection devices and disconnecting means that are permitted to be used for protecting and disconnecting high-voltage elements of electrical systems.

CIRCUIT BREAKERS
490.21(A); 490.22

Circuit breakers must be the type that will trip independently of the circuit breaker handle. With the handle closed, the breaker can trip internally without moving the handle. Thus, the breaker is "free to open in case it is closed on an overload." If a circuit breaker is used, it must be this type. Circuit breakers must have an interrupting capacity at least equal to the maximum short circuit current that can be delivered to its terminals. (**See Figure 10-13**)

ENCLOSED OVERCURRENT DEVICES
230.208(B)

Sections **215.2(B)(1)** through **(B)(3)** and **215.3** require derating of 600 volt or less feeder conductors and overcurrent devices in cases of "long continuous" loads. For services rated over 600 volts, the derating requirement for sizing the OCPD does not apply for continuous operated loads per **230.208(B)**.

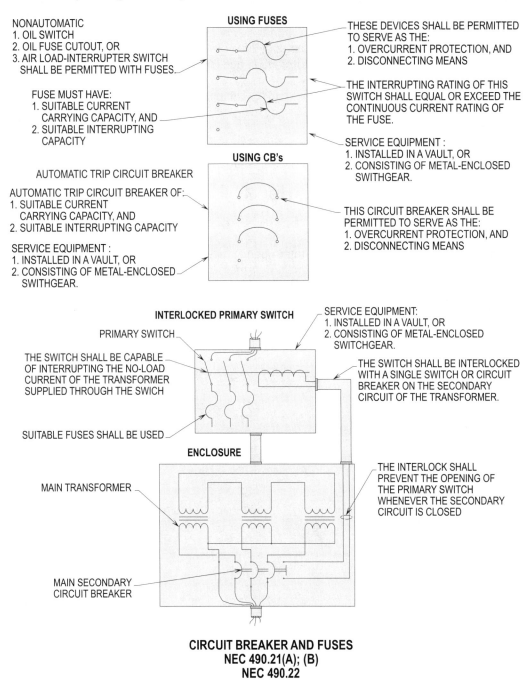

Figure 10-13. Using fuses and CB's in a special adaption of switch-disconnect and fused overcurrent protection for high-voltage electrical systems.

ARRESTERS (LIGHTNING ARRESTERS)
230.209; ARTICLE 280

Surge arresters installed in accordance with the requirements of **Article 280** must be installed on each ungrounded overhead service conductor on the supply side of the service equipment.

A surge arrester is defined as a protective device for limiting surge voltage by discharging or bypassing surge current. It also prevents continued flow of follow current while remaining capable of repeating these functions as needed.

Such surge arresters break down at voltages higher than the supply voltage, permitting the higher voltage and currents to flow to ground, thus protecting the equipment on the electrical system.

Once the surge passes to ground, the arrester heals itself, shutting off follow current from the supply system. (**See Figure 10-14**)

SERVICE EQUIPMENT - GENERAL PROVISIONS
230.210

Service equipment, including instrument transformers, shall conform to **Article 490, Part I** of the NEC.

METAL-ENCLOSED SWITCHGEAR
230.211; 490.46

Metal-enclosed switchgear must consist of a substantial metal structure with a sheet-metal enclosure. If properly built the above switchgear will protect personnel from energized parts of the service elements. Where installed over a wood floor, suitable protection must be provided, as required per **408.7**. This rule is intended to prevent the heat from the service elements from igniting combustible material and starting a fire. (**See Figure 10-15**)

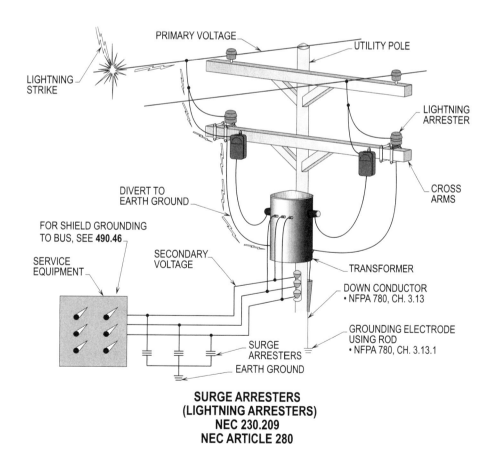

Figure 10-14. Dangerous high-voltage surges are bypassed by using an arrester.

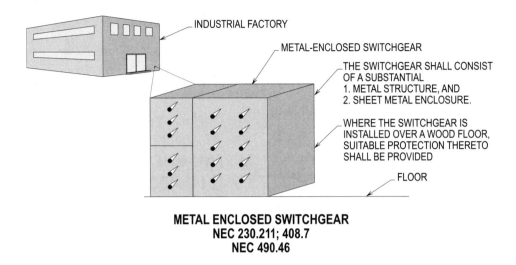

Figure 10-15. Metal-enclosed switchgear must be installed in a safe and reliable manner.

FEEDERS SUPPLYING TRANSFORMERS 215.2(B)(1)

Section **215.2(B)(1)** governs the rules and regulations for determining the size of feeder-circuit conductors supplying transformers rated over 600 volts. This rule requires the ampacity of such conductors to be not less than the sum of the nameplate ratings of the transformers supplied. **(See Figure 10-16)**

FEEDERS SUPPLYING TRANSFORMERS AND UTILIZATION EQUIPMENT 215.2(B)(2)

Section **215.2(B)(2)** governs the rules and regulations for determining the size of feeder-circuit conductors supplying transformers and utilization equipment rated over 600 volts. This rule requires such conductors to be not less than the sum of the nameplate ratings of the transformers plus 125 percent of the designed potential load of all utilization equipment that is capable of operating simultaneously. **(See Figure 10-17)**

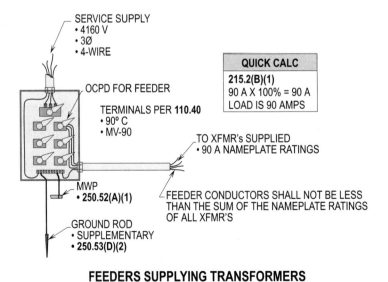

Figure 10-16. The above illustrates the procedure for calculating the load for a feeder supplying transformers.

Equiment Over 600 Volts

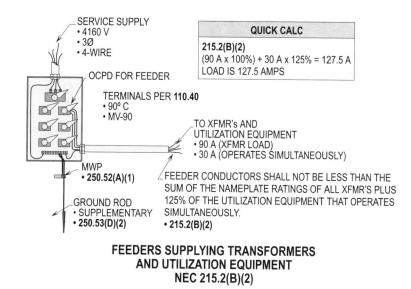

Figure 10-17. The above illustrates the procedure for calculating the load to supply utilization equipment and transformers.

SUPERVISED INSTALLATIONS
215.2(B)(3)

Section **215.2(B)(3)** governs the rules and regulations for determining the size of feeder-circuit conductors for supervised installations which such sizes are determined by qualified person(s) under engineering supervision. Supervised installations are considered those portions of a facility where the following conditions are complied with:

• Conditions of design and installation are provided under engineering supervision.

• Qualified persons with documented training and experience in over 600 volt systems, provide maintenance, monitoring and servicing of such systems.

(See **Figure 10-18** for a detailed description of this section.)

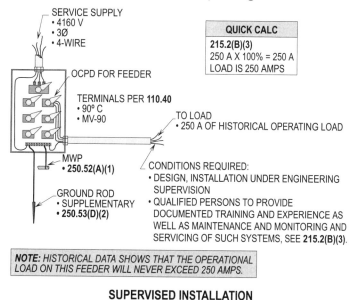

Figure 10-18. The above illustrates the procedure for calculating the load for a feeder supplying a supervised installation.

SERVICES
230.208

Service-entrance conductors must have a short-circuit protective device in each ungrounded conductor on the load side or be installed as an integral part of the service-entrance disconnect. The protective device must be capable of detecting and interrupting all values of current in excess of its trip setting or melting point which can develop at its terminals. A fuse rated in continuous amperes must not exceed three times the ampacity of the conductor, or a CB with a trip setting of not more than six times the ampacity of the conductors. Protected devices sized with these percentages are considered capable of providing short-circuit protection.

Such protection must be installed as follows:

(1) Either on load side of service disconnect.

(2) Or as an integral part of the service equipment.

Section **230.208(B)** indicates that there are overcurrent protection devices that are not required to be derated 80% of their rating where they are supplying continuous duty loads (three hours or more per **Article 100**).

Service-entrance conductors must be protected from dangerous overloads which can damage insulation from short-circuits and ground-faults.

To accomplish the above, there are three sections to review when calculating ampacities to size the OCPD(s) and conductors. These sections are **230.42, 230.90, 230.208** and **230.202(A)**.

See **Figure 10-19** for the maximum percentages permitted to be applied for selecting fuses and CB's for the protection of service-entrance conductors.

FEEDERS
240.100; 240.101(A)

For short circuit protection of high-voltage feeders, fuse rating may be up to three times the conductor ampacity. Circuit breaker setting may be up to six times conductor ampacity. These values apply for short circuit protection only, and are not meant to provide overload protection. An overload feature of a lower rating could be included in the same fuse or circuit breaker.
The melting time-current characteristics of fuse units, refill units, and links for power fuses must be determined as follows:

(1) The current-responsive element ratings 100 amperes or less must melt in 300 seconds at a rms current within the range of 200 to 250 percent of the continuous current rating of the fuse unit, refill unit, or fuse link.

(2) The current-responsive element with ratings above 100 amperes must melt in 600 seconds at a rms current within the range of 220 to 264 percent of the continuous current rating of the fuse unit, refill unit, or fuse link.

(3) The melting time-current characteristics of a power fuse at any current greater than the 200 to 240 or 264 percent listed in (1) or (2) above must be listed by the manufacturer's published time current curves, since the current-responsive element is a distinctive feature of each manufacturer.

(4) The maximum steady-state rms current must not exceed the minimum melting time by more than 20 percent.

Note: E rated fuses are assigned melting times at 200 percent or more of their continuous-current rating. The 300 percent for fuses times 200 percent is essential 600 percent which in effect is the same as circuit breakers. It is easily seen why **230.208** allows the rating of 300 percent for fuses and 600 percent for circuit breakers.

See **Figure 10-20** for the maximum percentages allowed to be applied for selecting fuses or CB's for the protection of feeder-circuit conductors.

Codes and Standards
NEC 240.100
NESC 161.A   (where applicable)

BRANCH-CIRCUITS
240.100

High-voltage branch-circuits must have a short circuit protective device in each ungrounded conductor. Or, as an alternative, if circuit breakers are used for protection, two overcurrent relays operated from two current transformers in each of two phases is allowed.

This rule is intended to provide protection of personnel from high-voltage electrical systems. The protective device must be capable of detecting and interrupting all values of current which might occur at their location in amounts exceeding their trip setting or melting point.

The ampacities of high-voltage conductors are listed in **Tables 310.77** through **310.86** of the NEC. **Table 310.77** is utilized for copper conductors in isolated conduit. Isolated is defined as not readily accessible for personnel unless special means for access are used.

Equiment Over 600 Volts

See Figure 10-21 for the maximum percentages allowed to be applied for selecting fuses and CB's for the protection of branch-circuit conductors.

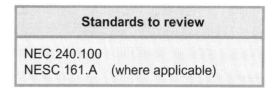

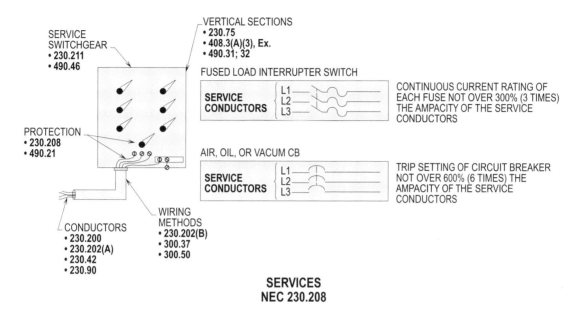

Figure 10-19. The maximum setting of fuses and CB's protecting service-entrance conductors is determined by multiplying the ampacity of the service conductors by percentages as permitted per **230.208**.

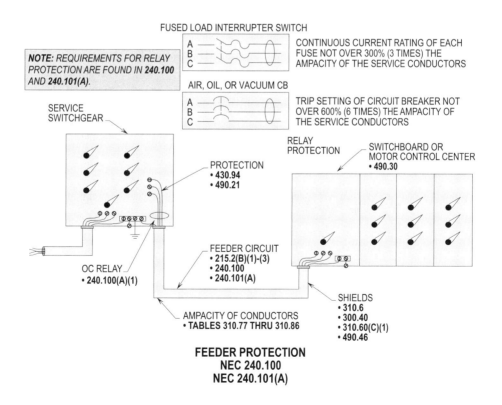

Figure 10-20. The maximum setting of fuses and CB's protecting feeder-circuit conductors is determined by multiplying the ampacity of the feeder conductors by percentages as permitted per **240.100** and **240.101**.

GROUNDING SPECIAL EQUIPMENT 250.180

Special consideration must be given to grounding alternating-current systems and circuits of 1 kV (1000 volts) and over. The reason for grounding high-voltage systems (over 600 V) is the same reasons as for grounding low-voltage systems (600 V or less). System and circuits are solidly grounded for several reasons and they are as follows:

(1) To limit the voltage due to lightning,

(2) To limit the voltage due to line surges,

(3) To limit the voltage due to unintentional contact of the supply with higher voltage lines,

(4) To stabilize the voltage to ground during normal operation,

(5) To facilitate the operation of the overcurrent device in case of ground fault.

Basically, it is up to the designer whether a high-voltage electrical system is grounded or not. General, it is the method in which the electrical system is utilized that really determines if it operates grounded or ungrounded. (**See Figure 10-22**) Note for substation grounding, see Figure 10-37 in this book.

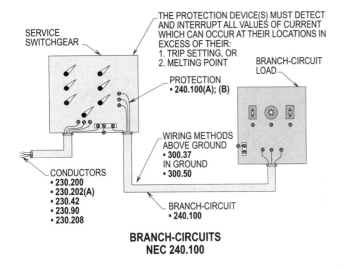

Figure 10-21. The general rule requires a high-voltage branch-circuit to be protected by a short-circuit protective device in each ungrounded conductor.

Note: When grounding high-voltage electrical systems, circuits and equiment, review **250.180** through **250.190** very carefully.

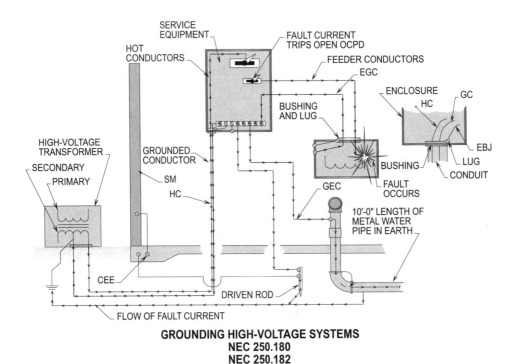

Figure 10-22. Solidly grounded electrical systems are grounded to protect from lighting surges, line surges, to regulate the voltage to ground, and help facilitate the operation of the OCPD's.

WIRING METHODS - COVERS REQUIRED
300.31

Any boxes, fittings, and similar enclosures must have covers suitable for preventing accidental contact with energized parts or physical damage to parts or insulation.

Sections **110.34(C)** and **314.72(E)** requires boxes to be closed by suitable covers and securely fastened in place. Underground box covers that weigh over one hundred pounds are considered as complying with this requirement.

Note: This rule can be applied to manhole covers. Covers for boxes must be permanently marked **"DANGER -HIGH VOLTAGE - KEEP OUT."** The marking must be on the outside of the box cover and is required to be readily visible. Letters must be block type at least 1/2 in. in height. See **110.34(C)** for rules pertaining to enclosures housing high-voltage conductors and elements. **(See Figure 10-23)**

Codes and Standards
NEC 300.31
OSHA 1910.269(u)(4)(iii)
OSHA 1910.269(w)(6)(ii)
NESC 441.D (where applicable)

WIRING METHODS - CONDUCTORS OF DIFFERENT SYSTEMS
300.32

Conductors of high-voltage and low-voltage systems shall not occupy the same wiring enclosure or pull and junction box. **(See Figure 10-24)**

Section **300.3(C)(2)** allows conductors of high-voltage (over 600 V) and low-voltage (600 V or less) systems to occupy the same wiring enclosure for motors, switchgear, control assemblies, and other similar enclosures. **(See Figure 10-25)**

Section **300.3(C)(2)** allows the mixing of conductors of such systems in manholes, if low-voltage conductors are separated from high-voltage conductors in such a manner acceptable to the AHJ. **(See Figure 10-26)**

Mixing conductors

The rules for mixing conductors of high-voltage (over 600 V) and low-voltage (600 V or less) together can be summed up as follows:

Over 600 volt conductors can not be installed in the same raceway or enclosure with conductors of 600 volts or less, except:

(1) For motor, switchgear, and control assemblies.

(2) In manholes.

For more detailed information on mixing conductors of low and high voltage systems, see **300.3(C)(2)(a)** through **(e)**.

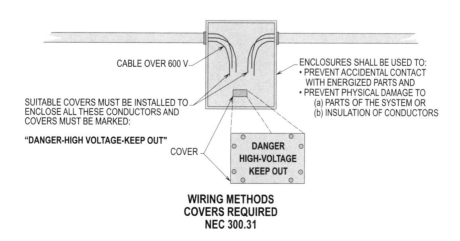

Figure 10-23. Properly marked covers are required to enclose high-voltage cables.

Stallcup's Electrical Design Book

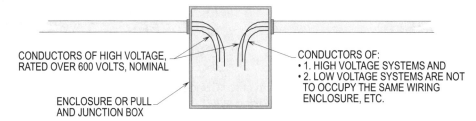

Figure 10-24. Conductors of different systems are not to occupy the same enclosures that are used to enclose the high-voltage conductors.

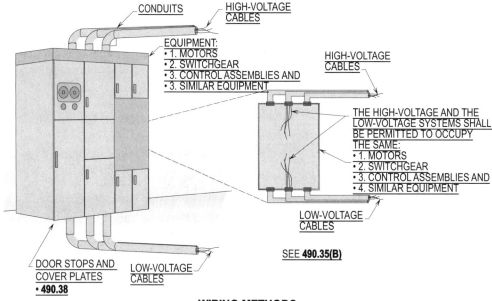

Figure 10-25. High-voltage and low-voltage cables can occupy the same enclosures, such as motors, switchgear, control assemblies, and other similar equipment.

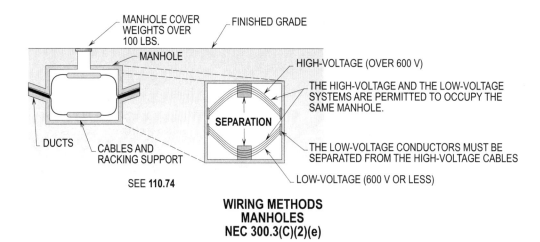

Figure 10-26. High-voltage and low-voltage cables are permitted to occupy the same manhole if they are properly separated.

10-20

WIRING METHODS - INSERTING CONDUCTOR IN RACEWAYS
300.18

Raceways, except those used for exposed work and having removable covers, shall first be installed as a complete raceway system without the conductors. Pull wires, if used, must not be installed until the raceway system is in place. Approved pulling compound may be used as a lubricant in inserting conductors in raceways. Cleaning agents or lubricants having a deleterious effect on conductor insulation and coverings must not be used.

Although this requirement is found in **300.18** it is still a good rule to apply even for high-voltage conductors. Carefully review the exception to **300.18**. **(See Figure 10-27)**

EXTRA PROTECTION

The following procedure will add extra protection to the deteriorating effect of pulling compounds and cleanness on the inside portion of raceways:

(1) It is recommended for ducts encased in concrete to have a wire brush, mandrel and cleaning rags pulled through them before pulling conductors.

(2) After conductors are installed in underground duct banks, unused ducts should be plugged and those with conductors should be sealed with an approved electrical sealing compound.

WIRING METHODS - CONDUCTOR BENDING RADIUS
300.34; 314.71

The bending radius of high-voltage conductors must comply with the following:

(1) Shielded or lead-covered conductors must not be bent to a radius of less than 12 times overall diameter.

(2) Other conductors must not be bent to a radius of less than 8 times overall diameter.

Section **314.71(A)** and **(B)** requires for straight pulls, the length of a box, to be 48 times the outside diameter of the largest cable or conductor entering the box. For angle or U pulls, minimum width and length are at least 36 times outside diameter of the largest cable or conductor plus the sum of the outside diameters, over sheath, of all other cables entering the same wall. **(See Figures 10-28(a) and 28(b))**.

Note: When pulling conductors off a reel be sure that the reels are placed so that the natural bends of the cable as it leaves the reels can enter the raceways without reversing their natural bends. Damage may occur to the shielding of shielded cables if precautions are not taken to prevent over bending the cable during the pulling process.

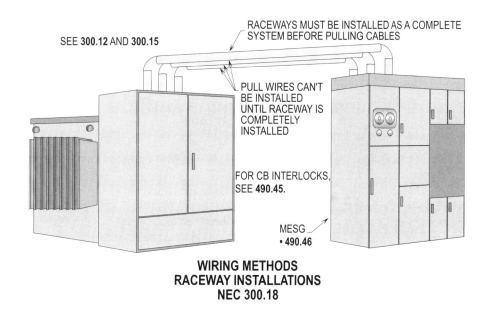

Figure 10-27. Raceway systems enclosing high-voltage cables must be installed completely between enclosures before pulling pull wires and calbes.

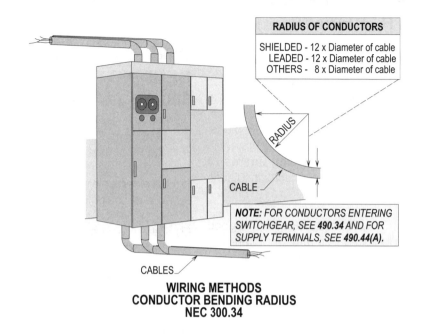

Figure 10-28(a). Conductors must not exceed a certain bending radius to prevent the damaging of insulation, shields, etc.

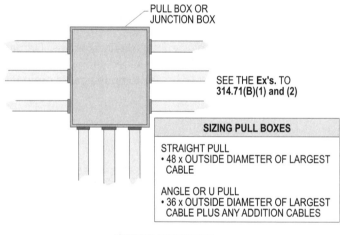

Figure 10-28(b). The above rules must be applied when sizing boxes to enclose high-voltage cables and conductors.

WIRING METHODS - PROTECTION AGAINST INDUCTION HEATING 300.35

AC circuits pulled in metal ducts can cause circulating currents in the wall of the duct. Induced heating of the ducts can be avoided by grouping all conductors of the same circuit together in the same conduit or cable, or binding single conductor cables together in circuit groups.

Where an alternating current flows in a conductor it sets up a varying magnetic field around the conductor. The magnetic field sets up induced currents in the surrounding metal, such as conduit. Where conductors of a circuit are grouped together in the same conduit, the field set up by one conductor cancels that of the other(s), for the currents are at all times in opposite directions. There will be no magnetic field, and therefore no induced currents in the metal wall of the conduit.

Equiment Over 600 Volts

If one phase of an AC circuit is enclosed in a conduit, there is no opposite field to cancel its field, and such current set up in the metal conduit might be of sufficient magnitude to raise the temperature of the conduit to a dangerous value. The NEC requires where AC conductors are pulled in a metal conduit or raceway, all conductors of a circuit must be grouped together in the same raceway.

Note: For nonmetallic and aluminum conduit this requirement does not apply, for induced currents will not set up in nonconducting material. Therefore, conductors can be grouped or run individual (same place) in raceways such as PVC as permitted in **300.5(I), Ex.2**.

Where a single AC conductor passes through a separate hole in the steel wall of a box or enclosure, heating of the steel will result due to hysteresis. To avoid such heating, all conductors of the same circuit must be grouped together, and enter through one bushing. However, if it is necessary for each conductor to enter through an individual hole, heating can be avoided by cutting slots in the metal between the individual holes. These slots will break the magnetic path in the steel. Enclosures that are made of nonferrous metals such as PVC, aluminum, brass, and bronze do not require slots when individual conductors pass through individual holes per **300.35** and **300.20**. (**See Figure 10-29**)

GROUNDING
250.20(C); 250.180; 250.184(B)

High-voltage wiring and equipment installations are required to be grounded in accordance with the provisions listed in Article 250 of the NEC. (**See Figure 10-30**)

Note: The supply transformer does not have to be grounded per **250.20(C)** and **250.130(B)**. However, metal enclosures enclosing electrical elements and conductors are required to be grounded with an equipment grounding means and should be sized per **250.122**.

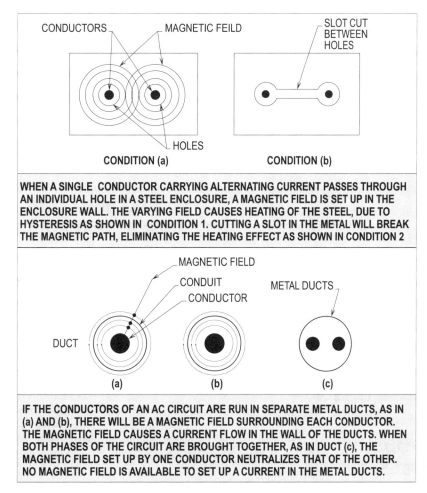

WIRING METHODS
PROTECTION AGAINST INDUCTION HEATING
NEC 300.35
NEC 300.20

Figure 10-29. Conductors and cables must be routed in raceways and ducts to prevent induction heating capable of damaging insulation.

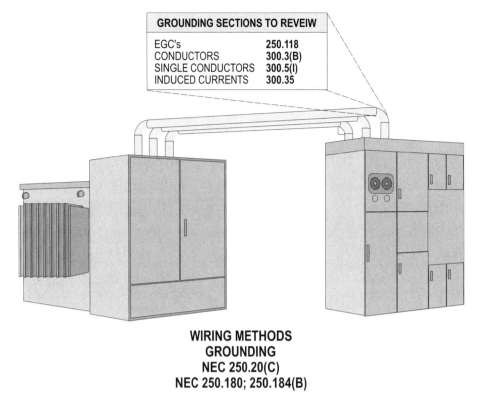

Figure 10-30. High-voltage electrical systems and wiring methods must comply with Article 250 of the NEC, as well as other pertinent sections.

WIRING METHODS - SIZE OF PULL AND JUNCTION BOXES
314.71

Pull and junction boxes are required to provide adequate space and dimensions for the installation of conductors and accessories.

Terminal housing supplied with motors are required to comply with the provisions listed in **430.12**. Motors that are provided with terminal housings per **314.71, Ex.**, are required to be of metal and of substantial construction.

When these terminal housings enclose wire-to-wire connections, they must have minimum dimensions and usable volumes in compliance with **430.12(C)** and Table **430.12(B)** of the NEC. **(See Figure 10-31)**

WIRING METHODS - STRAIGHT PULLS
314.71(A)

The length of the box must be not less than 48 times the outside diameter, over sheath, of the largest conductor or cable entering the box.

Note: 48 times is a minimum, as in many cases, one may wish to increase this length for ease in handling of the cables and for splicing. **(See Figure 10-32)**

WIRING METHODS - ANGLE OR U PULLS
314.71(B)

The distance between each cable or conductor entry inside the box and the opposite wall of the box must not be less than 36 times the outside diameter, over sheath, of the largest cable or conductor. This distance must be increased for additional entries by the amount of the sum of the outside diameters, over sheath of all other cables or conductor entries through the same wall of the box.

The distance between a cable or conductor entry and its exit from the box must be not less than 36 times the outside diameter, over sheath, of that cable or conductor.

EXCEPTION 1 TO 314.71(B)

Where a conductor or cable entry is in the wall of a box opposite to a removable cover and where the distance from that wall to the cover is in conformance with provisions of **300.34**. **(See Figure 10-33)**

Equiment Over 600 Volts

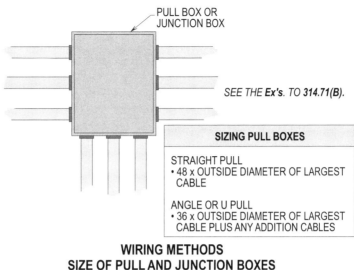

**WIRING METHODS
SIZE OF PULL AND JUNCTION BOXES
NEC 370.71(A)**

Figure 10-32. The above rules must be applied when sizing boxes to enclosed high-voltage cables and conductors.

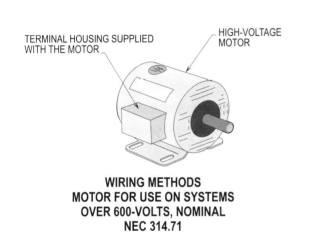

**WIRING METHODS
MOTOR FOR USE ON SYSTEMS
OVER 600-VOLTS, NOMINAL
NEC 314.71**

Figure 10-31. The terminal housing must comply with the provisions of **430.12**.

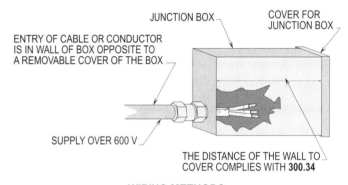

**WIRING METHODS
ENTRY IN WALL OPPOSITE REMOVABLE COVER
NEC 314.71(B), Ex. 1**

Figure 10-33. A certain distance must be maintain where a cable enters the wall of a junction box opposite to a removable cover.

Section **300.34** requires the conductor to be bent to a radius of not less than 8 times the overall diameter for nonshielded conductors or 12 times the diameter for shielded or lead-covered conductors during or after the installation has been made. (**See Figure 10-33**)

EXCEPTION 2 TO 314.71(B)

Where cables are nonshielded and not lead covered, the distance of thirty-six times the outside diameter shall be permitted to be reduced to twenty-four times the outside diameter. (Also see **314.71(B)(2)** and **Ex.**)

WIRING METHODS - REMOVABLE SIDES 314.71(C)

One or more sides of any pull box must be removable so personnel will have reasonable access to repair or service splices or conductors.

WIRING METHODS - CONSTRUCTION AND INSTALLATION REQUIREMENTS 314.72

Boxes enclosing high-voltage conductors are required to be constructed of suitable material to substantially support and hold such conductors. Boxes must be located in areas where the conductors inside can be maintained in a safe and reliable manner.

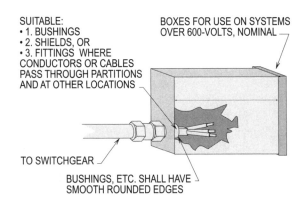

Figure 10-34. Protection of cables or conductors passing through partitions is required to prevent damaging insulation.

WIRING METHODS - CORROSION PROTECTION 314.72(A)

Boxes shall be made of material inherently resistant to corrosion or must be suitably protected, both internally and externally, by enameling, galvanizing, plating, or other effective means.

For example: The following wiring methods and equipment are considered protected against corrosion:

(1) Certain materials, like brass and aluminum, are corrosion-resistant in themselves.

(2) Equipment made of such materials does not require a corrosion-resistant coating.

(3) Steel or iron conduit and boxes must be galvanized or cadmium-coated.

(4) Boxes or cabinets marked "Raintight" or "Outdoor type" may be used outdoors.

(5) Enameled conduit and fittings may not be used outdoors, nor indoors in wet or corrosive locations.

(6) In wet indoor locations, conduit, cable, and fittings must be stooled off the surface at least 1/4 in.

PARTITIONS

The following methods will aid in helping prevent the spread of fire through partitions:

(1) When holes are made in fire-resistant walls, partitions, ceilings, the fire resistance is destroyed.

(2) When a hole must be made for a cable pass through, space around the cable must be plugged with a fire-resistant material.

(3) The material must be equal to the fire-resistant rating of the wall.

WIRING METHODS - COMPLETE ENCLOSURE 314.72(C)

Boxes shall provide a complete enclosure for the contained conductors or cables. Section **300.31** requires suitable covers to be installed on all enclosures and **300.11** requires enclosures to be properly secured.

WIRING METHODS - PASSING THROUGH PARTITIONS 314.72(B)

Suitable bushings, shields or fittings having smooth rounded edges shall be provided where conductors or cables pass through partitions and at other locations where necessary. The insulation of conductors must not be subjected to physical damage. **(See Figure 10-34)**

WIRING METHODS - WIRING IS ACCESSIBLE 314.72(D)

Boxes are required to be so installed that the wiring is accessible without removing any part of the building. Working space must be provided in accordance with the rules listed in **110.34**. **(See Figure 10-35)**

Equiment Over 600 Volts

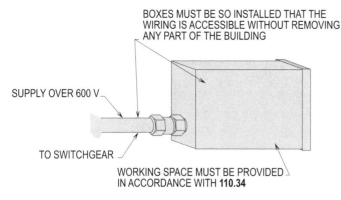

Figure 10-35. Wiring in boxes enclosing high-voltage cables and conductors must be installed in such a manner so wiring is accessible.

WIRING METHODS - SUITABLE COVERS 314.72(E)

Boxes are to be closed by a suitable cover securely fastened in place. Underground box covers that weigh over one hundred pounds shall be considered as meeting this requirement.

For example: This 100 lb. rule will satisfy the requirement for a manhole cover being securely fastened in place. Covers for boxes must be permanently marked **"DANGER - HIGH VOLTAGE - KEEP OUT."** The marking is required to be on the outside of the box cover and must be readily visible. Letters must be the block type at least 1/2 in. in height to be considered legible to read adequately. **(See Figure 10-36)**

WIRING METHODS - SUITABLE FOR EXPECTED HANDLING 314.72(F)

Boxes and their covers must be capable of withstanding the handling to which they may be subjected.

The recommended construction specifications for boxes and covers can be summed up as follows:

(1) Must be corrosion-resistant.

(2) Galvanized, cadmium-plated, or painted equipment will comply with this rule.

(3) Smaller sheet-metal boxes must be at least 16 gauge.

(4) Larger boxes (over 100 cu. in.) must comply to requirements for cabinets and cutout boxes. However, this rule does not apply for hinging.

(5) Malleable iron boxes, cast aluminum, brass, and bronze boxes must have a wall thickness of at least 3/32 in.

(6) Other cast metal boxes are required to have a wall thickness of at least 1/8 in.

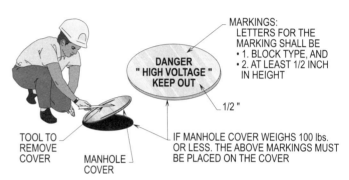

Figure 10-36. Manhole covers or underground box covers that weigh over 100 lbs. are considered accessible only to qualified personnel and do not require the above marking in the illustration.

GROUND GRID

A grid grounding scheme consists of buried conductors and driven ground rods which are interconnected to form a continuous grid network.

Ground grids are usually used for the grounding of lines and equipment located in the substation. The equipment and apparatus shall be bonded and grounded to the ground grid using large conductors to minimize the grounding resistance which limits the potential between equipment and ground surface.

Fences shall be constructed inside the ground grid and connected as needed to provide a safe bonding scheme. Sometimes large conductors are run underneath substations and spaced properly to create a very low resistance to ground. **(See Figure 10-37)**

DESIGNING FEEDER-CIRCUITS
215.2(B)(1) THRU (B)(3)

Feeder-circuits can be run from the switchgear installed in the building that is normally located in the substation. **(See Figure 10-38)**

When calculating the load for sizing feeder-circuits, see Figures 10-16 and 10-18 in this chapter.

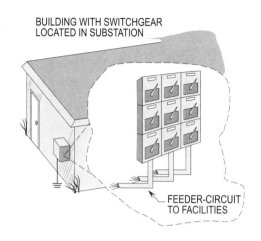

FEEDER-CIRCUITS ROUTED TO FACILITIES
NEC 215.2(B)(1) THRU (3)

Figure 10-38. Feeder-circuits can be run from the switchgear installed in the building that is located in the substation

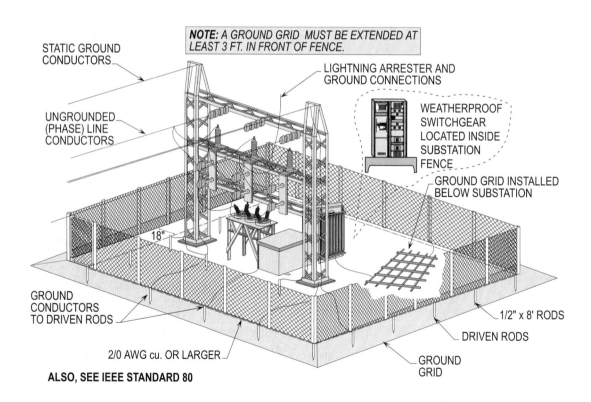

GROUND GRID
NFPA 70B - CH. 6-1.5 THRU 6-1.7

Figure 10-37. To create equipotental planes between ungrounded (phase) line conductors and equipment, elements are grounded to a counterpoise ground and driven rods or ground grid.

Name Date

Chapter 10. Equipment Over 600 Volts

Section	Answer		
_____	T F	1.	Isolating switches shall be accessible to unqualified and qualified personnel.
_____	T F	2.	The service disconnecting means shall simultaneously disconnect all ungrounded conductors.
_____	T F	3.	An overcurrent protection device shall be provided in each ungrounded conductor for service conductors operating over 600 volts.
_____	T F	4.	A load interrupters is a switch or circuit breaker that is intended to provide overload protection.
_____	T F	5.	Interrupting rating of fuses shall be at least equal to the maximum possible short circuit current available for the circuit.
_____	_____	6.	Service _____ is the point of connection between the facilities of the serving utility and the premises wiring.
_____	_____	7.	Open wire services and service wires in conduit shall be at least _____ AWG.
_____	_____	8.	When oil switches are used as the disconnecting means, an additional _____ switch shall be installed ahead of the oil switch.
_____	_____	9.	Isolating switches shall provide means for _____ grounding wiring on the premises.
_____	_____	10.	Circuit breakers shall have an interrupting capacity at least _____ to the maximum short circuit current that can be delivered to its terminals.
_____	_____	11.	Service-entrance conductors shall have a short-circuit protective device in each _____ conductor on the load side or be installed as an integral part of the service-entrance disconnect.
_____	_____	12.	For short circuit protection of high-voltage feeders, fuse rating may be up to _____ times the conductor ampacity.
_____	_____	13.	For short circuit protection of high-voltage feeders, circuit breaker rating may up to _____ times the conductor ampacity.
_____	_____	14.	Shielded or lead-covered high-voltage conductors shall not be bent to a radius of less than _____ times overall diameter.
_____	_____	15.	Other types of high-voltage conductors that are not shielded or lead-covered shall not be bent to a radius of less than _____ times overall diameter.

Grounding

Equipment and circuits that are grounded and bonded properly are protected from lightning storms and excessive surges of voltages. Property is also protected as well as safety for people is provided. There are three grounding schemes in a well designed and installed grounding system and they are as follows:

- Circuit and system grounding
- Equipment grounding
- Bonding supply and load side

There are types of conductors per the NEC, if sized and selected correctly, shall be permitted to be used as a network to ground and bond electrical systems for safety.

Before attempting to design and install the elements of an effective grounding scheme, the following definitions of ground, grounded, grounded conductor, grounding conductor, bonding (bonded) and equipment grounding conductor per **Article 100** and **250.2** must be reviewed very carefully. These definitions are related to the grounding and bonding of the electrical system and equipment.

Ground is defined as a conducting connection, whether intentional or accidental, between an electrical circuit or equipment and the earth or to some conducting body that serves in place in the earth.

Grounded is defined as connected to earth or to some conducting body that serves in place of the earth.

Grounded conductor is defined as a system or circuit conductor that is intentionally grounded.

Grounding conductor is a conductor used to connect equipment or the grounded circuit of a wiring system to a grounding electrode or electrodes.

Bonding (bonded) is defined as the permanent joining of metallic parts to form an electrically conductive path that ensures electrical continuity and the capacity to conduct safely any current likely to be imposed.

Equipment grounding conductor is a conductor used to connect the noncurrent-carrying metal parts of equipment, raceways and other enclosures to the system grounded conductor, the grounding electrode conductor or both, at the service equipment or at the source of a separately derived system.

When grounding and bonding electrical systems for safety, such grounding and bonding procedures shall be done at the following locations:

- Service equipment
- Separately derived system
- Separate buildings or structures
- Load side of OCPD

This Chapter deals with grounding and bonding of these locations as well as other methods of grounding.

GENERAL REQUIREMENTS FOR GROUNDING AND BONDING
250.4

The following types of systems have general requirements that are to be applied for grounding and bonding:

- Grounded systems
- Ungrounded systems

GROUNDED SYSTEMS
250.4(A)

The following general requirements for grounding and bonding shall be accomplished for grounded systems to protect property as well as provide safety for people:

- Electrical system grounding
- Grounding of electrical equipment
- Bonding of electrical equipment
- Bonding of electrically conductive materials and other equipment
- Effective ground-fault current path

ELECTRICAL SYSTEM GROUNDING
250.4(A)(1)

Systems and circuits shall be grounded to limit voltage due to lightning, line surges or unintentional contact with higher voltage lines. Systems and circuits shall be grounded to stabilize the voltage-to-ground during normal operation. Systems and circuits shall be solidly grounded to facilitate overcurrent device operation in case of ground-faults. **(See Figure 11-1)**

GROUNDING OF ELECTRICAL EQUIPMENT
250.4(A)(2)

Noncurrent-carrying conductive materials that enclose electrical conductors or equipment, or forming part of such equipment, shall be connected to earth so as to limit the voltage-to-ground on these materials.

Circuits and equipment shall be grounded to facilitate overcurrent device operation in case of insulation failure or ground-faults. To insure this, the grounding path from circuit equipment and metal enclosures shall be continuous and not subject to damage. This path shall be capable of safely handling fault currents that may be imposed on it. Impedance shall be sufficiently low enough to keep the voltage-to-ground at a minimum and facilitate the opening of overcurrent protection devices ahead of the circuit. **(See Figure 11-2)**

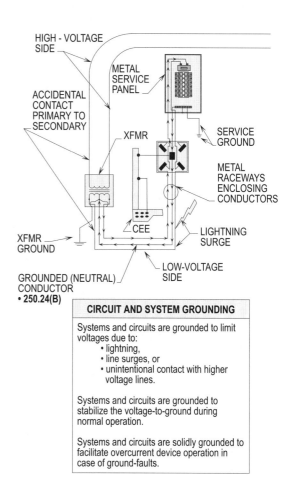

**ELECTRICAL SYSTEMS GROUNDING
NEC 250.4(A)(1)**

Figure 11-1. This illustration shows the benefits of grounding an electrical system for safety on the supply side.

BONDING OF ELECTRICAL EQUIPMENT
250.4(A)(3)

Noncurrent-carrying conductive materials that enclose electrical conductors or equipment, or forming part of such equipment, shall be connected together and to the electrical supply source in such a manner so as to establish an effective ground-fault current path. **(See Figure 11-3)**

BONDING OF ELECTRICALLY CONDUCTIVE MATERIALS AND OTHER EQUIPMENT
250.4(A)(4)

Electrically conductive materials, such as metal water piping, gas piping and structural steel members that are likely to become energized shall be bonded together and to the electrical supply source in such a manner that an effective ground-fault current path is established. **(See Figure 11-3)**

Grounding

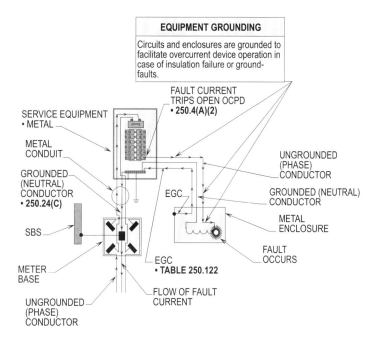

Figure 11-2. This illustration shows the benefits of equipment grounding on the load side.

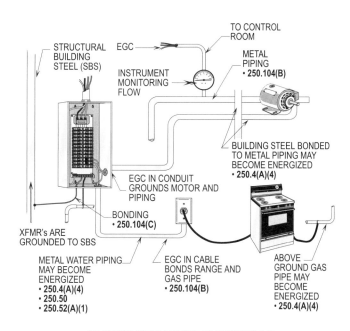

Figure 11-3. This illustration shows methods of how to bond electrically conductive materials (other than metal water piping) and other equipment.

11-3

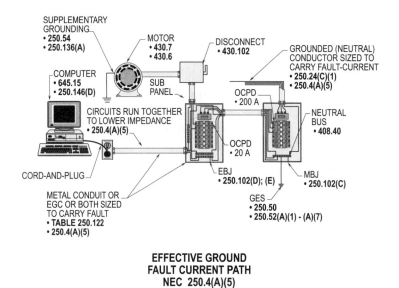

Figure 11-4. This illustration shows that an effective ground-fault current path from circuits, equipment and conductor enclosures shall be installed in a manner that creates a permanent, low-impedance circuit capable of safely carrying the maximum ground-fault current likely to be imposed on it from any point on the wiring system.

EFFECTIVE GROUND-FAULT CURRENT PATH
250.4(A)(5)

An effective ground-fault current path from circuits, equipment and conductor enclosures shall be installed in a manner that creates a permanent, low-impedance circuit facilitating the operation of the overcurrent device or ground detector for high-impedance grounded systems. It shall be capable of safely carrying the maximum ground-fault current likely to be imposed on it from any point on the wiring system. **(See Figure 11-4)**

Note: The earth shall not be used as the sole equipment grounding conductor or effective ground-fault current path.

UNGROUNDED SYSTEMS
250.4(B)

The following general requirements for grounding and bonding shall be accomplished for ungrounded systems to protect property as well as provide safety for people:

- Grounding electrical equipment
- Bonding of electrical equipment
- Bonding of electrically conductive materials and other equipment
- Path for fault-current

GROUNDING ELECTRICAL EQUIPMENT
250.4(B)(1)

Noncurrent-carrying conductive materials that enclose electrical conductors or equipment, or forming part of such equipment, shall be connected to earth in such a manner so as to limit the voltage imposed by lightning or unintentional contact with higher voltage lines and limit the voltage-to-ground on these materials. **(See Figure 11-5)**

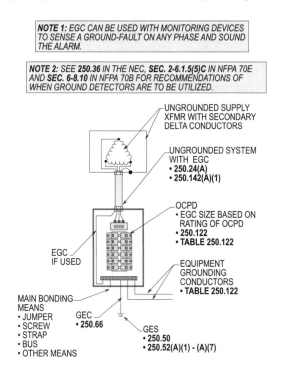

Figure 11-5. This illustration shows the grounding and bonding requirements for an ungrounded system.

BONDING OF ELECTRICAL EQUIPMENT
250.4(B)(2)

Noncurrent-carrying conductive materials that enclose electrical conductors or equipment, or forming part of such equipment, shall be connected together and to the supply system grounded equipment in such a manner so as to create a permanent, low-impedance path for ground-fault current that is capable of carrying the maximum fault current likely to be imposed on it.

BONDING OF ELECTRICALLY CONDUCTIVE MATERIALS AND OTHER EQUIPMENT
250.4(B)(3)

Electrically conductive materials, such as metal water piping, gas piping and structural steel members that are likely to become energized shall be bonded together and to the supply system grounded equipment in such a manner to create a permanent, low-impedance path for grounding current that is capable of carrying the maximum fault current likely to be imposed on it.

PATH FOR FAULT CURRENT
250.4(B)(4)

Electrical equipment, wiring and other electrically conductive material likely to become energized shall be installed in a manner that creates a permanent, low-impedance circuit from any point on the wiring system to the electrical supply source so as to facilitate the operation of overcurrent devices should a second fault develop on the wiring system.

Note: The earth shall not be used as the sole equipment grounding conductor or effective ground-fault current path.

OBJECTIONABLE CURRENT OVER GROUNDING CONDUCTORS
250.6

Unbalanced loads along with multiple grounding points may have objectionable current flowing in grounding conductors within the circuits of the wiring system. When the grounded (neutral) conductor is intentionally connected to earth ground at the supply transformer and the service equipment of the building, there will be two parallel paths present for unbalanced current to travel over. The two parallel paths of unbalanced current will return back to the utility supply transformer through the grounded (neutral) conductor and through the ground path of the earth soil. More current returns from the transformer to the service equipment over the faulted ungrounded (phase) conductor and trips open the OCPD of the faulted circuit conductor. The following conditions of use shall be considered for objectionable current over grounding conductors: **(See Figure 11-6)**

- Arrangement to prevent objectionable current
- Alterations to stop objectionable current
- Temporary currents not classified as objectionable currents
- Limitations to permissible alterations
- Isolation of objectionable direct-current ground currents

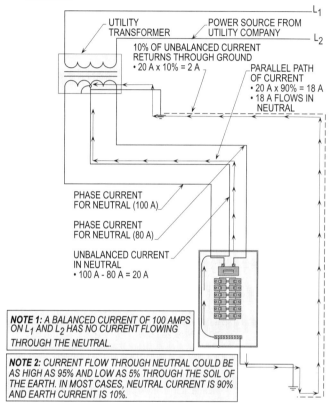

Figure 11-6. This illustration shows a condition where objectionable current flow in the earth will not be a problem.

ARRANGEMENT TO PREVENT OBJECTIONABLE CURRENT
250.6(A)

Depending on the resistance of the ground path of the grounded (neutral) conductor and earth soil, about 10 percent of the current will flow through the ground and approximately 90 percent will flow over the grounded (neutral) conductor. Balancing the load of the system as well as possible reduces the unbalanced current (objectionable) so that excessive amounts will not flow over the grounded (neutral) conductor. **(See Figure 11-7)**

Stallcup's Electrical Design Book

Design Tip: The grounded (neutral) conductor only carries the unbalanced current between the ungrounded (phase) conductors per **220.61(A)**. The grounded (neutral) conductor shall be permitted to be increased in size due to nonlinear loading on the ungrounded (phase) conductors per **220.61(C), FPN 2**. For transformer requirements pertaining to nonlinear loading, see **450.3, FPN 2**.

ALTERATIONS TO STOP OBJECTIONABLE CURRENT 250.6(B)

Objectionable current could occur from multiple grounds being installed on the same system. Objectionable current could flow from one ground connection to another, after such current has entered from one or more of the grounding points in the grounding system. If objectionable current is present, one or more of the following steps will help solve this problem:

- One or more grounding connections should be disconnected, but do not disconnect all of them.

- Placement of grounds should be altered.

- Break the conductive path of interconnecting grounding connections, so as to eliminate the continuity between grounding connections.

- Consult the authority having jurisdiction for suitable methods which are acceptable to help solve the problem. **(See Figure 11-8)**

Design Tip: There are conditions where a grounding conductor shall be permitted to be removed without disrupting the safety of the grounding system.

For example, a grounding conductor connecting the metal of an electrical enclosure to the structural steel for the diversion of static electricity build-up on such equipment.

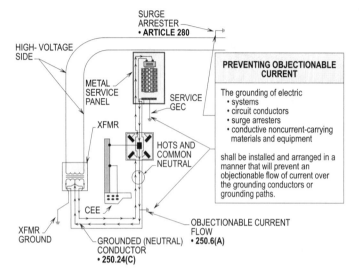

Figure 11-7. This illustration shows a connection to ground of the grounded (neutral) conductor that is not considered an objectionable current flow.

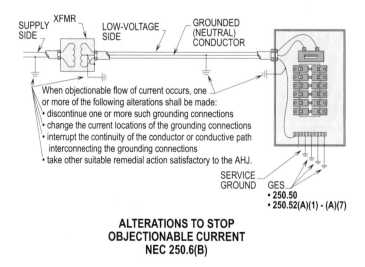

Figure 11-8. This illustration shows methods used to help eliminate objectionable current flow.

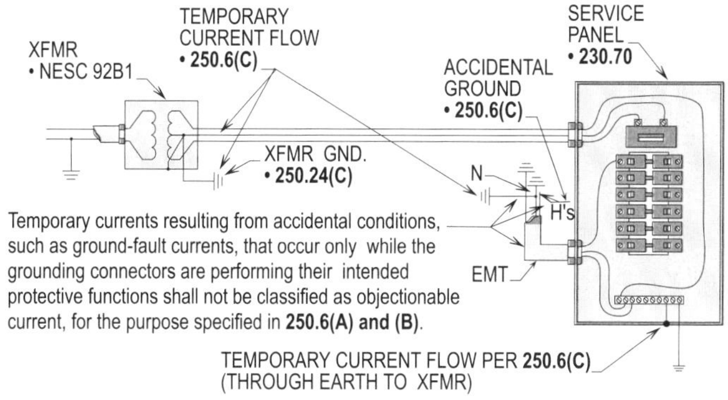

TEMPORARY CURRENTS NOT CLASSIFIED AS OBJECTIONABLE CURRENTS 250.6(C)

Temporary currents develop from accidental shorts such as ground-fault currents that will flow over grounding conductors and shall not be considered objectionable currents. **(See Figure 11-9)**

> **Design Tip:** Currents that flow over equipment grounding conductors, bonding jumpers and grounded (neutral) conductors shall be considered temporary currents and shall not be defined as objectionable current flow per **250.6(C)**, for they will only flow until the OCPD of the circuit trips open.

TEMPORARY CURRENTS NOT CLASSIFIED AS OBJECTIONABLE CURRENTS NEC 250.6(C)

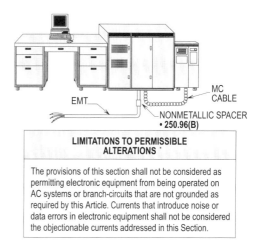

Figure 11-9. This illustration shows examples of temporary currents that are not considered objectionable currents.

LIMITATIONS TO PERMISSIBLE ALTERATIONS 250.6(D)

Currents generated from electronic equipment shall not be considered objectionable currents where such currents introduce noise or data errors. Currents that have such electrical noise or voltage fluctuations that produce dirty power conditions over the circuits shall also not be considered objectionable current flow. **(See Figure 11-10)**

> **Design Tip:** To correct noise problems for sensitive electronic equipment, it shall be permitted to isolate the supply conduit with a nonmetallic spacer. This spacer shall be located at the sensitive electronic equipment and shall not be permitted to take the place of the safety ground (equipment grounding conductor).

ISOLATION OF OBJECTIONABLE DIRECT-CURRENT GROUND CURRENTS 250.6(E)

Where isolation from undesirable DC ground current is required, such as in the area of cathodic protected systems, a listed solid state AC coupling/DC isolating device shall be permitted to be placed in the grounding path to provide an effective return path for AC ground-fault current and also blocking DC current. **(See Figure 11-11)**

LIMITATIONS TO PERMISSIBLE ALTERATIONS NEC 250.6(D)

Figure 11-10. This illustration shows a conduit system that is altered to help reduce electrial noise generated from the operation of the electronic equipment.

Note: It is necessary to restrict the flow of DC cathodic protection current to only the metallic parts to be protected. Such listed AC coupling and DC blocking devices permits flow of the AC current while blocking DC. These devices are evaluated for their ability to withstand the rated fault-currents. They are also evaluated for connections in the electrical grounding system in accordance with the instructions for the equipment.

> **Design Tip:** Equipment installed in series with the equipment grounding conductor shall be approved by the AHJ or be third party approved by a UL (type) 508 shop.

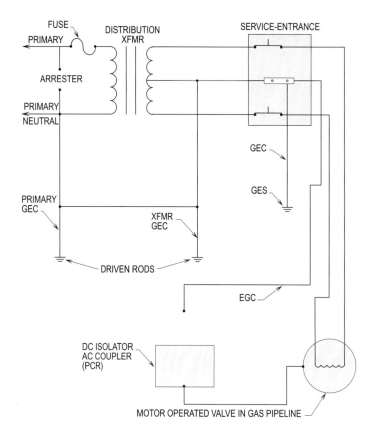

Figure 11-11. This illustration shows an AC coupling and DC blocking device used for cathodic protection.

CONNECTION OF GROUNDING AND BONDING EQUIPMENT
250.8

Pressure connectors, clips, clamps, lugs and devices shall be listed for the installation when used to connect grounding conductors to metal enclosures, electrodes and other types of equipment requiring grounding.

The grounding conductors and bonding jumpers shall be attached to circuits, conduits, cabinets, equipment, etc. which shall be grounded, by means of suitable lugs, listed pressure connectors, listed clamps, exothermic welding or other listed means. **(See Figure 11-12)**

Note: The device used shall be identified as being suitable for such purpose. Connection devices or fittings that depend solely on solder shall not be permitted to be used. Sheet-metal screws shall not be permitted to be used to connect grounding conductors or connection devices to enclosures.

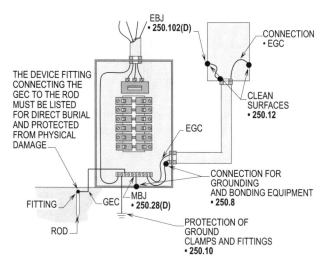

Figure 11-12. This illustration shows that the devices used to connect the grounding electrode conductor to the electrodes shall be listed for such use.

Grounding

PROTECTION OF GROUND CLAMPS AND FITTINGS
250.10

Unless approved for general use without protection, ground clamps and fittings shall be installed as follows:

- Be located so that they will not be subject to damage, or
- Be enclosed in metal, wood or equivalent protective covering.

With exothermic welding (cad-welding), checked for the resistance of the connection with a ductor. If a low-resistance is measured, the problems of a high-resistance connection should be eliminated. **(See Figure 11-13)**

AC SYSTEMS TO BE GROUNDED
250.20

The following AC systems shall be grounded:

- AC systems of less than 50 volts.
- AC systems of 50 to 1000 volts.
- AC systems of 1 kV and over.
- Separately derived systems.

See Figure 11-14 for a detailed discription when applying these requirements.

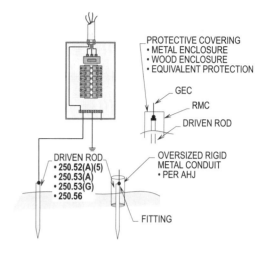

PROTECTION OF GROUND CLAMPS AND FITTINGS NEC 250.10

Figure 11-13. This illustration shows the methods that shall be permitted to be used to protect the grounding clamp from physical damage.

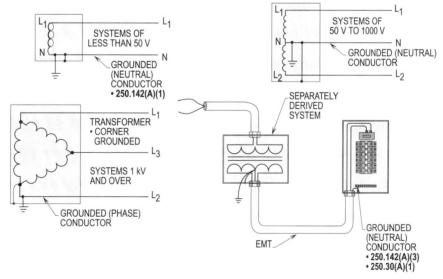

AC CIRCUITS AND SYSTEMS TO BE GROUNDED NEC 250.20

Figure 11-14. This illustration shows electrical systems which are required to be grounded under certain conditions of use.

AC SYSTEMS OF LESS THAN 50 VOLTS
250.20(A)

Circuits and equipment operating at less than 50 volts are found in **Article 720**. Class 1, 2 and 3 circuits including rules pertaining to their installation procedures are listed in **725.21(A)** and **725.41**.

AC systems of less than 50 volts shall be grounded under the following conditions:

- A transformer installed to supply low-voltage receives its supply from a transformer exceeding 150 volts-to-ground.

For example, a circuit of 277 volts-to-ground supplying the primary side of such a transformer is considered a circuit of over 150 volts-to-ground and its secondary shall be grounded.

- A transformer installed to receive its supply from a transformer with an ungrounded system. This condition includes a supply voltage obtained from a transformer which has an ungrounded secondary.

- Where low-voltage overhead conductors are installed outside and not inside.

For example, AC conductors of 50 volts which are run outside overhead shall have one conductor grounded.

See **Figure 11-15** for a detailed illustration for AC systems of less than 50 volts.

AC SYSTEMS OF 50 TO 1000 VOLTS
250.20(B)

AC systems of 50 to 1000 volts shall be grounded under any one of the following conditions:

- The maximum voltage-to-ground on the ungrounded (phase) conductors does not exceed 150 volts. This voltage-to-ground circuit is usually a 120 volt circuit derived from a 120/240 or 120/208 volt system.

- The system is nominally rated as 120/208 and 277/480 volt, three-phase, four-wire wye and is connected so that the grounded (neutral) conductor can be used as a circuit conductor. The voltage-to-ground and between phases for 50 to 1000 volts may be any level. Wye systems of 2400/4160 volts are also used to supply circuits requiring a higher voltage served from a wye hookup.

- The system is nominally rated as 240/120 volt, three-phase, four-wire delta connected in which the midpoint of one phase is used as a circuit conductor with one phase conductor having a higher voltage-to-ground than the other two. **(See Figure 11-16)**

The following types of AC systems which are rated 50 volts to 1000 volts shall be grounded:

- 120 volt, two-wire, single-phase
- 120/240 volt, three-wire, single-phase
- 120/208 volt, four-wire, three-phase wye
- 277/480 volt, four-wire, three-phase wye
- 480 volt corner grounded, three-phase delta
- 240 volt, three-wire, three-phase delta
- 480 volt, three-wire, three-phase delta
- 600 volt, three-wire, three-phase delta

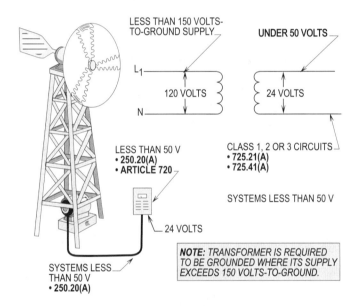

**AC SYSTEMS OF LESS THAN 50 VOLTS
NEC 250.20(A)**

Figure 11-15. This illustration shows examples of systems less than 50 volts.

AC SYSTEMS OF 1 kV AND OVER
250.20(C)

AC systems of 1 kV and over shall be grounded if supplying mobile or portable equipment per **250.188**. Other types of AC systems of 1 kV and over that are installed do not have to be grounded. Where such systems are grounded, they shall comply with the applicable provisions of **Article 250**. **(See Figure 11-17)**

Grounding

SEPARATELY DERIVED SYSTEM
250.20(D)

A separately derived system is derived from an generator, converter windings or transformer to reduce the voltage from high-voltage to low-voltage levels or vice versa in a building. Transformers of 4160 or 13,800 volts are voltages that may be reduced to utilization voltage of 480 volt or 120/208 volt to supply loads that are located at different floor levels within a building. (**See Figure 11-18**)

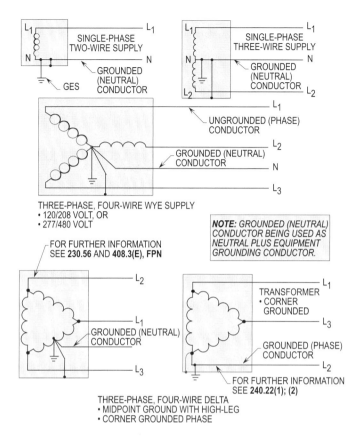

Figure 11-16. This illustration shows AC systems of 50 volts to 1000 volts that shall be grounded when the grounded (neutral) conductor or ungrounded (phase) conductor is used as an equipment grounding conductor during a ground-fault condition.

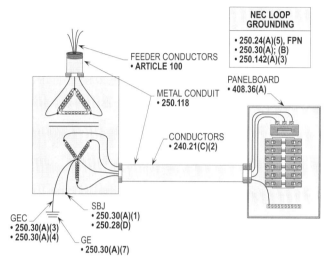

Figure 11-18. This illustration shows a separately derived system is derived from a generator, converter windings or transformers to reduce the voltage from high-voltage to low-voltage or vice versa in a building.

AC SYSTEMS OF 50 VOLTS TO 1000 VOLTS NOT REQUIRED TO BE GROUNDED
250.21

The following AC systems of 50 volts to 1000 volts shall not be required to be grounded:

- Circuits installed for industrial electric furnaces or any other means of heating metals for refining, melting or tempering.

- Separately derived systems used exclusively for rectifiers supplying only adjustable speed industrial drives. Such systems are used for speed control in industrial facilities and shall not be permitted to be utilized for anything else.

- Separately derived systems installed with primaries

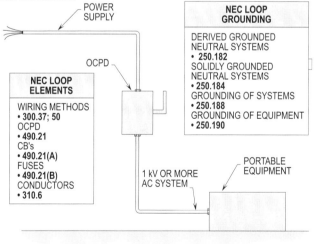

Figure 11-17. This illustration shows AC systems of 1 kV and over shall be grounded if supplying mobile or portable equipment.

not exceeding 1000 volts used exclusively for secondary control circuits where conditions of maintenance and supervision ensure that only qualified persons will service the installation. However, the continuity of control power is required with ground detectors installed on the control system to sound an alarm if one phase should become grounded and a ground-fault condition occurs. **(See Figure 11-19)**

CIRCUITS NOT TO BE GROUNDED 250.22

The following circuits shall not be grounded due to their conditions of use:

- Circuits for electric cranes operating over combustible fibers in Class III locations per **503.155**. Combustible fibers are easily ignited with sparking or arcing devices. This rule was designed to eliminate this problem and hazardarous condition.

- Circuits operating in health care facilities such as anesthetizing locations per **517.160(A)(2)**. In operating rooms, circuits shall not be permitted to be grounded to ensure against ground-faults.

- Circuits for electrolytic cells shall not be permitted to be grounded per **Article 668.**

- Secondary circuits for lighting systems shall not be permitted to be grounded per **411.5(A)**.

See Figures 11-20(a), (b) and **(c)** for a detailed illustration when applying these requirements.

SYSTEM GROUNDING CONNECTIONS 250.24(A)

The grounding electrode conductor shall be installed to bond and ground the service equipment terminal neutral bus to the grounding electrode system per **250.52(A)(1) through (A)(7)**. The grounding electrode conductor is installed to connect equipment grounding conductors, grounded (neutral) conductors and the service equipment enclosure to a reliable earth ground. Bonding jumpers shall be installed by using a wire, bus, screw or other suitable conductor per **250.28(A)**. Grounding electrodes bonded together for two or more services shall be considered a single grounding electrode system which will bond the metal enclosures of equipment of all sources together to create equipotential planes. Note that the head of a screw used to bond in the above items shall be green in color and be visible after the installation per **250.28(B)**. **(See Figure 11-21)**

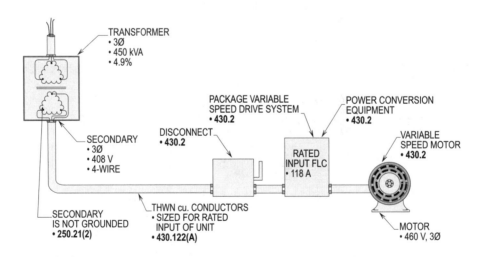

AC SYSTEMS OF 50 VOLTS TO 1000 VOLTS NOT REQUIRED TO BE GROUNDED
NEC 250.21

Figure 11-19. This illustration shows a system shall not be required to be grounded due to its supplying rectifiers powering and adjustable speed industrial drive.

Grounding

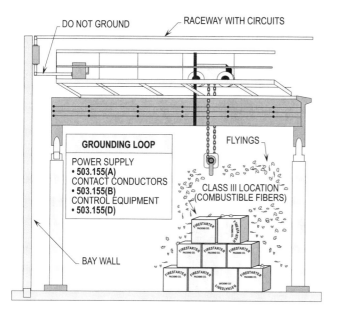

Figure 11-20(a). This illustration shows a circuit that is not permitted to be grounded.

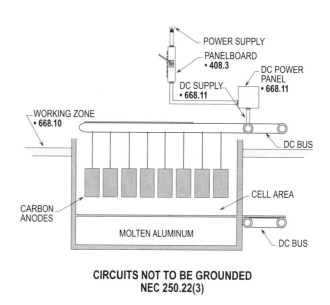

Figure 11-20(c). This illustration shows a circuit that is not permitted to be grounded.

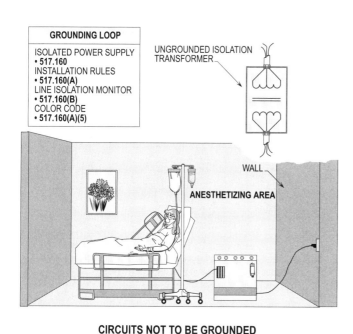

Figure 11-20(b). This illustration shows a circuit that is not permitted to be grounded.

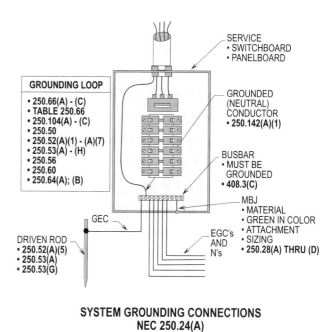

Figure 11-21. The grounding electrode conductor shall be installed in such a manner so as to ground the grounded busbar, the equipment grounding conductors and grounded (neutral) conductor to the grounding electrode system to form a single-point ground connected to earth ground.

11-13

GROUNDING SERVICE-SUPPLIED AC SYSTEMS
250.24(A)(1) THRU (A)(5)

AC grounded systems shall be grounded at each service by a grounding electrode conductor which is usually from the neutral bus terminal to a grounding electrode system which could be a cold water pipe, driven rod or other electrode, or any combination per **250.52(A)(1) through (A)(7)**. The grounding electrode conductor shall be connected to the grounded service conductor when applying one or more of the following grounding requirements:

- General
- Outdoor transformer
- Dual fed services
- Main bonding jumper as wire or busbar
- Load-side grounding connections

GENERAL
250.24(A)(1)

The grounding electrode conductor shall be installed to bond and ground the service equipment terminal neutral bus to the grounding electrode system per **250.52(A)(1) through (A)(7)**. The grounding electrode conductor shall be installed to connect equipment grounding conductors, grounded (neutral) conductors and the service equipment enclosure to a reliable earth ground. **(See Figure 11-22)**

OUTSIDE TRANSFORMER
250.24(A)(2)

Transformers which supply the service and are located outside the building, at least one additional grounding connection shall be installed from the grounded service conductor to a grounding electrode, either at the transformer or elsewhere outside the building. **(See Figure 11-22)**

DUAL FED SERVICES
250.24(A)(3)

A single grounding electrode connection to the tie point of the grounded circuit conductor(s) from each power source shall be permitted to be installed where services are dual fed (double ended) in a common enclosure or grouped together in service enclosures and employing a secondary tie. **(See Figure 11-23)**

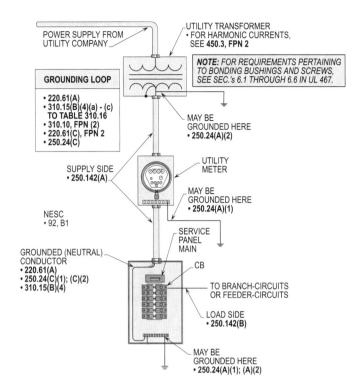

Figure 11-22. This illustration shows AC grounded systems shall be grounded at each service and the outside supply transformer.

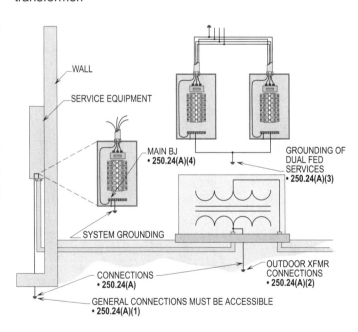

Figure 11-23. This illustration shows the requirements for grounding and bonding dual fed services from a service-entrance supply.

MAIN BONDING JUMPER AS WIRE OR BUSBAR
250.24(A)(4)

Where the main bonding jumper (wire or busbar) is installed from the grounded conductor terminal bar or bus to the equipment grounding terminal bar or bus in the service equipment, the grounding electrode conductor shall be permitted to be connected to the equipment grounding terminal bar or bus to which the main bonding jumper is connected.

LOAD SIDE GROUNDING CONNECTIONS
250.24(A)(5)

Unless otherwise permitted, a grounding connection shall not be permitted to be made to any grounded circuit conductor on the load side of the service disconnecting means. **(See Figure 11-24)**

> **Design Tip:** All equipment grounding conductors and equipment bonding jumpers on the load side are sized from **Table 250.122** based on the rating of the OCPD ahead of the circuit conductors.

CALCULATING FAULT CURRENT

By dividing the length of wire between supply and load by 1000 and multiplying by the resistance, the amount of fault current that will flow may be calculated. The resistance of each length is added together and divided into the voltage-to-ground to derive the fault current at the location of such short. **(See Figure 11-25)**

MAIN BONDING JUMPER
250.24(B)

For a grounded systems, an unspliced main bonding jumper shall be used to connect the equipment grounding conductor(s) and the service disconnect enclosure to the grounded conductor within the enclosure for each service disconnect in accordance with **250.38**.

GROUNDED CONDUCTOR BROUGHT TO SERVICE EQUIPMENT
250.24(C)

The grounded (neutral) conductors shall be run to each service disconnecting means and shall be bonded to each disconnecting means enclosure for AC systems which operate at less than 1000 volts and are grounded at any point. **(See Figure 11-26)**

> **Design Tip:** A good example of this rule is where six disconnects are installed at the service equipment and three are used with neutral connections and three are not. The three without neutral connections are still required to have the grounded (neutral) conductor run to each disconnect to provide a low-impedance path for the fault current to return over and trip the cutout fuses on the primary side of the power transformer.

The grounded (neutral) conductor(s) shall be installed in accordance with the following conditions of use:

- Sizing
- Routing
- Parallel conductors
- High impedance

GROUNDED CONDUCTORS BROUGHT TO ASSEMBLY LISTED FOR USED AS SERVICE EQUIPMENT
250.24(C), Ex.

Where more than one service disconnecting means is located in an assembly which is listed for use as service equipment, the grounded (neutral) conductor shall be permitted to be run to the assembly and the grounded (neutral) conductor shall be bonded to the assembly enclosure.

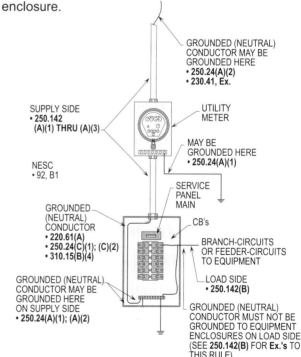

LOAD SIDE GROUNDING CONNECTIONS
NEC 250.24(A)(5)

Figure 11-24. This illustration shows the grounded circuit conductor shall not be permitted to have a grounding connection installed on the load side of the service disconnecting means.

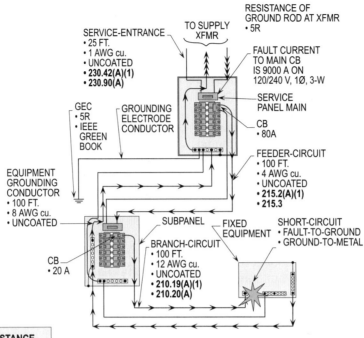

Figure 11-25. This illustration shows the procedures for calculating the fault current at the point of the fault.

Grounding

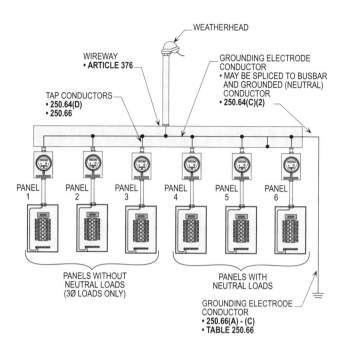

Figure 11-26. The grounded (neutral) conductor shall be run to each service disconnecting means, even if it's not used to supply single-phase loads with neutral connections.

ROUTING AND SIZING
250.24(C)(1)

The grounded (neutral) conductor shall be installed with the ungrounded (phase) conductors and shall not be permitted to be smaller than the required grounding electrode conductor per **Table 250.66**, but shall not be required to be larger than the largest ungrounded service (phase) conductor. The grounded (neutral) conductor shall not be permitted to be smaller than 12 1/2 percent of the area of the largest service-entrance (phase) conductor larger than 1100 KCMIL copper or 1750 KCMIL aluminum. **(See Figure 11-27)**

It is estimated that the amount of ground-fault current that will flow in the system is approximately 5 to 10 percent in the ground and 90 to 95 percent on the grounded (neutral) conductor between the supply transformer and service equipment. The rating of the grounded (neutral) conductor shall be calculated at least 12 1/2 percent of the largest ungrounded (phase) conductor. However, the equipment grounding conductor should be calculated at not less than 25 percent of the largest ungrounded (phase) conductor to ensure grounding conductors provide safe and dependable fault-ground paths. For an OCPD to clear a circuit safely, a fault current of at least 6 to 10 times its rating shall be available.

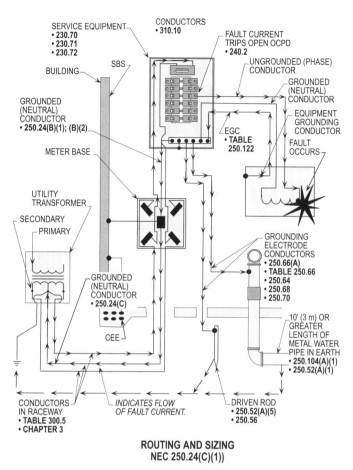

Figure 11-27. The grounded (neutral) conductor (fault-current return path) shall be installed with the ungrounded (phase) conductors and run to each service if the secondary of the utility's power transformer is grounded.

Note that fault-current will travel from the point of fault through the grounded (phase or neutral) conductor to the supply transformer and then will return through the phase which was faulted to ground and open the OCPD. The overcurrent protection will rapidly trip a faulted phase that has a fault current of 6 to 10 times its rating.

For example: What is the (appropriate) amount of fault current that is required to clear a 150 amp OCPD?
Step 1: Calculating percentage **250.24(C)** 150 A x 6 = 900 A 150 A x 10 = 1500 A **Solution: The minimum fault current is 900 amps and the maximum fault current is 1500 amps.**

11-17

Grounded (neutral) conductors shall be installed to provide an effective path for fault currents to travel over where phase-to-ground faults occur in the electrical system. The grounded (neutral) conductors shall be sized as large as the grounding electrode conductor per **250.66** and **Table 250.66**. The grounded (neutral) conductor shall be sized at least 12 1/2 percent of the area of the largest ungrounded (phase) conductor where the service conductors are installed larger than 1100 KCMIL copper or 1750 KCMIL aluminum.

Note: Two-phase or three-phase conductors shall not be permitted to be run to the service without installing a grounded (neutral) conductor when the utility company's secondary is grounded. An example of this rule is where all the service loads are three-phase, 480 volt and a step-down separately derived system is installed to supply single-phase loads of 120/208 volt.

For example: What size copper grounded (neutral) conductor is required based on the service-entrance conductors being rated 250 KCMIL THWN-THHN copper?

Step 1: Finding the grounded (neutral) conductor
250.142(A)(1); 250.24(C)(1); Table 250.66
250 KCMIL requires 2 AWG cu.

Solution: The size grounded (neutral) conductor is 2 AWG copper.

For example: In a paralleled three-phase, four-wire service with 4 - 300 KCMIL THWN copper conductors per phase, what size grounded (neutral) conductor is required when installed in a single conduit run, using RMC?

Step 1: Finding total KCMIL
250.24(C)(1)
300 KCMIL x 4 per phase = 1200 KCMIL

Step 2: Finding KCMIL to size grounded (neutral) conductor
250.24(C)(2)
1200 KCMIL x .125 = 150 KCMIL

Step 3: Finding CM rating
Table 8, Ch. 9
150 KCMIL requires 3/0 AWG cu.

Step 4: Sizing grounded (neutral) conductor for the conduit run
250.24(C)(2); Table 8, Ch. 9; 310.4

Solution: At least 3/0 THWN copper grounded (neutral) conductors are required per 250.24(C)(2) and 310.4.

PARALLEL CONDUCTORS
250.24(C)(2)

The grounded (neutral) conductor shall be based on the total circular mil area of the service-entrance conductors when installed in parallel. Grounded (neutral) conductors which are installed in two or more raceways shall be based on the size of the ungrounded service-entrance (phase) conductor in the raceway, but not smaller than 1/0 AWG.

Note: See **310.4** for grounded (neutral) conductors which are connected in parallel.

For example: In a paralleled three-phase, four-wire service with 3 - 700 KCMIL THWN copper conductors per phase, what size grounded (neutral) conductor is required in each run of conduit (3)?

Step 1: Finding total KCMIL
250.24(C)(1)
700 KCMIL x 1 = 700 KCMIL

Step 2: Finding KCMIL to size grounded (neutral) conductor
250.24(C)(2)
700 KCMIL requires 2/0 AWG cu.

Step 3: Dividing KCMIL in each conduit run
250.24(C)(2)
12 1/2% rule does not apply

Step 4: Sizing grounded (neutral) conductor for each conduit run
250.24(C)(2); Table 8, Ch. 9; 310.4
700 KCMIL requires 2/0 AWG cu.

Solution: At least 1 - 2/0 AWG THWN copper grounded (neutral) conductor is required in each conduit per 250.24(C)(2) and 310.4.

HIGH-IMPEDANCE
250.24(C)(3)

The grounded (neutral) conductor on a high-impedance grounded neutral system shall be grounded per **250.36**.

GROUNDING ELECTRODE CONDUCTOR
250.24(D)

A grounding electrode conductor shall be used to connect the equipment conductors, the service equipment enclosures, and where the system is grounded, the grounded service conductor to the grounding electrode or electrodes. **(See Figure 11-28)**

Where the three-phase, three-wire critical load is relatively large compared with loads that require a grounded (neutral) circuit conductor, a high-impedance grounded service power supply is sometimes used. This arrangement requires an on-site transformer for loads that require a grounded (neutral) circuit conductor.

Note that high-impedance grounded neutral system connections shall be made per **250.36**. High-impedance grounded systems should not be used unless they are equipped with ground-fault indicators or alarms, or both, and qualified persons are available to quickly locate and remove ground-faults. If ground-faults are not promptly removed, the service reliability will be reduced.

UNGROUNDED SYSTEM GROUNDING CONNECTIONS
250.24(E)

A premises wiring system that is supplied by an AC service supply that is ungrounded, shall have at the service, a grounding electrode conductor connected to the grounding electrode system. The grounding electrode conductor shall be connected to a metal enclosure of the service conductors at any accessible point from the load end of the service drop or service lateral to the service disconnecting means.

On an ungrounded system, none of the circuit conductors of the system are intentionally grounded. Note that the bonding together of all conductive enclosures and equipment in each circuit with equipment grounding conductors shall be done. These equipment grounding conductors shall be run with or enclose the circuit conductors, and they shall provide a permanent, low-impedance path for ground-fault currents to be detected. **(See Figure 11-29)**

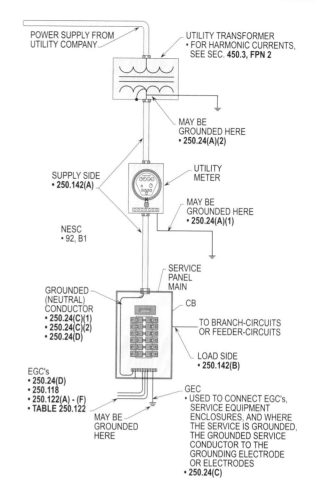

GROUNDING ELECTRODE CONDUCTOR NEC 250.24(D)

Figure 11-28. This illustration shows the procedure for using a grounding electrode conductor to connect equipment grounding conductors, the service equipment enclosures, and where the system is grounded, the grounded service conductor to the grounding electrode or electrodes.

CONDUCTOR TO BE GROUNDED - AC SYSTEMS
250.26

The grounded (neutral) conductor whether a grounded phase or neutral is usually installed with a white or gray insulation or otherwise identified. The following systems shall be grounded and have a grounded (phase) conductor or grounded (neutral) conductor: **(See Figure 11-30)**

- 120 volt, two-wire, single-phase - one neutral conductor
- 120/240 volt, three-wire, single-phase - the neutral conductor
- 120/208 volt, four-wire, three-phase wye - the common conductor
- 277/480 volt, four-wire, three-phase wye - the common conductor
- 240 volt, three-wire, three-phase delta - the common conductor

- 480 volt, three-wire, three-phase delta - the common conductor
- 600 volt, three-wire, three-phase delta - the common conductor
- 480 volt corner grounded delta - one phase conductor

MAIN BONDING JUMPER AND SYSTEM BONDING JUMPER 250.28

An unspliced main bonding jumper (for grounded systems) shall be installed to connect the equipment grounding conductors and the service disconnecting enclosure to the grounded (neutral) conductor of the system within the enclosure for each service disconnect. The following conditions of use shall be complied with when installing main bonding jumpers:

- Material
- Construction
- Attachment
- Size

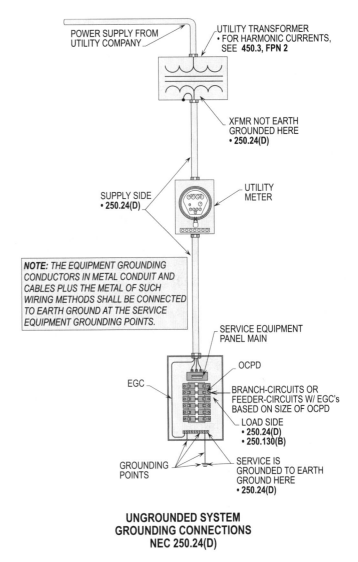

Figure 11-29. This illustration shows the procedures for installing an ungrounded service supply.

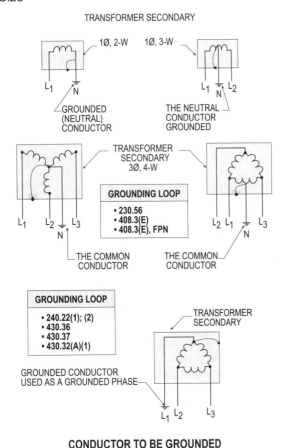

Figure 11-30. The grounded (neutral) conductor, whether a grounded phase or neutral is usually installed with a white or gray insulation or otherwise identified.

MAIN BONDING JUMPER - MATERIAL 250.28(A)

Unspliced main bonding jumpers shall be of copper or other corrosion-resistant material. Main bonding jumpers shall be installed as a wire, bus, screw or similar suitable conductor. **(See Figure 11-31)**

MAIN BONDING JUMPER - CONSTRUCTION 250.28(B)

When a screw is used as the main bonding jumper, it shall be identified with a green colored finish which is visible with the screw installed. **(See Figure 11-31)**

Grounding

MAIN BONDING JUMPER - ATTACHMENT
250.28(C)

Section **250.8** specifies the manner in which the main and equipment bonding jumpers shall be attached for electrical circuits and equipment. Section **250.70** specifies the manner by which grounding electrodes shall be attached together after sizing and selecting such grounding conductors. **(See Figure 11-31)**

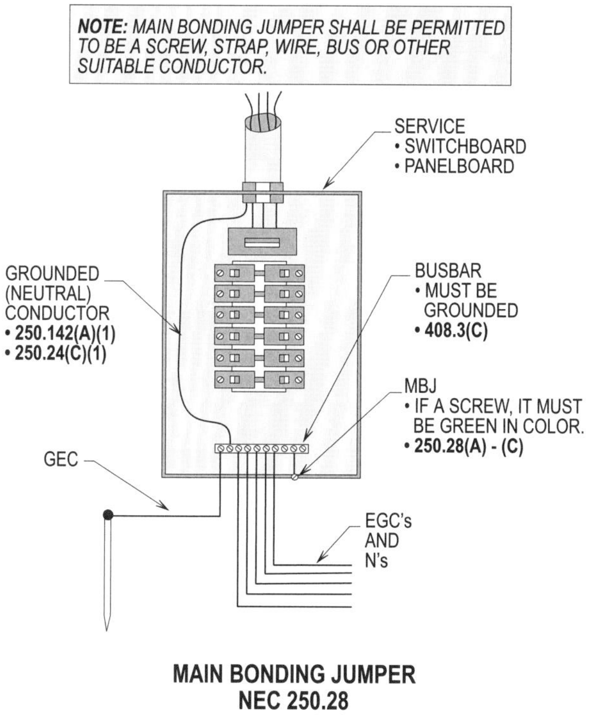

Figure 11-31. This illustration shows the items permitted to be used as a main bonding jumper.

For example: What size copper main bonding jumper is required based on the service-entrance conductors being rated at 250 KCMIL THWN-THHN copper?

Step 1: Finding the main bonding jumper
250.28(D); Table 250.66
250 KCMIL requires 2 AWG cu.

Solution: **The size of the main bonding jumper is required to be 2 AWG copper.**

For example: In a paralleled three-phase, four-wire service with 3 - 700 KCMIL THWN copper conductors per phase, what size main bonding jumper is required to ground the busbar to the service equipment enclosure?

Step 1: Finding total KCMIL
250.28(D)
700 KCMIL x 3 = 2100 KCMIL

Step 2: Finding KCMIL to size MBJ
250.28(D)
2100 KCMIL x .125 = 262.5 KCMIL

Step 3: Sizing MBJ
250.28(D); Table 8, Ch. 9
262.5 KCMIL requires 300 KCMIL

Solution: **The size of the main bonding jumper is required to be 300 KCMIL copper.**

MAIN BONDING JUMPER - SIZE
250.28(D)

The size of the service main bonding jumper, the system bonding jumper and the service equipment bonding jumper shall be based on the size of the service-entrance conductors and shall be sized per **Table 250.66**. These bonding jumpers shall be sized at least 12 1/2 percent of the area of the largest ungrounded (phase) conductor where the service conductors are installed larger than 1100 KCMIL copper or 1750 KCMIL aluminum.

GROUNDING SEPARATELY DERIVED AC SYSTEMS
250.30

Low-voltage and high-voltage feeder-circuits are sometimes installed from floor-to-floor in a high-rise building with transformers installed on each floor to reduce the voltage to 120/240 or 120/208 volts for general use lighting and receptacle loads in large building applications. Such grounding, since the 1978 NEC, may be installed either at the XFMR or at the load served which is connected and supplied from the secondary side per **240.21(C)(2), (C)(3)** and **240.92(B)(1) through (B)(3)**. **(See Figure 11-32)**

11-21

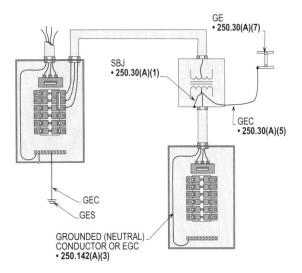

GROUNDING SEPARATELY DERIVED AC SYSTEMS NEC 250.30

Figure 11-32. The system bonding jumper and grounding electrode conductor is designed and installed based on the derived ungrounded (phase) conductors supplying the panel, switch or other equipment connected from the secondary side of the transformer.

GROUNDED SYSTEMS 250.30(A)

A separately derived system (SDS) that is grounded shall comply with the following:

- System bonding jumper
- Equipment bonding jumper size
- Grounding electrode conductor, single separately derived system
- Installation
- Bonding
- Grounding electrode
- Grounded conductor

SYSTEM BONDING JUMPER 250.30(A), Ex.

The grounding connection requirements for high-impedance grounded neutral systems shall be made per **250.36** and **250.186**.

SYSTEM BONDING JUMPER 250.30(A)(1)

The system bonding jumper (unspliced) shall be designed and installed based on the derived ungrounded (phase) conductors supplying the panel, switch or other equipment connected from the secondary side of the transformer and sized per **250.28(D)** and **250.102(C)**. The system bonding jumper shall be sized per **250.66** and **Table 250.66** from the ungrounded (phase) conductors up to 1100 KCMIL for copper and 1750 KCMIL for aluminum. The system bonding jumper shall be sized at least 12 1/2 percent (.125) of the area of the largest ungrounded (phase) conductor where the service conductors are installed larger than 1100 KCMIL copper or 1750 KCMIL aluminum. The system bonding jumper shall be installed and connected at any single point on the separately derived system from the source to the first system disconnecting means or overcurrent protection device. If the ungrounded (phase) conductors are larger than 1100 KCMIL for copper and 1750 KCMIL for aluminum, the system bonding jumper will normally be larger than the grounding electrode conductor.

For example: What size copper system bonding jumper is required to bond and ground the secondary of a separately derived system having 4 - 3/0 AWG THWN copper conductors connected to its secondary?

Step 1: Finding SBJ
250.30(A)(1); Table 250.66
3/0 AWG THWN cu. requires
4 AWG cu.

Solution: The size of the system bonding jumper is required to be 4 AWG copper.

For example: In a paralleled three-phase, four-wire service with 4 - 600 KCMIL THWN copper conductors per phase, what size system bonding jumper is required to bond and ground the separately derived system?

Step 1: Finding total KCMIL for SBJ
250.30(A)(1); 250.28(D)
600 KCMIL x 4 = 2400 KCMIL

Step 2: Finding KCMIL for SBJ
250.30(A)(1); 250.28(D)
2400 KCMIL x .125 = 300 KCMIL

Step 3: Sizing BJ
**250.30(A)(1); 250.28(D);
Table 8, Ch. 9**
300 KCMIL requires 300 KCMIL

Solution: The size of the system bonding jumper is required to be 300 KCMIL copper.

Grounding

DUAL FED (DOUBLE ENDED)
250.30(A)(1), Ex.1

A single system bonding jumper connection to the point of the grounded circuit conductors from each power source shall be permitted for separately derived systems that are dual fed (double ended) in a common enclosure and employing a secondary tie. **(See Figure 11-33)**

ADDITIONAL SYSTEM BONDING JUMPER
250.30(A)(1), Ex. 2

An additional system bonding jumper connection (more than one) shall be permitted to be made only where doing so will not create a parallel path for the grounded circuit conductor. The additional system bonding jumper shall not be permitted to be smaller than the other bonding jumper but shall not be required to be larger than the ungrounded (phase) conductor(s). **(See Figure 11-34)**

This section states that no parallel path may be formed when installing an additional system bonding jumper at equipment supplied by a separately derived system. The basic rule requires the system bonding jumper and the grounding electrode conductor connection to the grounded system conductor to be made at the same point. There is an Ex. to this rule, but only for cases, where there won't be a parallel path, such as a nonmetallic conduit (PVC) run back to the transformers source and so located to not share conductive contact with common structural elements or equipment grounding conductors. The purpose of this rule is to keep the grounded (neutral) conductor current confined and flowing over insulated electrical circuit conductors.

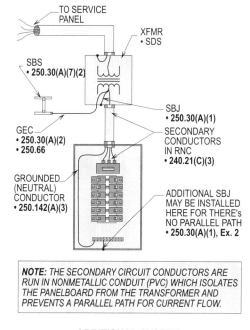

Figure 11-34. This illustration shows the grounding and bonding necessary when an additional bonding jumper is used.

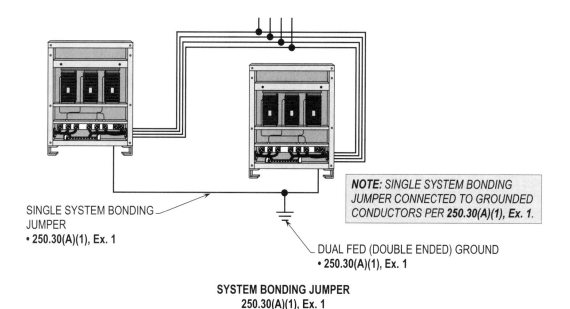

Figure 11-33. This illustration shows a single system bonding jumper connection for separately derived systems that are dual fed (double ended).

BONDING JUMPER - CLASS 1, CLASS 2 OR CLASS 3 CIRCUITS
250.30(A)(1), Ex. 3

The system bonding jumper shall be sized not smaller than the derived ungrounded (phase) conductors and shall not be smaller than 14 AWG copper or 12 AWG aluminum for systems which supplies a Class 1, Class 2 or Class 3 circuit, and is derived from a transformer which is rated not more than 1000 volt-amperes. **(See Figure 11-35)**

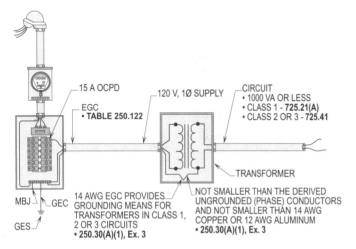

Figure 11-35. This illustration shows the bonding requirements for a small transformer using the equipment grounding conductor instead of installing a grounding electrode conductor and connecting it to a grounding electrode.

EQUIPMENT BONDING JUMPER SIZE
250.30(A)(2)

Where a bonding jumper (wire) is run with the derived phase conductors from the source of a separately derived system to the first disconnecting means, it shall be sized in accordance with **250.102(C)**, based on the size of the derived phase conductors.

GROUNDING ELECTRODE CONDUCTOR SINGLE SEPARATELY DERIVED SYSTEM
250.30(A)(3)

The grounding electrode conductor shall be designed and installed based on the derived ungrounded (phase) conductors supplying the panel, switch or other equipment connected from the secondary of the transformer and shall be sized per **Table 250.66**. The grounding electrode conductor shall be installed and connected at any point on the separately derived system from the source to the first system disconnecting means or overcurrent protection device. When the KCMIL rating is greater than 1100 for copper and 1750 for aluminum, the grounding electrode conductor will usually be smaller than the bonding jumper.

A common continuous grounding electrode conductor shall be permitted to be extended from the grounding electrode system and run through the building and the connection made at an accessible location near the separately derived system required to be grounded per **250.30(A)(4)**.

For example: What size copper grounding electrode conductor is required to bond and ground the secondary of a separately derived system having 4 - 3/0 AWG THWN copper conductors connected to its secondary?

Step 1: Finding size GEC
250.30(A)(3); Table 250.66
3/0 AWG THWN cu. requires
4 AWG cu.

Solution: The size of the grounding electrode conductor is required to be 4 AWG copper.

For example: In a paralleled three-phase, four-wire service with 4 - 600 KCMIL THWN copper conductors per phase, what size grounding electrode conductor is required to bond and ground the separately derived system?

Step 1: Finding total KCMIL for GEC
250.30(A)(3); 250.28(D)
600 KCMIL x 4 = 2400 KCMIL

Step 2: Sizing GEC
250.30(A)(3); Table 250.66
2400 KCMIL requires 3/0 AWG cu.

Solution: The size of the grounding electrode conductor is required to be 3/0 AWG copper.

SYSTEM BONDING JUMPER
250.30(A)(3), Ex. 1

The system bonding jumper (wire or busbar) specified in 250.30(A)(1) shall be permitted to connect the grounding electrode conductor to the equipment grounding terminal, bar or bus, provided the equipment grounding terminal, bar or bus is of sufficient size for the separately derived system. **(See Figure 11-36)**

GROUNDING ELECTRODE CONDUCTOR - LISTED EQUIPMENT
250.30(A)(3), Ex. 2

Where a separately derived system originates in listed equipment suitable as service equipment, the grounding electrode conductor from the service or feeder eqiupment to the grounding electrode conductor for the separtely derived system. Provided the grounding electrode conductor is of sufficients for the separately derived system, the grounding electrode connection for the separately derived system shall be permitted to be made to the bus where the equipment ground bus internal to the equipment is not smaller than the required grounding electrode conductor for the separately derived system. **(See Figure 11-37)**

GROUNDING ELECTRODE CONDUCTOR - CLASS 1, CLASS 2 OR CLASS 3 CIRCUITS
250.30(A)(3), Ex. 3

A Class 1, Class 2 or Class 3 remote-control or signaling transformer that is rated 1000 VA or less and has a grounded secondary conductor bonded to the metal case of the transformer, no grounding electrode conductor shall be required to be installed. However, the metal transformer case shall be properly grounded by a grounded metal raceway or equipment grounding conductor that supplies its primary or by means of an equipment grounding conductor that connects the case back to the grounding electrode for the primary system.

At least a 14 AWG copper conductor shall be used to bond and ground the transformer secondary to the transformer frame. This leaves the supply raceway, equipment grounding conductor or both to the transformer to provide the grounding return path back to the common service ground.

GROUNDED ELECTRODE CONDUCTOR MULTIPLE SEPARATELY DERIVED SYSTEM
250.30(A)(4)

A properly sized continuous common grounding electrode conductor shall be permitted to ground all separately derived systems on each floor of a high-rise building. The continuous common grounding electrode conductor shall be sized in accordance with **250.66** based on the total area of the derived ungrounded (phase) conductors smaller than 3/0 AWG copper or 250 KCMIL aluminum. **(See Figure 11-38)**

GROUNDING ELECTRODE CONDUCTOR SYSTEM BONDING JUMPER
250.30(A)(4), Ex. 1

The system bonding jumper (wire or busbar) specified in **250.30(A)(1)** shall be permitted to connect the grounding electrode conductor to the equipment grounding terminal, bar or bus, provided the equipment grounding terminal, bar or bus is of sufficient size for the separately derived system. **(See Figure 11-39)**

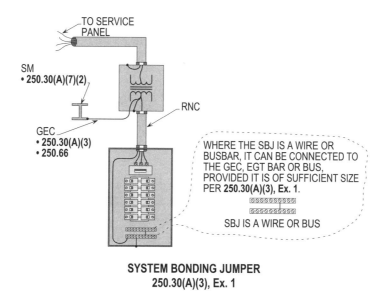

SYSTEM BONDING JUMPER
250.30(A)(3), Ex. 1

Figure 11-36. This illustration shows the requirements for grounding to the equipment grounding bus in the equipment listed as service equipment.

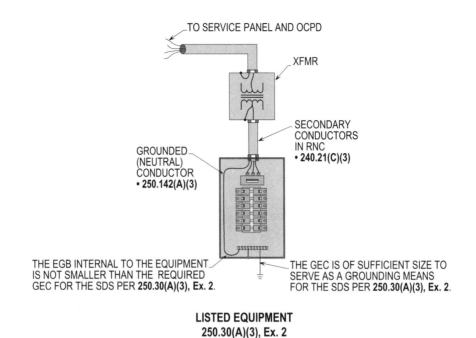

Figure 11-37. This illustration shows the requirements for grounding to the equipment grounding bus in the equipment listed as service equipment.

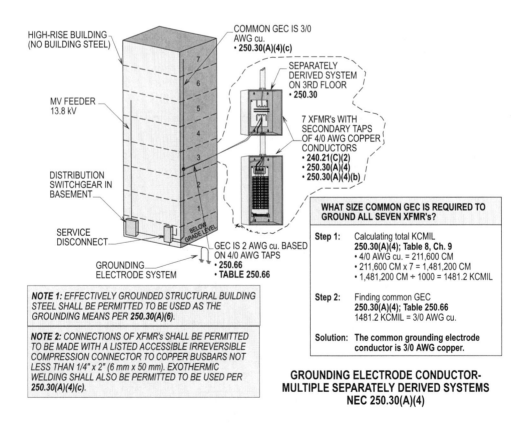

Figure 11-38. This illustration shows a properly sized continuous common grounding electrode conductor to serve as grounding means on each floor of a high-rise building.

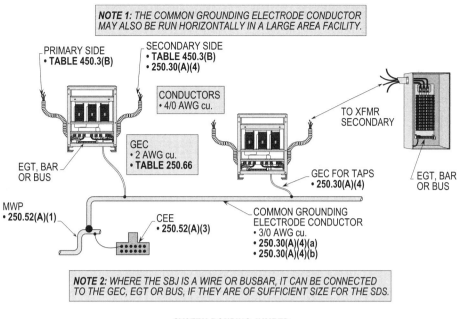

Figure 11-39. This illustration shows the requirements for connecting the system bonding jumper and grounding the separately derived system to the common GEC.

GROUNDING ELECTRODE CONDUCTOR - CLASS 1, CLASS 2 OR CLASS 3 CIRCUITS 250.30(A)(4), Ex. 2

A Class 1, Class 2 or Class 3 remote-control or signaling transformer that is rated 1000 VA or less and has a grounded secondary conductor bonded to the metal case of the transformer, no grounding electrode conductor shall be required to be installed. However, the metal transformer case shall be properly grounded by a grounded metal raceway or equipment grounding conductor that supplies its primary or by means of an equipment grounding conductor that connects the case back to the grounding electrode for the primary system.

At least a 14 AWG copper conductor shall be used to bond and ground the transformer secondary to the transformer frame. This leaves the supply raceway, equipment grounding conductor or both to the transformer to provide the grounding return path back to the common service ground. **(See Figure 11-40)**

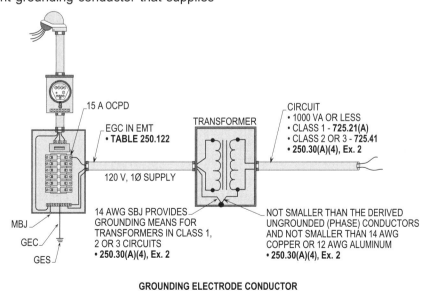

Figure 11-40. This illustration shows that no grounding electrode conductor is required to be installed.

COMMON GROUNDING ELECTRODE CONDUCTOR - SIZE
250.30(A)(4)(a)

The common grounding electrode conductor tap shall not be smaller than 3/0 AWG copper or 250 KCMIL minimum.

TAP CONDUCTOR SIZE
250.30(A)(4)(b)

Each tap conductor from the separately derived system to the continuous common grounding electrode conductor shall be sized per **250.66** based on the derived ungrounded (phase) conductors of the separately derived system that it supplies. **(See Figure 11-41)**

TAP CONDUCTOR SIZE – LISTED EQUIPMENT
250.30(A)(4)(b), Ex.

Where a separately derived system originates in listed equipment suitable as service equipment, the grounding electrode conductor from the service or feeder equipment to the grounding electrode conductor for the separtely derived system. Provided the grounding electrode conductor is of sufficients for the separately derived system, the grounding electrode connection for the separately derived system shall be permitted to be made to the bus where the equipment ground bus internal to the equipment is not smaller than the required grounding electrode conductor for the separately derived system. **(See Figure 11-42)**

CONNECTIONS
250.30(A)(4)(c)

Connections of the grounding electrode conductor from the separately derived system to the continuous common grounding electrode conductor shall be made at an accessible location by a connector listed for the purpose, listed connections to aluminum or copper busbars, not less than 1/4 in. x 2 in. (6 mm x 50 mm) or by the exothermic welding process.

INSTALLATION
250.30(A)(5)

The taps to each separately derived system and the continuous common grounding electrode conductor shall comply with **250.64(A), (B), (C)** and **(E)** which covers the installation requirements for grounding electrode conductors.

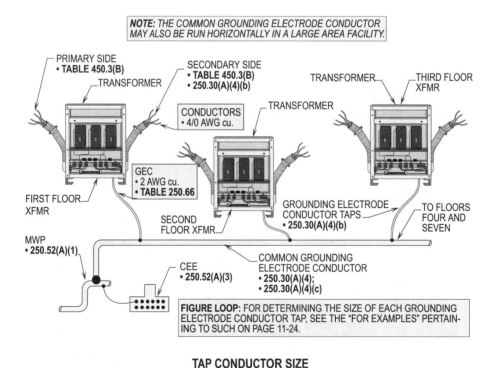

TAP CONDUCTOR SIZE
250.30(A)(4)(b)

Figure 11-41. This illustration shows that each tap conductor from the separately derived system to the continuous common grounding electrode conductor shall be sized per **250.66** based on the derived ungrounded (phase) conductors of the separately derived system that it supplies.

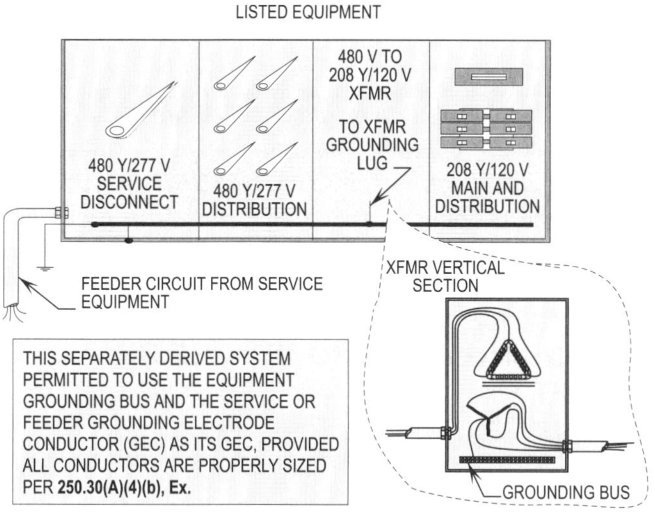

**GROUNDING ELECTRODE
USED IN SERVICE EQUIPMENT
NEC 250.30(A)(4)(b), Ex.**

Figure 11-42. The grounding electrode conductor shall be as near as possible and preferably in the same area as the grounding electrode conductor connection to the system.

BONDING
250.30(A)(6)

If the structural steel or interior metal piping is available and the continuous common grounding electrode conductor is also installed, these electrodes shall be bonded together as near as practicable to the location of the separately derived system.

GROUNDING ELECTRODE
250.30(A)(7)

The grounding electrode conductor shall be as near as possible and preferably in the same area as the grounding electrode conductor connection to the system. From the following choices, one shall be selected and installed in the order as they are listed:

- Nearest effectively grounded metal water pipe within 5 ft. (1.5 m) from the point of entrance into the building per **250.52(A)(1)**
- Nearest effectively grounded structural building steel per **250.52(A)(1)**

Note: Metal water pipe located in the area shall be bonded to the grounded conductor per **250.104(D)(1)**.

See Figure 11-43 for a detailed illustration pertaining to the installation of the grounding electrode conductor.

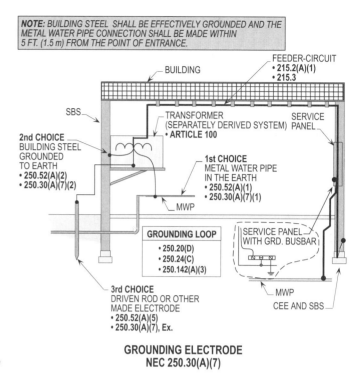

**GROUNDING ELECTRODE
NEC 250.30(A)(7)**

Figure 11-43. The grounding electrode conductor shall be as near as possible and preferably in the same area as the grounding electrode conductor connection to the system.

OTHER ELECTRODES
250.30(A)(7), Ex. 1

Any of the electrodes identified in **250.52(A)** can be used where the electrodes specified by **250.30(A)(7)** are not available.

The grounding electrode conductor shall not be required to be installed larger than 3/0 AWG for copper or 250 KCMIL for aluminum when connecting to the nearest building steel or nearest metal water pipe system. The grounding electrode conductor shall not be required to be installed larger than 6 AWG copper or 4 AWG aluminum when connecting to a driven rod or other made electrodes.

BUILDINGS OR STRUCTURES SUPPLIED BY FEEDER(S) OR BRANCH CIRCUIT(S)
250.32

These requirements apply to two or more buildings or structures supplied from one service. Basically, the grounded system of each building or structure shall have a grounding electrode that is also connected to the metal enclosure of the disconnecting means which is actually defined as a feeder-circuit. However, a grounding electrode shall be provided at each building or structure and connected to the disconnecting means as well as the grounded circuit conductor of the AC supply on the supply side of the building or structure disconnecting means. **(See Figure 11-44)**

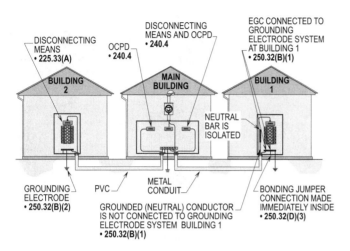

GROUNDING ELECTRODE
250.32(A)

Where two or more buildings or structures are supplied from one service by a feeder or branch-circuit, any of the grounding electrodes that are listed in **250.52(A)(1) through (A)(7)** shall be bonded together to form the grounding electrode system, if one or all are available. **(See Figure 11-45)**

GROUNDING ELECTRODE - ONE BRANCH-CIRCUIT
250.32(A), Ex.

A grounding electrode at a separate building or structure shall not be required to be installed where only one branch-circuit serves the building or structure and the branch-circuit includes an equipment grounding conductor for grounding noncurrent-carrying parts of all equipment.

For further information, see **250.32(D)** for applying the requirements when installing the disconnecting means per **225.32, Ex.'s 1 and 2**. **(See Figure 11-46)**

Figure 11-44. This illustration shows the use of the grounded (neutral) conductor and the equipment grounding conductor in a feeder-circuit supplying another building or structure.

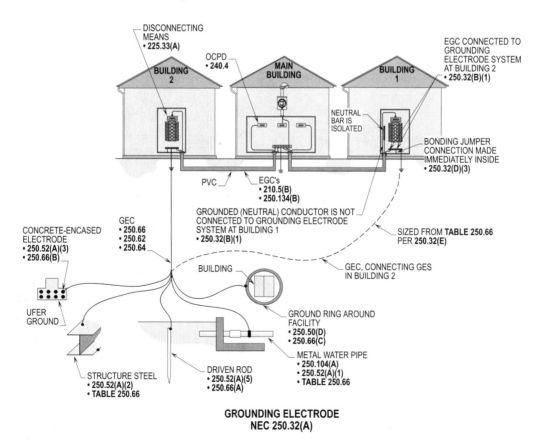

Figure 11-45. If any of the electrodes that are listed in 250.52(A)(1) through (A)(7) are available, they shall be bonded together to form the grounding electrode system.

Grounding

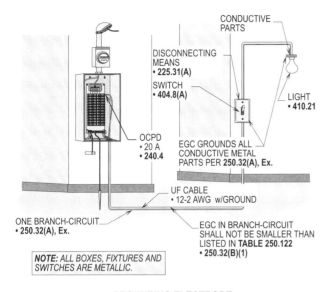

Figure 11-46. A grounding electrode shall not be required to be installed at a separate bulding or structure where only one branch-circuit is run and the branch-circuit includes an equipment grounding conductor for grounding noncurrent-carrying parts of all equipment.

TWO OR MORE BUILDINGS OR STRUCTURES - GROUNDED SYSTEMS
250.32(B)(1); (B)(2)

Where one or more buildings are supplied from a common AC grounded service, each panelboard at each building structure shall be separately grounded. The grounded (neutral) conductor or equipment grounding conductor run from the service panel of a building to a panel in a separate building or structure shall be at least the size specified in **Table 250.122**. The supply to each building or structure shall be disconnected by one to six disconnecting means per **225.33(A)** or the exception shall be permitted to applied under certain conditions. OCPD's shall comply with the rules of **Article 240**. See **240.4(A) through (G)** for such rules and regulations.

Note: The metal of a panelboard enclosure shall be grounded in the manner specified in **Article 250**, or **408.3(C)** and **408.40**. An approved terminal bar for equipment grounding conductors shall be provided and secured inside the enclosure for the attachment of all feeder and branch-circuit equipment grounding conductors when the panelboard is used with nonmetallic raceways, cable wiring or where separate grounding conductors are provided. The terminal bar for the equipment grounding conductors shall be bonded to the enclosure or panelboard frame (if it is metal), or connected to the grounding conductor that is run with the conductors which supplies the panelboard. Grounded (neutral) conductors shall not be permitted to be connected to a neutral bar, unless it is identified for such use. In addition, is shall be located at the connection between the grounded (neutral) conductor and the grounding electrode.

EQUIPMENT GROUNDING CONDUCTOR 250.32(B)(1)

Where a building or structure is supplied from a service in another building by more than one branch-circuit, a grounding electrode shall be installed at the additional building(s) or structure(s) being served. The equipment grounding conductor run to the other building or structure shall be sized per **Table 250.122**.

Where livestock is housed, the equipment grounding conductor shall be insulated or covered where routed with the feeder or branch-circuit and bonded to the metal case of the enclosure. Note that the neutral bus and grounded (neutral) conductor is isolated from the case. See **250.142(B)** and **408.40** for rules pertaining to this type of installation. **(See Figure 11-47)**

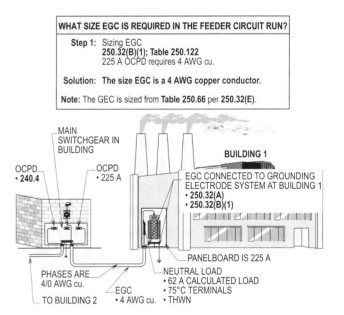

Figure 11-47. This type of installation is used to prevent the grounded (neutral) conductor and equipment grounding conductor from joining together at both ends, which if done, provides a parallel path for stray currents to travel over.

> **For example:** A 150 amp (75°C terminals) panelboard in building 1 is supplied by 3 - 1/0 AWG THWN copper conductors and the neutral calculated load is 48 amps. The feeder-circuit is protected by a 150 amp OCPD (75°C terminals). What size neutral and equipment grounding conductor is required using THWN copper conductors in the feeder-circuit?
>
> **Step 1:** Finding neutral in feeder
> **215.2(A)(1); Table 310.16**
> 48 A load requires 8 AWG cu.
>
> **Step 2:** Finding EGC in feeder
> **250.122(A); Table 250.122**
> 150 A OCPD requires 6 AWG cu.
>
> **Soution:** **The size of the neutral is 8 AWG copper and the equipment grounding conductor is 6 AWG copper.**

> **Design Tip:** An individual equipment grounding conductor shall be permitted to be installed to provide an effective return path for the fault current to travel over instead of using a grounded (neutral) conductor for such use. This particular installation is used to prevent the grounded (neutral) conductor and equipment grounding conductor from joining together at both ends, which if done, provides a parallel path for stray currents to travel over and this is not desirable for such an installation and causes problems for certain types of equipment.

> **For example:** A 225 amp (75°C terminals) panelboard in building 1 is supplied by 3 - 4/0 AWG THWN copper conductors and the neutral load is 50 amps. The feeder-circuit is protected by a 225 amp OCPD (75°C terminals). What size neutral is required to service as an equipment grounding conductor? (Loads are continuous)
>
> **Step 1:** Calculating neutal load
> **220.61(A); 225.3(B); 310.15(B)(4)(c)**
> 50 A x 125% = 62.5 A
>
> **Step 2:** Sizing neutral
> **250.32(B)(2); Table 310.16**
> 62.5 A requires 6 AWG cu.
>
> **Step 3:** Sizing neutral to be used as EGC
> **250.32(B)(2); Table 250.122**
> 225 A OCPD requires 4 AWG cu.
>
> **Soution:** **The grounded (neutral) conductor is required to be 4 AWG copper to service as an EGC.**

UNGROUNDED SYSTEMS
250.32(C)

A grounding electrode for an ungrounded system shall be connected only to the service equipment enclosure where installed at one or more buildings. The feeder or branch-circuit shall be grounded to an electrode at the other building being served from the service of the main building. All equipment grounding conductors at the separate building shall be connected to a grounding bus which is bonded to the enclosure. Such enclosure is bonded and grounded to the grounding electrode conductor and grounding electrode system.

GROUNDED CONDUCTOR
250.32(B)(2)

If a grounded (neutral) conductor, without an equipment grounding conductor, is routed in the circuit to a separate building, it shall be equal to the size required per **Table 250.122** to ensure the capacity to clear a ground-fault condition. See **250.142(A)(2)** for permission to use the grounded (neutral) conductor as a current-carrying conductor and equipment grounding conductor in a feeder-circuit. **(See Figure 11-48)**

PORTABLE AND VEHICLE-MOUNTED GENERATORS
250.34

Portable is defined as equipment that is easily carried from one location to another. Mobile equipment is capable of being moved, as on wheels or rollers, such as vehicle-mounted or placed on a trailer. Under certain conditions of use, the frame of a portable generator shall not be required to be connected to ground such as to a ground rod, water pipe, structural steel, etc.

Grounding

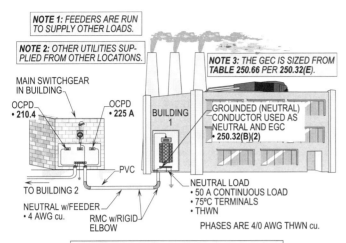

Figures 11-48. Where the grounded (neutral) conductor is used as an equipment grounding conductor plus a neutral, the raceway system that contains the ungrounded (phase) conductors shall be of the nonmetallic type.

PORTABLE GENERATORS
250.34(A)

The frame of a portable generator shall not be required to be connected to a grounding electrode as derived in **250.52**. if it supplies only the equipment on the generator or cord-and-plug-connected equipment connected to receptacles mounted on the generator, provided all the following conditions are complied with:

- An equipment grounding conductor is installed to bond the receptacles to the frame of the generator.
- The equipment grounding conductor in the cord is installed to bond the exposed noncurrent-carrying metal parts of the equipment to the frame of the generator.

See Figure 11-49 for a detailed illustration when applying these requirements.

VEHICLE-MOUNTED GENERATORS
250.34(B)

The frame of the vehicle shall not be required to be connected a grounding electrode as outlined in **250.52** for asystem supplied by a generator on the vehicle provided:

- The generator frame is bonded to the vehicle frame.
- The generator supplies only equipment mounted to the vehicle or if the generator supplies cord-and-plug connected equipment through receptacles mounted on the vehicle or both equipment located on the vehicle and cord-and-plug connected equipment through receptacles mounted on the vehicle or on the generator.
- Exposed metal parts of the equipment served are bonded to the generator frame either direct or through the receptacles.

See Figure 11-50 for a detailed illustration when applying these requirements

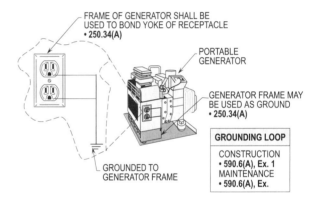

Figure 11-49. The frame of a portable generator shall not be required to be connected to a grounding electrode as defined in **250.52** if it supplies only the equipment on the generator or cord-and-plug connected equipment to receptacles mounted on the generator.

HIGH-IMPEDANCE GROUNDING
250.36

High-impedance grounding is accomplished by inserting a resistance or reactance between the grounding electrode conductor and the grounded (neutral) conductor. Ground-fault current is limited to a safe value which will not cause excessive damage to components and equipment should such a condition occur. Under ground-fault conditions, ground-fault current may be monitored using ground-fault relays and trip the system overcurrent protection devices open, only if necessary. Ungrounded systems generally have high-impedance grounding installed so the system has a reference to ground without deenergizing the circuits, if a ground-fault occurs on one of the ungrounded (phase) conductors.

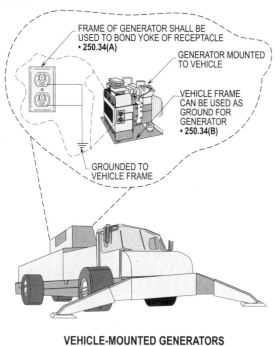

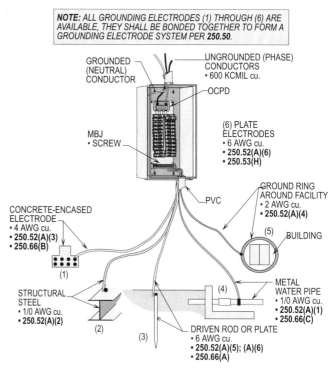

Figure 11-50. The frame of a generator shall not be required to be connected to a grounding electrode as specified in **250.52** if it supplies only the equipment supplied from the generator or cord-and-plug connected equipment connected to receptacles mounted and bonded to the frame of the vehicle.

Figure 11-51(a). This illustration shows the types of grounding electrodes that shall be bonded together to form the grounding electrode system if available.

GROUNDING ELECTRODE SYSTEM 250.50

All grounding electrodes that are present at each building or structure served, the following (first six) electrodes shall be bonded together to form the grounding electrode system:

- Metal underground water pipe
- Metal frame of the building or structure
- Concrete-encased electrode
- Ground ring
- Rod and pipe electrodes
- Plate electrodes

If one of the above electrodes are not available, one or more the following electrodes shall be installed and used:

- Ground ring
- Rod and pipe electrodes
- Plate electrodes
- Other local metal underground systems or structures

See Figures 11-51(a) and (b) for a detailed illustration when applying these requirements.

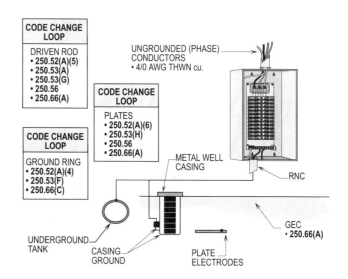

Figure 11-51(b). This illustration shows the types of grounding electrodes that shall be bonded together to form the grounding electrode system if available.

ELECTRODES PERMITTED FOR GROUNDING
250.52

The following types of electrodes shall be permitted for grounding:

- Metal underground water pipe
- Metal frame of the building or structure
- Concrete-encased electrode
- Ground ring
- Rod and pipe electrodes
- Plate electrodes
- Other local metal underground system or structures

METAL UNDERGROUND WATER PIPE
250.52(A)(1)

Metal water pipe with lengths of at least 10 ft. (3 m) long in the earth shall be connected to the grounded (neutral) bar in the service equipment enclosure. The grounded (neutral) bar shall be bonded to the service equipment enclosure, the grounded (neutral) conductor, the neutral conductors and the equipment grounding conductors by a bonding jumper sized per **Table 250.66**. Metal water piping shall be electrically continuous or made electrically continuous by bonding around insulated joints or sections.

The grounding connection to the metal water pipe shall be made at a point no more than 5 ft. (1.52 m) where the piping enters a building. It is recommended that the grounding conductor from the supplementary driven ground rod be terminated to the common grounded bar in the service equipment panel. **(See Figure 11-52)**

INDUSTRIAL AND COMMERCIAL BUILDINGS
250.52(A)(1), Ex.

The grounding electrode conductor shall be permitted to be installed further than 5 ft. (1.52 m) in industrial and commercial buildings where conditions of maintenance and supervision ensure that only qualified personnel will service the installation and the entire length of the interior metal water pipe that is being used for the conductor that is exposed. Under these conditions, such metal water pipes shall be permitted to be used as grounding electrode conductors to ensure continuity. **(See Figure 11-53)**

METAL FRAME OF THE BUILDING OR STRUCTURE
250.52(A)(2)

The metal frame of the building or structure shall be permitted to be used a grounding electrode where any of

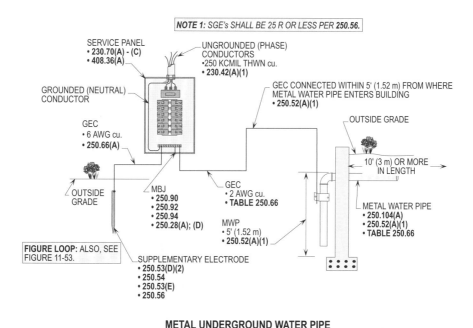

METAL UNDERGROUND WATER PIPE
NEC 250.52(A)(1)

Figure 11-52. In most all installation (existing), the grounding electrode conductor connects the service equipment grounded (neutral) conductor to a copper tubing or metal water piping system or a driven rod or all other electrodes, if available.

the following methods are used to a earth connection:

- A single structural metal member has 10 ft. (3 m) or more in direct contact with the earth or encased in concrete that is in direct contact with the earth.
- The structural metal frame is bonded to one or more of the grounding electrodes found in **250.52(A)(1), (A)(3)** or **(A)(4)**.
- The structural metal frame is bonded to one or more of the grounding electrodes in **250.52(A)(5)** or **(A)(6)** that comply with **250.56**.
- Other approved means of establishing a connection to earth.

(See Figure 11-54)

conductor installed properly usually provides a resistance of about 3 to 5 ohms resistance. **(See Figure 11-55)**

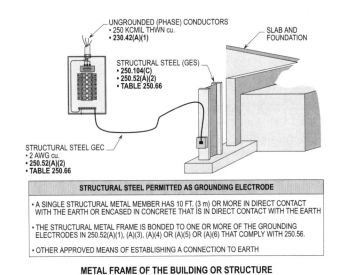

Figure 11-54. This illustration shows structural building steel used as a grounding electrode.

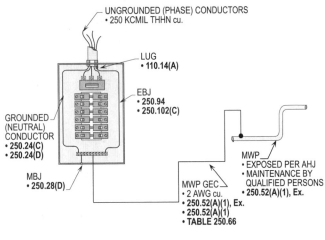

Figure 11-53. In industrial and commercial buildings, where the metal water pipe is exposed and maintained by qualified personnel, the grounding electrode conductor shall be permitted to be connected at any accessible location.

CONCRETE-ENCASED ELECTRODE 250.52(A)(3)

Concrete-encased electrodes consist of 1/2 in. x 20 ft. (13 mm x 6 m) lengths of reinforcing rebar located in the foundation. The 20 ft. (6 m) length of rebar shall be permitted to be in one continuous piece or many pieces spliced together to form a 20 ft. (6 m) or more continuous length. The metal reinforcing rebars shall be of the conductive type. A concrete-encased electrode shall also be permitted to be a 4 AWG bare copper conductor at least 20 ft. (6 m) long that is installed in the footing of the foundation. The 20 ft. (6 m) long 4 AWG conductor shall be located in at least 2 in. (50 mm) of concrete near the bottom of the foundation or footing. The reinforcing rebars or 4 AWG bare copper

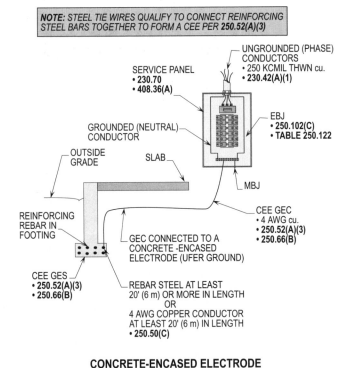

Figure 11-55. Concrete-encased electrodes consist of 1/2 in. x 20 ft. (13 mm x 6 m) lengths of reinforcing rebar located in the foundation. A concrete-encased electrode shall be permitted to be a 4 AWG bare copper conductor at least 20 ft. (6 m) long that shall be installed in the footing of the foundation.

GROUND RING
250.52(A)(4)

A bare copper conductor not smaller than 2 AWG and at least 20 ft. (6 m) shall be permitted to be installed in the earth to form a ground ring grounding system that encircles the building or structure. **(See Figure 11-56)**

> **Design Tip:** A more reliable and dependable ground may be obtained by burying the conductor about 3 ft. (900 mm) from the building and periodically driving ground rods and attaching them to the bare conductor by listed connecting methods.

system in the ground is nonmetallic (PVC) and converts to metal pipe or copper tubing above ground, a driven rod shall be permitted to be used to bond and ground the metal water pipe system and connect the electrical system to earth ground. **(See Figure 11-57)**

The metal water pipe system shall be bonded into the grounding electrode system, whether or not the piping in the ground is nonmetallic (PVC) or metal. The following are acceptable made electrodes which shall be permitted to be used:

- 1/2 in. x 8 ft. (13 mm x 2.5 m) copper rod
- 3/4 x 10 ft. (21 x 3 m) metal pipe or conduit
- 5/8 in. x 8 ft. (15.87 mm x 2.5 m) rebar
- 5/8 in. (16 mm) or larger stainless steel rod or 1/2 in. x 8 ft. (13 mm x 2.5 m) if listed for such use

PLATE ELECTRODES
250.52(A)(6)

Plate electrodes shall be a minimum of 2 sq. ft. (0.186 m sq. ft.), and if made of iron or steel they shall be at least 1/4 in. (6.4 mm) thick. If made of a nonferrous metal such as copper, they shall be 0.06 in. (1.5 mm) in thickness. **(See Figure 11-57)**

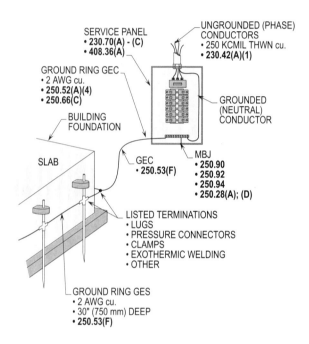

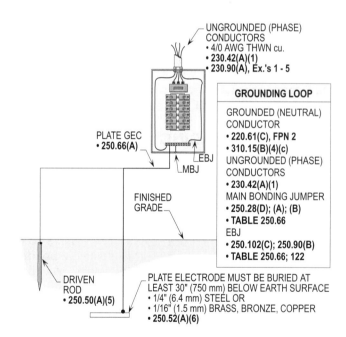

Figure 11-56. A bare copper conductor not smaller than 2 AWG and at least 20 ft. (6 m) shall be permitted to be installed in the earth to form a ground ring grounding system that encircles the building or structure.

ROD AND PIPE ELECTRODES
250.52(A)(5)

A driven electrode used to ground the service shall be permitted to be a 1/2 in. x 8 ft. (13 mm x 2.5 m) copper (brass) rod or a 3/4 x 10 ft. (21 x 3 m) galvanized pipe. A copper rod is usually the type of driven electrode that is installed. Driven rods are utilized to connect the grounded (neutral) bar terminal to earth ground. Where the water pipe

Figure 11-57. Rod and plate electrodes shall be installed when none of the electrodes found in **250.52(A)(1) through (A)(6)** are available.

OTHER LOCAL METAL UNDERGROUND SYSTEMS OR STRUCTURES
250.52(A)(7); FIGURE 11-51(b)

Other local metal underground systems and structures such as piping systems, underground tanks and underground metal well casings that are efficiently bonded to a metal water pipe.

GROUNDING ELECTRODE SYSTEM INSTALLATION
250.53

To complete the grounding electrode system, the following grounding electrodes have specific installation requirements applied:
- Rod, pipe and plate electrodes
- Electrode spacing
- Bonding jumper
- Metal underground water pipe
- Supplemental electrode bonding connection size
- Ground ring
- Rod and pipe electrodes
- Plate electrode

ROD AND PIPE ELECTRODES
250.53(A)

Rod and pipe electrodes shall be driven 8 ft. into the earth and plate electrodes shall be embedded below the permanent moisture level, where practicable. Rod, pipe and plate electrodes shall be free from nonconductive coatings such as paint and enamel.

ELECTRODE SPACING
250.53(B)

Where more than one rod, pipe or plate electrode is used, each electrode of one grounding system (including that used for air terminals) shall not be less than 6 ft. (1.83 m) from any other electrode of another grounding system. Two or more grounding electrodes that are effectively bonded together shall be considered a single grounding electrode system.

BONDING JUMPER
250.52(C)

Where bonding jumper(s) are used to connect the grounding electrodes together to form the grounding electrode system, the bonding jumper(s) shall be installed in accordance with **250.64(A), (B)** and **(E)**, they shall be sized in accordance with **250.66** and shall be connected in accordance with **250.70**.

METAL UNDERGROUND WATER PIPE
250.52(D)

The continuity of the grounding path for the bonding connection to the interior metal water piping shall not be permitted to rely on water meters, filtering devices or similar equipment. **(See Figure 11-58)**

> **Design Tip:** Many rural areas do not have water meters but may have water filtering or softening equipment that could be removed. Such removal would break or impair the grounding path or bonding connection of the metal interior water piping. Proper bonding and joining together of the metal piping can be accomplished by using correctly sized bonding jumpers.

The copper tubing or metal water pipe in a new installation may be installed in the earth or above grade. Either installation requires the copper or metal water pipe to be supplemented by an additional electrode. The additional grounding electrode to supplement the metal water pipe shall be permitted to be any of the electrodes listed in **250.52(A)(2) through (A)(6)**. **(See Figure 11-59)**

SUPPLEMENTAL ELECTRODE BONDING CONNECTION SIZE
250.52(E)

Where a rod, pipe or plate electrode is used as the supplemental electrode, that portion of the bonding jumper that is the sole connection to the supplemental grounding electrode shall not be required to be larger than 6 AWG copper wire or 4 AWG aluminum wire.

GROUND RING
250.52(F)

Where a ground ring is used as the grounding electrode, the ground ring shall be buried not less than 30 in. (750 mm) below the earth's surface.

ROD AND PIPE ELECTRODES
250.52(G)

Rod and pipe electrodes shall be installed where there is at least 8 ft. (2.44 m) of length in contact with the soil. Where a rock bottom is encountered, the electrode shall be permitted to be driven at an angle not to exceed 45 degrees from the vertical or shall be permitted to be buried at least 30 in. (750 mm) deep. **(See Figure 11-60)**

Grounding

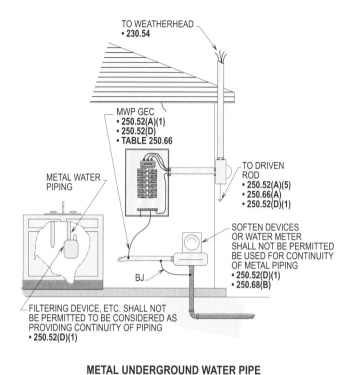

Figure 11-58. This illustration makes it clear that continuity of the grounding path for the bonding connection to the interior metal water piping shall not be permitted to rely on water meters, filtering devices or similar equipment.

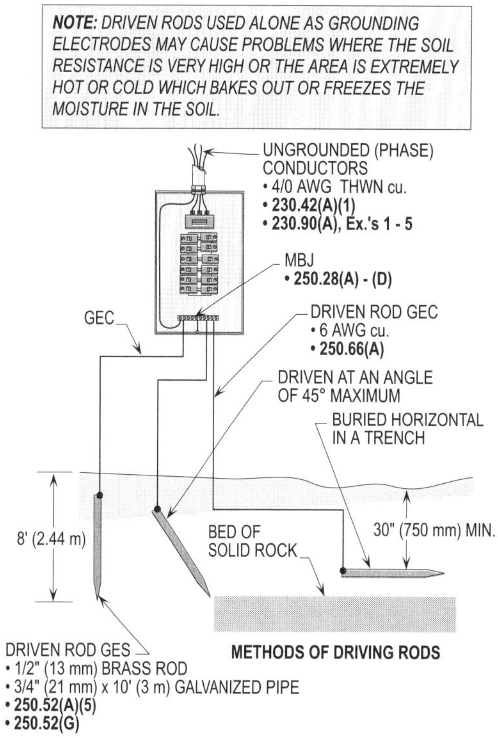

Figure 11-60. This illustration shows the methods in which rod and pipe electrodes are to be installed.

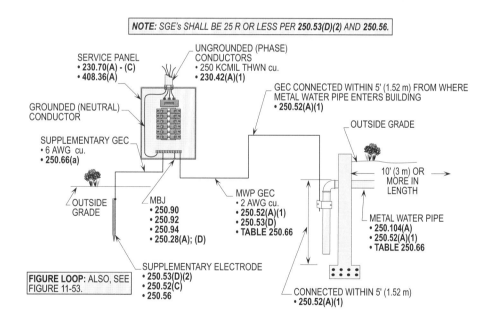

Figure 11-59. In most all installations (existing), the grounding electrode conductor connects the service equipment grounded (neutral) conductor to a copper tubing or a metal water pipe system which is supplemented by a driven rod or other electrodes (shall be permitted for this purpose), if available.

11-39

PLATE ELECTRODE
250.52(H)

Where a plate electrode is used as the grounding electrode, the plate electrode shall be buried not less than 30 in. (750 mm) below the earth's surface.

SUPPLEMENTARY GROUNDING
250.54

Supplementary grounding electrodes shall be permitted to be installed to augment the equipment grounding conductors specified in **250.118** and shall not be required to comply with the electrode bonding requirements of **250.50** or **250.53(C)** or the resistance requirements of **250.56**, but the earth shall not be permitted to be used as an effective ground-fault current path as specified in **250.4(A)(5)** and **250.4(B)(4)**. **(See Figure 11-61)**

Design Tip: This for instance, requires an equipment grounding conductor to be used, because the earth resistance in almost every case is too high to properly cause OCPD's to operate when a ground-fault occurs.

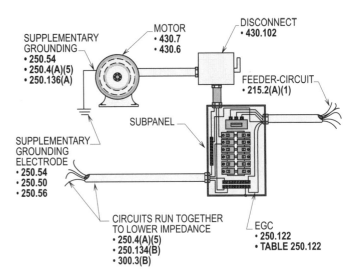

Figure 11-61. Supplementary grounding electrodes (SGE) shall be permitted to be used to supplement the equipment grounding conductor but are never to be used as the sole grounding means.

RESISTANCE OF ROD, PIPE AND PLATE ELECTRODES
250.56

Rod, pipe and plate electrodes shall have a resistance to ground of 25 ohms or less, wherever practicable. When the resistance is greater than 25 ohms, two or more electrodes shall be permitted to be connected in parallel or extended to a greater length. Note that such an electrode or supplementary grounding electrode that measures more than 25 ohms shall be augmented by one additional electrode of a type permitted by **250.52(A)(2) through (A)(7)**.

Continuous metal water piping systems usually have a ground resistance of less than 3 ohms. Metal frames of buildings normally have a good ground and usually have a resistance of less than 25 ohms. As pointed out in **250.52(A)(2)**, the metal frame of a building, if effectively grounded shall be permitted to be used as an electrode, per **250.104(C)**. Local metallic water systems and well casings make good grounding electrodes in most all types of installations. **(See Figure 11-62(a), (b) and (c))**

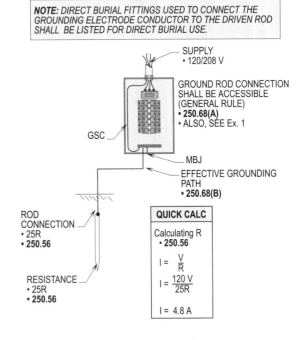

Figure 11-62(a). This illustration shows the procedure for calculating the amps to a driven rod based on 25 ohms.

Grounding

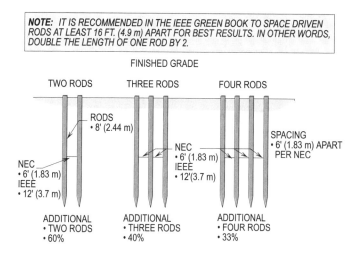

RESISTANCE OF ROD, PIPE AND PLATE ELECTRODES NEC 250.56

Figure 11-62(b). Rod, pipe and plate electrodes shall have a resistance to ground of 25 ohms or less, wherever practicable. When the resistance is greater than 25 ohms, two or more electrodes shall be permitted to be connected in parallel or extended to a greater length.

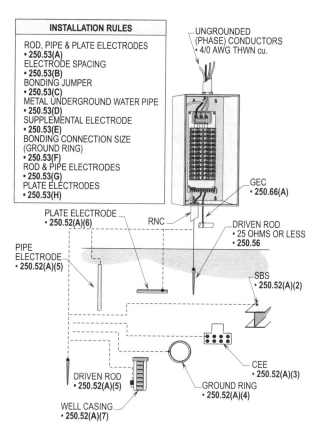

RESISTANCE OF ROD, PIPE AND PLATE ELECTRODES NEC 250.56

Figure 11-62(c). If a resistance of 25 ohms is not obtained, a rod, pipe or plate electrode or supplementary grounding electrode shall be augmented by one additional electrode per **250.52(A)(2) through (A)(7)**.

SOIL TREATMENT

When rod, pipe or plate electrodes are used, grounding may be greatly improved by the use of chemicals, such as magnesium sulfate, copper sulfate or rock salt. A trench, a basin or a doughnut-type hole system may be dug around the ground rod into which the chemicals are put. Rain and snow will dissolve the chemicals and allow them to penetrate the soil and saturate the rod, providing lower resistance. **(See Figure 11-63)**

USE OF AIR TERMINALS 250.60

Lightning down conductors and rod, pipe and plate electrodes, which are connected to lightning rods for grounding, shall not be permitted to be used in place of electrodes for grounding electrical wiring systems and equipment. However, they shall be permitted to be bonded together to limit the difference of potential that might appear between them. See **250.50** and **250.106**. **(See Figure 11-64)**

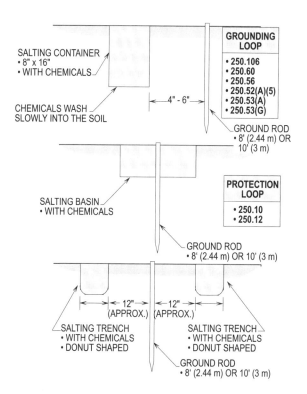

SOIL TREATMENT

Figure 11-63. When made electrodes are used, grounding may be greatly improved by the use of chemicals, such as magnesium sulfate, copper sulfate or rock salt.

11-41

Stallcup's Electrical Design Book

See **250.106, 800.100(D), 810.21(J)** and **820.100(D)** for required bonding and grounding of other interconnected systems. **(See Figure 11-65)**

Design Tip: Difference of potential between two different grounding systems on the same building, if tied together, will reduce the potential difference between them.

For example, lightning rods and the electrical systems grounding electrodes are bonded together for this reason.

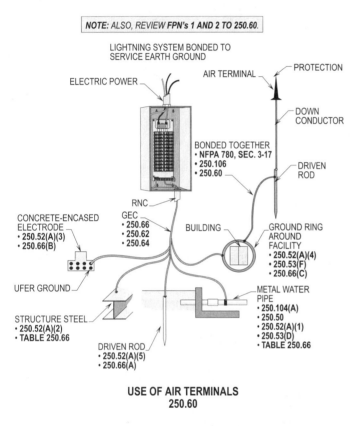

Figure 11-64. This illustration shows the requirements for installing and grounding lightning protection systems.

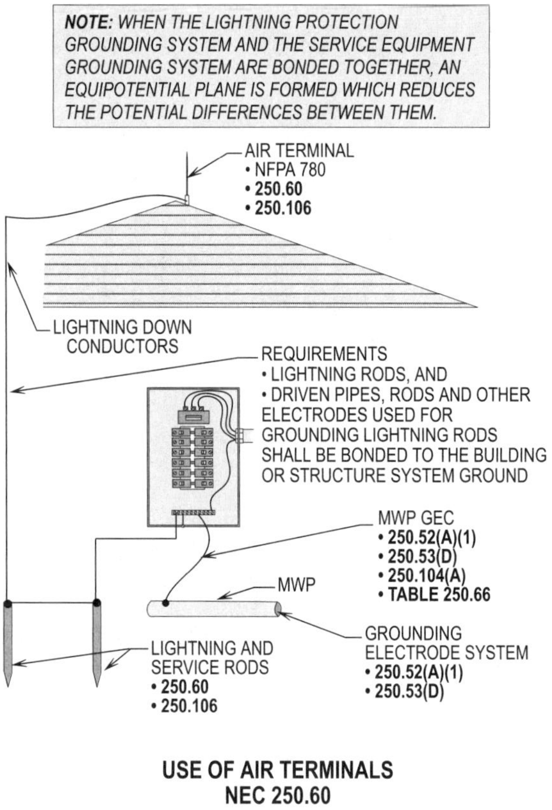

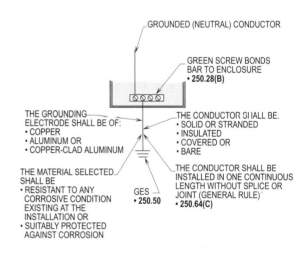

Figure 11-65. This illustration shows the requirements for installing and grounding lightning protection systems.

GROUNDING ELECTRODE CONDUCTOR MATERIAL
250.62

The grounding electrode conductor shall be of copper, aluminum or copper clad aluminum and shall be installed as solid or stranded, insulated, covered or bare. The grounding electrode conductor selected shall be resistant to any corrosive condition existing at the installation or be suitably protected against corrosion. **(See Figure 11-66)**

Figure 11-66. This illustration shows the materials that the grounding electrode may be made of and be used to ground the service equipment to a grounding electrode.

GROUNDING ELECTRODE CONDUCTOR INSTALLATION 250.64

The following wiring methods shall be permitted to be utilized for bonding, grounding and enclosing grounding electrode conductors:

- Aluminum or copper-clad aluminum conductors
- Securing and protection from physical damage
- Continuous
- Grounding electrode conductor taps
- Enclosures for grounding electrode conductors
- To electrode(s)

See Figure 11-67 for a detailed illustration when applying these requirements.

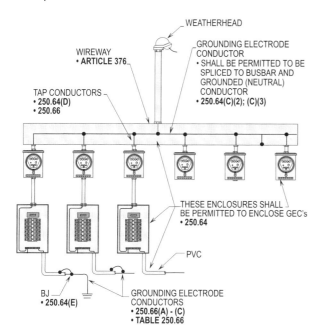

GROUNDING ELECTRODE CONDUCTOR INSTALLATION NEC 250.64

Figure 11-67. These enclosures shall be permitted to enclose grounding electrode conductors and also serve as a grounding and bonding means.

ALUMINUM OR COPPER-CLAD ALUMINUM CONDUCTORS 250.64(A)

Bare or aluminum or copper-clad aluminum conductors shall not be permitted to be used in the following locations:

- Where in direct contact with masonry or the earth
- Where subject to corrosive conditions

Aluminum or copper-clad aluminum grounding conductors shall not be permitted to be installed within 18 in. (450 mm) of the earth where such coductors are installed outside.

SECURING AND PROTECTION FROM PHYSICAL DAMAGE 250.64(B)

The following requirements shall be used to select and install the grounding electrode conductor that is used to connect the service enclosure to the grounding electrode system:

- Grounding electrode conductors 4 AWG or larger shall be securely fastened to the surface. Grounding electrode conductors not exposed to severe physical damage shall not require additional protection.

- A 6 AWG grounding electrode conductor shall be permitted to be run along the building surface construction without metal covering if not exposed to physical damage. Rigid metal conduit, intermediate metal conduit, rigid nonmetallic conduit, electrical metallic tubing or cable armor shall be permitted to be used where the grounding electrode conductor is exposed to physical damage.

- Rigid metal conduit, intermediate metal conduit, rigid nonmetallic conduit, electrical metallic tubing or cable armor shall be permitted to be used for installing a grounding electrode conductor smaller than 6 AWG. In most cases, a 8 AWG grounding electrode conductor is required to be installed per **Table 250.66** due to no OCPD ahead of service conductors.

See Figure 11-68 for a detailed illustration pertaining to these requirements.

CONTINUOUS 250.64(C)

The grounding electrode conductor shall be installed in one continuous length without a splice or joint, except as permitted in the following:

- Splicing shall be permitted only by irreversible compression-type connectors listed for the purpose or by the exothermic welding process. **(See Figure 11-69)**

- A grounding electrode conductor shall be permitted to be spliced in busbars if they are located as meter enclosures, etc. **(See Figure 11-70)**

- Bonding jumper(s) from grounding electrode(s) and

grounding electrode conductor(s) shall be permitted to be connected to an aluminum or copper busbar, not less than 1/4 in x 2 in. (6 mm x 50 mm). The busbar for such installation shall be securely fastened and shall be installed in an accessible location. These connections shall be made by a listed connector or by the exothermal welding process.

- Where aluminum busbars are used, the installation shall comply with **250.64(A)**.

GROUNDING ELECTRODE CONDUCTOR TAPS
250.64(D)

A grounding electrode conductor shall be permitted to be connect to taps in a service consisting of more than a single enclosure. The common grounding electrode conductor shall be sized per **250.66**, based on the sum of the circular mil area of the largest ungrounded (phase) conductor. The common grounding electrode conductor shall be sized in accordance with **Table 250.66**. Note that were more than one set of service entrance conductors are permitted per **230.40, Ex. 2**, connect directly to a service drop or lateral. The tap conductors shall be connected to the common grounding electrode conductor in such a manner that the grounding electrode conductor remains without a splice or joint. The tapped conductor shall be designed and connected to a common grounding electrode conductor per **Table 250.66** based on the largest ungrounded (phase) conductors. The tapped grounding conductor is then connected to an enclosure. **(See Figure 11-70)**

ENCLOSURES FOR GROUNDING ELECTRODE CONDUCTORS
250.64(E)

If a ferrous metal enclosure is used over the grounding electrode, it shall be electrically continuous from its origin to the grounding electrode. It shall be securely fastened to the grounding clamp so that there is electrical continuity from where it originates to the grounding electrode. If a ferrous metal enclosure is used for physical protection for the grounding electrode conductor and it is not continuous from the enclosure to the grounding electrode, it shall be bonded at both ends. The bonding jumper shall be the same size as, or larger than the required enclosed grounding electrode conductor. **(See Figure 11-71)**

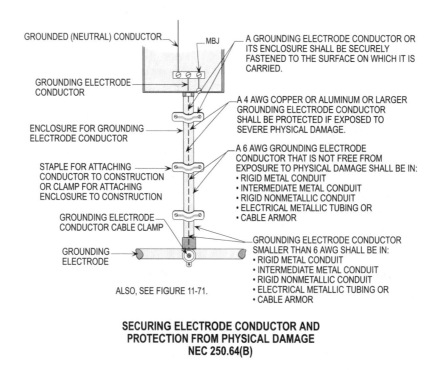

SECURING ELECTRODE CONDUCTOR AND PROTECTION FROM PHYSICAL DAMAGE NEC 250.64(B)

Figure 11-68. This illustration shows the requirements for installing and protecting the grounding electrode conductor with specific wiring methods.

Grounding

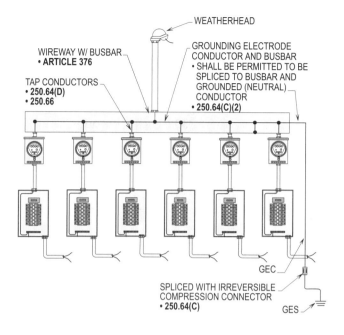

Figure 11-69. A grounding electrode conductor shall be permitted to be spliced in busbars located in meters, gutter, enclosures, wireways or with irreversible compression connectors.

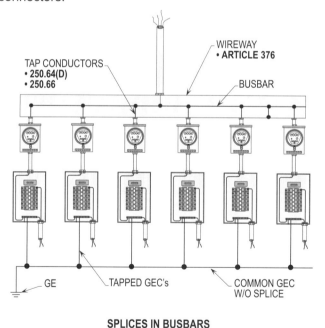

Figure 11-70. A grounding electrode conductor shall be permitted to be connected to taps in a service consisting of more than a single enclosure. The tap conductors shall be connected to the grounding electrode conductor in such a manner that the grounding electrode conductor remains without a splice or joint.

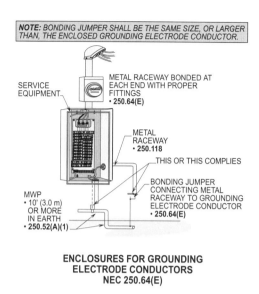

Figure 11-71. This illustration shows requirements for bonding a raceway containing the grounding electrode conductor.

TO ELECTRODE(S)
250.64(F)

A grounding electrode conductors shall be permitted to be run to any convenient grounding electrode that is available in the grounding electrode system or to one or more grounding electrode(s) individually or to the aluminum or copper busbar as permitted in **250.64(C)**. The grounding electrode conductor shall be sized for the largest grounding electrode conductor required among all the electrodes connected to it.

SIZE OF ALTERNATING-CURRENT GROUNDING ELECTRODE CONDUCTOR
TABLE 250.66

The following grounding electrodes shall be grounded by conductors sized and selected as listed in **Table 250.66** and **250.66(A) through (C)**.
• Connections to metal water pipe
 • Connections to structural steel
 • Connections to rod, pipe or plate electrodes
 • Connections to concrete-encased electrodes
 • Connections to ground rings

CONNECTIONS TO METAL WATER PIPE
250.66; TABLE 250.66; 250.104(A); 250.52(A)(1); 250.53(D)

The procedure for selecting the grounding electrode conductor to ground the service to a metal water pipe shall be determined by the size of the service-entrance conductors. **(See Figure 11-48)**

> **For example:** What size copper GEC is required to ground a service to a metal water pipe supplied by 250 KCMIL copper conductors?
>
> **Step 1:** Finding size GEC
> **Table 250.66**
> 250 KCMIL cu. = 2 AWG cu.
>
> **Solution: The size grounding electrode conductor is required to be 2 AWG copper.**

CONNECTIONS TO STRUCTURAL BUILDING METAL
250.66; TABLE 250.66; 250.104(C); 250.52(A)(2)

The procedure for selecting the grounding electrode conductor to ground the service to structual metal shall be determined by the size of the service-entrance conductors. **(See Figure 11-50)**

> **For example:** What size copper grounding electrode conductor is required to ground a service to structural metal supplied by 250 KCMIL copper conductors?
>
> **Step 1:** Finding the GEC
> **Table 250.66**
> 250 KCMIL cu. requires 2 AWG cu.
>
> **Solution: The size grounding electrode conductor is required to be 2 AWG copper.**

CONNECTIONS TO ROD, PIPE OR PLATE ELECTRODES
250.66(A)

The service equipment shall be permitted to be grounded with a rod, pipe or plate electrode where there are no other electrodes available in **250.52(A)(1) through (A)(4) and (A)(7)**. A driven rod or supplementary grounding electrode with a resistance of 25 ohms or less is considered low enough to allow the grounded system to operate safely and function properly. If the driven rod is used as a supplementary grounding electrode to the metal water pipe system, it should be connected to the grounded terminal bar in the service equipment panelboard. For further information, see **250.52(A)(5) through (A)(7)** and **250.53(A), (G)** and **(H)** for the application of this rule when using one of the electrodes listed there. The grounding electrode conductor shall not be required to be larger than 6 AWG copper or 4 AWG aluminum, where connected to electrodes such as driven rods. **(See Figure 11-56)**

> **For example:** What is the current flow in a 6 AWG copper grounding electrode conductor connecting the common grounded terminal bar in the service equipment to a driven rod? (The supply voltage is 120/208 volt, three-phase system)
>
> **Step 1:** Finding amperage
> **250.56**
> I = 120 V ÷ 25 R
> I = 4.8 A
>
> **Solution: The normal current flow is about 4.8 amps.**

CONNECTIONS TO CONCRETE-ENCASED ELECTRODES
250.66(B)

Lengths of rebar 1/2 in. (13 mm) in diameter and at least 20 ft. (6 m) long in length shall be permitted to be used as a grounding electrode to ground the service equipment. The size of the grounding electrode conductor shall be a 4 AWG or larger copper conductor per **250.66(B)** and **250.52(A)(3)**. The rebar system shall be permitted to be one length that is 1/2 in. x 20 ft. (13 mm x 6 m) long or a number of such that are spliced together to form a 20 ft. (6 m) length. This grounding method is known in the electrical industry as the UFER ground.

A concrete-encased electrode shall also be permitted to be a 4 AWG copper conductor instead of 1/2 in. x 20 ft. (13 mm x 6 m) rebar. The 4 AWG copper conductor shall be located within 2 in. (50 mm) of concrete located within or near the bottom of the foundation.

The size of the grounding electrode shall be at least a 4 AWG copper per **250.66(B)**. The 4 AWG copper grounding electrode conductor is selected based on the 4 AWG copper grounding electrode installed in the foundation. **(See Figure 11-51)**

CONNECTIONS TO GROUND RINGS
250.66(C)

A bare copper conductor not smaller than 2 AWG and at least 20 ft. (6 m) long shall be installed to create a ground ring that will encircle a building or structure. This ring shall be permitted to be utilized as the main grounding electrode if necessary or just bonded in as one per **250.50, 250.52(A)(4)** and **250.53(C)**. **(See Figure 11-52)**

GROUNDING ELECTRODE CONDUCTOR AND BONDING JUMPER CONNECTION TO GROUNDING ELECTRODES
250.68

The following methods shall be adhered to when connecting the grounding electrode conductor to grounding electrodes:

- Accessibility
- Effective grounding path

ACCESSIBILITY
250.68(A)

The 1987 NEC made it very clear that installers and inspectors shall work together to see that grounding clamps and fittings used to connect the grounding electrode conductor or bonding jumpers to water pipes or building steel are accessible. All installations built before the adoption of the 1978 NEC did not necessarily require the grounding clamps or fittings to be accessible. This revision in the 1978 NEC came about due to the grounding clamps or fittings being removed or damaged during construction and was never fixed or replaced. A missing ground clamp could cause the grounded (neutral) conductor to float and proper voltage regulation would not be accomplished. **(See Figure 11-72)**

ACCESSIBILITY NOT REQUIRED
250.68(A), Ex.

Grounding clamps, fittings or other approved methods that connect the grounding electrode conductor to concrete-encased electrodes, ground rings, plates or driven rods, are not required to be accessible. However, they shall be approved (listed) for direct burial.

EFFECTIVE GROUNDING PATH
250.68(B)

The connection of the grounding electrodes or bonding jumpers to the grounding electrode shall be in a manner that ensures a permanent and effective grounding path. When metal piping systems are used as a grounding electrode, effective bonding conductors of sufficient length shall be installed around insulated joints and sections and around any equipment that is likely to be disconnected for repairs or replacement. **(See Figure 11-73)**

METHODS OF GROUNDING AND BONDING CONDUCTOR CONNECTIONS TO ELECTRODES
250.70

When connecting grounding or bonding conductors to grounding fittings by suitable lugs, use pressure connectors, clamps or other listed means including exothermic welding. Soldering shall never be used. The ground clamps shall be of a material that is compatible for both the grounding electrode and the grounding electrode conductor. No more than one conductor shall be connected to an electrode unless the connector is listed for the purpose per **110.14(A)**. Grounded fittings used to connect the grounding conductor shall be pemitted to be any of the following:

- Exothermic welding
- Listed lugs
- Listed pressure connectors
- Listed clamps

Direct burial grounding clamps shall be utilized on grounding electrodes in the earth such as pipes, rods, rebar, encased electrodes, etc. **(See Figure 11-74)**

BONDING
250.90

Bonding is the permanent joining of metallic parts to form

Stallcup's Electrical Design Book

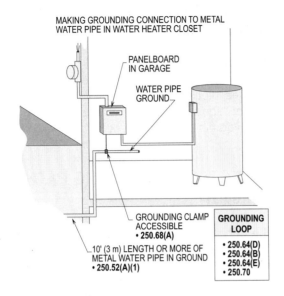

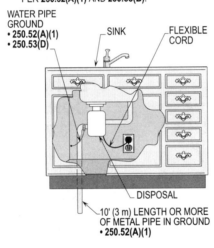

Figure 11-72. The 1987 NEC made it very clear that installers and inspectors shall work together to see that grounding clamps and fittings used to connect the grounding electrode conductor to metal water pipes or building steel are accessible. All installations built before the adoption of the 1978 NEC did not necessarily require the grounding clamps or fittings to be accessible.

an electrically conductive path that will assure electrical continuity and the capacity to conduct safely any current likely to be imposed. Such bonding jumpers shall be sized to handle these larger available fault currents when necessary to prevent damaging components and equipment.

SERVICES
250.92

The following bonding requirements shall be considered for services:
- Bonding of services
- Method of bonding at the service

Grounding

BONDING OF SERVICES
250.92(A)

The following noncurrent-carrying metal parts of service equipment shall be bonded together:

- Service raceways, cable trays, cablebus framework, auxiliary gutters or service cable armor or sheath except as permitted per **250.84**.

- All service enclosures containing service conductors, including meter fittings, meter base enclosure or CT cans, interposed in the service raceway or armor.

- Any metallic raceway or armor enclosing a grounding electrode conductor per **250.64(B)**. Bonding shall apply at each end and to all intervening raceways, boxes and enclosures between the service equipment and the grounding electrode.

Service-entrance raceways, meter base enclosures, service equipment enclosures and all other metal enclosures of the service equipment shall be bonded together to ensure a continuous metal-to-metal grounding system. **(See Figure 11-75)**

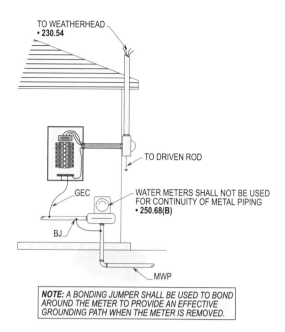

Figure 11-73. This illustration shows water meters and other devices shall not be permitted to be used for continuity of the metal water piping.

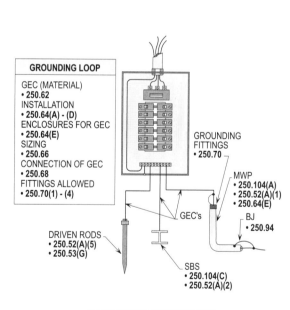

Figure 11-74. When connecting grounding conductors to grounding fittings by suitable lugs, use pressure connectors, clamps or other listed means including exothermic welding.

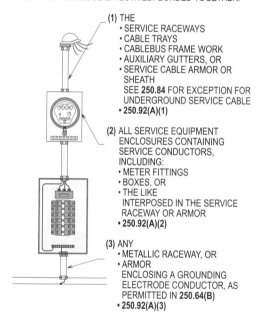

Figure 11-75. All metal raceways, metal-clad of cables, meter bases and metal equipment enclosures shall be bonded together for safety.

METHOD OF BONDING AT THE SERVICE 250.92(B)

Any one of the following five methods can be utilized for the bonding of service equipment:

- Grounded service conductor
- Threaded connections
- Threadless couplings and connectors
- Other listed devices such as bonding type locknuts, bushings and bushings with bonding jumpers
- Bonding jumpers

See Figure 11-76 for a detailed illustration when applying these requirements.

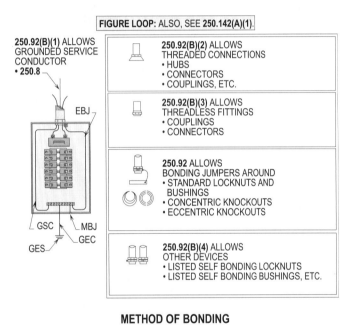

Figure 11-76. This illustration show the proper methods for bonding raceways at the service equipment

METHOD OF BONDING AT THE SERVICE 250.92(B)(1) THRU (B)(4)

A bonding conductor shall be used to connect and bond the raceway to the enclosure where smaller concentric or eccentric knockouts are removed, leaving larger ones. A metal bushing with a lug on the threaded raceway is used to terminate the bonding jumper to a lug connected to the enclosure or grounded busbar terminal. Any one of the following methods shall be permitted to be utilized for the bonding of service equipment:

- Grounded service conductor
- Threaded connections
- Threadless couplings and connectors
- Other listed devices

GROUNDED SERVICE CONDUCTOR 250.92(B)(1)

The grounded service conductor shall be permitted for bonding equipment by one of the following methods per **250.8**:

- Exothermic welding
- Listed pressure connectors
- Listed clamps
- Other listed means

See Figure 11-77 for a detailed illustration when applying these requirements.

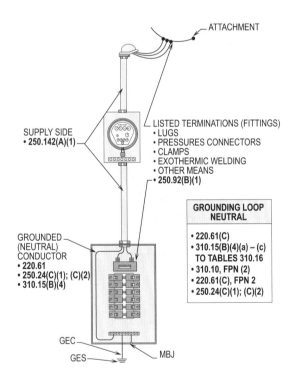

Figure 11-77. This illustration shows the fittings that shall be permitted to be used to bond and connect the grounded (neutral) conductor to the metal enclosure housing the components of the service.

Grounding

THREADED CONNECTIONS
250.92(B)(2)

Threaded connections, such as threaded couplings or threaded bosses, shall be permitted to be installed where the service equipment is threaded into the meter base or the service equipment enclosure. Such threaded connections shall be made up wrench-tight. **(See Figure 11-78)**

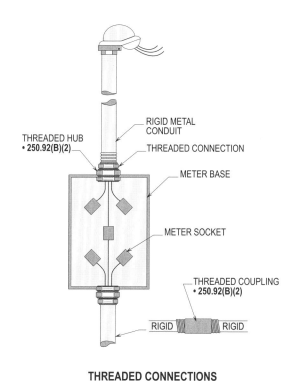

THREADED CONNECTIONS
NEC 250.92(B)(2)

Figure 11-78. This illustration shows a threaded hub used as the bonding means for grounding the rigid metal conduit to the metal of the meter base enclosure.

THREADLESS COUPLINGS AND CONNECTORS
250.92(B)(3)

Threadless couplings and connectors installed for rigid metal conduit, intermediate metal conduit and electrical metallic tubing shall be made up tight. Standard locknuts and bushings shall not be permitted to be installed for bonding the raceway to the enclosure. Only locknuts which are approved for bonding such as extra thick locknuts and bushings equipped with a set screw shall be permitted for this purpose. **(See Figure 11-79)**

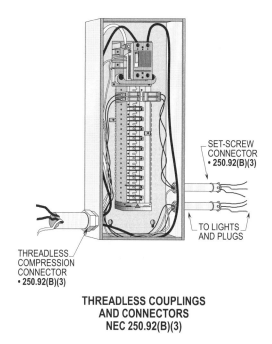

THREADLESS COUPLINGS AND CONNECTORS
NEC 250.92(B)(3)

Figure 11-79. This illustration shows threadless connectors used as a bonding means for grounding the electrical metallic tubing to the metal panelboard.

OTHER LISTED DEVICES
250.92(B)(4)

A wedge fitting (bushing) or locknut fitting is an application of other approved devices that shall be permitted to be used. A grounding wedge is equipped with a set screw that tightens the metal conduit and ensures proper bonding of the service equipment enclosure.

Design Tip: Two locknuts (one is self-bonding) with a bushing are an acceptable bonding means. However, two standard locknuts shall not be permitted to be used as a bonding means per **250.92(B)(3)**. Note that bonding jumpers shall be sized by the provisions listed in **250.102(C)** and **Table 250.66**. **(See Figure 11-80)**

BONDING FOR OTHER SYSTEMS
250.94

One of the following accessible means external to enclosures shall be provided at the service equipment and at the disconnecting means for any individual buildings or structures for connecting intersystem bonding and grounding electrode conductors:

- Exposed nonflexible metallic service raceways
- Exposed grounding electrode conductor
- Approved means for the external connection of a copper or other corrosion-resistant bonding and grounding conductor to the service raceway or equipment

See Figure 11-81 for a detailed illustration when applying these requirements.

Figure 11-80. This illustration shows self-bonding devices such as locknuts and bushings to bond and ground rigid metal conduit to the metal enclosure enclosing electrical components of equipment.

BONDING OTHER ENCLOSURES GENERAL 250.96(A)

Metal raceways, cable trays, cable armor, etc. shall be bonded to metal enclosures, boxes, cabinets, etc. Metal raceways shall be installed as a complete run from a metal box to metal box per **300.10**. Such wiring methods shall be terminated with listed connectors which comply with **300.12, 300.13** and **300.15** in the NEC. Note that when metal raceways, armor cables, cable sheaths, enclosures, frames and other metal noncurrent-carrying parts which serve as the equipment grounding conductors, whether using a supplementary grounding conductor or not, proper bonding shall be required to ensure the safe handling of such fault currents that may be imposed and to ensure electrical continuity. Nonconductive paint or similar coatings shall be removed at the threads or other contact points to ensure clean surfaces. Fittings or bonding jumpers are used for such a purpose. **(See Figure 11-82)**

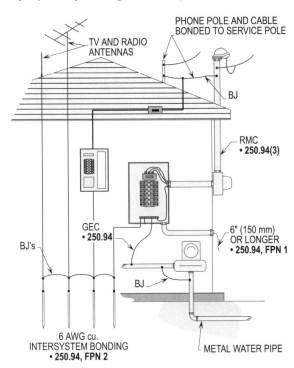

Figure 11-81. This illustration shows the procedures for bonding intersystems together such as communications lines, CATV lines, etc.

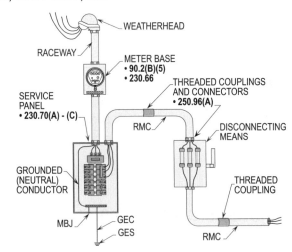

Figure 1-82. This illustration shows the metal weatherhead and raceway to the meter base, raceway between the panelboard, raceway to the disconnecting means and the raceway from the disconnect supplying the load is properly connected together to form a continuous unbroken grounding path.

ISOLATED GROUNDING CIRCUITS 250.96(B)

A nonmetallic spacer or fitting shall be permitted to be installed between the metal raceway supplying sensitive electronic equipment at the point of connection. The equipment grounding conductor shall be isolated and the metal raceway grounded. Electromagnetic interference that creates unwanted noise may be reduced in sensitive electronic equipment when this type of installation is applied. **(See Figure 11-83)**

BONDING FOR OVER 250 VOLTS 250.97

When on the load side, one of the following wiring methods shall be used for bonding circuits over 250 volts-to-ground, not including services:

- Install the following wiring methods between sections of metal conduit, intermediate metal conduit, electrical metallic tubing or other type of metal raceways.
    (a) Threaded couplings
    (b) Threadless couplings
- Install the following wiring methods between metallic raceways, metallic boxes, cabinet enclosures, etc.
    (a) Jumpers
    (b) Two locknuts and bushing
    (c) Threaded (bosses) hubs
    (d) Other approved devices

Metal raceways and conduits enclosing conductors at more than 250 volts-to-ground (other than service conductors) shall have electrical continuity assured by one of the methods outlined in **250.92(B)(2) through (B)(4)**. Note that the grounded (neutral) conductor shall not be permitted to be used (shall be permitted for the supply side of services) except as permitted for separate buildings and separately derived systems per **250.24(A)(5), FPN, 250.92(B)(1)** and **250.142(A)(1), (A)(2) and (A)(3). (See Figure 11-84)**

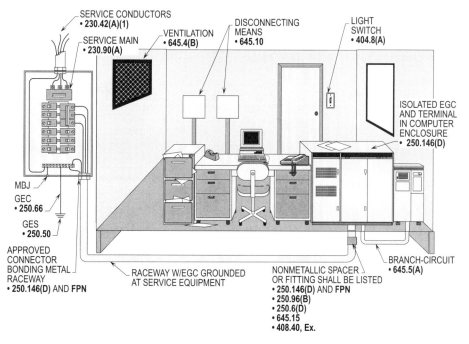

**ISOLATED GROUNDING CIRCUITS
NEC 250.96(B)**

See Figure 11-83. Reduction of electromagnetic interference that creates unwanted noise in sensitive electronic equipment may be reduced by installing a nonmetallic spacer or fitting in the metal raceway at the point of connection to such equipment.

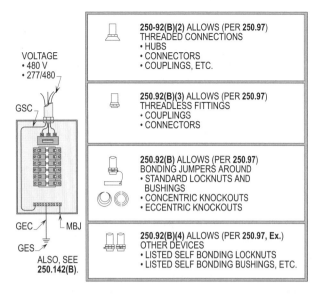

BONDING FOR OVER 250 VOLTS
NEC 250.97

Figure 11-84. This illustration shows the methods that shall be permitted to be used to bond and ground metal conduits to metal enclosures.

ECCENTRIC OR CONCENTRIC KNOCKOUTS
250.97, Ex.

If oversized eccentric or concentric knockouts are not encountered, the following items shall be permitted to be utilized:

- Threadless fittings
- Two locknuts
- One locknut
- Listed fittngs

See Figure 11-85 for a detailed illustration when applying these requirements.

THREADLESS FITTINGS
250.97, Ex. (a)

Threadless couplings and connectors shall be permitted to be used with metal sheath cables. However, they shall be listed as required per **300.15 and 300.15(F)**.

TWO LOCKNUTS
250.97, Ex. (b)

Two locknuts shall be permitted to be used with rigid metal conduit or intermediate metal conduit, one locknut on the outside and one on the inside of the enclosure when entering boxes or cabinets. Appropriate bushings inside the enclosure, etc. shall be used to protect the insulation of the conductors. Grounding connections and circuit conductors that are not made-up tight are the cause of many arcing faults and electrical fires.

ONE LOCKNUT
250.97, Ex. (c)

One locknut on the inside of the boxes or cabinets provides proper bonding where the fitting shoulders seat firmly on the box or cabinet. These include electric metallic tubing connectors, cable connectors and flexible metal conduit connectors, etc.

LISTED FITTINGS
250.97, Ex. (d)

A wedge fitting (bushing) or locknut fitting is an application of other listed devices that shall be permitted to be used. A grounding wedge is equipped with a set-screw that tightens the metal conduit and ensures proper bonding of the equipment enclosure.

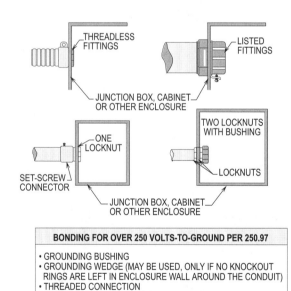

ECCENTRIC OR CONCENTRIC KNOCKOUTS
NEC 250.97, Ex.

Figure 11-85. This illustration shows fittings that shall be permitted to be used to bond and ground the raceways and enclosures together to form a continuous grounding path.

Grounding

SIZE - EQUIPMENT BONDING JUMPER ON SUPPLY SIDE OF SERVICE
250.102(C)

There has always been considerable confusion as to when to use **Table 250.66** and **Table 250.122**. By reviewing the headings of each Table there should be no confusion. **Table 250.66** is used for grounding electrode conductors for AC systems and sized based on the fact that no OCPD is ahead of the conductors.

Table 250.122 is used for equipment grounding conductors for grounding raceways and equipment and sized based on an OCPD ahead of the circuit conductors.

For example: What size copper equipment bonding jumper is required to bond the service raceway (having 3 - 250 KCMIL THWN copper conductors) to the grounded busbar in a panelboard?

Step 1: Sizing EBJ on supply side
250.102(C); Table 250.66
250 KCMIL THWN cu. requires 2 AWG cu.

Solution: **The size equipment bonding jumper required on the supply side is 2 AWG copper.**

For example: What size individual equipment bonding jumper is required to bond each service raceway (RMC) containing 3 - 700 KCMIL THWN copper conductors paralleled three times per phase?

Step 1: Sizing EBJ on supply side
250.102(C); Table 250.66
700 KCMIL THWN cu. requires 2/0 AWG cu.

Solution: **The size equipment bonding jumper required for each conduit on the supply side is 2/0 AWG copper.**

For example: What size equipment bonding jumper (single conductor) is required to bond all service raceways (RMC) containing 3 - 700 KCMIL THWN copper conductors paralleled three times per phase?

Step 1: Sizing EBJ on supply side
250.102(C)
700 KCMIL x 3 = 2100 KCMIL

Step 2: Finding KCMIL to size EBJ
2100 KCMIL x .125 = 262.5 KCMIL

Step 3: Sizing EBJ
Table 8, Ch. 9
262.5 KCMIL requires 300 KCMIL

Solution: **The size equipment bonding jumper required for all service raceways on the supply side is 300 KCMIL copper.**

SIZE - EQUIPMENT BONDING JUMPER ON LOAD SIDE OF SERVICE
250.102(D)

Table 250.122 lists the size of the equipment bonding jumpers on the load side of the service OCPD's. The bonding jumper shall be a single conductor sized per **Table 250.122** based on the largest OCPD that supplies the electrical circuits. If the bonding jumper supplies two or more raceways or cables, the same rules apply and proper size conductors shall be selected. Note that it is not necessary to size the equipment bonding jumper larger than the unground (phase) conductors, but in no case shall it be smaller than 14 AWG.

For example: What size equipment bonding jumper is required to bond the raceway of a feeder-circuit having a 225 amp OCPD ahead of its conductors?

Step 1: Sizing EBJ on load side
250.102(D); Table 250.122
225 A OCPD requires 4 AWG cu.

Solution: **The size equipment bonding jumper required on the load side is 4 AWG copper.**

INSTALLATION
250.102(E)

It shall be permitted to install the equipment bonding jumper either inside or outside of the raceway or enclosure. When it is installed outside of the raceway or enclosure, it shall not be over 6 ft. (1.8 m) in length and shall be routed with raceway or enclosure. If the equipment bonding jumper is inside the raceway, refer to **310.12(B), 250.119** and **250.148** for the conductor identification requirements. Note that if such equipment bonding jumper is routed on the outside of the raceway, listed grounding fittings shall be used to terminate raceway and equipment bonding jumper.

BONDING OF PIPING SYSTEMS AND EXPOSED STRUCTURAL STEEL
250.104

The rules of this section requires that certain piping systems and structural steel to be bonded and grounded properly to ensure the safety of the electrical system.

METAL WATER PIPING - GENERAL
250.104(A)(1)

Regardless of whether the water piping is supplied by nonmetallic pipe to the building or structure, proper bonding of the interior metal water piping to the service equipment enclosure shall be required. Metal water piping that is installed in or attached to a building or structure shall be bonded to the service equipment enclosure, the grounded (neutral) conductor at the service, the grounding electrode conductor where of sufficient size or to the one or more grounding electrodes used. The bonding jumper to metal water piping shall be sized per **Table 250.66**, except as permitted in **250.104(A)(2)** and **(A)(3)**. **(See Figure 11-86)**

METAL WATER PIPING - BUILDINGS OF MULTIPLE OCCUPANCY
250.104(A)(2)

In multiple occupancies if the interior piping is metal and the piping systems of all occupancies are not tied together, the metal piping shall be isolated from all other occupancies due to the use of nonmetallic pipe as a supply line. Each individual metal water pipe system shall be bonded separately to the panelboard or switchboard enclosure in each apartment. Bonding jumper points of attachment shall be accessible and sized per **Table 250.122**. This ensures that the metal piping system in each occupancy is at ground potential and the threat of electrical shock and fires are eliminated. **(See Figure 11-87)**

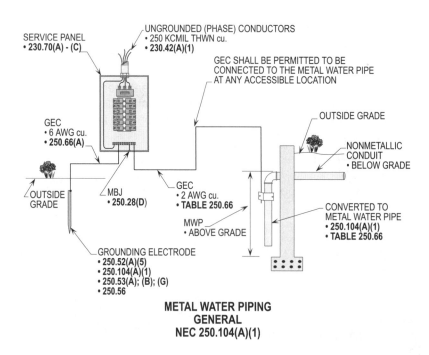

Figure 11-86. Regardless of whether the water piping is supplied by nonmetallic pipe to the building or structure, proper bonding of the interior metal water piping to the service equipment enclosure shall be required.

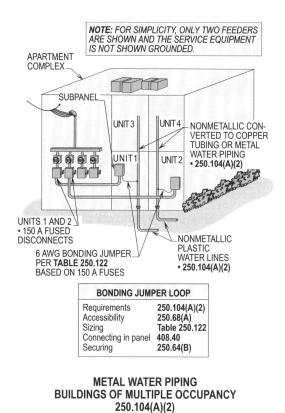

METAL WATER PIPING
BUILDINGS OF MULTIPLE OCCUPANCY
250.104(A)(2)

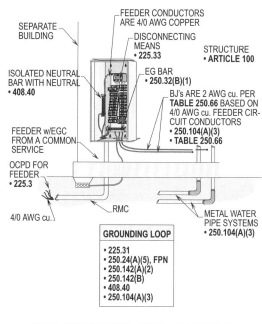

METAL WATER PIPING - MULTIPLE BUILDINGS
OR STRUCTURES SUPPLIED BY A
FEEDER(S) OR BRANCH CIRCUIT(S)
250.104(A)(3)

Figure 11-87. This illustration shows the procedure for sizing the bonding jumper to bond and ground the copper tubing or metal water piping to the grounded busbar in the subpanel in apartment units 1 and 2.

Figure 11-88. This illustration shows the procedure for sizing the bonding jumper to connect the equipment grounding conductors (run with the feeder-circuit conductors) and the grounded busbar to the concrete-encased electrode of the separate building supplied from a common servce.

METAL WATER PIPING - MULTIPLE BUILDINGS OR STRUCTURES SUPPLIED BY A FEEDER(S) OR BRANCH CIRCUIT(S) 250.104(A)(3)

The interior metal water piping system shall be bonded to the building or structure disconnecting means enclosure where located at the building or structure, or to the equipment grounding conductor run with the supply conductors or to the one or more grounding electrodes used. The bonding jumper shall be sized per **Table 250.66** based on the size of the feeder or branch-circuit conductors that supply the building. The bonding jumper shall not be required to be larger than the largest ungrounded (phase) feeder or branch-circuit conductors supplying the building. **(See Figure 11-88)**

OTHER METAL PIPING
250.104(B)

All metal piping systems and gas piping that is likely to be subjected to being energized shall be bonded to the service equipment ground at the service enclosure to the common grounding conductor, if it is of sufficient size, or to one or more of the other grounding electrodes. This bonding jumper shall be sized per **Table 250.122**. **(See Figure 11-89)**

Design Tip: With circuit conductors that are likely to energize metal piping, it is permitted to use the equipment grounding conductor run with the circuit conductors to bond and ground such piping system.

STRUCTURAL METAL
250.104(C)

Exposed building steel frames that are not intentionally grounded shall be bonded to either the service enclosure, the neutral bus, the grounding electrode conductor or to a grounding electrode. The bonding conductor shall be sized per **Table 250.66**, and installed per **250.64(A)**, **(B)** and **(E)**. **(See Figure 11-90)**

Design Tip: For additional safety, it is recommended to bond all metal air ducts and metal piping systems on the premises to reduce voltage differences.

SEPARATELY DERIVED SYSTEM 250.104(D)

The following shall be bonded to separately derived systems:
- Metal water piping system(s)
- Structural metal

METAL WATER PIPING(S) 250.104(D)(1)

When there is a separately derived system utilizing a grounding electrode near an available point of metal water pipe which is in the area supply by such derived system, the metal water pipe system shall be bonded to the derived systems grounded (neutral) conductor. Each bonding jumper shall be sized per **Table 250.66** based on the largest ungrounded conductor of the separately derived system. **(See Figure 11-91)**

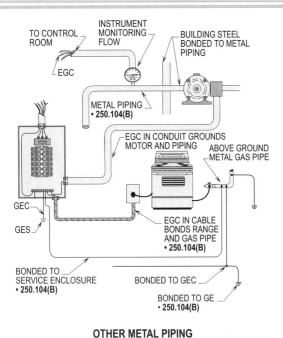

Figure 11-89. Metal piping and gas piping that may be energized shall be bonded and grounded with the equipment grounding conductor in the supplying branch-circuit or feeder that could energize such piping.

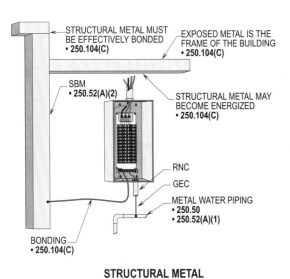

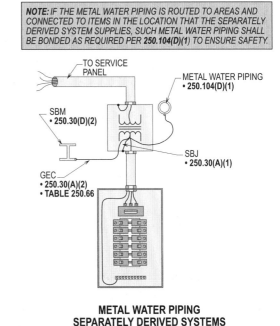

Figure 11-91. This illustration shows that metal water pipe located in the area of the separately derived system shall be bonded and grounded to the grounded (neutral) conductor located in the transformer's grounding point.

STRUCTURAL METAL 250.104(D)(2)

Exposed structural metal shall be bonded to the grounded conductor of each separately derived system where it is interconnected to form the building frame (exists in the area).

Figure 11-90. Exposed interior building steel frames that are not intentionally grounded and is likely to become energized, shall be bonded to either the service enclosure, the neutral bus, the grounding electrode conductor or to a grounding electrode.

The connection to the separately derived system shall be made at the same point where the grounding electrode conductor is connected. Each bonding jumper shall be sized per **Table 250.66**, based on the largest ungrounded (phase) conductor of the separately derived system.

TYPES OF EQUIPMENT GROUNDING CONDUCTORS
250.118

The equipment grounding conductor run with or enclosing the circuit conductors shall be one or more or a combination of the following:

- A copper, aluminum or copper-clad aluminum conductor
- Rigid metal conduit (RMC)
- Intermediate metal conduit (IMC)
- Electrical metallic tubing (EMT)
- Listed flexible metal conduit (FMC)
- Listed liquidtight flexible metal conduit (LTFC)
- Flexible metal tubing (FMT)
- Type AC cable (BX)
- Mineral-insulated sheathed cable (MI)
- Type MC cable (MC)
- Cable trays
- Cablebus framework
- Other listed electrically continuous metal raceways
- Surface metal raceways listed for grounding

See Figure 11-92 for a detailed illustration pertaining to the various types of wiring methods that are permitted to be used as equipment grounding conductors. **Sections 250.118(6)a through d** and **(7)a through e** allows FMC and LTFC to be used as follows:

The maximum length that shall be permitted to be used in any combination of flexible metal conduit, flexible metallic tubing and liquidtight flexible metal conduit in any grounding run is 6 ft. (1.8 m) or less.

The circuit conductors in a 6 ft. (1.8 m) run shall be protected by overcurrent devices not to exceed 20 amperes, but shall also be permitted to be less than 20 amps. The flexible metal conduit or tubing shall be terminated only in fittings listed for grounding. **(See Figure 11-93)**

Liquidtight flexible metal conduit shall be permitted to be used in sizes up to 1 1/4 in. (35) that have a total length not to exceed 6 ft. (1.8 m) in the ground return path. Fittings for terminating the flexible conduit shall be listed for grounding. Liquidtight flexible metal conduit 3/8 (12) and 1/2 in. (16) trade size shall be protected by overcurrent devices rated at 20 amps or less, and sizes of 3/4 in. (21) through 1/ 1/4 in. (35) trade size shall be protected by overcurrent devices rated at 60 amps or less. **(See Figure 11-94)**

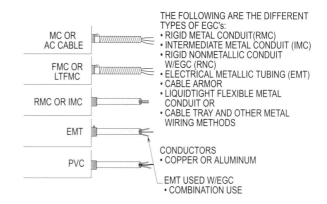

**TYPES OF EQUIPMENT
GROUNDING CONDUCTORS
NEC 250.118**

Figure 11-92. The various types of wiring methods permitted to be used as equipment grounding conductors. (See **250.118(A)(1) through (14)** for a complete list of wiring methods.

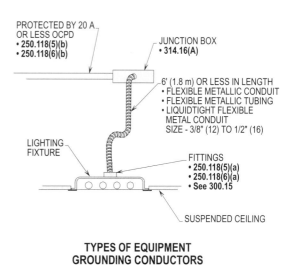

**TYPES OF EQUIPMENT
GROUNDING CONDUCTORS
NEC 250.118**

Figure 11-93. This illustration shows the maximum length permitted for listed flexible metal conduit and listed liquidtight flexible metal conduit to be used as an equipment grounding means.

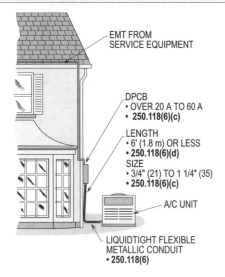

TYPES OF EQUIPMENT GROUNDING CONDUCTORS
NEC 250.118

Figure 11-94. This illustration shows the maximum length that listed liquidtight flexible metal conduit shall be permitted to be used as a grounding means without an equipment grounding conductor routed with the circuit conductors inside the flex.

IDENTIFICATION OF EQUIPMENT GROUNDING CONDUCTORS 250.119

Equipment grounding conductors shall be permitted to be installed as bare, covered or insulated, unless otherwise permitted elsewhere in the NEC. Equipment grounding conductors smaller than 6 AWG copper or aluminum shall have a continuous outer finish that is either green or green with one or more yellow stripes, unless otherwise permitted elsewhere in the NEC.

CONDUCTORS LARGER THAN 6 AWG 250.119(A)

Equipment grounding conductors that are insulated or covered shall be permitted to be installed larger than 6 AWG copper or aluminum if color coded at each end and at every point where the conductor is accessible. The following methods shall be permitted to be used to identify an equipment grounding conductor larger than 6 AWG:

- Stripping the insulation or covering from the entire length exposed.
- Coloring the exposed insulation or covering green.
- Marking the exposed insulation or covering with green colored tape or green adhesive labels.

Note: Equipment grounding conductors that are larger than 6 AWG shall not be reuiqred to be marked in conduit bodies that contain no splices or unused hubs.

See Figure 11-95 for a detailed illustration when applying these requirements.

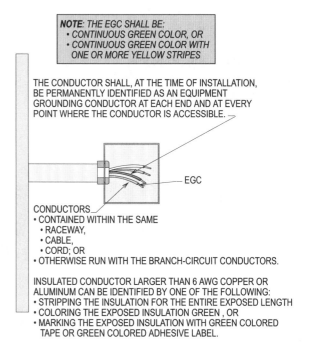

CONDUCTORS LARGER THAN 6 AWG
NEC 250.119(A)

See Figure 11-95. This illustration shows the methods of identifying and using the equipment grounding conductor when its insulation is other than green in color.

EQUIPMENT GROUNDING CONDUCTOR INSTALLATION 250.120(A) THRU (C)

When an equipment grounding conductor is used in a raceway, cable tray, cable armor, cablebus framework or cable sheath, and where it is routed in a raceway or cable, the installation shall conform to the applicable provisions in the NEC based upon this type wiring system.

All terminations and joints shall be listed and approved for such use. All joints and fittings shall be made up tight by using tools suitable for the purpose.

Sections 250.130(A) and **(B)** provides for using the equipment grounding conductor as a separate conductor.

Equipment grounding conductors shall be permitted to be bare, insulated aluminum or copper-clad aluminum. Bare conductors shall not be permitted to come in direct contact with masonry or the earth or where subject to corrosive conditions. Aluminum or copper-clad aluminum conductors shall not be permitted to be terminated within 18 in. (450 mm) of the earth.

If the grounding conductors are routed in the hollow spaces of a building that is protected, sizes smaller than 6 AWG shall not be required to be in raceways.

SIZING OF EQUIPMENT GROUNDING CONDUCTORS
250.122; TABLE 250.122

When an equipment grounding conductor is used in a raceway, cable tray, cable armor or cable sheath, and where it is routed in a raceway or cable, the installation and size of the conductor shall conform to the applicable provisions in the NEC. The size of the equipment grounding conductor shall be based on the size of the overcurrent protection device (OCPD) protecting the circuit conductors per **Table 250.122**. **(See Figure 11-96)**

The equipment grounding conductor is the conductor used to ground the noncurrent-carrying metal parts of equipment. The function of the EGC is to keep the equipment elevated above ground, at zero potential, and provides a path for the ground-fault current. Equipment grounding conductors protect elements of circuits and equipment and also ensure the safety of personnel from electrical shock. The size of the equipment grounding conductor shall be based on the size of the overcurrent protection device protecting the circuit conductors.

> **For example:** What size copper equipment grounding conductor is required to supply a subpanel, if the OCPD protecting the feeder-circuit is rated at 400 amps?
>
> **Step 1:** Sizing EGC
> **Table 250.122**
> 400 A OCPD requires 3 AWG cu.
>
> **Solution: The size equipment grounding conductor is required to be 3 AWG copper.**

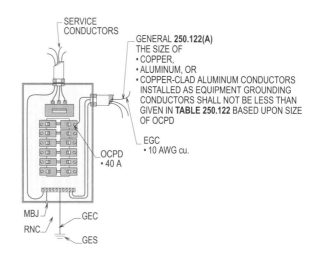

SIZING OF EQUIPMENT GROUNDING CONDUCTORS NEC 250.122; TABLE 250.122

Figure 11-96. This illustration shows the equipment grounding conductor shall be sized based on the OCPD ahead of the circuit conductors supplying the electrical equipment.

GENERAL
250.122(A)

Equipment grounding conductors 6 AWG and smaller shall have their insulation for its entire length colored green or green with one or more yellow stripes. Note that the equipment grounding conductor shall be permitted to be bare under certain conditions of use. The insulation or covering shall be stripped from its entire length or length exposed. (For this requirement, see **250.119(A)**.

Equipment grounding conductors that are insulated or covered shall be permitted to be installed larger than 6 AWG copper or aluminum if color coded at each end and at every point where the conductor is accessible. The following methods shall be permitted to be used to identify an equipment grounding conductor larger than 6 AWG.

- Stripping the insulation or covering from the entire length exposed.

- Coloring the exposed insulation or covering green.

- Marking the exposed insulation or covering with green colored taps or green adhesive labels.

See Figure 11-97 for a detailed illustration when applying these requirements.

Stallcup's Electrical Design Book

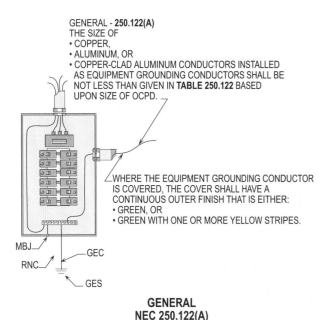

Figure 11-97. This illustration shows marking techniques that shall be permitted to be used when identifying the equipment grounding conductor routed through a raceway or cable.

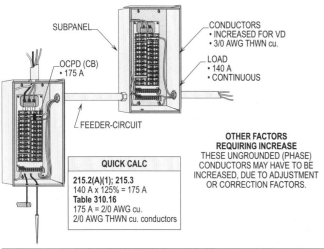

Figure 11-98. This illustration shows the equipment grounding conductor being increased in size proportionately according to the circular mil area of the ungrounded (phase) conductors.

INCREASED IN SIZE
250.122(B)

Where ungrounded (phase) conductors are increased in size, equipment grounding conductors (if installed) shall be increased in sized proportionately according to the circular mil area of the ungrounded (phase) conductors. **(See Figure 11-98)**

MULTIPLE CIRCUITS
250.122(C)

Single equipment grounding conductors that are run with multiple circuits in the same raceway shall be permitted to be sized based on the largest OCPD protecting the circuit conductors in the raceway or cable.

> **For example:** What size equipment grounding conductor is required for multiple circuits in the same raceway with the following known values:
> - 3 - 10 AWG copper conductors (30 A OCPD)
> - 2 - 12 AWG copper conductors (20 A OCPD)
> - 3 - 14 AWG copper conductors (15 A OCPD)
>
> **Step 1:** Sizing EGC
> **250.122(C)**
> 30 A OCPD requires 10 AWG cu.
>
> **Solution: A 10 AWG copper equipment grounding conductor is required for multiple circuits in the same raceway.**

MOTOR CIRCUITS
250.122(D)

If the OCPD is a motor short-circuit protector or an instantaneous trip breaker, see **430.52** and **Table 430.52** for sizing such device. The size of the equipment grounding conductor in such cases shall be permitted to be based upon the rating of the protective device. In any design, the equipment grounding conductor never has to be larger than the largest ungrounded (phase) conductor supplying the motor. **(See Figure 11-99)**

FLEXIBLE CORD AND FIXTURE WIRE
250.122(E)

The equipment grounding conductor in a flexible cord, with the largest circuit conductor 10 AWG or smaller and equipment grounding conductors that are part of flexible cords or used with fixture wires per **240.5** shall not be permitted to be smaller than 18 AWG copper and not smaller than the circuit conductors. The equipment grounding conductor shall be sized per **Table 250.122** where larger than 10 AWG.

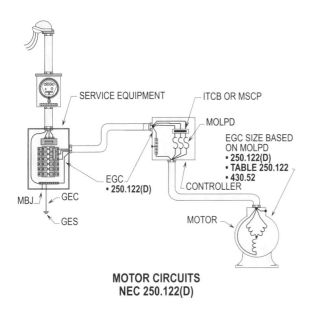

MOTOR CIRCUITS
NEC 250.122(D)

Figure 11-99. This illustration shows that under certain design conditions, the size of the equipment grounding conductor shall be permitted to be determined by the rating of the motor overload protective device (MOLPD) in the controller.

CONDUCTORS IN PARALLEL
250.122(F)

Circuit conductors that are connected in parallel and routed through separate conduits shall have a separate equipment grounding conductor run in each conduit. The equipment grounding conductor shall be sized per **Table 250.122**, based on the rating of the OCPD protecting the circuit.

> **For example:** What size copper equipment grounding conductors are required to ground the metal parts of a piece of equipment supplied by 3 - 3/0 AWG THWN copper conductors per phase connected to a 600 amp OCPD in the panelboard?

> **Step 1:** Sizing EGC's in each conduit
> **250.122(F); Table 250.122**
> 600 A OCPD requires 1 AWG cu.
>
> **Solution:** The size equipment grounding conductors are required to be 1 AWG copper for conductors installed in parallel.

EQUIPMENT GROUNDING CONDUCTOR CONNECTIONS 250.130

Equipment grounding conductor connections for a separately derived system and for service equipment shall be made in accordance with **250.30(A)(1)**. The equipment grounding conductor connections shall be made at service equipment and separately derived systems as follows:

- For grounded systems
- For ungrounded systems

FOR GROUNDED SYSTEMS
250.130(A)

The equipment grounding conductor shall be bonded to the grounded (neutral) conductor and the grounding electrode conductor at the service equipment neutral busbar terminal. The grounded conductor (may be a neutral) shall be installed and connected to the grounding electrode system per **250.50** and **250.52(A)(1) through (A)(7)**. Note that the grounded (neutral) conductor shall be connected to the busbar where the grounding electrode conductor is terminated per **250.24(A)**. **(See Figure 11-100)**

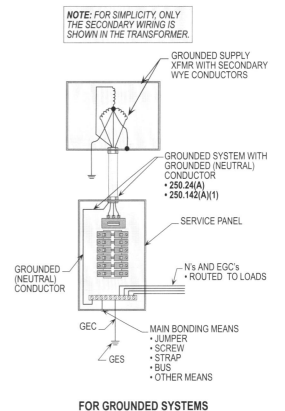

FOR GROUNDED SYSTEMS
NEC 250.130(A)

Figure 11-100. This illustration shows the utility transformer and the panelboard both connected to earth ground for safety.

FOR UNGROUNDED SYSTEMS
250.130(B)

The equipment grounding conductor shall be bonded to the grounding electrode conductor at the service equipment neutral busbar terminal. A grounded (neutral) conductor is not present to be connected to ground. Equipment grounding conductors installed per **250.118** shall be permitted to be used to bond and connect all metal noncurrent-carrying parts of the wiring system to the grounding electrode conductor at the service equipment. **(See Figure 11-101)**

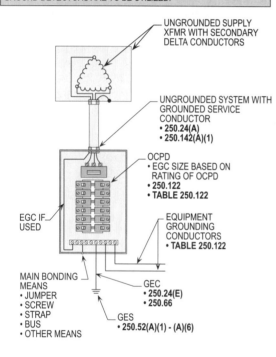

Figure 11-101. This illustration shows an ungrounded transformer being used to supply uninterrupted power to an industrial plant.

NONGROUNDING RECEPTACLE REPLACEMENT OR BRANCH-CIRCUIT EXTENSIONS
250.130(C)(1) THRU (C)(5)

The basic rule for replacement of receptacles is that grounding-type receptacles shall be used to replace existing receptacles where there is an equipment grounding conductor routed with the branch-circuit. A nongrounding-type receptacle shall be used with a branch-circuit that has no equipment grounding conductor or grounding means. The equipment grounding conductor shall connect the green grounding terminal of the receptacle to the grounding electrode conductor or to the closest metal water pipe per **250.104(A)**, **250.50**, **250.52(A)(1)** and **250.53(D)**. **(See Figure 11-102)**

The NEC allows five methods by which a nongrounding receptacle shall be permitted to be replaced with a circuit without an equipment grounding conductor and they are as follows:

- Install a nongrounding type receptacle at each receptacle location.
- A GFCI receptacle protecting each outlet at receptacle location.
- A GFCI receptacle protecting a single outlet and additional outlets downstream.
- An equipment grounding conductor routed from the receptacle to a metal water pipe per **250.53(D)** in the NEC.
- A GFCI CB protecting all outlets.

See Figure 11-103 for a detailed illustration when applying these requirements.

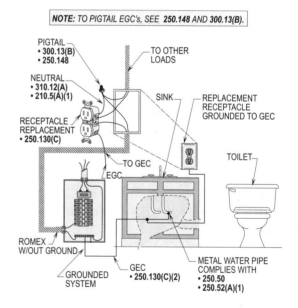

Figure 11-102. This illustration shows one method in which a grounding type receptacle shall be permitted to be grounded to the grounding electrode conductor. Note that the branch-circuit is not equipped with an equipment grounding conductor.

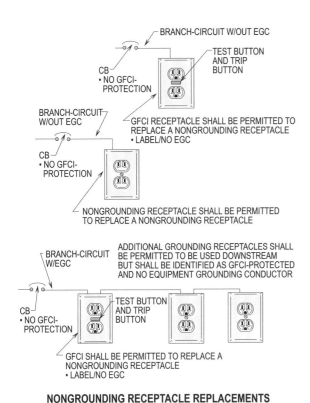

NONGROUNDING RECEPTACLE REPLACEMENTS OR BRANCH-CIRCUIT EXTENSIONS
NEC 250.130(C)(1) THRU (C)(5)

Figure 11-103. This illustration shows methods that shall be permitted to be used to replace and protect receptacles when there is no equipment grounding conductor available in the branch-circuit wiring method.

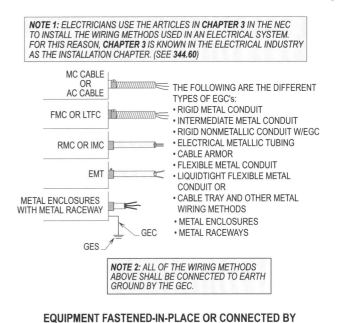

EQUIPMENT FASTENED-IN-PLACE OR CONNECTED BY PERMANENT WIRING METHODS (FIXED) - GROUNDING
NEC 250.134

Figure 11-104. This illustration shows wiring methods that shall be permitted to be used to earth ground equipment that is fastened-in-place.

EQUIPMENT FASTENED-IN-PLACE OR CONNECTED BY PERMANENT WIRING METHODS (FIXED) - GROUNDING
250.134

Noncurrent-carrying metal parts of equipment, raceways, cables and other enclosures shall be grounded by one of the following methods where required to be grounded for the safety of circuits, equipment and personnel:
- Equipment grounding conductors
- Metal raceways
- Metal clad cables, etc.

See Figure 11-104 for a detailed illustration when applying these requirements.

EQUIPMENT GROUNDING CONDUCTOR TYPES
250.134(A)

The following wiring methods shall be permitted to be installed as the equipment grounding conductor to ground electrical enclosures and equipment cases:
- Rigid metal conduit (RMC)
- Electrical metallic tubing (EMT)
- Type AC cable
- Approved metal conduit or cables

See Figure 11-105 for a detailed illustration when applying these requirements.

WITH CIRCUIT CONDUCTORS
250.134(B)

An equipment grounding conductor contained within the same raceway, cable or cord shall be permitted to be carried along with circuit conductors and installed to ground electrical enclosures and equipment. Equipment grounding conductors shall be permitted to be bare, covered or insulated. Insulated equipment grounding conductors shall be identified green or green with one or more yellow stripes. For more detailed information on this subject (with circuit conductors), see **300.3(B)**, **300.5(I)** and **300.20**. **(See Figure 11-106)**

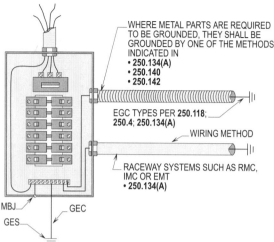

Figure 11-105. This illustration shows wiring methods that shall be permitted to be used to earth ground the noncurrent-carrying metal parts of electrical related equipment and enclosures.

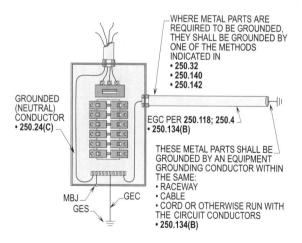

Figure 11-106. This illustration shows the methods by which the equipment grounding conductors shall be routed to ensure a low-impedance path for clearing ground-faults.

CORD-AND-PLUG CONNECTED EQUIPMENT
250.138

The metal cases of portable cord-and-plug connected stationary equipment shall be permitted to be grounded by any one of the following three wiring methods:
- By means of metal enclosure
- By means of an equipment grounding conductor
- By means of a separate flexible wire or strap

Metal enclosures housing conductors shall be permitted to be used as a grounding means if an approved grounding type attachment plug is utilized with one fixed member making contact for the purpose of bonding and grounding the metal enclosure. However, the attachment plug shall be of a type which is approved for grounding from a grounding type receptacle and such cord is required to carry an equipment grounding conductor. One end shall be attached to the grounding terminal of the attachment plug and the other end shall terminate to the frame of the portable equipment. For proper grounding, it is important that the metal enclosure of such conductors be attached to the attachment plug and to the equipment by approved connectors.

BY MEANS OF AN EQUIPMENT GROUNDING CONDUCTOR
250.138(A)

The equipment grounding conductor in a cord is used to connect the grounding prong of a plug to ground the equipment. Equipment grounding conductors shall be permitted to be bare or insulated with a green color or green with one or more yellow stripes. Such grounding conductors shall be permitted to be routed in the cable or flexible cord assembly, provided that it terminates in an approved grounding type attachment plug having a fixed grounding type contact member. **(See Figure 11-107)**

The **Ex. to 250.138(A)** allows a movable self-restoring type such as an approved attachment plug that is equipped with a hinged grounding prong with a spring. Note that this spring may be folded out of the way and still restore itself back to the normal position on the grounding type receptacle.

BY MEANS OF A SEPARATE FLEXIBLE WIRE OR STRAP
250.138(B)

A separate flexible wire or strap shall be permitted to be used to ground portable equipment. This strap should be protected from physical damage as well as practical to ensure proper bonding and grounding.

Grounding

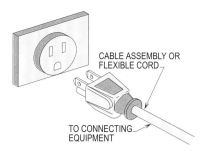

Figure 11-107. This illustration shows a flexible cord with an attachment cap and equipment grounding conductor used to ground the connected equipment to the receptacle and its supply.

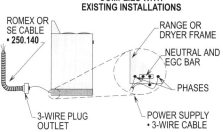

Figure 11-108. This illustration shows the procedure for installing a range, cooktop, oven or dryer on existing and new branch-circuit installations.

FRAMES OF RANGES AND CLOTHES DRYERS
250.140

A new branch-circuit installation for ranges, cooktops, ovens, clothes dryers, including junction boxes or outlet boxes that are part of the circuit shall be permitted to be bonded and grounded with an equipment grounding conductor. However, an isolated grounded (neutral) conductor shall also be installed. For further information concerning this type of installation, see **250.114, 250.134** and **250.138**. **(See Figure 11-108)**

Existing branch-circuit installations for the frames of ranges, cooktops, ovens, clothes dryers and junction boxes or outlet boxes that are part of the circuit shall be permitted to be bonded and grounded with the grounded (neutral) conductor under the following conditions:

- The supply circuit is 120/240 volt, single-phase, three-wire or 208Y/120 volt derived from a three-phase, four-wire, wye-connected system.
- The grounded (neutral) conductor installed is no smaller than 10 AWG copper or 8 AWG aluminum.
- The grounded (neutral) conductor is insulated and part of NM cable (romex) or PVC conduit.
- Grounding contacts of receptacles furnished as part of the equipment are bonded to the equipment.

USE OF GROUNDED CIRCUIT CONDUCTOR FOR GROUNDING EQUIPMENT
250.142

Under certain conditions, all metal parts of enclosures used to install the service equipment shall be permitted to be grounded by the grounded (neutral) conductor on the supply side of the system. The service weatherhead, service raceway, service meter base and the service equipment enclosure are included when using this type of grounded system.

The grounded (neutral) conductor shall be installed and isolated from the other system circuit conductors and from metal enclosures or metal conduits when installed on the load side of the system.

SUPPLY SIDE EQUIPMENT
250.142(A)

The grounded (phase or neutral) conductor shall be permitted to be used as a current-carrying conductor and grounding means on the supply side of the service disconnecting means and secondary side of a separately derived system as follows:

- On the supply side of service equipment per **250.142(A)(1)** and **250.24(C)(1)**.

- On the supply side of the main service disconnecting means for separate buildings and structures per **250.142(A)(2)** and **250.32(B)(1)** and **(B)(2)**.

- On the supply side of the disconnect or overcurrent protection device of a separately derived system per **250.142(A)(3)** and **250.30(A)(1)**.

See Figure 11-109 for a detailed illustration when applying these requirements.

- Frame of ranges, wall-mounted ovens, counter-mounted cooking units and clothes dryers per **250.140**. (Only for existing branch-circuits)

- Where one or more buildings or structures are supplied from a common AC grounded service, each grounded service at each individual building or structure shall be separately grounded per **250.32**.

- If no service ground-fault protection is provided, all meter socket enclosures located near the service disconnecting means shall be permitted to be grounded by the grounded (neutral) conductor on the load side of the service disconnect. The grounded (neutral) conductor shall not be permitted to be smaller than the size specified in **Table 250.122**.

- DC systems shall be permitted to be grounded at the first disconnecting means or overcurrent protection device per **250.164(B)(2)**.

See Figure 11-110 for a detailed illustration when applying these requirements.

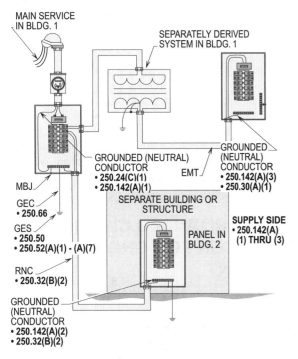

Figure 11-109. This illustration shows three installations where the grounded (neutral) conductor shall be permitted to be used as a current-carrying neutral plus equipment grounding conductor.

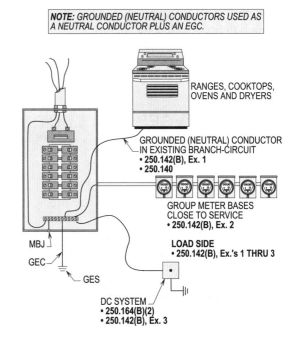

Figure 11-110. This illustration shows an installation where the grounded (neutral) conductor is used as a neutral conductor plus an equipment grounding conductor.

LOAD-SIDE EQUIPMENT
250.142(B)

The grounded (neutral) conductor shall not be permitted to be used as an equipment grounding conductor on the load side except as follows:

CONNECTING RECEPTACLE GROUNDING TERMINAL TO BOX
250.146

A good connection to ground shall be made by connecting the receptacle grounding terminals to the box. A bonding jumper shall be permitted to be used to connect the grounding terminal of a grounding-type receptacle to a metal grounded box. The following are four exceptions when a connection shall be permitted to be made without using a bonding jumper:

- Surface-mounted boxes
- Contact devices or yokes
- Floor boxes
- Isolated receptacles

SURFACE-MOUNTED BOXES
250.146(A)

If a surface-mounted box such as a handy box complies with **250.146(B)**, then a bonding jumper providing metal-to-metal contact is not required. In other words, the 6-32 screws on the yoke of the device shall be considered an acceptable means of bonding the box and yoke together. Note that a bonding jumper shall be required if metal-to-metal contact is not provided by the connection above. **(See Figure 11-111)**

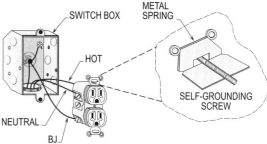

CONTACT DEVICES OR YOKES
NEC 250.146(B)

Figure 11-112. This illustration shows a self-bonding device used to make metal-to-metal contact between switch box and receptacle yoke.

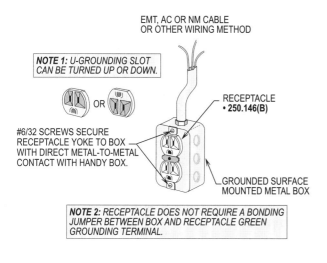

SURFACE MOUNTED BOX
NEC 250.146(A)

Figure 11-111. This illustration shows the surface box and the receptacle yoke making metal-to-metal contact.

CONTACT DEVICES OR YOKES
250.146(B)

A flush-mounted box shall be permitted to be installed with self-grounding screws to ground receptacles with a bonding jumper. Note that these contact devices and yokes are specifically listed for proper grounding when used in conjunction with supporting screws to ensure adequate grounding between the yoke and flush-type mounted boxes, receptacles or switches. **(See Figure 11-112)**

FLOOR BOXES
250.146(C)

The metal yoke of the receptacle requires no bonding jumper to be installed for floor boxes. Floor boxes are designed to provide a good contact between the metal box. However, such floor boxes shall be specifically designed and listed for such purpose. **(See Figure 11-113)**

ISOLATED RECEPTACLES
250.146(D)

An isolated equipment grounding conductor shall be used to connect the receptacle isolated ground terminal back to the grounded connection at the service equipment or separately derived system to reduce electromagnetic interference.

This rule applies, when electrical noise known as electromagnetic interference occurs in the grounding circuit, an insulated equipment grounding conductor shall be permitted to be run to the isolated terminal with the circuit conductors. This insulated grounding conductor shall be permitted to pass through one or more panelboards in the same building without being connected thereto as allowed in **408.40, Ex.** Note that it still has to terminate at the equipment grounding terminal at a separately derived system or service panel. Also, see **250.96(B)**. **(See Figure 11-114)**

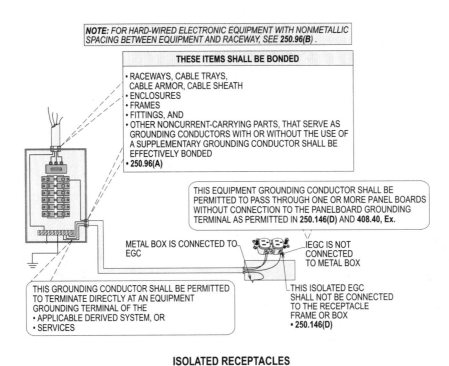

ISOLATED RECEPTACLES
NEC 250.146(D)

Figure 11-114. This illustration shows the techniques for bonding and grounding the yoke of an isolation receptacle to a metal box.

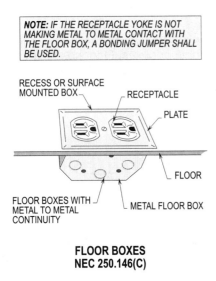

FLOOR BOXES
NEC 250.146(C)

Figure 11-113. This illustration shows the technique for bonding and grounding a receptacle yoke to a metal floor plate.

CONTINUITY AND ATTACHMENT OF EQUIPMENT GROUNDING CONDUCTORS TO BOXES
250.148

Where more than one equipment grounding conductor enters a box, all such conductors shall be spliced within the box with devices suitable for the purpose. Equipment grounding conductors shall be connected and bonded to metal boxes to ensure grounding continuity. In other words, all grounding conductors entering the box shall be made electrically and mechanically secure, and the pigtail from them serves as the grounding connection to the device being installed. If the device is removed, the continuity of the equipment grounding conductors will not be disrupted. **(See Figure 11-115)**

COMPUTER GROUNDING USING A SPACER
645.15; 250.96(B)

Problems often occur with computers due to electrical noise within the electrical circuits. Special grounding methods shall be installed to minimize the electrical noise. A nonmetallic spacer shall be permitted to be installed between sensitive electronic equipment and the service panel or source per **250.6(D)** and **250.96(B)**. An insulated equipment grounding conductor shall be installed with circuit conductors. **(See Figure 11-116)**

Note that when installing a grounding electrode conductor to a driven rod or other type of electrode which is not connected to the grounding electrode system of the building, the computer is not completely isolated and this creates a

Grounding

grounding hazard. The grounding electrode conductor has no low-impedance return path for the fault-current to return to the power source and trip open the overcurrent protection device. This is due to the high resistance of the earth.

RADIAL GROUNDING
645.15

More grounding noise is sensed when there is not an equipotential plane between the computer and other grounded metal.

To eliminate such noise, the computer shall be grounded with a radial system which clears the ground system of ground loops. Ground loops produce paths for circulation of the ground currents and these currents create unwanted electrical noise that cause computers to malfunction.

Grounding for a computer shall be installed in accordance with **Article 250**. The grounding electrode conductor at the service equipment or subfed transformer shall be connected to all exposed noncurrent-carrying parts of the computer enclosure. The equipment grounding conductor is used to provide this function of grounding to the enclosure of the computer. **(See Figure 11-117)**

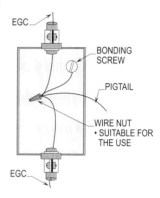

CONTINUITY AND ATTACHMENT OF EQUIPMENT GROUNDING CONDUCTORS TO BOXES
NEC 250.148

Figure 11-115. This illustration shows that equipment grounding conductors shall be bonded and grounded to boxes, enclosures, etc. with a method that will ensure uninterrupted continuity.

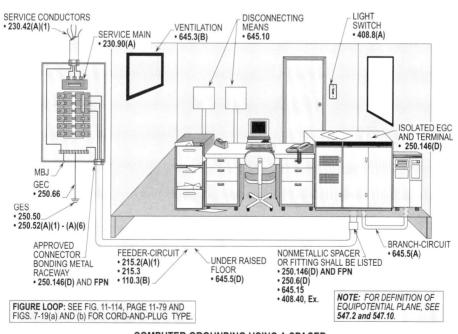

COMPUTER GROUNDING USING A SPACER
NEC 645.15
NEC 250.96(B)

Figure 11-116. A nonmetallic spacer shall be permitted to be installed between sensitive electronic equipment and the service panel or source. However, an equipment grounding conductor shall be run to such equipment.

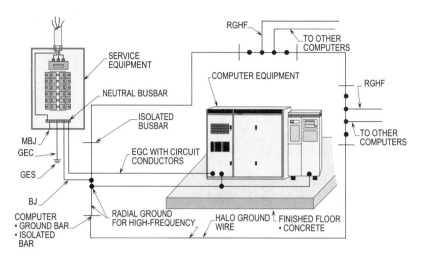

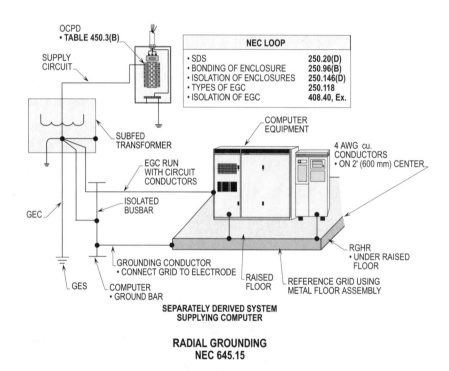

Figure 11-117. Ground loops produce paths for circulation of the ground currents, and these currents create unwanted electrical noise that cause computers to malfunction.

HIGH-FREQUENCY REFERENCE GRID
645.15

A high-frequency reference grid system shall be permitted to be connected to a computer grounding point. A single ground point shall be grounded radially and tied to a high-frequency reference grid for computer equipment. The computer shall be grounded by the grid at the same ground point. The grounding electrode is then grounded to the computer ground point. This type of grounding scheme is installed to help minimize electrical noise and complies with the rules and regulations of the NEC.

The power conductors, grounds and grounded (neutral) conductor shall be installed in the same conduit or cable except for the reference grid ground, if need be, it shall be permitted to be routed away from these circuit conductors.

SWIMMING POOLS
ARTICLE 680, PART II

Swimming pools and decorative fountains shall be grounded and bonded to prevent electrical shock to people wading and swimming in the pool water. Proper size bonding and grounding conductors are essential to ensure the safety of persons having a good time in or around the wet areas of the swimming pool.

BONDING
680.26(A); (B); (C)

The bonding of all metal parts and grounding of the elements keeps the interconnected system and steel at the same potential which creates an equipotential plane. Stray currents moving in the water, metal and steel of the pool that are not cleared by the overcurrent protection device are kept at zero or at earth potential. This prevents electrical shock hazards to persons in the pool water or the area surrounding the pool, such as the deck where there may be stray currents in the steel or mesh in the concrete or gunite.

A 8 AWG or larger solid copper conductor shall be used to bond together all metal parts to the common bonding grid which is the foundation steel of the pool. The reinforcing bars in the pool and metal in the walls of the pool are used to form the common grid system. The bonding conductor shall not be permitted to be run to the service equipment and be connected to the grounded terminal bar in the service equipment enclosure per **680.26, FPN**. Note that this connection could allow more current from the service ungrounded (neutral) conductor to flow to the pool.

The equipment and metal parts of a swimming pool required to be bonded are as follows:

- All fixed parts located within 5 ft. (1.5 m) of the walls of the pool or they shall be separated by a permanent barrier.
- All metal parts and reinforcing steel in the pool or deck.
- All forming shells housing wet-niche or dry-niche fixtures.
- All metal parts and fittings of recirculating motor, etc.
- All fixed metal parts, such as conduit, pipes, cables, equipment, etc. located within 5 ft. (1.5 m) of the pool and not separated by a permanent barrier, and within 12 ft. (3.7 m) vertically of the maximum water level. **(See Figure 11-118)**

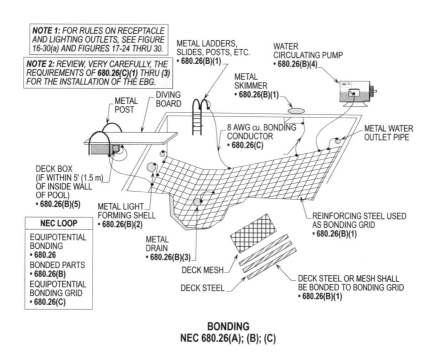

Figure 11-118. The bonding of all metal parts and grounding of the elements keeps the interconnected system and steel at the same potential, which creates an equipotential plane.

GROUNDING
680.6

All metal parts around the pool shall be bonded and grounded to ground potential and thus reduce the threat of shock hazard. The equipment and parts that shall be grounded includes the following items:

- Through-wall lighting assemblies and underwater luminaires (lighting fixtures), other than those low-voltage systems listed for the application without a grounding conductor.
- All electrical equipment located within 5 ft. (1.5 m) of the inside walls of the pool.
- All electrical equipment associated with the recirculating pump.
- Junction boxes.
- Transformer enclosures.
- Ground-fault circuit-interrupters
- Panelboards that are not part of the service equipment and they supply electrical equipment associated with the pool.

The equipment grounding conductor used to ground the metal parts of equipment, junction boxes, etc. shall be sized from **Table 250.122**, based on the rating of the overcurrent protection device protecting the branch-circuit conductors.

Grounding of the junction box and wet-niche fixtures shall be accomplished by a 12 AWG copper insulated equipment grounding conductor. The grounding conductor shall be sized according to the rating of the overcurrent protection device per **Table 250.122**. It shall not be smaller than 12 AWG and shall be unspliced unless the exception is applied.

The grounding conductor shall be installed and run with the circuit conductors in rigid metal conduit, intermediate metal conduit or rigid nonmetallic conduit from the panelboard to the deck box. The main function of the 12 AWG equipment grounding conductor is to clear the overcurrent protection device in case of a ground-fault. **(See Figure 11-119)**

Flexible cords used to ground wet-niche fixtures shall be provided with a 12 AWG or larger insulated copper equipment grounding conductors connected to a terminal in the supply deck box. The type of connection provides continuity when the luminaire (lighting fixture) is removed for servicing.

> **Design Tip:** The 12 AWG equipment grounding conductor is used to clear a ground-fault while the 8 AWG bonding jumper is utilized to bond the metal parts of the pool to the grid system.

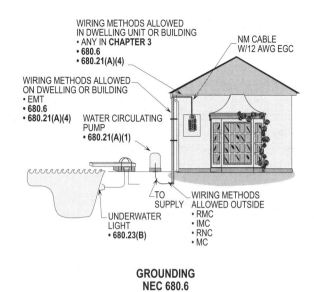

Figure 11-119. This illustration shows the types of wiring methods permitted to be installed for swimming pools.

HOT TUBS, SPAS AND HYDROMASSAGE TUBS
ARTICLE 680, PARTS IV AND VII

Hot tubs, spas and hydromassage tubs shall be bonded and grounded for the same reasons as swimming pools for the safety of those using them. Such tubs shall be permitted to be located inside or outside of the dwelling unit or building.

BONDING
680.42(B)

Bonding by metal-to-metal mounting on a common frame or base shall be permitted. The metal band or hoops that are used to secure wooden staves shall not be required to bonded per **680.26**. **(See Figure 11-120)**

GROUNDING
680.43(F)

The following electrical equipment shall be grounded:

- All electrical equipment located within 5 ft. (1.5 m) of the inside wall of the spa or hot tub.
- All electrical equipment associated with the circulating system of the spa or hot tub.

Grounding

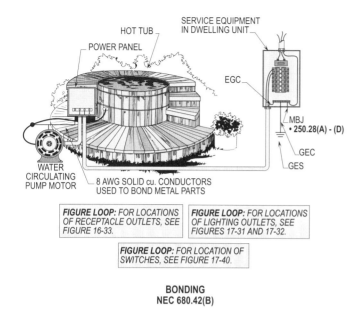

BONDING
NEC 680.42(B)

Figure 11-120. A 8 AWG solid copper conductor shall be used to ground all noncurrent-carrying parts associated with the hot tub including metal piping, etc. located within 5 ft. (1.5 m) of the tub.

HYDROMASSAGE BATHTUBS
680.70; 680.71; 680.72

Hydromassage bathtubs are treated in the same manner as conventional bathtubs. The wiring methods for hydromassage bathtubs shall comply with **Chapters 1 through 4** in the NEC. All elements for hydromassage bathtubs shall be supplied by GFCI-protected circuits. See **410.4(D)** for hanging luminaires (fixtures) over and around the tub. Location of switches for luminaires (lighting fixtures) and receptacle outlets are treated in the same manner as a regular bathtub. **(See Figure 11-121)**

Design Tip: See **404.4** and **406.8(C)** for locating and installing switches and receptacles which are not associated with the hydromassage tub. Access to the circulating motor for servicing is required per **Article 100, 430.14(A)** and **680.73**.

GROUNDING ANTENNA SYSTEMS
PER NEC AND UL

Satellite dishes, CATV and antenna systems shall be grounded for safety. Inspectors find that these systems are not bonded and grounded properly when making an inspection of such systems. Many times they are not grounded into the grounding electrode system, but are grounded independently with different potentials to ground. **(See Figure 11-122)**

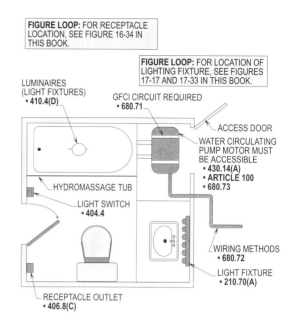

HYDROMASSAGE BATHTUBS
NEC 680.70 THRU 74

Figure 11-121. Hydromassage bathtubs are treated in the same manner as regular bathtubs except they shall be supplied with a GFCI-protected circuit to protect the users from electrical shock. Luminaires (lighting fixtures) shall comply with **410.4(D)**. Switches and receptacles shall meet the provisions of **404.4** and **406.8(C)**.

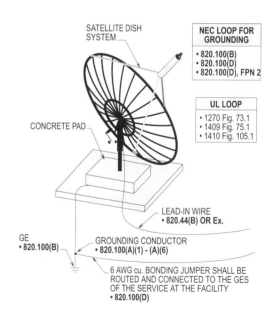

GROUNDING ANTENNA SYSTEMS
PER NEC AND UL

Figure 11-122. The illustraton shows the proper grounding of a satellite dish system.

Name   Date

Chapter 11. Grounding

Section	Answer		
_____	T	F	1. All AC circuits of less than 50 volts shall not be required to be grounded.
_____	T	F	2. AC systems of 1 kV and over shall be grounded if supplying mobile or portable equipment.
_____	T	F	3. The frame of a portable generator shall be grounded if it supplies only the equipment on the generator or cord-and-plug connected equipment connected receptacles mounted on the generator.
_____	T	F	4. AC grounded systems shall be grounded at each service and outside transformer if the system has a grounded conductor.
_____	T	F	5. Where one or more buildings or structures are supplied from a common AC grounded service, each grounded service at each individual building or structure shall be separately grounded.
_____	T	F	6. The equipment grounding conductor shall be bonded to the grounding electrode conductor at the service equipment neutral busbar terminal for a grounded system.
_____	T	F	7. A GFCI receptacle shall be permitted to be used to protect a single outlet and additional outlets downstream when replacing nongrounding-type receptacles.
_____	T	F	8. The metal cases of portable cord-and-plug connected stationary equipment cannot be grounded by means of metal enclosures and building steel.
_____	T	F	9. New branch-circuit installations for grounding the frames of ranges, cooktops, ovens, clothes dryers, and junction boxes or outlet boxes that are part of the circuit shall be permitted to bonded and grounded with an equipment grounding conductor and a grounded neutral conductor shall be installed.
_____	T	F	10. The grounded phase or neutral conductor shall not be permitted to be used as a current-carrying conductor and grounding means on the supply side of the service disconnecting means and secondary side of a separately derived system.
_____	T	F	11. Service-entrance raceways, meter base enclosures, service equipment enclosures, and all other metal enclosures of the service equipment shall not be required to be bonded together.
_____	T	F	12. The metal gas piping shall be electrically continuous and bonded to the grounding electrode system. (General Rule)
_____	T	F	13. The size of equipment bonding jumpers and the service equipment main bonding jumper shall be based on the size of the service-entrance conductors.

Section	Answer		
_____	T F	14.	It is permitted to install the equipment bonding jumper either inside or outside of the raceway or enclosure in 6 ft. or less lengths. (using flex)
_____	T F	15.	The grounding electrode system shall consist of one electrode only.
_____	T F	16.	The grounding electrode conductor shall be installed within 5 ft. where a metal water pipe enters a commercial building.
_____	T F	17.	Lightning rod conductors and made electrodes which are connected to lightning rods for grounding shall not be used in place of made electrodes for grounding electrical wiring systems and equipment.
_____	T F	18.	Grounding electrode conductors shall only be permitted to be installed as copper.
_____	T F	19.	All electrical equipment located within 5 ft. of the inside walls of the pool shall be bonded and grounded.
_____	T F	20.	Satellite dishes, CATV, and antenna systems shall be grounded.

21. Circuits for electric cranes operating over _____ fibers in Class III locations shall not be required to be grounded.

22. Unbalanced loads along with _____ grounding points may have objectionable current flowing in grounding conductors within the circuits of the wiring system.

23. Temporary currents develop from _____ shorts such as ground-fault currents that flow over grounding conductors are not considered objectionable currents.

24. The grounded conductor whether a grounded phase or neutral is usually installed with a _____ or natural gray insulation or otherwise identified.

25. The bonding jumper shall be sized at least _____ percent of the area of the largest conductor where the service conductors are installed larger than 1100 KCMIL copper or 1750 KCMIL aluminum.

26. High-impedance grounding is accomplished by inserting a grounding _____ between the grounding electrode conductor and the grounded conductor.

27. A _____ flexible wire or strap shall be permitted to be used to ground portable equipment.

28. Existing branch-circuit installations for frames or ranges, cooktops, ovens, clothes dryers, and junction boxes or outlet boxes that are part of the circuit shall be permitted to be bonded and grounded with grounded (neutral) conductor provided it is installed no smaller than _____ copper.

29. Bonding is the permanent joining of _____ parts to form an electrically conductive path that will assure electrical continuity.

30. Threadless couplings and connectors installed for rigid metal conduit, intermediate metal conduit, and electrical metallic tubing shall be made up _____.

	Section	Answer

31. A flush-mounted box may be installed with self-grounding _____ to ground receptacles with a bonding jumper.

32. An _____ equipment grounding conductor shall be used to connect the receptacle isolated ground terminal back to the grounded connection at the service equipment.

33. A _____ spacer fitting shall be permitted to be installed between the metal raceway and sensitive electronic equipment at the point of connection to reduce electrical noise.

34. At the service equipment, a screw or wire shall be permitted to be used as the _____ bonding jumper.

35. When an equipment bonding jumper is installed outside of the raceway or enclosure, it shall not be over _____ ft. in length. (using flex)

36. Exposed interior building steel frames that are not _____ grounded shall be bonded to either the service enclosure, the neutral bus, the grounded electrode conductor, or to a grounding electrode.

37. A grounding electrode conductor shall be permitted to be _____ in busbars such as meter enclosures, etc.

38. Liquidtight flexible metal conduit 3/8 and 1/2 in. trade sizes shall be protected by overcurrent devices rated at _____ amps or less.

39. The metal frame of a building or structure shall have _____ ft. or more of a single structural metal member in direct contact with the earth.

40. Grounding of the junction box and wet-niche fixtures for a pool is accomplished by installing a _____ AWG copper insulated equipment grounding conductor.

41. Which of the following AC circuits and systems shall be grounded?
    (a) AC circuits of 50 to 1000 volts
    (b) Separately derived systems
    (c) None of the above
    (d) All of the above

42. The rating of the grounded conductor shall be installed at least _____ percent of the largest ungrounded phase conductor.
    (a) 10
    (b) 12 1/2
    (c) 15 1/2
    (d) 25

43. Which of the following methods can be utilized for the bonding of service quipment?
    (a) Grounded service conductor
    (b) Threaded connections
    (c) Bonding jumpers
    (d) All of the above

Section	Answer

44. The interior metal water piping, located more than _____ ft. from the point of entrance to the building, shall not be permitted to be used as part of the grounding electrode system? (Splicing other GEC's to it)
    (a) 5
    (b) 6
    (c) 10
    (d) 20

45. Metal water pipe lengths of at least _____ ft. long in the earth shall be connected to the grounded neutral bar in the service equipment enclosure.
    (a) 5
    (b) 10
    (c) 15
    (d) 20

46. A concrete-encased electrode no smaller than _____ AWG bare copper conductor at least 20 ft. long shall be permitted to be installed in the footing of the foundation.
    (a) 8
    (b) 4
    (c) 2
    (d) 1

47. A bare copper conductor not smaller than 2 AWG and at least 20 ft. in length shall be installed at least _____ ft. below the earth to create a ground ring grounding system for encircling the building or structure.
    (a) 2 1/2
    (b) 4 1/2
    (c) 5 1/2
    (d) 6 1/2

48. A driven electrode used to ground the service can be a _____ in. x _____ ft. copper (brass) rod.
    (a) 1/2 x 8
    (b) 1/2 x 10
    (c) 3/4 x 8
    (d) 3/4 x 10

49. Made electrodes shall have a resistance to ground of _____ ohms or less.
    (a) 10
    (b) 20
    (c) 25
    (d) 50

50. Liquidtight flexible metal conduit can be used in sizes up to 1 1/4 in. which has a total length not to exceed _____ ft. from the ground return path and with fittings for terminating the flexible conduit.
    (a) 3
    (b) 5
    (c) 6
    (d) 10

51. What size copper bonding jumper and grounding electrode conductor is required to bond the secondary side of a separately derived system having 2/0 AWG THWN copper conductors?

	Section	Answer

52. What size copper main bonding jumper is required for service equipment supplied with phase conductors of 500 KCMIL copper?

53. What size copper equipment bonding jumper is required where three raceways enter a switchgear with each raceway individually bonded to the enclosure supplied with phase conductors of 250 KCMIL copper?

54. What size copper EBJ is required where three raceways enter a switchgear with only one bonding jumper used to bond the raceway? Service conductors are 250 KCMIL copper, paralleled 3 times. (Supply Side)

55. What size copper grounding electrode conductor is required to be installed to a metal water pipe supplied with phase conductors of 2/0 AWG THWN copper?

56. What size copper grounding electrode conductor is required to be installed for a concrete-encased electrode supplied with phase conductors of 3/0 AWG THWN copper?

57. What size copper equipment grounding conductor is required to be installed for an A/C unit supplied by an 80 amp circuit breaker?

12

Designing Wiring Methods

Boxes and conduits shall be sized and selected to accommodate the number of conductors required to supply power to various pieces of electrical equipment in residential, commercial and industrial locations. Boxes, in addition to housing conductors, shall have enough fill area to provide sufficient space for receptacles, switches, dimmers and combination devices without damaging the conductor's insulation. When boxes and conduits are not sized properly, conductors are sometimes packed in undersized boxes and are pinched by large size devices that intrude into the box's fill area. Therefore, conductors, devices, fittings and clamps shall be counted and the appropriate box selected. Boxes and conduits shall be sized based on the number of conductors being either the same size and having the same characteristics or different sizes and having different characteristics.

BOX FILL CALCULATIONS
314.16(B)

The size box selected shall be based on the number and size of conductors, devices and fittings that are contained within the box. The designer, installer and inspector shall count the number of items that are in the box and select the size box from **Table 314.16(A)** based upon the maximum number of conductors or the minimum cubic inch capacity rating. The column for the maximum number of conductors shall be used when the conductors are the same size. The appropriate cubic inch capacity column shall be applied for combination conductors in the same box.

CONDUCTOR FILL
314.16(B)(1)

Conductors passing through the box unbroken and not pulled into a loop or spliced together with scotchlocks shall be counted as one conductor. Conductors passing through boxes are utilized to supply power to boxes supporting luminaires (lighting fixtures), receptacles, etc. Spliced conductors shall be counted as one conductor for each.

> **For example:** What is the count used for ten conductors passing straight through an octagon box and the count for ten that are spliced?
>
> **Step 1:** Counting conductors
> 314.16(B)(1)
> Ten passing through counted as 10
>
> **Solution:** Ten conductors are counted for the ten conductors passing through the box.
>
> **Step 2:** Counting conductors
> 314.16(B)(1)
> Ten spliced x 2 count as 20
>
> **Solution:** Twenty conductors are counted for the ten spliced conductors in the box.

See Figures 12-1(a) and (b) for counting the number of conductors passing through or spliced in a box.

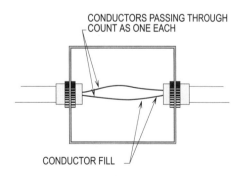

CONDUCTOR FILL
NEC 314.16(B)(1)

Figure 12-1(a). Conductors passing straight through the box shall be counted as one each.

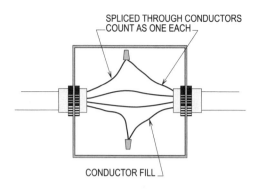

CONDUCTOR FILL
NEC 314.16(B)(1)

Figure 12-1(b). Spliced conductors shall be counted as one for each conductor spliced in the box.

CONDUCTOR FILL
314.16(B)(1)

A conductor that is not entirely within the box shall not be required to be counted as a conductor used toward the box fill. A pigtail from a splice within the box to a receptacle or a switch mounted to the box or a bonding jumper from the receptacle yoke bonded to the box is an example of such a conductor. **(See Figure 12-2)**

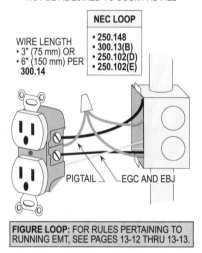

CONDUCTOR FILL
NEC 314.16(B)(1)

Figure 12-2. A pigtail or bonding jumper shall not be required to be counted toward the fill space of the box.

CONDUCTOR FILL
314.16(B)(1), Ex.

Wires from luminaires (fixtures) shall be counted in determining the size of the box. Luminaire (fixture) wire sizes from 18 AWG to 12 AWG or larger shall be counted to determine the size box used to support the luminaire (fixture). The **Ex. to 314.16(B)(1)** permits the canopy of a luminaire (fixture) to be used to house conductors or four or less luminaire (fixture) wires in sizes smaller than 14 AWG plus an equipment grounding conductor. It is not necessary to count these luminaire (fixture) wires toward the fill of the ceiling box supporting the luminaire (fixture) or ceiling fan where there's a canopy.

Review and verify both the luminaire (fixture) box and luminaire (fixture) canopy can be used for fill area. If so, a smaller ceiling box shall be permitted to be installed to support the luminaire (fixture).

Design Tip: Until the 1984 NEC was published, luminaire (fixture) wires were not counted in determining the fill space of a box.

Note: The 1984 NEC required luminaire (fixture) wires 14 AWG and larger to be counted and the 1987 NEC required luminaire (fixture) wire in sizes 18 AWG and 16 AWG to be counted. However, these requirements shall not be permitted to be retroactive to existing installations which would require larger boxes.

For example: What is counted toward the fill space of two boxes where there are 2 - 14 AWG luminaire (fixture) wires without the use of a canopy and 3 - 16 AWG luminaire (fixture) wires plus an equipment grounding conductor utilizing a canopy?

Step 1:   Counted conductors
          (luminaire (fixture) wire only)
          314.16(B)(1), Ex.
          2 - 14 AWG luminaire (fixture) wires
          count as 2

Solution: **Two luminaire (fixture) wires are counted as 2 conductors in the box.**

Step 1:   Counted conductors
          (luminaire (fixture) wire only)
          314.16(B)(1), Ex.
          3 - 16 AWG luminaire (fixture) wires
          count is 0

Solution: **The luminaire (fixture) wires plus the equipment grounding conductor do not count toward the fill of the box.**

See Figure 12-3 for the procedures used to count luminaire (fixture) wires.

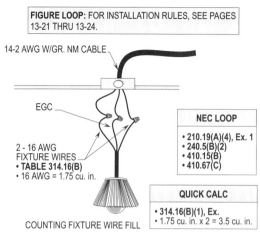

**CONDUCTOR FILL
NEC 314.16(B)(1), Ex.**

Figure 12-3. Luminaire (fixture) wires shall be counted in determining the size box to support a luminaire (fixture).

Note: Luminaire (fixture) wires smaller than 14 AWG which are inside a luminaire (fixture) canopy shall not be required to be counted toward the fill space for sizing a box.

CLAMP FILL
314.16(B)(2)

Boxes with one or more cable clamps that are installed to support cables in the box shall have one conductor added toward the fill computation to determine the size box. The conductor used to represent the one or more cable clamps shall be based on the largest conductor in any one cable entering the box. The outer covering of the cable that is used to protect the insulation of the conductors shall be extended 1/4 in. (6 mm) past the clamp in the box to prevent the clamp from damaging the insulation of the conductor. When the clamp is tightened to support the cable, this extended covering beyond the clamp will prevent the insulation of the conductor from being pinched or nicked. The arcing or sparking of a bare conductor to metal that is grounded may cause a fire hazard. **(See Figure 12-4)**

Note: See the next design tip below "support fitting fill" as to the time when the NEC required cable clamps, luminaire (fixture) studs and hickeys to be counted as one conductor.

SUPPORT FITTINGS FILL
314.16(B)(3)

Boxes that contain luminaire (fixture) studs or hickeys shall have one conductor added for each fitting. The conductor for each fitting is selected and based on the largest conductor entering the box. By adding a conductor for each fitting in the box, a larger box is required to be selected and helps prevent an undersized box. Conductors in boxes shall have sufficient room in the box per **110.7** and **314.16** to breathe and dissipate heat and prevent the overheating of its insulation per **310.10, FPN's (1) through (4)**.

For example, a box contains cables of 14-2 AWG and 12-2 AWG w/ground and is equipped with a luminaire (fixture) stud and hickey. The size conductors used for the count of the stud and hickey shall be determined by each one being counted as a 12 AWG. In other words, the two fittings count as 12 AWG conductors and shall be based on the largest conductor entering the box which is the 12-2 AWG w/ground cable. **(See Figure 12-5)**

Design Tip: Luminaire (fixture) studs, hickeys and cable clamps were counted as one conductor by the 1978 NEC and previous editions. It did not matter how many were present, the total number counted as one. The 1981 NEC required luminaire (fixture) studs and hickeys to be counted as one each. However, these rules are not retroactive to installations before the publication of the 1981 NEC. The designer must be careful when routing larger conductors with smaller conductors through or in the same box, for a larger box may be needed due to such protection.

conductor that is connected to the receptacle or switch. One strap or yoke could have one, two or three receptacles mounted on it and still counted as two conductors. Article 100 defines a receptacle as a single receptacle having a contact device with no other contact device on the same strap. A multiple receptacle is a single device with two or more receptacles mounted to it. A duplex device has two receptacles and despar device may have three or four receptacles mounted on it. The same rules apply for switches. **(See Figure 12-6)**

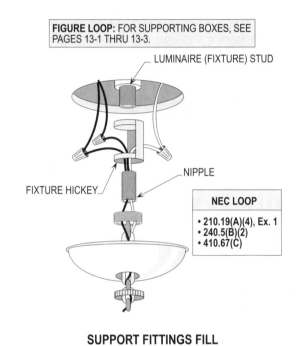

SUPPORT FITTINGS FILL
NEC 314.16(B)(3)

Figure 12-5. Boxes containing luminaire (fixture) studs or hickeys shall have one conductor added for each fitting.

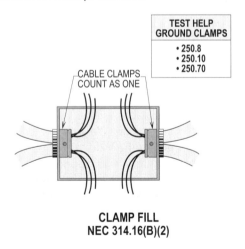

CLAMP FILL
NEC 314.16(B)(2)

Figure 12-4. Boxes with one or more cable clamps shall have one conductor counted toward the fill space of the box.

DEVICE OR EQUIPMENT FILL
314.16(B)(4)

A receptacle or switch that is mounted on a strap or yoke shall be counted as two conductors in determining the fill of the box. The count of two shall be based on the size

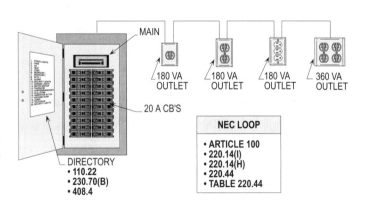

DEVICE OR EQUIPMENT
NEC 314.16(B)(4)

Figure 12-6. The yoke or strap with mounted devices shall be counted as two conductors based on the size conductors connecting to its terminals.

EQUIPMENT GROUNDING CONDUCTOR FILL
314.16(B)(5)

Equipment grounding conductors passing through or spliced together in the box count shall be counted as one. It doesn't matter how many equipment grounding conductors are installed in a box. All the equipment grounding conductors added together shall be counted as one. It is a good recommendation to place the equipment grounding conductors into the first box. The equipment grounding conductors may be insulated or bare but shall be bonded to the box if the box is the metal type. Should the bare copper of one of the equipment grounding conductors come in contact with the metal of the box, it would not create a problem because it is bonded to the box anyway per **250.148(A)**.

For example: What is the count toward the fill of the box where there are six equipment grounding conductors in the box that are used to ground the metal enclosures of the equipment.

Step 1: Counting conductors
314.16(B)(5)
Six EGC's count as 1

Solution: One conductor is counted for the 6 equipment grounding conductors in the box.

The only time that equipment grounding conductors in a box count more than one is where an isolation equipment grounding conductor (IEGC) is run to an isolation receptacle and used to ground the metal enclosure of a PC. The isolation equipment grounding conductor is routed with the equipment grounding conductor in the same conduit and bypasses the metal yoke (strap) of the receptacle and connects directly to the metal enclosure of the PC. All the regular equipment grounding conductors shall be counted as one plus one for the isolation equipment grounding conductor for a total of two. See **250.146(D)** and **408.40, Ex.** for the requirements of installing the isolation equipment grounding conductor.

Design Tip: An isolation equipment grounding conductor shall be permitted to be used to ground the metal of any type of cord-and-plug connected sensitive electronic equipment to reduce electromagnetic noise.

For example: How many conductors are counted for one equipment grounding conductor to a metal box and one isolation equipment grounding conductor for a PC?

Step 1: Counting conductors
314.16(B)(5)
1 EGC counts as 1
1 IEGC counts as 1

Solution: There are 2 conductors counted for the equipment grounding conductor and isolation equipment grounding conductor in the box.

See Figures 12-7(a) and (b) for a detailed illustration of installing equipment grounding conductors and isolation grounding conductors in boxes.

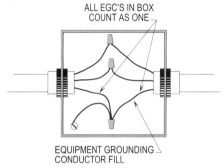

EQUIPMENT GROUNDING CONDUCTOR FILL
NEC 314.16(B)(5)

Figure 12-7(a). Equipment grounding conductors passing through or spliced together in the box shall be counted as one

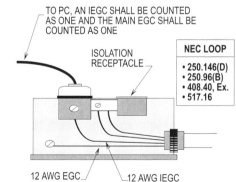

EQUIPMENT GROUNDING CONDUCTOR FILL
NEC 314.16(B)(5)

Figure 12-7(b). One additional conductor shall be added to the count where an isolation equipment grounding conductor is contained in the box.

BOXES
314.16(A), Table 314.16(A)

Boxes are usually required to be mounted to support luminaires (fixtures), devices, fans, etc. Boxes are available in four main types from **Table 314.16(A)**. The four types most used are octagonal, round, square or device. The octagonal or round box is used to mount luminaires (lighting fixtures), ceiling fans or for splicing conductors. Square boxes are used with plaster rings to mount and support luminaires (lighting fixtures), ceiling fans or devices. They are also used for the splicing of conductors.

Device boxes are used to support the yoke (strap) of receptacles, switches or combinations of receptacles, switches, pilot lights, etc. They are also used to splice conductors. **Table 314.16(A)** lists the sizes of these boxes in cubic inches (cu. in.) and the size conductors of the same size in rating from 18 AWG through 6 AWG. See **314.28(A)(1)** and **(A)(2)** to size boxes for conductors sized 4 AWG and larger.

Based upon the number of the same size conductors or cubic inch ratings of combination conductors (18 AWG through 6 AWG) the size box shall be determined from **Table 314.16(A)**.

The requirements of **314.16(A)(1)** and **(B)(1) through (B)(5)** shall be applied to calculate the number of conductors or cubic inch value of each conductor used to select the size box from **Table 314.16(A)**.

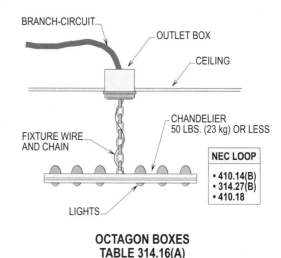

Figure 12-8(a). Outlet boxes shall not be permitted to support luminaires (lighting fixtures) weighing more than 50 lb (23 kg).

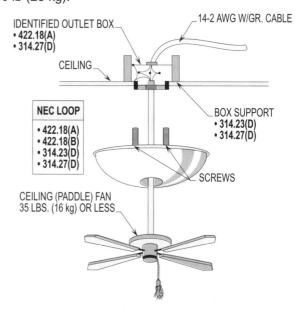

Figure 12-8(b). Outlet boxes shall not support ceiling (paddle) fans weighing more than 35 lb (16 kg).

OCTAGON BOXES
Table 314.16(A)

Octagon boxes are used mainly to mount and properly support luminaires (lighting fixtures). They are also utilized for splicing the conductors of branch-circuits or feeder-circuits supplying power to various pieces of electrical apparatus such as ranges, cooktops, ovens, heating units, AC units, etc. Special octagon boxes that are listed shall be permitted to support ceiling fans. Octagon boxes that are installed to support luminaires (lighting fixtures) shall not be permitted to support luminaires (lighting fixtures) that weigh more than 50 lb (23 kg) per **314.27(B)**. Ceiling fans that are supported to listed octagon boxes shall weigh 35 lb (16 kg) or less per **422.18(A)** and **314.27(D)**. (See Figure 12-8(a); (b))

Design Tip: Reference **314.23** and **314.27** for the rules pertaining to the proper methods of supporting boxes.

SAME SIZE CONDUCTORS
314.16(B)(1) through (B)(5), Table 314.16(A)

Boxes containing conductors that are the same size shall have their total fill space determined by adding the number of conductors plus additional conductors for each fitting or device. The cubic inch value listed in **Table 314.16(B)** for the largest conductor in the box shall be used to determine the fill capacity for the fittings. For devices, use the size conductor that is connected to its terminals.

Designing Wiring Methods

For example: What size octagon box is required to support a luminaire (lighting fixture) supplied with 2 - 14-2 AWG w/ground nonmetallic sheathed cables (romex) with a luminaire (fixture) stud, hickey and 2 - 14 AWG luminaire (fixture) wires? Neutrals are spliced with a pigtail and romex connectors are used instead of romex clamps.

Step 1: Same size conductors
314.16(B)(1) through (B)(5)
(Conductor fill)
314.16(B)(1)
2 - 14 AWG hots = 2
2 - 14 AWG neutrals = 2
(EGC fill)
314.16(B)(5)
2 - 14 AWG EGC's = 1
(Support fitting fill)
314.16(B)(3)
1 fixture stud = 1
1 hickey = 1
(Conductor fill)
314.16(B)(1)
1 pigtail = 0
(Clamp fill)
314.16(B)(2)
2 romex connectors = 0
(Fixture wire fill)
314.16(B)(1), Ex.
2 - 14 AWG fixture wires = 2
Total = 9

Step 2: Selecting box
Table 314.16(A)
9 - 14 AWG conductors requires
4" x 2 1/8" (100 mm x 54 mm) box

Solution: A 4 in. x 2 1/8 in. (100 mm x 54 mm) octagon box is required.

See Figure 12-9 for a step by step procedure on sizing and selecting octagon boxes where the conductors are the same size.

Problem: What size octagon box is required to support a luminaire (lighting fixture) with a canopy supplied with 2 - 12-2 AWG w/ground nonmetallic sheathed cables (romex) with a luminaire (fixture) stud, hickey and 4 - 18 AWG luminaire (fixture) wires, plus EGC's?

Step 1: Counting conductors
314.16(B)(1) thru (5)
(Conductor fill)
2 - 12 AWG hots = 2
2 - 12 AWG neutrals = 2
(EGC fill)
2 - 12 AWG EGC's = 1
(Support and fitting fill)
1 fixture stud = 1
1 hickey = 1
(Fixture wire fill)
3 - 18 AWG fixture wires = 0
plus EGC = 0
Total = 7

Step 2: Selecting box
Table 314.16(A)
7 - 12 conductor requires a
4" x 2 1/8" (100 mm x 54 mm) box

Solution: A 4 in. x 2 1/8 in. (100 mm x 54 mm) octagon box is required.

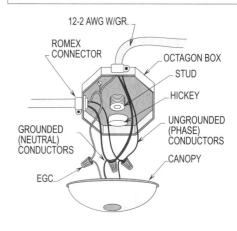

**SAME SIZE CONDUCTORS
NEC 314.16(B)(1) THRU (5)
NEC 314.16(A)**

Figure 12-9. Sections **314.16(B)(1) through (B)(5)** and **Table 314.16(A)** are utilized to size and select an octagon box based on the box containing the same size conductors. **Note:** The canopy is used for the fill of the luminaire (fixture) wires.

DIFFERENT SIZE CONDUCTORS
314.16(B)(1) through (B)(5); Table 314.16(B)

Boxes containing conductors that are not the same size shall be calculated based on the cubic inch rating of each conductor listed in **Table 314.16(B)**. The cubic inch rating of each conductor shall be added together plus the number of fittings or devices. The cubic inch rating used for each fitting shall be based on the largest conductor entering the box. The cubic inch rating for yokes or straps shall be based on the size conductor that is connected to the switch or receptacle.

Many times the luminaire (fixture) wires are 18 AWG or 16 AWG which allows more flexibility for terminating such wires to the branch-circuit conductors, the fill of the box shall be determined by applying the combination calculation per **Table 314.16(B)**.

> **For example:** What size octagon box is required to support a luminaire (lighting fixture) with 2 - 12-2 AWG w/ground nonmetallic sheathed cables (rope) with a luminaire (fixture) stud, two cable clamps, two pigtails and 2 - 18 AWG fixture wires? (No luminaire (fixture) canopy)

> **Step 1:** Combination conductors
> **314.16(B)(1) through (B)(5);**
> **Table 314.16(B)**
> **(Conductor fill)**
> **314.16(B)(1)**
> 2 - 12 AWG Hots
> 2.25 cu. in. x 2 = 4.5 cu. in.
> 2 - 12 AWG Neutrals
> 2.25 cu. in. x 2 = 4.5 cu. in.
> **(EGC fill)**
> **314.16(B)(5)**
> 2 - 12 AWG EGC's
> 2.25 cu. in. x 1 = 2.25 cu. in.
> **(Supporting fitting fill)**
> **314.16(B)(3)**
> 1 Fixture Stud
> 2.25 cu. in. x 1 = 2.25 cu. in.
> **(Cable clamp fill)**
> **314.16(B)(2)**
> 2 Cable Clamps
> 2.25 cu. in. x 1 = 2.25 cu. in.
> **(Conductor fill)**
> **314.16(B)(1)**
> 2 Pigtails = 0 cu. in.
> **(Fixture wire fill)**
> **314.16(B)(1), Ex.**
> 2 - 18 AWG luminaire (fixture) wires
> 1.5 cu. in. x 2 = 3.0 cu. in.
> Total = 18.75 cu. in.
>
> **Step 2:** Selecting Box
> **Table 314.16(A)**
> 18.75 cu. in. requires 4" x 2 1/8"
> (100 mm x 54 mm) box
>
> **Solution: A 4 in. x 2 1/8 in. (100 mm x 54 mm) octagon box is required.**

There are installations where different size conductors are spliced in boxes and fed out of the boxes to supply power to loads. The boxes are used as junction points for the branch-circuits between the panelboard and loads.

> **For example:** What size octagon box with cable clamps is required for 2 - 14 AWG and 2 - 12 AWG w/ground romex cables that are spliced in the box between the panelboard and the loads?

> **Step 1:** Combination conductors
> **314.16(B)(1) through (B)(5);**
> **Table 314.16(B)**
> **(Conductor fill)**
> **314.16(B)(1)**
> 2 - 12 AWG Hots
> 2.25 cu. in. x 2 = 4.5 cu. in.
> 2 - 12 AWG Neutrals
> 2.25 cu. in. x 2 = 4.5 cu. in.
> 2 - 14 AWG Hots
> 2 cu. in. x 2 = 4 cu. in.
> 2 - 14 AWG Neutrals
> 2 cu. in. x 2 = 4 cu. in.
> **(EGC fill)**
> **314.16(B)(5)**
> 2 - 12 AWG EGC's
> 2.25 cu. in. x 1 = 2.25 cu. in.
> 2 - 14 AWG EGC's
> 2 cu. in. x 0 = 0
> **(Cable clamp fill)**
> **314.16(B)(2)**
> 2 Cable Clamps
> 2.25 cu. in. x 1 = 2.25 cu. in.
> Total = 21.5 cu. in.
>
> **Step 2:** Selecting box
> **Table 314.16(A)**
> 21.5 cu. in. requires 4" x 2 1/8"
> (100 mm x 54 mm)
>
> **Solution: A 4 in. x 2 1/8 in. (100 mm x 54 mm) octagon box is required**

See Figure 12-10 for a step by step procedure on sizing and selecting octagon boxes where the conductors are different sizes.

Designing Wiring Methods

Problem: What size octagon box is required for a 12-2 AWG w/ground cable and 3 - 16 AWG luminaire (fixture) wires?

Step 1: Counting conductors
314.16(B)(1) thru (5); Table 314.16(B)
(Conductor fill)
1 -12 AWG hot   2.25 cu. in. x 1 = 2.25
1 -12 AWG neutral   2.25 cu. in. x 1 = 2.25
(EGC fill)
1 -12 AWG EGC   2.25 cu. in. x 1 = 2.25
1 -16 AWG EGC   = 0
(Fixture wire fill)
2 -16 AWG fix. wires 1.75 cu. in. x 2 = 3.5
Total   = 10.25

Step 2: Selecting box
Table 314.16(A)
10.25 cu. in. requires 4" x 1/4" box
(100 mm x 32 mm)

Solution: A 4 in. x 1 1/4 in. (100 mm x 32 mm) octagon box is required.

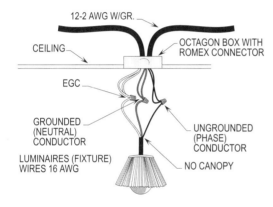

DIFFERENT SIZE CONDUCTORS
NEC 314.16(B)(1) THRU (B)(5)
TABLE 314.16(B)

Figure 12-10. Sections **314.16(B)(1) through (B)(5)** and **Table 314.16(B)** are utilized to size and select the box based on an octagon box containing different size conductors. The cubic inch of each conductor times the number derives the size box.

SQUARE BOXES
Table 314.16(A)

Square boxes have greater cubic inch ratings than octagon boxes. Square boxes are usually installed where more than four romex cables enter and leave the box. Square boxes may be used with plaster rings that are designed to accommodate the mounting of a receptacle, switch, light, etc.

The rise on the plaster ring may be 1/4, 1/2 or 3/4 inches to provide more space inside to mount receptacles and switches. This extra space in the box due to the rise of the plaster ring permits a receptacle or switch with a greater depth to be installed. Plaster rings are also used to extend the surface of the box inside the wall or ceiling to the face of the wall or ceiling where a receptacle (device) or luminaire (lighting fixture) may be mounted. Square boxes are utilized to splice circuit conductors to supply power to various electrical loads in dwelling units or buildings.

SAME SIZE CONDUCTORS
314.16(B)(1) through (B)(5); Table 314.16(A)

Section **314.16(A)** lists the items in or out of the box and the number of conductors that shall be added for each item based on the condition of use. Where all conductors are the same size, each item in or out of the box shall be counted based on these same size conductors.

For example: What size square box is required for 2 - 12-2 AWG w/ground romex cables connecting to a receptacle and 4 - 12-2 AWG w/ground romex cables that pass through the box? (Use an extension ring.)

Step 1: Same size conductors
314.16(B)(1) through (B)(5)
(Conductor fill)
314.16(B)(1)
2 - 12 AWG Hots   = 2
2 - 12 AWG Neutrals   = 2
4 - 12 AWG Hots (through)
4 x 2   = 8
4 - 12 Neutrals (through)
4 x 2   = 8
(EGC fill)
314.16(B)(5)
10 - 12 AWG EGC's   = 1
Total   = 21

Step 2: Selecting box
Table 314.16(A)
21 - 12 AWG requires 4 11/16" x 1 1/4" (120 mm x 32 mm) with ring

Solution: A 4 11/16 in. x 1 1/4 in. (120 mm x 32 mm) square box with extension ring is required. Note: Extension ring allows 22 - 12 AWG conductors to be installed per 314.16(A).

See Figure 12-11 for a step by step procedure on sizing and selecting square boxes where the conductors in the box are the same size.

12-9

Stallcup's Electrical Design Book

> **Problem:** What size square box is required to contain 4 -12-2 AWG w/ground cables which are supported within in the box with cable clamps?
>
> **Step 1:** Counting conductors
> 314.16(B)(1) thru (B)(5)
> (Conductor fill)
> 4 -12 AWG hots = 4
> 4 -12 AWG neutrals = 4
> (EGC fill)
> 4 12 AWG EGC's = 1
> (Clamp fill)
> 2 cable clamps = 1
> Total = 10
>
> **Step 2:** Selecting box
> Table 314.16(A)
> 10 -12 AWG conductors requires 4 11/16" x 1 1/4" (120 mm x 32 mm)
>
> **Solution:** A 4 11/16 in. x 1 1/4 in. (120 mm x 32 mm) square box is required.

> **For example:** What size square box is required where 4 - 14 AWG and 4 - 12 AWG romex cables w/grounds are spliced and routed to loads in the dwelling unit or building?
>
> **Step 1:** Combination conductors
> 314.16(B)(1) through (B)(5);
> Table 314.16(B)
> (Conductor fill)
> 314.16(B)(1)
> 4 - 14 AWG Hots
> 2 cu. in. x 4 = 8 cu. in.
> 4 - 14 AWG Neutrals
> 2 cu. in. x 4 = 8 cu. in.
> 4 - 12 AWG Hots
> 2.25 cu. in. x 4 = 9 cu. in.
> 4 - 12 AWG Neutrals
> 2.25 cu. in. x 4 = 9 cu. in.
> (EGC fill)
> 314.16(B)(5)
> 4 - 12 AWG EGC's
> 2.25 cu. in. x 1 = 2.25 cu. in.
> Total = 36.25 cu. in.
>
> **Step 2:** Selecting box
> Table 314.16(A)
> 36.25 cu. in. requires
> 4 11/16" x 2 1/8" (120 mm x 54 mm)
>
> **Solution:** A 4 11/16 in. x 2 1/8 in. (120 mm x 54 mm) square box is required.

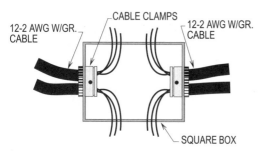

**SAME SIZE CONDUCTORS
NEC 314.16(B)(1) THRU (5)
TABLE 314.16(A)**

Figure 12-11. Sections **314.16(B)(1) through (B)(5)** and **Table 314.16(B)** are utilized to size and select a square box based on the box containing the same size conductors.

See **Figure 12-12** for a step by step procedure for sizing and selecting square boxes where the conductors are different sizes.

DEVICE BOXES
Table 314.16(A)

Device boxes are used to mount switches to control luminaires (lighting fixtures) and other electrical loads. They also are used to mount receptacles for the cord-and-plug connection of electrical appliances. The depth space (fill) in the box shall be sized to accommodate the size device installed. A dimmer switch (device) will take up more depth (space) in the box than a regular sized toggle switch. A GFCI receptacle in most cases takes up greater depth (space) for installation than other size devices. The designer and installer shall select the proper size device box to accommodate not only conductors and fittings but the depth of the device, how much fill it requires shall also be considered.

DIFFERENT SIZE CONDUCTORS
314.16(B)(1) through (B)(5); Table 314.16(B)

Section **314.16(B)** refers to **Table 314.16(B)** which requires the cubic inch rating of each conductor in the box to be used for determining the fill space in cubic inches. This total rating shall be used for selecting the box size. Section **314.16(B)** refers to **314.16(A)** for selecting the cubic inch rating of each fitting, cable clamp, etc. in the box, based on the larger conductor entering the box. The cubic inch rating for each yoke shall be based on the size conductor that is connected to the device terminal.

Designing Wiring Methods

Problem: What size square box with cable clamps is required for 2 - 14 AWG and 2 - 12 AWG w/grounds romex cables that are spliced in the box between the panelboard and the loads?

Step 1: Counting conductors
314.16(B)(1) thru (B)(5); Table 314.16(B)
(Conductor fill)
2 - 12 AWG hots   2.25 cu. in. x 2 = 4.5 cu. in.
2 - 12 AWG neutrals 2.25 cu. in. x 2 = 4.5 cu. in.
2 - 14 AWG hots   2 cu. in. x 2   = 4. cu. in.
2 - 14 AWG neutrals 2 cu. in. x 2   = 4. cu. in.
(EGC fill)
2 - 12 AWG EGC's   2.25 cu. in. x 1 = 2.25 cu. in.
2 - 14 AWG EGC's   2. cu. in. x 1   = 0
(Clamp fill)
2 cable clamps   2.25 cu. in. x 1 = 2.25 cu. in.
Total   = 21.5 cu. in.

Step 2: Selecting box
Table 314.16(A)
21.5 cu. in. requires 4 11/16" x 1 1/4" (120 mm x 32 mm)

Solution: **A 4 11/16 in. x 1 1/4 in. (120 mm x 32 mm) square box is required.**

For example: What size device box with cable clamps is required for 2 - 14-2 AWG w/ground romex cables and 1 duplex receptacle?

Step 1: Same size conductors
314.16(B)(1) through (B)(5)
(Conductor fill)
314.16(B)(1)
2 - 14 AWG Hots   = 2
2 - 14 AWG Neutrals   = 2
(EGC fill)
314.16(B)(5)
2 - 14 AWG EGC's   = 1
(Device fill)
314.16(B)(4)
1 receptacle   = 2
(Clamp fill)
314.16(B)(2)
2 cable clamps   = 1
Total   = 8

Step 2: Selecting box
Table 314.16(A)
8 - 14 requires 3" x 2" x 3 1/2" (75 mm x 50 mm x 90 mm)

Solution: **A 3 in. x 2 in. x 3 1/2 in. (75 mm x 50 mm x 90 mm) device box is required.**

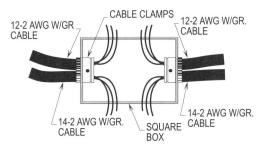

DIFFERENT SIZE CONDUCTORS
NEC 314.16(B)(1) THRU (B)(5)
TABLE 314.16(B)

Figure 12-12. Section **314.16(B)** and **Table 314.16(B)** is utilized to size and select a square box based on the box containing different size conductors.

See Figure 12-13 for a step by step procedure for sizing and selecting device boxes where the conductors are the same size in the box.

SAME SIZE CONDUCTORS
314.16(B)(1) through (B)(5); Table 314.16(A)

Section **314.16(A)** contains specific rules that shall be applied where the device box has the same size conductors with the various types of fittings, devices, etc. all present in the box. Devices shall be counted as two conductors to allow more fill (space) to accommodate the different size devices that are available in the electrical industry today. The count of two conductors shall be based upon the size conductors that are connected to the terminals of the device mounted on the yoke (strap).

DIFFERENT SIZE CONDUCTORS
314.16(B)(1) through (B)(5); Table 314.16(B)

Section **314.16(B)** requires the cubic inch rating of each conductor from **Table 314.16(B)** to be utilized where conductors are not the same size in the box. The cubic inch rating of each conductor shall be multiplied by the number in the box and the total cubic inch value shall be used to determine the size device box from **Table 314.16(A)**.

12-11

For example: What size device or square box with cable clamps is required for 1 - 14-2 AWG w/ground passing through, 1 - 10-2 AWG w/ground passing through and 4 - 12-2 AWG w/ground that are spliced with pigtails and connected to a dimmer switch?

Step 1: Sizing box with different size conductors
314.16(B)(1) through (B)(5);
Table 314.16(B)
(Conductor fill)
314.16(B)(1)
2 - 14 AWG Hots (spliced through)
2   cu. in. x 2   = 4   cu. in.
2 - 14 AWG Neutrals (spliced through)
2   cu. in. x 2   = 4   cu. in.
(EGC fill)
314.16(B)(5)
2 - 14 AWG EGC's
(spliced through)   = 0
(Conductor fill)
314.16(B)(1)
4 - 12 AWG Hots (spliced)
2.25 cu. in. x 4   = 9   cu. in.
4 - 12 AWG Neutrals (spliced)
2.25 cu. in. x 4   = 9   cu. in.
(EGC fill)
314.16(B)(5)
4 - 12 AWG EGC's (spliced)
2.25 cu. in. x 0   = 0
(Conductor fill)
314.16(B)(1)
2 - 10 AWG Hots (spliced)
2.5 cu. in. x 2   = 5   cu. in.
2 - 10 AWG Neutrals (spliced)
2.5 cu. in. x 2   = 5   cu. in.
(EGC fill)
314.16(B)(5)
2 - 10 AWG EGC's (spliced)
2.5 cu. in. x 1   = 2.5 cu. in.
(Device fill)
314.16(B)(4)
dimmer switch (device)
2.25 cu. in. x 2   = 4.5 cu. in.
Total   = 43   cu. in.

Step 2: Selecting box with different size conductors
Table 314.16(A)
43 cu. in. requires a 4 11/16" x 1 1/4" (120 mm x 32 mm) square box with extension ring and plaster ring

Solution: **A 4 11/16 in. x 1 1/4 in. (120 mm x 32 mm) square box w/extension ring and plaster ring is required.**

See **Figure 12-14** for a step by step procedure for sizing and selecting device boxes where the conductors in the box are different sizes.

Problem: What size device box is required for 2 - 12 AWG hots, 2 - 12 AWG neutrals, a receptacle, 1 bonding jumper and 1 pigtail?

Step 1: Counting conductors
314.16(B)(1) thru (B)(5)
(Conductor fill)
Device   = 2
2 - 12 AWG hots   = 2
2 - 12 AWG neutrals   = 2
2 - 12 AWG pigtails   = 0
1 - 12 AWG EGC (BJ)   = 0
Total   = 6

Step 2: Selecting box
Table 314.16(A)
6 - 12 AWG conductors requires 3" x 2" x 2 3/4" (75 mm x 50 mm x 70 mm)

Solution: **A 3 in. x 2 in. x 2 3/4 in. (75 mm x 50 mm x 70 mm) device box is required.**

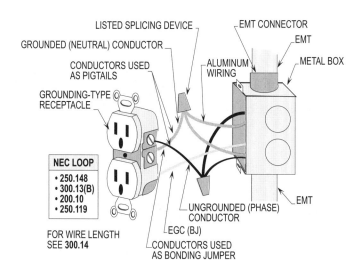

Figure 12-13. Sections **314.16(B)(1) through (B)(5)** and **Table 314.16(A)** are utilized to size and select a device box based on the box containing the same size conductors.

OTHER BOXES
314.16(A)(2)

Other boxes are boxes that are not listed in **Table 314.16(A)**. They are junction boxes with greater fill (space) area that allows more conductors to be routed straight through or spliced in the box. Junction boxes come in many sizes and are built to house the different size conductors in **Table 314.16(A)**. If a box size selected from **Table 314.16(A)** with an extension ring will not accommodate the number of conductors ranging from 18 AWG to 6 AWG installed in the box then a junction box shall be used. See the chart in the Appendix for the standard size junction boxes available to contain branch-circuit and feeder-circuit conductors.

SAME SIZE CONDUCTORS
314.16(B)(1) through (B)(5); 314.16(A)

Because there are not any tables available with the number of conductors permitted in junction boxes for conductors that are the same size, a calculation applying the cubic inch value of each conductor shall be done. By multiplying the cubic inch rating of each conductor by the number in the box and using this total cubic inch rating, the proper size junction box may be selected. The cubic inch rating of each conductor shall be found in **Table 314.16(B)** for conductors 18 AWG through 6 AWG. The cubic inch fill space of a junction box shall be found by multiplying the dimensions of the box.

For example, a 4" x 4" x 4" junction box has a cubic inch (cu. in.) rating of 64 cu. in. (4" x 4" x 4" = 64 cu. in.). This box will contain any arrangement of conductors not exceeding 64 cu. in. in total rating.

When an existing panelboard in a remodeling job in a dwelling unit is used for a junction box, it shall comply with **240.24(D)** and **(E)** which prohibits a panelboard with OCPD's in a clothes closet around combustible material. Section **230.70(A)** prohibits the service disconnecting means to be located in a bathroom of dwelling units, commercial or industrial locations.

Problem: What size square box is required to contain 2 - 10 AWG conductors passing through, 4 - 12 AWG conductors and 1 receptacle? There are 2 pigtails and 1 bonding jumper in the box.

Step 1: Counting conductors
314.16(B)(1) thru (B)(5); Table 314.16(B)
(Conductor fill)
2 - 10 AWG hots   2.5 cu. in. x 2 = 5 cu. in.
2 - 12 AWG hots   2.25 cu. in. x 2 = 4.5 cu. in.
2 - 12 AWG neutrals 2.25 cu. in. x 2 = 4.5 cu. in.
(Device fill)
1 receptacle   2.25 cu. in. x 2 = 4.5 cu. in.
(Conductor fill)
2 pigtails   = 0
1 bonding jumper   = 0
Total   = 18.5 cu. in.

Step 2: Selecting box
Table 314.16(A)
18.5 cu. in. requires 4" x 1 1/2" (100 mm x 38 mm) square box

Solution: **A 4 in. x 1 1/2 in. (100 mm x 38 mm) square box with plaster ring is required.**

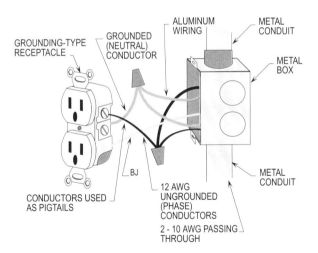

DIFFERENT SIZE CONDUCTORS
NEC 314.16(B)(1) THRU (B)(5)
TABLE 314.16(B)

Figure 12-14. Sections **314.16(B)(1) through (B)(5)** and **Table 314.16(B)** are utilized to size and select a device box based on the box containing different size conductors. The box shall be sized based upon the number of cu. in. of each conductor.

For example: What size junction box is required to contain 6 - 12-2 AWG w/ground, 4 - 12-2 AWG w/ ground and 2 - 12 AWG without ground cables where all the conductors in the cables are spliced?

Step 1: Sizing box with the same size conductors
314.16(A)(2)
(Conductor fill)
314.16(B)(1)
12 - 12 AWG Hots (spliced)
2.25 cu. in. x 12   = 27   cu. in.
12 - 12 AWG Neutrals (spliced)
2.25 cu. in. x 12   = 27   cu. in.
(EGC fill)
314.16(B)(5)
12 - 12 AWG EGC's (spliced)
2.25 cu. in. x 1   = 2.25 cu. in.
(Conductor fill)
314.16(B)(1)
8 - 12 AWG Hots (spliced)
2.25 cu. in. x 8   = 18   cu. in.
8 - 12 AWG Neutrals (spliced)
2.25 cu. in. x 8   = 18   cu. in.
(EGC fill)
314.16(B)(5)
8 - 12 AWG EGC's (spliced)
2.25 cu. in. x 0   = 0
(Conductor fill)
314.16(B)(1)
4 - 12 AWG Hots (spliced)
2.25 cu. in. x 4   = 9   cu. in.
4 - 12 AWG Neutrals (spliced)
2.25 cu. in. x 4   = 9   cu. in.
(EGC fill)
314.16(B)(5)
4 - 12 AWG EGC's (spliced)
2.25 cu. in. x 0   = 0
Total   = 110.25 cu. in.

Step 2: Selecting box for the same size conductors
314.16(B), Chart
6" x 6" x 4" box = 144 cu. in.
144 cu. in. will contain
110.25 cu. in.

Solution: A 6 in. x 6 in. x 4 in. junction box is required.

Problem: What size junction box is required to contain 4 raceways with 8 - 12 AWG in each and 2 raceways with 4 - 12 AWG in each?

Step 1: Counting sq. in.
314.16(B)(1) thru (B)(5); Table 314.16(B)
(Conductor fill using other boxes)
8 - 12 AWG conductors   2.25 cu. in. x 8   = 18
8 - 12 AWG conductors   2.25 cu. in. x 8   = 18
8 - 12 AWG conductors   2.25 cu. in. x 8   = 18
8 - 12 AWG conductors   2.25 cu. in. x 8   = 18
4 - 12 AWG conductors   2.25 cu. in. x 4   = 9
4 - 12 AWG conductors   2.25 cu. in. x 4   = 9
Total   = 90

Step 2: Selecting box
Box chart
6" x 4" x 4" box = 96 cu. in.
96 cu. in. will contain 90 cu. in.

Solution: A 6 in. x 4 in. x 4 in. junction box is required.

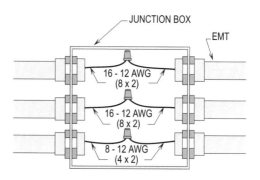

Figure 12-15. Sizing and selectng a junction box to contain the same size conductors.

See **Figure 12-15** for a step by step procedure for sizing and selecting junction boxes with the same size conductors.

DIFFERENT SIZE CONDUCTORS
314.16(A)(2)

To select the proper size junction box housing different size conductors, the cubic inch rating of each conductor from **Table 314.16(B)** shall be selected and multiplied by the number of conductors for each cubic inch rating. The total calculation of the cubic inch ratings shall be used to select the proper size junction box.

For example: What size junction box is required to house the following different size conductors?

Step 1: Sizing box with different size conductors
314.16(B)(1) through (B)(5);
Table 314.16(B)
(Conductor fill)
314.16(B)(1)
8 - 10 AWG Hots (spliced)
2.5 cu. in. x 8   = 20 cu. in.
8 - 10 AWG Neutrals (spliced)
2.5 cu. in. x 8   = 20 cu. in.
(EGC fill)
314.16(B)(5)
 8 #10 EGC's (spliced)
2.5 cu. in. x 0   = 0
(Conductor fill)
314.16(B)(1)
20 - 14 AWG Hots (spliced)
2 cu. in. x 20   = 40 cu. in.
20 - 14 AWG Neutrals (spliced)
2 cu. in. x 20   = 40 cu. in.
(EGC fill)
314.16(B)(5)
20 - 14 AWG EGC's (spliced)
2 cu. in. x 0   = 0
(Conductor fill)
314.16(B)(1)
12 - 12 AWG Hots (spliced)
2.25 cu. in. x 12   = 27 cu. in.
12 - 12 AWG Neutrals (spliced)
2.25 cu. in. x 12   = 27 cu. in.
(EGC fill)
314.16(B)(5)
12 - 12 AWG EGC's (spliced)
2.25 cu. in. x 0   = 0
(Conductor fill)
314.16(B)(1)
4 - 6 AWG Hots (spliced)
5 cu. in. x 4   = 20 cu. in.
4 - 6 AWG Neutrals (spliced)
5 cu. in. x 4   = 20 cu. in.
(EGC fill)
314.16(B)(5)
4 - 6 AWG EGC's (spliced)
5 cu. in. x 1   = 5 cu. in.
Total   = 219 cu. in.

Step 2: Selecting box with different size conductors
Chart for junction boxes
8" x 8" x 4" box = 256 cu. in.
256 cu. in. will contain 219 cu. in.

Solution: An 8 in. x 8 in. x 4 in. junction box is required.

See Figure 12-16 for a step by step procedure for sizing and selecting junction boxes with different size conductors.

Problem: What size junction box is required to contain the following different size conductors?

Step 1: Counting conductors
314.16(B)(1) thru (B)(5); Table 314.16(B)
(Conductor fill using other boxes)
16 - 10 AWG conductors 2.5 cu. in. x 16 = 40
20 - 14 AWG conductors 2   cu. in. x 20 = 40
24 - 12 AWG conductors 2.25 cu. in. x 24 = 54
10 - 6 AWG conductors   5   cu. in. x 10 = 50
Total   = 184

Step 2: Selecting box
Box chart
8" x 6" x 4" = 192 cu. in.
192 cu. in. will contain 184 cu. in.

Solution: A 8 in. x 6 in. x 4 in. junction box is required.

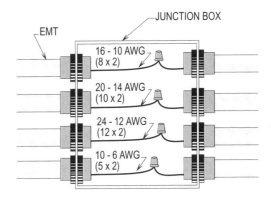

**DIFFERENT SIZE CONDUCTORS
NEC 314.16(B)(1) THRU (B)(5)
TABLE 314.16(B)**

Figure 12-16. Sizing and selecting a junction box to contain different size conductors.

PLASTER RINGS AND EXTENSION RINGS
314.16(A)(1)

Plaster rings and extension rings are accessories that are used in conjunction with octagon, round, square or device boxes to extend the face of the box to mount luminaires (fixtures), ceiling fans, devices or to provide more fill space in the box.

For example, the face of an octagon box needs to be extended 1/4 in., 1/2 in. or 3/4 in. to the surface of a sheet-rock ceiling. A 1/2 in. plaster ring with provisions for mounting a luminaire (lighting fixture) may be used to extend the face of the octagon box to the surface edge of the sheet-rock.

Extension rings are used to extend the face of boxes a greater distance than plaster rings because plaster rings only come with limited extension height to extend the face of a box to a given height.

For example, one or two extension rings may be used to extend the face of a box to the surface edge of a wall or ceiling where a plaster ring is mounted and a device, luminaire (fixture), etc. is mounted to the plaster ring. Most designers, installers and inspectors interpret **314.22** to permit one extension ring to be used to provide additional fill space in the box. Check with the local inspector for his or her interpretation of this rule.

The reason for this requirement of limiting the number of extension rings for fill space is to allow access to the inside box area for servicing the spliced conductors. **(See Figure 12-17)**

CONDUIT BODIES
314.16(C)

Conduit bodies containing 6 AWG conductors and smaller shall not be permitted to be less than twice the cross-sectional area of the largest conduit to which it is connected. The number of conductors that are permitted in conduit bodies shall be determined by the square inch area per **Table 5** and **Table 4 of Chapter 9**.

> **For example:** What is the cross-sectional area of an LB (conduit body) that is connected to a 3/4 in. conduit?
>
> **Step 1:** Finding cross-sectional area
> **Table 4, Ch. 9**
> 3/4" EMT conduit = .533 sq. in.
>
> **Step 2:** Calculating size LB
> **314.16(C)**
> .533 sq. in. x 2 = 1.066 sq. in.
>
> **Solution:** The LB (conduit body) has to have a cross-sectional area of 1.066 sq. in.

> **Design Tip:** A larger LB may be used with reducing bushings to accomplish this requirement where needed.

Conduit bodies with less conduit entries shall not be permitted to contain splices, taps or devices unless they comply with **314.16(B)** and are supported properly. **(See Figure 12-18)**

> **Problem:** What size square box with extension ring is required to contain 21 - 12 AWG conductors? (use min. size)
>
> **Step 1:** Counting conductors
> **314.16(B)(1) thru (B)(5); Table 370.16(A)**
> (Using extension rings for greater fill area)
> 21 - 12 AWG conductors requires
> 4 11/16" x 1 1/4" (120 mm x 32 mm) with extension ring
>
> **Solution:** Use a 4 11/16 in. x 1 1/4 in. (120 mm x 32 mm) square box with an extension ring. Note: With a ring, 22 - 12 AWG conductors may be housed.

Figure 12-17. Increasing the fill space using an extension ring.

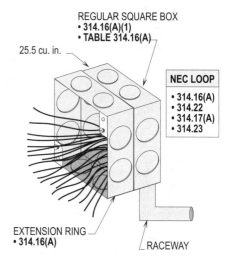

PLASTER RINGS AND EXTENSION RINGS
NEC 314.16(A)

Designing Wiring Methods

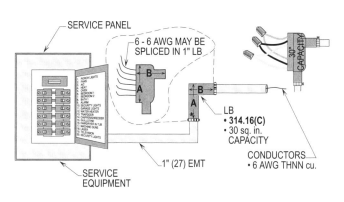

Figure 12-18. Conduit bodies containing 6 AWG or smaller conductors shall be at least twice the cross-sectional area of the largest conduit that it is connected to. This rule is designed to provide proper room in the LB.

JUNCTION BOXES
314.28

Junction boxes housing conductors 4 AWG and larger that are pulled through raceways shall be sized based upon a straight pull or an angle pull.

STRAIGHT PULLS
314.28(A)(1)

A straight pull (straight through) is where the raceway is connected to one side of the box and another raceway is connected to the opposite side with the junction box used as a junction between the raceways. Junction boxes utilized in a straight pull shall be sized by multiplying the largest raceway in the run by not less than the multiplier 8 per **314.28(A)(1)**.

For example: What is the minimum length for a straight pull consisting of 1 run of 2 in. (53) raceways?

Step 1: Finding the multiplier
314.28(A)(1)
Multiplier = 8

Step 2: Calculating length
314.28(A)(1)
2" (53) raceway x 8 = 16"

Solution: The minimum length of the junction box is 16 in..

Design Tip: The minimum length of the junction box is 16 in. long while the width of the wall that the raceways are connected to shall be sized to accommodate the number of raceways plus the locknuts and bushings. The proper length and width is usually provided when a standard size junction box is selected.

Sometimes junction boxes will have more than two raceways connected to its sides (walls). Junction boxes with more than two raceways connected to their sides shall be sized by multiplying the largest raceway by 8.

For example: What is the minimum length for a junction box with a straight pull that has 4 in. (103), 2 in. (53) and 1 in. (27) raceways connected to its sides?

Step 1: Finding the multiplier
314.28(A)(1)
Multiplier = 8

Step 2: Calculating length
314.28(A)(1)
4" (103) x 8" = 32"

Solution: The minimum length is 32 in.

See Figure 12-19 for a step by step procedure for sizing junction boxes for straight pulls.

12-17

Problem: What is the minimum length for a junction box with a straight pull that has 4 in. (103) raceway on opposite sides and a second box that has 3 1/2 in. (91), 2 in. (53) and 1 in. (27) raceways connected to its sides?

Step 1: Finding the multiplier
314.28(A)(1)
Multiplier = 8

Step 2: Calculating length (3 1/2" Raceway)
314.28(A)(1)
3 1/2" (91) x 8 = 28"

Step 3: Calculating length (4" Raceway)
314.28(A)(1)
4" (103) x 8 = 32"

Solution: The minimum length is 28 in. and 32 in. respectively for each box.

For example: What is the minimum length of a junction box for an angle pull that has one run of 4 in. (103) raceways?

Step 1: Finding the multiplier
314.28(A)(2)
Multiplier = 6

Step 2: Calculating length
314.28(A)(2)
4" (103) x 6 = 24"

Solution: The minimum length is 24" x 24".

Junction boxes having more than one run of raceways connecting to their walls to form an angle pull shall be sized by multiplying the largest raceway by not less than 6 and adding the remaining raceways to this value.

For example: What is the minimum length of a junction box for an angle pull that has a 4 in. (103), 3 in. (78), 2 in. (53) and 1 in. (27) raceway which is connected to the right wall and bottom wall?

Step 1: Finding the multiplier
314.28(A)(2)
Multiplier = 6

Step 2: Calculating length
314.28(A)(2)
4" (103) x 6 = 24"
3" (78)
2" (53)
1" (27)
30"

Solution: The minimum length is 30" x 30".

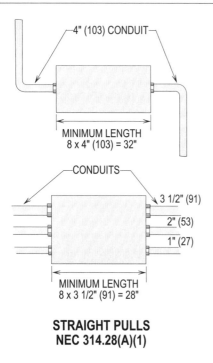

Figure 12-19. The size junction box for a straight pull shall be found by multiplying the largest conduit by not less than 8.

See **Figure 12-20** for a step by step procedure for sizing junction boxes for angle pulls.

ANGLE PULLS
314.28(A)(2)

An angle pull is where the raceways enter and leave the junction box from the top wall to one of the side walls or from one of the side walls to the bottom wall. Junction boxes in angle pulls shall be sized by multiplying the largest raceway in the run by the multiplier 6 per **314.28(A)(2)**. A U pull is where the raceway leaves the same wall and forms a U pull configuration.

SIZING CONDUITS
ANNEX C

Conduits are used to enclose conductors to supply power to various types of electrical equipment. The same size conductors with the same type of insulation may be pulled through conduits. They may also be routed through conduits

Designing Wiring Methods

with different sizes and types of insulation. **Tables in Chapter 9** shall be used to size conduits based on the types of insulation and size of conductors that are pulled through the conduit system.

The procedure for selecting the size of the conduit is to take the size of the conductors and align them up in one column of the correct Tables and select the proper size in the appropriate column based on the type insulation and number of conductors. This is the easiest and fastest method to use in determining the size conduit housing the same size conductors.

Problem: What is the minimum length of a junction box for an angle pull that has a 3 in. (78), 3 in. (78), 2 in. (53) and 1 in. (27) raceway which is connected to the right wall and bottom wall?

Step 1: Finding the multiplier
314.28(A)(2)
Multiplier = 6

Step 2: Calculating length
314.28(A)(2)
3" x 6 = 18"
3"
2"
1"
24"

Solution: The minimum length is 24 in. x 24 in.

For example: What size EMT conduit is required for 10 - 12 AWG XHHW copper conductors routed through the conduit to a pull box?

Step 1: Finding size conduit
Table C1, Appendix C
10 - 12 AWG requires 3/4" (21)

Solution: A 3/4 in. (21) conduit is required.

See Figure 12-21 for a step by step procedure for sizing conduits enclosing the same size conductors.

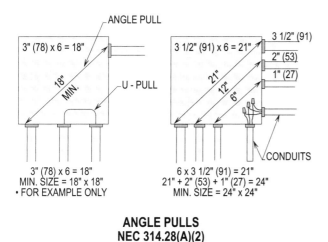

ANGLE PULLS
NEC 314.28(A)(2)

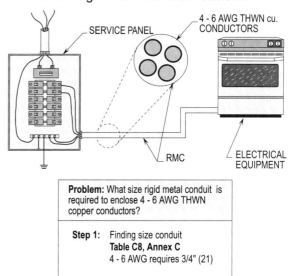

Problem: What size rigid metal conduit is required to enclose 4 - 6 AWG THWN copper conductors?

Step 1: Finding size conduit
Table C8, Annex C
4 - 6 AWG requires 3/4" (21)

Solution: A 3/4 in. (21) RMC is required.

QUICK CALC	
TABLE 4, CH. 9	= 40%
3/4" (21)	= .220
TABLE 5, CH. 9	
6 AWG	= .0507 sq. in.
.220 (40%) ÷ .0507 sq. in. = 4.339	
4.339	= 4 conductors
3/4" (21)	= 4 conductors

ENCLOSING SAME
SIZE CONDUCTORS
NEC TABLE C1 THRU C12A

Figure 12-20. The size junction box for a angle pull shall be found by multiplying the largest conduit by not less than 6 and adding this number to the remaining conduits. Calculation will determine the size junction box.

ENCLOSING SAME SIZE CONDUCTORS
TABLES C1 THROUGH C12A, ANNEX C

Conduits with conductors that have the same type of insulation and are the same size shall be permitted to be selected by using **Tables C1 through C12A, Appendix C**.

Figure 12-21. Tables C1 through C12A in Annex C shall be permitted to be used to size the conduit enclosing the same size conductors.

ENCLOSING DIFFERENT SIZE CONDUCTORS
TABLES 5, TO CH. 9

Conduits enclosing different size conductors shall be sized by selecting the square inch area of each conductor per **Table 5 in Chapter 9** and multiplying by the number. This total is used to select the size conduit per **Table 4 in Chapter 9**.

For example: What size EMT conduit is required to enclose 2 - 4 AWG, 1 - 6 AWG and 1 - 8 AWG THWN copper conductors?

Step 1: Finding sq. in. area
Table 5, Ch. 9
4 AWG = .0824 sq. in.
6 AWG = .0507 sq. in.
8 AWG = .0366 sq. in.

Step 2: Calculating sq. in. area
Table 5, Ch. 9
.0824 sq. in. x 2 = .1648 sq. in.
.0507 sq. in. x 1 = .0507 sq. in.
.0366 sq. in. x 1 = .0366 sq. in.
Total = .2521 sq. in.

Step 3: Selecting size conduit
Table 4, Ch. 9
.2521 sq. in. requires .346 sq. in.

Solution: A 1 in. (27) conduit is required.

See **Figure 12-22** for an illustrated procedure for sizing conduits with different size conductors.

SIZING NIPPLES
NOTE (4) TO TABLES, CH. 9

The difference in a nipple and a conduit is that a nipple is 24 in. (600 mm) or less in length. A conduit system is any run over 24 in. (600 mm) in length.

A conduit system with more than two conductors shall be permitted to have a 40 percent fill per **Table 4, Chapter 9**, while a nipple shall be permitted to have a 60 percent fill per **Note (4) to Chapter 9**.

Nipples shall be sized by finding the square inch area of each conductor and multiplying by the number. The total 100 percent fill in square inches in **Table 5, Chapter 9** shall be selected based upon the size conduit and this value shall be multiplied by 60 percent. If the total square inch area of all the conductors is less or equal to the total of the square inch area produced by applying the 60 percent factor, the size of the nipple shall be selected based upon that conduit size used in the calculation.

For example: What size EMT nipple is required to enclose 36 - 12 AWG THWN copper conductors that are installed between a panelboard and junction box?

Step 1: Finding sq. in. area
Table 5, Ch. 9
12 AWG = .0133 sq. in.

Step 2: Calculating sq. in. area
Table 5, Ch. 9
.0133 sq. in. x 36 = .4788 sq. in.

Step 3: Finding conduit at 100% total fill
Table 4, Ch. 9
1" (27) conduit that has a 100% total of .864 sq. in.
(diameter squared x
.7854 = sq. in. area)
(1.049 x 1.049 x .7854 = .864 sq. in.)

Step 4: Applying 60% fill
Note 4 to Ch. 9
sq. in. area x 60% = fill area
.864 x 60% = .518 sq. in.

Step 5: Selecting the nipple
Table 4, Ch. 9
.518 sq. in. is greater than .4788 sq. in.

Solution: The size nipple is 1 in. (27) Note: The derating rules of Table 310.15(B)(2)(a) do not apply to nipples per Ex. 3 to 310.15(B)(2)(a).

See **Figure 12-23** for an illustrated procedure for sizing nipples applying the 60 percent fill rule.

Designing Wiring Methods

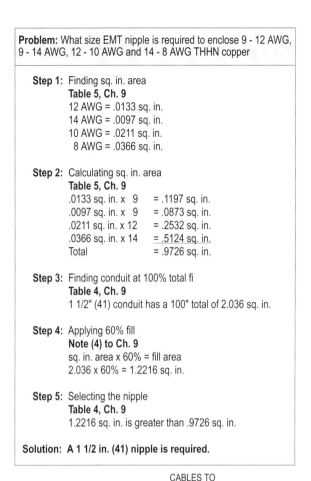

Problem: What size EMT nipple is required to enclose 9 - 12 AWG, 9 - 14 AWG, 12 - 10 AWG and 14 - 8 AWG THHN copper

Step 1: Finding sq. in. area
Table 5, Ch. 9
12 AWG = .0133 sq. in.
14 AWG = .0097 sq. in.
10 AWG = .0211 sq. in.
8 AWG = .0366 sq. in.

Step 2: Calculating sq. in. area
Table 5, Ch. 9
.0133 sq. in. x 9 = .1197 sq. in.
.0097 sq. in. x 9 = .0873 sq. in.
.0211 sq. in. x 12 = .2532 sq. in.
.0366 sq. in. x 14 = .5124 sq. in.
Total = .9726 sq. in.

Step 3: Finding conduit at 100% total fi
Table 4, Ch. 9
1 1/2" (41) conduit has a 100" total of 2.036 sq. in.

Step 4: Applying 60% fill
Note (4) to Ch. 9
sq. in. area x 60% = fill area
2.036 x 60% = 1.2216 sq. in.

Step 5: Selecting the nipple
Table 4, Ch. 9
1.2216 sq. in. is greater than .9726 sq. in.

Solution: A 1 1/2 in. (41) nipple is required.

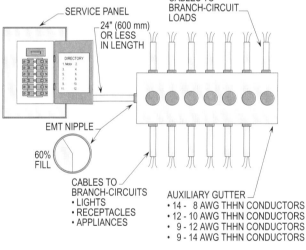

**SIZING NIPPLES
NOTE (4) TO CH. 9 TABLES**

Figure 12-23. Table 5, Ch. 9 and **Table 4, Ch. 9** and **Note (4) to Ch. 9 Tables** shall be permitted to be used to calculate the size of nipples. Nipples are 24 in. (600 mm) or less in length and a raceway is over 24 in. (600 mm) in length.

Problem: What size EMT is required to enclose 12 - 14 AWG, 4 - 12 AWG and 4 - 10 AWG THWN copper conductors?

Step 1: Finding sq. in. area
Table 5, Ch. 9
14 AWG = .0097 sq. in. area
12 AWG = .0133 sq. in. area
10 AWG = .0211 sq. in. area

Step 2: Calculating sq. in. area
Table 5, Ch. 9
.0097 sq. in. x 12 = .1164 sq. in.
.0133 sq. in. x 4 = .0532 sq. in.
.0211 sq. in. x 4 = .0844 sq. in.
Total = .254 sq. in.

Step 3: Selecting size EMT
Table 4, Ch. 9
.254 sq. in. requires .346 sq. in.

Solution: A 1 in. (27) EMT is required.

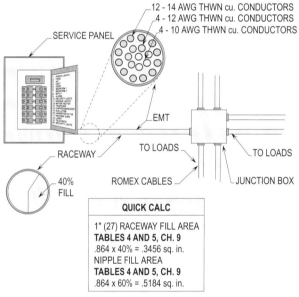

**ENCLOSING DIFFERENT SIZE CONDUCTORS
NEC TABLE 5, CH. 9
NEC TABLE 4, CH. 9**

Figure 12-22. Table 5, Ch. 9 and **Table 4, Ch. 9** shall be permitted to be used to size the conduit enclosing different size conductors.

GUTTER SPACE
312.7; 366.56

Gutter space is used to enter or leave enclosures that enclose conductors which are terminated to serve the electrical equipment or tapped or spliced to serve other loads. Certain clearances and space requirements to protect the conductors and prevent overcrowding is essential per **312.7** and **366.56**.

AUXILIARY GUTTERS
366.22

The number of current-carrying conductors permitted in an auxiliary gutter without derating per **Table 310.15(B)(2)(a) to Table 310.16** is 30 or less per **366.6(A)**. The size of an auxiliary gutter shall determined by dividing the total square inch area of the conductors by 20 percent fill area for installing conductors. The total square inch area of the conductors shall be found by multiplying the square inch area per **Table 5, Ch. 9** of each conductor which is based on the size and insulation of each conductor placed in the auxiliary gutter.

For example: What size auxiliary gutter is required to house 3 - 350 KCMIL THWN copper conductors (feeder-circuit) that have 3 - 1/0 AWG, 3 - 1 AWG and 3 - 2 AWG THWN copper conductors spliced to them?

Step 1: Finding sq. in. area
Table 5, Ch. 9
350 KCMIL = .5242
1/0 AWG = .1855
1 AWG = .1562
2 AWG = .1158

Step 2: Calculating sq. in. area
Table 5, Ch. 9
.5242 sq. in. x 3 = 1.5726 sq. in.
.1855 sq. in. x 3 = .5565 sq. in.
.1562 sq. in. x 3 = .4686 sq. in.
.1158 sq. in. x 3 = .3474 sq. in.
Total = 2.9451 sq. in.

Step 3: Sizing gutter
366.22(A)
sq. in. area divided 20% = total fill
2.9451 divided 20% = 14.7255 sq. in.

Step 4: Selecting gutter
Chart
4" x 4" = 16"

Step 5: Applying 75% fill for splices
sq. in. of gutter x 75% = fill area
16" x 75% = 12 sq. in.

Solution: A 4 in. x 4 in. auxiliary is required with only 12 in. of the gutter space used for splicing conductors.

See Figure 12-24 for a detailed procedure for designing, sizing and selecting auxiliary gutters.

Problem: What size auxiliary gutter is required to house 3 - 250 KCMIL THHN copper conductors that have 3 - 4/0 AWG and 3 - 1/0 AWG THHN conductors spliced to them?

Step 1: Finding sq. in. area
Table 5, Ch. 9
250 KCMIL THHN = .397 sq. in.
4/0 AWG THHN = .3237 sq. in.
1/0 AWG THHN = .1855 sq. in.

Step 2: Calculating sq. in. area
Table 5, Ch. 9
.397 sq. in. x 3 = 1.191
.3237 sq. in. x 3 = .9711
.1855 sq. in. x 3 = .5565
Total = 2.7186

Step 3: Sizing gutter
366.22(A)
sq. in. area ÷ 20% = total fill
2.7186 sq. in. ÷ 20% = 13.593

Step 4: Selecting gutter
Chart
13.593 sq. in. requires 16 sq. in.
4" x 4" = 16"

Solution: A 4 in. x 4 in. auxiliary gutter is required.

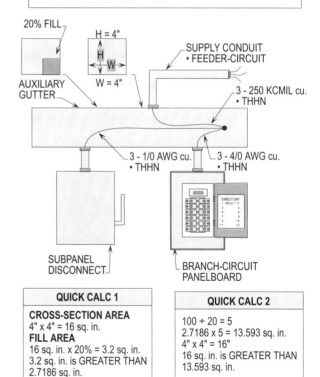

QUICK CALC 1	QUICK CALC 2
CROSS-SECTION AREA 4" x 4" = 16 sq. in. **FILL AREA** 16 sq. in. x 20% = 3.2 sq. in. 3.2 sq. in. is GREATER THAN 2.7186 sq. in.	100 ÷ 20 = 5 2.7186 x 5 = 13.593 sq. in. 4" x 4" = 16" 16 sq. in. is GREATER THAN 13.593 sq. in.

AUXILIARY GUTTERS
NEC 366.22

Figure 12-24. **Table 5, Ch. 9** and **Table 4, Ch. 9** and **366.22** shall be used to calculate and size auxiliary gutters.

PANELBOARDS
408.3(F)

Panelboards shall comply with the provisions of **408.3(F)** which requires the minimum gutter space in the panel to meet the clearances in **Table 312.6(A)** for L-bends and **Table 312.6(B)** for S- or Z-bends.

Electricians shall enter panelboards using an L-, S- or Z-bends so that there is adequate room to terminate the conductors without damaging the insulation which later could create a ground-fault condition and cause a power failure.

> **For example:** What are the minimum clearances between the terminating lugs and the side of a panelboard employing an L-bend using 4/0 AWG THWN copper conductors?
>
> **Step 1:** Finding the minimum clearances
> Table 312.6(A)
> 1 - 4/0 AWG per lug = 4" (102 mm)
>
> **Solution:** The minimum clearance required is 4 in. (102 mm).

Panelboards utilizing S- or Z-bends require greater clearance between lugs and the sides (walls) where conductors are terminated. S- or Z-bends are harder to handle, bend, shape and make ready for termination. L-bends are a lot easier to handle and connect than S or Z-bends.

> **For example:** What is the minimum clearance between the terminating lugs and the sides of the panelboard employing 4/0 AWG THHN copper conductors?
>
> **Step 1:** Finding the minimum clearance
> Table 312.6(B)
> 1 - 4/0 AWG per lug = 7" (178 mm)
>
> **Solution:** The minimum clearance required is 7 in. (178 mm).

If the lugs in the panelboard are removable, the clearance shall be permitted to be reduced from the dimension in **Table 312.6(B)** based on the number and size of conductors connected to each lug.

> **For example:** What is the minimum clearance for 4/0 AWG THHN copper conductors terminating to the lugs in a panelboard entering from the bottom wall? (Lugs removable)
>
> **Step 1:** Finding the minimum clearance
> Table 312.6(B)
> 1 - 4/0 AWG per lug
> 7" - 1" = 6"
>
> **Solution:** The minimum clearance required is 6 in.

See Figure 12-25 for an illustrated procedure for determining the minimum clearance of conductors between lugs and sides (walls) in a panelboard.

CABLE TRAYS
ARTICLE 392

Cable trays shall be permitted to be used for the installation of a number of cables that are installed in a single run from one location to another. Steel or aluminum cable tray sections are used with the ends bolted together to form a single run of the required length. Where installing and routing cables in cable trays, the cables are laid on racks, troughs or hangers.

MULTICONDUCTOR CABLES
392.9

Where sizing and installing multiconductor cables in ladder, ventilated-trough, solid-bottom or ventilated channel-type trays, the requirements of **392.9** shall be complied with.

The diameter of all cables 4/0 AWG and larger shall be added together and the total of each cable shall not be permitted to exceed the width of ladder or ventilated-trough trays. All cables 4/0 AWG and larger shall be installed side-by-side. Cables smaller than 4/0 AWG shall not be required to be installed side-by-side per **Table 392.9, Column 1**. When installing cables rated larger than 4/0 AWG and smaller than 4/0 AWG, the requirements per **Table 392.9, Column 2** shall be used. **(See Figure 12-26)**

Note that cables 1/0 AWG through 4/0 AWG shall be permitted to be stacked in a angular and square configuration.

The total cross-sectional areas of all cables in cable trays with an inside depth of 6 in. (150 mm) or less containing control and/or signal cables shall not be permitted to exceed 50 percent of the cross-sectional area of the cable tray.

The diameter of all cables that are rated 4/0 AWG and larger shall not be permitted to exceed 90 percent of the solid-bottom tray width. Cables smaller than 4/0 AWG shall meet the requirements of **Table 392.9, Column 3**. A combination of cables that are rated 4/0 AWG and larger or less than 4/0 AWG shall meet the requirements of **Table 392.9, Column 4**.

Control and/or signal cables installed in cable trays with an inside depth of 6 in. (150 mm) or less, the total cross-sectional areas of the cables shall not be permitted to exceed 40 percent of the cross-sectional area of the tray.

The total cross-sectional areas of all cables shall not be permitted to exceed 1.3 square inches (850 sq. mm) for 3 in. (75 mm) wide trays, 2.5 square inches (1600 sq. mm) for 4 in. (100 mm) wide trays, or 3.8 square inches (2450 sq. mm) for 6 in. (150 mm) wide trays when installing ventilated channel-type trays.

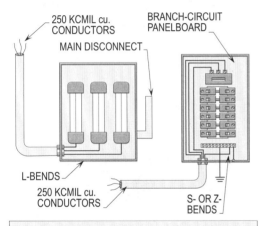

PANELBOARDS
NEC 408.3(F)

Problem: Determine the size cable tray required for the copper conductor.

Step 1: Finding sq. in. area
392.9(A)(3)
4/0 AWG and smaller = 22 sq. in.
4/0 AWG and larger = 12.6 in. dia.

Step 2: Calculating sq. in. area
392.9(A)(3)
4/0 AWG and smaller = 22.00 sq. in.
4/0 AWG and larger (12.6" x 1.2") = 15.12 sq. in.
Total = 37.12 sq. in.

Step 3: Selecting size cable tray
Table 392.9, Col. 2
37.12 sq. in. requires 36 in.
36 in. wide tray = 42 in.

Solution: A 36 in. (900 mm) wide tray is required.

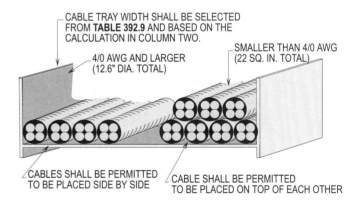

MULTICONDUCTOR CABLES
NEC 392.9; NEC 392.9(A)(3)

Figure 12-25. The distance between lugs and enclosures walls for L, S or Z-bends shall be permitted to be determined in **Tables 312.6(A)** and **312.6(B)**. These tables based upon bends provide safe terminating space for conductors.

Figure 12-26. The diameter of all cables 4/0 AWG and larger shall be added together and the total of each cable shall not be permiited to exceed the width of ladder or ventilated-trough trays.

Designing Wiring Methods

SINGLE CONDUCTOR CABLES
392.10

When designing and installing single-conductor cables in ladder, ventilated-trough or ventilated channel-type trays the requirements in **392.10** shall be applied.

When installing cables in ladder or ventilated-trough trays, the total diameter of all cables 1000 KCMIL and larger shall not be permitted to exceed the width of the cable tray. Cables smaller than 1000 KCMIL shall be installed per **Table 392.10(A), Column 1**.

A combination of cables that are rated 1000 KCMIL and larger or less than 1000 KCMIL shall meet the requirements of **Table 392.10(A), Column 2**. **(See Figure 12-27)**

The total diameter of all cables installed in ventilated channel-type trays shall not be permitted to exceed the inside width of the channel.

CONTROL AND SIGNALING CABLES
392.9(B); (D)

Ladder or ventilated trough trays shall be permitted to be filled with control or signaling or both up to 50 percent of the cross-sectional area of their trays up to 6 in. (150 mm) deep. Solid bottom trays shall be calculated using the same procedures except 40 percent is utilized as the allowable multiplier. **(See Figure 12-28)**

DETERMINING AMPACITIES OF CONDUCTORS

The derating factors of **Table 310.15(B)(2)(a) to Tables 0 through 2000 volts** covers and applies only to multiconductor cables that have more than three current-carrying conductors. Derating is based upon the number of conductors in the cable and not on the number of conductors in the cable tray.

Problem: Determine the size cable tray required for the copper conductors?

Step 1: Finding sq. in. area
Table 5, Ch. 9
300 KCMIL THHN = .4608 sq. in.
1000 KCMIL THHN = 1.3478 sq. in.

Step 2: Calculating sq. in. area
392.10(A)(2); Table 392.10(A), Col. 1
.4608 sq. in. x 14 = 6.4512 sq. in.
1.3478 sq. in. x 10 = 13.478 sq. in.
Total = 19.9292 sq. in.

Step 3: Selecting size cable tray
Table 392.10(A), Col. 1
19.9292 sq. in. requires 24 in. width
24 in. wide tray requires 26 sq. in.

Solution: A 24 in. (600 mm) wide cable tray has 26 sq. inches per Table 392.10(A).

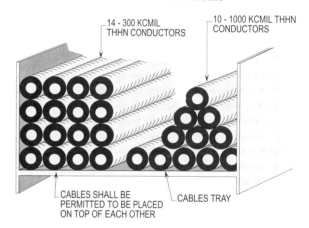

SINGLE CONDUCTOR CABLES
NEC 392.10(A)(2)
TABLE 392.10(A), COL.'s 1 & 2

Figure 12-27. When installing cables in ladder or ventilated-trough trays, the total diameter of all cables 1000 KCMIL and larger shall not be permitted to exceed the width of the cable tray.

MULTICONDUCTOR CABLES
392.11(A)

Tables **310.16** and **310.18** covers the allowable ampacities that apply to multiconductor cables nominally rated 2000 volts or less, provided they are installed accordingly to the requirements of **392.9**.

Ampacities shall be cut to 95 percent of the ampacity in **Tables 310.16** and **310.18** where cable trays are continuously covered for more than 6 ft. (1.8 m) with solid cover that is not ventilated per **392.11(A)(2)**.

When multiconductor cables in uncovered trays are installed in a single layer and maintain a spacing of one conductor diameter between cables, their ampacity shall be determined according to **310.15(B)(2)(a)** as required per **392.11(A)(3)** (ambient corrected ampacities of multiconductor cables rated 0-2000 volts). Refer also to **Table B.310.3** in the Annex for another ampacity Table to be used under certain conditions. **(See Figures 12-29(a) and (b)).**

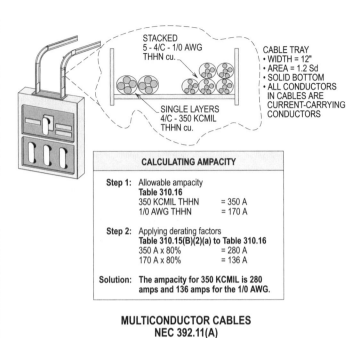

Figure 12-29(a). Calculating the ampacity of multiconductor cables supported by a cable tray.

SIZING CASE 1	SIZING CASE 2
Step 1: Calculating cable tray for case 1 **392.9(B)** 18 x .35 sq. in. = 6.3 sq. in. 35 x .15 sq. in. = 5.25 sq. in. Total = 11.55 sq. in.	**Step 1:** Calculating solid bottom tray **392.9(B)** 16 x .170 = 2.72 sq. in.
Step 2: Applying multiplier **392.9(B)** 11.55 sq. in. ÷ 50% = 23.1 sq. in.	**Step 2:** Selecting cable tray **Table 392.9** 2.72 sq. in. requires 6 in.
Step 3: Applying depth of tray **392.9(B); Table 392.9** 23.1 sq. in. ÷ 4" = 5.775 in.	**Solution:** The size cable tray required is 6 in. (150 mm) per the NEC.
Solution: A cable tray with a 6 in. (150 mm) width is required.	
Note: In the case of a solid bottom tray, the multiplier is 40% and the results would be 11.55 ÷ 40% = 28.88 ÷ 4" = 7.2 in. in which a 12 in. (300 mm) cable tray is required per **Table 392.9**.	

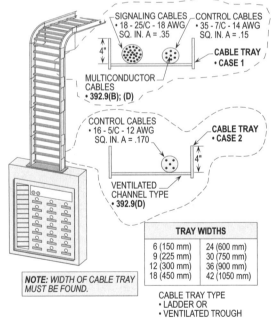

CONTROL AND SIGNALING CABLES
TABLE 392.9; NEC 392.9(B); (D)

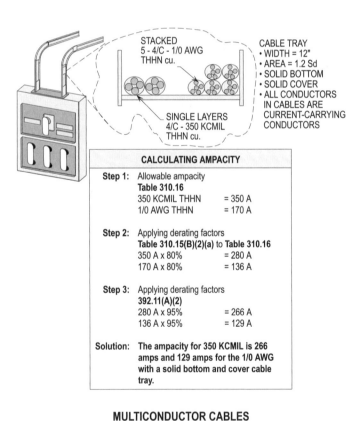

Figure 12-29(b). Calculating the ampacity of multiconductor cables supported by a solid bottom cable tray with a solid cover.

Figure 12-28. The above is the procedure for calculating and sizing cable trays for supporting control and signaling cables using a solid or open cable tray system.

SINGLE CONDUCTOR CABLES
392.11(B)(1); (B)(2)

The following will apply to the ampacity of single conductor cables or single conductor cables that are twisted or bound together, such as triplex or quadraplex, etc. Note that this applies to nominally rated cables of 2000 volts or less.

If 600 KCMIL or larger single conductor cables are installed per **392.10** and in uncovered cables trays, the ampacities of the single conductor cables shall not exceed 75 percent of the ampacities as shown in **Tables 310.17** and **310.19**. Where cable trays are continuously covered for more than 6 ft. (1.8 m) and unventilated covers are used, ampacities for 600 KCMIL and larger cables shall not exceed the ampacity of 70 percent of the level ampacities shown in **Tables 310.17** and **310.19**. **(See Figures 12-30(a) and (b))**

When cables are installed according to **392.10**, single conductor cables from 1/0 AWG through 500 KCMIL installed in uncovered cable trays shall have their ampacities calculated at 65 percent of the allowable ampacities in **Tables 310.17** and **310.19**. If the cable trays are continuously covered for more than 6 ft. (1.8 m) with solid unventilated covers, for 1/0 AWG through 500 KCMIL single conductor cables, the ampacities shall not be permitted to exceed 60 percent of the allowable ampacities as covered in **Tables 310.17** and **310.19**. **(See Figures 12-31(a) and (b))**

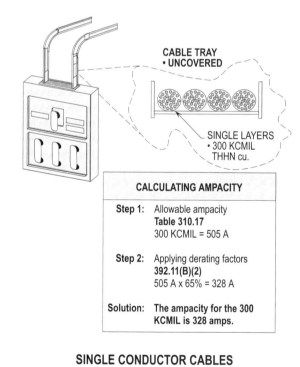

Figure 12-30(b). Single conductors that are 600 KCMIL or larger shall have their ampacities in **Table 310.17** derated by 70 percent to obtain allowable ampacities.

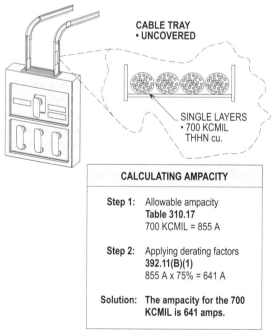

Figure 12-30(a). Single conductors that are 600 KCMIL or larger shall have their ampacities in **Table 310.17** derated by 75 percent to obtain allowable ampacities.

Figure 12-31(a). Single conductors that are 1/0 AWG through 500 KCMIL shall have their ampacities in **Table 310.17** derated by 65 percent to obtain allowable ampacities.

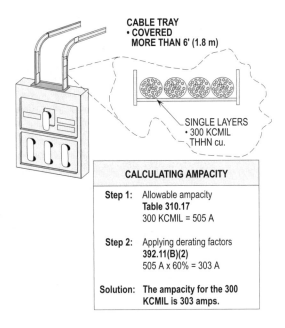

SINGLE CONDUCTOR CABLES
NEC 392.11(B)(2)

Figure 12-31(b). Single conductors that are 1/0 AWG through 500 KCMIL shall have their ampacities in **Table 310.17** derated by 60 percent to obtain allowable ampacities.

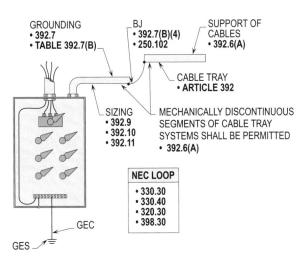

GROUNDING AND BONDING CABLE TRAYS
NEC 392.5(E); NEC 392.6(A)

Figure 12-32(a). Discontinuous segments of cable trays such as for changes of direction or elevation shall be permitted.

GROUNDING AND BONDING CABLE TRAYS
392.5(E); 392.6(A)

Cable trays shall be permitted to have mechanically discontinuous segments between cable tray runs and/or between cable tray runs and equipment. Such discontinuous segments shall be bonded and grounded for safety to protect wiring methods, equipment and personnel from electrical shock and fire hazards.

In industrial facilities where conditions of maintenance and supervision ensure that only qualified people will service the installation, cable trays that are designed to support cables and other loads shall be permitted to be used to support such systems.

For conduits, cables, etc., that bonds to cable trays, a listed clamp or adapter shall be used. If a listed cable tray clamp or adapter is used, a nearby support, such as a support of 3 ft. (900 mm) shall not be required. **(See Figures 12-32(a), (b), and (c))**.

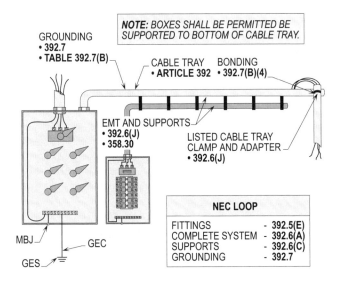

GROUNDING AND BONDING CABLE TRAYS
NEC 392.6(J)

Figure 12-32(b). The procedure by which conduits and cables may be run, connected and supported to or by cable tray systems.

EXPANSION FITTINGS
352.44

Where expansion and contraction, due to temperature difference might be encountered, approved rigid nonmetallic conduit expansion joints shall be used. The substances of PVC expand and contract with temperature changes, which will put undue strain on the conduit. Properly designed expansion joints absorb this expansion and contraction, and therefore, avoiding damage to the conduit and fittings. **(See Figure 12-33)**

For Example: RNC is run between two panels. It runs 40 ft. up the wall, then 90's into the attic and runs 16 ft.. How many expansion joints with 6 in. expansion capacity is needed for the run, if the temperature for the run over the year has an average of 100°F?

Step 1: Finding number of joints
352.44; Table 352.44(A)
100°F = 4.06

Step 2: Finding total charge
Table 352.44(A) and text
• 200 ft. run /100 ft.   = 2
• 4.06 x 2   = 8.12
• 8.12 ÷ 6   = 1.35

Solution: **Two expansion joints should be used and installed at about 100 ft. intervals.**

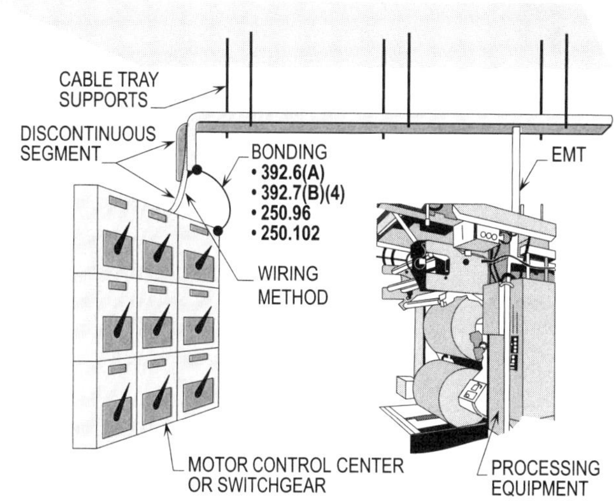

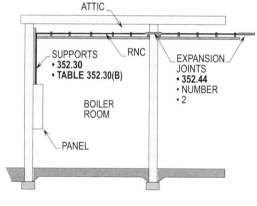

GROUNDING AND BONDING CABLE TRAYS
NEC 392.5(E); NEC 392.6(A)

Figure 12-32(c). This illustration shows discontinuous segments of the cable tray and wiring methods shall be permitted.

EXPANSION FITTINGS
NEC 352.44

Figure 12-33. Installing expansion joints in a run of RNC.

Name
Date

Chapter 12. Designing Wiring Methods

Section	Answer		
_____	T F	1.	Boxes and conduits shall be sized and selected to accommodate the number of conductors required to supply power to various pieces of electrical equipment.
_____	T F	2.	Conductors passing through the box unbroken and not pulled into a loop or spliced together shall be counted as two each in determining the fill of the box.
_____	T F	3.	Equipment grounding conductors passing through or spliced together in the box shall be counted as one in determining the fill of the box. (General Rule)
_____	T F	4.	A conductor that is not entirely within the box shall be required to be counted as a conductor used toward the box fill.
_____	T F	5.	A receptacle or switch that is mounted on a strap or yoke is counted as one conductor in determining the fill of the box.
_____	T F	6.	Boxes with one or more cable clamps that are installed to support cables in the box shall have one conductor added in determining the fill of the box.
_____	T F	7.	Boxes that contain fixture studs and hickeys shall have three conductors added in determining the fill of the box.
_____	T F	8.	Boxes are usually required to be mounted to support luminaires (lighting fixtures), fans, etc.
_____	T F	9.	Octagon boxes, unless listed for such, shall not support luminaires (lighting fixtures) that weigh more than 50 lbs.
_____	T F	10.	Octagon boxes shall not support ceiling fans that weigh more than 35 lbs. or less.
_____	T F	11.	Square boxes shall not be used with plaster rings for the purpose of mounting a receptacle or switch.
_____	T F	12.	Device boxes are used to mount switches to control lighting and other electrical loads.
_____	T F	13.	Boxes and junction boxes used to support luminaires (lighting fixtures), ceiling fans, devices, shall be mounted in such a manner to adequately support such items.
_____	T F	14.	The number of conductors that are permitted in conduit bodies are determined by the square inch area per **Table 2** and **Table 3 of Chapter 9**.
_____	T F	15.	Junction boxes housing conductors 4 AWG and larger that are pulled through raceways shall be sized based upon a straight pull or an angle pull.

12-31

Section	Answer		

_____ T F **16.** A conduit system is any run over 24 in. long in length.

_____ T F **17.** The size of an auxiliary gutter shall be determined by dividing the total square inch area of the conductor by 40 percent fill area for installing conductors.

_____ T F **18.** Cable trays can be used for the installation of a number of cables that are installed in a single run from one location to another.

_____ T F **19.** Ladder or ventilated trough trays may be filled with control or signaling cables or both up to 60 percent of the cross-sectional area of their trays up to 6 in. deep.

_____ T F **20.** When installing cables in ladder or ventilated trough trays, the total diameter of all cables 1000 KCMIL and larger shall not exceed the width of the cable tray.

_____ _____ **21.** Junction boxes in angle pulls shall be sized by multiplying the largest raceway in the run by the multiplier _____.

_____ _____ **22.** Junction boxes in straight pulls shall be sized by multiplying the largest raceway in the run by the multiplier _____.

_____ _____ **23.** A nipple shall be permitted to have an _____ percent fill area.

_____ _____ **24.** Panelboards utilizing _____ or _____ bends require greater distance between lugs and the sides (walls) where conductors are terminated.

_____ _____ **25.** A nipple is _____ in. or less in length.

_____ _____ **26.** Ceiling (paddle) fans that are supported to octagon boxes (identified) shall not weigh more than:
    (a) 25 lbs.
    (b) 35 lbs.
    (c) 40 lbs.
    (d) 50 lbs.

_____ _____ **27.** Luminaires (lighting fixtures) that are supported by a octagon box (General Rule) shall not weigh more than:
    (a) 25 lbs.
    (b) 35 lbs.
    (c) 40 lbs.
    (d) 50 lbs.

_____ _____ **28.** The rise on the plaster ring can be:
    (a) 1/4 in.
    (b) 1/2 in.
    (c) 3/4 in.
    (d) All of the above

_____ _____ **29.** Junction boxes utilized in a straight pull shall be sized by multiplying the largest raceway in the run by the multiplier:
    (a) 4
    (b) 8
    (c) 10
    (d) 12

	Section	Answer

30. The fill area for installing conductors in an auxiliary gutter shall be determined by dividing the total square inch area by:
    (a)   20 percent
    (b)   40 percent
    (c)   60 percent
    (d)   80 percent

31. What is the count used for eight conductors passing through an octagon box and the count for eight that are spliced?

32. What is the count toward the fill of the box where there are eight equipment grounding conductors in the box that are used to ground the metal enclosures of the equipment?

33. How many conductors are counted toward the fill for one equipment grounding conductor to a metal box and one isolation equipment grounding conductor for a PC?

34. What size octagon box (w/canopy) is required to support a lighting fixture supplied with 2 - 14-2 AWG w/ground nonmetallic sheathed cables with a fixture stud, hickey and 4 - 18 AWG fixture wires, plus EGC's?

35. What size octagon box, with cable clamps, is required for 2 - 12 AWG and 2 - 14 AWG w/ground romex cables that are spliced in the box between the paneboard and the loads?

36. What size square box is required for 2 - 14-2 AWG w/ground romex cables that are supported within in the box with two cable clamps?

37. What size square box, with cable clamps, is required for 2 - 12 AWG and 2 - 10 AWG w/grounds romex cables that are spliced in the box between the panelboard and the loads?

38. What size device box is required for 1 - 10 AWG straight through conductor, 2 - 10 AWG hots, 2 - 10 AWG neutrals, 2 EGC's, 1 bonding jumper and 1 pigtail? (Romex is the wiring method)

39. What size device box is required for 2 - 12 AWG conductors (Hot and N), 4 - 10 AWG conductors (Hots and N's) and 1 - 30 amp receptacle? There is 1 pigtail and 1 bonding jumper in the box. (EMT is the wiring method)

40. What size junction box is required for 4 raceways with 10 - 12 AWG in each and 2 raceways with 8 - 12 AWG in each that are spliced together?

41. What size junction box is required for 2 raceways with 24 - 14 AWG, 2 raceways with 18 - 12 AWG, 2 raceways with 14 - 10 AWG and 2 raceways with 8 - 6 AWG conductors that are spliced together?

42. What size square box, with extension ring, is required for 26 - 12 AWG conductors? (Use minimum size)

43. What is the cross-sectional area of an LB (conduit body) that is connected to a 2 in. EMT?

Section	Answer

44. What is the minimum length for a junction box, with a straight pull, that has 4 in., 2 in., and 1 1/2 in. raceways connected to its sides?

45. What is the minimum length of a junction box for an angle pull that has a 3 1/2 in., 2 in., and 1 in. raceway that is connected to the right wall and bottom walls?

46. What size EMT conduit is required for four 8 AWG THWN copper conductors?

47. What size EMT conduit is required for ten 14, six 12, and 10 AWG copper conductors?

48. What size EMT nipple is required for ten 14, ten 12, eight 10, and twelve 8 AWG THWN copper conductors?

49. What size auxiliary gutter is required to house three 500 KCMIL THWN copper conductors that have three 4/0 and three 2/0 AWG THWN conductors spliced to them?

50. What are the minimum clearances between the terminating lugs and the side of a panelboard employing and L-bend using 500 KCMIL copper conductors?

51. What is the minimum clearance for 4/0 AWG THHN copper conductors terminating to the lugs in a panelboard, entering from the bottom wall? (S or Z- bends with non-removable lugs)

52. What is the ampacity for four 2/0 AWG THHN multiconductor copper conductors and four 300 KCMIL THHN multiconductor copper conductors supported by a solid bottom cable tray?

53. What is the ampacity for four - 2/0 AWG THHN multiconductor copper conductors and four - 300 KCMIL THHN multiconductor copper conductors supported by a solid bottom cable tray with a solid cover?

54. What is the ampacity for single layer 600 KCMIL THHN copper conductors installed in a cable tray that is uncovered?

55. What is the ampacity for single layer 600 KCMIL THHN copper conductors installed in a cable tray that is covered?

56. What is the ampacity for single layer 500 KCMIL THHN copper conductors installed in a cable tray that is uncovered?

57. What is the ampacity for single layer 500 KCMIL THHN copper conductors installed in a cable tray that is covered?

13

Installing Wiring Methods

Wiring methods used to wire in residential, commercial and industrial locations are cable systems, conduit systems and tubing systems which enclose and protect the conductors. Wiring methods installed by the provisions of the *National Electrical Code* whether in walls, ceilings, floors or attics assure safety. Good workmanship and proper inspection procedures work together in ensuring safe installations. For rules and regulations pertaining to designing wiring methods, see Chapter 12 in this book.

SUPPORTING BOXES
314.23

Boxes and junction boxes used to support luminaires (lighting fixtures), ceiling fans and devices, shall be mounted in such a manner to adequately support the box. Section **314.23(A)** requires the box to be fastened to the surface to which it is mounted unless the surface does not provide an adequate means of support. Section **314.23(B)** permits enclosures to be structurally mounted to a framing member of the dwelling unit or building. The box shall be permitted to be directly mounted to the framing member or by using a metal or wooden brace. Any of the methods used shall provide rigid support where luminaires (fixtures) and devices are mounted to the box.

Metal braces shall be made of at least .020 inches (.51 mm) of metal and wooden braces are shall not be less than 1 in. by 2 in. (25 mm x 50 mm) in cross-section per **314.23(B)(2)**.

Nails shall be permitted to be used to mount boxes when the nails pass through the inside of the box within 1/4 in. (6 mm) of the back or ends of the box. This provision of **314.23(B)(1)** indicates that screws otherwise shall be used to support a box to framing members.

See Figures 13-1(a) and (b) for methods of mounting boxes to framing members.

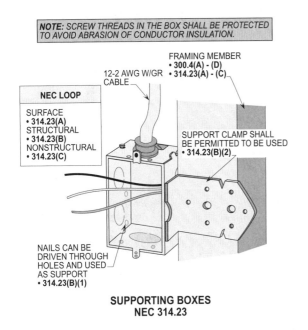

Figure 13-1(a). Mounting methods that shall be permitted to be used to support boxes to the framing members.

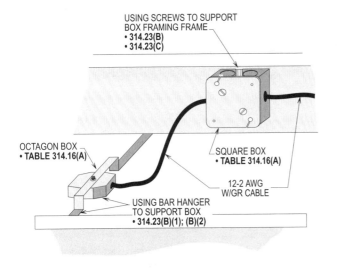

Figure 13-1(b). Using a bar hanger and screw to support boxes to framing members.

DEVICE BOXES
Table 314.16(A)

Device boxes are used to support receptacles, switches or a combination of receptacles and switches, etc. mounted on a single yoke (strap). Device boxes are available in different sizes to accommodate the many sizes of switches and dimmers or receptacles and GFCI receptacles. **Table 314.16(A)** has a list of device boxes with different depths that are designed to contain conductors, receptacles, switches, GFCI receptacles, dimmers, etc. with oversized dimensions.

The size of device boxes along with the proper depth shall be chosen to house the width and depth of the receptacle or switch. The device box not only shall have the adequate space for the device mounted to it, but it shall provide space for the number of conductors entering and leaving the box.

OTHER BOXES
314.16(A)(2)

Boxes 100 cubic inches or less in size, conduit bodies with more than two conduit entries and boxes that are nonmetallic shall be durably and legibly marked with their cubic inch fill area. The fill capacity of each box shall be determined by multiplying each conductor by the cubic inch rating of each conductor from **Table 314.16(B)**. The cubic inch fill area of a junction box shall be found by multiplying the dimensions of the box.

For example, a box with dimensions of 4 in. x 4 in. x 4 in. has a cubic inch fill area of 64 sq. in. (4" x 4" x 4" = 64").

BOXES USED AS SUPPORTS
314.23(D); (E)

Boxes with threaded enclosures or hubs shall be permitted to be supported with two or more conduits that are threaded wrench tight into the enclosure or hubs. Boxes with threaded enclosures or hubs shall be permitted to support devices or luminaires (fixtures) without additional supporting where the conduit is supported properly. **(See Figure 13-2)**

WITHOUT SUPPORTING DEVICES OR FIXTURES
314.23(E)

Boxes with threaded sides or hubs shall be permitted to be used to support devices or luminaires (fixtures) without the boxes being supported from within to the frame, wall, pole, structure, etc. However, for this method of supporting to be applied, the conduit shall be threaded wrench tight into the box and be supported within 18 in. (450 mm) of the box. The box is limited to 100 cubic in. (1650 cubic meters) or less in size.

For example, a box with threaded sides and with a threaded conduit entering and leaving is placed underneath the diving board of a swimming pool. The threaded conduit shall be supported within 18 in. (450 mm) of the box and has a toggle switch mounted to it to control the pool light. The box does not have to be independently supported because the conduit with the 18 in. (450 mm) supports shall be permitted to serve as the required supporting means. **(See Figure 13-3)**

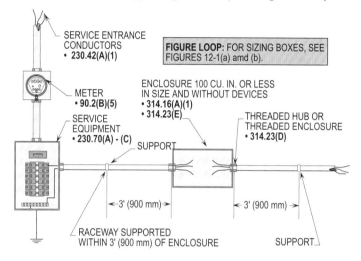

Figure 13-2. If supported within 3 ft. (900 mm) on two or more sides, enclosures without devices do not require additional supports.

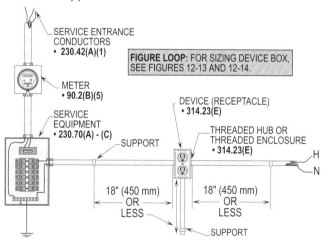

Figure 13-3. If supported within 18 in. (450 mm) on two or more sides, enclosures with devices do not require additional supports. (**Note:** Threaded hubs shall be identified.)

CABLE SYSTEMS
ARTICLES 320 THROUGH 340

Multiconductor cables are flexible assemblies with protective coverings for protection of the conductors. Armored cable (BX), metal-clad cable (MC), nonmetallic sheathed cable (romex) and service-entrance cable are four types of multiconductor cables which are used as wiring methods for the wiring of modern day electrical systems.

ARMORED CABLE: TYPE AC
ARTICLE 320

Type AC cable (BX) is a fabricated assembly of insulated conductors in a flexible metallic enclosure. To assure a reliable path for fault-current to return from the point of fault to the grounded (neutral) conductor connected to the grounded busbar in the main service panelboard, a bonding conductor or spinal is in contact with the armor. When installing cables in wet locations, a special type of armored cable (ACL) shall be used, but the cable shall not be permitted to be buried in the ground.

USES PERMITTED
320.10

Type AC cable shall be permitted to be installed in the following locations:

- In both exposed and concealed work
- In cable trays
- In cable where identified for such use
- In dry locations
- Embedded in plaster finish on brick or other masonry, except in damp in wet locations
- Where run or fished in air voids of masonry block or tile walls where such walls are not exposed or subject to excessive moisture or dampness.

Note: This is not an all inclusive list.

USES NOT PERMITTED
320.12

Type AC cable shall not be permitted to be installed in the following locations:

- Where subject to physical damage
- In damp or wet locations
- In air voids of masonary block or tile walls where such walls are exposed to excessive moisture or dampness
- In areas with corrosive fumes or vapors
- Embedded in plaster finish or brick or other masonry in damp or wet locations.

> **Design Tip:** Check area to see if only flexible connections are permitted for the wiring connection to flexible equipment.

IN ACCESSIBLE ATTICS
320.23

When attics are accessible without stairways or pull-down ladders, armored cable shall be protected by guard strips only when installed within 6 ft. (1.8 m) of the scuttle-hole opening. Guard strips shall not be required for AC cable that is routed along the sides of the joists or rafters or studded through drilled holes.

Guard strips shall be used to protect cables that are installed below 7 ft. (2.1 m) in an attic. These strips are normally 1 in. x 2 in. (25 mm x 50 mm) wooden strips nailed directly to the structure. Guard strips shall be used to protect cables run on top of the floor joists. Guard strips shall not be required for cables installed above 7 ft. (2.1 m) or cables run on the side of floor joists, rafters and studs. **(See Figures 13-13(a) and (b))**

SECURING AND SUPPORTING
320.30

Type AC cable shall be supported and secured by staples, cable ties, straps, hangers or small fittings. Type AC cable shall be secured with 12 in. (300 mm) and secured and supported at intervals not exceeding 4 1/4 ft. (1.4 m).

Type AC cable shall be permitted to be unsupported in the following locations

- Where flexibility is needed, lengths not more than 2 ft. (600 mm)
- Lengths not more than 6 ft. (1.8 m) within an accessible ceiling
- Installed in cable trays

BOXES AND FITTINGS
320.40

When AC cable is terminated in a box, cabinet or piece of equipment, the conductors shall be protected from abrasion by a connector. An insulating bushing shall be used with the connector. This bushing slips over the conductor and under the armor of the cable for additional protection.

METAL-CLAD CABLE: TYPE MC
ARTICLE 330

Type MC cable is a factory assembly of one or more insulated circuit conductors with or without optical-fiber members enclosed in a metallic sheath of interlocking tape, or a smooth or corrugated metallic sheath. The following are these types of MC cable:

- Interlocked metal,
- Corrugated metallic sheath and
- Smooth sheath.

> **Design Tip:** All three types of MC cable are intended for above ground installation except when marked for direct burial.

USES PERMITTED
330.10

When installing Type MC cable and where not subject to physical damage, the following installations shall be permitted:

- For services, feeders and branch-circuits
- For power, lighting, control and signal circuits
- Indoors or outdoors
- Where exposed or concealed
- Direct buried where identified for such use
- In cable trays where identified for such use
- In any raceway
- As aerial cable on a messenger
- In hazardous (classified) locations as permitted
- In dry locations and embedded in plaster finish on brick or other masonry except in damp or wet locations
- Wet locations where any of the following conditions are complied with:

  (a) metallic covering is impervious to moisture
  (b) lead-sheath or moisture-impervious jacket is provided under the metal covering
  (c) insulated conductors under the metallic covering are listed for use in wet locations

- Where single conductor cables are used, all phase conductors and, where used, the grounded (neutral) conductor shall be grouped together to minimize induced voltage on the sheath

Installing Wiring Methods

USES NOT PERMITTED
330.12

Type MC cable shall not be permitted to installed where exposed to distinctive conditions such as:

- Where subject to physical damage
- Direct burial in the earth
- In concrete
- Where exposed to cinder fills, strong chlorides, caustic alkalis or vapors of chlorine or of hydrochloric acids

> **Design Tip:** Type MC cable shall be permitted where the metallic sheath is suitable for the conditions or is protected by material suitable for the conditions.

SECURING AND SUPPORTING
330.30

Type MC cable shall be supported and secured by staples, cable ties, straps, hangers or similar fittings or other approved means.

Type MC cable shall be secured at intervals not exceeding 6 ft (1.8 m). MC cable containing four or fewer conductors sized no larger than 10 AWG shall be secured within 12 in. (300 m). MC cable shall be supported at intervals not exceeding 6 ft. (1.8 m).

Type MC cable shall be permitted to be unsupported in the following locations:

- Cable that is fished between access points
- Lengths not more than 6 ft. (1.8 m) within an accessible ceiling

NONMETALLIC-SHEATHED CABLE: TYPES NM, NMC AND NMS
ARTICLE 334

Romex or rope is the term used in the field when installing nonmetallic-sheathed cable. This cable has two, three or four conductors with a green insulated or bare conductor used for grounding equipment. The conductors are protected by an outer nonmetallic jacket. There are three types of nonmetallic sheathed cable and they are as follows:

- NM cable
- NMC cable
- NMS cable

Type NM cable has a flame-retardant and moisture-resistant outer jacket and is restricted to be installed for inside locations. Type NMC cable has a flame-retardant and moisture-resistant outer jacket and also is fungus-resistant and corrosion-resistant and shall be permitted to be installed for outside locations. Type NMS cable is a factory assembly of insulated power, communications and signaling conductors enclosed within a common sheath of moisture-resistant, flame-retardant, nonmetallic material.

> **Design Tip:** Nonmetallic-sheathed cable shall be permitted to be used for economical wiring of single-family dwellings, multifamily dwellings and small commercial buildings.

USES PERMITTED
334.10

Type NM, NMC and NMS shall be permitted to be installed in the following locations:

- In one and two-family dwellings
- In multifamily dwellings of Types III, IV and V construction except as prohibited per **334.12**
- In other structures of Types III, IV and V construction except as prohibited per **334.12**. Cables shall be concealed within walls, floors or ceilings that provide a thermal barrier of at least 15-minute finish rating.
- Cable trays in structures permitted to be types II IV, or V, where the cables are identified for such use.

Type NM cable shall be permitted for the following locations:

- For both exposed and concealed work in normally dry locations
- Where installed or fished in air voids in masonry block or tile walls

Type NMC cable shall be permitted for the following locations:

- For both exposed and concealed work in dry, moist, damp or corrosive locations
- In outside and inside walls of masonry block or tile
- In a shallow chase in masonry, concrete or adobe protected against nails or screws by a steel plate at least 1/16 in. (1.59 mm) thick and covered with plaster, adobe or similar finish

Type NMS cable shall be permitted for the following locaions:

- For both exposed and concealed work in normally dry locations
- Where installed or fished in air voids in masonry block or tile walls
- Where listed and identified for closed-loop and programmed power distribution.

USES NOT PERMITTED
334.12

Type NM, NMC and NMS shall not be permitted to be installed in the following locations:

- In any dwelling unit not specifically in 334.10(1) - (3)
- Exposed in dropped or suspended ceilings in other than one- and two-family and multifamily dwellings
- As service-entrance cable
- In commercial garages having hazardous (classified) locations
- In theaters and similar locations
- In motion picture studios
- In storage battery rooms
- In hoistways or on elevators or escalators
- Embedded in poured cement, concrete or aggregate
- In hazardous (classified) locations

Type NM and NMS shall not be permitted to be installed in the following locations:

- Where exposed to corrosive fumes or vapors
- Where embedded in masonry, concrete, adobe, fill or plaster
- In a shallow chase in masonry, concrete or adobe and covered with plaster, adobe or similar finish
- Where exposed or subject to excessive moisture or dampness

EXPOSED WORK
334.15

Exposed cable systems shall be run along the surface of finished areas or running boards shall be provided and cables shall closely follow such surfaces and be supported per **334.30**.

Nonmetallic-sheathed cable shall be protected from physical damage by the following:

- Conduit
- Electrical metallic tubing
- Schedule 80 PVC rigid nonmetallic conduit
- Pipe
- Guard strips
- Listed surface metal or nonmetallic raceway
- Other means

Where nonmetallic-sheathed cable passes through a floor, the cable shall be enclosed at least 6 in. (150 mm) above the floor by one of the following means:

- Rigid metal conduit
- Intermediate metal conduit
- Electrical metallic tubing
- Schedule 80 PVC rigid nonmetallic conduit
- Listed surface metal or nonmetallic raceway
- Other metal pipe

Cable systems mentioned above that are installed in unfinished basements shall be permitted to be routed through the center of framing members (studs) without additional supports. Cable systems that are not run through the center of the framing members shall be pemitted to be placed in a cut notch in the framing member and protected by a 1/16 in. (1.59 mm) thick steel plate. Cable systems smaller than 8-3 AWG and 6-2 AWG shall be secured to the face or placed and supported on running boards.

THROUGH OR PARALLEL TO FRAMING MEMBERS
334.17

Framing members that are notched and have the above raceways installed in the notched area, shall not be required to have a 1/16" (1.52 mm) thick plate placed over them for protection.

SECURING AND SUPPORTING
334.30

Nonmetallic sheathed cable shall be supported and secured every 4 1/2 ft. (1.4 m) and within 12 in. (300 mm) of every outlet or fitting. Cables installed within 6 ft. (1.8 m) of a scuttle hole or trap door with a ladder shall not be permitted to be installed unless the cable is run along the sides of floor joists, rafters or studs. A guard strip shall be required when the cable is installed on the top of joists or rafters. **(See Figures 13-7 and 13-13(a) and (b))**

SERVICE-ENTRANCE CABLE: TYPES SE AND USE
ARTICLE 338

The following are three types of service-entrance cable that can be used as a wiring method for the wiring of modern day electrical systems and they are as follows:

- SE
- USE
- ASE

Type SE cable has a flame-retardant, moisture-resistant covering that is unarmored and will not withstand severe mechanical abuse. Type USE cable has a moisture-resistant covering that is unarmored and aids in preventing deterioration of the cable when it is buried in the earth. Type ASE cable has an armored covering to provide additional protection of the cable.

Type SE cable is available in sizes ranging from 12 AWG to 4/0 AWG. The cable may be installed with two insulated conductors and one bare conductor or with three insulated conductors and one bare conductor.

USES PERMITTED - SERVICE-ENTRANCE CONDUCTORS
338.10(A)

Service-entrance cable shall be permitted to be used for service-entrance wiring but shall be installed as required per **Article 230**. Type USE used for service laterals shall be permitted to emerge aboveground outside at terminations in meter bases or other enclosures where protected in accordance with **300.5(D)**.

USES PERMITTED - BRANCH-CIRCUITS OR FEEDERS
338.10(B)

Service-entrance cable shall be permitted to be used for general interior wiring but the grounded (neutral) conductor shall be insulated when installed with the type of insulation approved for the purpose, such as rubber or thermoplastic. When installing the grounded (neutral) conductor in service-entrance cable, the grounded (neutral) conductor shall be permitted to be insulated or bare where the uninsulated conductor is installed as an equipment grounding conductor and not as a circuit conductor or grounded (neutral) conductor.

The temperature limitations for the insulation on the conductors is rated either:

- 60°C
- 75°C
- 90°C

For example, TW insulation has a 60°C rating and THWN insulation has a 75°C rating. These ratings are applied per **Table 310.16** and matched to the terminal rating per **110.14(C)(1)** and **110.14(C)(2)**.

Conductors with 60°C rated insulation are used on most cooking appliances such as ranges, cooktops and ovens. When installing conductors with 90°C rated insulation, the load shall be limited to the current rating capacity of conductors rated with 60°C rated ampacities. The 90°C ampacities are only used for derating purposes due to equipment adjustment or correction factors or both.

> **Design Tip:** Table 310.15(B)(2)(a) to Ampacity Tables 0 to 2000 volts and correction factors at the bottom of the Tables shall be observed when designing and installing service-entrance conductors.

METAL CONDUITS
ARTICLES 342 THRU 358

Metal conduit provides excellent protection for conductors. When designing wiring systems utilizing metal conduit, the manufacturer's suggestions and applicable sections of the NEC should be followed. Depending on the type used, metal conduit shall be permitted to be installed outdoors or indoors in damp or dry areas.

INTERMEDIATE METAL CONDUIT: TYPE IMC
ARTICLE 342

Intermediate metal conduit (IMC) is a steel threadable raceway of circular cross section design for the physical protection and routing of conductors and cables and for use as an equipment grounding conductor when installed with its integral or associated coupling and appropriate fittings.

USES PERMITTED
342.10

Intermediate metal conduit (IMC) shall be listed and shall be permitted to be used for all atmospheric conditions and occupancies.

Intermediate metal conduit, elbows, couplings and fittings shall be permitted to be installed in the following:

- Where installed in concrete
- Where installed in direct contact with the earth
- Where installed in areas subject to severe corrosive influences

Rigid metal conduit shall be protected by an encased concrete layer at least 2 in. (50 mm) thick to be buried in a cinder fill or it shall be buried at least 18 in. (450 mm) under the cinder fill to ensure protection.

All suports, bolts, straps, screws, etc. shall be of corrosion-resistant materials or protected against corrosion by corrosion-resistant materials.

SIZE
342.20

Intermediate metal conduit shall be permitted to be installed in a minimum size of 1/2 in. (16) and a maximum size of 4 in. (103).

REAMING AND THREADING
342.28

When cutting and threading rigid metal conduit in the field, a standard cutting die with a 3/4 in. taper per foot shall be used. To protect the wire insulation from abrasion all cut conduit ends shall be reamed to remove rough edges. For more details on this requirement, see the standard for pipe threads, general purpose which is ANSI B.1.20.1.

SUPPORTS
342.30

Rigid metal conduit shall be supported every 10 ft. (3 m) and within 3 ft. (900 mm) of each outlet box, junction box, device box, etc. Where structural members do not permit fastening within 3 ft. (900 mm), the distance shall be supported at 5 ft. (1.5 m). Rigid metal conduit shall not be required to be supported in the following locations:

- Where approved, a distance of 3 ft. (900 mm) shall not be required at the service head for above-the-roof termination of a service mast.

- When installing threaded couplings for straight runs, the support shall be permitted to be increased more than 10 ft. (3 m) for larger sizes of conduit connected together with threaded couplings and threaded hubs.

- A distance of 20 ft. (6 m) between supports shall be permitted be used for vertical rises from machines, motors, etc. if the conduit is made up with threaded couplings, the conduit is firmly supported and securely fastened at the top and bottom of the riser and no other means of intermediate support is readily available.

- Horizontal runs supported by openings through framing members at a distance not greater than 10 ft. (3 m) and within 3 ft. (900 mm) of termination points shall be permitted.

COUPLINGS AND CONNECTORS
342.42

Threadless couplings and connectors shall be made tight where installed with intermediate metal conduit. Threadless couplings and connectors shall not be permitted to be used on threaded ends unless listed for the purpose. Where buried in masonry or concrete and installed in wet locations the following types shall be used:

- Concretetight
- Raintight

Running threads shall not be permitted to be used for connection at couplings. The galvanized coating would be removed and the conduit would rust if running threads where installed at the connection of couplings. The conduit at the joint would also be weaken if installed with running threads and subjected to breakage if stepped on or run over with a wheelbarrow full of concrete during the pouring of the slab.

BUSHINGS
342.46

A bushing shall be provided to protect the wire from abrasion where a conduit enters a box, fitting or other enclosure unless the box, fitting or enclosure design provides equivalent protection.

RIGID METAL CONDUIT: TYPE RMC
ARTICLE 344

Rigid metal conduit is a threadable raceway of circular cross section designed for the physical protection and routing of conductors and cables and for use as an equipment grounding conductor when installed with its integral or associated coupling and appropriate fittings. RMC is generally made of steel (ferrous) with protective coatings or aluminum (nonferrous). Special use types are silicon bronze and stainless steel. **(See Page 13-12 and Figure 13-14)**

USES PERMITTED
344.10

Rigid metal conduit (RMC) shall be listed and shall be permitted to be used for all atmospheric conditions and occupancies.

Ferrous raceways and fittings shall be permitted to be installed only indoors and in occupancies not subject to severe corrosive influences where protected from corrosion solely by enamel.

Rigid metal conduit, elbows, couplings and fittings shall be permitted to be installed in the following:

- Where installed in concrete
- Where installed in direct contact with the earth
- Where installed in areas subject to severe corrosive influences

Rigid metal conduit shall be protected by an encased concrete layer at least 2 in. (50 mm) thick to be buried in a cinder fill or it shall be buried at least 18 in. (450 mm) under the cinder fill to ensure protection.

All suports, bolts, straps, screws, etc. shall be of corrosion-resistant materials or protected against corrosion by corrosion-resistant materials.

SIZE
344.20

Rigid metal conduit shall be permitted to be installed in a minimum size of 1/2 in. (16) and a maximum size of 6 in. (155). The following exception permits the conduit size to be installed smaller than 1/2 in. (16):

- A rigid metal conduit shall be permitted to enclose leads to connect a motor that is separated from the motor box per **430.145(B)**.

REAMING AND THREADING
344.28

When cutting and threading rigid metal conduit in the field, a standard cutting die with a 3/4 in. taper per foot shall be used. To protect the wire insulation from abrasion all cut conduit ends shall be reamed to remove rough edges.

SECURING AND SUPPORTING
344.30

Rigid metal conduit shall be supported every 10 ft. (3 m) and within 3 ft. (900 mm) of each outlet box, junction box, device box, etc. Where structural members do not permit fastening within 3 ft. (900 mm), the distance shall be supported at 5 ft. (1.5 m). Rigid metal conduit shall not be required to be supported in the following locations:

- Where approved, a distance of 3 ft. (900 mm) shall not be required at the service head for above-the-roof termination of a service mast.

- When installing threaded couplings for straight runs, the support shall be permitted to be increased more than 10 ft. (3 m) for larger sizes of conduit connected together with threaded couplings and threaded hubs.

- A distance of 20 ft. (6 m) between supports shall be permitted be used for vertical rises from machines, motors, etc. if the conduit is made up with threaded couplings, the conduit is firmly supported and securely fastened at the top and bottom of the riser and no other means of intermediate support is readily available.

- Horizontal runs supported by openings through framing members at a distance not greater than 10 ft. (3 m) and within 3 ft. (900 mm) of termination points shall be permitted.

COUPLINGS AND CONNECTORS
344.42

Threadless couplings and connectors shall be made tight where installed with rigid metal conduit. Where buried in masonry or concrete and installed in wet locations the following types shall be used:

- Concretetight
- Raintight

Running threads shall not be permitted to be used for connection at couplings. The galvanized coating would be removed and the conduit would rust if running threads where installed at the connection of couplings. The conduit at the joint would also be weaken if installed with running threads and subjected to breakage if stepped on or run over with a wheelbarrow full of concrete during the pouring of the slab.

BUSHINGS
344.46

A bushing shall be provided to protect the wire from abrasion where a conduit enters a box, fitting or other enclosure unless the box, fitting or enclosure design provides equivalent protection.

FLEXIBLE METAL CONDUIT: TYPE FMC
ARTICLE 348

Flexible metal conduit is a raceway of circular cross section made of helically wound, formed, interlocked metal strip.

USES PERMITTED
348.10

Flexible metal conduit shall be listed and shall be permitted to be installed in both exposed and concealed locations.

USES NOT PERMITTED
348.12

Flexible metal conduit shall not be permitted to be installed in the following locations:

- In wet locations unless installed with a W-rated weatherproof insulation.

- In hoistways, except for the installation of control circuits and short extensions on elevator cars.

- In storage-battery rooms.

- In hazardous (classified) locations, except as permitted in **501.4(B)** and **504.20**.

- In oil or gasoline areas without the installation of wire insulation is approved for such locations.

- In embedded in poured concrete or aggregate or installed underground.

- Where subject to physical damage.

SIZE
348.20

Flexible metal conduit shall be permitted to be installed in a minimum size of 1/2 in. (16) and a maximum size of 4 in. (103). Flexible metal conduit in 3/8 in. (12) sizes shall be permitted to be used as a wiring method for the following installations:

- To enclose the motor leads to connect a motor that is separated from the motor box per **430.145(B)**.

- Part of a listed assembly does not exceed 6 ft. (1.8 m) in length for tap connections to luminaires (lighting fixtures) per **410.67(C)** or for utilization equipment.

- For installing manufactured wiring systems per **604.6(A)**.

- In hoistways per **620.21(A)(1)**.

- Part of a listed assembly to connect wired luminaire (fixture) sections per **410.77(C)**.

SECURING AND SUPPORTING
348.30

Flexible metal conduit shall be supported within 12 in. (300 mm) of each box, cabinet, etc. and supported at intervals not exceeding 4 1/2 ft. (1.4 m). Flexible metal conduit shall not be required to be supported in the following locations:

- Where flexible metal conduit is fished.

- Where flexibility is needed, lengths shall not exceed the following
    - 3 ft. (900 mm) for trade sizes 1/2 thru 1 1/4 (metric designations 16 thru 25)
    - 4 ft. (1200 mm) 1 1/2 thru 2 (41 thru 58)
    - 5 ft. (1500 mm) 2 1/2 (63) and larger

- Part of a listed assembly does not exceed 6 ft. (1.8 m) in length for tap connections to luminaires (lighting fixtures) per **410.67(C)**.

GROUNDING AND BONDING
348.60

Flexible metal conduit shall be permitted to be installed as a equipment grounding conductor per **250.118(5)**. Lengths of 6 ft. (1.8 m) or less and protected at 20 amps or less may be installed for grounding of equipment per **348.60**. An equipment grounding conductor shall be installed to connect equipment where flexibility is needed. Where required, an equipment bonding jumper shall be installed around flexible metal conduit per **250.102(D)** and **(E)** and such bonding jumper shall be sized per **Table 250.122** based upon the size of the branch-circuit OCPD.

LIQUIDTIGHT FLEXIBLE METAL CONDUIT: TYPE LFMC
ARTICLE 350

Liquidtight flexible metal conduit is a raceway of circular cross section having an outer liquidtight, nonmetallic, sunlight-resistant jacket over an inner flexible metal core with associated couplings, connectors and fittings for the installation of electric conductors.

USES PERMITTED
350.10

Liquidtight flexible metal conduit shall be permitted to be installed in both exposed or concealed locations as follows:

- Where conditions of installation, operation or maintenance requires flexibility or protection from liquids, vapors or solids.

- For installations permitted by **501.4(B), 502.4(A)(2), 502.4(B)(2), 503.3(A)** and **504.20** and in other hazardous (classified) locations where specifically approved, and by **553.7(B)**.

- Where listed and marked for direct burial.

USES NOT PERMITTED
350.12

Liquidtight flexible metal conduit shall not be permitted to be installed where:

- Subject to physical damage.

- An operating temperature will be produced in excess of that for which the material is approved for any combination of ambient and conductor temperatures.

SIZE
350.20

Liquidtight flexible metal conduit shall be permitted to be installed in a minimum size of 1/2 in. (16) and a maximum size of 4 in. (103). The exception allows 3/8 in. (12) conduit to be installed per **348.20(A)**.

SECURING AND SUPPORTING
350.30

Liquidtight flexible metal conduit shall be supported within 12 in. (300 mm) of each box, cabinet, etc. and supported at intervals not exceeding 4 1/2 ft. (1.4 m). Horizontal runs shall be supported through framing members at intervals not exceeding 4 1/2 ft. (1.4 m) and within 12 in. (300 mm) of each termination point. Liquidtight flexible metal conduit shall not be required to be supported in the following locations:

- Where liquidtight flexible metal conduit is fished.

- Lengths not exceeding 3 ft. (900 mm) at terminals where flexibility is needed.

- Part of a listed assembly does not exceed 6 ft. (1.8 m) in length for tap connections to luminaires (lighting fixtures) per **410.67(C)**.

RIGID NONMETALLIC CONDUIT: TYPE RNC
ARTICLE 352

Schedule 40 and Schedule 80 are the two types of rigid nonmetallic conduit normally used in the industry. Rigid nonmetallic conduit is known in the industry as PVC plastic conduit. A rigid nonmetallic conduit is a nonmetallic raceway of circular cross section, with integral or associated couplings, connectors and fittings for the installation of electrical conductors. **(See Figures 13-10(a) and (b))**

SCHEDULE 40

Nonmetallic materials such as fiber, polyvinyl chloride or polyethylene are used so as to make Schedule 40 PVC waterproof, rustproof and rotproof.

SCHEDULE 80

Schedule 80 PVC is durable and constructed of a thicker and heavier plastic material than Schedule 40 PVC and shall be permitted to be used above and below grade.

USES PERMITTED
352.10

Listed rigid nonmetallic conduit shall be permitted to be installed in any one of the following locations:

- Concealed in walls, floors and ceilings.

- Where subject to severe corrosive influences per **300.6** and where subject to chemicals.

- In cinder fill.

- In wet locations. Corrosion-resistant materials such as supports, bolts, straps, etc. shall be installed or be protected by approved corrosion resistant materials.

- Dry and damp locations are not prohibited per **352.12**.

- Exposed work not subjected to physical damage if identified for such use.

> **Design Tip:** Schedule 80 PVC shall be permitted be installed and utilized in exposed locations subjected to physical damage per **300.5(D)**. However, Schedule 40 PVC shall not be permitted be used for this particular installation.

- Underground installations. See **300.5** and **300.50(B)** in the NEC. Conduits listed for the purpose shall be permitted to be installed underground in continuous lengths from a reel.

- Rigid nonmetallic conduit shall be permitted to support nonmetallic conduit bodies not larger than the largest trade size of an entering raceway.

USES NOT PERMITTED
352.12

Rigid nonmetallic conduit (general rule) shall not be permitted to be installed in the following locations:

- Hazardous (classified) locations. See **503.3(A), 504.20, 514.8, 515.8** and **501.4(B), Ex.**

- Support of luminaires (fixtures) or other equipment.

- Subject to physical damage unless identified for such use.

- For ambient temperatures in excess of 50°C (122°F) unless listed otherwise.

- For insulation temperature limitations not exceeding the listing for the conduit.

- Theaters and similar locations, except as provided per **Articles 518.4** and **520.5**.

SIZE
352.20

Rigid nonmetallic conduit shall be permitted to be installed in a minimum size of 1/2 in. (16) and a maximum size of 6 in. (155).

SECURING AND SUPPORTING
352.30

Rigid nonmetallic conduit shall be supported within 3 ft. (900 mm) of each outlet box, junction box, device box, etc. Rigid nonmetallic conduit shall be supported so that movement from thermal expansion or contraction will be permitted. When installing rigid nonmetallic conduit that is larger than 1 in. (27), it shall be supported per **Table 352.30(B)**.

> **Design Tip:** Listed rigid nonmetallic conduit shall be permitted to be supported in accordance to the listing other than the spacing requirements per **Table 352.30(B)**. For expansion fittings, see **352.44** and page 12-29.

TUBING
ARTICLES 358 THRU 362

Tubing provides protection for conductors. When designing wiring systems utilizing tubing, the manufacturer's suggestions and applicable sections of the NEC should be followed.

ELECTRICAL METALLIC TUBING
ARTICLE 358

Electrical metallic tubing is an unthreaded thin-wall raceway of circular cross section designed for the physical protection and routing of conductors and cables and for use an equipment grounding conductor when installed utilizing appropriate fittings. Electrical metallic tubing is generally made of steel (ferrous) with protective coatings or aluminum (nonferrous).

USES PERMITTED
358.10

Electrical metallic tubing shall be permitted to be installed in both exposed and concealed locations.

Electrical metallic tubing, elbows, couplings and fittings shall be permitted to be installed in the following:

- Where installed in concrete
- Where installed in direct contact with the earth
- Where installed in areas subject to severe corrosive influences

All suports, bolts, straps, screws, etc. shall be of corrosion-resistant materials or protected against corrosion by corrosion-resistant materials.

USES NOT PERMITTED
358.12

Electrical metallic tubing shall not be permitted to be installed in the following locations:

- Where subject to severe physical damage.
- Protected from corrosion solely by enamel.
- Protected by a encased noncinder concrete layer at least 2 in. (50 mm) thick to be buried in a cinder concrete or cinder fill or it shall be buried at least 18 in. (450 mm) under the cinder fill.
- In any hazardous (classified) locations.
- For the support of luminaire (fixtures) or other equipment except conduit bodies no larger than the largest trade size of the tubing.
- Where practicable, dissimilar metals in contact anywhere in the system shall be avoided to eliminate the possibility of galvanic action.

SIZE
358.20

Electrical metallic tubing shall be permitted to be installed in a minimum size of 1/2 in. (16) and a maximum size of 4 in. (103). The following exception permits the conduit size to be installed smaller than 1/2 in. (16):

- A electrical metallic tubing shall be permitted to enclose leads to connect a motor that is separated from the motor box per **430.145(B)**.

SECURING AND SUPPORTING
358.30

Electrical metallic tubing shall be supported every 10 ft. (3 m) and within 3 ft. (900 mm) of each outlet box, junction box, device box, etc. Horizontal runs supported by openings through framing members at a distance not greater than 10 ft. (3 m) and within 3 ft. (900 mm) of termination points shall be permitted. Electrical metallic tubing shall not be required to be supported in the following locations:

- Where structural members do not permit fastening within 3 ft. (900 mm), the distance shall be supported at 5 ft. (1.5 m)
- Electrical metallic tubing shall be permitted to be fished for unbroken lengths.

COUPLINGS AND CONNECTORS
358.42

Threadless couplings and connectors shall be made tight where installed with electrical metallic conduit. Where buried in masonry or concrete and installed in wet locations the following types shall be used:

- Concretetight
- Raintight

ROUGH-IN

Rough-in wiring methods are used in slabs, walls, ceilings, attics and floors between the service panelboard and boxes which are used to support devices and luminaires (fixtures) or contain spliced conductors.

Rough-in is the stage of construction where all the wiring methods are left open in ditches, walls, ceilings, attics, floors, etc. for inspection before being covered up. Wiring methods in the rough-in stage should never be insulated over or sheetrocked until the inspector leaves a card of approval which is usually green or blue while a card of disapproval or rejection is normally red. **(See Figure 13-4)**

Boxes and other enclosures are mounted to the framing members. Cable or conduit systems are routed from the service panel and connected properly per **300.15**. Cable and conduit systems are to be continuous and comply with **300.10** and **300.12**.

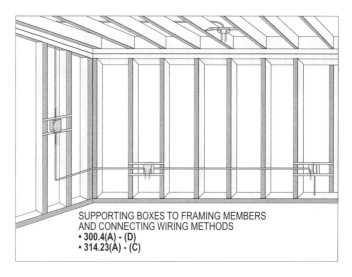

Figure 13-4. Wiring methods installed in walls, ceilings, attics and floors are not allowed to be covered up with insulation or sheetrock until the rough-in inspection has been made by the AHJ.

TEMPORARY POLE

Temporary poles (construction poles) consists of the meter base and panelboard with receptacle outlets mounted beneath the panelboard. The temporary pole must be installed in the earth at a depth (usually 3 ft. (900 mm)) that will hold the pole with mounted equipment substantially. Braces shall be installed to support the pole and the service drop between the pole and the utility pole. The braces shall be arranged so that they are opposite in direction from the point of attachment of the service drop conductors. Areas that have service conductors underground, rather than overhead, will utilize a temporary pole or pedestal. A pedestal has the meter and the OCPD's installed within, in other words, it is self contained. **(See Figures 13-5(a) and (b))**

Procedures for design, installation and inspections of temporary power poles are as follows:

- Check support of pole.
- Check bracing of pole.
- Check grounding of pole.
- Check for main(s) in panelboard.
- Check for GFCI-protection of receptacles.

Stallcup's Electrical Design Book

- Check point of attachment for service drop conductors.
- Check size of service-entrance conductors.
- Check for weatherproof equipment.

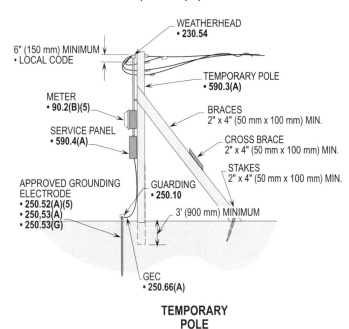

Figure 13-5(a). The above is a recommended procedure for installing an overhead temporary power pole to be used on a constuction site.

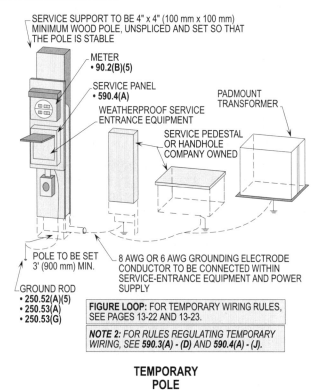

Figure 13-5(b). The above is a recommended procedure for installing an underground temporary pole (pedestal) to be used on construction sites. **Note:** For temporary wiring rules, see pages 13-22 and 13-23.

SLAB
TABLE 300.5

Raceways are installed in slabs between the service equipment panelboard and the outlet boxes used to support devices such as receptacles and switches. Raceways are installed in the slab to connect subpanels to the service panelboard and allow the capacity for more circuits to be installed. Raceways are sometimes routed in the slab to supply power to special pieces of equipment such as heating units, AC units, etc. For future additions, such as a detached garage or a small tool house out back, a raceway shall be installed in the slab between the panelboard and the foundation and extended outside for later use. Raceways installed in the slab shall be considered supported where they are tied to the rebar reinforcing steel. Boxes and enclosures shall be supported to the framing members per **314.23(B)** and **(C)**. **(See Figure 13-6)**

Procedures for design, installation and inspection of wiring methods are as follows:

- Check that raceways are tied to the steel and supported.
- Check floor boxes for support by raceways and tie wire.
- Check ends of raceways for concrete caps.
- Check raceways to verify they are complete and fittings are tight.
- Check that raceways are supported where they stub-up in the walls.

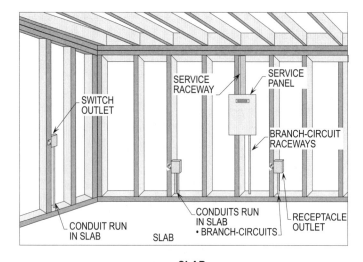

Figure 13-6. Conduits shall be permitted to be run in the slab to connect outlets and other enclosures to the service equipment panel.

WALLS
300.4

Wiring methods installed horizontally or run parallel to framing members (studs) shall be supported. Wiring methods run horizontally through drilled holes shall be considered properly supported by the framing members. Wiring methods run parallel to framing members shall be supported by the provisions of the NEC pertaining to each type of wiring method.

CABLE SYSTEMS
ARTICLES 320, 330, 334 AND 338

Cable systems shall be supported and securely fastened where they are routed through stud, joists or similar wood members per **320.17, 330.17, 334.17** and **338.10(B)(4)**. Exposed cable systems running along the surface of finished areas or running boards shall closely follow such surfaces and shall be supported per **320.17, 330.17, 334.17** and **338.10(B)(4)**. AC cable, nonmetallic sheathed cable and service-entrance cable shall be supported at intervals not exceeding 4 1/2 ft. (1.4 m) and within 12 in. (300 mm) of each cabinet, box or fitting. **(See Figure 13-7)**

Cable systems mentioned above that are installed in unfinished basements shall be permitted to be routed through the center of framing members (studs) without additional supports. Cable systems that are not run through the center of the framing members shall be permitted to be placed in a cut notch in the framing member and protected by a 1/16 in. (1.6 mm) thick steel plate. Cables systems smaller than 8-3 AWG and 6-2 AWG shall be secured to the face or placed and supported on running boards. **(See Figure 13-8)**

Procedures for design, installation and inspection of wiring methods are as follows:

- Check supports that are supporting cables.

- Check spacing of supports.

- Check to verify that proper fittings for terminating cables are utilized.

- Check loose fitting supports of cables at boxes, equipment, etc.

- Check cables for exposure to damage.

- Check type of installation:
  (a) Cables run through drilled holes in wall studs.
  (b) Cables run through notched cuts on the edge of wall studs.
  (c) Cables run on the sides of wall studs.
  (d) Cables run on the bottom of wall studs.
  (e) Cables run on the top of wall studs.

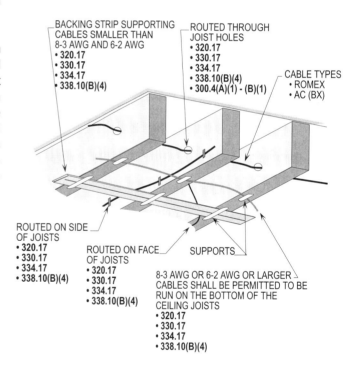

CABLE SYSTEMS
ARTICLES 320, 330, 334 AND 338

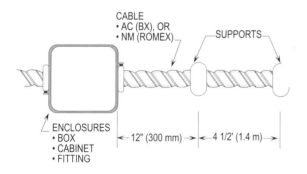

CABLE SYSTEMS
ARTICLES 320, 330, 334 AND 338

Figure 13-7. Nonmetallic sheathed cable and AC cable shall be supported within 12 in. (300 mm) of each box cabinet or fitting and at intervals not exceeding 4 1/2 ft. (1.4 m).

Figure 13-8. The method in which cables are routed through framing members in an unfinished basement shall be determined by the size and number of conductors in the cable.

RACEWAY SYSTEMS
ARTICLES 342, 344, 348, 350 AND 352

Raceway systems run horizontally through drilled holes in the framing members shall be considered supported. Those that are run along the surface, on top, on bottom or parallel to the framing members shall be supported per **342.30, 344.30, 348.30, 350.30** and **350.30**.

Raceway systems such as rigid metal conduit and IMC are required to be supported and securely fastened every 10 ft. (3 m) and within 3 ft. (900 mm) of each box, enclosure and fitting for sizes 1/2 in. (16) and 3/4 in. (21) per **Table 344.30(B)(2)**. See **Table 344.30(B)(2)** for spacing of supports for larger raceways using threaded connections.

> **For example:** What is the maximum distance between supports for 1 in. (27) IMC using threaded connections?
>
> **Step 1:** Finding distance for supports
> **342.30(B)(2); Table 344.30(B)(2)**
> 1" (27) IMC = 12' (3.7 m)
>
> **Solution: Supports are required at 12 ft. (3.7 m) intervals.**

Framing members that are notched and have the above raceways installed in the notched areas, shall not be required to have a 1/16 in. (1.6 mm) thick plate placed over them for protection. **(See Figure 13-9)**

Rigid nonmetallic conduit shall be be supported and securely fastened every 3 ft. (900 mm) and within 3 ft. (900 mm) of every box, cabinet or fitting per **352.30(A)**. Schedule 40 or 80 rigid nonmetallic conduit shall be supported and securely fastened within 3 ft. (900 mm) of each box, enclosure and fitting and at intervals per **352.30(B)** and **Table 352.30(B)**.

> **For example:** What is the maximum distance between supports for 1 1/2 in. (41) rigid nonmetallic conduit?
>
> **Step 1:** Finding distance for supports
> **352.30(B); Table 352.30(B)**
> 1 1/2" (41) RNC = 5' (1.5 m)
>
> **Solution: Supports are required at 5 ft. (1.5 m) intervals.**

See Figures 13-10(a) and (b) for the installation requirements for rigid nonmetallic conduit.

Procedures for design, installation and making inspections are as follows:

• Check supports supporting raceways.

• Check spacing of supports.

• Check and verify that proper fittings for terminating raceways are utilized.

• Check loose fittings and support of raceways at boxes, equipment, etc.

• Check type of installation such as:

  (a) Raceways run through drilled holes in wall studs.
  (b) Raceways run through notched cuts on the edge of the wall studs.
  (c) Raceways run on the sides of the wall studs.
  (d) Raceways run on the bottom of wall studs.
  (e) Raceways run on the top of wall studs.

CEILINGS
300.4

Wiring methods run through the center of the ceiling rafters shall be considered supported. When they are routed on the top, bottom or face they shall be supported by the provisions in the NEC. Wiring methods supported by the NEC shall be considered to be safe and reliable.

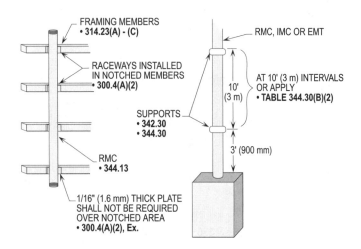

**RACEWAY SYSTEMS
ARTICLES 342, 344, 348, 350 AND 352**

Figure 13-9. Installation requirements and supporting rules pertaining to metal raceways.

Installing Wiring Methods

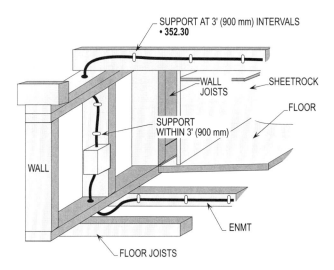

Figure 13-10(a). Rigid nonmetallic conduit shall be supported and securely fastened at intervals not exceeding 3 ft. (900 mm) and within 3 ft. (900 mm) of each box, cabinet, fitting, etc.

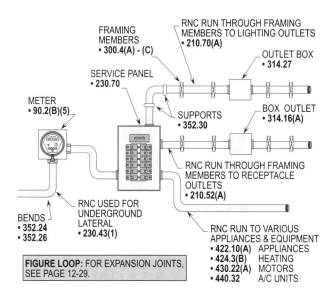

Figure 13-10(b). Rigid nonmetallic conduit (RNC) shall be permitted to be used to wire residential, commercial and industrial locations.

CABLE SYSTEMS
ARTICLES 320, 330, 334 AND 338

Cable systems shall be permitted to be run through the center of rafters and joists or they may be run on the surface (face). Cables shall be considered supported where the center of the rafter or ceiling joist is drilled and the cable is pulled through each drilled joist. Cables routed on the face shall be supported per **320.23(A), 330.23, 334.23** and **338.10(B)(4)**.

AC cable, nonmetallic sheathed cable and service-entrance cable shall be supported and securely fastened within 12 in. (300 mm) of every box, cabinet or fitting and at intervals not exceeding 4 1/2 ft. (1.4 m).

Some inspectors, by special permission, allow the cables to be supported at both ends, if they are pulled tight and supported in the middle of the cable run.

Design Tip: This method of supporting is only allowed where the cable is installed on the top of the ceiling studs. The inspector shall always be consulted if supports are installed other than by the provisions in the NEC.

Ceiling joists shall be permitted to be notched and the cables installed in the notch if protected by a 1/16 in. (1.6 mm) steel plate. Sometimes notching the ceiling joist is easier than drilling a hole in the center of the joist. **(See Figure 13-11)**

Procedures for design, installation and inspection of cables are as follows:

- Check supports supporting cables.
- Check spacing of supports.
- Check and verify that proper fittings for terminating cables are utilized.
- Check loose fittings and supports of cables at boxes, equipment, etc.
- Check cables for exposure to damage.
- Check type of installation such as:

  (a) Cables run through drilled holes in ceiling joists.
  (b) Cables run through notched cuts on the edge of the ceiling joist.
  (c) Cables run on the sides of the ceiling joists.
  (d) Cables run on the bottom of the ceiling joists.
  (e) Cables run on the top of the ceiling joists.

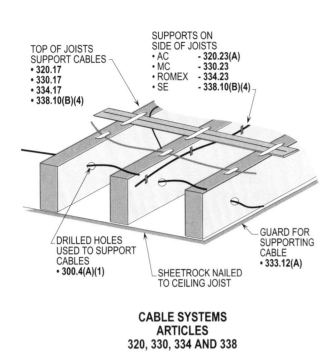

CABLE SYSTEMS
ARTICLES 320, 330, 334 AND 338

Figure 13-11. The above includes the rules for installing cables in attics.

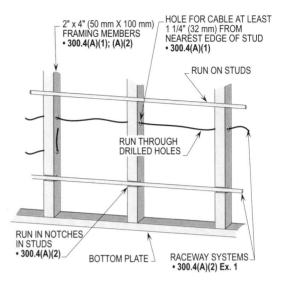

RACEWAY SYSTEMS
ARTICLES 342, 344, 348, 350 AND 352

Figure 13-12. Raceway systems may be run on the studs, in notches in the studs or through drilled holes in the studs.

RACEWAY SYSTEMS
ARTICLES 342, 344, 348, 350 AND 352

Raceway systems are usually installed in notches that are cut into the face of the framing members. It is much easier to run the raceway system in the notched area than to drill and run the raceway through the center of the joist. Where the construction will permit, it is simpler to route the raceway system on the top or bottom of the ceiling joist. The raceway systems shall be supported as previously mentioned. **(See Figure 13-12)**

ATTICS
320.23

There are two height levels where cable systems shall be protected when they are installed in attics.

For example, cables installed above 7 ft. (2.1m) shall not be required to be protected. However, cables installed at or below 7 ft. (2.1 m) shall be protected under certain conditions of use. Raceway systems shall be considered protected from physical damage in most all types of installations.

CABLE SYSTEMS
ARTICLES 320, 330, 334 AND 338

Cable systems installed in attics shall be considered inaccessible unless stairways, pulldown ladders or scuttle hole openings are provided.

Cable systems run within 6 ft. (1.8 m) of a scuttle hole shall have guard strips if they are not routed on the sides of the ceiling joists. Cable systems installed at or less than 7 ft. (2.1 m) and run on top of the ceiling joists shall have guard strips installed to protect the cables if exposed to physical damage. Cables that are studded through drilled holes or run on the sides of the ceiling joists shall not be required to be protected with guard strips even when they are at or below the 7 ft. (2.1 m) limitation.

Cables installed above 7 ft. (2.1 m) shall not be required to be protected with guard strips. The proper supports are required as mentioned above to protect the cables from being damaged.

Procedures for design, installation and inspection of cables are as follows:

- Check supports supporting cables.
- Check spacing of supports.
- Check to verify that proper fittings to terminate cables are utilized.

- Check loose fittings and supports of cables at boxes, equipment, etc.
- Check cables for exposure to damage.
- Check the type of installation such as:

  (a) Cables run through drilled holes in ceiling joists.
  (b) Cables run in cuts (notches) in the edge of ceiling joists.
  (c) Cables run on the sides of ceiling joists.
  (d) Cables run on the bottom of ceiling joists.
  (e) Cables run on the top of ceiling joists.

See Figures 13-13(a) and (b) for cable systems installed in attics with installation requirements.

RACEWAY SYSTEMS
ARTICLES 342, 344, 348, 350 AND 352

Raceway systems installed in attics, and on the top of a joist or on the side shall not be required to be protected. They shall be supported according to the NEC as mentioned above for walls and ceilings.

Inspectors usually require electrical nonmetallic conduit to be protected in the attic in the same manner as nonmetallic sheathed cable. See **320.23** and **334.23**.

See Figure 13-14 for raceways installed in attics with installation requirements listed.

Some cities require raceways to enclose the conductors instead of cables so that arcs and sparks will be contained. If a short-circuit or ground-fault should occur in the electrical system, such arcing and sparking is contained in the raceway system which will help prevent fires.

Procedures for design, installation and inspection of raceway systems are as follows:

- Check supports supporting raceways.
- Check spacing of supports.
- Check to verify that proper fittings for connecting raceways are utilized.
- Check loose fittings and supports of raceways at boxes, equipment, etc.
- Check raceways for exposure to damage.
- Check type of installation such as:
  (a) Raceways run through drilled holes in ceiling joists.
  (b) Raceways run through notched cuts on the edge of the ceiling joists.
  (c) Raceways run on the sides of ceiling joists.
  (d) Raceways run on the bottom of the ceiling joists.
  (e) Raceways run on the top of the ceiling joists.

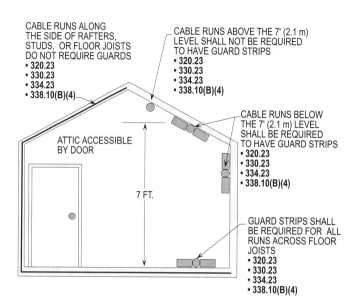

Figure 13-13(a). Requirements for installing cables in accessible attics where using a door or pulldown stairs.

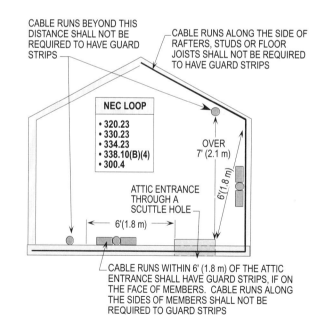

Figure 13-13(b). Requirements for installing cables in accessible attics where using a scuttle hole.

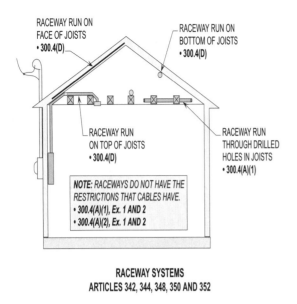

Figure 13-14. Requirements for installing raceways in attics where using raceway systems.

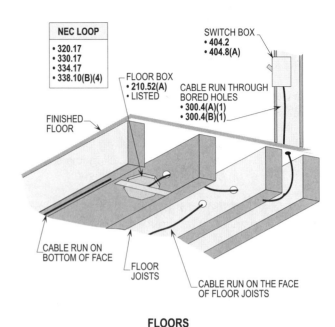

Figure 13-15. Requirements for installing cables in floors, where cables are used as the wiring method.

FLOORS
300.4

Cable systems are only installed in floor areas where the foundation is the pier and beam type. The cables shall be permitted to be pulled through drilled holes in the center of the floor joists or stapled to the sides or routed on the bottom. Guard strips shall only be required if the cable is exposed to physical damage.

The supporting requirements for cables installed under the floor on the sides, bottoms or drilled holes of floor joists are the same as mentioned for walls or ceilings. **(See Figure 13-15)**

Procedures for making inspections are as follows:

- Check supports supporting cables.

- Check spacing of supports.

- Check loose fittings and supports of cables at boxes, equipment, etc.

- Check cables that are exposed to physical damage.

- Check type of installation such as:

    (a) Cables run through drilled holes in floor joists.
    (b) Check cables routed in cut notches in floor joists.
    (c) Check cables routed on sides of floor joists.
    (d) Check cables routed on bottom of floor joists.

FLEXIBLE CORDS AND EXTENSION CORDS
400.7, 400.8 AND ARTICLE 305

Flexible cords shall be permitted to be used for wiring certain types of utilization equipment. Flexible cords shall not be permitted as a permanent wiring method to supply power to receptacles, luminaires (lighting fixtures), etc. Extension cords shall be permitted to be used only on a temporary basis and shall not be permitted to be substituted as a permanent wiring method. **(See Figure 13-16)**

FLEXIBLE CORDS
400.7 AND 400.8

Flexible cords are approved to be attached to luminaires (lighting fixtures), appliances and similar equipment. Luminaires (lighting fixtures) and appliances shall not be permitted to be permanently wired into the circuit with flexible cords. Flexible cords for portable equipment shall be equipped with caps and plugs and plugged into a receptacle. There are three types of flexible cords used and they are as follows:

- Not permitted for hard usage
- Permitted for hard usage
- Permitted for extra-hard usage

Installing Wiring Methods

Types of flexible cords and their permitted use are listed in **Table 400.4**. The ampacities of the different types of flexible cords are listed in **Table 400.5**.

> **Design Tip:** Flexible cords are to be used very carefully where they are used for temporary power reasons. After their use, the AHJ will require such cords to be removed due to the threat of fire. They are only to be used for temporary use and not for hard-wired systems.

> **For example:** What type of three conductor flexible cord and what ampacity rating is required where the cord is subject to abrasion and diggings. The load on the cord is 16.5 amps and is the hard service type (noncontinuous load)?
>
> Step 1: Finding type of cord
> **Table 400.4**
> Type S, SE or SEO may be used
>
> **Solution: Flexible cord is Type S, SE or SEO.**
>
> Step 1: Finding the ampacity of cord
> **Table 400.5**
> 16.5 A requires a No. 12 cord
>
> **Solution: A No. 12-3 w/ground flexible cord is required.**

Flexible cord shall be permitted to be used to cord-and-plug connect electrical apparatus and they are as follows:

- Pendants
- Luminaires (lighting fixture)
- Appliance or portable lamp connections
- Elevator cables
- Cranes and hoists
- Stationary equipment requiring frequent interchange
- Noise or vibration utilization equipment
- Appliances requiring maintenance
- Connection of moving parts
- Where specifically permitted elsewhere in the NEC

Flexible cords shall not be permitted to be used as a permanent wiring method under certain conditions of use. They are not to be permanently attached at both ends, concealed or pass through walls, ceilings, floors or openings in doors. They are never to be used as a substitute for permanent wiring methods.

Flexible cord shall not be permitted as a wiring method and they are as follows:

- Not permitted as a substitute for building wiring.
- Not permitted to pass through walls, ceilings or floors.
- Not permitted to pass through doorways, windows or similar openings.
- Not permitted to be permanently attached to building surfaces.
- Not permitted to be concealed behind walls, ceilings or floors.
- Where subject to physical damage

See Figure 13-17 for a detailed description on how flexible cords shall not be permitted to be used.

> **Design Tip:** Temporary power taps per **UL white book, 90.7** and **110.3(B)** shall be permitted for the cord-and-plug connection to sensitive electronic equipment. Temporary power taps allow several such pieces of equipment such as a PC hard disk, printer, etc. to be cord-and-plug connected into one self protected multioutlet strip which is plugged into a receptacle outlet.

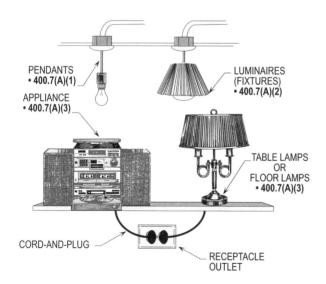

FLEXIBLE CORDS AND EXTENSION CORDS NEC 400.7

Figure 13-16. The above are approved installations using flexible cord.

See Figure 13-18 for a detailed illustration of a temporary power tap being used for cord and plug connections when connecting a PC and its accessories.

Stallcup's Electrical Design Book

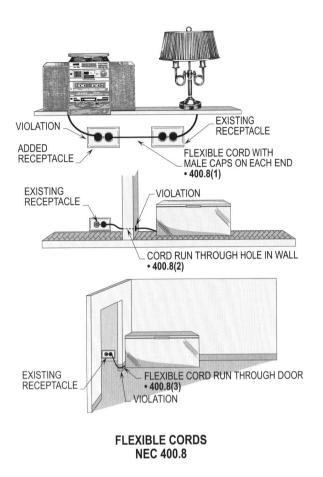

FLEXIBLE CORDS
NEC 400.8

Figure 13-17. The above are installations where flexible cords shall not be permitted to be used.

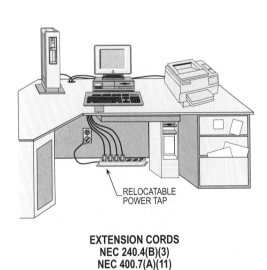

EXTENSION CORDS
NEC 240.4(B)(3)
NEC 400.7(A)(11)

Figure 13-18. A number of sensitive electronic pieces of equipment shall be permitted to be plugged into a temporary power tap per **UL White Book, 90.7 and 110.3(B)**.

EXTENSION CORDS
240.5(B)(3)

This section allows electricians to make up extension cords in the field using flexible cords, if separately listed cord bodies and components are utilized.

Note that listed manufactured cords shall be permitted to be used as before. The NEC now agrees with OSHA that extension cords shall be permitted to be made up in the field. **(See Figure 13-19)**

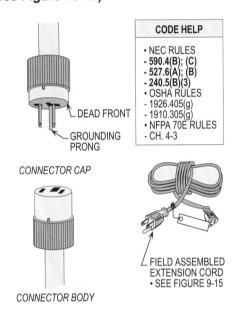

EXTENSION CORDS
NEC 240.4(B)(3)
NEC 400.7(A)(11)

Figure 13-19. Extension cords shall be permitted to be field wired using flexible cord, if listed bodies and components are utilized.

TEMPORARY WIRING
ARTICLE 590

Temporary wiring shall be permitted to be used on construction sites and for remodeling, maintenance, repair or demolition activities. The following requirements give the authorities having jurisdiction something to use for inspecting such wiring.

FEEDER-CIRCUITS
590.4(B)

Temporary feeders shall comply with **Article 240**. They shall also originate in an approved method for the distribution of

Installing Wiring Methods

power. Cord and cable assemblies shall be permitted to be used if they are of a type identified in **Table 400.4** for hard usage or extra-hard usage.

For the purpose of this section, Type NM and Type NMC cables (romex) shall be permitted to be used in any dwelling, building or structure without any height limitations.

BRANCH-CIRCUITS
590.4(C)

All branch-circuits shall originate in an approved power outlet or panelboard. Conductors shall be permitted within cable assemblies, or within multiconductor cords or cables, of a type identified in **Table 400.4** for hard usage or extra-hard usage. All conductors shall be protected as provided in **240.4, 240.5** and **240.100**. For the purposes of this section, Type NM and Type NMC cables shall be permitted to be used in any dwelling, building or structure without any height limitations.

Branch-circuits installed for the purposes specified in **590.3(B)** or **(C)** shall be permitted to be run as single insulated conductors. Where the wiring is installed in accordance with **590.3(B)**, the voltage-to-ground shall not be permitted to exceed 150 volts, the wiring shall not be permitted to be subject to physical damage and the conductors shall be supported on insulators at intervals of not more than 10 ft. (3 m) or for festoon lighting, the conductors shall be arranged so that excessive strain is not transmitted to the lampholders. **(See Figure 13-20)**

Note: For GFCI-protection of receptacles and assured equipment grounding conductor program, see pages 16-20 of chapter 16 of this book.

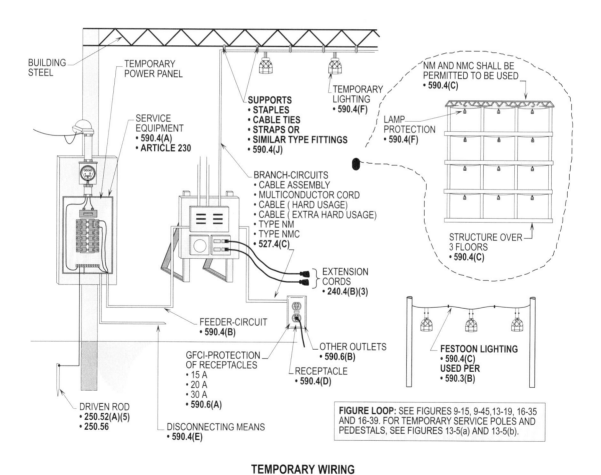

Figure 13-20. The illustration shows the rules pertaining to temporary wiring.

Name Date

Chapter 13. Installing Wiring Methods

Section	Answer			
_____	T	F	1.	Nails shall be permitted to be used to mount boxes when the nails pass through the inside of the box within 1/4 in. of the back or ends of the box.
_____	T	F	2.	Nonmetallic boxes are not required to be marked with their cubic inch fill area.
_____	T	F	3.	Boxes with threaded enclosures or hubs shall be considered supported with two or more conduits that are threaded wrench tight into the enclosure or hubs.
_____	T	F	4.	If supported within 36 in. on two or more sides, enclosures without devices do not require additional supports.
_____	T	F	5.	If supported within 36 in. on two or more sides, enclosures with devices do not require additional supports.
_____	T	F	6.	Type NM cable shall not be permitted for both exposed and concealed work in normally dry locations.
_____	T	F	7.	UF cable is designed and made to be used outside and underground.
_____	T	F	8.	Service-entrance cable shall be permitted to be used inside or outside. (General Rule)
_____	T	F	9.	Type AC cable shall not be permitted to wire branch-circuits and feeder circuits in both concealed and exposed work.
_____	T	F	10.	Rigid metal conduit and IMC are the most rugged of the raceway systems and are used to protect wiring methods subjected to physical damage.
_____	T	F	11.	Rigid nonmetallic conduit (RNC) consists of two main types which are known as Schedule 40 or Schedule 80.
_____	T	F	12.	Nonmetallic sheathed cable (romex) and AC cable (BX) shall be supported within 18 in. of each box, cabinet or fitting and at intervals not exceeding 4 1/2 ft.
_____	T	F	13.	Wiring methods run through the center of ceiling rafters shall be considered supported.
_____	T	F	14.	Cables installed in attics above 7 ft. shall not be required to be protected.
_____	T	F	15.	Cable systems run within 5 ft. of a scuttle hole shall have guard strips if they are not routed on the sides of the ceiling joists.
_____ _____ _____			16.	Rigid metal conduit shall be permitted to be installed in a minimum size of _____ in. and a maximum size of _____ in.

13-25

Section	Answer
_____	_____

17. When cutting and threading rigid metal conduit in the field, a standard cutting die with a _____ in. taper per foot shall be used.

18. _____ threads shall not be used for connection at couplings when installing rigid metal conduit.

19. Intermediate metal conduit shall be permitted to be installed in a minimum size of _____ in. and a maximum size of _____ in.

20. EMT shall be supported every _____ ft. and within _____ ft. of each outlet box, junction box, device box, etc.

21. Electrical metallic tubing shall be permitted to be installed in a minimum size of _____ in. and a maximum size of _____ in.

22. Flexible metal conduit and liquidtight flexible metal conduit shall be supported within _____ in. of each box, cabinet, etc. and supported at intervals not exceeding _____ ft.

23. Rigid nonmetallic conduit shall be permitted to be installed in a minimum size of _____ in. and a maximum size of _____ in.

24. Temporary poles must be installed in the earth at a depth usually _____ ft. that will hold the pole with mounted equipment substantially.

25. NM cable run _____ through drilled holes shall be considered properly supported by the framing members.

26. Schedule _____ is durable and shall be permitted to be exposed to physical damage.

27. AC cable, nonmetallic sheathed cable and service-entrance cable shall be supported within _____ in. of every box, cabinet or fitting and at intervals not exceeding _____ ft.

28. _____ strips are only required if the cable is exposed to physical damage.

29. Types of flexible cords and their permitted use are listed in Table _____.

30. Flexible cords are prohibited as a _____ wiring method to supply power receptacles, luminaires (lighting fixtures), etc.

31. Cable systems that are not run through the center of the framing members can be placed in a cut notch in the framing member and protected by a steel plate of:
    - (a) 1/16 in.
    - (b) 1/8 in.
    - (c) 3/16 in.
    - (d) 1/4 in.

Branch-circuits

Branch-circuit conductors extend between the final overcurrent protection device protecting the circuit conductors supplying power to equipment and outlets. The equipment supplied is either permanently installed and hard-wired or cord-and-plug connected to properly design and select receptacle outlets.

Branch-circuits in this Chapter are designed to supply the following type of occupancies:

- Residential
- Commercial
- Industrial

RESIDENTIAL

Branch-circuits utilized in residential occupancies shall be calculated differently than those for commercial and industrial locations. The general-purpose circuits are calculated at 3 VA per square foot of the dwelling unit.

For example, a dwelling unit of 2000 sq. ft. has a volt-amp rating of 6000 VA (2000 sq. ft. x 3 VA = 6000 VA). This VA rating shall be used to supply all the general-purpose lighting and receptacle outlets in the dwelling unit.

Another example would be the 2 - 20 amp small appliance circuits which are used to supply countertop receptacles and other wall outlets in the kitchen, pantry, dining room and breakfast room respectively. Individual circuits such as a 12 kW range circuit shall be calculated as 8 kW per **Table 220.55**. Other loads in the dwelling unit shall be permitted to be reduced in VA due to their operation and use.

GENERAL-PURPOSE CIRCUITS
220.12; TABLE 220.12

The calculations for loads in various occupancies shall be based on VA (volt-amperes) per square foot. In computing the VA per square foot, the outside dimensions of the building shall be used. They do not include the area of open porches and attached garages with dwelling unit occupancies. However, if there is an unused basement, it should be assumed that it will be finished later, so it shall be included in the calculation so that the capacity of the wiring system will be adequate to serve such loads at a later date.

The load values used in **Table 220.12** shall be considered at 100 percent power factor. If less than 100 percent power factor, equipment is installed with sufficient capacity. Such equipment shall be figured in to take care of the additional higher current values.

Design Tip: The number of 15 or 20 amp branch-circuits for a dwelling unit shall be determined by multiplying the square footage by 3 VA per sq. ft. per **220.12** and **Table 220.12** and dividing by the size overcurrent protection device times the voltage of the circuit. **(See Figure 14-1)**

GENERAL-PURPOSE CIRCUITS
TABLE 220.12
NEC 220.12

Figure 14-1. Determining the number of 15 amp, 2-wire circuits to supply power to the general-purpose lighting and receptacle outlets throughout the dwelling unit.

SMALL APPLIANCE CIRCUITS
220.52; 210.11(C)(1); 210.52(B)(1)

A minimum of 2 - 20 amp, 1500 VA small appliance circuits shall be required to supply receptacle outlets that are located in the kitchen, pantry, breakfast room and dining room. The 2 small appliance circuits shall be routed to the kitchen countertop(s) and the outlets proportioned among the 2 circuits as evenly as possible to prevent unbalanced loading of the circuits. Unbalanced loading may trip open the overcurrent protection device due to, too many portable appliances being plugged into the same small appliance circuit. **(See Figure 14-2)**

LAUNDRY CIRCUIT
220.52B); 210.11(C)(2); 210.52(F)

At least 1 - 20 amp, 1500 VA laundry circuit shall be required to supply receptacle outlets in the laundry room. All laundry equipment shall be located within 6 ft. (1.8 m) of the receptacle outlet per **210.50(C)**. Sometimes a 20 amp duplex receptacle is used to cord-and-plug connect a washing machine and gas dryer. No other outlets shall be permitted to be supplied by this 20 amp, 2-wire small appliance circuit. **(See Figure 14-3)**

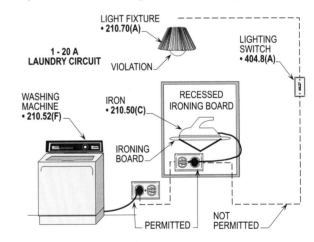

LAUNDRY CIRCUIT
NEC 220.52(B)
NEC 210.11(C)(2)
NEC 210.52(F)

Figure 14-3. At least 1 - 20 amp, 1500 VA laundry circuit shall be required to supply receptacle outlets in the laundry room.

Branch-circuits

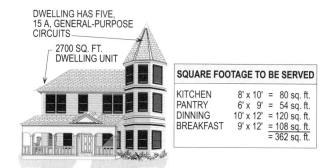

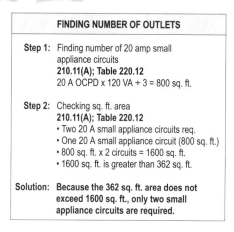

Figure 14-2. Determining the number of outlets that are permitted on a 20 amp small appliance circuit.

INDIVIDUAL CIRCUITS
210.19(A)(1); 220.18(B)

Individual branch-circuits in dwelling units shall be calculated at 100 percent or 125 percent of full-load amps of the appliance served or by applying demand factors based upon the type of appliance.

RESIDENTIAL COOKING EQUIPMENT
220.55; TABLE 220.55

The procedure for calculating the branch-circuit loads for ranges, cooktops and ovens shall be determined by applying the demand factors in **Table 220.55**. The demand factors listed in Columns A, B and C are based on the size of the range, cooktop or oven. When the kW rating exceeds 12 or there's more than one unit, one of the **Footnotes** shall be used in conjunction with **Table 220.55**.

COLUMN A
TABLE 220.55

Column A in Table 220.55 shall be used for cooking equipment rated less 3.5 kW. These units vary in rating and have limited use.

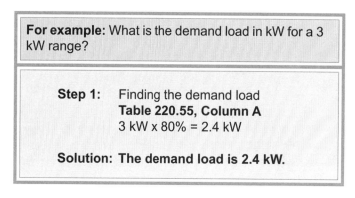

See **Figure 14-4** for calculating cooking equipment loads per **Column A in Table 220.55**.

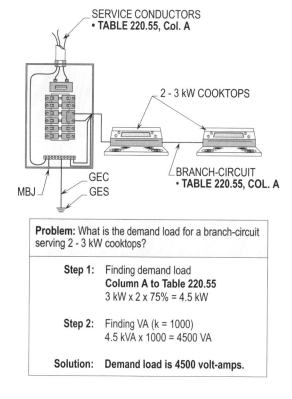

Figure 14-4. The maximum demand load shall be determined by the number of cooking units times the percentage factor applied in **Table 220.55, Column A**.

14-3

COLUMN B
TABLE 220.55

Column B in Table 220.55 shall be used for cooking equipment rated from 3.5 kW to 8.75 kW. These size units are usually the types in use today.

> **For example:** What is the demand load in kW for an 8.75 kW range?
>
> **Step 1:** Finding the demand load
> Table 220.55, Column B
> 8.75 kW x 80% = 7 kW
>
> **Solution:** The demand load is 7 kW.

See **Figure 14-5** for calculating cooking equipment loads per **Column B in Table 220.55**.

are usually rated 9 to 12 kW respectively.

> **For example:** What is the demand load in kW for a 12 kW range?
>
> **Step 1:** Finding the demand load
> Table 220.55, Column C
> 12 kW range = 8 kW
>
> **Solution:** The demand load is 8 kW.

See **Figure 14-6** for calculating cooking equipment loads per **Column C in Table 220.55**.

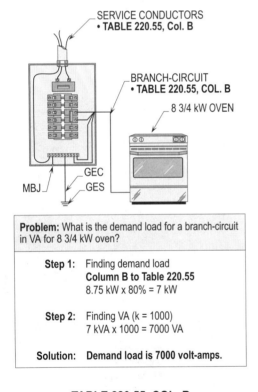

TABLE 220.55, COL. B

Figure 14-5. The maximum demand load for a branch-circuit shall be determined by the number of cooking units times the percentage factors applied in **Table 220.55, Column B**.

COLUMN C
TABLE 220.55

Column C in Table 220.55 shall be used for cooking equipment rated from over 8.75 kW to 12 kW. These units

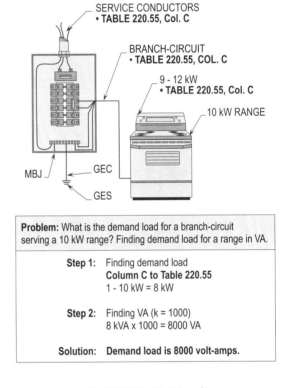

TABLE 220.55, COL. C

Figure 14-6. The demand load in kW for a branch-circuit serving cooking equipment in **Column C** is already determined per **Table 220.55**.

NOTE 1
TABLE 220.55

Note 1 to Table 220.55 shall be applied where the kW rating of the cooking equipment is over 12 kW but not over 27 kW. All kW ratings that exceed 12 kW shall be multiplied by 5 percent. The kW rating of 1 range is listed in **Column C in Table 220.55** and shall be multiplied by this total demand.

Units of sizes larger than 12 kW are usually combination units found in kitchens with limited space.

> **For example:** What is the demand load in kW for a 18 kW range?
>
> **Step 1:** Finding percentage
> **Note 1 to Table 220.55**
> 18 kW - 12 kW = 6 kW
> 6 kW x 5% = 30%
>
> **Step 2:** Finding the demand load
> **Table 220.55, Col. C**
> 8 kW x 130% = 10.4 kW
>
> **Solution: The demand load is 10.4 kW.**

See **Figure 14-7** for calculating cooking equipment loads per **Note 1 to Table 220.55**.

NOTE 2
TABLE 220.55

For cooking equipment of unequal values rated over 12 kW to 27 kW in **Column C**, **Note 2** shall be applied and calculated by adding the kW ratings of all units and dividing by the number of units, all ranges below 12 kW shall be computed at 12 kW. When an average rating is found, the number of units shall be increased by 5 percent for each kW exceeding 12 kW to derive the allowable kW.

See **Figure 14-8** for demand factors to be applied per Table **220.55, Column C, Note 2**. **Note:** The demand factor selected from **Table 220.55, Column C** shall be based upon the number of units.

> **Design Tip:** It is permissible per **Note 3 to Table 220.55** to add all pieces of cooking equipment with ratings over 1 3/4 kW through 8 3/4 kW together and multiply by the percentage of **Columns A** or **B**, whichever produces the smaller kW rating shall be permitted to be used. This calculation shall be permitted to be applied by permission of the AHJ if it provides the smaller kW rating of all the methods available in **Table 220.55** and **Notes**.

NOTE 4
TABLE 220.55

Note 4 to Table 220.55 shall be permitted to be applied to a counter-installed cooktop with one or two wall-mounted ovens. The total kW of each cooking unit shall be totaled and all kW exceeding 12 kW shall be multiplied by 5 percent. The kW rating of 1 range not 3 as listed in **Column C in Table 220.55** shall be multiplied by this total. The advantage of this rule is allowing 1 branch-circuit to be run, instead of 3 individual circuits, 1 to each unit.

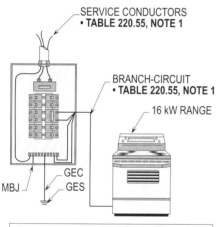

> **Problem:** What is the demand load in VA for a branch-circuit serving a 16 kW range?
>
> **Step 1:** Finding demand load
> **Note 1 to Table 220.55**
> 16 kW - 12 kW = 4 kW
> 4 kW x 5% = 20%
>
> **Step 2:** Finding VA
> **Table 220.55, Col. C**
> 8 kVA x 1000 x 120% = 9600 VA
>
> **Solution:** Demand load is 9600 volt-amps.

TABLE 220.55, NOTE 1

Figure 14-7. For cooking equipment rated over 12 kW to 27 kW in **Column C**, **Note 1** requires an increase of 5 percent for each kW over 12 kW. This percentage times 8 kW will determine the demand load to be used to size the elements.

> **For example:** What is the demand load in kW for a 10 kW cooktop, and 8 kW oven and a 6 kW oven connected to a 240 volt, single-phase, branch-circuit?
>
> **Step 1:** Finding percentage
> **Note 4 to Table 220.55**
> 10 kW + 8 kW + 6 kW = 24 kW
> 24 kW - 12 kW = 12 kW
> 12 kW x 5% = 60%
>
> **Step 2:** Finding the demand load
> **Table 220.55, Col. C**
> 8 kW x 160% = 12.8 kW
>
> **Solution: The demand load is 12.8 kW.**

See **Figure 14-9** for calculating cooking equipment loads for taps per **Ex. 1 to 210.19(A)(3)**.

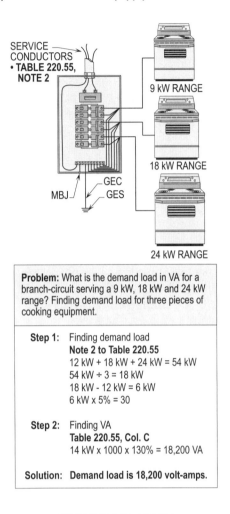

TABLE 220.55, NOTE 2

Figure 14-8. Cooking equipment of unequal values rated over 12 kW to 27 kW in **Column C**, **Note 2** shall be computed by adding the kW ratings of all units and dividing by the number of units, all ranges below 12 kW shall be computed at 12 kW. When an average rating is determined, the number of units shall be increased by 5 percent for each kW exceeding 12 kW.

SIZING TAPS
210.19(A)(3), Ex. 1

A tap to connect a cooktop and ovens shall be permitted to be made from a 50 amp branch-circuit when the tap conductors are sized from the kW rating of each piece of cooking equipment per **Note 4 to Table 220.55**. Anytime taps are made from larger conductors with smaller conductors, the tap shall comply with **240.21(A)**.

For example, can a 14 AWG tap be made from a 12 AWG branch-circuit and be down sized and used as a switch leg? Naturally, the answer is no, because the tap cannot meet the provisions of **240.21(A) through (G)** for making an approved tap. However, **240.21(A)** and **210.19(A)(3), Ex. 1** when applied together permits a smaller tap to be made for cooking equipment circuits tapped from conductors of a larger sized branch-circuit.

See **Figure 14-10** for calculating cooking equipment loads for taps per **Ex. 1 to 210.19(A)(3)**.

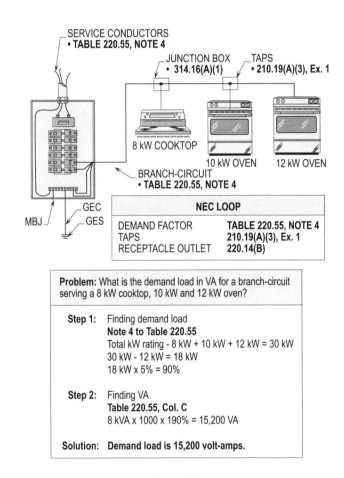

TABLE 220.55, NOTE 4

Figure 14-9. The demand load for a cooktop and 2 or less wall-mounted ovens shall be determined by finding the amperage rating of each unit in kW. Each kilowatt that exceeds 12 kW shall be increased by 5 percent to obtain the multiplier. The multiplier times the demand for 1 unit in **Column C in Table 220.55** derives the demand load for the elements of the circuit.

Branch-circuits

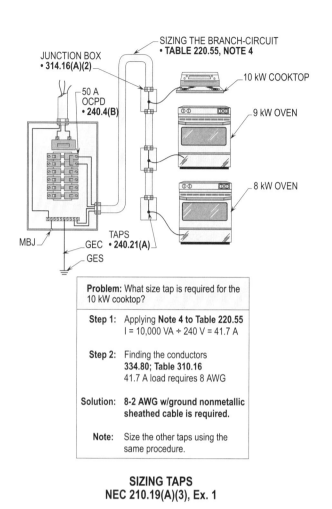

Figure 14-10. Determining the load in kW and amps to size the elements for a branch-circuit from which taps to cooking equipment are made.

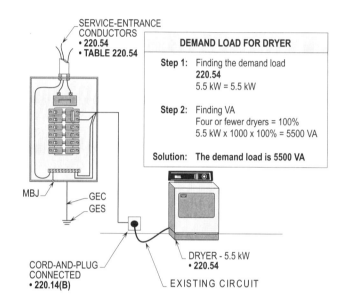

Figure 14-11. The demand load for household dryers shall be figured at 5 kVA or the nameplate rating, whichever provides the greater rating. Four or fewer dryers shall be calculated at 100 percent of the nameplate rating. Five or more dryers shall be permitted to have a percentage applied based on the number of units per **Table 220.54**. The Table is based on five dryers being used at different times and the load limited to about 80 percent of total. **Table 220.54** is used for dwelling units in apartment complexes.

ELECTRIC CLOTHES DRYERS
220.54

Dryer equipment loads shall be at least 5000 VA or the nameplate rating, whichever is larger. This value is used for sizing the branch-circuit load. When installing four or fewer dryers the load shall be calculated at 100 percent. When installing five or more dryers the load shall be calculated by the percentages listed in **Table 220.54** based on the number of dryers being installed. When calculating the load for a 4500 VA dryer, the load shall be calculated at 5000 VA for the branch-circuit that is used to cord-and-plug connect the unit. Dryers installed in dwelling units shall comply with provisions of **220.14(B)** and **220.54**. Dryers shall be calculated at 100 percent for noncontinuous operation and 125 percent for continuous operation if they are installed as commercial dryers.

Note that an existing branch-circuit in a dwelling unit can be used as a wiring method to supply an electric dryer. If NM cable is used, the neutral shall be insulated. However, the neutral shall be permitted to be uninsulated if SE cable is the existing wiring method.

14-7

RATINGS
210.3

Branch-circuits shall be classified by the rating or setting of the overcurrent protection device protecting the circuit. Two or more outlets shall be permitted to be protected by the following overcurrent protection devices based upon the computed load per **210.19(A)**:

- 15 amp
- 20 amp
- 30 amp
- 40 amp
- 50 amp

PERMISSIBLE LOADS
210.23

The rating or setting of the overcurrent protection device shall not be permitted to be exceeded by the load on an individual branch-circuit. If the load is continuous, the load shall be multiplied by 125 percent. A fastened-in-place appliance shall be permitted to be connected to a general-purpose circuit if its amp rating does not exceed 50 percent of the branch-circuit.

15- AND 20-AMPERE BRANCH-CIRCUITS
210.23(A)

A 15 or 20 amp branch-circuit shall be permitted to supply luminaires (lighting fixtures) and/or utilization equipment in residential, commercial or industrial locations.

For example: What is the total VA rating for a 15 or 20 amp, 120 volt, 2-wire branch-circuit supplying a noncontinuous load?

Finding VA of 20 amp branch-circuit

Step 1: Finding VA
         210.23(A)
         VA = 15 A x 120 V
         VA = 1800 VA

Solution: The total VA is 1800.

Step 1: Finding VA
         210.23(A)
         VA = 20 A x 120 V
         VA = 2400 VA

Solution: The total VA is 2400.

For example: What is the total VA rating for a 15 or 20 amp, 240 volt, 3-wire branch-circuit supplying noncontinuous loads?

Step 1: Finding VA
         210.23(A)
         VA = 15 A x 240 V
         VA = 3600

Solution: The total VA is 3600.

Step 1: Finding VA
         210.23(A)
         VA = 20 A x 240 V
         VA = 4800

Solution: The total VA is 4800.

The rating of any one cord-and-plug connected utilization equipment shall not be permitted to exceed 80 percent of the branch-circuit rating if connected to a general-purpose circuit supplying 2 or more outlets.

For example: What size overcurrent protection device is required to be installed for a 7 amp, 120 volt, single-phase compactor that is cord-and-plug connected per **422.16(B)(2)**?

Step 1: Finding usable amperage of 15 or
         20 amp branch-circuit
         210.23(A)
         (If continuous)
         A = 15 A x 80%
         A = 12 A
         A = 20 A x 80%
         A = 16 A

         210.23(A)
         (If noncontinuous)
         A = 15 A x 100%
         A = 15 A
         A = 20 A x 100%
         A = 20 A

Solution: The OCPD of 15 amp shall be permitted to be used.

The 7 amp full-load current rating of the compactor does not exceed the (15 A x 80% = 12 A) loading range of the 15 amp overcurrent protection device. The 7 amp full-load current rating of the compactor does not exceed the (20 A x 80% = 16 A) loading range of the 20 amp overcurrent protection device. Therefore, the 20 amp overcurrent protection device shall be permitted to be used to supply the 7 amp compactor. **Note:** The AHJ may require the 15 amp OCPD to be used.

Fixed appliances (fastened-in-place) shall be permitted to draw up to 50 percent of the rating of a branch-circuit supplying 2 or more general purpose outlets which serve lighting and receptacle loads.

For example: Can an 8 amp, 120 volt, single-phase A/C window unit be connected to an existing branch-circuit?

Step 1: Finding A of BC
210.23(A)
A = 50% of 20 A OCPD
A = 10 A

Step 2: Calculating A for A/C unit
210.23(A); 440.62(B); (C); 440.32
A = 8 A x 125%
A = 10 A

Step 3: Verifying permissive A
210.23(A); 440.62(B); (C); 210.3
A = 20 A OCPD x 80%
A = 16 A

Solution: Yes, the A/C window unit rated at 8 amps shall be permitted to be connected to the 20 amp branch-circuit.

A 20 amp overcurrent protection device shall be permitted to protect a fastened-in-place appliance with a rating of 10 amps or less after applying the 50 percent rule. The remaining 50 percent of the overcurrent protection device is used to supply lighting and/or cord-and-plug connected appliances. **(See Figure 14-12)**

30-AMPERE BRANCH-CIRCUITS
210.23(B)

A 30 amp branch-circuit shall be permitted to be installed to supply fixed lighting units with heavy-duty lampholders in other than a dwelling unit(s) or utilization equipment in any occupancy. The rating of any individual cord-and-plug connected appliance shall not be permitted to exceed 80 percent of the branch-circuit rating. The rating of any individual cord-and-plug connected appliance shall not be permitted to draw more than 24 amps (30 A x 80% = 24 A) when connected to an 30 amp overcurrent protection device. **(See Figure 14-13)**

For example: What is the load for a 23 amp dishwasher used at continuous duty?

Step 1: Finding amperage
220.12
A = 23 A x 125%
A = 28.75

Step 2: Finding branch-circuit
210.23(B)
28.75 A requires 30 A

Solution: The branch-circuit load is 28.75 amps.

A 30 amp branch-circuit shall be permitted to supply a single appliance that is used at continuous operation in any type occupancy.

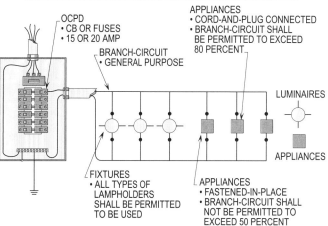

15 - AND 20 - AMPERE BRANCH-CIRCUITS NEC 210.23(A)

Figure 14-12. A 20 amp overcurrent protection device shall be permitted to protect a fastened-in-place appliance with a rating of 10 amps or less after applying the 50 percent rule. The remaining 50 percent of the overcurrent protection device is used to protect lighting and/or cord-and-plug connected appliances.

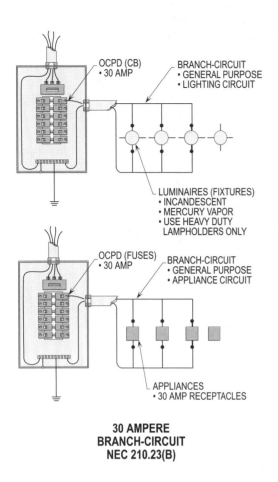

Figure 14-13. A 30 amp branch-circuit shall be permitted to be installed to supply fixed lighting units with heavy-duty lampholders in other than a dwelling unit(s) or utilization equipment in any occupancy.

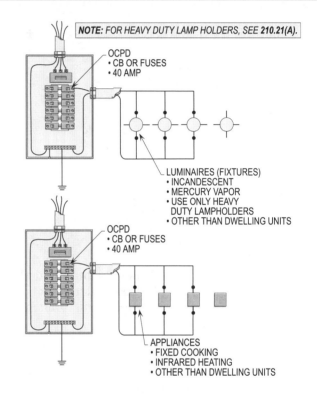

For example: What is the load for a 37 amp water heater used at continuous operation in a commercial building?

Step 1: Finding amperage
422.13
A = 37 A x 125%
A = 46.25

Solution: The branch-circuit load is 46.25 amps.

Figure 14-14. A 40 or 50 amp branch-circuit shall be permitted to supply cooking appliances that are fastened-in-place in any occupancy.

40- AND 50-AMPERE BRANCH-CIRCUITS
210.23(C)

A 40 or 50 amp branch-circuit shall be permitted to supply cooking appliances that are fastened-in-place in any occupancy. Fixed lighting units with heavy-duty lampholders or infrared heating units shall be permitted for such circuits except for other than dwelling units. Equipment such as a water heater, dryer or heating unit shall be permitted to be supplied by a 40 or 50 amp branch-circuit. **(See Figure 14-14)**

BRANCH-CIRCUITS LARGER THAN 50 AMPERES
210.23(D)

Branch-circuits larger than 50 amps shall supply only nonlighting outlet loads. A maximum load of 50 amps shall be used for multioutlet branch-circuits installed for lighting.

A combination of loads exceeding 50 amps shall be permitted be used for multioutlet branch-circuits that are not connected to lighting units such as plugs for welders.

CONDUCTORS
ARTICLE 310

Branch-circuit conductors used for general wiring shall be rated for the following insulations per **Table 310.16**:

- 60°C
- 75°C
- 90°C

The temperature rating of conductors and their conditions of use are listed in **Table 310.13**. Not more than 3 current-carrying conductors in a raceway or cable run in an ambient temperature of not more than 30°C or 86°F shall be used for branch-circuit conductors based on **Table 310.16**. The types of insulation available are listed in **Table 310.16** for copper and aluminum conductors. Conductor ampacities shall be determined by condition of use and by the terminal ratings of OCPD's and equipment per **110.14(C)(1)** and **(C)(2)**. **(See page 8-3 of Chapter 8)**

60°C CONDUCTORS
TABLE 310.16, COLUMN 2; 110.14(C)(1)

The following type of conductors shall be permitted to be installed when using **Table 310.16, Column 2** for 60°C ampacities:

- TW
- UF

The temperature rating of the conductors are rated at 60°C per **Table 310.13** and **Table 310.16** and shall be terminated to 60°C terminals and 60°C ampacities shall be used.

Design Tip: All conductors with the "W" (60°C or 75°C) rating insulation shall be permitted to be installed in dry, damp or wet locations per **310.8(B)** and **(C)**.

Overcurrent protection devices rated at 100 amps or less shall be permitted to be terminated with conductors 14 AWG through 1 AWG with 60°C ampacities per **110.14(C)(1)**. The maximum size of 1 AWG is listed so that a 60°C conductor ampacity will match to terminals of OCPD's and equipment rated at 100 amps.

For example: What size THHN copper conductor is required to supply a piece of equipment operating at 90 amps? (Load already computed at continuous or noncontinuous operation).

Step 1: Finding load
Table 310.13; Table 310.16
Load = 90 A

Step 2: Finding conductor A at 60°C
Table 310.16
95 A = 2 AWG THHN

Solution: **The conductor is required to be 2 AWG THHN.**

The allowable ampacity for each conductor shall be rated at 60°C when installing nonmetallic-sheathed cable (romex or rope) per **334.112**. However, these conductors have insulation rated at 90°C which permits such conductors to be used for derating purposes. **(See Figure 14-15)**

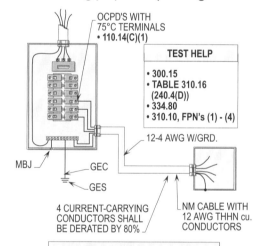

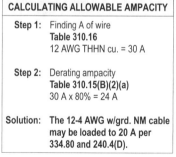

Figure 14-15. The allowable ampacity for each conductor is rated at 60°C, when installing nonmetallic-sheathed cable (romex or rope) per **334.80** and **334.112**. However, these conductors have insulation rated at 90°C which permits such conductors to be used for derating purposes.

75°C CONDUCTORS
TABLE 310.16, COLUMN 3; 110.14(C)(2)

The following types of conductors shall be permitted to be installed when using **Table 310.16, Column 3** for 75°C ampacities:

- FEPW
- RH
- RHW
- THHW
- THW
- THWN
- XHHW
- USE
- ZW

Overcurrent protection devices rated over 100 amps shall be permitted to be terminated with conductors larger than 1 AWG with 75°C ampacities per **110.14(C)(2)**. Conductors shall be permitted to be installed with higher temperature ratings if the ampacities are matched to the terminals of the overcurrent protection device and equipment. The terminals shall be marked by one of the following temperature ratings:

- 60°C
- 60°C/75°C

Note that 60°C ampacities shall only be permitted to be applied to terminals marked 60°C. 75°C ampacities shall only be permitted to be applied to terminals marked 75°C. Note that 60°C ampacities shall only be permitted be applied using 75°C ampacities where conductors 14 AWG through 1 AWG are installed. 75° terminals shall be permitted be used and connected with any size conductor listed in the 75°C column.

> **For example:** What size amperage rating is allowed for a 4 AWG copper conductor using the ampacities of the 60°C column and 75°C column respectively?

> **Step 1:** Finding amperage
> Table 310.16, Columns 2 and 3
> 60°C = 70 A
> 75°C = 85 A
>
> **Solution:** The allowable amperage rating is 70 amp for 60°C and 85 amps for 75°C. Note: The 60°C terminal rating reduces the 4 AWG THHN copper conductor to only 70 amps and not 85 or 95 amps.

90°C CONDUCTORS
TABLE 310.16, COLUMN 4

The following types of conductors shall be permitted to be used to wire modern day electrical systems. See **Table 310.16, Column 4** and **Table 310.13** for ampacities and conditions of use for 90°C rated conductors:

- TBS
- SA
- SIS
- FEP
- FEPB
- MI
- RHH
- RHW-2
- THHN
- THHW
- THW-2
- THWN-2
- USE-2
- XHH
- XHHW
- XHHW-2
- ZW-2

These conductors with the 90°C rated insulation shall be permitted to be connected to terminals rated at 60°C, 75°C and 90°C. Devices and equipment with 90°C terminals shall be used for connecting higher ampacity conductors that are rated 90°C. However, such OCPD's and equipment that mated (matched) are not available as of today.

> **For example:** What is the allowable ampacity required for a 6 AWG THWN copper conductor connected to a 60°C OCPD installed in a panelboard?

> **Step 1:** Finding amperage and condition of use
> Table 310.16, Column 4;
> Table 310.13
> A = 75
>
> **Step 2:** Finding allowable ampacity
> Table 310.16, Column 2
> A = 55
>
> **Solution:** The allowable ampacity is limited 55 amps because of the 60°C terminals.

> **Design Tip:** Overcurrent protection devices rated at 100 amps or less shall be terminated with conductors 14 AWG through 1 AWG with 60°C ampacities per **110.14(C)(1)** if not otherwise marked. Overcurrent protection devices rated over 100 amps shall be permitted to be terminated with conductors larger than 1 AWG with 75°C ampacities per **110.14(C)(2)**,

The following ampacity values shall be used for the temperature ratings of terminals using a 1/0 AWG THHN copper conductor:

Branch-circuits

- 60°C = 125 amps
- 75°C = 150 amps
- 90°C = 170 amps

A load of 125 amps or less on 60°C terminals shall be permitted to be served by a 1/0 AWG THHN copper conductor. A load of 150 amps or less on 75°C terminals shall be permitted to be served by a 1/0 AWG THHN copper conductor. A load of 170 amps or less on 90°C terminals shall be permitted to be served by a 1/0 AWG THHN copper conductor. The ampacities shall be matched to the terminals of the overcurrent protection device and equipment. **(See Figure 14-16)**

AMBIENT TEMPERATURES
310.15(B)(2)(a)

Where there is not more than 3 current-carrying conductors in a raceway or cable, the allowable ampacities listed in **Table 310.16** shall be used. If four or more current-carrying conductors or surrounding temperature exceeding 86°F are present, derating factors shall be applied, based upon their conditions of use. See the top of **Table 310.16** for these conditions that shall be applied before selecting the allowable ampacities of such conductors.

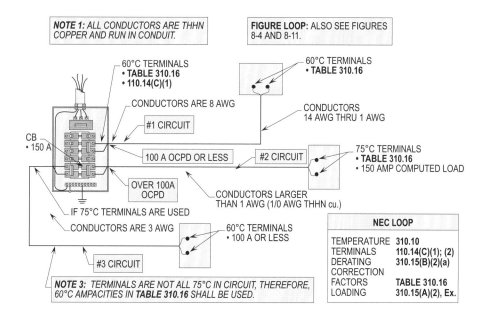

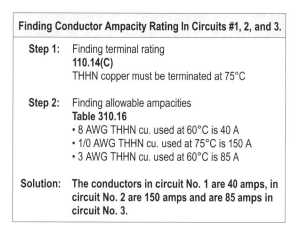

TEMPERATURE LIMITATIONS
NEC 110.14(C)(1); (C)(2)

Figure 14-16. Determining the allowable ampacity rating of terminals and sizing conductors based on 75°C, or 75°C ampacities from **Columns 2** and **3 in Table 310.16**. This rule prevents terminals from being overheated.

14-13

For example: What is the allowable ampacity for 6 - 12 AWG THHN cu. conductors that are all current-carrying? (**Note:** Adjustment factors shall be applied).

Step 1: Calculating ampacity
Table 310.16
12 AWG THHN cu. = 30 A

Step 2: Applying derating factors
310.15(B)(2)(a); 310.10, FPN (4)
30 A x 80% = 24 A

Solution: **The allowable ampacity is 24 amps.**

For example: What is the allowable ampacity for 4 - 12 AWG THHN cu. conductors (three current-carrying) that are in an ambient temperature of 102°F? (**Note:** Correction factors shall be applied).

Step 1: Calculating ampacity
Table 310.16
12 AWG THHN cu. = 30 A

Step 2: Applying derating factors
Asterisk to Table 310.16
30 A x 91% = 27.3 A

Solution: **The allowable ampacity is 27.3 amps.**

For example: What is the allowable ampacity for 4 - 12 AWG THHN cu. conductors that are all current-carrying and routed through an ambient temperature of 102°F?

Step 1: Calculating ampacity
Table 310.16
12 AWG THHN cu. = 30 A

Step 2: Applying adjustment factors
310.15(B)(2)(a)
30 A x 80% = 24 A

Step 3: Applying correction factors
240.4(D); Table 310.16
24 A x 91% = 21.8 A

Solution: **The allowable ampacity is 21.8 amps.**

For example: What size overcurrent protection device is required to serve a computed continuous load of 70.5 amps using 4 AWG THHN cu. conductors with an ampacity of 85 amps?

Step 1: Finding ampacity
210.19(A)(1); Table 310.16
70.5 A requires 80 A OCPD
4 AWG = 85 A = 80 A OCPD

Solution: **A 80 amp overcurrent protection device is required.**

Design Tip: An 80 amp overcurrent protection device shall be permitted to be installed, because 56.4 amps x 125% is 70.5 amps and **240.4(B)** permits the next size OCPD above 70.5 amps to be used. **(See Figure 14-17)**

The ampacity of conductors shall be derated at least three times when applying the following rules for branch-circuits and determining the allowable ampacities:

- More than 3 current-carrying conductors
  - Adjustment factors
- Ambient temperature exceeds 86°F
  - Correction factors
- Continuous duty loads
  - FLA x 125 percent

More than 3 current-carrying conductors in a raceway shall be derated by the percentages listed in **310.15(B)(2)(a)**. Conductors routed through ambient temperatures exceeding 86°F shall be derated by the percentages according to the Ampacity Correction Factors of **Table 310.16**. Continuous operation (three hours or more) shall be multiplied by 125 percent per **210.19(A)(1)**.

BRANCH-CIRCUIT VOLTAGE LIMITATIONS
210.6

The following voltage limitations are divided into three categories and each category of voltage ratings are designed to supply certain loads:

- 120 volts between conductors
- 277 volts-to-ground
- 600 volts between conductors

Branch-circuits

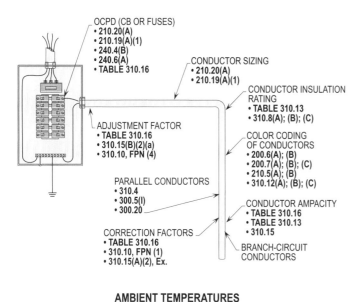

AMBIENT TEMPERATURES
NEC 310.15(B)(2)(a)

Figure 14-17. More than 3 current-carrying conductors in a raceway shall be derated by the percentages listed in **Table 310.16** and **Table 310.15(B)(2)(a)**. Conductors routed through ambient temperatures exceeding 86°F shall be derated by the percentages according to the **Ampacity Correction Factors of Table 310.16**. Continuous operation (three hours or more) shall be multiplied by 125 percent per **210.19(A)**.

120 VOLTS BETWEEN CONDUCTORS
210.6(B)

120 volts between conductors is not a restricted voltage and shall be permitted to be used to supply the following loads in any type of occupancy:

- Terminals of lampholders applied within their voltage ratings.
- Ballasts for fluorescent or high intensity discharge (HID) luminaires (lighting fixtures).
- Cord-and-plug connected or permanently connected appliances.

Cord-and-plug connected or permanently (hard-wired) connected appliances rated over 1440 VA are usually supplied by individual circuits. These appliances are generally connected to general purpose circuits that supply more than one outlet, etc.. Such appliances shall be permitted to be any one of the following types of equipment: **(See Figure 14-18)**

- Heating units
- A/C units
- Welders
- Water heaters
- Processing machine
- Etc.

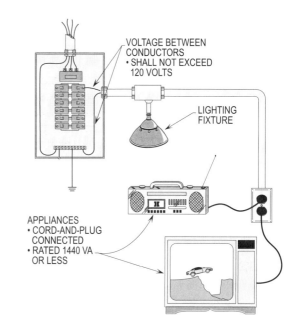

120 VOLTS BETWEEN CONDUCTORS
NEC 210.6(B)

Figure 14-18. Cord-and-plug connected or permanently (hard-wired) connected appliances rated over 1440 VA are usually supplied by individual circuits.

277 VOLTS-TO-GROUND
210.6(C); 225.7(C)

Circuits exceeding 120 volts between conductors and not exceeding 277 volts-to-ground shall be permitted to supply any one of the following types of electrical apparatus: **(See Figures 14-19(a) through (d))**

- Listed electric-discharge luminaires (lighting fixtures).

- Listed incandescent luminaires (lighting fixtures).

- Mogul-base screw-shell lampholders.

- Other than screw-shell type lampholders applied within their voltage ratings.

- Ballasts for fluorescent or high intensity discharge (HID) luminaires (lighting fixtures).

- Cord-and-plug connected or permanently connected appliances or other utilization equipment.

14-15

Design Tip: Listed electric-discharge or incandescent luminaires (lighting fixtures) are no longer required to be installed at a minimum height of 8 ft. (2.5 m) above finished grade when supplied by 480/277 volt, three-phase, four-wire system.

Luminaires (lighting fixtures) for illumination shall be permitted to be installed for outdoor areas of industrial establishments, office buildings, schools, stores and other commercial or public buildings where the luminaires (lighting fixtures) are supplied by 480/277 volt circuits per **225.7(C)**. However, luminaires (fixtures) shall not be permitted to be located within 3 ft. (900 mm) from windows, platforms, fire escapes, etc.

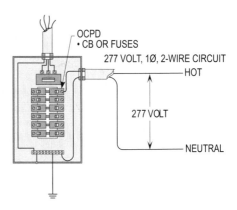

277 VOLTS - TO - GROUND
NEC 210.6(C)
NEC 225.7(C)

Figure 14-19(c). The above is a diagram of a 277 volt, single-phase, two-wire circuit.

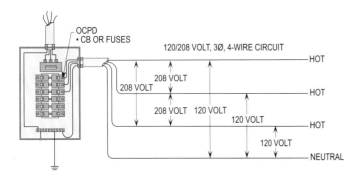

277 VOLTS - TO - GROUND
NEC 210.6(C)
NEC 225.7(C)

Figure 14-19(a). The above is a diagram of a 120/208 volt, three-phase, four-wire circuit.

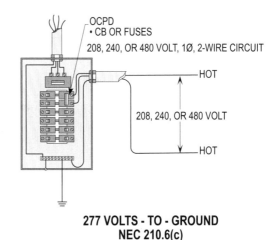

277 VOLTS - TO - GROUND
NEC 210.6(c)
NEC 225.7(c)

Figure 14-19(d). The above is a diagram of a 208, 240 or 480 volt, single-phase, two-wire circuit.

600 VOLTS BETWEEN CONDUCTORS 210.6(D); 225.7(D)

Circuits exceeding 277 volts-to-ground and not exceeding 600 volts between conductors shall be permitted to supply the following types of electrical apparatus:

- Ballasts for electric-discharge luminaires (fixtures) where mounted by one of the following methods:

  (a) At a height not less than 22 ft. (6.7 m) on poles or similar structures for the illumination of outdoor areas, such as highways, roads, bridges, athletic

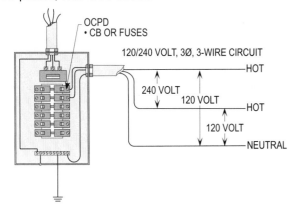

277 VOLTS - TO - GROUND
NEC 210.6(C)
NEC 225.7(C)

Figure 14-19(b). The above is a diagram of a 120/240 volt, single-phase, three-wire circuit.

fields or parking lots.

(b) At a height not less than 18 ft. (5.5 m) on other structures, such as tunnels.

- Utilization equipment that is cord-and-plug connected or permanently connected.

These circuits are usually derived by the following types of electrical systems:

- Ungrounded 480 volt, three-wire systems
- Corner grounded 480 volt, three-wire systems

Ungrounded 480 volt, three-wire systems will have 480 volts-to-ground should one of the legs be accidentally grounded, the power is not lost. Corner grounded 480 volt, three-wire systems will have 480 volts-to-ground because one leg is intentionally grounded. Circuits exceeding 277 volts and not exceeding 600 volts between conductors shall be permitted to supply the auxiliary equipment of electric-discharge lamps as permitted in **225.7(D)**.

DETERMINING AMPERAGE

The amperage for single-phase branch-circuits shall be determined by dividing the VA by the supply voltage (I = VA ÷ V). The amperage for three-phase branch-circuits shall be determined by dividing the VA by the supply voltage times the square root of 3 (I = VA ÷ V x 1.732). The 1.732 is determined by taking the √3. These amperage ratings shall be used to select the conductors and overcurrent protection devices and other elements of branch-circuits.

FINDING AMPERAGE
SINGLE-PHASE CIRCUITS

The amperage for single-phase branch-circuits shall be determined by dividing the VA rating of electrical equipment by the supply voltage of the circuit. **(See page 4-2 of Chapter 4)**

> **For example:** What is the amperage rating of an 2,400 VA computed load connected to a 120 volt, single-phase branch-circuit?
>
> **Step 1:** Finding amperage
> I = VA ÷ V
> I = 2400 VA ÷ 120 V
> I = 20 A
>
> **Solution:** The branch-circuit amperage is 20 amps.

FINDING AMPERAGE
THREE-PHASE CIRCUITS

The amperage for three-phase branch-circuits shall be determined by dividing the VA rating for the electrical equipment by the supply voltage times 1.732. The amperage is evenly (as possible) distributed on legs 1, 2 and 3 when dividing the VA by the voltage times 1.732. **(See page 4-3 of Chapter 4)**

In some cases, the values shown on the three-phase chart is used when determining the three-phase amperage instead of multiplying the supply voltage by 1.732. However, other calculations with the full square root of 3 times the voltage will be used. **(See page 4-4 of Chapter 4)**

> **For example:** What is the amperage rating of an 8,960 VA calculated load connected to a 208 volt, three-phase branch-circuit?
>
> **Step 1:** Finding amperage
> I = VA ÷ V x 1.732
> I = 8960 VA ÷ 360 V
> I = 25 A
>
> **Solution:** The branch-circuit amperage is 25 amps.

COMMERCIAL AND INDUSTRIAL

Branch-circuits used in commercial and industrial locations shall be computed differently than those in residential dwelling units. Most loads in commercial and industrial locations shall be used continuous for three hours or more without being interrupted. Therefore, such branch-circuits supplying these loads shall be calculated at 125 percent of their rating. However, there are loads that operate at noncontinuous operation and some are thermostatically controlled and these do not fall under such rules.

LIGHTING LOADS
ARTICLE 220, PART III

Incandescent or electric-discharge luminaires (lighting fixtures) shall be permitted to be installed in or on a premise. Such loads shall be calculated at noncontinuous operation (100 percent) or continuous operation (125 percent). Noncontinuous operated loads shall be calculated at 100 percent when used for less than three hours at any given time. Continuous operated loads shall be calculated at 125 percent when used for more than three hours without employing an OFF or cycle period.

NONCONTINUOUS OPERATION
210.19(A)(1); 210.20(A)

Noncontinuous lighting shall be calculated at 100 percent of the total VA or amperage rating of the branch-circuit. Luminaires (lighting fixtures) and cord-and-plug connected table lamps or floor lamps used for various periods of time shall be permitted to be installed or connected to these circuits. The branch-circuit will never be overloaded during their time of use if proper calculations are made and the correct size elements selected.

For example: How many outlets (1.5 per outlet) are permitted to be connected to a 20 amp branch-circuit used at noncontinuous operation?

Step 1: Finding amperage of outlets
210.11(A); 210.19(A)(1); 210.20(A)
A = 180 VA x 100% ÷ 120 V
A = 1.5

Step 2: Finding number of outlets
210.11(A)
# = 20 A OCPD ÷ 1.5 A
# = 13

Solution: The number of noncontinuous outlets permitted on a 20 amp branch-circuit is 13.

Design Tip: Applying the same procedure (15 A OCPD ÷ 1.5 A = 10), 10 outlets shall be permitted on a 15 amp branch-circuit. The limitation of outlets on branch-circuit alleviates the nuisance tripping of the OCPD.

CONTINUOUS OPERATION
210.19(A)(1); 210.20(A)

Continuous lighting loads shall be calculated at 125 percent of the total VA or amperage rating of the branch-circuit. Incandescent or electric-discharge luminaires (lighting fixtures) shall be permitted to be installed in or on commercial or industrial buildings for continuous operated lighting loads.

For example: How many lighting outlets are permitted to be installed on a 20 amp branch-circuit used at continuous operation?

Step 1: Finding amperage of outlets
210.11(A); 210.19(A)(1); 210.20(A)
A = 180 VA x 125% ÷ 120 V
A = 1.875

Step 2: Finding number of outlets
210.11(A)
# = 20 A OCPD ÷ 1.875 A
# = 10

Solution: The number of continuous operated lighting outlets permitted on a 20 amp branch-circuit is 10.

Design Tip: Applying the same procedure (15 A OCPD ÷ 1.875 A = 8), 8 outlets shall be permitted on a 15 amp branch-circuit. The limitation of outlets on branch-circuit alleviates the nuisance tripping of the OCPD.

OTHER LOADS
210.20(A)

The rating of the branch-circuit overcurrent device serving continuous loads, such as store lighting and similar loads, shall not be less than the noncontinuous load plus 125 percent of the continuous load.

Design Tip: The minimum branch-circuit conductor size, without the application of any adjustment or correction factors, shall have an allowable ampacity equal to or greater than the noncontinuous load plus 125 percent of the continuous load. **(See Figure 14-20)**

For example: What is the maximum continuous operated load that can be connected to a 20 amp branch-circuit?

Step 1: Finding amperage
210.19(A)(1); 210.20(A)
A = 20 A x 80%
A = 16 A

Solution: The branch-circuit load is limited to 16 amps.

Branch-circuits

For example: What size overcurrent protection device is permitted to serve a continuous operated load of 14 amps and noncontinuous load of 2.5 amps?

Step 1: Finding amperage
210.20(A); 210.19(A)(1)
14 A x 125% = 17.5 A
2.5 A x 100% = 2.5 A
Total   = 20   A

Solution: A 20 amp overcurrent protection device is required.

Design Tip: The grounded (neutral) conductor shall not be required to be considered a current-carrying conductor if it is designed to carry only the unbalanced load from resistive loads connected to the ungrounded (phase) conductors.

ELECTRIC DISCHARGE LOADS
220.4(B); 210.20(A); 310.10, FPN (2)

Branch-circuit overcurrent protection devices shall be sized at 125 percent of the VA or amperage rating of each ballast installed to supply the ballast of electric discharge lighting units. Such electric discharge lighting units and loads are as follows:

- Fluorescent
- Mercury vapor
- High pressure sodium
- Low pressure sodium
- Metal halide

The total wattage of each bulb or lamp shall not be permitted to be calculated at 125 percent to determine the total lighting load. This type of calculation does not comply with the NEC and usually requires a larger lighting load than the VA rating of each ballast times the number times 125 percent.

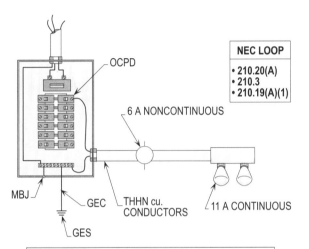

NEC LOOP
- 210.20(A)
- 210.3
- 210.19(A)(1)

What size OCPD and THHN cu. conductors are required based upon an ambient temperature of 155°F?

Step 1: Finding load
210.19(A)(1); 210.20(A)
Conductors   OCPD
11 A x 125% = 13.75 A
 6 A x 100% = 6.00 A
17 A   19.75 A

Step 2: Applying correction factors per **Table 310.16** using 12 AWG THHN cu.
30 A x 58% = 17.4 A

Step 3: Finding conductor and OCPD
Table 310.16; 240.4; 240.6(A)
17.4 A supplies 17 A load
19.75 A requires 20 A OCPD
20 A is the next size above 17.4 A

Solution: 12 AWG THHN cu. conductors and 20 amp OCPD is required.

OTHER LOADS
NEC 210.20(A)

Figure 14-20. Both the OCPD and conductors are calculated at 100 percent for noncontinuous loads and 125 percent for continuous loads. These values are added together to obtain the total load.

Design Tip: The grounded (neutral) conductor shall be considered a current-carrying conductor if it is installed for electric discharge lighting loads unless the loading does not comply with **310.15(B)(4)(c)**.

For example: What is the lighting load for a branch-circuit supplying 12 - 1.5 amp ballast's that serve 24 - F25 CW lamps used at continuous operation?

Step 1: Finding amperage
220.18(B)
I = ballast A x # of outlets
I = 1.5 A x 12
I = 18 A

Step 2: Applying percentage
210.19(A)(1); 210.20(A)
I = 18 A x 125%
I = 22.5 A

Solution: The branch-circuit load is 22.5 amps

OUTLETS
220.14(D); (E); (L)

Branch-circuits of 120 volts shall be calculated at 180 VA (1.5 A) per outlet where the VA rating of ballasts or the wattage of incandescent bulbs is unknown. Heavy-duty lampholders of 120 volts shall be calculated at 600 VA or 5 A. Lampholders connected to a branch-circuit having a rating in excess of 20 amps shall have a rating of not less than 660 watts or 5.5 amps if of the admedium type and not less than 750 watts or 6.25 amps if of any other type per **210.21(A)**. Outlets shall be calculated at noncontinuous operation (100 percent) and continuous operation (125 percent) times their VA rating to determine the branch-circuit load. Overcurrent protection devices shall be calculated at 125 percent. **(See Figure 14-21)**

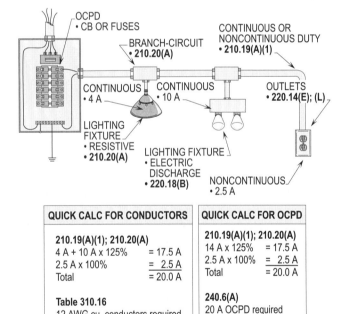

OUTLETS
NEC 220.14(D); (E); (L)

Figure 14-21. Outlets shall be calculated at noncontinuous operation (100 percent) and continuous operation (125 percent) times their VA rating to determine the branch-circuit load. Overcurrent protection devices shall be calculated at 125 percent.

For example: What size overcurrent protection device and THHN copper conductors (branch-circuits) are required for 11 outlets used at noncontinuous operation to supply luminaires (lighting fixtures)?

Step 1: Finding VA
210.19(A)(1); 210.20(A)
VA = # of outlets x VA x 100%
VA = 11 x 180 VA x 100%
VA = 1980

Step 2: Finding amperage
210.11(A)
I = VA / V
I = 1980 VA / 120 V
I = 16.5

Step 3: Finding conductor and OCPD
Table 310.16 and Asterisk;
240.4(B); 240.6(A); 240.4(D)
12 AWG THHN cu. = 30 A
Protected by 20 A OCPD

Solution: The branch-circuit conductors are sized 12 AWG and the overcurrent protection is selected at 20 amps.

For example: What size overcurrent protection device and THHN copper conductors (branch-circuit) are required for 9 outlets used at continuous operation to supply luminaires (lighting fixtures)?

Step 1: Finding VA
210.19(A)(1); 210.20(A)
VA = # of outlets x VA x 125%
VA = 9 x 180 VA x 125%
VA = 2025

Step 2: Finding amperage
210.11(A)
I = VA ÷ V
I = 2025 VA ÷ 120 V
I = 16.9

Step 3: Finding conductor and OCPD
Table 310.16 and Asterisk;
240.4(B); 240.6(A); 240.4(D)
12 AWG cu. = 30 A

Solution: The branch-circuit conductors are sized 12 AWG and the overcurrent protection device is 20 amps per Asterisk to Table 310.16.

SHOW WINDOWS
220.14(G); 220.43(A)

Show window lighting loads shall be calculated at a minimum of 180 VA per outlet for branch-circuits if the VA is unknown. The total VA rating for show window lighting loads shall be multiplied by 125 percent for each branch-circuit. Show window lighting loads shall be multiplied by 125 percent because they can operate for a period of three hours or more to provide illumination for advertisement of goods.

> **For example:** What is the number of outlets permitted to be installed on a 20 amp branch-circuit supplying lighting loads in a show window?
>
> **Step 1:** Finding # of outlets
> **220.14(G); 210.19(A)(1); 210.20(A)**
> # of outlets =
> device x 120 V ÷ VA x 125%
> # of outlets =
> 20 A x 120 V ÷ 180 VA x 125%
> # of outlets = 10.7
>
> **Solution:** The number of outlets permitted to be installed on a 20 amp branch-circuit is 10.7. Note: Inspectors usually permit 180 VA to be used if the load for such units are unknown.

See Figure 14-22 for calculating the load for show windows based upon either linear foot or individual outlets.

TRACK LIGHTING
220.43(B)

Lighting track loads shall be calculated at 150 VA for each 2 ft. (600 mm) of track to determine the load of branch-circuits. To properly balance the load, the VA rating is divided between the number of circuits supplying the length of lighting track.

> **For example:** What is the VA rating for 30 ft. of lighting track?
>
> **Step 1:** Finding VA
> **220.43(B)**
> VA = Track length ÷ 2' x 150 VA
> VA = 30' ÷ 2' x 150 VA
> VA = 2250
>
> **Solution:** The total VA rating to be used for computing the service or feeder loads is 2250 VA.

The load on multicircuit lighting tracks shall be balanced as adequately as possible using one of the following methods:

Calculating 2 circuits
2 branch-circuits
150 VA for each 2'
VA = 150 VA ÷ 2
VA = 75

Calculating 3 circuits
3 branch-circuits
150 VA for each 2'
VA = 150 VA ÷ 3
VA = 50

> **Design Tip:** Use the actual load on the track if possible.

The 150 VA multicircuit track lighting shall not apply to dwelling units per **220.43(B)**. The load for track lighting would be included in the 3 VA per sq. ft. listed in **Table 220.12(A)**. **(See Figure 14-23)**

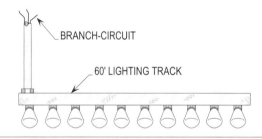

What is the VA rating of 60 ft. of lighting track and for 2 or 3 circuits with the load divided?

Finding VA of track

Step 1: Calculating VA
220.43(B)
VA = Length ÷ 2' x 150 VA
VA = 60' ÷ 2' x 150 VA
VA = 4500

Solution: VA = 4500

Dividing loads

Step 1: Calculating loads
2 circuits   3 circuits
4500 VA ÷ 2 = 2250 VA   4500 VA ÷ 3 = 1500 VA

Solution: For 2-circuits 2250 VA is used and for 3-circuits 1500 VA is used.

TRACK LIGHTING
NEC 220.43(B)

Figure 14-23. The 150 VA for each 2 ft. (600 mm) of lighting track shall be used to calculate the load for service or feeder-circuit loads. The load of the track shall be divided as evenly as possible on each circuit supplying the track.

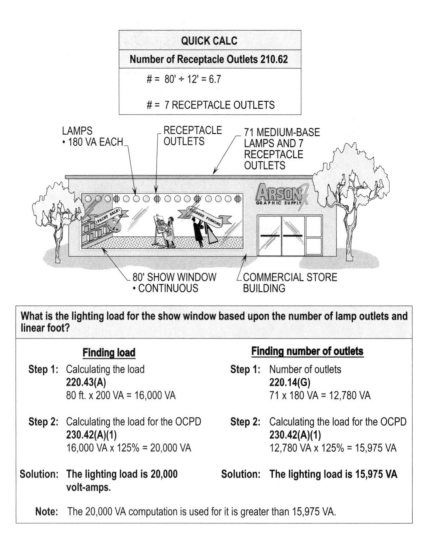

Figure 14-22. Calculating the load for a show window using largest load, either the number of outlets or the linear foot shall be used, whichever is greater.

SIGN LOADS
ARTICLE 600, PART I

Each commercial building and each commercial occupancy accessible to pedestrians shall be provided with a sign circuit that is accessible at each entrance for pedestrians from a sidewalk, street, etc. Signs shall be considered to be continuous operation (three hours or more) if installed and used in commercial occupancies.

REQUIRED BRANCH-CIRCUIT
600.5(A)

Each commercial building and each commercial occupancy shall be required to be installed with at least one 20 amp branch-circuit to supply an outlet for a sign or outline lighting that is located in an accessible location. This 20 amp branch-circuit shall not be permitted to have no other loads connected to the overcurrent protection device protecting such circuits.

RATING
600.5(B)(1); (B)(2)

Branch-circuits that supply signs and outline lighting systems containing incandescent and fluorescent forms of illumination shall be limited to 20 amps or less. Branch-circuits that supply transformers for neon tubing installations shall be limited to 30 amps or less.

For example, transformers installed for channel letters and ballasts for electric discharge lamps shall be rated 20 amps or less for branch-circuits. Transformers installed only to connect branch-circuits for neon or channel letters shall be rated 30 amps or less.

COMPUTED LOAD
220.14(F)

When sizing the service or feeder calculation for a sign calculated at a minimum of 1200 VA, the load shall be multiplied by 125 percent for continuous operation if the sign burns for three hours or more. When sizing the load for the overcurrent protection device, the load shall be multiplied by 125 percent to obtain the load which is based upon continuous operation.

For example: What is the total load allowed for a 16 amp wall sign supplied by a 20 amp, 120 volt branch-circuit?

Step 1: Finding amperage
600.5(A)
OCPD = 20 A

Step 2: Calculating amperage
220.14(F); 210.20(A)
20 A x 80% = 16 A
16 A x 125% = 20 A

Step 3: Selecting OCPD
220.14(F); 210.20(A)
16 A load is permitted

Solution: **The size overcurrent protection device required is 20 amps for a sign that burns three hours or more at 16 amps or less.**

See **Figure 14-24** for computing the load for a sign.

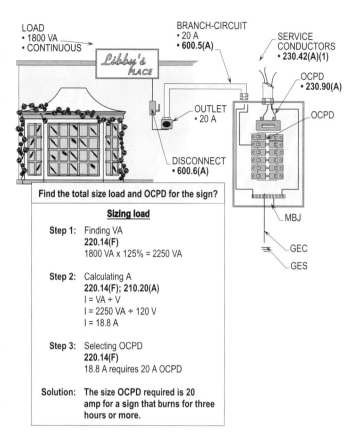

Find the total size load and OCPD for the sign?

Sizing load

Step 1: Finding VA
220.14(F)
1800 VA x 125% = 2250 VA

Step 2: Calculating A
220.14(F); 210.20(A)
I = VA ÷ V
I = 2250 VA ÷ 120 V
I = 18.8 A

Step 3: Selecting OCPD
220.14(F)
18.8 A requires 20 A OCPD

Solution: **The size OCPD required is 20 amp for a sign that burns for three hours or more.**

COMPUTED LOAD
NEC 220.14(F)

Figure 14-24. Sign loads that operate for three hours or more shall be calculated at 125 percent and elements selected from this value.

RECEPTACLE LOADS
ARTICLE 220, PART III

Receptacle outlets shall be permitted to be installed for cord-and-plug connected appliances, utilization equipment and table or floor lamps in dwelling units, apartments, condominiums, townhouses and commercial or industrial locations. Cord-and-plug connected loads shall be connected in an arrangement not to load up the branch-circuit. Cord-and-plug connected loads are general-purpose serving more than one outlet or individually serving one outlet from a branch-circuit. Cord-and-plug connected items shall be installed and located in a manner to prevent the use of extension cords except for temporary use.

RECEPTACLE OUTLETS
220.14(I)

A general purpose branch-circuit shall be calculated at 180 VA for each outlet where supplying more than one outlet utilizing cord-and-plug connected items. The number of outlets shall be calculated at 180 VA times noncontinuous operation at 100 percent and continuous operation at 125 percent. Note that overcurrent protection devices and conductors shall be sized at 125 percent. **(See Figure 14-25)**

For example: What is the load for 8 receptacle outlets that are supplying cord-and-plug connected loads used at noncontinuous operation?

Step 1: Finding VA
Load = # of outlets x 180 VA x 100%
Load = 8 x 180 VA x 100%
Load = 1440 VA

Solution: The load is 1440 VA.

For example: How many noncontinuous duty receptacle outlets can be connected to a 20 amp general purpose branch-circuit?

Step 1: Finding amperage of outlets
220.14(I)
180 VA ÷ 120 V = 1.5 A

Step 2: Finding number of outlets
220.14(I); 210.11(A)
20 A OCPD ÷ 1.5 A x 100% = 13

Solution: The number of outlets permitted on a 20 amp OCPD is 13.

For example: How many continuous duty receptacle outlets can be connected to a 20 amp general purpose branch-circuit?

Step 1: Finding amperage of outlets
220.14(I)
180 VA ÷ 120 V = 1.5 A

Step 2: Finding number of outlets
220.14(I); 210.11(A)
20 A OCPD ÷ 1.5 A x 125% = 10

Solution: The number of outlets permitted on a 20 amp OCPD is 10.

INDIVIDUAL
210.19(A)(1)

Individual cord-and-plug connected loads for branch-circuits shall be determined by multiplying the noncontinuous operated load by 100 percent and continuous operated load by 125 percent. The amount of amperage for each cord-and-plug connected load shall be found by applying the power formula. The following shall apply when using the power formula:

I = amps   I = VA ÷ V
P = volt-amps   VA = I x V
E = voltage   V = VA ÷ I

Design Tip: When applying calculations in this book, the variation of the power formula is used, since the NEC recognizes volt-amps (VA) for load calculations per the **Examples in Annex D**.

For example: What size THWN copper conductors and overcurrent protection device is required for an individual branch-circuit to a hot tub having a nameplate current rating of 42 amps?

Step 1: Finding amperage
680.9
42 A is the circuit current rating

Step 2: Calculating load
680.9
42 A x 125% = 52.5 A

Step 3: Selecting conductors
Table 310.16
52.5 A requires 6 AWG THWN cu.

Step 4: Selecting OCPD
240.4(B); 680.9
52.5 A load requires 60 A OCPD
65 A conductor allows 70 A OCPD

Solution: A 60 amp OCPD and 6 AWG THWN copper conductor is permitted.

See Figure 14-26 for calculating the load to an individual cord-and-plug connected load.

Branch-circuits

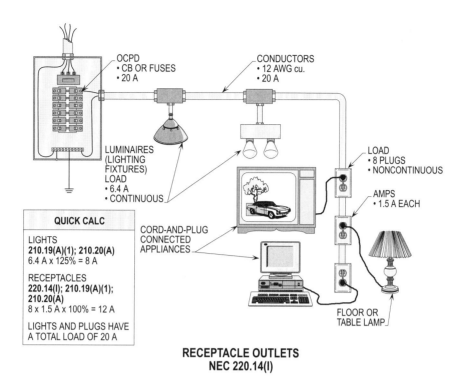

Figure 14-25. Receptacle outlets shall be calculated at 180 VA times noncontinuous operation at 100 percent and continuous operation at 125 percent.

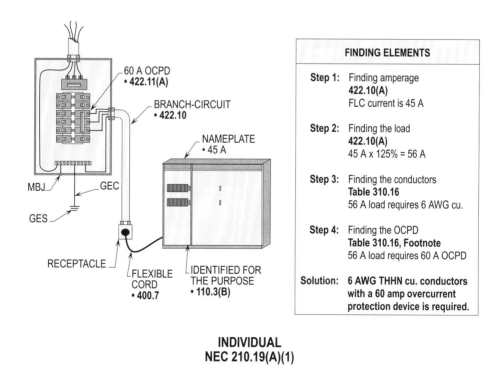

Figure 14-26. Individual loads shall be calculated at 100 percent for noncontinuous and 125 percent of continuous times the nameplate FLA of the appliance.

14-25

MULTIWIRE BRANCH-CIRCUITS
210.4

Multiwire branch-circuits are used to supply power to line-to-neutral loads only when more than one ungrounded (phase) conductor is sharing the grounded (neutral) conductor. All circuits shall originate from the same panelboard or similar distribution equipment and shall be switched by individual circuit breakers. Multiwire branch-circuits shall be supplied by double pole circuit breakers or individual circuit breakers with handle ties to provide safety from electrical shock while serving the circuits or equipment.

Section **300.13(B)** makes it mandatory that grounded (neutrals) conductors be pigtailed to prevent different voltage levels between the ungrounded (phase) conductors should the grounded (neutral) conductor become loose due to bad connections, etc.

Multiwire branch-circuits shall be permitted to serve power to one outlet for an individual piece of equipment with the other circuit(s) supplying power to a number of outlets that are used for receptacles, lights and other cord-and-plugged equipment. **(See Figure 14-27)**

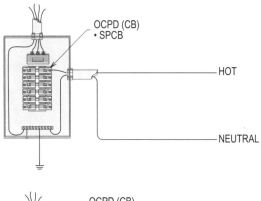

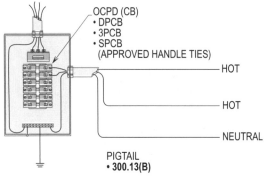

**MULTIWIRE BRANCH-CIRCUITS
NEC 210.4**

Figure 14-27. Multiwire branch-circuits are used to supply power to line-to-neutral loads only when more than one ungrounded (phase) conductor is sharing the grounded (neutral) conductor.

COMMERCIAL COOKING EQUIPMENT
210.19(A)(1); 210.20(A)

Commercial cooking equipment shall be calculated at noncontinuous operation (100 percent) or continuous operation (125 percent) to determine the branch-circuit load. When installing more than one piece of cooking equipment, the demand factors shall be selected from **Table 220.56**.

For example: What is the load for a 10 kW cooking unit that is supplied by a 240 volt, single-phase branch-circuit used at continuous operation?

Step 1: Finding kW for OCPD
210.20(A)
10 kW x 125% = 12.5 kW

Step 2: Finding kW for conductor
210.19(A)(1)
10 kW x 125% = 12.5 kW

Solution: The noncontinuous operated load is 12.5 kW and the continuous operated load is 12.5 kW.

Design Tip: The overcurrent protection device shall be calculated at 125 percent and the conductors shall be calculated at 125 percent. The OCPD shall be permitted to to be increased and decreased in size per **240.4** or **240.4(E)**.

See Figure 14-28 for calculating the load for a commercial cooking unit.

WATER HEATER LOADS
ARTICLE 422, PART III

Water heaters are designed with elements to heat water at different stages of use. For larger amounts of hot water, more of the elements shall be connected into the circuit to heat the water to replace that which was used. When smaller amounts of hot water is needed, fewer elements are used in the circuit to heat the water.

CONDUCTORS
422.13; 422.10(A)

Storage-type water heaters having a capacity of 120 gallons (450 L) or less shall have a rating not less than 125 percent of the nameplate rating to size the branch-circuit conductors.

Branch-circuits

> **For example:** What size conductors are required to supply power to a 240 volt, single-phase water heater pulling 5000 VA?
>
> **Step 1:** Finding amperage
> 5000 VA ÷ 240 V = 21 A
>
> **Step 2:** Finding the loads
> 422.13; 422.10(A)
> 21 A x 125% = 26 A
>
> **Step 3:** Finding the conductors
> 334.80; Table 310.16
> 26 A load requires 10 AWG cu.
>
> **Solution:** The size conductors required are 10 AWG.

OVERCURRENT PROTECTION
422.13; 422.11(A)

Overcurrent protection devices shall be sized not less than 125 percent of the heating load to prevent tripping open the OCPD and disconnecting all the elements connected into the circuit. Approximately 100 percent of the water heaters connected load is pulled from resistance heating elements.

> **For example:** What size overcurrent protection device is required to supply power to a 240 volt, single-phase water heater pulling 5000 VA (21 A) at continuous operation?
>
> **Step 1:** Finding load for OCPD
> 422.13; 422.11(A)
> 21 A x 125% = 26 A
>
> **Step 2:** Finding the OCPD
> 422.13; 240.4(B)
> 26 A load requires a 30 A OCPD
>
> **Solution:** The overcurrent protection device is 30 amps.

DISCONNECTING MEANS
422.31(B)

A disconnecting means shall not be required to be installed at the water heater when the overcurrent protection device is readily accessible. The overcurrent protection device shall be permitted to be installed as the disconnect where installed in a service panel that is readily accessible and is located outside or inside the building. An accessible cord-and-plug shall be permitted to serve as the disconnecting means for cord-and-plug connected water heaters where readily accessible per **422.33(A)**. However, such units shall be listed for cord-and-plug connection. **(See Figure 14-29)**

HEATING LOADS
ARTICLE 424, PART I

Heating elements in heating units are rated at 5 kW each. The elements are stacked in the heating unit to provide the rated kW for a particular size occupancy. Two stacked elements provide a 10 kW heating unit. Three stacked elements provide a 15 kW heating unit and so forth. Branch-circuit conductors and overcurrent protection devices shall be sized and selected based on the kW rating of each heating unit plus the blower motor per **424.3(B)**.

Heating units with a number of elements shall be subdivided so that they may be more easily supplied from a local panelboard or from the service equipment. Individual branch-circuits or a feeder-circuit shall utilized in routing power to the heating unit.

CONDUCTORS
424.3(B)

Heating units shall be calculated at 125 percent of the heating element load plus 125 percent of the blower motor load when present for sizing the branch-circuit conductors. The heating elements and blower motor load shall be at least 125 percent when sizing the ampacity of conductors.

> **For example:** What size THWN cu. conductors are required for a 25 kW, 240 volt, single-phase heating unit with a 4 amp blower motor?
>
> **Step 1:** Finding amperage
> 25 kVA x 1000 ÷ 240 V = 104 A
>
> **Step 2:** Finding the load
> 424.3(B)
> 104 A + 4 A x 125% = 135 A
>
> **Step 3:** Finding the conductors
> Table 310.16
> 150 A load requires 1/0 AWG THWN cu.
>
> **Solution:** 1/0 AWG THWN copper conductors are required.

Stallcup's Electrical Design Book

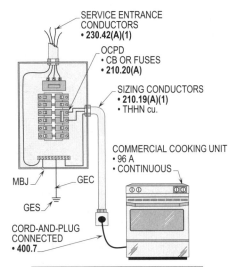

Figure 14-28. Commercial cooking equipment shall be calculated at noncontinuous operation (100 percent) or continuous operation (125 percent) to determine the branch-circuit load.

OVERCURRENT PROTECTION
424.3(B)

Heating units shall be calculated at 125 percent of the heating element load plus 125 percent of the blower motor load when present for sizing the overcurrent protection device. The next higher size rating overcurrent protection device shall be permitted to be installed if it does not correspond to this rating per **240.4(B)**. The next higher size overcurrent protection device prevents tripping open when the heating unit requires all the 5 kW rated elements to satisfy the heating load called for by the thermostat.

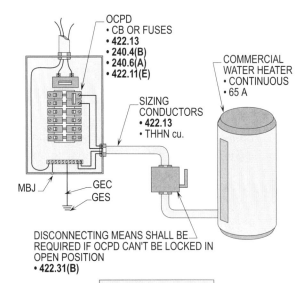

Figure 14-29. A disconnecting means shall not be required to be installed at the water heater when the overcurrent protection device is readily accessible.

14-28

DISCONNECTING MEANS
424.19

A disconnecting means shall be installed for a self-contained heating unit with a controller that energizes the circuits to the heating elements and blower motor. A fused or nonfused disconnect, an automatic breaker or a nonautomatic circuit breaker used as the disconnecting means for a heating unit shall be located within sight and within 50 ft. (15 m) of the heating unit per **Article 100, 424.19(A)** and **430.102**. **(See Figure 14-30)**

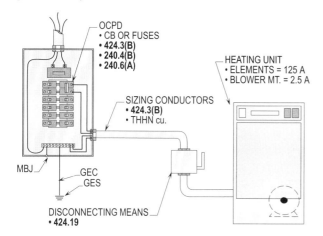

DISCONNECTING MEANS
NEC 424.19

Figure 14-30. A disconnecting means shall be installed for a self-contained heating unit with a controller that energizes the circuits to the heating elements and blower motor.

AIR-CONDITIONING LOADS
ARTICLE 440, PART III

Air-conditioning units are rated in Btu. or tons. One ton contains 12,000 Btu. A five ton compressor in a central air-conditioning unit has 60,000 Btu. (5 x 12,000 Btu. = 60,000 Btu.). The air-conditioner load shall be calculated by the square footage and construction of the premises. The size air-conditioner rated in Btu. shall be selected from this calculation. The conductors shall be sized large enough to carry the load of the compressor and the overcurrent protection device shall be selected to allow the compressor to start without tripping open, due to inrush current from high head pressures.

CONDUCTORS
440.32

Branch-circuit conductors supplying a single motor-compressor shall have an ampacity not less than 125 percent of either the motor-compressor rated load current or the branch-circuit selection current, whichever is greater.

The branch-circuit conductors shall be sized large enough to prevent damage to the insulation of the conductor caused by an overload. Overload relays are installed and are usually adjusted to trip open on an overload exceeding 140 percent of full-load current rating of the compressor. This type of condition shall be compensated for by sizing and selecting the conductors at 125 percent of the compressors full-load current rating. The condenser motor load shall be added at 100 percent when present.

> **For example:** What size conductors are required to supply power to a 240 volt, single-phase air-conditioning unit with a compressor rating of 6000 VA plus a 3 amp condenser motor? (Use THHN copper conductors terminated to 60°C terminals)
>
> **Step 1:** Finding amperage of compressor
> 6000 VA ÷ 240 V = 25 A
>
> **Step 2:** Finding the total load
> **440.32**
> 25 A x 125% + 3 A = 34 A
>
> **Step 3:** Finding the conductors
> **Table 310.16**
> 34 A load requires 8 AWG cu.
>
> **Solution:** **8 AWG THHN copper conductors are required.**

OVERCURRENT PROTECTION
440.22(A)

To allow the compressor to start and run, the overcurrent protection device shall be sized properly. The overcurrent protection device shall have a rating or setting not exceeding 175 percent of the motor-compressor rated full-load current or the branch-circuit selection current, whichever is greater to determine the minimum rating. The overcurrent protection device shall be permitted to be increased up to 225 percent of the full-load current if an air-conditioner unit will not start using 175 percent. Condenser motors shall be calculated at 100 percent and added to this total and the OCPD selected accordingly.

When air-conditioning units are installed on the roof or on the outside of the premises where the sun goes down, the air-conditioning unit may have trouble starting and operating during hot summer months. Overcurrent protection devices shall be sized by the listing on the nameplate of the unit. Fuses and circuit breakers (HACR) shall be installed where listed on the nameplate of the unit. For further information, see Underwriters Laboratory book, Electrical Appliance and Utilization Equipment Directory.

Note that OCPD's for A/C units installed on the roofs are already calculated and sized at the maximum value of 225 percent per **110.3(B)** and **440.22(A)**.

Conductors supplying A/C units on roofs are calculated already per **440.4(C)** and **440.32** and the nameplate will list the size conductors or circuit needed.

For example: What minimum and maximum size overcurrent protection device is required to supply power to a 240 volt, single-phase air-conditioning unit with a compressor rating of 6000 VA plus a 3 amp condenser motor?

Step 1: Finding load for OCPD
440.22(A)
25 A x 175% + 3 A = 46.7 A

Step 2: Finding OCPD
240.4(G); 240.6(A)
45 A is the next lower standard size

Solution: The lower size OCPD is 45 amps.

Step 1: Finding load for OCPD
440.22(A)
25 A x 225% + 3 = 59.25 A

Step 2: Finding OCPD
240.4(G); 240.6(A)
50 A is the next higher size permitted

Solution: The higher size OCPD is 50 amps.

DISCONNECTING MEANS
440.14

The disconnecting means shall be located within 50 ft. (15 m) and within sight from the air-conditioning unit per **Article 100**. The disconnecting means shall be permitted to be installed on or within the air-conditioning unit. The disconnecting means shall be permitted to be installed where capable of being locked in the open position where conditions of maintenance and supervision ensure that only qualified personnel will service the equipment. Note that **Ex. 1 to 440.14** only applies to industrial compressors and not A/C units. A disconnecting means shall not be required where within sight of such equipment. **(See Figure 14-31)**

MOTOR LOADS
ART. 430, PARTS II AND IV

Motors are rated in horsepower (HP). The amount of work that a motor can perform depends upon its HP rating class, code letter or design letter. A motor with a high horsepower rating may do more work than a motor with a low horsepower rating. The overcurrent protection devices and conductors used to supply power to motors shall be sized and based on the horsepower, voltage and number of phases. The starting current of a motor shall be based on its code letter per **Table 430.7(B)** or design letter per **Tables 430.251(A)** and **(B)**.

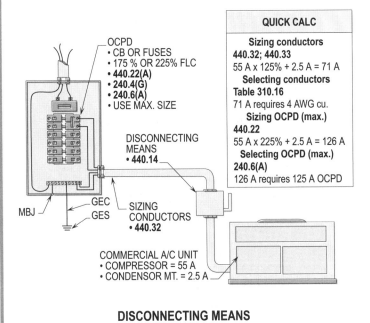

Figure 14-31. The disconnecting means shall be located within 50 ft. (15 m) and within sight from the air-conditioning unit per **Article 100**. Also, see Figures 19-9 and 19-16.

Branch-circuits

CONDUCTORS
430.22(A)

Branch-circuit conductors supplying a single motor shall have an ampacity not less than 125 percent of the motor full-load current rating. Single-phase motors shall have their full-load current ratings selected from **Table 430.48** per **430.6(A)(1)**. Three-phase motors shall have their full-load current rating selected from **Table 430.250** per **430.6(A)(1)**.

> **For example:** What size conductors are required to supply power to a 10 HP, 208 volt, single-phase motor? (Using THWN cu. conductors)
>
> Step 1: Finding amperage
>   430.6(A)(1); Table 430.248
>   10 HP = 55 A
>
> Step 2: Finding the load
>   430.22(A)
>   55 A x 125% = 68.75 A
>
> Step 3: Finding the conductors
>   Table 310.16
>   68.75 A load requires 4 AWG cu.
>
> **Solution: The branch-circuit conductors required are 4 AWG copper.**

> **For example:** What size conductors are required to supply power to a 10 HP, 208 volt, three-phase motor? (Using THWN cu. conductor)
>
> Step 1: Finding amperage
>   430.6(A)(1); Table 430.250
>   10 HP = 30.8 A
>
> Step 2: Finding the load
>   430.22(A)
>   30.8 A x 125% = 38.5 A
>
> Step 3: Finding the conductors
>   Table 310.16
>   38.5 A load requires 8 AWG cu.
>
> **Solution: The branch-circuit conductors required are 8 AWG copper.**

OVERCURRENT PROTECTION
430.52; TABLE 430.52

When power is applied to the windings of a motor and it starts, the motor has high inrush currents. The amount of current required to drive a load is called running current. Inrush currents of the motor shall be held by an overcurrent protection device sized large to allow the motor to accelerate the driven load. The inrush current of most motors is four to six times the running current of the motor, based on its code letter per **Table 430.7(B)** or design letter per **Table 430.52**. Note that Design E motors can have 8 1/2 to 15 times FLA.

Overcurrent protection devices shall not be permitted to exceed the percentages listed in **Table 430.52**. Section **430.52(C)(1)** requires the overcurrent protection device to based on the percentages for the type of device used in the columns of **Table 430.52**. If the percentage do not correspond to a standard device listed in **240.6(A)**, the higher standard size rating shall be permitted to be used per **430.52(C)(1), Ex. 1**. If the motor will not start and run, the percentage shall be permitted to be increased per **430.52(C)(1), Ex.'s 2(a) through (c)**. These percentages for the overcurrent protection device are as follows:

- Nontime-delay fuses -
  shall not be permitted to exceed 400 percent of the full-load current for fuses rated 600 amps or less per **430.52(C)(1), Ex. 2(a)**.

- Time-delay fuses -
  shall not be permitted to exceed 225 percent of the full-load current per **430.52(C)(1), Ex. 2(b)**.

- Circuit breakers -
  shall not be permitted to exceed 400 percent of the full-load current for ratings of 100 amps or less, or shall not be permitted to exceed 300 percent of the full-load current for ratings over 100 amps per **430.52(C)(1), Ex. 2(c)**.

- Instantaneous trip circuit breakers -
  shall not be permitted to exceed 1300 or 1700 percent of the full-load current per **430.52(C)(3), Ex. 1**.

> **For example:** What size nontime-delay fuse, time-delay fuse, circuit breaker and instantaneous circuit breaker are required for a 20 HP, 230 volt, three-phase motor with a full-load current rating of 54 amps per **Table 430.250**. (Nameplate amps is 49 amps)

Nontime-delay fuses

Step 1: Finding percentage
430.52(C)(1); Table 430.52
300%

Step 2: Finding amperage
430.52(C)(1)
54 A x 300% = 162 A

Step 3: Finding NTDF (round down size)
430.52(C)(1); 240.4(G); 240.6(A)
162 A requires 150 A OCPD

Step 4: Finding NTDF (round up size)
430.52(C)(1), Ex. 1; 240.4(G); 240.6(A)
The next higher standard size above 162 A is 175 A OCPD

Step 5: Finding NTDF (max. size)
430.52(C)(1), Ex. 2(a)
52 A x 400% = 208 A
The largest size below 208 is 200 A OCPD

Solution: The next higher size nontime-delay fuses required are 175 amps.

Time-delay fuses

Step 1: Finding percentage
430.52(C)(1); Table 430.52
175%

Step 2: Finding amperage
430.52(C)(1)
54 A x 175% = 94.5 A

Step 3: Finding TDF (round down size)
430.52(C)(1); 240.4(G); 240.6(A)
94.5 A requires 90 A OCPD

Step 4: Finding TDF (round up size)
430.52(C)(1), Ex. 1; 240.4(G); 240.6(A)
The next standard size above 94.5 A is 100 A OCPD

Step 5: Finding TDF (max. size)
430.52(C)(1), Ex. 2(b)
54 A x 225% = 121.5 A
The largest size below 121.5 A is 110 A OCPD

Solution: The next higher size time-delay fuse required is 100 amp.

Branch-circuits

Circuit breaker

Step 1: Finding percentage
430.52(C)(1); Table 430.52
250%

Step 2: Finding amperage
430.52(C)(1)
54 A x 250% = 135 A

Step 3: Finding CB (round down size)
430.52(C)(1); 240.4(G); 240.6(A)
135 A requires 125 A OCPD
The next size below 135 A is 125 A OCPD

Step 4: Finding CB (round up size)
430.52(C)(1), Ex. 1; 240.4(G); 240.6(A)
The next standard size above 135 A is 150 A OCPD

Step 5: Finding CB (max. size)
430.52(C)(1), Ex. 2(c)
54 A x 400% = 216 A
The largest size below 216 A is 200 A OCPD

Solution: The next higher size circuit breaker required is 150 amp.

Instantaneous circuit breaker

Step 1: Finding percentage
430.52(C)(3); Table 430.52
800%

Step 2: Finding amperage
430.52(C)(3)
54 A x 800% = 432 A

Step 3: Finding CB
430.52(C)(1); 240.4(G); 240.6(A)
The minimum setting is 432 A

Step 4: Finding CB
430.52(C)(3), Ex. 1
54 A x 1300% = 702 A
The maximum setting is 702 A

Solution: The maximum size instantaneous circuit breaker setting is 702 amps.

DISCONNECTING MEANS
430.102; 430.107

A disconnecting means shall be located within sight from the controller and shall disconnect the controller to allow personnel to service the motor without the danger of the branch-circuit accidentally being energized. The disconnecting means shall be permitted to be installed by one of the following methods to disconnect a motor circuit:

• The disconnecting means shall be located adjacent to the controller within sight and located within 50 ft. (15 m). Same rule shall be applied for the motor.

• The disconnecting means shall be capable of being locked in the open position if located adjacent to the controller. Under this rule, the motor shall not be required to be located within sight.

• An additional disconnecting means shall be provided within 50 ft. (15 m) of the motor and within sight if the disconnecting means by the controller cannot be locked in the open position and the motor is located out of sight. **(See Figure 14-32)**

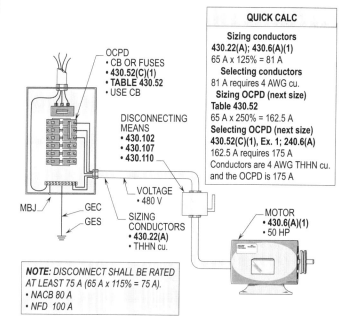

DISCONNECTING MEANS
NEC 430.102
NEC 430.107

Figure 14-32. A disconnecting means shall be located within sight from the controller and shall disconnect the controller to allow personnel to service the motor without the danger of the branch-circuit accidentally being energized. Also, see Figures 18-1, 18-2 and 18-24(a) thru (d).

WELDER LOADS
ARTICLE 630, PARTS II AND IV

There are three types of welders used in modern day welding shops. The type used determines how the circuit elements shall be designed.

The procedure for calculating the full-load amps for welders is to obtain the duty-cycle factor and select the multiplier. The primary amps of the welder shall be multiplied by the multiplier to derive full-load amps to size the elements of the branch-circuit. The following are the three types of welders that shall be permitted to be used:

- AC/DC arc welders
- Motor-generator arc welders
- Resistance welders

AC/DC ARC WELDERS
ARTICLE 630, PART II

Arc welders are used to perform welding operations through the medium of an arc drawn between the weld and a metal rod. The metal from the rod (electrode) is added to the weld which makes the weld substantially strong.

CONDUCTORS
630.11(A)

When sizing the branch-circuit conductors for AC/DC arc welders, the current-carrying capacity shall not be permitted to be less than the primary current of the welder times a duty cycle factor listed in **Table 630.11(A)**.

> **For example:** What size THWN cu. conductor are required to supply an AC transformer and DC rectifier arc welder rated at 68 amps with a 50 percent duty cycle?
>
> Step 1: Finding FLC
> 630.11(A)
> Welder = 68 A
>
> Step 2: Finding multiplier
> 630.11(A)
> 50% = .71
>
> Step 3: Calculating amps
> 630.11(A)
> 68 A x 71% = 48.28 A
>
> Step 4: Selecting conductors
> Table 310.16
> 48.28 A requires 8 AWG cu.
>
> **Solution: The size THWN copper conductors are 8 AWG.**

OVERCURRENT PROTECTION
630.12(A)

The welder's primary full-load current rating listed on the nameplate shall be selected at not more than 200 percent for sizing the overcurrent protection device. Branch-circuit conductors shall be protected at a rating not exceeding 200 percent of their allowable ampacities per **630.12(B)**. (See **Figure 14-33**)

> **For example:** What size OCPD is required for the conductors supplying an AC transformer and DC rectifier arc welder rated at 68 amps with a 50 percent duty cycle?
>
> Step 1: Finding FLC
> 630.12(A); 630.12(B); Table 310.16
> Welder = 68 A x 71% = 48 A
> Conductors = 50 A
>
> Step 2: Finding multiplier
> 630.12(A); 630.12(B)
> Multiplier = 200%
>
> Step 3: Calculating amps
> 630.12(A); 630.12(B)
> 68 A x 200% = 136 A
>
> Step 4: Selecting OCPD
> 240.4(G); 240.6(A); 630.12
> 136 A requires 125 A
>
> Step 5: Protecting conductors
> 630.22(B)
> 8 AWG THWN cu. = 50 A
> 50 A x 200% = 100 A
> 100 A requires 100 A
>
> **Solution: The size OCPD required is 100 amps based upon amps of conductors times 200 percent. The OCPD at the welder is 125 amps.**

MOTOR-GENERATOR ARC WELDERS
ARTICLE 630, PART II

The arc welding principles are used to perform work for motor-generator arc welders. Two metal parts are welded together by a main electrode used to strike an arc which melts the electrode and supplies the metal necessary to join the metal parts together.

CONDUCTORS
630.11(A)

When sizing the branch-circuit conductors for motor-generator arc welders, the current-carrying capacity shall not be permitted to be less than the rated primary current of the welder times a duty cycle factor listed in **Table 630.11(A)**.

> **For example:** What size THWN cu. conductors are required to supply a motor-generator arc welder rated at 76 amps having a 90 percent duty cycle?

Step 1:	Finding FLC 630.11(A) Welder = 76 A
Step 2:	Finding multiplier 630.11(A) 90% = .96
Step 3:	Calculating amps 630.11(A) 76 A x 96% = 72.96 A
Step 4:	Selecting conductors Table 310.16 72.96 A requires 4 AWG cu.
Solution:	**The size THWN copper conductors are 4 AWG.**

OVERCURRENT PROTECTION
630.12(A)

The welder's primary full-load current rating listed on the nameplate shall be selected at not more than 200 percent for sizing the overcurrent protection device. Branch-circuit conductors shall be protected at a rating not exceeding 200 percent of their allowable ampacities per **630.12(B)**. **(See Figure 14-34)**

> **For example:** What size overcurrent protection device is required for the conductors to supply a motor-generator arc welder rated at 76 amps having a 90 percent duty cycle?

Step 1:	Finding FLC 630.12(A); 630.12(B); Table 310.16 Welder = 76 A x 96% = 73 A Conductors = 85 A (4 AWG cu.)
Step 2:	Finding multiplier 630.12(A); 630.12(B) Multiplier = 200%
Step 3:	Calculating amps 630.12(A); 630.12(B) 76 A x 200% = 152 A
Step 4:	Selecting OCPD for welder 240.4(G); 240.6(A); 630.12(A) 152 A requires 150 A
Step 5:	Selecting OCPD for conductors 630.12(B) 85 A x 200% = 170 A
Solution:	**The size OCPD required for the conductors is 150 amps.**

RESISTANCE WELDERS
ARTICLE 630, PART III

Resistance welders use a heavy current which flows through the small area of material in contact with such material at a particular time. This type of welding is accomplished by pressing two metal parts together as they reach the molten state. Resistance welders do not add metal to the weld. Resistance welders may have high inrush current while welding together certain materials.

CONDUCTORS
630.31(A)

When sizing the branch-circuit conductors for resistance welders, the current-carrying capacity shall not be permitted to be less than the primary current of the welder times a duty cycle factor listed in **Table 630.31(A)**.

Stallcup's Electrical Design Book

For example: What size THWN cu. conductors are required to supply a resistance welder rated at 91 amps having a 40 percent duty cycle?

Step 1: Finding FLC
630.31(A)
Welder = 91 A

Step 2: Finding multiplier
630.31(A)
40% = .63

Step 3: Calculating amps
630.31(A)
91 A x 63% = 57.33 A

Step 4: Selecting conductors
Table 310.16
57.33 A requires 6 AWG cu.

Solution: The size THWN copper conductor are 6 AWG.

OVERCURRENT PROTECTION 630.32(A)

The welder's primary full-load current rating listed on the nameplate shall be selected at not more than 300 percent for sizing the overcurrent protection device. Branch-circuit conductors protected at a rating not exceeding this value shall be considered protected from overloading per **630.32(B). (See Figure 14-35)**

For example: What size overcurrent protection device is required for the conductors supplying power to a resistance welder rated at 91 amps having a 40 percent duty cycle?

Step 1: Finding FLC
630.32(A)
Welder = 91 A

Step 2: Finding multiplier
630.32(A)
Multiplier = 300%

Step 3: Calculating amps
630.32(A)
91 A x 300% = 273 A

Step 4: Selecting OCPD
240.4(G); 240.6(A); 630.12
273 A requires 250 A

Step 5: Protecting conductors
630.32(B)
6 AWG THWN cu. = 65 A
65 A x 300% = 195 A
195 A requires 175 A

Solution: The size OCPD required to protect the conductors is 175 amps. The size OCPD for the welder is 250 amps.

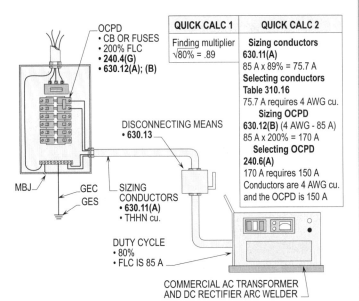

OVERCURRENT PROTECTION NEC 630.12(A)

Figure 14-33. The welder's primary full-load current rating listed on the nameplate shall be selected at not more than 200 percent for the overcurrent protection device to be installed. Branch-circuit conductors shall be protected at a rating not exceeding 200 percent of their allowable ampacities.

Branch-circuits

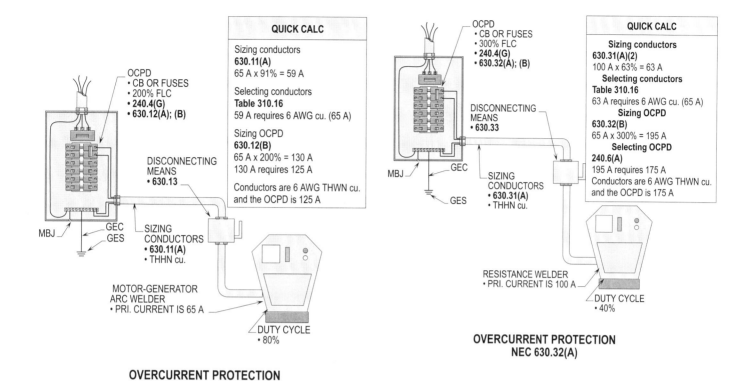

Figure 14-34. The welder's primary full-load current rating listed on the nameplate shall be selected at not more than 200 percent for the overcurrent protection device. Branch-circuit conductors shall be protected at a rating not exceeding 200 percent of their allowable ampacities.

Figure 14-35. The welder's primary full-load current rating listed on the nameplate shall be selected at not more than 300 percent for sizing the overcurrent protection device. Branch-circuit conductors protected at a rating not exceeding this value shall be considered protected from overloading.

14-37

Name   Date

Chapter 14. Branch-circuits

Section	Answer		
_____	T F	1.	A minimum of three 20 amp, 1500 VA small appliance circuits are required to supply receptacle outlets that are located in the kitchen, pantry, breakfast room, and dining room.
_____	T F	2.	At least one 20 amp, 1500 VA laundry circuit is required to supply receptacle outlets in the laundry room.
_____	T F	3.	**Column B to Table 220.19** is used for cooking equipment rated from 3.5 kW to 8.75 kW.
_____	T F	4.	A tap to connect a cooktop and ovens can be made from a 60 amp branch-circuit when the tap conductors are sized from the kW rating of each piece of cooking equipment per **Note 4 to Table 220.19**.
_____	T F	5.	Dryer equipment loads shall be at least 5000 VA or the nameplate rating, whichever is larger, when selecting a VA for the branch-circuit load.
_____	T F	6.	A 30 amp branch-circuit shall be permitted to be installed to supply fixed lighting units with heavy-duty lampholders in other than a dwelling unit(s) or utilization equipment in any occupancy.
_____	T F	7.	Conductors with the 90°C rated insulation can be connected to terminals rated at 60°C, 75°C or 90°C.
_____	T F	8.	Continuous lighting loads in commercial buildings shall be calculated at 125 percent of the total VA or amperage rating of the branch-circuit.
_____	T F	9.	Branch-circuit overcurrent protection devices shall be sized at 100 percent of the VA or amperage rating of each ballast installed to supply discharge lighting units that are used at continuous duty.
_____	T F	10.	Show window lighting loads shall be calculated at a minimum of 200 VA per outlet for branch-circuits if the VA is unknown.
_____	_____	11.	Note 1 to **Table 220.19** is applied where the kW rating of the cooking equipment is over _____ kW but not over 27 kW.
_____	_____	12.	When installing four or fewer dryers, the load shall be calculated at _____ percent.
_____	_____	13.	A fastened-in-place appliance shall be permitted to be connected to a general-purpose circuit if its amp rating does not exceed _____ percent of the branch-circuit.
_____	_____	14.	A _____ or _____ amp branch-circuit shall be permitted to supply cooking appliances that are fastened-in-place in any occupancy.
_____	_____		

14-39

Section	Answer

_____ _____ 15. Overcurrent protection devices rated over _____ amps are allowed to be terminated with conductors larger than 1 AWG with 75°C ampacities.

_____ _____ 16. General-purpose circuits shall be calculated at _____ VA per sq. ft. based on the square footage of the dwelling unit.
    (a)   2
    (b)   3
    (c)   3.5
    (d)   4

_____ _____ 17. The rating of any one cord-and-plug connected utilization equipment shall not exceed _____ percent of the branch-circuit rating if connected to a general-purpose circuit supplying two or more outlets.
    (a)   50
    (b)   75
    (c)   80
    (d)   100

_____ _____ 18. Branch-circuits larger than _____ amps shall supply only nonlighting outlet loads.
    (a)   20
    (b)   30
    (c)   40
    (d)   50

_____ _____
_____ _____ 19. 60°C ampacities shall only be applied using 75°C ampacities where conductors _____ AWG through _____ AWG are installed.
    (a)   14 - 2
    (b)   14 - 1
    (c)   12 - 2
    (d)   12 - 1

_____ _____ 20. Cord-and-plug connected or permanently (hard-wired) connected receptacles rated over _____ VA are usually supplied by individual circuits.
    (a)   1440
    (b)   1660
    (c)   2250
    (d)   2440

_____ _____ 21. Lighting track loads shall be calculated at _____ VA for each 2 ft. of track to determine the load for branch-circuits.
    (a)   150
    (b)   180
    (c)   200
    (d)   220

_____ _____ 22. How many 15 amp circuits are allowed to supply power to the general-purpose lighting and receptacle outlets in a 3000 sq. ft. dwelling unit?

_____ _____ 23. What is the demand load in VA for a branch-circuit serving a 9 kW range?

_____ _____ 24. What is the demand load in VA for a branch-circuit serving 2 - 3 kW cooktops?

_____ _____ 25. What is the demand load in VA for a branch-circuit serving a 8.5 kW oven?

	Section	Answer

26. What is the demand load in VA for a branch-circuit serving a 18 kW range?

27. What is the demand load in VA for a service or feeder serving a 10 kW, 18 kW and 20 kW range?

28. What is the demand load in VA for a branch-circuit serving a 9 kW cooktop, 8 kW and 14 kW oven?

29. What size nonmetallic-sheathed cable tap is required for a 12 kW cooktop?

30. What is the demand load in VA for a branch-circuit serving a 4500 VA dryer?

31. What is the demand load in VA for a branch-circuit serving a 6500 VA dryer?

32. What is the total VA rating for a 15 amp, 120 volt, two-wire branch-circuit supplying a continuous load?

33. What is the total VA rating for a 15 amp, 240 volt, two-wire branch-circuit supplying a continuous load?

34. What is the load in amps for a 24 amp dishwasher used at continuous duty?

35. What is the load in amps for a 32 amp water heater used at continuous operation in a commercial building?

36. What size amperage rating is allowed for a 2 AWG copper conductor using the ampacities of the 60°C column and 75°C column?

37. What is the allowable ampacity for 8 - 10 AWG THHN copper conductors that are all current-carrying?

38. What is the allowable ampacity for 3 - 10 AWG THHN copper conductors that are in an ambient temperature of 105°F?

39. What is the allowable ampacity for 6 - 10 AWG THHN copper conductors that are all current-carrying that are in an ambient temperature of 105°F?

40. What is the amperage rating of an 2800 VA computed load connected to a 120 volt, single-phase branch-circuit?

41. What is the amperage rating of an 9250 VA calculated load connected to a 208 volt, three-phase branch-circuit?

42. How many outlets are permitted to be connected to a 20 amp branch-circuit used at noncontinuous operation?

43. How many outlets are permitted to be connected to a 20 amp branch-circuit used at continuous operation?

44. What size overcurrent protection device is allowed to serve a continuous operated load of 13.5 amps and noncontinuous load of 3 amps in an ambient temperature of 155°F?

45. What is the lighting load in amps for a branch-circuit supplying ten 1.5 amp ballasts that serve 24 - F25, CW lamps used at continuous operation?

14-41

Section	Answer
_____	_____

46. What size overcurrent protection device and THHN copper conductors are required for 12 outlets used at noncontinuous operation to supply lighting fixtures used at 120 volts?

47. What size overcurrent protection device and THHN copper conductors are required for 10 outlets used at continuous operation to supply lighting fixtures?

48. How many outlets are permitted to be installed on a 20 amp branch-circuit-supplying lighting loads in a show window?

49. What is the lighting load in VA for 70 ft. of show window? (Lighting is used at continuous duty)

50. What is the show window lighting load in VA for 70 medium-base lamps? (Lamps are used at continuous duty)

51. What is the VA rating for 25 ft. of lighting track? (Lighting will be used at non-continuous duty)

14-42

15

Feeder-circuits

Feeder-circuits are all circuit conductors between the service equipment, the source of a separately derived system or other power supply source and the final overcurrent protection devices protecting the branch-circuit conductors and equipment. Feeder-circuits may be installed to supply subpanels in dwelling units, apartments and commercial and industrial occupancies. Feeder-circuits may also be installed in commercial and industrial occupancies and routed through the plant, where taps are sometimes made to these conductors and connected to supply panelboards, disconnect switches, control centers and other types of electrical equipment.

LOADS
ARTICLE 220, PART III

The rules and regulations of **Article 220, Part III** shall be used to size and select the feeder-circuit conductors, overcurrent protection devices and other pertinent elements. Demand factors shall be permitted to be applied to loads that are grouped and percentages shall be permitted to be applied that will reduce the load. These percentages shall be based on how the equipment is used in the electrical system.

VOLTAGE
220.5(A)

The ampacity of electrical components for feeder-circuits shall be determined by the voltage ratings listed in **220.5(A)**. Unless another voltage is specified, the following voltage ratings shall be used to compute feeder-circuit loads:

- 120 volt
- 120/240 volt
- 208 Y/120 volt
- 240 volt
- 347 volt
- 480 Y/277 volt
- 480 volt
- 600 Y/347 volt
- 600 volt

UNGROUNDED (PHASE) CONDUCTORS 215.2(A)(1); 215.3

Ungrounded (phase) conductors that are used in a feeder-circuit shall be sized and selected based on the total amps or the total VA divided by the voltage of the circuit. The total amps or VA of the feeder-circuit loads shall be calculated at continuous or noncontinuous operation or a combination of both. Also, loads with demand factors shall be calculated at percentages less than those used for continuous or noncontinuous loads and such percentages shall be derived for how the loads are used on the feeder-circuit. The smallest size feeder-circuit shall not be less than 30 amps per **215.2(A)(2)**. To comply with this minimum rating, 10 AWG copper or 8 AWG aluminum conductor shall be used per asterisk below **Table 310.16** and **240.4(D)**.

Ungrounded (phase) conductors shall be considered current-carrying conductors for derating purposes when applying adjustment or correction factors per **310.15(B)(2)(a)** and **Table 310.16**.

COMPUTING AMPS

If the VA is known, amps may be found by dividing the VA by the configuration of voltage for the feeder-circuit.

For example: What is the ampacity for a feeder-circuit load of 18,400 VA which is supplied by a 120/240 volt, single-phase system?

Step 1: Finding amperage
I = VA ÷ A
I = 18,400 VA ÷ 240 V
I = 77 A

Solution: The feeder-circuit ampacity is 77 amps.

For example: What is the ampacity for a feeder-circuit load of 18,400 VA which is supplied by a 120/208 volt, three-phase, four-wire system?

Step 1: Finding amperage
I = VA ÷ V x 1.732
I = 18,400 VA ÷ 360 V
I = 51.1 A

Solution: The feeder-circuit ampacity is 51 amps. Note: The .1 amp is dropped per 220.5(B).

The feeder-circuit components shall be selected from the 51 amps per phase. The 51 amps per phase shall be connected to Phase A, Phase B and Phase C. This calculation shall be verified by dividing the VA by the voltage per phase and dividing the total by three.

For example: What is the ampacity for a feeder-circuit load of 18,400 VA which is supplied by a 120/208 volt, three-phase system?

Step 1: Finding amperage
I = VA ÷ V
I = 18,400 VA ÷ 120 V
I = 153.3
I = 153.3 A ÷ 3
I = 51 A

Solution: The feeder-circuit ampacity is 51 amps.

See Figure 15-1 for calculating amps based on volt-amps and voltage.

POWER FACTOR

Power factor (PF) is the ratio of actual power used in a circuit to the apparent power drawn from the line. Actual power is the true power used to produce heat or work. Actual power is also known in the industry as real, true or useful power.

Terms:

- Power factor = PF
- Watts = W
- Volts = V
- Amps = I

Feeder-circuits

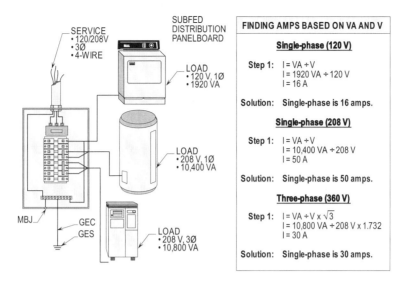

COMPUTING AMPS

Figure 15-1. Finding the amps of a circuit based upon volt-amps and voltage.

ACTUAL POWER SINGLE-PHASE

The actual power in watts in a pure resistance circuit shall be found by multiplying volts x amps.

> **For example:** A 240 volt, single-phase feeder-circuit with a 51 amp load has a load of 12,240 watts.

> **Step 1:** W = V x I
> W = 240 V x 51 A
> W = 12,240
>
> **Solution: Actual power equals 12,240 watts.**

If the circuit above has a poor power factor of 75 percent because of inductive loads, the load in amps shall be computed as follows:

> **For example:** A 240 volt, single-phase feeder-circuit of 51 amps having a power factor of 75 percent has a load of 9180 watts.

> **Step 1:** W = V x I x PF
> W = 240 V x 51 A x 75%
> W = 9180
>
> **Solution: Actual power equals 9180 watts.**

ACTUAL POWER THREE-PHASE

The actual power in watts in a three-phase circuit with inductive loads shall be found by multiplying V x $\sqrt{3}$ x I x PF.

> **For example:** The actual power in watts for a three-phase circuit with a load of 51 amps that is supplied by a 208 volt circuit with a power factor of 75 percent is 13,770 watts. (Using 360 V)

> **Step 1:** W = V x $\sqrt{3}$ x I x PF
> W = 208 V x 1.732 x 51 A x 75%
> W = 13,770
>
> **Solution: Actual power equals 13,770 watts.**

Note: The actual power is 18,360 watts (208 V x 1.732 x 51 A = 18,360 W) if the power factor is 100 percent instead of 75 percent.

See Figure 15-2 for calculating amps on a feeder-circuit having poor power factor.

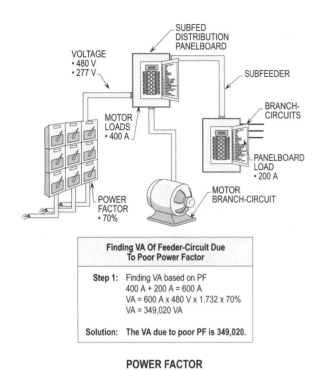

Figure 15-2. Calculating amps for a feeder-circuit based upon poor power factor.

NONCONTINUOUS OPERATED LOADS
215.2(A)(1); 215.3

Noncontinuous operated loads shall be calculated at 100 percent or where appropriate demand factors shall be permitted to be applied to specific loads. Based upon their condition of use, the total VA rating for such loads may be derived. Noncontinuous loads are classified as loads that operate for a period of time less than three hours per **Article 100**. Portable or fixed cord-and-plug connected or permanently connected loads may be classified noncontinuous loads.

For example: What is the VA rating for a connected load of 34,500 VA to size the elements of the feeder-circuit at noncontinuous operation?

Step 1: Finding VA
215.2(A)(1); 215.3
VA = VA x 100%
VA = 34,500 VA x 100%
VA = 34,500

Solution: The total VA for the feeder-circuit is 34,500 VA.

CONTINUOUS OPERATED LOADS
215.2(A)(1); 215.3

Continuous operated loads are classified as loads that operated for a period of three hours or more. Continuous loads do not operate at varying or intermittent operation. A continuous operated load shall be supplied continuously for a period of three hours or more without the circuit being interrupted. An industrial processing machine used in a facility to perform a work task for a work day of 8 hours falls under such use and classification.

The total VA rating for sizing conductors to such a machine shall be obtained by multiplying the continuous load by 125 percent. Feeder-circuit conductors shall be increased in size due to voltage drop, ambient temperature or too many current-carrying conductors in a raceway or cable.

For example: What is the VA rating for a continuous operated processing machine with a connected load of 34,500 VA ?

Step 1: Finding VA
215.2(A)(1); 215.3
VA = VA x 125%
VA = 34,500 VA x 125%
VA = 43,125 VA

Solution: The total VA for the feeder-circuit is 43,125 VA.

See Figure 15-3 for calculating the load in amps based upon continuous or noncontinuous operation.

DEMAND FACTORS
ARTICLE 220, PART II

Demand factors shall be permitted to be applied to the VA rating for specific loads, depending on their conditions of use. The percentages for which demand factors shall be permitted to be applied are listed in the NEC. Feeder-circuits supplying loads under specific condition of use shall be permitted to be reduced by applying demand factors to obtain the total rating. Loads shall be permitted to be reduced by applying a demand factor are as follows:

Types of loads per NEC:
- Lighting - **Table 220.42**
- Receptacles - **Table 220.44; 220.14(I)**
- Dryers - **Table 220.54**
- Ranges - **Table 220.55**

- Kitchen equipment — **Table 220.56**
- Dwelling units — **220.82(B); (C)**
- Existing dwelling units — **Table 220.83**
- Multifamily dwelling units — **Table 220.84**
- Schools — **Table 220.86**
- Existing installations — **220.87**
- Restaurants — **Table 220.88**
- Farms — **Table 220.102; Table 220.103**

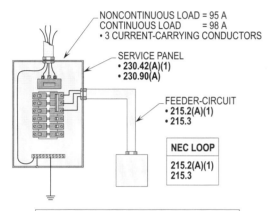

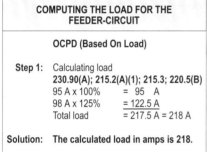

Figure 15-3. Calculating the load in amps for a feeder-circuit based upon continuous and noncontinuous operation.

APPLYING DEMAND FACTORS FOR RECEPTACLE LOADS
220.44; TABLE 220.44; TABLE 220.42

The VA rating for receptacle loads shall be calculated at 100 percent for the first 10,000 VA and the remaining VA shall be calculated at 50 percent per **Table 220.44**. **Table 220.44** shall only be permitted to be applied to receptacles that are used at noncontinuous operation. The demand factors of **Table 220.42** shall be permitted to be used determine the load for receptacle outlets that are figured at 180 VA and then added to the lighting load.

For example: What is the load for a number of general-purpose receptacles used at noncontinuous operation that are utilized to supply a connected load of 24,800 VA?

Step 1: Finding VA
Table 220.44
First 10,000 VA at 100%
10,000 VA x 100% = 10,000 VA
Remaining VA at 50%
14,800 x 50% = 7,400 VA
Total load = 17,400 VA

Solution: The total VA load is 17,400 VA.

See **Figure 15-4** for calculating the load in amps for a feeder-circuit with loads having demand factors.

GROUNDED (NEUTRAL) CONDUCTOR
ARTICLE 220, PART III; ARTICLE 310

The grounded (neutral) conductor shall be intentionally grounded at the service equipment. The grounded (neutral) conductor shall be sized to carry the maximum unbalanced current. The largest load between the grounded (neutral) conductor and any one ungrounded (phase) conductor is the maximum unbalanced current that shall be carried. A demand factor shall be permitted be applied to the grounded (neutral) conductor, under specific conditions of use.

SIZING
220.61(A) THRU (C)

The feeder grounded (neutral) conductor load shall be the maximum unbalanced load connected between the grounded (neutral) conductor and any one ungrounded (phase) conductor. The first 200 amps of neutral current shall be calculated at 100 percent. All resistive loads on the grounded (neutral) conductor exceeding 200 amps shall be permitted to have a demand factor of 70 percent applied and this value added to the first 200 amps taken at 100 percent. All inductive neutral current shall be calculated at 100 percent with no demand factor applied. The feeder grounded (neutral) conductor load shall be 70 percent of the demand load for cooking equipment or a dryer load which, for example, are installed in dwelling units. **(See Figure 15-5)**

The neutral current in amps for a three-wire, two-phase or five-wire, two-phase system shall be multiplied by 140 percent which is derived by taking the square root of 2. By multiplying the grounded (neutral) conductor amps by 140 percent ($\sqrt{2}$ = 141%), the unbalanced current in the grounded (neutral) conductor will be collected from all ungrounded (phase) conductors. The grounded (neutral) conductor shall be multiplied by 140 percent instead of 141 percent (rounded down) based on the ungrounded (phase) conductor with the highest ampacity per **220.61(A), Ex**. Basically, if calculated using this procedure, the grounded (neutral) conductors are not overloaded because 120 volt loads being switched in and out on the circuits at different intervals of time.

For example: What is the grounded (neutral) conductor load for a five-wire, two-phase feeder-circuit with each phase rated at 150 amps?

Step 1: Finding amperage
220.61(A), Ex.
150 A x 140% = 210 A

Solution: The grounded (neutral) conductor load is 210 amps.

For example: What is the neutral load for a feeder-circuit with an inductive load of 400 amps?

Step 1: Calculating neutral load
220.61(C); 310.15(B)(4)(c)
First 200 A x 100% = 200 A
Remaining 200 A x 100% = 200 A
Total load = 400 A

Solution: The total inductive neutral load is 400 amps.

For example: What is the neutral load for a feeder-circuit with a resistive load of 400 amps?

Step 1: Calculating neutral load
220.61(B); 310.15(B)(4)(c)
First 200 A x 100% = 200 A
Remaining 200 A x 70% = 140 A
Total load = 340 A

Solution: The total resistive neutral load is 340 amps.

For example: What is the neutral load for a 9 kW range?

Step 1: Finding demand
Table 220.54, Col. C
9 kW = 8 kW

Step 2: Calculating neutral load
220.61(B)
8 kW x 70% = 5.6 kW

Solution: The total neutral range load is 5.6 kW.

SIZING
NEC 220.61(A) - (C)

Figure 15-5. Calculating the feeder grounded (neutral) conductor load for an inductive, resistive and range load.

What is the demand load for the 82 kW cooking equipment load?

Finding demand load

Step 1: Calculating percentage
Table 220.56
17 pieces allows 65%

Step 2: Applying demand factors
Table 220.56: 220.56
82 kW x 65% = 53.3 kW

Solution: The demand load is 53.3 kV.

Note: 53.3 kW is greater than the two largest loads of 18 kVA.

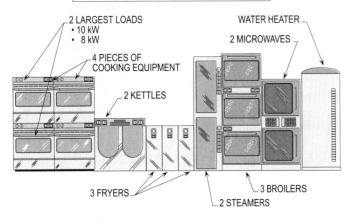

TOTAL KITCHEN EQUIPMENT LOAD IS 82 kW

NEC 220.56
TABLE 220.56

Figure 15-4. Calculating the load in amps for a feeder-circuit with 17 pieces of commercial cooking equipment.

UTILIZING THE GROUNDED (NEUTRAL) CONDUCTOR 310.15(B)(4)

Section **310.15(B)(4)** is divided into three main Subdivisions to explain the loading conditions and use of the grounded (neutral) conductors. The following are three Subdivisions to be used to derive loading conditions based on use as follows:

• Subdivision (a) - unbalanced current
• Subdivision (b) - common conductor carries
• Subdivision (c) - to determine if current-carrying

Feeder-circuits

SUBDIVISION (a)
310.15(B)(4)(a)

The grounded (neutral) conductor shall not be considered a current-carrying conductor when carrying only the unbalanced current from other ungrounded (phase) conductors. When circuits are properly balanced, the grounded (neutral) conductor carries very little current.

When sizing the load for a two-wire circuit, the grounded (neutral) conductor carries the same amount of current as the ungrounded (phase) conductor. This type of installation has no unbalanced load, therefore the grounded (neutral) conductor carries full current.

For example: What is the grounded (neutral) conductor load for a single-phase, 120 volt, two-wire circuit supplying a load of 14 amps?

Step 1: Finding amperage
220.61; 310.15(B)(4)(a)
Ungrounded (phase) conductor = 14 A
Grounded (neutral) conductor = 14 A

Solution: The grounded (neutral) conductor carries a load of 14 amps.

When sizing the load for a three-wire circuit, the grounded (neutral) conductor carries the unbalanced load of the two ungrounded (phase) conductors. This type of installation has an unbalance load unless both ungrounded (phase) conductors pull the same amount of current on each ungrounded (phase) conductor.

For example: What is the unbalanced grounded (neutral) conductor load for a three-wire circuit carrying 64 amps and 52 amps on the ungrounded (phase) conductors?

Unbalanced condition

Step 1: Finding amperage
220.61, 310.15(B)(4)(a)
Ungrounded (phase) conductor
Phase A   = 64 A
Ungrounded (phase) conductor
Phase B   = - 52 A
Total   = 12 A

Balanced condition

Step 1: Finding amperage
220.61; 310.15(B)(4)(a)
Ungrounded (phase) conductor
Phase A   = 64 A
Ungrounded (phase) conductor
Phase B   = - 64 A
Total   = 0 A

Solution: The grounded (neutral) conductor load is 12 amps for the unbalanced condition and zero amps for the balanced condition.

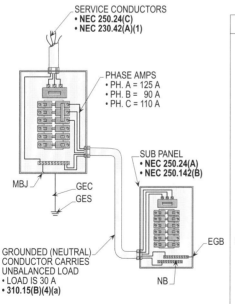

FINDING GROUNDED (NEUTRAL) CONDUCTOR LOAD

Step 1: Calculating grounded (neutral) conductor load
220.61; 310.15(B)(4)(a)

$$A = \sqrt{PH\,A^2 + PH\,B^2 + PH\,C^2 - (PH\,A \times PH\,B) - (PH\,B \times PH\,C) - (PH\,C \times PH\,A)}$$

$$A = \sqrt{125^2 + 90^2 + 110^2 - (125\,A \times 90\,A) + (90\,A \times 110\,A) + (110\,A \times 125\,A)}$$

$$A = \sqrt{15{,}625\,A + 8{,}100\,A + 12{,}100\,A - 11{,}250\,A + 9{,}900\,A + 13{,}750\,A}$$

$$A = \sqrt{35{,}825\,A - 34{,}900\,A}$$

$$A = \sqrt{925\,A}$$

$$A = 30\,A$$

Solution: The grounded (neutral) conductor load is 30 amps.

Note: If phases A and B were lost, the grounded (neutral) conductor current would be 110 amps and if Phase C were lost the grounded (neutral) conductor current would be 35 amps
(PH A = 125 A - PH B = 90 A = 35 A).

**CALCULATING NEUTRAL CURRENT
FOR THREE-PHASE CIRCUITS**

Figure 15-6. Calculating grounded (neutral) conductor current for three-phase circuits.

15-7

CALCULATING NEUTRAL CURRENT FOR THREE-PHASE CIRCUITS

A specific formula shall be used to calculate the grounded (neutral) conductor current for three-phase feeder-circuits. Where currents on Phases A, B and C are of different values, the grounded (neutral) conductor current shall be computed as shown in **Figure 15-6**.

SUBDIVISION (b)
310.15(B)(4)(b)

The grounded (neutral) conductor of a three-wire, 120/208 volt feeder-circuit shall be the same size as the ungrounded (phase) conductors for a feeder-circuit derived from a four-wire, 120/208 volt system. The reason for this is the grounded (neutral) conductor of a three-wire circuit consisting of two-phase conductors. A three-wire system carries approximately the same amount of current as the ungrounded (phase) condcutor. Therefore, a reduction in ampacity is not allowed.

For example: What is the grounded (neutral) conductor load for a 120/208 volt, single-phase circuit taken from a four-wire wye, three-phase system with 190 amps on phase A, 170 amps on phase B and 90 amps for the neutral?

Step 1: Finding ampacity of neutral
220.61; 310.15(B)(4)(b)
Ungrounded (phase) conductor
A = 190 A
Ungrounded (phase) conductor
B = 170 A
Grounded (neutral) conductor
N = 190 A

Solution: The grounded (neutral) conductor load is 190 amps based on the largest ungrounded (phase) conductor.

SUBDIVISION (c)
310.15(B)(4)(c)

The grounded (neutral) conductor for a four-wire, three-phase system supplying nonlinear loads shall be the same size as the ungrounded (phase) conductors. The grounded (neutral) conductor shall be considered a current-carrying conductor due to the harmonic currents generated by these loads.

For example, harmonic loads generated by ballasts are approximately 25 percent per phase in a four-wire wye, three-phase system. Such harmonic currents are collected by the grounded (neutral) conductor from the ungrounded (phase) conductors in a four-wire system because they are all in phase. Because these loads produce third harmonics, harmonic currents present in the grounded (neutral) conductor add up to 75 percent which is derived from 25 percent of such current on each ungrounded (phase) conductor. Under this condition of use, the grounded (neutral) conductor shall not be permitted to be reduced by 70 percent for current from 120 volt loads exceeding 200 amps per **220.61(C)**.

HARMONIC CURRENT
310.10, FPN (2)

Harmonic currents usually appear in power distribution equipment that supplies nonlinear loads. There are two basic types of nonlinear loads - single-phase and three phase. Single-phase nonlinear loads are found in offices, while three-phase nonlinear loads are dominant in industrial plants. However, there are other nonlinear loads that present problems to the electrical system because of their harmonic content.

EFFECTS ON NEUTRAL
220.61(C)(2), FPN 2

In electrical systems with many single-phase nonlinear loads, the grounded (neutral) conductor current may actually exceed the ungrounded (phase) current. Excessive overheating may occur because there is no overcurrent protection device in the grounded (neutral) conductor to limit the current to a safe value.

In a four-wire system with single-phase nonlinear loads, there are certain odd-numbered harmonics called triplens. These odd multiples of harmonic current are known as the 3rd, 9th, 15th, etc. and they do not cancel, but add together in the grounded (neutral) conductor of a wye connected circuit.

Design Tip: Under normal conditions of use for a balanced linear load, the fundamental 60 Hz portion of the ungrounded (phase) conductor currents will cancel in the grounded (neutral) conductor. Therefore, there is not an excessive heating problem in the grounded (neutral) conductor.

SIZING GROUNDED (NEUTRAL) CONDUCTOR ELEMENTS

Grounded (neutral) conductors, busbars and connecting lugs shall be sized to carry the full rating of ungrounded (phase) conductor currents. Such elements easily become overloaded due to grounded (neutral) conductors being overloaded with the triplen harmonics.

> **Design Tip:** Triplen harmonics are in addition to the normal grounded (neutral) conductor current. (See **FPN 2 to 220.61(C)(2)**)

DEMAND FACTORS
220.61(B)

A demand factor of 70 percent shall be applied to all grounded (neutral) conductor loads exceeding 200 amps for nonlinear loads. Nonlinear related loads shall be computed at 100 percent.

> **For example:** What is the load for the grounded (neutral) conductor if it exceeds 200 amps and has more than 50 percent of its load harmonically related? The ungrounded (phase) conductors are carrying a total grounded (neutral) conductor load of 275 amps respectively.
>
> Step 1: Finding amperage
> 310.15(B)(4)(c)
> Ungrounded (phase) conductors
> • 275 amps
>
> Step 2: Calculating amperage
> 220.61(C)
> First 200 A x 100%   = 200 A
> Next 75 A x 100%   = 75 A
> Total   = 275 A
>
> **Solution: The grounded (neutral) conductor load is 275 amps.**

The grounded (neutral) conductor shall be considered a current-carrying conductor because of the harmonic currents generated by these loads and **310.15(B)(2)(a)** shall be applied for four or more current-carrying conductors in a conduit, cable, etc.

> **For example:** What is the grounded (neutral) conductor load for 120 volt loads having harmonic currents of 400 amps per phase?
>
> Step 1: Finding amperage
> 310.15(B)(4)(c)
> Ungrounded (phase) conductors
> • 400 A
>
> Step 2: Calculating amperage
> 220.61(C)
> 400 A x 100% = 400 A
>
> **Solution: The grounded (neutral) conductor load is 400 amps. Note: No reduction of ampacity shall be permitted due to harmonic currents.**

VOLTAGE DROP
ARTICLE 215

Conductors are sometimes increased in size to prevent excessive voltage drop (VD) due to long runs between the OCPD's and the load served. Due to the long runs of feeder-circuit conductors, they shall be increased in size to compensate for poor voltage drop. The voltage drop on the feeder conductors should not exceed 2 to 3 percent at the farthest outlet supplying power to the loads. The voltage drop on the feeder-circuit conductors should not exceed 5 percent overall.

SINGLE-PHASE CIRCUITS
215.2(A)(3), FPN 2

The following formula shall be applied when calculating the voltage drop in a two-wire or three-wire, single-phase or three-phase feeder-circuit:

Resistive method
• $VD = 2 \times R \times L \times I \div 1000$
Circular-mill method
• $VD = 2 \times R \times L \times I \div CM$

The following values shall be used when calculating the voltage drop:

VD = voltage drop
R = resistivity for conductor material
Use Chapter 9, Table 8 Columns 6 or 8 (uncoated ohm/MFT) or use 12 for cu. and 18 for alu.
L = one-way length of circuit conductor in feet

I = current in conductor in amperage
CM = conductor area in circular mils
See **Chapter 9, Table 8**
1000 = length of conductors based on **Table 8, Chapter 9**

Conductors routed at greater lengths will have the resistance of each conductor increased which will oppose the flow of current. By running larger conductors, the diameter of each conductor is increased in size, creating a greater path for the flow of current and less opposition to the movement of electrons which keeps the voltage high at the conductor end.

As listed above, there are two methods that are used by designers and installers to calculate voltage drop in a feeder-circuit. One is the resistivity concept which is considered the most accurate. The second is the circular mil (CM) method that has been used for many years. The resistivity method is utilized by selecting the resistivity of the conductors. The CM method consists of selecting the CM rating of the conductors from **Table 8, Ch. 9** in the NEC. These values, whichever chosen, is inserted into the formula with other data and the voltage drop computed. **(See Figures 15-7(a) and (b))**

THREE-PHASE CIRCUITS 215.2(A)(3), FPN 2

The voltage drop for three-phase circuits shall be calculated by multiplying the voltage drop by .866. The same formula for finding voltage drop in single-phase circuits shall be used to determine the voltage drop in three-phase circuits. Voltage drop multiplied by .866 is found by dividing $\sqrt{3}$ by 2 (1.732 ÷ 2 = .866). The .866 is basically produced from the additional conductor (third conductor) which is derived from a three-phase system instead of a two-phase system. **(See Figure 15-8)**

FINDING THE SIZE EQUIPMENT GROUNDING CONDUCTOR 250.122(B)

Where current-carrying conductors have to be larger to compensate for voltage drop, the equipment grounding conductors shall be adjusted proportionally to the circular-mil area.

For example: Using the resistivity method to determine the voltage drop and size elements of a single-phase feeder-circuit having the following characteristics:

- VD is held to 3%
- Voltage is 240
- Length is 300 ft.
- Load is 180 amps
- OCPD is 200 amps
- Conductors are 3/0 AWG THWN copper

Finding the VD using the resistivty method

Step 1: Selecting percentage
215.2(A)(3), FPN 2
Feeder-circuit = 3%

Step 2: Calculating VD
215.2(A)(3), FPN 2; Table 8, Ch. 9
VD = 2 x R x L x I ÷ 1000
VD = 2 x .0766 x 300 x 180 ÷ 1000
VD = 8.2728

Step 3: Calculating allowable VD
VD = supply V x 3%
VD = 240 V x 3%
VD = 7.2 V

Step 4: Checking percentage
215.2(A)(3), FPN 2
% = VD ÷ V
% = 8.2728 ÷ 240
% = .03447 or 3.447

Solution: The voltage drop rating of 8.2728 is greater than 7.2 V and a larger conductor shall be used to reduce the 3.447% to 3% or less.

Lowering VD using larger 4/0 AWG conductor

Step 1: Selecting percentage
215.2(A)(3), FPN 2
Feeder-circuit = 3%

Step 2: Calculating VD
215.2(A)(3), FPN 2; Table 8, Ch. 9
VD = 2 x R x L x I ÷ 1000
VD = 2 x .0608 x 300 x 180 ÷ 1000
VD = 6.5664

Step 3: Checking percentage
215.2(A)(3), FPN 2
% = 6.5664 V ÷ 240 V
% = .02736 or 2.736

Solution: The voltage drop rating of 6.5664 volts is less than 7.2 volts, which is well below the 3% limit. The 4/0 awg conductors are large enough to reduce the voltage drop to 3% or less.

FIGURE LOOP: FOR SIZING GROUNDED (NEUTRAL) CONDUCTOR AND EGC TO CORRECT VD, SEE FIGURES 9-35 AND 15-9.

SINGLE-PHASE CIRCUITS NEC 215.2(A)(3), FPN 2

Figure 15-7(a). Calculating the VD using the resistivity method and using larger conductor.

Feeder-circuits

Design Tip: If a single equipment grounding conductor is run with multiple circuits in the same raceway, it shall be sized for the largest overcurrent device protecting the conductors in the raceway or cable.

See **Figure 15-9** for a detailed procedure for calculating and selecting the size equipment grounding conductor to be installed.

FINDING THE SIZE GROUNDED (NEUTRAL) CONDUCTOR 240.23

Where a change in the size of the ungrounded (phase) conductors occur due to voltage drop, a similar change may have to be made in the size of the grounded (neutral) conductor.

For example, if the ungrounded (phase) conductor supplying a 120 volt load of 175 amps were increased from 2/0 AWG THWN copper to 3/0 AWG THWN copper due to voltage drop problems, the grounded (neutral) conductor would have to be increased from 2/0 AWG copper to 3/0 AWG copper also.

SIZING CONDUCTORS BASED ON VD 215.2(A)(3), FPN 2

Sometimes, it is desirable to size and select the conductors based on the percentage of VD needed for the circuit to operate properly. Such conductors may be found by dividing R (12 for cu. and 18 for alu.) x L (length) x I (amps) by VD percent.

These rules and examples are also used for branch-circuits to determine the amount of voltage drop where there are excessive long runs of conductors between the protection devices and load served. (**See Figure 15-10**)

For example: Using the CM method to determine the voltage drop and size elements of a single-phase feeder-circuit having the following characteristics:

- VD is held to 3%
- Voltage is 240
- Length is 300 ft.
- Load is 180 amps
- OCPD is 200 amps
- Conductors are 3/0 AWG THWN copper

Finding the VD using the CM method

Step 1: Selecting percentage
215.2(A)(3), FPN 2
Feeder-circuit = 3%

Step 2: Calculating VD
215.2(A)(3), FPN 2; Table 8, Ch. 9
VD = 2 x R x L x I ÷ CM
VD = 2 x 12 x 300 x 180 ÷ 167,800
VD = 7.7235

Step 3: Calculating allowable VD
VD = supply V x 3%
VD = 240 V x 3%
VD = 7.2 V

Step 4: Checking percentage
215.2(A)(3), FPN 2
% = VD ÷ V
% = 7.7235 ÷ 240
% = .03218 or 3.218

Solution: The voltage drop rating of 7.7235 is greater than 7.2 V and a larger conductor shall be used to reduce the 3.218% to 3% or less.

Lowering VD using larger 4/0 AWG conductor

Step 1: Selecting percentage
215.2(A)(3), FPN 2
Feeder-circuit = 3%

Step 2: Calculating VD
215.2(A)(3), FPN 2; Table 8, Ch. 9
VD = 2 x R x L x I ÷ 211,600
VD = 2 x 12 x 300 x 180 ÷ 211,600
VD = 6.125

Step 3: Checking percentage
215.2(A)(3), FPN 2
% = 6.125 V ÷ 240 V
% = .0255 or 2.55

Solution: The voltage drop rating of 6.125 volts is less than 7.2 volts, which is well below the 3% limit. The 4/0 AWG conductors are large enough to reduce the voltage drop to 3% or less.

FIGURE LOOP: FOR SIZING GROUNDED (NEUTRAL) CONDUCTOR AND EGC TO CORRECT VD, SEE FIGURES 9-35 AND 15-9. FOR CALCULATING VD IN 3Ø CIRCUITS, SEE FIGURE 15-8

SINGLE-PHASE CIRCUITS
NEC 215.2(A)(3), FPN 2

Figure 15-7(b). Calculating the VD using the CM method and using larger conductor.

For example: Using the resistivity method to determine the voltage drop and size elements of a three-phase feeder-circuit having the following characteristics:

- VD is held to 3%
- Voltage is 480/277
- Length is 700 ft.
- Load is 180 amps
- OCPD is 200 amps
- Conductors are 4/0 AWG THWN copper

Finding the VD using the resistivity method

Step 1: Selecting percentage
215.2(A)(3), FPN 2
Feeder-circuit = 3%

Step 2: Calculating VD
215.2(A)(3), FPN 2; Table 8, Ch. 9
VD = 2 x R x L x I ÷ 1000
VD = 2 x .0608 x 700 x 180 ÷ 1000
VD = 15.3216

Step 3: Applying .866 for three-phase
VD = VD x .866
VD = 15.3216 x .866
VD = 13.269

Step 4: Calculating allowable VD
VD = supply V x 3%
VD = 480 V x 3%
VD = 14.4 V

Step 5: Checking percentage
215.2(A)(3), FPN 2
% = VD ÷ V
% = 13.269 ÷ 480
% = .0276 or 2.76

Solution: The voltage drop rating of 13.269 is less than 14.4 V, therefore the 4/0 AWG THWN copper shall be used.

THREE-PHASE CIRCUITS
NEC 215.2(A)(3), FPN 2

Figure 15-8. Calculating the VD using the resistivity method for a three-phase feeder-circuit.

For example: Using the percentage of VD, to determine the size conductors needed to correct the VD problem for a single-phase, feeder-circuit having the following characteristics:

- VD is held to 3%
- Voltage is 240
- Length is 125 ft.
- Load is 82.5 amps
- Conductors are THWN copper

Finding the size conductors based on VD

Step 1: Selecting VD percentage
215.2(A)(3), FPN 2
Feeder-circuit = 3%

Step 2: Calculating CM
215.2(A)(3), FPN 2; Table 8, Ch. 9
CM = 2 x K x L x I ÷ VD
CM = 2 x 12 x 125' x 82.5 ÷ 240 x 3%
CM = 34,375

Solution: The CM rating of 34,375 requires 4 AWG THWN copper conductors.

Note: When the circuit is three-phase, the voltage drop shall be multiplied by .866 to derive the total voltage drop.

SIZING CONDUCTORS BASED ON VD
NEC 215.2(A)(3), FPN 2

Figure 15-10. Calculating the size conductors based on VD to size conductors needed to correct the VD problem for a feeder-circuit.

See Figure 15-11 for calculation procedures pertaining to a feeder-circuit with combination loads.

Design Tip: For more detailed rules on sizing taps, sizing conductors and overcurrent protection schemes, see the appropriate Chapter in this book pertaining to such subjects. Also, see the calculation chapters for a variety of calculations pertaining to various types of buildings and loads.

CALCULATION OF A FEEDER-CIRCUIT
215.2(A)(1); 215.3

A feeder-circuit may consist of loads such as lighting, receptacles, appliances, heating, air-conditioning and motors. Each load shall be evaluated and calculated at continuous or noncontinuous operation.

If loads with demand factors are present, they shall be separated into individual loads and demand factors applied based on their condition of use in the electrical system.

OPTION CALCULATION FOR A FEEDER-CIRCUIT
220.87, Ex.

If the maximum demand data for a one-year period is not available, the calculated load shall be permitted to be based on the maximum demand (measure of average power

Feeder-circuits

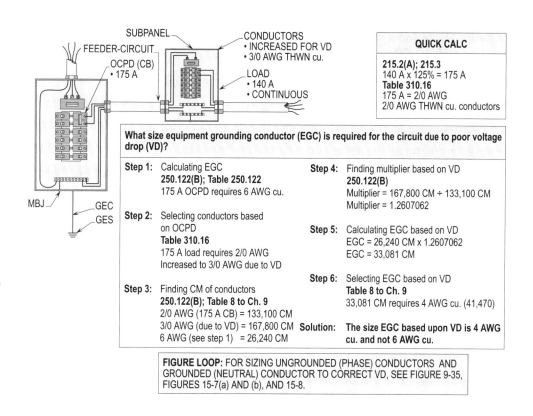

FINDING THE SIZE EGC
NEC 250.122(B)

Figure 15-9. Calculating the size equipment grounding conductor based upon poor VD in the feeder-circuit due to excessive length of conductors.

demand over a 15 minute period) continuously recorded over a minimum 30-day period using a recording ammeter or power meter connected to the highest loaded ungrounded (phase) conductor of the feeder or service, based on the initial loading at the start of the recording. The recording shall reflect the maximum demand of the feeder or service being taken when the building or space is occupied and shall include by measurement or calculation, the larger of the heating or cooling equipment load and other loads that may be periodic in nature due to seasonal or similar conditions.

Design Tip: The existing demand at 125 percent plus the new load calculated at continuous and noncontinuous operation shall not be permitted to exceed the ampacity of the feeder-circuit rating. The feeder-circuit shall have overcurrent protection provided as required by **240.4** and **230.90**. (See Figure 15-12)

OPTIONAL CALCULATION FOR A SERVICE
220.87

The optional calculation shall be permitted to be used to determine if additional loads may be added to existing feeder or service conductors. This rule allows the actual maximum demand figures to be utilized when the following conditions are complied with.

- The maximum demand data is available for one year.

- The existing demand at 125 percent plus the new load does not exceed the ampacity of such service.

- OCPD's comply with **240.4** and **230.90** respectfully.

See Figure 15-13 for a detailed illustration for applying such a procedure.

Stallcup's Electrical Design Book

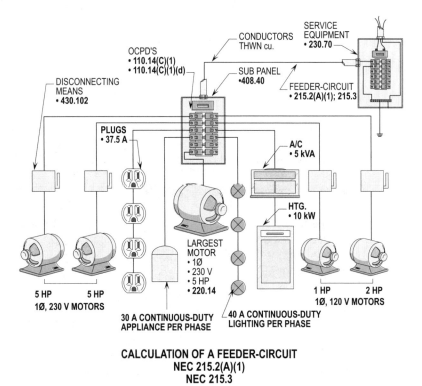

SIZING FEEDER-CIRCUIT OCPD BASED ON AMPACITY OF CONDUCTORS	

Step 1:	Calculating loads
	215.2(A)(1); Table 430.248; 220.50; 430.63
	• Lighting load
	   40 A x 125% = 50 A
	• Receptacle load
	   37.5 A x 100% = 37.5 A
	• Appliance load
	   30 A x 125% = 37.5 A
	• Heat or A/C load
	   10 kW x 1000 x
	   100% ÷ 240 V = 42 A
	• Motor load; Table 430.248
	   5 HP = 28 A x 100% = 28 A
	   5 HP = 28 A x 100% = 28 A
	   1 HP = 16 A x 100% = 16 A
	   2 HP = 24 A x 100% = 24 A
	• Largest motor load
	   28 A x 25% = 7 A
	Total load = 270 A

Step 2:	Selecting conductors 310.10, FPN (2); Table 310.16 270 A requires 300 AWG KCMIL THWN cu.
Step 3:	Selecting OCPD based on conductors 215.3; 240.4(B); 430.63; 240.6(A) 300 KCMIL THWN cu. = 285 A 285 A = 300 A OCPD
Solution:	The size OCPD is allowed to be a 300 amp CB, based upon amps of conductors.
Note:	If the largest OCPD for any motor of the group is calculated and added to the remaining loads, the same OCPD is produced as in Step 3 above.

SIZING FEEDER-CIRCUIT OCPD BASED ON LARGEST OCPD FOR ANY MOTOR OF THE GROUP	

Step 1:	Calculating loads
	215.3; Table 430.248
	• Lighting load
	   40 A x 125% = 50 A
	• Receptacle load
	   37.5 A x 100% = 37.5 A
	• Appliance load
	   30 A x 125% = 37.5 A
	• Heat or A/C load
	   10 kW x 1000 x
	   100% ÷ 240 V = 42 A
	• Motor load
	430.62(A); 430.52(C)(1); Table 430.52
	   5 HP = 28 A x 250% = 70 A
	   5 HP = 28 A x 100% = 28 A
	   1 HP = 16 A x 100% = 16 A
	   2 HP = 24 A x 100% = 24 A
	Total load = 305 A

Step 2:	Selecting OCPD 430.62(A); 240.6(A) 300 A is the next standard size below 305 A
Solution:	The size OCPD is a 300 amp.
Note:	The OCPD for the feeder-circuit is the same size (300 A) whether sized from the ampacity of the conductors or largest OCPD for any one motor of the group.

Note: For simplicity, a one line diagram is shown.

CALCULATION OF A FEEDER-CIRCUIT
NEC 215.2(A)(1)
NEC 215.3

Figure 15-11. Calculating a feeder-circuit with loads such as lighting, receptacle, appliances, heating, air-conditioning and motors with continuous and noncontinuous loads.

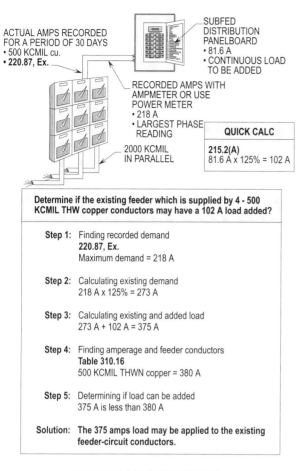

Figure 15-12. Calculating the load in amps for a feeder-circuit using the optional calculation.

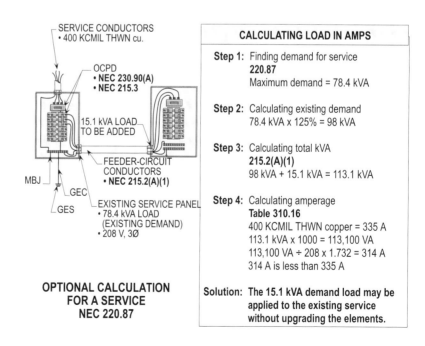

Figure 15-13. The above illustration is a calculation for a service to verify if a load may be added.

Name   Date

Chapter 15. Feeder-circuits

Section	Answer		

_____   T   F   **1.** Continuous loads are classified as loads that operate for a period of two hours or more?

_____   T   F   **2.** All inductive neutral current, even if noncontinuous shall be calculated at 125 percent with no demand factor applied.

_____   T   F   **3.** The neutral conductor shall not be considered current-carrying when carrying only the unbalanced current from other ungrounded phase conductors. (120/240 V)

_____   T   F   **4.** Harmonic currents usually appear in power distribution equipment that supplies non-linear loads.

_____   T   F   **5.** The voltage drop on the feeder-circuit conductors should not exceed 3 percent overall, if the VD on the branch-circuit does not exceed 2 percent.

_____   T   F   **6.** Noncontinuous operated loads shall be calculated at _____ percent or demand factors can be applied to specific loads under certain conditions.

_____   T   F   **7.** The VA rating for receptacle loads shall be calculated at 100 percent for
_____   T   F   the first _____ VA and the remaining VA shall be calculated at _____ percent per **Table 220.13**.

_____   T   F   **8.** The feeder neutral load shall be calculated at 100 percent for the first
_____   T   F   _____ amps and the remaining shall have a demand factor of _____ percent applied.

_____   T   F   **9.** The voltage drop on feeder conductors should not exceed 2 percent to _____ percent at the farthest outlet supplying power to the loads.

_____   T   F   **10.** If the maximum demand data for a one-year period is not available, the calculated load shall be permitted to be based on the maximum demand (measure of average power demand over a 15 minute period) continuously recorded over a minimum _____ day period using a recording ammeter or power meter connected to the highest loaded phase of the feeder or service, based on the initial loading at the start of the recording.

_____ _____   **11.** What is the ampacity for a feeder-circuit load of 20,800 VA which is supplied by a 120/240 volt, single-phase system?

_____ _____   **12.** What is the ampacity for a feeder-circuit load of 20,800 VA which is supplied by a 120/208 volt, three-phase system?

_____ _____   **13.** What is the ampacity for a 240 volt, single-phase feeder-circuit with a 58 amp load?

Section	Answer

14. What is the ampacity for a 240 volt, single-phase feeder circuit with a 58 amp load with a power of 75 percent?

15. What is the ampacity for a 208 volt, three-phase feeder-circuit with a 58 amp load with a power factor of 75 percent?

16. What is the VA rating for a connected load of 32,800 VA to size the elements of the feeder-circuit at noncontinuous operation?

17. What is the VA rating for a for a continuous operated processing machine with a connected load of 32,800 VA?

18. What is the load in VA for a number of general-purpose receptacles used at noncontinuous operation that are utilized to supply a connected load of 28,400 VA?

19. What is the neutral load for a feeder-circuit with a inductive load of 300 amps?

20. What is the neutral load for a feeder-circuit with a resistive load of 300 amps?

21. What is the neutral load in kW for a 10 kW range?

22. What is the voltage drop using the resistivity method with the following characteristics: (Use larger conductor if necessary)

    - VD is held to 3%
    - Voltage is 240
    - Length is 300 ft.
    - Load is 180 amps
    - OCPD is 200 amps
    - Conductors are 3/0 AWG THWN copper

23. What is the voltage drop using the CM method with the following characteristics: (Use larger conductor if necessary)

    - VD is held to 3%
    - Voltage is 240
    - Length is 300 ft.
    - Load is 180 amps
    - OCPD is 200 amps
    - Conductors are 3/0 AWG THWN copper

24. What size feeder-circuit OCPD is required for the following loads based on the ampacity of conductors:

    - Lighting load = 30 A
      (continuous operation)
    - Receptacle load = 35 A
    - Appliance load = 30 A
      (continuous operation)
    - Heating load = 10 kW
    - A/C load = 6 kVA

    Motor loads
    - 5 HP, 230 V, single-phase
    - 5 HP, 230 V, single-phase
    - 2 HP, 120 V, single-phase
    - 1 HP, 120 V, single-phase

25. Determine if the existing service which is supplied by 4 - 400 KCMIL THWN copper conductors can have an 86 amp load added to a service with a 198 amp recorded amperage rating for a period of 30 days?

15-18

16

Receptacle Outlets

The rules for installing receptacle outlets in one and two family dwelling units are more stringent than for commercial or industrial occupancies. Where there is not a proper number of receptacles installed in a dwelling unit, the amount of power required to serve the number of appliances in use today is not available. Therefore, in dwelling units only, a specific number shall be installed. However, this rule does not apply for commercial and industrial locations. They are installed as needed to cord-and-plug electrical items. Extension cords used improperly create a fire hazard. For this reason, the NEC requires outlets in all types of occupancies to be installed in such a manner as to cord-and-plug connect appliance loads without the use of extension cords.

GFCI-protection of 15 or 20 amp, 125 volt receptacle outlets shall be required in specified areas both indoors and outdoors to protect personnel from electric shock while using cord-and-plug connected electric hand tools, radios, TV's, stereos, etc. This rule applies for certain areas in dwelling units, commercial and industrial locations.

AFCI-protection of 15 or 20 amp, 125 volt outlets installed in bedrooms of dwelling units shall be required. Note that a listed AFCI shall protect the entire branch-circuit. Section **210.12(B)** must be reviewed carefully before installing AFCI protection for these outlets in bedrooms.

GROUNDING RECEPTACLES
406.3(A)

Receptacles installed on 15 or 20 amp general purpose branch-circuits shall be the grounding type. They are required to be the three-wire type with a brass terminal for the ungrounded (phase) conductor, silver terminal for the grounded (neutral) conductor and a green terminal for the equipment grounding conductor. The brass terminal is the short slot and represents the connection for the ungrounded (phase) conductor. The U slot is the equipment grounding conductor terminal on the receptacle while the long slot represents the grounded (neutral) conductor terminal. Any color conductor but white, gray, green, green with a yellow stripe or bare shall be permitted to be connected to the ungrounded (phase) brass terminal per **310.12(C)**. Only white or gray shall be connected to the grounded (neutral) silver terminal per **200.6(A), 210.5(A)** and **310.12(A)**. A green or green with yellow stripe conductor shall be the only color permitted to be connected to the green equipment grounding conductor terminal per **250.134(B), 210.5(B), 250.119(A)** and **310.12(B)**. The insulation of the equipment grounding conductor shall be permitted to be stripped per **250.119(A)**. **(See Figure 16-1)**

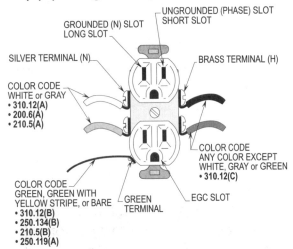

GROUNDING RECEPTACLES
NEC 406.3(A)

Figure 16-1. Identifying the components and correct color coding for receptacles.

Receptacles shall be rated for the voltage and current based upon the size of the general-purpose branch-circuit to which they are connected.

Receptacles shall be permitted to be connected to an individual branch-circuit, a general purpose branch-circuit, a general-purpose branch-circuit with two or more receptacles or a combination of lighting and receptacle outlets.

A receptacle shall be permitted to be connected to an individual branch-circuit, but to do so, it shall be rated not less than the ampacity of the branch-circuit. The size of the OCPD shall be utilized in determining the rating of the branch-circuit.

For example, a 20 amp OCPD (CB or Fuse) classifies the branch-circuit as 20 amp even if the conductors of the circuit are 10 AWG per **210.19(A)(1), 210.20(A)** and **210.23**. Sometimes conductors are increased for voltage drop correction, adjustment or correction factors which creates a larger conductor than normally required. **(See Figure 16-2)**

An individual branch-circuit supplies power to a single load that is cord-and-plug connected to a single receptacle. To determine the size of the load permitted to be connected to the receptacle, refer to **Table 210.21(B)(2)** which lists the maximum cord-and-plug connected load that shall be permitted to be plugged into a receptacle.

For example: What is the maximum size load that may be cord-and-plug connected to a receptacle supplied by a general purpose branch-circuit?

Step 1: Max. load allowed
Table 210.21(B)(2)
Col. 1 Branch-circuit = 15 or 20 A
Col. 2 Receptacle = 15 A
Col. 3 Max. load = 12 A

Solution: Max. load is 12 amp.

Table 210.21(B)(3) lists the amperage rating of a receptacle that shall be permitted be connected to a branch-circuit based on the rating of the OCPD and conductors.

For example: What is the minimum and maximum size receptacle in amps that may be connected to a 20 amp general purpose branch-circuit?

Step 1: Max. size receptacle allowed
Table 210.21(B)(3)
Col. 1 Branch-circuit = 20 A
Col. 2 Receptacle = 15 or 20 A

Solution: Max. size receptacle is 15 or 20 amp.

Design Tip: If the branch-circuit is an individual branch-circuit, only a 20 amp receptacle shall be used. A 15 amp receptacle would not be permitted. A general purpose branch-circuit supplies power to both receptacle outlets and hard-wired appliances.

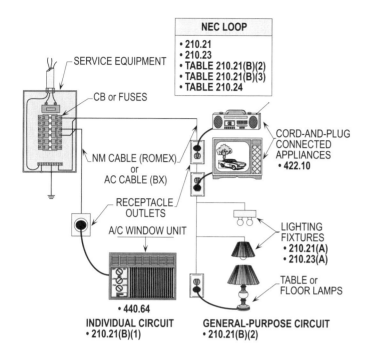

Figure 16-2. An individual branch-circuit has only one outlet and a single load whereas a general-purpose branch-circuit has more than one outlet and load which may be cord-and-plug connected or hard-wired.

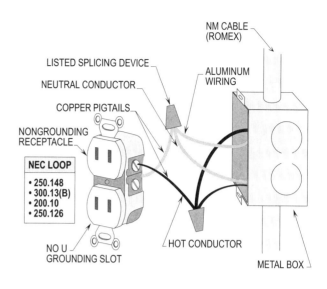

Figure 16-3. Nongrounding receptacles are used where the branch-circuit is not provided with an equipment grounding conductor. If the branch-circuit wiring method is aluminum, a copper pigtail shall be used to terminate the aluminum conductor to the terminals of the receptacle. **Note:** The splicing device shall be listed.

NONGROUNDING RECEPTACLES 406.3(A), Ex.

The general rule requires receptacles of the grounding type to be installed in existing dwelling units. Conductors of copper should be used to connect the terminals of the receptacle to the branch-circuit conductors. Otherwise, copper and aluminum type receptacles shall be installed listed for such use.

Aluminum conductors of a branch-circuit should be pigtailed to a 6 in. (150 mm) copper jumper and connected to the terminals of the receptacle. This is required because the existing terminals of receptacles are not usually rated for aluminum connections. Without certain rules are applied, nongrounding receptacles shall be connected to branch-circuits not having an equipment grounding conductor.

If an existing branch-circuit is wired with an AC cable and the existing receptacle outlet is the nongrounding type, a grounding receptacle shall be used to replace the existing nongrounding receptacle. This is due to AC listed cable being permitted to be used as an equipment grounding means per **406.3(D)** and **250.118(8)**. Only one conductor shall be placed under the bending screw on the terminal of the receptacle unless it is approved for more than one conductor per **110.3(B)** and **110.14(A)**. **(See Figure 16-3)**

REPLACEMENT OF RECEPTACLES 250.130(C)

The basic rule for replacement of receptacles is that grounding type receptacles shall be used to replace existing receptacles. There is no problem in applying this requirement as long as the branch-circuit is equipped with an equipment grounding conductor or grounding means that complies with **250.118**. A nongrounding type receptacle shall be used with a branch-circuit that has no equipment grounding conductor or grounding means. There are two conditions in the NEC that allows a GFCI receptacle to be used. Either a GFCI receptacle shall be permitted to be used to replace and protect a single receptacle outlet or a GFCI feed through receptacle shall be permitted to be used to protect additional outlets downstream. The receptacle outlets downstream shall be permitted to be of the grounding type but shall be identified as being protected by a GFCI circuit with no ground. **Section 250.130(C)** permits an equipment grounding conductor sized from the OCPD of the branch-circuit per **Table 250.122** to be utilized. The equipment grounding conductor shall connect the green grounding terminal of the receptacle to the closest metal water pipe as required by **250.104(A), 250.52(A)(1)** and **250.53(D)**. The equipment grounding conductor shall not be permitted to be connected to a driven rod.

> **Design Tip:** Check with the inspector for his or her interpretation of the point of connection to the metal water pipe. The authority having jurisdiction may require the connection of the equipment grounding conductor to be made within 5 ft. (1.52 m) of where the metal water pipe enters the dwelling unit per **250.52(A)(1)**. However, with the metal water pipe in the earth and entering the premises from T's from more than one point, the AHJ may allow connection to anyone of such entries.

The NEC allows five methods by which a nongrounding receptacle shall be permitted to be replaced and they are as follows:

- A nongrounding receptacle.
- A GFCI receptacle protecting a single outlet.
- A GFCI receptacle protecting a single outlet and additional outlets downstream.
- An equipment grounding conductor routed from the receptacle to a metal water pipe per **250.52(A)(1)**.
- A GFCI CB protecting all outlets.

See Figure 16-4 for a detailed illustration on methods used for replacing nongrounding receptacles on branch-circuits without an equipment grounding conductor.

NUMBER ON A CIRCUIT
210.11(A)

Section 210.11(A) is used to determine the number of outlets permitted on a branch-circuit. **Section 210.11** refers to **220.12** and **Table 220.12** which requires 3 VA per sq. ft. for a dwelling unit to determine the total VA to select the number of branch-circuits.

For example, a 3000 sq. ft. dwelling unit has a total of 9000 VA (3000 sq. ft. x 3 VA = 9000 VA) to supply general-purpose receptacle outlets. Using this concept, there are 4 - 120 volt, 20 amp branch-circuits provided to supply general-purpose receptacle outlets. (See below)

To determine the number of outlets permitted on a branch-circuit, the OCPD shall be multiplied by 120 volts per **210.11(B)** and divided by 3 VA per **Table 220.12**. This number may be verified as follows:

\# = 3,000 sq. ft. x 3 VA = 9000 VA ÷ 20 A OCPD x 120 V
\# = 3.75
3.75 requires 4 - 20 amp branch-circuits

> **For example:** How many receptacle outlets in a dwelling unit may be connected to a 20 amp general purpose branch-circuit?
>
> **Step 1:** Finding VA
> **210.11(B)**
> 20 A OCPD x 120 V = 2400 VA
>
> **Step 2:** Finding Sq. Ft.
> **Table 220.12**
> 2400 VA ÷ 3 VA = 800 Sq. Ft.
>
> **Step 3:** Finding number of outlets allowed
> **220.14(J); 220.12; 210.11(B)**
> Allows any number to be installed
>
> **Solution:** There is no limit to the number of outlets permitted in the 800 sq. ft. area. Note that local codes may limit the number of outlets that can be connected to a branch-circuit.

In commercial and industrial locations, the number of receptacle outlets permitted on a general-purpose branch-circuit are computed differently than those for dwelling units.

> **For example:** In a commercial or industrial area, the limited number of thirteen outlets on a general-purpose branch-circuit is based on the following procedure.
>
> **Step 1:** Finding amperage of outlets
> **220.14(J)**
> 180 VA ÷ 120 V = 1.5 A
>
> **Step 2:** Finding number of outlets
> **220.14(J)**
> 20 A OCPD ÷ 1.5 A = 13
> A 20 A circuit is limited to 13
>
> **Solution:** The number of outlets permitted on a 20 amp branch-circuit is 13.

> **Design Tip:** Using the same procedure, 10 outlets shall be permitted on a 15 amp OCPD. The limitation of outlets presents nuisance tripping of the OCPD.

Municipal Electrical Boards may amend the NEC and allow different numbers of outlets on a 15 or 20 amp branch-circuit. The two methods listed in determining the number of receptacle outlets on a branch-circuit in dwellings, commercial and industrial locations are the most common methods used. **(See Figure 16-5)**

Receptacle Outlets

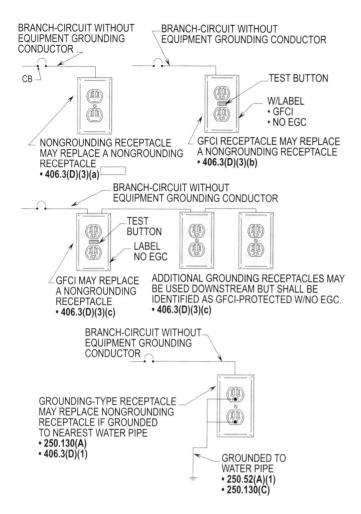

Figure 16-4. Methods for replacing nongrounding receptacles on branch-circuits without an equipment grounding conductor.

DWELLING UNIT LOCATIONS 210.52(A) THRU (H)

The provision for installing the required number of receptacle outlets in dwelling units are listed in **210.52(A) through (H)**. The basic rule requires receptacle outlets to be installed where flexible cords are used to connect electrical appliances and equipment. This Section, with **Subdivisions (A) through (H)**, lists the exact location and number to be installed. The number of receptacle outlets required are different based upon the location either inside or outside the dwelling unit.

Design Tip: Commercial and industrial installations of receptacle outlets shall be permitted to be installed either inside or outside and located where needed. Note that there are not a specific number required.

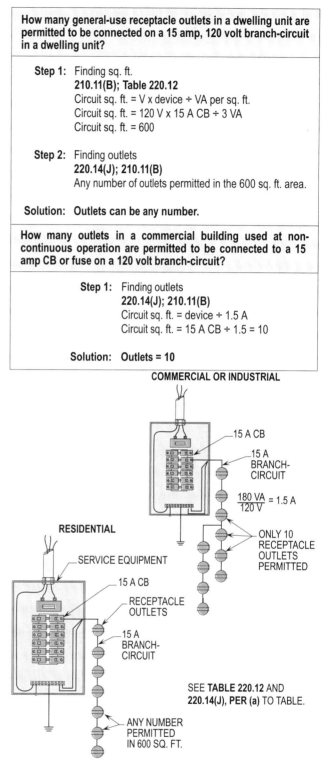

Figure 16-5. For determining the number of receptacle outlets on a 15 amp branch-circuit using the any number permitted method or the limited method is at the discretion of the AHJ. **Note:** The number per circuit may vary from city to city or state to state based upon the electrical ordinance of each individual area.

16-5

WALL RECEPTACLE OUTLETS
210.52(A)(1); (A)(2)

Section 210.52(A) lists the requirements for installing receptacle outlets in dwelling units, including single family dwellings, duplexes; single family dwellings in apartment complexes, townhouses and condominiums. A dwelling unit per **Article 100** in the NEC is defined as one or more rooms with permanent provisions for:

- Living
- Sleeping
- Sanitation
- Cooking

See **Figure 16-6** for a detailed illustration of a dwelling unit.

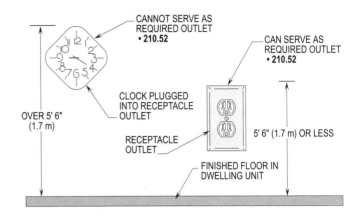

WALL RECEPTACLE OUTLETS
NEC 210.52

Figure 16-7. Receptacle outlets located over 5 ft. 6 in. (1.7 m) from the floor in a dwelling unit shall not serve as one of the required outlets listed in **210.52**.

Section 210.52 gives permission for a receptacle to be installed in a listed factory-assembled baseboard heater so as to prevent extension cords from being plugged into the receptacle outlet above the baseboard heater. Heat from the electric heater elements may over heat the insulation of the cord and cause a fire hazard from the arcing and sparking that can occur. The **FPN** recommends that baseboard heaters be installed by the instructions accompanying the heater unit per **90.7** and **110.3(B)**.

> **Design Tip:** This rule applies strictly to electric baseboard heater with heating elements. **(See Figure 16-8)**

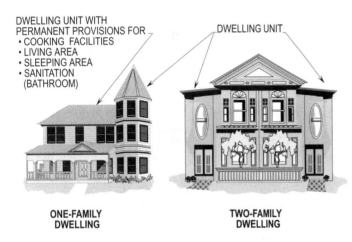

ARTICLE 100

Figure 16-6. A dwelling unit in a one-family house or duplex contains provisions for permanent cooking and sanitation as well as facilities for living and sleeping. These provisions shall be permitted to be located in one or more rooms to be classified as dwelling unit per **Article 100**.

Receptacles that are an integral part of a luminaire (lighting fixture), appliance or cabinet shall not be counted as one of the required receptacle outlets. Those outlets installed over 5 ft. 6 in. (1.7 m) above the finished floor are not permitted to be counted. In other words, they shall not be counted as one of the required outlets for normal use. In these cases, additional outlets shall be installed to comply with the number of outlets required per **210.52(A)(1)**. **(See Figure 16-7)**

Receptacle outlets shall be installed on walls in every inhabitable room, including the hallway, of every dwelling unit, except the bathroom. The procedure for properly spacing the receptacles along the floor line of an unbroken wall is to provide an outlet within 6 ft. (1.8 m) of the cord-and-plug connected equipment.

The first outlet shall be installed 6 ft. (1.8 m) from the door entering the room. The measurement for installing this outlet shall be permitted to be made on either side of the door. From the outlet measured at 6 ft. (1.8 m), additional outlets measured at 12 ft. (3.7 m) intervals along the unbroken wall shall be installed. Outlets installed in this manner allow electrical appliances with 6 ft. (1.8 m) cords to be cord-and-plug connected without the use of an extension cord. The first outlet installed at 6 ft. (1.8 m) and one every 12 ft. (3.7 m) thereafter complies with this requirement and makes it possible to cord-and-plug connect appliances without creating a fire hazard. Any wall space of 2 ft. (600 mm) or

Receptacle Outlets

more requires an outlet for a receptacle to be installed.

Design Tip: The wall space of 2 ft. (600 mm) is based on trimmed out walls, etc. and not from a 2 ft. (600 mm) rough in measurement of framing members.

A wall space is a wall unbroken along the floor line by a door, bookcase, fireplace or window that extends all the way to the floor line. A floor receptacle outlet located close to the wall shall be permitted to be counted as one of the outlets required by **210.52(A)(3)**. **(See Figure 16-9)**

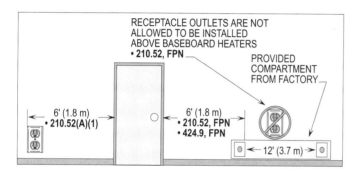

Figure 16-8. Electric baseboard heaters shall not be located below a receptacle outlet. Partitions from the factory, for installing receptacles, shall be permitted to be put in the baseboard heater to comply with the 12 ft. (3.7 m) requirement of **210.52(A)(1)**.

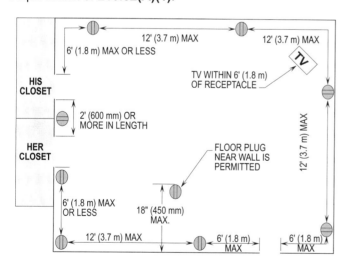

Figure 16-9. Receptacle outlets shall be installed so that there is no point on the wall greater than 6 ft. (1.8 m) from a receptacle outlet.

Railings used for room dividers, where furniture with floor or table lamps are backed up to the railing, shall have receptacle outlets installed for cord-and-plug connections such as table or floor lamps. This rule is to prevent the use of extension cords which are usually run through walls, doors, under carpets, etc. per **400.8(1) through (7)**. **(See Figure 16-10(a))**

The sliding portion of an exterior door shall not be considered wall space. Fixed panels, including the fixed portion of a sliding glass door unit in an exterior wall shall be considered wall space for the purpose of spacing the required number of receptacle outlets per **210.52(A)(2)(2)**. The NEC allows a floor receptacle to be used to provide the required number of receptacle outlets per **210.52(A)(3)**. **(See Figure 16-10(b))**

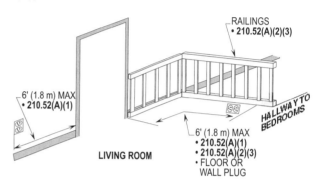

Figure 16-10(a). Railings used as room dividers shall have receptacle outlets installed per **210.52(A)(2)(3)** to cord-and-plug connect floor lamps, table lamps, etc.

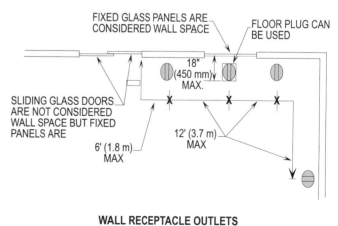

Figure 16-10(b). Sliding glass doors or panels shall not be considered wall space when spacing receptacles to comply with **210.52(A)**. However, fixed panels shall be considered wall space.

RECEPTACLE OUTLETS FOR SMALL APPLIANCE CIRCUITS
210.52(B); 210.52(C); 210.11(C)(1); 220.52(A)

A minimum of 2 - 20 amp, 120 volt, 1500 VA small appliance circuits shall be required to supply receptacle outlets that are located in the kitchen, pantry, breakfast room and dining room. The 2 small appliance circuits shall be routed to the kitchen countertop(s) and the outlets proportioned among the 2 circuits as evenly as possible to prevent unbalanced loading of the circuits. Unbalanced loading may trip open the overcurrent protection device due to too many portable appliances being plugged into the same small appliance circuit. **(See Figure 16-11)**

The 2 - 20 amp small appliance circuits wired with 12-2 AWG w/ground, nonmetallic-sheathed cable (romex or rope) shall only supply receptacle outlets located in the kitchen, pantry, breakfast room and dining room. All other receptacle outlets shall be served by the general purpose branch-circuits or individual branch-circuits per **Table 220.12** and **210.23(A)** or **210.19(A)(1)**. Switched receptacle outlets connected to the general purpose branch-circuits per **210.52(B)(1), Ex. 1** and **210.70(A)(1), Ex. 1** to serve swag lights, table lamps, floor lamps, etc. shall also be allowed. These are in addition to those required in **210.52(B)**. All receptacle outlets shall be installed so that no point along the floor line of an unbroken wall is further than 6 ft. (1.8 m) from an outlet per **210.52(A)(1)**.

Exception 2 to 210.52(B)(1) allows a receptacle outlet to be installed from an individual branch-circuit to supply a compressor motor for refrigeration equipment.

Exception 2 to 210.52(B)(2) permits receptacle outlets to be installed on the small appliance circuits to supply loads such as timers, clocks, burner ignition systems on gas ovens, cooktops and ranges.

> **Design Tip:** This rule makes it clear that such mentioned motor loads shall be permitted to be supplied by an individual 15 amp branch-circuit per **210.19(A)(1)**, general purpose motor branch-circuit per **430.53(A)**.

There is an **Ex. to 210.52(B)(2)** which permits another outlet to be supplied by the small appliance circuits. The **Ex. 1** permits a clock outlet to be served by any one of the small appliance circuits. To apply this exception, the clock outlet shall be located in the kitchen, pantry, dining room or breakfast room. **(See Figure 16-12)**

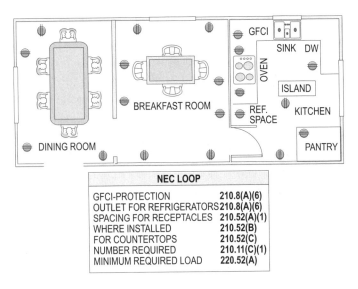

RECEPTACLE OUTLETS FOR SMALL APPLIANCE CIRCUITS NEC 210.52(B)

Figure 16-11. A minimum of 2 - 20 amp small appliance circuits shall be provided to serve the receptacle outlets located in the kitchen, pantry, breakfast room and dining room.

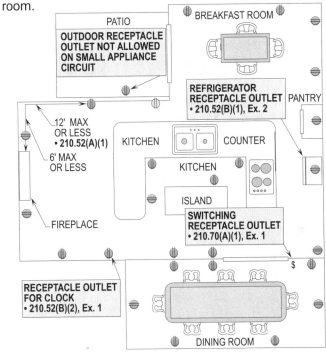

RECEPTACLE OUTLETS FOR SMALL APPLIANCE CIRCUITS NEC 210.52(B)(1), Ex. 1 NEC 210.70(A)(1), Ex. 1

Figure 16-12. A clock outlet shall be permitted to be installed on the small appliance circuits. In addition, a switched receptacle outlet shall be permitted to be installed on the general-purpose branch-circuit.

RECEPTACLE OUTLETS OVER COUNTERTOPS
210.52(C)

There are three types of countertops that may be present in dwelling units and they are as follows:
- countertop with a wall behind it per **210.52(C)(1)**
- island countertops per **210.52(C)(2)**
- countertop with no wall behind it (peninsular) per **210.52(C)(3)**

WALL TYPE COUNTERTOPS
210.52(C)(1)

A receptacle outlet shall be provided at each countertop space 12 in. (300 mm) or wider in the kitchen and dining room areas. No point along the wall shall be permitted to be more than 24 in. (600 mm) from a receptacle outlet. Each countertop shall be treated separately in providing the number and spacing of outlets. Countertops divided by a sink, a range, a cooktop, an oven or a refrigerator shall have receptacle outlets installed on each side where there is 12 in. (300 mm) or more of countertop space along the wall.

The procedure for laying out the location of these outlets is to measure 24 in. (600 mm) and then at 4 ft. (1.2 m) intervals until the last receptacle outlet is no further than 2 ft. (600 mm) in any direction from an outlet. Receptacle outlets will be accurately provided for each countertop space by applying these measurements. Any receptacle outlets rendered inaccessible by the position of installed appliances such as a refrigerator shall not be considered as one of the required outlets. **(See Figure 16-13)**

ISLAND COUNTERTOPS
210.52(C)(2)

An island countertop stands alone with no wall behind or beside it. Island countertops with a short dimension of 12 in. (300 mm) or greater or a long dimension of 24 in. (600 mm) or greater shall have at least one receptacle installed. **(See Figure 16-14)**

Design Tip: See **210.52(B)(2)** which lists the requirements for installing the small appliance circuits for countertops in the kitchen and dining room areas. Refer to **210.8(A)(6)** for countertop receptacle outlets requiring GFCI-protection.

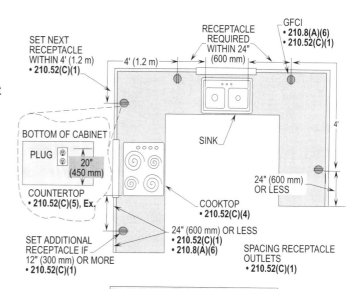

WALL TYPE COUNTERS
NEC 210.52(C)(1)

Figure 16-13. Receptacle outlets installed at kitchen countertops shall be located so not point along the countertop space is more than 24 in. (600 mm), measured horizontally from a receptacle outlet.

PENINSULAR COUNTERTOPS
210.52(C)(3)

Receptacle outlets shall be installed at each peninsular countertop. A peninsular is a countertop without a wall on either side. Receptacle outlets shall be installed for each peninsular with a long dimension of 24 in. (600 mm) or greater and a short dimension of 12 in. (300 mm) or greater. A peninsular countertop shall be measured from the connecting edge and receptacle outlets installed accordingly. The receptacle outlets shall be installed above or within 12 in. (300 mm) below the countertop or a tombstone type receptacle shall be permitted to be installed on the countertop.

Design Tip: If the countertop has an overhang of more than 6 in. (150 mm), a receptacle outlet shall not be permitted to be installed below the countertop per **210.52(C)(5), Ex. (See Figure 16-14)**

RECEPTACLE OUTLETS IN BATHROOMS
210.52(D)

A bathroom is not just a room but is an area including a basin(s) with a toilet, a tub or shower per **Article 100**.

A utility sink or wet bar sink shall not be considered a bathroom because no other fixtures mentioned are present.

> **Design Tip:** The key in defining a bathroom is that a basin and at least one or more of the other fixtures shall be present.

Section 210.52(D) requires at least one receptacle outlet to be installed in bathrooms within 3 ft. (900 mm) of the outside edge of each basin in a dwelling unit. The receptacle outlet shall be located on a wall or partition that is adjacent to the basin or basin countertop. **Section 210.8(A)(1)** requires all 15 or 20 amp, 125 volt receptacles to be GFCI-protected for the safety of personnel using electric grooming tools. **(See Figure 16-15)**

> **Design Tip:** Each basin is required to have a receptacle outlet installed within 3 ft. (900 mm) and located in such a manner so as to serve the sink area safely without the use of extension cords.

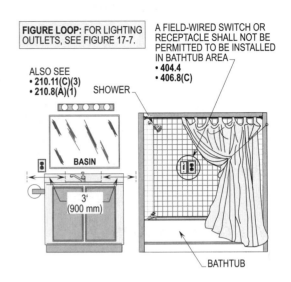

RECEPTACLE OUTLETS IN BATHROOMS NEC 210.52(D)

Figure 16-15. At least one receptacle outlet to serve the bathroom area shall be installed within 3 ft. (900 mm) of the outside edge of each basin in a dwelling unit.

RECEPTACLE OUTLETS OUTDOORS 210.52(E)

At least 2 receptacle outlets shall be installed outdoors for one and two family dwellings. There shall be at least one receptacle outlet installed in the front and the back of such units. **Section 210.8(A)(3)** requires such outdoor receptacle outlets to be GFCI-protected to protect personnel from electrical shock when using such tools as drill motors, lawn mowers, hedge clippers, etc.

All outdoor receptacles with direct grade level access shall be GFCI-protected. Direct grade level access is defined as being located 6 ft. 6 in. (2 m) or less from grade level and readily accessible to the user. Outdoor receptacle outlets shall be located on outside walls, open porches or fed out of the ground serve as such required outlets. Receptacle outlets located in the eve of dwellings to connect Christmas lights or on balconies to cord-and-plug connect radios, VCR's, stereos or TV's shall be GFCI-protected even if they are located over 6 ft. 6 in. (2 m) from direct grade level.

Section 406.8(B)(1) requires unattended receptacle outlets to have rain tight covers to protect the components. Unattended outlets are those serving items such as outdoor Christmas lights, water pumps, etc. For receptacle outlets located outdoors for townhouses and condominiums with zero lot lines, see **210.8(A)(3)** and **210.52(E)**. **(See Figure 16-16)**

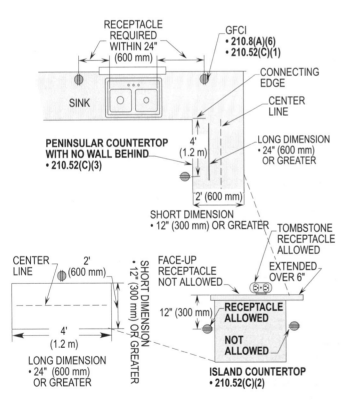

NEC 210.52(C)(2); (C)(3)

Figure 16-14. Installation requirements for receptacle outlets on wall, island or peninsular countertops.

Receptacle Outlets

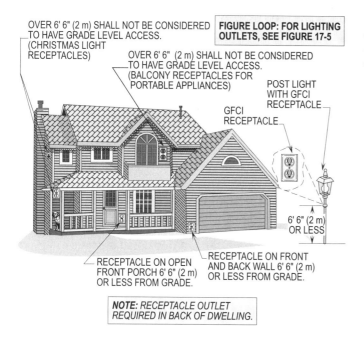

RECEPTACLE OUTLETS OUTDOORS
NEC 210.52(E)

Figure 16-16. Receptacle outlets mounted on a wall or located on an open front porch or stoop of a dwelling unit shall be GFCI-protected. **Note:** All the receptacles above shall be GFCI protected.

LAUNDRY CIRCUIT
210.52(F)

All laundry receptacle outlets shall be supplied by a 20 amp circuit utilizing a 12-2 AWG nonmetallic-sheathed cable with ground. This cable is usually nonmetallic-sheathed cable or armored cable.

> **Design Tip:** A conduit system shall be permitted to be used with individual conductors pulled in after the conduit is installed. For the installer to use this wiring method, a local code usually requires it.

At least one receptacle outlet shall be installed for the laundry area. All laundry equipment is required to be located within 6 ft. (1.8 m) of the receptacle outlet per **210.50(C)**. Sometimes a 20 amp duplex receptacle is used to cord-and-plug connect a washing machine and a gas dryer. No other outlets shall be supplied by this 20 amp, 2-wire small appliance circuit. Where the laundry area is located in the garage or basement, GFCI-protection shall not be required for the receptacle outlets per **210.8(A)(2), Ex. 2** and **210.8(A)(5) Ex. 2**.

For example, a 20 amp small appliance laundry circuit (20 A OCPD x 80% = 16 A) should be loaded no more than 16 amps, if used for long periods of time such as three hours or longer. **(See Figure 16-17)**

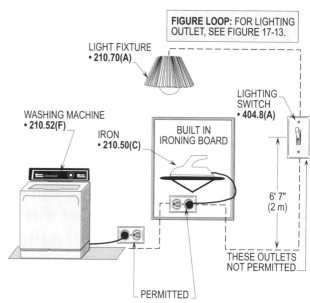

LAUNDRY CIRCUIT
NEC 210.52(F)

Figure 16-17. The 20 amp laundry circuit shall be permitted to be used only to serve laundry equipment. The laundry circuit shall be permitted to be routed to the garage, basement, utility room or wherever the laundry room is located.

RECEPTACLE OUTLETS IN BASEMENTS AND GARAGES
210.52(G)

GFCI-protected receptacles shall be installed in basements and garages to protect people from electric shock when using electric hand tools. GFCI-protected circuit breakers or receptacles shall be used to provide this protection. GFCI stands for ground-fault circuit interrupter and has been designed to protect individuals from fatal electric shock. For example, it's trip open point is 4 to 6 milliamps respectively.

RECEPTACLES IN BASEMENTS
210.52(G)

At least one GFCI-protected receptacle outlet per **210.8(A)(5)** shall be installed in the basement, when it is left unfinished. An unfinished basement means that the basement is used for storage or a workshop. If more than one receptacle outlet is installed, then all of the receptacle outlets shall be GFCI-protected. This receptacle outlet or outlets are to protect people from electric shock when using electrical hand tools such as electric drills, saws, sanders, etc. If the basement is finished into one or more habitable rooms, each separate unfinished portion of the basement shall have a receptacle outlet installed.

There are three **Ex.'s to 210.8(A)(5)**, where receptacles shall not be required to be GFCI-protected. **Exception 1** does not require receptacles to be GFCI-protected where they are not readily accessible. **Exception 2** does not require cord-and-plug connected appliances such as refrigerators or freezers to be GFCI-protected. **Exception 3** does not require GFCI-protection where a receptacle supplies only a permanently installed fire alarm or burgular alarm system.

Exception 2 does not require the sump pump receptacle to be GFCI-protected. When any one of the **Ex.'s to 210.8(A)(5)** are applied, the receptacle shall be of the single type. The 15 or 20 amp general purpose receptacle shall be permitted to be the duplex type, if it is GFCI-protected.

At least one GFCI-protected receptacle per **210.8(A)(4)** shall be installed in a crawl space at or below grade level. This receptacle shall be permitted to be used for serving HVAC equipment or be used for a drop light, hand tool, etc.

See Figure 16-18 and **210.70(A)(3)** for the requirements of lighting outlets.

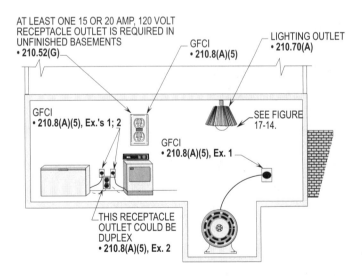

RECEPTACLES IN BASEMENTS
NEC 210.52(G)

Figure 16-18. At least one 15 or 20 amp, 125 volt receptacle shall be installed in unfinished basements. See the three exceptions for receptacles, which shall not be required to be GFCI-protected.

RECEPTACLES IN GARAGES
210.52(G)

At least one GFCI receptacle per **210.52(G)** and **210.8(A)(2)** shall be installed in attached or unattached garages. There are two **Ex.'s to 210.8(A)(2)** where GFCI-protection shall not be required to be provided.

Exception 1 does not require receptacles that are not readily accessible to be GFCI-protected. Such receptacle outlets are those used to cord-and-plug connect a motor that opens and closes an overhead garage door. **Exception 2** permits any dedicated appliance that is cord-and-plug connected to be exempted from GFCI-protection. A refrigerator or a freezer shall be considered a dedicated appliance.

A GFCI receptacle outlet shall not be required in a detached garage unless electric power is routed to the garage. Refer to **210.70(A)(2)(a)** for lighting outlet requirements. **(See Figure 16-19)**

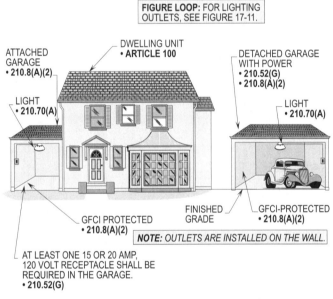

RECEPTACLES IN GARAGE
NEC 210.52(G)

Figure 16-19. At least one 15 or 20 amp receptacle outlet shall be installed in the garage at 5 ft. 6 in. (1.7 m) or less from the finished floor to be considered a required outlet.

RECEPTACLES IN HALLWAYS
210.52(H)

Hallways in dwelling units that are 10 ft. (3 m) or more in length without passing through a doorway shall have a receptacle outlet installed. This receptacle outlet may be used for the connection of plugged-in appliances such as table lamps, floor lamps, vacuum cleaners, etc. The length of the hallway shall be measured along its center line to determine if a receptacle outlet is required. The receptacle outlet should be located so it serves the hallway in the most convenient manner. **(See Figure 16-20)**

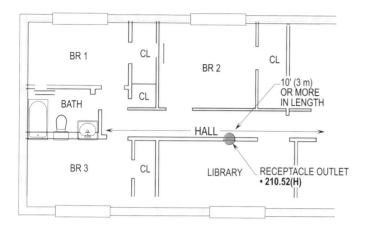

RECEPTACLES IN HALLWAYS
NEC 210.52(H)

Figure 16-20. At least one receptacle outlet shall be installed in dwelling unit hallways that are 10 ft. (3 m) or more in length without passing through a doorway.

GFCI-PROTECTION OF RECEPTACLES
210.8(A)(1) THROUGH (A)(8)

A GFCI circuit breaker or receptacle detects any imbalance current in the circuit such as current leaking to ground through the body of a person using an electric hand tool. This imbalance of current trips open the GFCI unit at about 4 to 6 milliamps plus or minus 1 milliamp. The action of the GFCI prevents any harm to the person using the receptacle under adverse conditions of use.

It only takes a relatively small amount of current flowing through the body to be fatal. Electrocution may occur at about 380 milliamp. GFCI-protection protects people from such hazards while working with electrical hand tools.

Design Tip: GFCI-protection of receptacles shall be required in bathrooms, garages, outdoors, basements, crawl spaces, kitchens, wet bars and boathouses. **(See Figure 16-21)**

PROTECTION IN BATHROOMS
210.8(A)(1)

All 15 or 20 amp, 125 volt receptacles shall be provided with GFCI-protection, if located in the bathroom. (For the definition of bathroom, see **Article 100**). The receptacles shall be supplied by a circuit protected by a GFCI circuit breaker, by a GFCI receptacle or be served by a fed through GFCI-protected receptacle.

The NEC requires at least one receptacle outlet to be installed within 3 ft. (900 mm) of the outside edge of each sink per **210.52(D)** and **210.8(A)(1)** and also requires this receptacle and all others, if present, to be GFCI-protected. **Sections 404.4** and **406.8(C)** does not permit a switch outlet to be installed in the bathtub area where the elements of such devices may be sprayed with water.

Design Tip: If a 20 amp, 120 volt laundry receptacle is installed in the bathroom, it is required to be GFCI-protected per **210.8(A)(1)**. There are no exceptions to the rule as there are for garages per **210.8(A)(2), Ex. 2** and basements per **210.8(A)(5), Ex. 2**. **(See Figure 16-22)**

PROTECTION IN GARAGES
210.8(A)(2)

All 15 or 20 amp, 125 volt receptacles installed in garages shall have GFCI-protection for the safety of personnel. Since the concrete slab is in direct contact with the earth, the garage shall be considered a hazardous area for people using electric hand tools. Leakage current from a faulty electric hand tool may flow through the human body, through the concrete slab and through the earth to complete the circuit.

The NEC requires only one 15 or 20 amp, 125 volt receptacle outlet to be installed. If there are more installed, such as over a workbench, they shall be GFCI-protected.

There are two exceptions where GFCI-protection shall not be required for receptacle outlets in the garage.

- Receptacles that are not readily accessible such as an overhead garage door, etc. per **210.8(A)(2), Ex. 1**

- Receptacles that supply power to dedicated appliances such as a freezer, refrigerator, gas dryer, etc. per **210.8(A)(2), Ex. 2**

See Figure 16-23 for a detailed illustration pertaining to receptacles installed in garages.

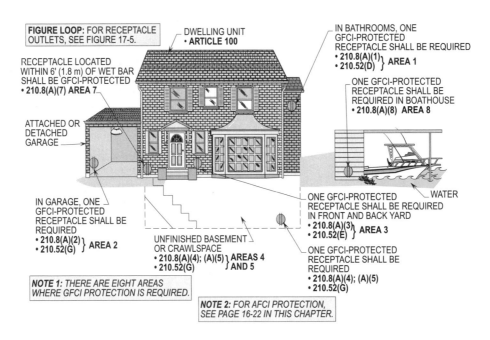

GFCI - PROTECTION OF RECEPTACLES
NEC 210.8(A)(1) THROUGH (A)(8)

Figure 16-21. One or more GFCI-protected receptacles shall be required for the bathroom, kitchen, garage, crawlspace, unfinished basement and the outdoors (front and back) of a dwelling unit. GFCI-protected receptacles are also required within 6 ft. (1.8 m) of the wet bar sink. At least one GFCI-protected receptacle outlet shall be required for a boathouse.

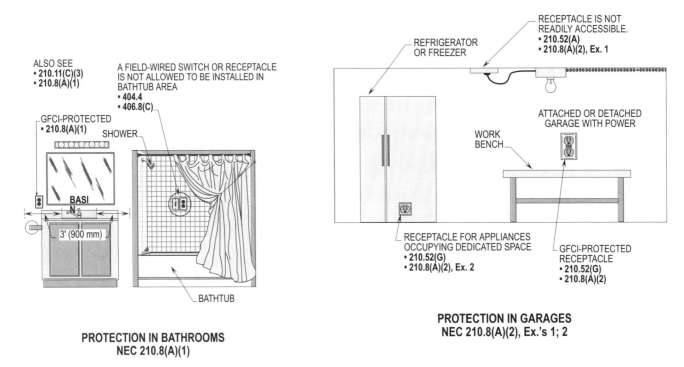

PROTECTION IN BATHROOMS
NEC 210.8(A)(1)

PROTECTION IN GARAGES
NEC 210.8(A)(2), Ex.'s 1; 2

Figure 16-22. At least one GFCI receptacle outlet shall be installed within 3 ft. (900 mm) of each basin in a bathroom.

Figure 16-23. All 15 or 20 amp, 125 volt receptacles installed in garages shall be GFCI-protected except those not readily accessible or those serving dedicated appliances. **Note:** At least one shall be installed.

Receptacle Outlets

PROTECTION IN ACCESSORY BUILDINGS
210.8(A)(2)

Accessory buildings (unfinished or finished) for a dwelling unit not intended to be used as a habitable room and limited to storage areas, work areas or similar purposes, shall have GFCI-protection for all 125 volt, single-phase, 15 and 20 amp receptacles. **(See Figure 16-24)**

PROTECTION OUTDOORS
210.8(A)(3)

All 15 or 20 amp, 125 volt receptacle outlets located outdoors that are installed 6 ft. 6 in. (2 m) or less from direct grade level shall be GFCI-protected. Direct grade level is defined as being readily accessible and located 6 ft. 6 in. (2 m) or less from grade level.

Receptacles at 6 ft. 6 in. (2 m) or less from grade and located in the front and back yard of houses shall be GFCI-protected. Receptacles located in post lights at 6 ft. 6 in. (2 m) or less shall be GFCI-protected. Receptacles outlets located above 6 ft. 6 in. (2 m) such as in the eves for the connection of Christmas lights or on balconies shall also be GFCI-protected. **(See Figure 16-25)**

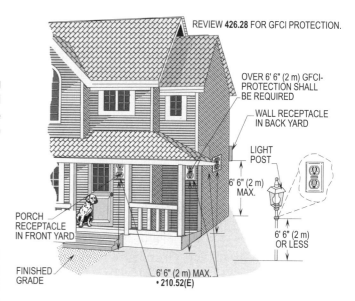

PROTECTION OUTDOORS
NEC 210.8(A)(3)

Figure 16-25. At least one receptacle outlet located outdoors in the front and back shall be installed 6 ft. 6 in. (2 m) or less from finished grade and be readily accessible to the user. These outlets shall be GFCI-protected.

PROTECTION IN BASEMENTS AND CRAWLSPACES
210.8(A)(4); 210.8(A)(5)

All 15 or 20 amp, 125 volt receptacle outlets installed in crawl spaces at or below grade level to serve electrical equipment shall be GFCI-protected per **210.8(A)(4)**. Most mechanical codes require a receptacle outlet to be installed for the servicing of mechanical and related equipment.

See Figure 16-26 for rules pertaining to the installation and protection of receptacles in crawlspaces and basements.

At least one 15 or 20 amp, 125 volt receptacle outlet shall be installed in unfinished basements of dwelling units. These basements are unfinished and are not habitable. Habitable is the case in which they are finished as a bedroom, den, game room, etc. and occupied by people. An unfinished basement is not trimmed out and may be utilized as a workshop area or storage area. There are three exceptions where receptacles in the basement shall not be required to be GFCI-protected.

• **Exception 1 to 210.8(A)(5)** does not require the sump pump receptacle to be GFCI-protected.

• **Exception 2 to 210.8(A)(5)** does not require cord-and-plug connected appliances, such as refrigerators or freezers to be plugged into GFCI-protected receptacles.

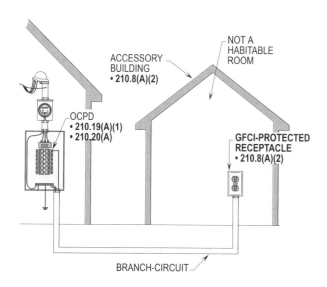

PROTECTION IN ACCESSORY BUILDINGS
NEC 210.8(A)(2)

Figure 16-24. GFCI-protection shall be provided for all 125 volt, single-phase, 15 and 20 amp receptacles for accessory buildings (unfinished or finished) for a dwelling unit.

• **Exception 2 to 210.8(A)(5)** does not require the laundry circuit receptacle to be GFCI-protected.

• **Exception 3 to 210.8(A)(5)** does not require a permanently installed fire alarm or burgular alarm system receptacle to be GFCI-protected.

PROTECTION OVER COUNTERTOPS 210.8(A)(6) AND (A)(7)

All 15 or 20 amp, 125 volt receptacle outlets installed over kitchen countertops shall be GFCI-protected. Receptacle outlets included are those located over or below the countertop and on islands or peninsulars which are supplied by the small appliance circuits per **210.11(C)(1), 210.52(B)** and **210.52(C). (See Figure 16-27)**

Design Tip: If a pull chain luminaire (lighting fixture) with a receptacle outlet is installed over the kitchen sink, it shall be GFCI-protected per **210.8(A)(6)**.

Receptacle outlets located over utility sinks shall not be required to be GFCI-protected because the sink does not have an additional fixture such as a toilet, a tub or shower per **Article 100**. Receptacle outlets installed within 6 ft. (1.8 m) of laundry, utility, and wetbar sinks shall be GFCI-protected to protect personnel using blenders and other drink related appliances per **210.8(A)(7)**.

Utility sinks are usually used to clean mops, etc. while wetbar sinks are utilized to entertain guests at parties and other social related functions in the home.

Design Tip: A receptacle used to cord-and-plug a refrigerator shall not be required to be GFCI-protected per **210.8(A)(6). (See Figure 16-28)**

PROTECTION FOR BOATHOUSES 210.8(A)(8)

All receptacle outlets rated at 15 or 20 amp, 125 volt that are installed in boathouses to provide power for electric hand tools to work on boats shall be GFCI-protected. Personnel using electric hand tools are exposed to wet locations due to the presence of water. GFCI-protected circuits protect personnel from electrical shock while using electricity around such hazardous conditions. See **555.19(B)(1)** for such protection pertaining to marinas and boatyards. **(See Figure 16-29)**

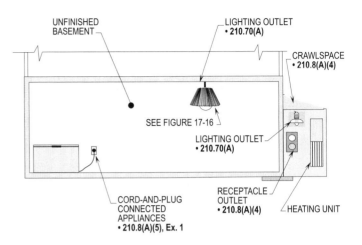

Figure 16-26. All 15 or 20 amp, 125 volt receptacle outlets installed in crawl spaces at or below grade level to serve electrical equipment shall be GFCI-protected.

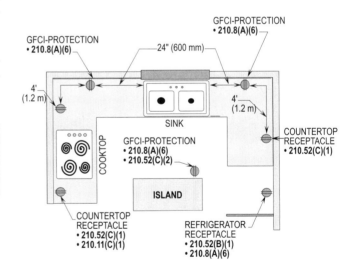

Figure 16-27. All 15 or 20 amp, 125 volt receptacles serving kitchen countertop surfaces located above or below shall be GFCI-protected. This rule is to protect the user.

Receptacle Outlets

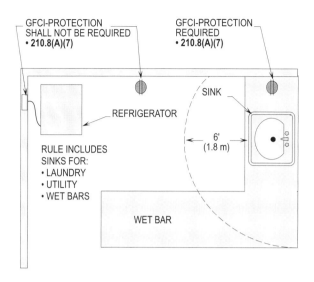

PROTECTION OVER COUNTERTOPS
NEC 210.8(A)(7)

Receptacle outlets installed near swimming pools and fountains shall be properly located to minimize the threat of shock hazards to personnel in or around the pool. Receptacles shall not be permitted to be located within 10 ft. (3 m) of the inside walls of the pool per **680.22(A)(2)** and **(3)**.

At least one 15 or 20 amp, 125 volt receptacle outlet shall be installed at a minimum of 10 ft. (3 m) from and not more than 20 ft. (6 m) from the inside walls of the pool or fountain per **680.22(A)(2), (3)** and **(5)**. The 10 ft. (3 m) rule is to prevent users from plugging in portable radios, televisions, stereos, etc. into the receptacle and being too close to the water. Receptacle outlets which are installed behind enclosed rooms with hinged or sliding doors, windows or other barriers are permitted within the 10 ft. (3 m) limitation of a swimming pool per **680.22(A)(6)**. The receptacle outlet located within the 20 ft. (6 m) boundary of the swimming pool shall be GFCI-protected per **680.22(A)(5)**. **(See Figures 16-30 (a) and (b))**

Figure 16-28. All 15 or 20 amp, 125 volt receptacle outlets located within 6 ft. (1.8 m) of a wet bar sink shall be GFCI-protected except for a properly located refrigerator plug.

> **Design Tip:** See **680.22(A)(1)** which allows a single locking receptacle outlet to be located not less than 5 ft. (1.5 m) from the inside walls of the pool. This receptacle outlet shall be GFCI-protected per **680.22(A)(1)**. A cord-and-plug connected recirculating pool pump can easily be removed in winter. **(See Figure 16-31)**

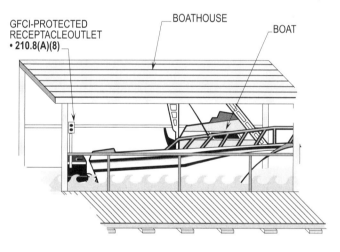

PROTECTION FOR BOATHOUSES
NEC 210.8(A)(8)

Figure 16-29. All 15 or 20 amp, 125 volt receptacle outlets shall be GFCI-protected when installed in boathouses.

PROTECTION AROUND SWIMMING POOLS
680.22(A)(2); (3)

The following requirements for protection around swimming pools apply to residential, commercial and industrial locations.

PROTECTION FOR STORABLE POOLS
680.32

Storable pools shall be installed at least 10 ft. (3 m) from any receptacle outlet per **680.22(A)(1)**. Electrical equipment associated with storage pools shall be provided with GFCI-protection for the safety of personnel using the pool. See **90.7** and **110.3(B)**.

> **Design Tip:** The cord supplying power from the receptacle outlet to the pool equipment shall also have GFCI-protection. The supply cord may be longer than 3 ft. (900 mm) per **680.7(A)**. There are UL listed supply cords of 25 ft. (7.5 m) in length and approved to be used to cord-and-plug connect filter pumps to storable pools. The 3 ft. (900 mm) limitation does not apply to cords used to connect filter pumps to storable pools. The GFCI-protection shall be permitted to be provided with a GFCI-protected receptacle or a remote GFCI-protected receptacle or circuit breaker per **680.32**. **(See Figure 16-32)**

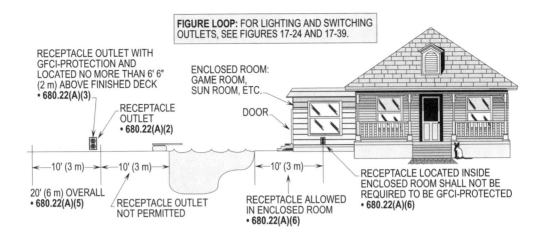

PROTECTION AROUND SWIMMING POOLS
NEC 680.22(A)(5)

Figure 16-30(a). At least one GFCI-protected receptacle shall be located in an area within 10 ft. (3 m) to 20 ft. (6 m) from the inside wall of the pool. Receptacles located inside enclosed rooms and within 10 ft. (3 m) of the inside walls of the pool shall not be required to be GFCI-protected. Personnel inside the enclosed room are not exposed to the pool area.

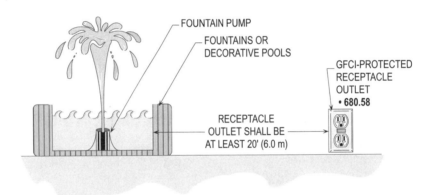

PROTECTION AROUND FOUNTAINS
NEC 680.58

Figure 16-30(b). Receptacle outlets for fountains or decorative pools shall be located at least 20 ft. (6.0 m) from the receptacle to the inside walls. **Note:** This receptacle is not required, but if installed, shall be GFCI-protected.

PROTECTION FOR SPAS OR HOT TUBS
680.43(A)(1); (A)(2)

Receptacle outlets shall be located at least 5 ft. (1.5 m) from the inside walls of a spa or hot tub to ensure the safety of the user. Receptacle outlets installed 5 ft. (1.5 m) to 10 ft. (3 m) from the inside walls of the spa or hot tub shall be GFCI-protected by a GFCI-protected receptacle or CB.

At least one receptacle outlet shall be provided in this area for the protection of personnel using cord-and-plug connected radios, TV's, stereos, etc. Receptacle outlets supplying power to a spa or hot tub shall be GFCI-protected per **680.43(A)(3)**. **(See Figure 16-33)**

Design Tip: See **680.42** for the installation requirements of spas or hot tubs that are located outdoors and apply **Parts I and II of Article 680** which pertain to swimming pools.

Receptacle Outlets

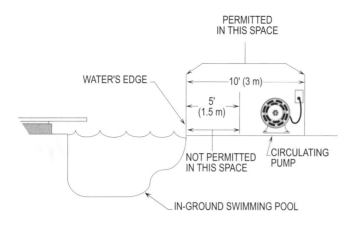

PROTECTION AROUND SWIMMING POOLS
NEC 680.22(A)(1)

Figure 16-31. A single locking type GFCI-protected receptacle outlet shall be permitted to be installed between 5 ft. (1.5 m) to 10 ft. (3 m) from the inside walls of a swimming pool or fountain to supply power to a water pump motor.

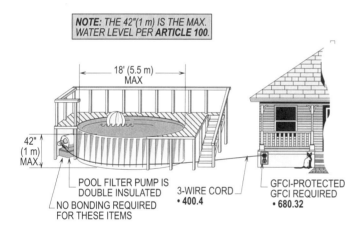

PROTECTION FOR STORABLE POOLS
NEC 680.32

Figure 16-32. Electrical equipment that is cord-and-plug connected shall be served by a GFCI-protected receptacle outlet or circuit.

PROTECTION FOR HYDROMASSAGE TUBS
680.71

Receptacle outlets installed by hydromassage tubs shall be provided with GFCI-protection per **210.8(A)(1)**. Hydromassage tubs are treated just like regular bathtubs per **680.72**. All electrical elements for hydromassage tubs shall be supplied by GFCI circuits per **680.71**. The wiring methods shall comply with the provisions of **Chapters 1 through 4** in the NEC, which are to be applied generally. Circulating motors shall comply with the requirements of **430.14(A)** and **680.73** for accessibility, so that proper maintenance and adequate ventilation may be provided. **(See Figure 16-34)**

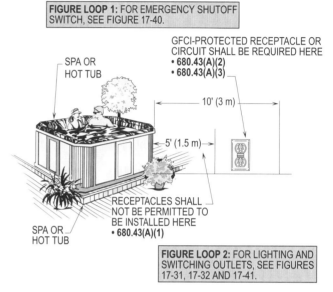

PROTECTION FOR SPAS OR HOT TUBS
NEC 680.43(A)(2)

Figure 16-33. Receptacle outlets installed around a spa or hote tub that are located inside the dwelling unit. For spa or hot tubs located outside, see **680.40**.

RECEPTACLES ON CONSTRUCTION SITES
590.6

All 15, 20 and 30 amp, 125 volt receptacles used by personnel during construction, remodeling, maintenance, repair or demolition purposes shall be GFCI-protected. GFCI-protection shall be provided so that workers are protected from electrical shock. There are three methods by which this may be accomplished.

- A GFCI circuit breaker (CB) shall be permitted to be used to supply power to all receptacle outlets. The

GFCI CB protects all receptacle outlets down stream.

- A GFCI receptacle shall be permitted to be installed at each outlet. This provides protection at each location of use and may be reset immediately when tripped open due to a faulty hand tool, etc.

- A portable GFCI receptacle shall be permitted to be used by workers which is protected by a regular CB. Each worker carries a portable GFCI and plugs all electric hand tools into a plain receptacle outlet. If the GFCI receptacle trips for any reason, it can be reset at the work location without workers losing time. **(See Figure 16-35)**

The authority having jurisdiction (AHJ) has the authority to approve a written procedure that requires three-wire extension cords with equipment grounding conductors. Personnel shall be designated to enforce the program utilizing the following procedures:

- All equipment grounding conductors shall be tested for continuity and shall be electrically continuous.

- Each receptacle and attachment plug shall be tested for correct attachment of the equipment grounding conductor.

- All required tests shall be performed as follows:

  (a) Before first use on site
  (b) When there is evidence of damage
  (c) Before equipment is returned to service following any repairs
  (d) At intervals not exceeding 3 months

Note: The assured equipment grounding conductor program only applies for industrial sites where conditions of maintenance and supervision ensure that only qualified personnel are involved per **590.6(B)(2)** and **590.6(A), Ex.**

Receptacle outlets shall not be connected to any branch-circuits supplying power to lighting outlets per **590.4(D)**. The reason for this rule is to prevent a ground-fault on an electric hand tool from opening the OCPD on the circuit and putting the workman in the dark, which could be dangerous.

Design Tip: Section **590.6(B)(2)(1)** and **OSHA 1926, Subpart K** must be studied carefully and applied properly to protect the users of such cords and provide a safe workplace for employees that are building or remodeling dwelling units.

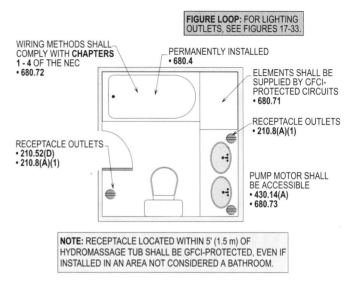

PROTECTION FOR HYDROMASSAGE TUBS
NEC 680.71

Figure 16-34. Hydromassage bathtubs are treated as regular bathtubs and shall comply with **680.70, 680.71, 680.72** and **680.73** and other pertinent Sections of the NEC.

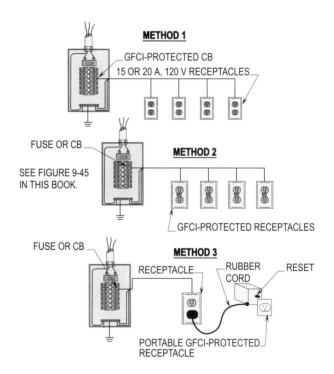

RECEPTACLES ON CONSTRUCTION SITES
NEC 590.6(A)

Figure 16-35. Three methods of providing GFCI-protection for construction sites to protect personnel.

Receptacle Outlets

RECEPTACLES INSTALLED IN BATHROOMS OF COMMERCIAL AND INDUSTRIAL LOCATIONS
210.8(B)(1)

In the bathrooms of commercial and industrial locations, GFCI's shall be installed for the protection of personnel. This will include all 125 volt, 15 and 20 ampere circuits in the bathroom. Note that such receptacle outlets shall not be required to be installed, but if they are, they shall be GFCI-protected. By definition, a bathroom is an area including a basin and at least one of the following: a tub, a toilet, a shower or a combination. **(See Figure 16-36)**

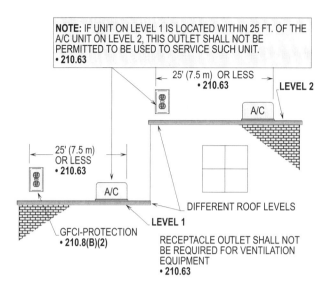

RECEPTACLES INSTALLED ON ROOFTOPS OF COMERCIAL AND INDUSTRIAL BUILDINGS
NEC 210.8(B)(2)

Figure 16-37. All 125 volt, 15 or 20 amp receptacles installed on roofs shall have ground-fault protection for personnel.

RECEPTACLE INSTALLED IN KITCHENS OF COMMERCIAL AND INDUSTRIAL BUILDINGS
210.8(B)(3)

All 125, 15 or 20 amp receptacles installed in kitchens of commercial and industrial buildings shall have ground-fault protection for personnel. **(See Figure 16-38)**

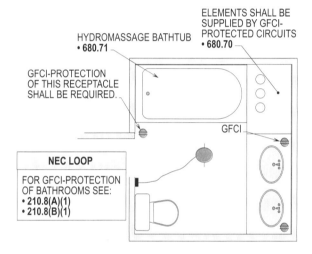

RECEPTACLES INSTALLED IN BATHROOMS OF COMMERCIAL AND INDUSTRIAL LOCATIONS
NEC 210.8(B)(1)

Figure 16-36. In the bathrooms of commercial and industrial locations, GFCI's shall be installed for the protection of personnel.

RECEPTACLES INSTALLED ON ROOFTOPS OF COMMERCIAL AND INDUSTRIAL BUILDINGS
210.8(B)(2)

All 125 volt, 15 or 20 amp receptacles installed on roofs shall have ground-fault protection for personnel. Note that this does not apply to receptacles installed on the roofs of dwelling units. **(See Figure 16-37)**

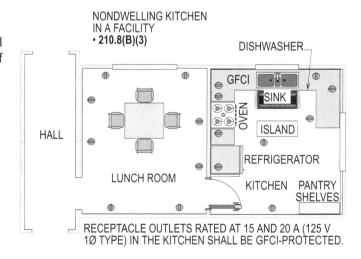

Figure 16-38. All 125, 15 or 20 amp receptacles installed in kitchens of commercial and industrial buildings shall have ground-fault protection for personnel.

RECEPTACLE USED FOR MAINTENANCE ACTIVITIES 590.6(A)

All 125 volt, 15, 20 and 30 amp receptacles used for electric hand tools, etc. to pull maintenance on electrical apparatus and equipment shall be GFCI-protected. **(See Figure 16-39)**

AFCI-PROTECTED OUTLETS 210.12(A); (B)

Section **210.12(B)** requires all branch-circuits that supply 125 volt, single-phase, 15 and 20 amp outlets installed in dwelling unit bedrooms to be protected by a listed arc-fault circuit interrupter(s) that is capable of providing protection of the entire branch-circuit. The requirements to protect all 125 volt, single-phase, 15 and 20 ampere outlets makes better sense and allows more flexibility and protection by placing all outlets on the same branch-circuit equipped with an AFCI protection scheme. Note that an AFCI circuit breaker can be used to provide such protection. **(See Figure 16-40)**

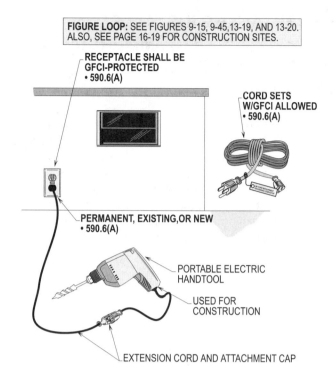

Figure 16-39. Employees using electric hand tools to pull maintenance on electrical equipment shall be fully protected by GFCI-protection of power circuit and elements.

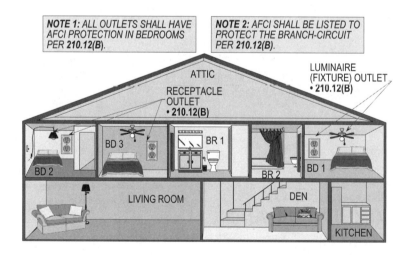

Figure 16-40. Certain outlets in bedrooms of dwelling units shall be AFCI protected.

Name   Date

Chapter 16. Receptacle Outlets

Section   Answer

_____   T   F   1. Receptacles installed on 15 or 20 amp general-purpose branch-circuits with an EGC shall be of the grounding type.

_____   T   F   2. An individual branch-circuit shall not supply power to a single load that is cord-and-plug connected to a single load.

_____   T   F   3. When two or more branch circuits supply devices or equipment on the same yoke, a means to simultaneously disconnect the ungrounded conductors shall be provided.

_____   T   F   4. Grounding type receptacles shall never be used to replace existing receptacles.

_____   T   F   5. Receptacle outlets shall be installed where flexible cords are used to connect electrical appliances and equipment in dwelling units.

_____   T   F   6. Receptacle outlets shall not be installed in the hallway of dwelling units.

_____   T   F   7. Receptacles that are an integral part of a lighting fixture, appliance or cabinet shall not be counted as one the required receptacle outlets in dwelling units.

_____   T   F   8. A maximum of 2 - 20 amp, 1500 VA small appliance circuits are required to supply receptacle outlets in dwelling units that are located in the kitchen, pantry, breakfast room and dining room.

_____   T   F   9. Receptacle outlets shall be installed so that there is no point on the wall greater than 6 ft. from a receptacle outlet.

_____   T   F   10. Sliding glass doors or panels shall be considered wall space when spacing receptacles in dwelling units.

_____   _____ 11. Receptacles shall be permitted to be installed above an electric baseboard heater in a dwelling unit.

_____   _____ 12. A clock shall be permitted to be installed on the small appliance circuit.

_____   _____ 13. An island countertop with a short dimension of 10 in. or greater shall have a receptacle outlet for each 4 ft. in a dwelling unit.

_____   _____ 14. A bathroom is an area including a basin(s), with a toilet, tub or shower.

_____   _____ 15. Residential outdoor outlets are not required to be GFCI-protected.

_____   _____ 16. All 15 or 20 amp, 125 volt receptacles shall not be required to be provided with GFCI-protection if located in the bathroom.

16-23

Section	Answer

17. GFCI-protection shall be provided for all receptacles in the kitchen to serve countertop surfaces.

18. A receptacle outlet shall be installed, at a minimum of 10 ft. from and not more than 25 ft. from the inside walls of a pool. (General Rule)

19. Receptacle outlets installed by a hydromassage tub shall be provided with GFCI-protection if they are located within 5 ft..

20. GFCI-protection shall be installed for all 120 volt receptacle outlets on construction sites which are used for electric hand tools.

21. The NEC allows a floor plug to be used to provide the required number of receptacle outlets in a _____ unit.

22. Receptacle outlets shall be installed for each peninsular countertop with a long dimension of _____ in. or greater and a short dimension of _____ in. or greater.

23. Grade level access shall be considered _____ ft. _____ in. or less from finished grade.

24. At least _____ GFCI-protected receptacle shall be installed in a basement.

25. Receptacles located inside enclosed rooms and within _____ ft. of the inside walls of the pool shall not be required to be GFCI-protected.

26. Receptacle outlets shall be installed so that there is no point on the wall is greater than:
    (a) 4 ft.
    (b) 5 ft.
    (c) 6 ft.
    (d) 10 ft.

27. Receptacle outlets in a dwelling unit shall not serve as one of the required outlets if located over:
    (a) 4 ft. 6 in. from the floor
    (b) 5 ft. 6 in. from the floor
    (c) 6 ft. 6 in. from the floor
    (d) 7 ft. 6 in. from the floor

28. What is considered wall space when spacing receptacles:
    (a) Sliding glass doors
    (b) Sliding panels
    (c) Fixed panels
    (d) None of the above

29. Which of the following countertops are present in a dwelling unit:
    (a) Wall-type
    (b) Peninsular
    (c) Island
    (d) All of the above

	Section	Answer

30. All receptacle outlets in a dwelling unit installed in kitchens shall be GFCI-protected if located within:
    - (a) 4 ft. 0 in. of the sink
    - (b) 5 ft. 6 in. of the sink
    - (c) 6 ft. 0 in. of the sink
    - (d) None of the above

31. What is the minimum wall space for receptacle outlets to be installed in a dwelling unit:
    - (a) 1 ft.
    - (b) 2 ft.
    - (c) 6 ft.
    - (d) 12 ft.

32. When installing receptacles in a dwelling unit for hallways, at least one receptacle outlet shall be installed at a minimum of:
    - (a) 2 ft. or more in length
    - (b) 6 ft. of more in length
    - (c) 10 ft. or more in length
    - (d) 12 ft. or more in length

33. Receptacle outlets shall be GFCI-protected if they are located from the inside walls of an indoor spa or hot tub at:
    - (a) 2 ft. to 10 ft.
    - (b) 5 ft. to 10 ft.
    - (c) 10 ft. to 20 ft.
    - (d) 10 ft. to 25 ft.

17

Lighting and Switching Outlets

Lighting outlets can be installed in specified locations to ensure the proper illumination for residential, commercial and industrial locations. Lighting shall be provided by luminaires (lighting fixtures), controlled by a wall switch or by table lamps, floor lamps, swag lamps, etc., which are cord-and-plug connected into wall switch controlled receptacles. Pull-chain lighting fixtures may be installed without wall switches in some locations and under certain conditions of use.

Luminaires (lighting fixtures) supported by metallic and nonmetallic boxes are either ceiling mounted or wall mounted. For the convenience of the user, wall-switched receptacle outlets can be mounted to boxes installed in the wall, baseboard, or floor and may be used to cord-and-plug connect table or floor lamps.

Switches may be installed to operate luminaires (fixtures) and receptacles from more than one location in a residential, commercial and industrial location. Switching outlets are mounted on the wall at convenient locations and heights to switch the lighting outlets or receptacle outlets on and off. Only in residential occupancies are lighting outlets and switches mandated by the NEC to be installed in specific locations to switch and illuminate certain areas.

Note that where the rules of the 2005 NEC apply, the term luminaire(s) is used before fixture(s) in the text.

GROUNDING LUMINAIRES
410.18(A); 410.21

In residential, commercial, and industrial locations, the exposed metal parts of luminaires (fixtures) shall be grounded with an equipment grounding conductor (EGC) if the branch-circuit is provided with an EGC. The EGC shall be selected from any of the wiring methods listed in **250.118** and sized per **Table 250.122**, based upon the size OCPD. Section 410-92 of the 1971 NEC and early editions required luminaires (fixtures), with metal parts, to be used with metallic wiring systems such as metal conduit or metal clad cables.

The metal clad AC cables (BX) and metal clad cables (MC) were used to ground the exposed metal parts. The metal of metal conduits such as rigid metal conduit and EMT was mostly used as a grounding means. Copper or aluminum conductors were also pulled in conduits and utilized as an additional grounding means. Because nonmetallic wiring systems were not always equipped with an EGC, the 1975 NEC in **410.18(A)** required the exposed metal parts of fixtures to be grounded with an approved grounding means per **250.118**. Section **250.148(A)** required the metal of boxes supporting fixtures to be grounded with the metal of the conduit or cable or with an EGC. Since the 1975 NEC, luminaires (fixtures) with exposed metal parts are required to be grounded in new or existing installations. **(See Figure 17-1)**

UNGROUNDED LUMINAIRES (FIXTURES)
410.18(B)

The branch-circuit wiring in older sites of residential, commercial, and industrial locations does not have an EGC in the nonmetallic-sheathed cable (Romex) or in knob-and-tube wiring systems to ground the exposed metal parts of lighting fixtures. Lighting fixtures in older facilities were not required to be grounded with an EGC. Lighting fixtures were not required by the NEC to be grounded with an EGC until the 1975 edition of the code.

Part R and 410-91 through 410-93 in the 1971 NEC (grounding lighting fixtures) was deleted and relocated to Part E and 410-18(a) and (b) of the 1975 NEC.

The Ex. to Sec. 410-93 in the 1971 NEC permitted lighting fixtures with metal parts to be wired with nonmetallic raceways and nonmetallic-sheathed cables. If a metal cable was used as a wiring method, it had to have a listed grounding means. To accomplish this rule, the cable had to be AC (BX) or metal clad (MC) of the grounding type.

In those days, nonmetallic wiring systems were systems that did not have an equipment grounding means. Such wiring systems were knob-and-tube, nonmetallic raceways, or nonmetallic-sheathed cable (Romex).

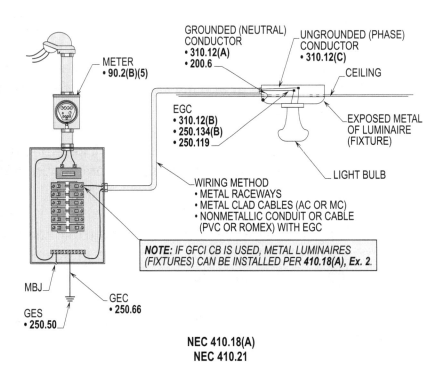

Figure 17-1. The metal parts of luminaires (fixtures) have been required to be grounded in new or existing work since the 1975 NEC. For exceptions to the rule, see **Ex.'s 1** and **2 to 410.18(A)**.

Lighting and Switching Outlets

The Ex. to Sec. 410-93 in the 71 NEC allowed metal fixtures and boxes that were mounted on nonconducting ceilings or walls and located not less than 8 ft. vertically or 5 ft. horizontally from grounded surfaces to be installed without grounding the metal of the fixtures and boxes. **(See Figure 17-2)**

> **Design Tip:** The 1971 NEC and earlier editions required metal boxes, that were not grounded, to be located 5 ft. from a bathtub or shower. It was the metal box and not the snap switch that required either grounding or to be located 5 ft. from the bathtub or shower. (See **404.4** of the 2002 NEC)

The 1975 NEC required the metal boxes to be grounded with an EGC which had to be installed in nonmetallic cables from the factory, or the use of metal conduits or the metal clad of cables has to provide such grounding. Therefore, the 5 ft. requirement was deleted and no longer required due to a grounding means being provided in the wiring method. See **404.4** and **406.8(C)** for rules pertaining to field wiring switches and receptacle outlets in the bathtub area.

> **Design Tip:** Boxes were not required for the mounting of lighting fixtures until the 1928 edition of the NEC. Because of this requirement, old existing occupancies may not have boxes installed for the mounting and supporting of lighting fixtures and switches.

REPLACEMENT LUMINAIRES (FIXTURES) 410.18(B)

Luminaires with exposed metal parts shall not be used to replace fixtures on existing wiring systems that are not equipped with an equipment grounding means. Section **410.18(B)** in the 1975 NEC as well as present editions require a lighting fixture with an insulated material to be used to replace an existing metal fixture in an existing wiring system. If luminaires (fixtures) with exposed metal parts are used for replacements, a branch-circuit with an EGC shall be provided per **410.18(B)**. **(See Figure 17-3)**

Care shall be taken when replacing an existing fixture with a new one. If a luminaire (fixture) with exposed metal parts is installed as a replacement, use an insulating nonmetallic type or any insulating type that isolate the metal of the fixture. The branch-circuit shall have a wiring method that has an EGC or installed, as permitted by the one of the exceptions to **410.18(B)**. See **250.118, 250.122** and **250.134(B)** for the selection, routing, and termination of the EGC. Note that the above rules and regulations apply to residential, commercial and industrial locations.

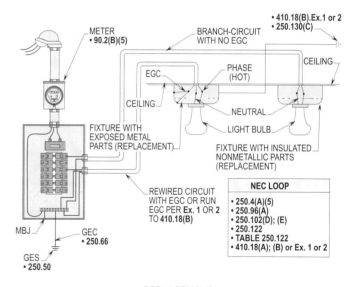

REPLACEMENT
NEC 410.18(B) OR Ex. 1 OR 2

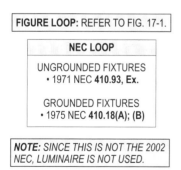

NEC 410.18(B)

Figure 17-2. The metal exposed parts of lighting fixtures were not required to be grounded by the 1971 NEC and earlier editions. The 1975 NEC and later editions do require exposed metal parts of lighting fixtures to be grounded.

Figure 17-3. Since the publication of the 1975 NEC, replacement fixtures installed on branch-circuits without an EGC must be of the insulated or nonmetallic type. The parts of fixtures that are not insulated to isolate the exposed metal parts of the lighting fixtures shall be wired with a branch-circuit having an EGC to ground all metal parts or an EGC can be installed, as permitted by **410.18(B), Ex. 1** or **2**.

NUMBER ON A CIRCUIT
210.11(A)

In residential dwelling units, the number of lighting outlets that may be connected to a 15 or 20 amp general purpose branch-circuit can be determined by multiplying the rating of the branch-circuit by 120 volts and dividing by 3 VA per square foot.

> **For example:** The number of lighting outlets allowed on a 15 amp branch-circuit in a dwelling unit is determined by the following procedure.
>
> **Step 1:** Finding VA
> **210.11(A)**
> VA = V x OCPD
> VA = 120 V x 15 A
> VA = 1800
>
> **Step 2:** Finding number of outlets
> No. = VA ÷ 3 VA sq. ft.
> No. = 1800 ÷ 3
> No. = 600 sq. ft.
>
> **Solution: In a residential dwelling unit, there is no limit to the number of lighting outlets permitted in the 600 sq. ft. area.**

The concept of the number of outlets allowed on a 15 or 20 amp general purpose branch-circuit is more easily understood by referencing **220.12**, **Table 220.12** and **210.11(A) and (B)**. **Table 220.12** requires 3 VA per sq. ft. times the sq. ft. area of a dwelling unit to determine the VA rating to supply the number of outlets for general purpose lighting and receptacle outlets. Notice that the reference ª by the dwelling units, **Table 220.12,** refers to the footnote and verifies this procedure for determining the number of outlets on a branch-circuit.

Inspection authorities that disagree with this concept and want to limit the number, usually apply the 1.5 amp (180 VA ÷ 120 V = 1.5) method per **220.4(I)**. The rating of the OCPD of 15 amps is divided by 1.5 amp to determine the number of lighting outlets permitted on a 15 or 20 amp branch-circuit. A 15 amp OCPD divided by 1.5 amps (15 A OCPD ÷ 1.5 A = 10) permits ten lighting outlets to be connected to a 15 amp general purpose branch-circuit.

The Authority Having Jurisdiction (AHJ) may permit any number of lighting outlets high or low by the local electrical ordinance to be connected to a 15 or 20 amp general purpose branch-circuit.

For example, by applying such local codes, 15 amp general purpose branch-circuit is permitted to have ten outlets installed for lighting and a 20 amp circuit is permitted to have only thirteen outlets for lighting. **(See Figure 17-4(a))**

> **For example:** Finding the number of lighting outlets on a 20 amp general purpose branch-circuit.
>
> **Step 1:** Finding VA
> **210.11(B)**
> VA = V x OCPD
> VA = 120 x 20
> VA = 2400
>
> **Step 2:** Finding # of outlets
> # = VA ÷ 3 VA per sq. ft.
> # = 2400 ÷ 3 VA
> # = 800 sq. ft.
>
> **Solution: In a residential dwelling unit, there is not a limit to the number of lighting outlets permitted in the 800 sq. ft. area.**

Design Tip: There are cases where the AHJ wishes to limit the number of outlets on a circuit, see text for methods used to determine the number of lighting outlets allowed on a general purpose branch-circuit.

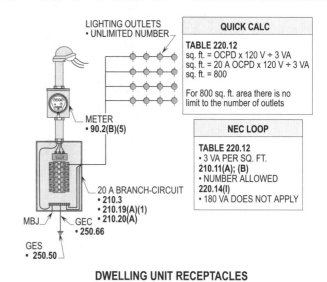

DWELLING UNIT RECEPTACLES
NEC 210.11(B)

Figure 17-4(a). Section **210.11(B)** can be used to determine the number of lighting outlets permitted on a general-purpose branch-circuit for a dwelling unit.

NUMBER ON A CIRCUIT
210.11(A)

In commercial and industrial locations, the outlets on a general-purpose branch-circuit shall be computed at 180 VA each or the load rating, whichever is greater. The number of outlets times 180 VA times 100 percent is used to compute the load of a branch-circuit supplying outlets of noncontinuous operation. The number of outlets times 180 VA times 125 percent is used to compute the load of a branch-circuit supplying outlets of continuous operation. OCPD's and conductors shall be computed at 125 percent and selected per **240.4** and **240.6(A)** for OCPD's and **Table 310.16** for conductors.

For example: What is the VA rating for thirteen receptacle outlets supplying cord-and-plug connected loads used at noncontinuous operation?

Noncontinuous operation

Step 1: Finding VA
220.14(I); (11); 210.19(A)(1)
VA = # of outlets x 180 VA x 100%
VA = 13 x 180 VA x 100%
VA = 2340

Solution: The VA is 2340 and the number of outlets at noncontinuous operation are limited to 13 as computed.

For example: What is the VA rating for ten receptacle outlets supplying cord-and-plug connected loads used at continuous operation?

Continuous operation

Step 1: Finding VA
220.14(I); (11); 210.19(A)(1)
VA = # of outlets x 180 VA x 125%
VA = 10 x 180 VA x 125%
VA = 2250

Solution: The VA is 2250 for ten outlets uses at continuous operation.

See **Figure 17-4(b)** for calculating the number of outlets allowed on a branch-circuit for commercial and industrial locations.

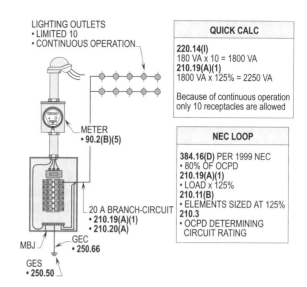

Figure 17-4(b). Receptacles in commercial and industrial locations must be calculated at continuous or noncontinuous operation based on each outlet at a minimum of 180 VA.

LIGHTING OUTLETS IN DWELLING UNITS
210.70(A)

At least one wall switch controlled lighting outlet is required in all habitable rooms, halls, stairways, attached garages or detached garages with electric power, bathrooms, and outdoor exits and entrances to the dwelling unit. Kitchens and bathrooms shall have a lighting outlet on the ceiling or wall that is controlled by a wall switch. In addition, a lighting fixture shall be installed in an unfinished or finished basement, attic, or crawl space used for storage or for air handling equipment, etc. Note that in some cases, these rules pertaining to lighting outlets may be applied to commercial and industrial locations as well as residential. Such will be noted in the text when appropriate. **(See Figure 17-5)**

LIGHTING OUTLETS IN HABITABLE ROOMS
210.70(A)(1)

At least one lighting outlet shall be installed to provide lighting for the illumination of habitable rooms. Habitable rooms are rooms in the dwelling unit such as the bedroom, living room, den, dining room, breakfast room, etc. The lighting outlets may be installed in the ceiling or on the wall, if the location in which they are mounted provides proper lighting. Section **210.70(A)(1), Ex. 1** allows wall-switched receptacle outlets to be mounted on the wall at a height so floor lamps or table lamps may be cord-and-plug connected to provide proper lighting. The only rooms in a dwelling unit not allowed to have a wall-switched receptacle outlet to provide lighting is the kitchen and bathroom(s). The kitchen is required to have at least one lighting outlet mounted to the ceiling or wall that is switched by a wall switch. A pull chain lighting outlet installed over the sink or a lighting outlet in the vent-a-hood is in addition to and shall not be counted as the required lighting outlet per **210.70(A)(1)**. **(See Figure 17-6)**

Design Tip: Ex. 2 to 210.70(A)(1) allows lighting outlets to be controlled by occupancy sensors, listed for such use. This rule allows an occupancy sensor to control lighting outlets in habitable rooms of dwelling units which includes bathrooms, hallways, stairways, garages, and at each outdoor entrance and exit. However, a manual override that will allow the sensor to function as a wall switch shall be provided.

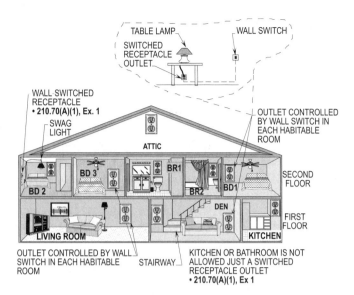

SWITCHED OUTLETS
NEC 210.70(A)(1), Ex. 1

Figure 17-6. Habitable rooms in a dwelling unit shall be provided with a wall or ceiling lighting outlet switched by a wall switch. Rooms, except the kitchen and bathroom(s), are permitted to have a switched receptacle outlet with cord-and-plug connected table lamps or floor lamps to provide the necessary lighting.

LIGHTING OUTLETS IN BATHROOMS
210.70(A)(1)

Lighting outlets are required in bathrooms to provide lighting for bathing and personal care. The lighting outlets may be installed in the ceiling or on the wall above the mirror. Lighting outlets are sometimes installed over bathtubs or in showers to prevent shadows due to the location of the required lighting outlets per **210.70(A)(1)**. Luminaires (lighting fixtures) installed over bathtubs or in showers are usually surface gasket or recessed type. At least one lighting outlet shall be provided in the bathroom and it shall be wall-switched. **(See Figure 17-7)** Check with the inspector for the type that is allowed.

Design Tip: The lighting outlet could be a combination vent/fan/heater/lighting fixture which complies with **210.70(A)(1)**.

WALL-SWITCHED RECEPTACLE
LIGHTING OUTLETS
210.70(A)(1), Ex. 1

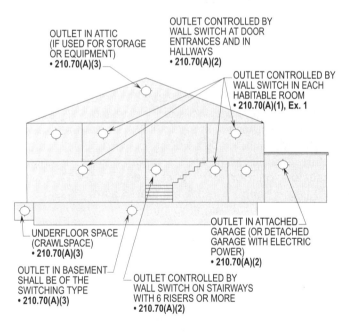

HABITABLE ROOMS
NEC 210.70(A)

Figure 17-5. At least one lighting outlet shall be installed in these locations to provide proper lighting for safety, pleasure, etc. For receptacle outlets, see **Figure 16-9**.

A wall-switched receptacle outlet may be used in lieu of a wall-switched lighting outlet in habitable rooms other than kitchens and bathrooms. A wall-switched receptacle with cord-and-plug connected table lamps or floor lamps that are used to provide lighting per **210.70(A)(1), Ex. 1** may be used. **(See Figure 17-8)**

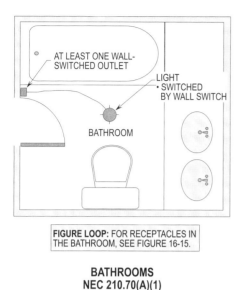

Figure 17-7. Bathrooms are required to have a wall-switched lighting outlet. Others are in addition to this switch.

LIGHTING OUTLETS IN HALLWAYS 210.70(A)(2)

The hallways in dwelling units shall have a wall-switched ceiling or wall mounted lighting outlet to provide proper lighting. Section **210.70(A)(2), Ex.** allows remote, central, or automatic control of lighting outlets installed in hallways.

The control method used to switch lighting outlets in hallways shall turn the lighting outlets on and off as needed to provide the necessary lighting. **(See Figure 17-9)**

LIGHTING OUTLETS IN STAIRWAYS 210.70(A)(2)

Lighting outlets shall be installed in interior stairways for illumination and a wall switch shall be provided at each level to control the lighting outlets. Where there is a difference between floor levels of six risers or more, a wall switch to control the lighting outlet or outlets shall be provided at each level. A lighting outlet at a door on a landing in a stairway that provides the proper lighting and switching complying with **210.70(A)(2)** is also required. **(See Figure 17-10)**

For remote, central, or automatic control of lighting in multifamily dwellings for hallways and stairways, it may be desirable to locate switches or use time clocks where they may not be intentionally or inadvertently turned to the "off" position.

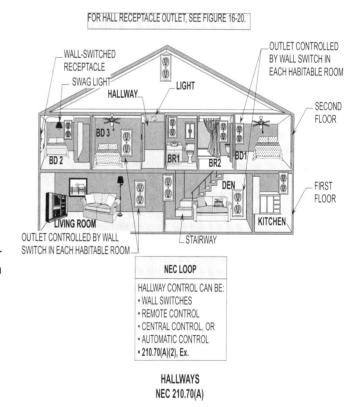

Figure 17-9. Hallways in dwelling units shall have a lighting outlet installed that is switched by a wall switch.

Stallcup's Electrical Design Book

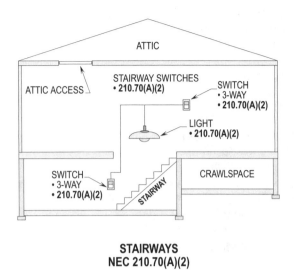

STAIRWAYS
NEC 210.70(A)(2)

Figure 17-10. Stairways in dwelling units shall have wall-switched lighting outlets installed to provide proper illumination. Floor levels with six or more risers between them are considered different levels.

LIGHTING OUTLETS IN GARAGES
210.70(A)(2)

One lighting outlet shall be installed in the garage to provide lighting for parking vehicles. If the utility room is in the garage, there shall be lighting to see how to wash and dry clothes. One or more of the lighting outlets shall be controlled by a wall-mounted switch.

A lighting outlet is not required at a vehicle door in an attached or unattached garage because it is not considered an outdoor entrance per **210.70(A)(2)**.

A detached garage with power routed to it requires a lighting outlet. If the detached garage has a walkway between the dwelling unit and garage, a lighting outlet is usually installed with a set of three-way switches to control the lighting outlet at either the dwelling or garage. **(See Figure 17-11)**

LIGHTING OUTLETS AT
OUTSIDE DOORS
210.70(A)(2)

A lighting outlet is required at each outside door that is classified as an entrance or exit. This lighting outlet must be installed in a location that provides lighting at the door and steps, to prevent people from accidentally falling due to darkness. The lighting outlet may be mounted on the ceiling or wall and shall be wall-switched. The wall switch should be located by the door in a location so that the control of the lighting outlet and fixture can easily be found. **(See Figure 17-12)**

Design Tip: The switch should not be located behind a closing door because of safety as well as easy access for switching purposes by the user.

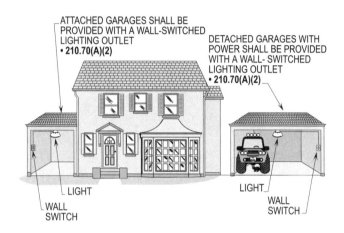

GARAGES
NEC 210.70(A)(2)

Figure 17-11. Attached garages shall be provided with a wall-switched lighting outlet. Detached garages with power are also required to have lighting outlets controlled by a wall switch. These switches are usually the three-way type. (For receptacle outlets see Figure 16-19)

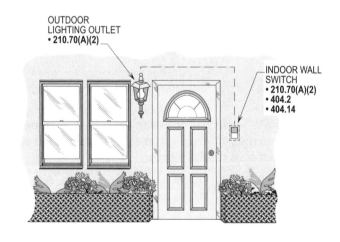

OUTDOORS
NEC 210.70(A)(2)

Figure 17-12. Wall switches installed in the dwelling unit shall control a lighting outlet that is located at each door used as an entrance or exit.

Lighting and Switching Outlets

LIGHTING OUTLETS IN UTILITY ROOMS
210.70(A)(3)

One wall-switched lighting outlet shall be required in utility rooms. The wall switch shall be located at the entry of the utility room. At least one lighting outlet is required, if the utility room is used for storage or equipment that needs serving. If the washing machine and clothes dryer are located in the utility room, lighting outlets are required to see how to wash and dry clothes and service the machines. **(See Figure 17-13)**

Design Tip: The lighting outlet located in an utility room does not necessarily require a wall switch to switch the lighting unit on or off. In other words, a properly located pull chain could serve as such lighting outlet.

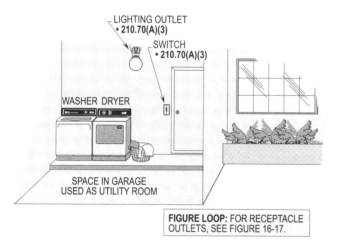

**UTILITY ROOMS
NEC 210.70(A)(3)**

Figure 17-13. A lighting outlet controlled by a pull chain or wall switch shall be installed in a utility room. The utility room may be located in the dwelling unit or garage.

The lighting outlet installed in the utility room may be surface or recess mounted. It could be cove lighting or any other wall-switched type lighting. The type luminaire (fixture) may be incandescent or fluorescent, whichever type the designer wants to install. The type really depends on the foot candles needed to illuminate the area.

LIGHTING OUTLETS IN BASEMENTS
210.70(A)(3)

At least one lighting outlet shall be required for the illumination of basements where there is storage or equipment installed that requires servicing. Such equipment can be air-handling equipment, refrigeration equipment, air-conditioning equipment, etc. Table saws, routers, sanders, etc. may be located in the basement for the purpose of a workshop. More than one lighting outlet may be required, in this case, to provide the proper lighting. Lighting to service the sump pump should be provided. The sump pump, with its equipment, is usually considered equipment requiring servicing due to being located in a pit. **(See Figure 17-14)**

Section **210.70(A)(3)** requires the lighting outlet to be switched. However, it does not specifically state that it must be a wall-switched lighting outlet. If **210.70(A)(3)** required this lighting outlet to be wall-switched, it wouldn't comply with the various mechanical codes which allows it to be a pull chain type under certain conditions of use.

Design Tip: The Uniform Mechanical Code requires a wall switch while the Standard Mechanical Code requires only a switched lighting outlet which could be a pull-chain type luminaire (fixture).

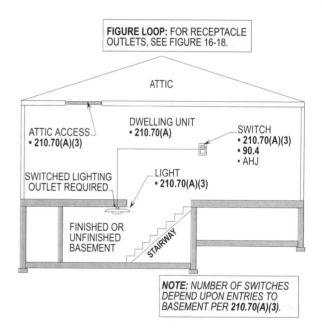

**BASEMENTS
NEC 210.70(A)(3)**

Figure 17-14. A switched lighting outlet shall be required in the basement of a dwelling unit to provide lighting for safe entrance, exiting, and servicing of equipment. This lighting outlet may be controlled by a pull chain or wall switch.

LIGHTING OUTLETS IN ATTICS
210.70(A)(3)

At least one lighting outlet shall be required in attics that are floored and used for storage or for electric equipment that requires servicing. A switch is required to turn the

17-9

lighting outlet on and off. The switch may be incorporated into the luminaire (lighting fixture) or be controlled by a wall switch at the point of entry into the attic. Some inspectors permit a wall switch only, while others allow a pull-chain at the point of entry. Either complies, depending on which mechanical code is used. The pull-chain is usually controlled by an extended string from the luminaire (lighting fixture) at the point of entry. **(See Figure 17-15)**

Design Tip: See Uniform Mechanical Code, Southern Mechanical Code or other appropriate codes for requirements concerning the switching rules for lighting outlets installed in attics to service the HVAC.

Note that **210.70(C)** requires a wall-switched lighting outlet for such use in commercial and industrial locations.

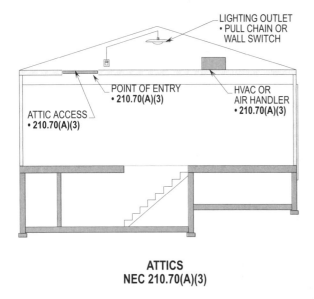

ATTICS
NEC 210.70(A)(3)

Figure 17-15. Attic space used for storage or equipment shall have a pull chain or wall-switched lighting outlet. **Note:** See local mechanical code for the type required and receptacle outlet requirements.

LIGHTING OUTLETS IN UNDERFLOOR SPACES
210.70(A) AND (C)

A switched lighting outlet shall be required at underfloor spaces or crawlspaces where the space is used for storage or for equipment that requires servicing. Crawlspaces are located at or below grade level. Underfloor spaces are usually located under pier and beam type constructed dwelling units or those constructed on the side of a hill. Lighting outlets shall not be required where there is no storage or equipment needing service. **(See Figure 17-16)**

Note that **210.70(C)** requires a wall-switched lighting outlet for such use in commercial and industrial locations.

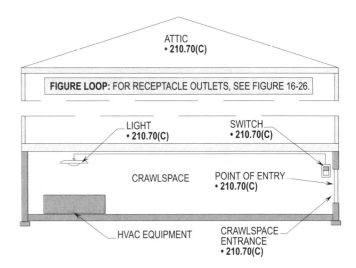

UNDERFLOOR SPACES
NEC 210.70(C)

Figure 17-16. Underfloor spaces such as crawlspaces shall be provided with a lighting outlet that is controlled by a pull chain or wall switch. The lighting outlet shall not be required, if the space is not used for storage or HVAC.

LIGHTING OUTLETS OVER BATHTUBS
410.4(D)

Hanging luminaires (fixtures), track lighting, and ceiling (paddle) fans are not permitted to be hung over bathtubs. There is a hazard to electrical shock when changing light bulbs and danger to electrocution due to grabbing the hanging luminaire (fixture) for support, if the bather should slip when stepping from the tub. Due to these hazardous conditions, hanging units shall be installed at least 8 ft. (2.5 m) vertically and shall be located at least 3 ft. (900 mm) horizontally from the tub or shower threshold in all directions.

The authority having jurisdiction usually requires lumiaries (lighting fixtures) to be surface mounted with a gasket or be recessed, if the bathtub is used for bathing and showering. Bathtubs used just for bathing are usually permitted to have regular surface mounted luminaires (fixtures) installed. (Check with AHJ) **(See Figure 17-17)**

Design Tip: The bathroom is not classified as a wet location and these rules apply for hydromassage tubs and bathtubs where lighting fixtures are concerned.

Lighting and Switching Outlets

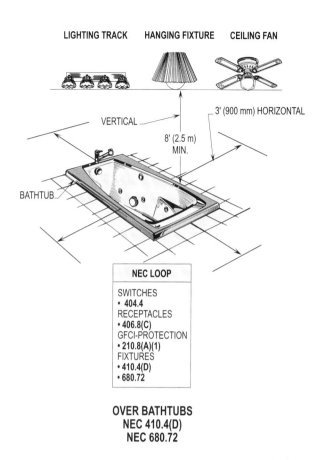

Figure 17-17. Hanging luminaires (fixtures), ceiling fans, and track lighting are permitted to be hung over bathtubs if installed at least 8 ft. (2.5 m) over the tub or shower threshold and located at least 3 ft. (900 mm) from the tub or shower threshold in all directions.

LIGHTING OUTLETS IN CLOTHES CLOSETS
410.8(A) THRU (D)

In residential, commercial, and industrial locations, **210.70(A)** does not require a lighting outlet to be installed in clothes closets. If a lighting outlet is installed in a clothes closet, **410.8** lists the requirements for locating and installing the lighting outlets. Section **410.8** is divided into four subdivisions. Subdivision **(A)** deals with the definitions of storage space. Subdivision **(B)** lists the types of luminaires (fixtures) that are permitted to be used. Subdivision **(C)** lists the types of luminaires (fixtures) that are not permitted. Subdivision **(D)** deals with the location in the closet where the luminaire (fixture) can be mounted.

STORAGE SPACE
410.8(A)

Section **410.8(A)** gives the definition of the storage space for the purpose of measuring and positioning luminaires (lighting fixtures). The storage space is the area for the storage of clothes, shoes, hats, etc.

Storage space above the rod shall be at least the shelf width or 12 in., whichever is greater. Storage space below the rod is 24 in. to the rod or 6 ft. above the floor, whichever is greater. The location of lighting fixtures, depending on the type, are measured and positioned from these boundaries.

TYPES PERMITTED
410.8(B)(1) AND (B)(2)

Section **410.8(B)** lists the types of luminaires (lighting fixtures) that are permitted in clothes closets. Surface-mounted or recessed fluorescent luminaires (fixtures) are allowed. Incandescent lamps, depending on wattage rating, get very hot and the elements, if accidentally broken, can fall into stored combustible materials and cause fires. The NEC requires incandescent luminaires (fixtures) to be provided with a lens, that covers the luminaire (fixture) completely to prevent such accidents.

Fluorescent lamps operate much cooler than incandescent bulbs and therefore, can be mounted closer to the storage area than the incandescent type.

TYPES NOT PERMITTED
410.8(C)

Section **410.8(C)** does not permit incandescent luminaires (fixtures) with open or partially enclosed light bulbs (pull-chain or keyless) to be installed. Pendant luminaires (fixtures) or lampholders such as rosettes or the hanging type are not permitted.

LOCATION
410.8(D)(1) THRU (D)(4)

Section **410.8(D)** lists the dimensions that luminaires (lighting fixtures) shall be positioned from the storage area. The luminaires (lighting fixtures) permitted are grouped into two types. They are of the surface mounted and recessed type. Surface-mounted luminaires (fixtures) are mounted to a ceiling box and recessed luminaires (lighting fixtures) are recessed in the ceiling with an approved can (Hat) listed for the purpose. Surface-mounted luminaires (fixtures) of the incandescent type shall be mounted at least 12 in. (300 mm) from the storage space (area). The incandescent luminaire (fixture) may be mounted on the wall above the door or ceiling per **410.8(D)(1)**.

17-11

Surface-mounted fluorescent luminaires (fixtures) shall have a clearance of at least 6 in. (150 mm) from the storage area. The luminaire (fixture) may be mounted on the wall above the door or on the ceiling in the same way as the incandescent luminaire (fixture) per **410.8(D)(2)**.

Recessed luminaires (lighting fixtures) are available as an incandescent or fluorescent type. The minimum clearance of 6 in. (150 mm) or less shall be maintained from the storage area whether incandescent or fluorescent recessed fixtures are installed per **410.8(D)(3) and (4)**. **(See Figure 17-18)**

crawlspaces. The switch shall be installed at the entry point to the attic or crawlspace. This rule provides a wall-switched outlet at the point of entry so the user or maintenance person does not have to search for such switch which could cause a safety problem. **(See Figure 17-20)**

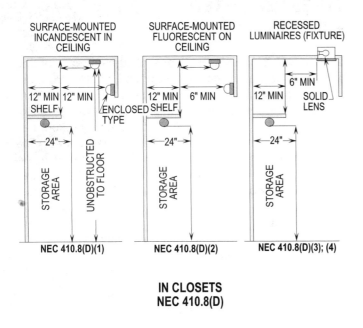
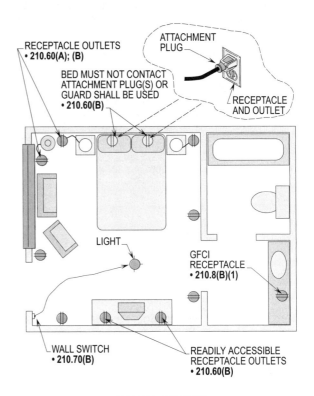

Figure 17-18. Installation requirements for luminaires (lighting fixtures) located in clothes closets.

GUEST ROOMS IN HOTELS AND MOTELS
210.70(B)

This rule requires at least one wall-switched controlled lighting outlet or receptacle for a table, floor, or hanging lamp to be installed in the guest rooms of hotels, motels, etc. **(See Figure 17-19)**

COMMERCIAL AND INDUSTRIAL ATTICS AND CRAWLSPACES
210.70(C)

One or more switch controlled lighting outlets shall be installed near equipment requiring service, such as heating or air-conditioning equipment installed in attics or

Figure 17-19. At least one wall-switched outlet shall be provided to switch a lamp in the guest rooms of hotels and motels.

ILLUMINATION FOR ELECTRICAL EQUIPMENT IN COMMERCIAL AND INDUSTRIAL LOCATIONS
110.26(D)

Illumination shall be provided for all working spaces about service equipment, switchboards, panelboards, or motor control centers installed indoors. Additional luminaires (lighting fixtures) shall not be required where the workspace is illuminated by an adjacent light source. In electrical equipment rooms, the illumination shall not be controlled by automatic means only. **(See Figure 17-21)**

Lighting and Switching Outlets

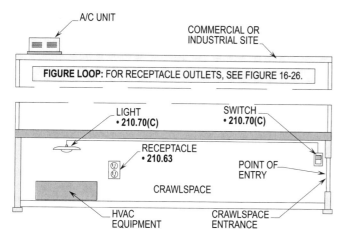

Figure 17-20. At least one wall-switched outlet shall be provided at crawlspace entry to switch light for storage and equipment service area.

RECESSED LUMINAIRES (LIGHTING FIXTURES) 410.65(C)

In residential, commercial and industrial locations, recessed luminaires (lighting fixtures) are found in new or existing installations. In such premises, per the 1987 NEC, all recessed luminaires (lighting fixtures) were required to be equipped with a thermal protector to prevent overheating. There have been two types allowed since the 1987 NEC and they are thermal protected (TP) and insulation covered (IC) with thermal protection. The thermal protected type shall be clear of all insulation by 3 in. (75 mm) on the top and sides per **410.66(B)**.

The luminaire (fixture) can (Hat) shall have a clearance of at least 1/2 in. (13 mm) from combustible materials such as wooden rafters per **410.66(A)(1)**. The IC recessed listed luminaire (fixture) may be covered with insulation and set against the wooden rafter for support per **410.66(A)(2)**.

In existing facilities with hung (suspended) ceilings, a suspended recessed luminaire (fixture) may be used per the 1984 NEC. All other types of recessed incandescent luminaires (lighting fixtures) must have a clearance of at least 3 in. (75 mm) from the top to the sides and be located at 1/2 in. (13 mm) from combustible material. Fluorescent recessed luminaires (fixtures) are also permitted to be installed. **(See Figure 17-22)**

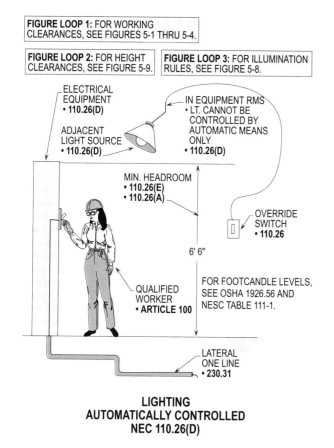

Figure 17-21. Lighting outlets providing illumination over electrical equipment may be controlled automatically. However, a regular switch for overriding purposes shall be provided.

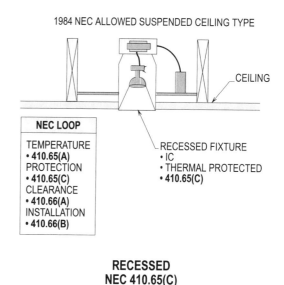

Figure 17-22. Recessed luminaires (lighting fixtures) may be utilized with certain installation requirements being applied.

DISCHARGE LIGHTING
410.80(B)

Discharge lighting systems requiring open secondary voltage of more than 1000 volts are not permitted to be installed inside or outside dwelling units. Open secondary voltage is the secondary output side of a ballast or transformer which supplies power to the electric discharge lighting unit. The primary is the input side and is supplied by a 120 volt branch-circuit which is protected by an OCPD per **210.20(A)**. **(See Figure 17-23)**

Design Tip: This rule has been interpreted to apply only to discharge lighting units installed inside. It clearly allows them to be installed both inside or outside of such premises.

Lighting outlets with luminaires (fixtures) or ceiling (paddle) fans shall be located at least 12 ft. (3.7 m) above the maximum water level of the pool per **680.22(B)(1)**. A swimming pool built at an existing premise may have luminaires (lighting fixtures) located in the area above the 5 ft. (1.5 m) horizontal boundary per **680.22(B)(3)**. The luminaire (fixture) shall be existing (not new) and rigidly attached to the structure. Luminaires (lighting fixtures) may be installed between 5 ft. (1.5 m) and 10 ft. (3 m) horizontally with GFCI-protected circuits per **680.22(B)(3)**.

Luminaires (lighting fixtures) located at safe heights over and around the pool protects personnel while they are servicing or changing lamps. Persons swimming in or playing near the pool will be protected from contacting live parts in a luminaire (fixture), if it is located at these required heights. **(See Figure 17-24)**

Lighting outlets and ceiling (paddle) fans shall be installed over swimming pools located inside such structures. The lighting outlets supporting the luminaires (fixtures) or ceiling fans shall be located at least 7 ft. 6 in. (2.3 m) above the maximum water level of the swimming pool. The branch-circuit supplying power to these fixtures shall be GFCI-protected per **680.6(B)(4)**. **(See Figure 17-25)**

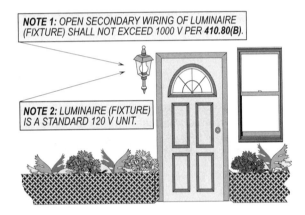

OPEN SECONDARY WIRING
NEC 410.80(B)

Figure 17-23. The open-circuit voltage of ballast and transformers supplying electric discharge lighting units shall not exceed 1000 volts when installed outside or inside dwelling units.

LIGHTING OUTLETS AT SWIMMING POOLS
680.22(B)

In residential, commercial, and industrial locations, lighting outlets used to support luminaires (fixtures) and ceiling fans shall be located at specific locations over or around swimming pools. The positioning and location of the lighting outlets are measured from the inside walls of the pool. This basic rule prohibits lighting outlets to be installed 5 ft. (1.5 m) horizontally or 5 ft. (1.5 m) vertically from the inside walls of the pool per **680.22(B)(3)**.

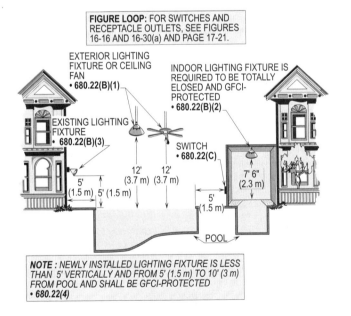

OUTDOORS
NEC 680.22(B)(1)

Figure 17-24. Location of luminaires (lighting fixtures) around a swimming pool shall be a certain height and distance from the inside walls of the pool. This rule protects personnel from electrical hazards and shock.

Lighting and Switching Outlets

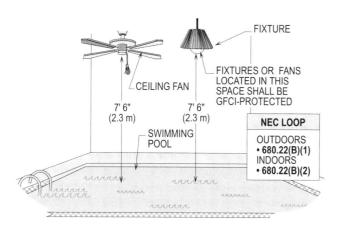

Figure 17-25. Installation requirements for installing lighting units and fans over swimming pools located inside.

UNDERWATER LIGHTING FIXTURES 680.23(A)(1) THRU (A)(8)

In residential, commercial and industrial locations, underwater luminaires (lighting fixtures) shall be installed in such a manner so as to prevent electrical shock to persons swimming in the pool. Luminaires (lighting fixtures) installed underwater shall be located at least 18 in. below the normal water level per **680.23(A)(5)**. Section **680.23(A)(5)** allows specially designed underwater luminaires (lighting fixtures) to be installed not less than 4 in. (100 mm) below the normal water level per **90.7** and **110.3(B)**.

Underwater luminaires (fixtures) shall be inherently equipped with thermal protection to protect against overheating when the fixture relies on submersion in water for safe operation. The luminaire (fixture) shall be supplied by a branch-circuit with 150 volts or less between conductors. The branch-circuit shall include a grounded neutral conductor. There are three types and they are as follows:

- Wet-niche
- Dry-niche
- No-niche

WET-NICHE 680.23(B)(1) THRU (B)(5)

Where used in residential, commercial and industrial locations, wet-niche luminaires (fixtures) shall be equipped with a metal or plastic forming shell that is approved for installation in swimming pool walls. The forming shell should be equipped with threaded entries to connect rigid metal conduit, IMC, LFNC or RNC, and be made with metallic or nonmetallic material. The conduit shall extend from the shell to a junction box on the deck or yard area. The deck box shall be located under the diving board or in another protected area used for this purpose. The deck box shall be located at least 4 ft. (1.2 m) from the inside walls of the pool and be at least 4 in. (100 mm) from the deck to the inside bottom of the deck box per **680.23(A)(5)** and not less than 8 in. (200 mm) from the maximum water level. The measurement that produces the greater elevation is naturally chosen.

For example, if 4 in. (100 mm), measured from the deck to the inside bottom of the box is greater than 8 in. (200 mm) from the maximum water level, the deck box shall be installed using the 4 in. (100 mm) measurement. **(See Figure 17-26)**

> **Design Tip:** The deck box may be located in a flower bed at the side of the facility where an approved conduit is routed all the way between the forming shell and deck box. If rigid metal brass conduit is used, it shall be routed as a complete system. In the same manner, if PVC is used, the 8 AWG solid or stranded copper conductor shall be pulled unbroken from the shell to the deck box, wherever it is located.

The 4 ft. (1.2 m) minimum distance of the deck box may be reduced with an effective barrier between the deck box and inside walls of the pool.

For example: A solid fence may be located 2 ft. from the pool with the deck box located on the opposite side of such barrier and pool. **(See Figure 17-27)**

Where RNC is used to connect the shell to the junction box (deck box), an 8 AWG solid or stranded insulated copper conductor shall be run through the RNC and connected to the forming shell. The termination of the 8 AWG conductor shall be potted to prevent corrosion. The cord end and terminals within the wet-niche luminaire (fixture) shall be sealed to prevent the entry of water per **680.23(B)(5)**. The EGC in the cord is used to ground the wet-niche luminaire (fixture) when it is removed from the shell to be serviced. **(See Figure 6-28)**

DRY-NICHE 680.23(C)

In the walls of pools installed in residential, commercial, and industrial locations, dry-niche luminaires (fixtures) may be installed instead of wet-niche luminaires (fixtures). This type of luminaire is installed outside the walls of the pool in closed recesses which provides for drainage of water that might accumulate. Dry-niche luminaires shall be wired with approved rigid metal conduit, IMC or RNC conduit with an EGC for each conduit entry. The luminaire shall have adequate provisions for drainage of water per **680.23(C)(1)**. **(See Figure 17-29)**

Stallcup's Electrical Design Book

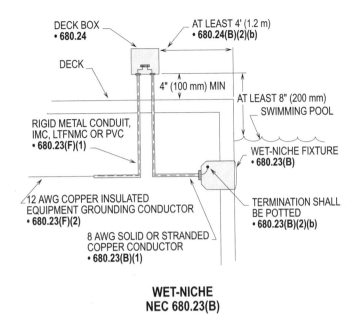

WET-NICHE
NEC 680.23(B)

Figure 17-26. The deck box shall be located at least 4 ft. (1.2 m) from the inside wall of the pool and have a height of 4 in. (100 mm) measured from the deck or 8 in. (200 mm) measured from the water, whichever is greater.

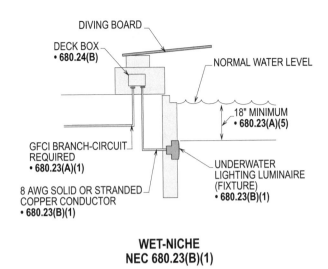

WET-NICHE
NEC 680.23(B)(1)

Figure 17-28. An 8 AWG insulated copper conductor shall be routed in RNC conduit to ground to the metal forming shell.

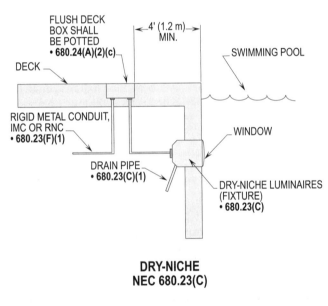

DRY-NICHE
NEC 680.23(C)

Figure 17-29. Installation requirements for dry-niche luminaires in swimming pool walls shall be carefully designed.

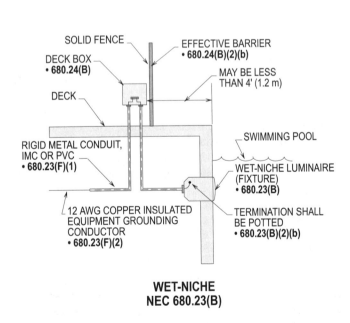

WET-NICHE
NEC 680.23(B)

Figure 17-27. The deck box located behind a solid permanent barrier may be located less than 4 ft. (1.2 m) from the inside walls of the pool. The barrier prevents easy access if properly designed and isolates personnel between box and pool.

NO-NICHE
680.23(D)

No-niche fixtures may be installed instead of wet or dry-niche fixtures. No-niche fixtures have no exposed metal parts, and they contain impact-resistant polymeric lenses. If installed in residential, commercial, and industrial pools, they shall be required to be of the listed type per **680.23(D)**, **90.7** and **110.3(B)**. **(See Figure 17-30)**

Lighting and Switching Outlets

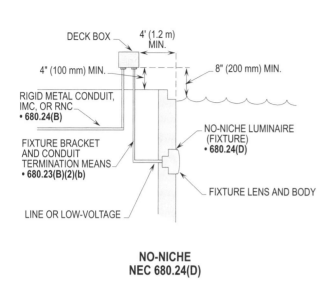

Figure 17-30. No-niche fixtures are now available in both line and low-voltage type. **Note:** All grounding connections are to be made to the mounting bracket to ensure grounding.

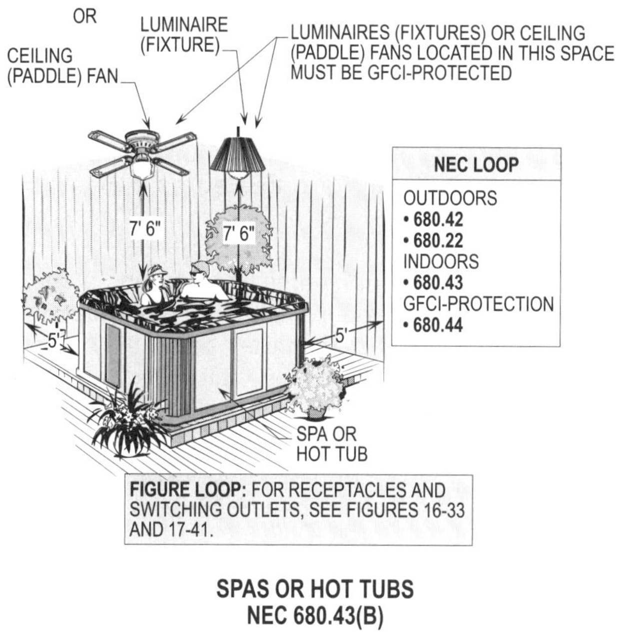

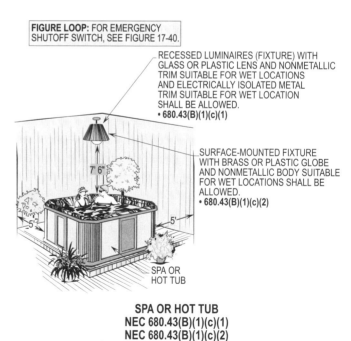

Figure 17-31. GFCI-protected lighting outlets with luminaires or ceiling (paddle) fans shall be located at least 7 ft. 6 in. from the maximum water level of the spa or hot tub. **Note:** Luminaires or ceiling (paddle) fans located over 12 ft. above the maximum water level are not required to be GFCI-protected.

LIGHTING OUTLETS OVER SPAS OR HOT TUBS
680.43(B)

Lighting outlets with luminaires and ceiling (paddle) fans located over spas or hot tubs shall have a clearance of 7 ft. 6 in. above the maximum water level. This clearance is required within 5 ft. in all directions. The supplying branch-circuit shall be GFCI-protected per **680.43(B)(1)(b)**. If luminaires and ceiling fans are located 12 ft. above spas or hot tubs, the supplying branch-circuit is not required to be GFCI-protected per **680.43(B)(1)(a)**. **(See Figure 17-31)**

Recessed and surface-mounted luminaires may be hung over a spa or hot tub, under certain conditions. Recessed luminaires with glass or plastic lens and nonmetallic trim suitable for wet locations are permitted to be hung less than 7 ft. 6 in. per **680.43(B)(1)(b); (c)**. Luminaires installed in this manner will protect users of spas and hot tubs from serious electrical shock. Surface-mounted luminaires with glass or plastic globes and nonmetallic bodies suitable for wet locations per **680.43(B)(c)(1); (2)** may also be used. Note that the above rules applies for residential, commercial, and industrial installed spas and hot tubs. **(See Figure 17-32)**

Figure 17-32. Location and installation requirements for recessed surface-mounted luminaires over hot tubs or spas.

LIGHTING OUTLETS OVER HYDROMASSAGE BATHTUBS
680.72

In residential, commercial, and industrial locations, hydromassage bathtubs are treated as conventional bathtubs.

Lighting outlets and luminaires shall be installed per **410.4(D)**. All the elements for hydromassage bathtubs shall be wired with GFCI-protected circuits whether they are cord-and-plug connected or permanently hard wired per **680.72**. Receptacle outlets shall be GFCI-protected per **210.8(A)(1)**. Motors used to circulate the water shall be accessible for maintenance and service by a removable cover or trap door per **430.14** and **680.73**. **(See Figure 17-33)**

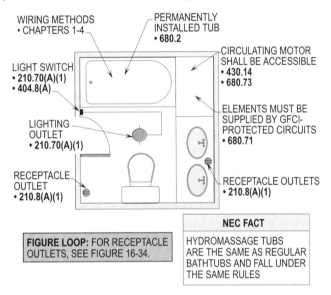

Figure 17-33. Installation requirements for lighting outlets over a hydromassage bathtub.

SWITCHING OUTLETS
404.2

In residential, commercial, and industrial locations, switches are used to control wiring systems, they provide easy access to complete circuits for energizing and deenergizing electrical equipment. A variety of types and styles is available to meet the various requirements and locations. Switches are designed so that they are ON when they are in the up position and OFF in the down position. This position indicates, in a precise manner, whether they are ON or OFF.

Switches shall disconnect only the ungrounded hot phase conductors to an electrical load. They shall not disconnect the grounded neutral conductor unless the hot and neutral conductors are disconnected simultaneously. Switches shall disconnect all hot phase conductors from the terminals of a screw shell type lampholder. Switches shall be located and mounted at specific heights that are acceptable to the user per **404.8(A)**.

TYPES OF SWITCHES
404.2(A)

Single-pole switches have two terminals and are designed to switch only one hot phase and return conductor. A single conductor must not be routed between the switch and the fixture. Where nonmetallic-sheathed cable (Romex) or armored cable (BX) can be used as the wiring method, the white or natural gray conductor in the cable may be used as a switch leg while the black conductor in the cable may be used as the return leg (switch leg) per **200.7(C)(2)**. The switch leg supplies power to the line side of the switch and the return line from the load side of the switch connects power to the luminaire or other type of load. The conductor of the other side (return) of the circuit shall be run with the switched (leg) conductor to prevent induced currents. Induced currents are only a problem where metal conduits or metal-clad cables are used as the wiring method. Since current in one conductor is equal to and opposite in direction to current in the other conductor, the inductive effect is canceled. **(See Figure 17-34)**

> **Design Tip:** Always run the switch leg and return leg in the same conduit or cable. See **300.3(B)** and **300.20** for more information on methods used to prevent inductive heating.

Three-way switches have three terminals to connect conductors. The odd color terminal can be used for the connection of the hot supply or switch leg conductor. The other two of the same color can be used for the traveler conductors. Four-way switches have the same color terminals identified for the travelers. Travelers for three-way and four-way switches may be routed alone in metal conduits or metal clad cables, where used as wiring methods in residential, commercial, and industrial locations. The grounded conductor (neutral) does not have to be run with the travelers. A set of three-ways (two switches) may be used to switch one or more luminaires from two locations. A set of three-ways and a four-way may control one or more luminaires from three locations. For example, a set of three-ways with three four-ways will switch ON and OFF one or more luminaires in five different locations.

Lighting and Switching Outlets

Design Tip: With each four-way switch added, the luminaire(s) may be controlled at additional locations. **(See Figure 17-35)**

SWITCHING OUTLET HEIGHTS 404.8(A)

In residential, commercial, and industrial locations, switches shall be located at a height accessible to the user. To comply with this requirement, switches should be located at 6 ft. 7 in. (2 m) or less from the toggle of the switch in the ON position. (Most inspectors apply this rule.) Switching outlet boxes are usually installed at 56 in. to the center of the box to provide a height accessible to users. Lesser heights may be used where needed. **(See Figure 17-36)**

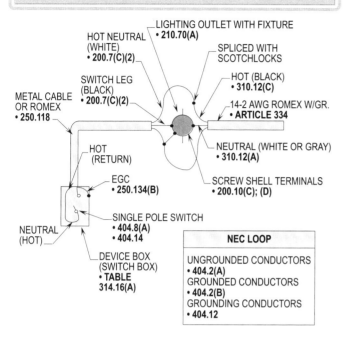

Design Tip: Switches may be located at any height below the 6 ft. 7 in. (2 m) to the toggle in the ON position as long as they comply with **404.8(A)**. This rule is usually applied by electrical inspectors to ensure a maximum height.

Figure 17-34. The white or gray conductor in a romex cable (nonmetallic-sheathed cable) can be used to run the hot conductor to a switch and return as a black conductor (switch leg) to the luminaire.

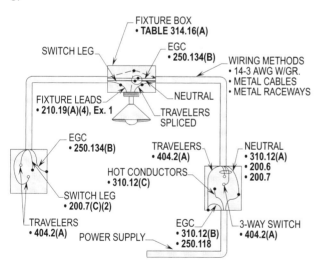

THREE-WAY SWITCHES
NEC 404.2(A)

Figure 17-35. Three-way switches may be used to switch a luminaire from two different locations by using a set of travelers.

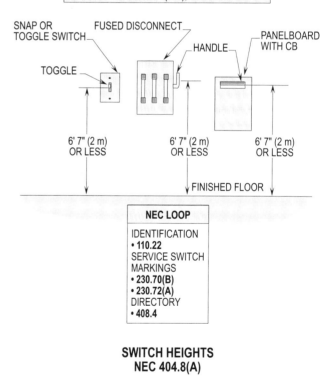

SWITCH HEIGHTS
NEC 404.8(A)

Figure 17-36. Enclosure with switching devices such as (toggle) snap switches, disconnects with fuses, or panelboards with CB's may be located to a maximum height of 6 ft. 7 in. (2 m) from the finished grade to the center of the toggle, disconnect handle, or CB handle in the ON position.

17-19

LOADING SWITCHES
404.14

General-use switches are basically divided into two types for energizing and deenergizing loads. AC general-use switches may be used to control resistive loads (incandescent lights) or electric discharge lighting (fluorescent lights). The switch rating shall be at least equal to the load it controls.

For example, a load for luminaires of 15 amps may be connected to a 15 amp AC rated switch. A load of 20 amps may be connected to a 20 amp AC rated switch.

AC general use switches controlling motor loads shall be computed at 125 percent of the full load current of the motor or limited to not more than 80 percent of the switch rating.

For example, a motor load of 12 amps requires a 15 amp AC toggle switch (12 A x 125% = 15 A). Note that 80 percent of a 15 amp AC toggle switch (15 A x 80% = 12 A) allows a 12 amp load to be switched. **(See Figure 17-37)**

AC-DC general use switches may be used to turn ON and OFF incandescent or inductive lighting loads. The rating of an AC-DC switch shall be twice the amperage rating of the inductive load at 50 percent of the amperage rating of the switch at the applied voltage.

For example, a 15 amp AC-DC toggle switch is required to supply a lighting load of 7.5 amps (7.5 A x 2 = 15 A). Note that 50 percent of a 15 amp AC-DC toggle switch (15 A x 50% = 7.5 A) allows a load of 7.5 amp to be switched. The above rules for switches apply to residential, commercial, and industrial locations where switches are used for such use. **(See Figure 17-38)**

SWITCHING OUTLETS BY SWIMMING POOLS
680.22(C)

Switching devices installed in residential, commercial, and industrial locations shall be located at least 5 ft. (1.5 m) from the inside walls of swimming pools. If the swimming pool and switching devices are separated by a permanent barrier, than this rule does not apply. Switching devices such as toggle switches, safety switches, meters, OCPD's in panelboards, etc. shall not be permitted within 5 ft. (1.5 m) of the inside walls of the pool. A distance of 5 ft. (1.5 m) or less shall be permitted with proper barriers between switches and pool. **(See Figure 17-39)**

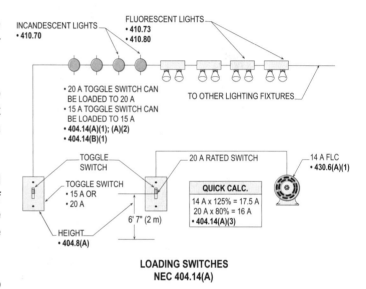

Figure 17-37. The size load that a toggle switch will supply and disconnect is determined by the load served. Switches supplying lighting loads can be loaded to 100 percent of their rating. Switches supplying motors are sized at 125 percent of the FLC.

EMERGENCY SWITCH FOR SPAS AND HOT TUBS
680.41

A clearly labeled emergency shutoff switch for control of the recirculation system and jet system shall be installed at least 5 ft. (1.5 m) away, adjacent to and within sight of the spa or hot tub. Note that this requirement does not apply to single family dwellings. For a maintenance disconnect, see **680.12**. **(See Figure 17-40)**

Lighting and Switching Outlets

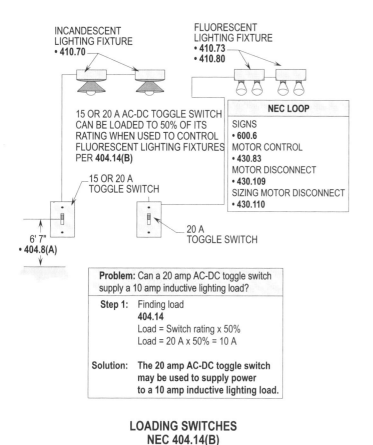

**LOADING SWITCHES
NEC 404.14(B)**

Figure 17-38. A 20 amp AC-DC toggle switch may be used at 100 percent of its rating to control incandescent lighting loads. It is limited to only 50 percent of its amp rating to control inductive lighting loads.

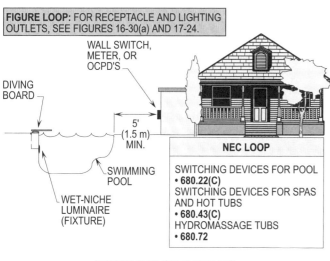

**SWITCHING OUTLETS BY
SWIMMING POOLS
NEC 680.22(C)**

Figure 17-39. Switching devices are not allowed to be installed within 5 ft. (1.5 m) of the inside walls of swimming pools. Switching devices include electrical apparatus such as wall switches, meter bases with meters, and panelboards with OCPD's, etc.

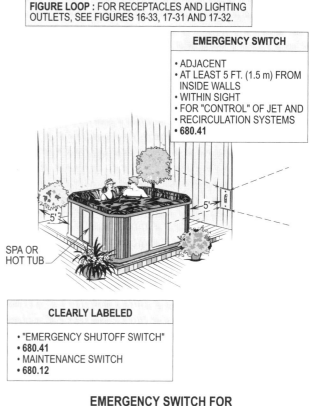

**EMERGENCY SWITCH FOR
SPAS AND HOT TUBS
NEC 680.41**

Figure 17-40. A clearly labeled emergency shutoff switch for control of the recirculation system and jet system shall be installed at least 5 ft. (1.5 m) away, adjacent to and within sight of the spa or hot tub.

SWITCHING OUTLETS BY SPAS OR HOT TUBS 680.43(C)

Switching outlets with toggle switches installed in residential, commercial, and industrial locations shall be located at least 5 ft. (1.5 m) from the inside walls of spas or hot tubs. This is to prevent someone from turning a lighting load ON or OFF and receiving an electrical shock from a faulty switch. Switches and controls associated with a spa or hot tub may be closer than 5 ft. (1.5 m), where used as an approved (listed) assembly. **(See Figure 17-41)**

GROUNDED OR UNGROUNDED 404.12

Switching outlets and boxes that are used to support switches shall be grounded if they are metal per **250.148(A)**.

17-21

Outlets that are nonmetallic are not required to be grounded per **250.148(B)**. A means shall be provided to ensure grounding continuity of the branch-circuit. The EGC may be used for this purpose. Toggle switches, without metal plates and used to control one or more luminaires, are not required to be grounded per **110.3(B)** and **404.12**. **(See Figure 17-42)**

> **Design Tip:** See **410.18(A)** for the grounding of metal boxes, if they are used. Nonmetallic boxes are not required to be grounded per **410.18(B)** for lighting outlets. See **250.96** and **404.12** for the grounding of device yokes and metal switch plates.

Note that the rules above shall be required for switching outlets and boxes used to support switches that are installed in residential, commercial, and industrial locations.

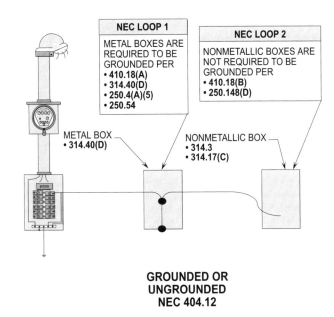

GROUNDED OR UNGROUNDED NEC 404.12

Figure 17-42. Metal boxes are required to be grounded with an EGC or other grounding means. Nonmetallic boxes are not required to be grounded with a grounded means. See **250.2(D)** and **250.54** for further explanation.

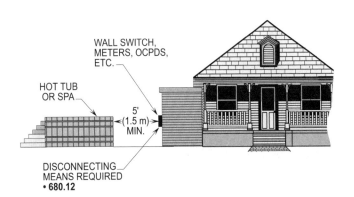

SWITCHING OUTLETS BY SPAS OR HOT TUBS NEC 680.43(C)

Figure 17-41. Switching devices are not permitted to be installed within 5 ft. (1.5 m) of hot tubs or spas. Switching devices include electrical apparatus such as wall switches, meter bases with meters per utility, and panelboards with OCPD's, etc. Check with local utility for requirements.

Name Date

Chapter 17. Lighting and Switching Outlets

Section   Answer

_____   T F   1. Luminaires with exposed metal parts shall not be required to be grounded in new or existing installations.

_____   T F   2. A replacement luminaire installed on branch-circuits without an EGC shall be of the insulated or nonmetallic type or grounded per **250.130(C)**.

_____   T F   3. The authority having jurisdiction (AHJ), by local codes, can limit the number of outlets to be connected to a 15 or 20 amp general-purpose branch-circuit.

_____   T F   4. At least one lighting outlet shall be installed to provide lighting for the illumination of habitable rooms in dwelling units. (General Rule)

_____   T F   5. At least one lighting outlet shall be provided in the bathroom and it does not have to be wall-switched if located in dwelling units.

_____   T F   6. Hanging luminaires shall be permitted over bathtubs in dwelling units.

_____   T F   7. One lighting outlet shall be installed in the garage to provide lighting for parking vehicles in dwelling units.

_____   T F   8. A lighting outlet in dwelling units may be controlled by a pull-chain or wall switch when installed in attics.

_____   T F   9. A lighting outlet shall not be required at each outside door that is classified as an entrance or exit in dwelling units.

_____   T F   10. At least one lighting outlet shall be required in attics that are floored and used for storage or for electric equipment that requires servicing in dwelling units.

_____ _____   11. A wall-switched receptacle outlet cannot be used in lieu of a wall-switched lighting outlet in habitable rooms other than kitchens and bathrooms in dwelling units.

_____ _____   12. Lighting outlets and ceiling fans shall be permitted to be installed over swimming pools located inside of dwelling units.

_____ _____   13. Underwater luminaires shall be installed in such a manner as to prevent electric shock to personnel swimming in the pool.

_____ _____   14. No-niche luminaires can be installed instead of wet or dry-niche in swimming pools.

_____ _____   15. Lighting outlets with luminaires or ceiling (paddle) fans located 12 ft. above spas or hot tubs shall not be required to be GFCI-protected.

Section	Answer

_____ _____ **16.** Hydromassage bathtubs are not treated as conventional bathtubs.

_____ _____ **17.** AC general-use switches controlling motor loads shall be calculated at 100 percent of the full-load current (in amps) of the motor or not more than 80 percent of the switch rating.

_____ _____ **18.** Toggle switches supplying motors shall be sized at 125 percent of the FLC (in amps).

_____ _____ **19.** A distance of less than 5 ft. is not permitted, even if a proper barrier between switches and pool is installed.

_____ _____ **20.** Metal boxes shall be grounded with an EGC or other grounding means.

_____ _____ **21.** Hanging luminaires shall be installed at least _____ ft. over the tub and
_____   shall be located at least _____ ft. from the tub in all directions in dwelling units.

_____ _____ **22.** The lighting outlets supporting luminaires or ceiling fans shall be located at least
_____   _____ ft. _____ in. above the maximum water level of the swimming pool, if installed indoors.

_____ _____ **23.** Specially designed underwater lighting luminaires shall be installed not less than _____ in. below the normal water level.

_____ _____ **24.** What are three types of underwater lighting luminaires installed in swimming
_____   pools _____, _____ and _____.
_____

_____ _____ **25.** A _____ AWG insulated copper conductor shall be routed in RNC conduit to ground to the metal forming shell.

_____ _____ **26.** Habitable rooms are such rooms in the dwelling units as:
   (a) Bedroom
   (b) Living room
   (c) Dining room
   (d) All of the above

_____ _____ **27.** Surface-mounted luminaires of the incandescent type shall be mounted in dwelling units from the storage space (area) at least:
   (a) 6 in.
   (b) 10 in.
   (c) 12 in.
   (d) 24 in.

_____ _____ **28.** Surface-mounted luminaires of the fluorescent type shall be mounted in dwelling units from the storage space (area) at least:
   (a) 6 in.
   (b) 12 in.
   (c) 18 in.
   (d) 24 in.

	Section	Answer

29. Underwater lighting luminaires shall be located below the normal water level at least:
  (a) 6 in.
  (b) 12 in.
  (c) 18 in.
  (d) 24 in.

30. A wall switch to control the lighting outlets in dwelling units shall be provided at each level, where there is a difference between floor levels of:
  (a) 4 risers or more
  (b) 5 risers or more
  (c) 6 risers or more
  (d) 10 risers or more

18
Motors

There are three currents that must be found before designing and selecting the elements to make-up circuits supplying power to motors. The first current that must be found is the full-load amps (FLA) from **Table 430.248** for single-phase and **Table 430.250** for three-phase. This current rating in amps, per **430.6(A)(1),** is used to size all the elements of the circuit except the overload protection. The second current to be determined is the nameplate amps on the motor per **430.6(A)(2)**. This current rating in amps is used to size the overloads (OL's) to protect the motor windings and conductors. The third current, per **430.7(A)(9)**, is the locked-rotor current (LRC), in amps, from **Tables 430.251(A)** and **(B)**. The OCPD must be sized large enough to hold this current rating (LRC) in amps and permit the motor to start-and-run. When using the code letter to determine the locked rotor current (starting current), see **Table 430.7(B)**.

Electrical systems containing AC and DC motors must have their circuits and elements designed per **Article 430** of the NEC. AC and DC motors are available in various types and sizes. To protect such motors and circuits and still allow them to operate, the conductors, controllers, starters, protection devices, and disconnecting means must be designed and installed properly.

BRANCH-CIRCUIT AND FEEDER-CIRCUIT CONDUCTORS
430.1

The following branch-circuit and feeder-circuit elements of a motor system are designed and installed based upon the characteristics of the motor involved:

- Motor branch-circuit and feeder conductors
- Motor branch-circuit and feeder OCPD's
- Motor overload protection
- Motor control circuits
- Motor controllers
- Motor disconnecting means

SIZING CONDUCTORS FOR SINGLE MOTORS
430.22(A)

Branch-circuit conductors supplying a single motor shall have an ampacity not less than 125 percent of the motor full-load current rating, in amps, per **Tables 430.247, 430.248, 430.249,** and **430.250** respectively. For example, a 20 HP, 208 volt, three-phase motor per **Table 430.250** has a full-load current of 59.4 amps. The full-load amps (FLA) for sizing the conductors is determined by multiplying 59.4 A x 125% which equals 74.25 amps.

A motor will normally have a starting current of four to six times the full-load current of the motor's FLA for motors marked with code letters A through G and 8 1/2 to 15 times for NEMA B, high-efficiency motors. Design B, C and D motors have a starting current of about 4 to 6 times their full-load amps when starting and driving a motor load.

There are heating effects that develop on the conductors when motors are starting and accelerating the driven load. To eliminate such effects, the conductor's current-carrying capacity is increased by taking 125 percent of the motor's full-load current rating in amps. For example, a motor with a FLC rating of 42 amps must have conductors with a current-carrying capacity of at least 52.5 amps (42 A x 125% = 52.5 A) to safely carry the load when starting and also protect insulation due to overload conditions.

SIZING CONDUCTORS FOR SINGLE-PHASE MOTORS
430.22(A)

Section 430.6(A)(1) of the NEC requires the full-load current in amps for single-phase motors to be obtained from **Table 430.248**. This FLC rating in amps is then multiplied by 125 percent per **430.22(A)** to derive the total amps to select the conductors from **Table 310.16** to supply power to the motor windings. **(See Figure 18-1)**

SIZING CONDUCTORS FOR THREE-PHASE MOTORS
430.22(A)

Section **430.6(A)(1)** of the NEC requires the full-load current in amps for three-phase motors to be obtained from **Table 430.250**. This FLC rating in amps is multiplied by 125 percent per **430.22(A)** to derive the total amps to select the conductors for the motor windings. **(See Figure 18-2)**

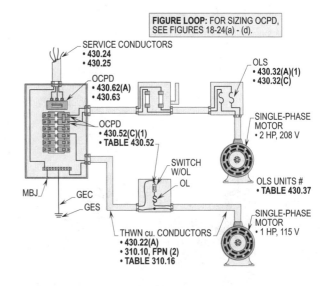

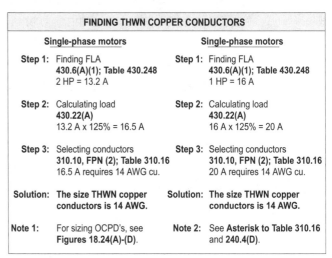

Figure 18-1. Determining the size branch-circuit conductors to supply single-phase motors.

SIZING CONDUCTORS FOR MULTISPEED MOTORS
430.22(B)

The circuit conductors for multispeed motors must be sized large enough, to the controller, to supply the highest nameplate full-load current rating of the multispeed motor winding involved. A single OCPD is allowed to serve each speed for a multispeed motor per **430.22(B)**. The speed with the greater amps is used to size the OCPD and conductors. Overload protection shall be provided for each speed to protect each winding from excessive current during an overload condition. **(See Figure 18-3)**

Motors

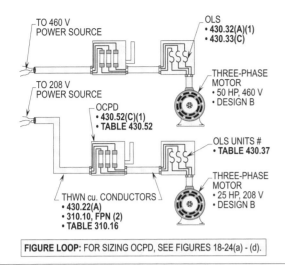

sized with a current-carrying capacity of 125 percent of the motor's full-load current, in amps. **Table 430.22(E)** allows the conductors to be sized with a percentage times the nameplate current rating, in amps, based upon the duty cycle classification of the motor.

When sizing conductors to supply individual motors that are used for short-time, intermittent, periodic, or varying duty the requirements of **Table 430.22(E)** shall apply. Varying heat loads are produced on the conductors by the starting and stopping duration of operation cycles which permits conductor sizing changes. In other words, such conductors are never subjected to continuous operation due to ON and OFF periods and therefore conductors are never fully loaded for long intervals of time. For this reason, conductors can be down sized. **(See Figure 18-5)**

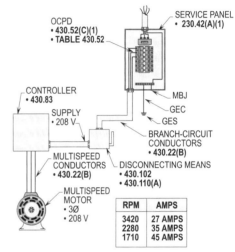

Figure 18-2. Determining the size branch-circuit conductors to supply three-phase motors.

SIZING CONDUCTORS FOR WYE START AND DELTA RUN MOTORS 430.22(C)

The branch-circuit conductors for wye-start and delta run connected motors shall be selected based on the full-load current on the line side of the controller. The selection of conductors between the controller and the motor must be based on 58 percent (1 ÷ 1.732 = .58) of the motor's full-load current, in amps, times 125 percent for continuous use. **(See Figure 18-4)**

SIZING CONDUCTORS FOR DUTY CYCLE MOTORS 430.22(E)

Conductors for a motor used for short-time, intermittent, periodic, or varying duty do not require conductors to be

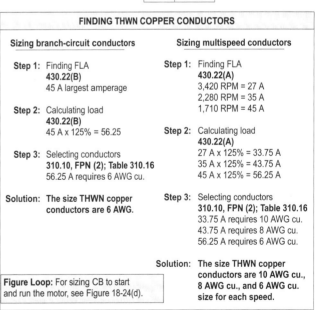

Figure 18-3. Determining the size branch-circuit conductors to supply multispeed motors.

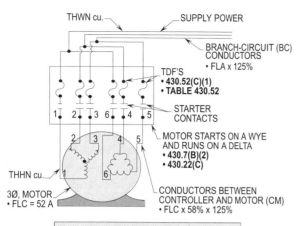

Figure 18-4. Determining the size conductors to supply motors starting on a wye and running on delta.

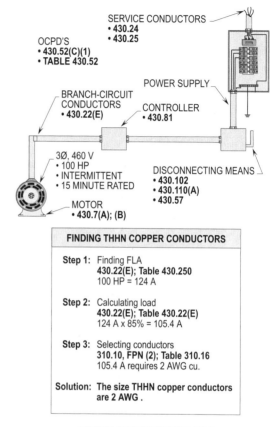

Figure 18-5. Determining the size conductors to supply duty cycle related motors.

SIZING CONDUCTORS FOR ADJUSTABLE SPEED DRIVE SYSTEMS
430.122(A); 430.2

Power conversion equipment, when supplied from a branch-circuit includes all elements of the adjustable speed drive system. The rating in amps is used to size the conductors that are based upon the power required by the conversion equipment. When the power conversion equipment provides overcurrent protection for the motor, no additional overload protection is required.

The disconnecting means can be installed in the line supplying the conversion equipment, and the rating of the disconnect shall not be less than 115 percent of the input current rating of the conversion unit.

Power conversion equipment requires the conductors to be sized at 125 percent of the rated input of such equipment.

Note: For more information on adjustable speed drives, see **430.120** through **430.128** at the end of this chapter.

SIZING CONDUCTORS FOR PART-WINDING MOTORS
430.22(D); 430.4

Induction or synchronous motors that have a part-winding start are designed so that at starting they energize the primary armature winding first. After starting, the remainder of the winding is energized in one or more steps. The purpose of this arrangement is to reduce the initial inrush current until the motor accelerates to its running speed.

The inrush current at start is locked-rotor current and at times can be quite high. A standard part-winding-start induction motor is designed so that only half of its winding is energized at start; then, as it comes up to speed, the other half is energized, so both halves are energized and carry equal current to drive the load.

Separate overload devices must be used on a standard part-winding-start induction motor to protect the windings from excessive damaging currents. This means that each half of the motor winding has to be individually provided with overload protection. These requirements are covered in **430.32** and **430.37**. Each half of the windings has a trip current value that is one half of the specified running current. As required by **430.52(C)(1)**, each of the two motor windings shall have branch-circuit, short-circuit, and ground-fault protection that is to be selected at not more than one half the percentages listed in **430.52(C)(1)** and **Table 430.52**. **(See Figure 18-7)**

Design Tip: Section **430.3, Ex.** allows a single device with this one half rating, for both windings, provided that it will permit the motor to start and run. If a time-delay (dual element) fuse is used as a single device for both windings, its rating is permitted if it does not exceed 150 percent of the motor's full-load current.

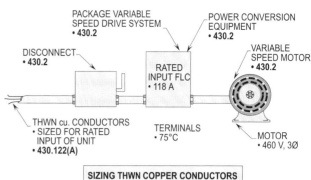

Figure 18-6. Determining the size conductors to supply power conversion equipment.

WOUND-ROTOR SECONDARY
430.23

Wound-rotor motors are three-phase motors that are installed with two sets of leads. The main leads to the motor windings (field poles) is one set and the secondary leads to the rotor is the other set. The secondary leads on one end connect to the rotor through slip rings, and the other end of the leads connects through a controller and a bank of resistors. The speed of the motor varies when the amount of resistance in the motor circuit is varied. The rotor will turn slower when the resistance is greater in the rotor, and faster when such resistance is lowered.

SIZING CONDUCTORS FOR CONTINUOUS DUTY
430.23(A)

The motor shall have an ampacity not less than 125 percent of the full-load secondary current of the motor where secondary leads are installed between the controller and the motor. The secondary full-load current rating, in amps, is obtained from the manufacturer or found on the nameplate of the motor.

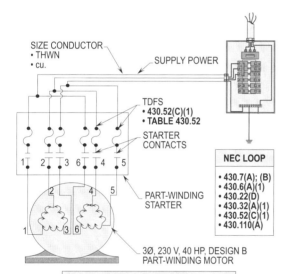

Figure 18-7. Determining the size conductors to supply part-winding motors.

18-5

SIZING CONDUCTORS FOR OTHER THAN CONTINUOUS DUTY
430.23(B)

When installing a motor to be used for short-time, intermittent, periodic, or varying duty, the secondary conductors shall be sized not less than 125 percent of the secondary current per **Table 430.22(E)**. The classification of service determines the correct percentages to select and apply, when sizing the conductors, which is based upon the cycles of the motor.

SIZING CONDUCTORS FOR RESISTORS, SEPARATED FROM CONTROLLER
430.23(C)

Where the secondary resistor is separate from the controller, the ampacity of the conductors between the controller and resistor shall not be less than the resistor duty classification percentages listed in **Table 430.23(C)**. **(See Figure 18-8)**

SIZING CONDUCTORS FOR SEVERAL MOTORS
430.24

The full-load current rating of the largest motor shall be multiplied by 125 percent to select the size of conductors for a feeder supplying a group of two or more motors. The remaining motors of the group must have their full-load current ratings added to this value and this total amperage is then used to size the conductors. **(See Figure 18-9)**

SIZING CONDUCTORS FOR DUTY CYCLE MOTORS
430.24, Ex. 1

Table 430.22(E) shall be used for sizing the amperage rating for a motor that is classified as either short-time, intermittent, periodic, or varying duty. The amperage rating shall be based on 100 percent of the full-load current rating on the motor's nameplate if rated for continuous operation.

The full-load current rating, in amps, of the largest motor shall be multiplied by 125 percent to select the size conductors for a feeder-circuit supplying a group of two or more motors. The remaining motors of the group must have their total full-load current ratings added to the computed amps of the duty cycle motor or the amps of the 125 percent motor, whichever is greater. The feeder conductors shall be sized by this total full-load current, in amperes. **(See Figure 18-10)**

SIZING CONDUCTORS SUPPLYING MOTORS AND OTHER LOADS
430.25

The motor load shall be calculated per **430.22(A)** or **430.24** when designing combination loads that consist of one or more motor loads on the same circuit with lights, receptacles, appliances, or any combination of such loads. For other than motor loads **Art. 220** and other applicable articles shall be used to compute such loads. The ampacity required for the feeder-circuit conductors shall be equal to all the total loads involved. The OCPD's used to protect conductors and elements from short-circuits and ground-faults are sized per **430.62(A)** and **430.63**. **(See Figure 18-11)**

SIZING THE BRANCH-CIRCUIT PROTECTIVE DEVICE
TABLE 430.52; 430.52(C)(1); (C)(3)

The motor branch-circuit overcurrent device shall be capable of carrying the starting current of the motor. Short-circuit and ground-fault current is considered as being properly taken care of when the overcurrent protection device does not exceed the values in **Table 430.52,** as permitted by the provisions of **430.52(C)(1)** with exceptions.

Different percentages are selected for particular devices based upon one of the four columns listed in **Table 430.52**. The percentages are used to size and select the proper size overcurrent protection device to allow a certain type of motor to start and run. The motor has a momentary starting current that is necessary for the motor to have power to start and drive the connected load at the driven equipment. Note that the OCPD sized per **430.52(C)(1)** provides protection from short-circuits and ground-faults. Overload protection must be provided for conductors and motor windings per **430.32(A)(1)** and **430.32(C)**. **(See Figure 18-12)**

> **Design Tip:** In cases where the values for branch-circuit protective devices determined by **Table 430.52** do not correspond to the standard sizes or ratings of fuses, nonadjustable circuit breakers, or thermal devices, or possible settings of adjustable circuit breakers adequate to carry the starting currents of the motor, the next higher size rating or setting may be used.

Motors

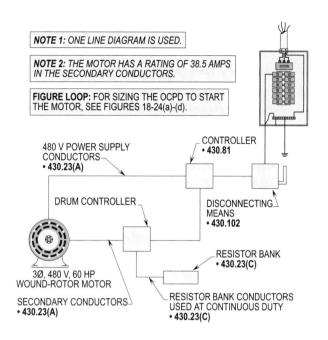

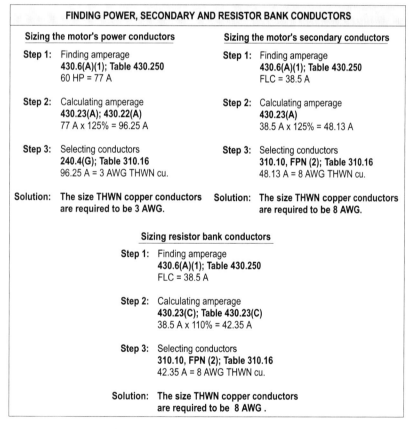

Figure 18-8. Determining the size conductors to supply wound-rotor motors.

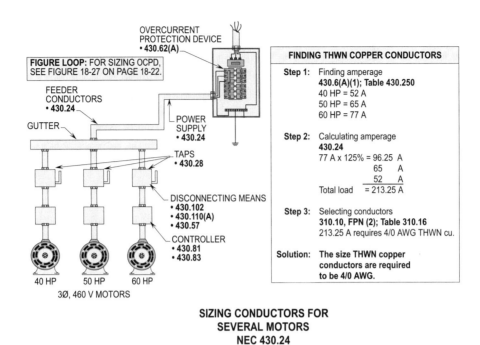

Figure 18-9. Determining the size conductors for a feeder-circuit to supply several motors.

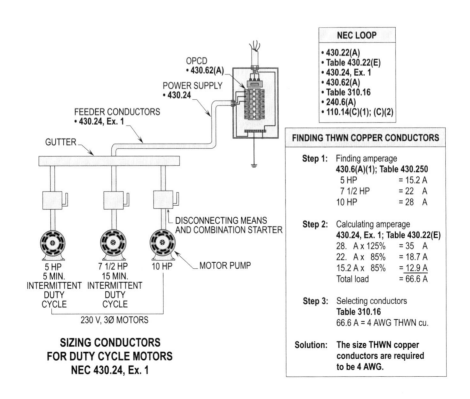

Figure 18-10. Determining the size conductors for a feeder-circuit to supply duty cycle related motors.

Motors

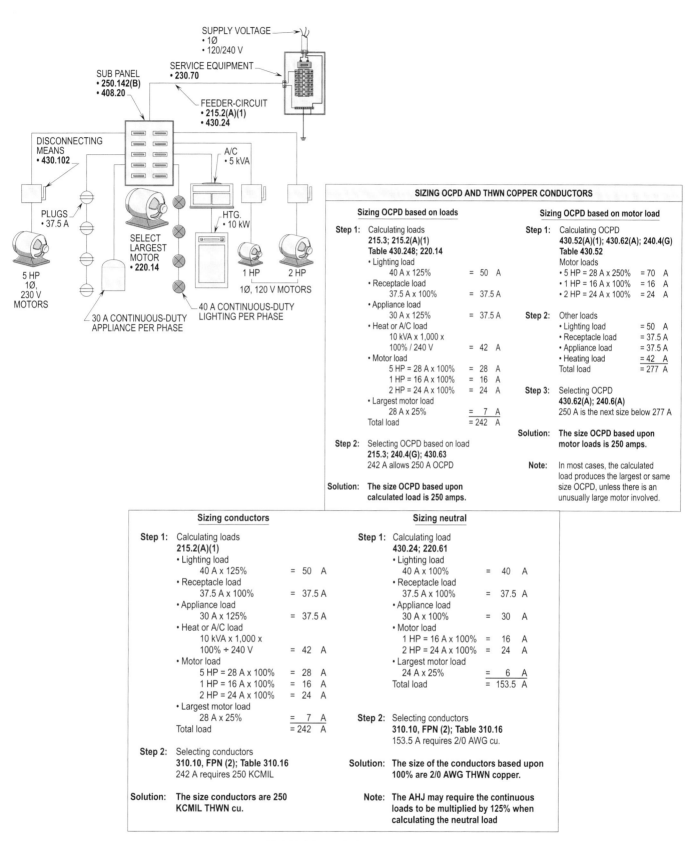

Figure 18-11. Determining the size conductors for motors and other loads supplied by a feeder-circuit.

APPLYING THE EXCEPTIONS
430.52(C)(1), Ex. 1 AND Ex. 2

There are Exceptions that permit larger OCPD's to be used where the overcurrent protection device, as specified in **Table 430.52**, will not take care of the starting current to allow the motor to start and run. Where the motor fails to start and run because of excessive inrush starting currents, one of the following exceptions can be applied:

APPLYING Ex. 1

If the values of the branch circuit, short-circuit, and ground-fault protection devices determined from **Table 430.52** do not conform to standard sizes or ratings of fuses, nonadjustable circuit breakers, or possible settings on adjustable circuit breakers, it does not matter if they are capable or not capable of adequately carrying the load involved, the next higher setting or rating will be permitted. In other words, you can round up or round down the size of the overcurrent protection device automatically by choice. **(See Figure 18-13)**

APPLYING Ex. 2

If the ratings listed in **Table 430.52** and **Ex. 1** to **430.52(C)(1)** are not sufficient for the starting current of the motor. The following OCPD's with percentages shown can be used to start and run motors having high inrush starting currents: **(See Figure 18-14(a))**

Anytime the percentages of **Table 430.52** and **Ex. 1** to **430.52(C)(1)** won't allow the motor to start and run its driven load. The following percentages can be applied for different types of OCPD's:

- When nontime delay fuses are used and they do not exceed 600 amperes in rating, it shall be permitted to increase the fuse size up to 400 percent of the full-load current, but never over 400 percent is to be used.

- Time-element fuses (dual-element) shall not exceed 225 percent of the full-load current, in amps, but they may be increased to up this percentage.

- Inverse time-element breakers are permitted to be increased in rating. However, they shall not exceed:

  (a) 400 percent of full-load current of the motor for 100 amperes or less, or
  (b) They may be increased to 300 percent where a full-load current is greater than 100 amperes.

See Figure 18-14(b) for a detailed illustration on selecting percentage for sizing OCPD's.

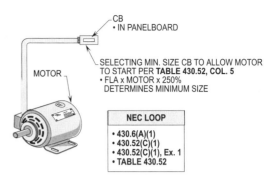

Figure 18-12. Selecting the percentages to size the minimum (rounding down), next size (rounding up), and maximum size circuit breaker to allow a motor to start and run.

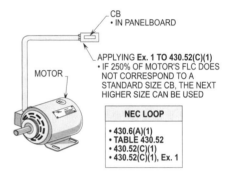

Figure 18-13. Where the percentages of **Table 430.52** times the full-load current of motor in amps does not correspond to a standard size OCPD, the next higher size above this percentage may be used.

USING INSTANTANEOUS TRIP CB's
430.52(C)(3), Ex. 1

An instantaneous trip circuit breaker shall be used only if it is adjustable, and it is a part of a combination controller that has overcurrent protection in each controller. Such combination when used has to be approved. An instantaneous trip circuit breaker is allowed to have a damping device, to limit the inrush current when the motor is started.

If the specified setting in **Table 430.52** is not sufficient for the starting current of the motor, the setting on an instantaneous trip circuit current may be increased, provided that in no instance it exceeds 1300 percent of the motor's full-load current rating, in amps, for motors marked as Class B, C, or D.

> **Design Tip:** For Design E and Class B, NEMA motors, the setting on the INST. CB can be adjusted up to 1700 percent to allow the motor to start and run.

See **Figure 18-15** for adjusting the maximum trip settings on instantaneous trip circuit breakers to allow motors to start and accelerate their driven load.

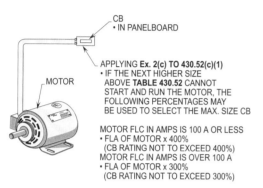

SIZING THE BRANCH-CIRCUIT
PROTECTIVE DEVICE
NEC TABLE 430.52
NEC 430.52(C)(1)
NEC 430.52(C)(1), Ex. 1
NEC 430.52(C)(1), Ex. 2(c)

Figure 18-14(a). When the percentages of **Table 430.52** and **430.52(C)(1), Ex. 1** won't allow the motor to start and run the driven load, the maximum size CB of **430.52(C)(1), Ex. 2(c)** may be used.

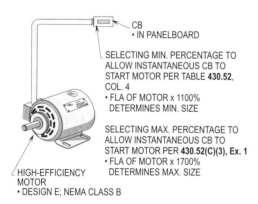

USING INSTANTANEOUS TRIP CB'S
NEC TABLE 430.52
NEC 430.52(C)(3), Ex. 1

Figure 18-15. Determining the minimum and maximum size setting on an instantaneous trip circuit breaker to allow a motor to start and run a driven load. Note that after the minimum size OCPD has been determined, a smaller size can be selected.

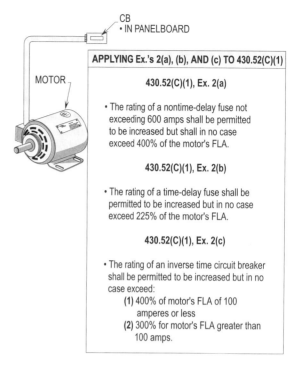

SIZING THE BRANCH-CIRCUIT
PROTECTIVE DEVICE
NEC 430.52(C)(1), Ex. 2(a), (b), AND (c)

Figure 18-14(b). When the percentages of **Table 430.52** and **430.52(C)(1), Ex. 1** won't allow the motor to start and run, the maximum percentages of **430.52(C)(1), Ex.'s 2(a), (b),** and **(c)** may be applied.

TYPES OF MOTORS
TABLE 430.52

The following are five types of motors to be considered when sizing OCPD's to allow motors to start and run:
- Single-phase AC squirrel-cage
- Three-phase AC squirrel-cage
- Wound-rotor
- Synchronous
- DC

SINGLE-PHASE AC SQUIRREL-CAGE MOTORS

Squirrel-cage motors are known in the electrical industry as induction motors. An induction motor operates on the same principles as the primary and secondary windings of a transformer. When power energizes the field windings they serve as the primary by inducing voltage into the rotor which serves as the secondary windings. Squirrel-cage motors have two windings on the stator, one winding is the run winding and the other is the starting winding. This additional

18-11

starting winding on the stator is required for split-phase, single-phase, induction motors to have the capacity to start and run. The starting winding has a higher resistance to ground than the running winding which creates a phase displacement between the two windings. It is this phase displacement between the two windings that gives split-phase motors the power to start.

The phase displacement is about 18 degrees to 30 degrees in angular phase displacement which provides enough starting torque (twist or force) to start the motor. The motor operates on the running winding when the rotor starts turning and has established a running speed at about 75 percent to 80 percent of the motor's synchronous speed. The starting winding is disconnected by a centrifugal switch that is installed in the circuit of the starting winding. (**See Figure 18-16**)

to the motor windings (field poles) and the other set is the secondary leads to the rotor. The secondary leads are connected to the rotor through the slip rings, while the other end of the leads are connected through a controller and a bank of resistors. The speed of the motor varies with the amount of resistance added in the motor circuit. The rotor will turn slower when the resistance is greater in the rotor, and vice versa. The resistance may be incorporated in the controller or the resistor banks may be separate from the motor. (**See Figure 18-18**)

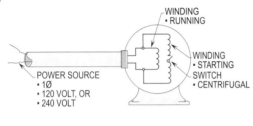

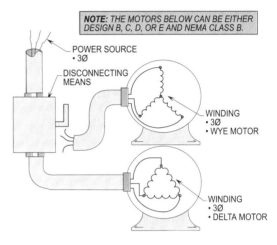

Figure 18-16. The above is an example of a single-phase squirrel-cage induction motor which is listed in **Table 430.52** of the NEC.

Figure 18-17. The above is an example of a three-phase squirrel-cage induction motor which is listed in **Table 430.52** of the NEC.

THREE-PHASE SQUIRREL-CAGE MOTORS

Three-phase induction motors have three separate windings per pole on the stator that generate magnetic fields that are 120 degrees out-of-phase with each other. An additional starting winding is not required for three-phase motors to start and run. An induction motor will always have a peak phase of current. This is due to alternating current reversing its direction of flow. In other words, when alternating current of one phase reverses its direction of flow, a peak current will be developed on one phase and as current reverses direction again, a second phase will peak, etc. Three-phase motors provide a smooth and continuous source of power once they are started and driving their load. (**See Figure 18-17**)

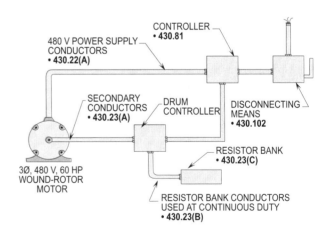

WOUND-ROTOR MOTORS

Wound-rotor motors are classified as three-phase induction motors. Wound-rotor motors are similar in design to squirrel-cage induction motors. Wound-rotor motors are three-phase motors having two sets of leads. One set is the main leads

Figure 18-18. The above is an example of a three-phase wound-rotor motor which is listed in **Table 430.52** of the NEC.

SYNCHRONOUS MOTORS

The following are two types of synchronous motors that are available:
- Nonexcited
- Direct-current excited

Synchronous motors are available in a wide range of sizes and types which are designed to run at designed speeds. A DC source is required to excite a direct-current excited synchronous motor. The torque required to turn the rotor for a synchronous motor is produced when the DC current of the rotor field locks in with the magnetic field of the stator AC current. **(See Figure 18-19)**

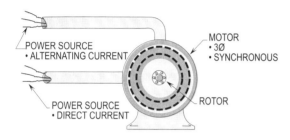

SYNCHRONOUS MOTORS
NEC TABLE 430.52

Figure 18-19. The above is an example of a three-phase synchronous motor which is listed in **Table 430.52** of the NEC.

DC MOTORS

Direct-current only is used to operate DC related motors. A DC motor is designed with the following two main parts:
- The stator
- The rotor

The stationary frame of the motor is called the stator. The armature mounted on the drive shaft is known as the rotor. By applying direct-current to the rotor, the speed may be adjusted for a DC motor which drives the driven load at a specific speed.. **(See Figure 18-20)**

CODE LETTERS
430.7(B); TABLE 430.7(B)

Code letters are installed on motors by manufacturers for calculating the locked-rotor current (LRC) in amps based upon the kVA per horsepower which is selected from the motor's code letter. Overcurrent protection devices shall be set above the locked-rotor current of the motor to prevent the overcurrent protection device from opening when the rotor of the motor is starting. The following two methods can be used to calculate and select the locked-rotor current of motors:

- Utilizing code letters to determine LRC
- Utilizing horsepower to determine LRC

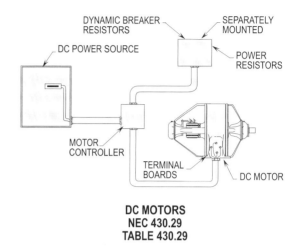

DC MOTORS
NEC 430.29
TABLE 430.29

Figure 18-20. The above is an example of a DC motor which is listed in **Table 430.52** of the NEC.

UTILIZING CODE LETTERS TO FIND LRC
430.7(B); TABLE 430.7(B)

Code letters shall be marked on the nameplate and such letters are used for designing locked-rotor current. Locked-rotor current for code letters are listed in **Table 430.7(B)** in kVA (kilovolt-amps) per horsepower which is based upon a particular code letter.

For example: What is the locked-current rating for a three-phase, 208 volt, 20 horsepower motor with a code letter B marked on the nameplate of the motor?

Step 1: Finding LRC amps
Table 430.7(B)
A = kVA per HP x 1000 ÷ V x 1.732
A = 3.54 x 20 x 1000 ÷ 208 V x 1.732
A = 70,800 ÷ 360
A = 197

Solution: The locked-rotor current is 197 amps. Note that Table 430.7(B) must be used to find LRC's of the motor, based on their code letters per the 1996 NEC and earlier editions.

LOCKED-ROTOR CURRENT UTILIZING HP TABLES 430.251(A) AND (B)

The locked-rotor current of a motor may be found in **Tables 430.251(A)** and **(B)**. The locked-rotor current for single-phase and three-phase motors are selected from one of these Tables, based upon the phases, voltage, and horsepower rating of the motor. For motors with code letters A through G, round the nameplate current in amps up to an even number (unit of ten) and multiply by 6 to obtain the LRC of the motor. Note that code letters are not found in **Tables 430.251(A)** and **(B)** because they will not be listed on the motor's nameplate anymore. Motors will be marked either as Design B, C, D, or E letter to indicate which locked-rotor currents are to be selected from **Tables 430.251(A)** and **(B)** based on horsepower, phases, and voltages.

For example: What is the locked-rotor current rating for a three-phase, 460 volts, 50 horsepower, Design B motor?

Table method using Design letter

Step 1: Finding LRC amps
Table 430.251(B)
50 HP requires 363 A

Solution: The locked-rotor current is 363 amps.

For example: Consider a motor with a nameplate current of 63 amps and determine the LRC of the motor based upon code letter A through G?

Rule of thumb method using code letter

Step 1: Finding even number (unit of ten)
Table 430.7(B)
Round up 63 A to 70 A

Step 2: Calculating LRC
Table 430.7(B)
70 A x 6 = 420 A

Solution: The locked-rotor current is 420 amps. Note: This method can only be used for code letters A thru G.

See **Figures 8-21(a)** and **(b)** for calculating and selecting the locked-rotor current of a motor.

Design Tip: Engineers and electricians must select the locked-rotor current rating from **Tables 430.251(A) and (B)** when using Design B, C, D, or E motors. The overcurrent protection device must be set above the locked-rotor current of the motor so the motor can start and run.

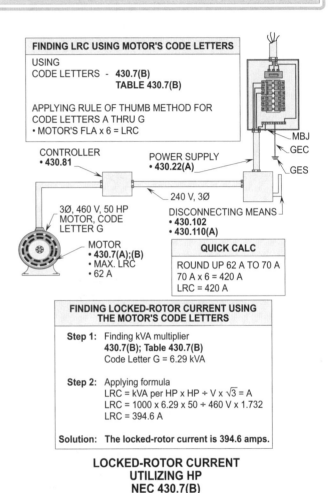

Figure 18-21(a). For motors having code letters instead of Design letters, the LRC must be calculated per **Table 430.7(B)** using the code letter of the motor.

SIZING AND SELECTING OCPD'S TABLE 430.52, COLUMNS 2, 3, 4, & 5

The overcurrent protection device must be sized for the starting current of the motor and selected to allow the motor to start and run. The OCPD per **Table 430.52** must protect the branch-circuit conductors from short-circuits and ground-faults. The following four overcurrent protection devices selected from **Table 430.52** will start most motors under normal starting conditions:

- Nontime-delay fuses per Column 2
- Time-delay fuses per Column 3
- Instantaneous trip circuit breakers per Column 4
- Inverse-time circuit breakers per Column 5

NONTIME-DELAY FUSES
TABLE 430.52, COLUMN 2

Nontime-delay fuses are installed with instantaneous trip features to detect short-circuits and thermal characteristics to sense slow heat build-up in the circuit. A nontime-delay fuse will hold 5 times (500 percent) its rating for approximately 1/4 to 2 seconds based upon the type used.

> **For example:** What is the holding time in amps for a nontime-delay fuse of 150 amps?
>
> **Step 1:** Finding holding amps
> A = fuse rating x 500%
> A = 150 A x 500%
> A = 750
>
> **Solution: The holding time in amps of a nontime-delay fuse is 750 amps. Note: This fuse will blow in 1/4 to 2 seconds so the motor will have to start and accelerate the load quickly.**

See **Figure 18-22(a)** for a detailed illustration of sizing nontime-delay fuses to allow motors to start and run.

TIME-DELAY FUSES
USING MAXIMUM SIZE
TABLE 430.52, COLUMN 3

Time-delay fuses are also equipped with instantaneous trip features to detect short-circuits and thermal characteristics to sense slow heat build-up in the circuit. Time-delay fuses are used because of their time-delay action to allow a motor to start. Time-delay fuses will hold 5 times (500 percent) of their rating which will permit most motors to start and accelerate the driven load. Note that time-delay fuses which are sized at 125 percent or less of the motor's FLC rating can provide overload protection for the motor.

A time-delay fuse will hold 5 times its rating for ten seconds and this delayed action provides more acceleration time to allow the motor to start without tripping the OCPD.

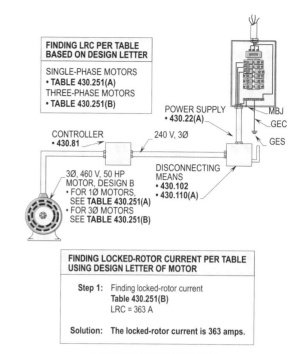

Figure 18-21(b). Tables 430.251(A) and (B) can be used to determine the LRC in amps for motors with Design letters.

> **For example:** What is the holding power in amps for a time-delay fuse of 150 amps?
>
> **Step 1:** Finding holding amps
> A = fuse x 500%
> A = 150 A x 500%
> A = 750
>
> **Solution: The rating of a time-delay fuse is 750 amps. Note: This fuse holds five times its rating for ten seconds without blowing and opening the circuit.**

See **Figure 18-22(b)** for sizing time-delay fuses to hold the motors locked-rotor current, in amps.

INSTANTANEOUS TRIP CIRCUIT BREAKERS
TABLE 430.52, COLUMN 4

Instantaneous trip circuit breakers are installed with instantaneous values of current to respond from short-circuits only. Thermal protection is not provided for instantaneous trip circuit breakers. Instantaneous trip circuit breakers will hold about three times their rating on the low setting and five times their next setting, seven times their next setting, approximately ten times their rating on the high setting. Certain types allow such settings to be adjusted from 0 to 1700 percent. **(See Figure 18-23(a))**

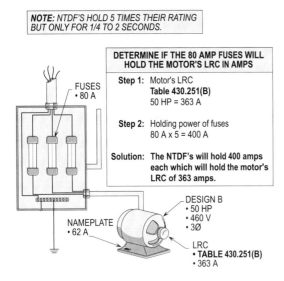

NONTIME-DELAY FUSES
NEC TABLE 430.251(B)

Figure 18-22(a). Nontime-delay fuses will hold five times their rating and when this rating is above the locked-rotor current it should allow the motor to start and run based upon LRC.

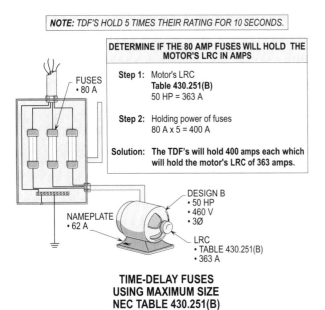

TIME-DELAY FUSES
USING MAXIMUM SIZE
NEC TABLE 430.251(B)

Figure 18-22(b). Time-delay fuses will hold five times their rating and when this rating is above the locked-rotor current, in amps, it should allow the motor to start and run based upon LRC.

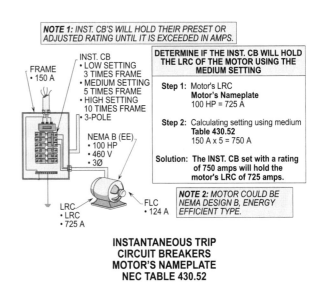

INSTANTANEOUS TRIP
CIRCUIT BREAKERS
MOTOR'S NAMEPLATE
NEC TABLE 430.52

Figure 18-23(a). An instantaneous circuit breaker with its rating set above the locked-rotor current, in amps, of a motor will allow the motor to start and run.

INVERSE-TIME CIRCUIT BREAKERS TABLE 430.52, COLUMN 5

Inverse-time circuit breakers are designed with instantaneous trip features to detect short-circuits and thermal characteristics to sense slow heat build-up in the circuit. If heat should occur in the windings of the motor, the instantaneous values of current will be detected by the thermal action of the circuit breaker and will trip open the circuit if it is sized properly. The magnetic action of the circuit breaker will clear the circuit if short-circuits or ground-faults should occur on the circuit elements or equipment served.

> **Design Tip:** Inverse-time circuit breakers will hold about three times their rating for different periods of time based upon their frame size. For example, a motor with a full-load current of 585 amps can be started with a 200 amp circuit breaker.

This can be verified by multiplying the 200 amp circuit breaker by 3 which is equal to 600 amps or 585 amps divided by 3 is equal to 195 amps. By rounding up to the next size circuit breaker per **430.52(C)(1), Ex. 1**, the size circuit breaker is 200 amps, per **240.6(A)**. Note that this size circuit breaker allows the motor to start and run. **(See Figure 18-23(b))**

Motors

Figure 18-23(b). Circuit breakers sized at least three times their rating provides an amp rating above the locked-rotor current of the motor and will hold such LRC.

NOTE: CB'S WILL HOLD THREE TIMES THEIR RATING FOR PERIODS OF TIME, BASED UPON THEIR FRAME SIZE.

DETERMINE IF THE CB WILL HOLD THE LRC OF THE MOTOR
Step 1: Motor's LRC
Motor's Nameplate
100 HP = 725 A
Step 2: Calculating size CB
250 A x 3 = 750 A
Note: CB size is 250 A
Solution: The circuit breaker with a holding power of 750 amps will hold the motor's LRC of 725 amps.

- CB
- FRAME
  - 250 A
  - 3-POLE
- LRC
  - 725 A
- DESIGN E
  - 100 HP
  - 460 V
  - 3Ø
- FLC
  - 124 A

QUICK CALC
- LRC = 725 A
- CB HOLDS 3 × ITS RATING
- HOLDING A = LRC ÷ 3
  A = 725 A ÷ 3
  A = 242 (NEXT SIZE)
  CB = 250 A

INVERSE-TIME CIRCUIT BREAKERS
MOTOR'S NAMEPLATE
NEC TABLE 430.52

SIZING MAXIMUM OCPD
430.52(C)(1), Ex.'s 2(a) thru (c)

Where the rating specified in **Table 430.52** is not sufficient for the starting current of the motor the following ratings (percentages) shall be applied:

- Nontime-delay fuses (400 percent)
- Time-delay fuses (225 percent)
- Inverse time circuit breakers (400 and 300 percent)
- Instantaneous trip circuit breakers (0-1700 percent based on Design letter or Code letter)

SIZING OCPD'S TO ALLOW MOTORS TO START AND RUN
430.52(C)(1); TABLE 430.52

The branch-circuit protection for a motor may be a fuse or circuit breaker located in the line at the point where the branch-circuit originates. The fuse or circuit breaker is located either at a service cabinet or distribution panel or in the motor control center. When there is only one motor on a branch-circuit, the fuse or circuit breaker is sized according to **Table 430.52** and **430.52(C)(1)**.

To use the table properly will require explanation. There is the matter of "Design letters." A Design letter provides certain electrical characteristics of a particular motor which are needed to size the OCPD to permit the motor to start and accelerate its load. To apply **Table 430.52**, it is necessary to take the following steps:

- Select the phase of the motor
  (a) Single-phase
  (b) Three-phase (poly-phase)

- Select type of motor
  (a) Squirrel-cage induction
  (b) Wound-rotor
  (c) DC
  (d) Synchronous

Note that motors can be single-phase or three-phase types.

- Select the Design letter of the motor
  (a) Design B
  (b) Design C
  (c) Design D
  (d) Design E

- Select the type OCPD
  (a) Column 2 is for NTDF's
  (b) Column 3 is for TDF's
  (c) Column 4 is for CB's with instantaneous trip settings or adjustments
  (d) Column 5 is for CB's with both instantaneous trip settings and thermal trip characteristics

See Figures 18-24(a) through (d) for sizing and selecting the size OCPD's per **Table 430.52** to allow motors to start and run their driven loads. Note that the minimum (rounded down) and next size OCPD will be sized for a particular size (rounded up) motor.

NONTIME-DELAY FUSES USING THE MAXIMUM SIZE
430.52(C)(1), Ex. 2(a)

If the minimum or next size OCPD does not allow the motor to start and run, the maximum size rating of a nontime-delay fuse not exceeding 600 amps shall be permitted to be increased but shall in no case exceed 400 percent of the FLA of the motor. **(See Figure 18-25(a))**

18-17

Stallcup's Electrical Design Book

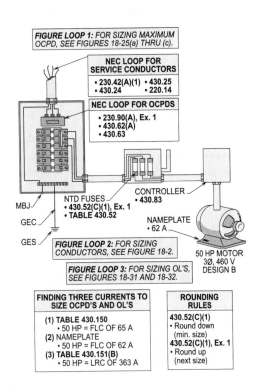

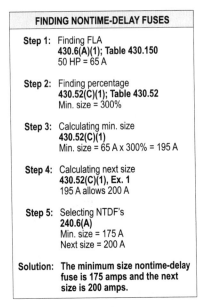

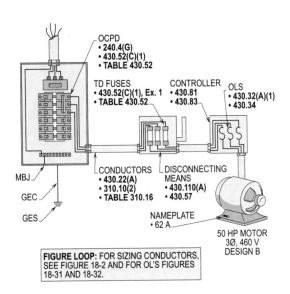

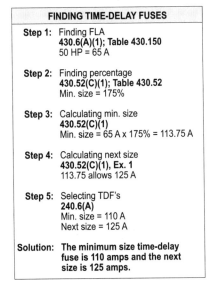

Figure 18-24(a). Determining the minimum and next size nontime-delay fuses per **Table 430.52** and **430.52(C)(1)** and **Ex. 1** to start and run a motor. Note that a smaller OCPD than the minimum size (rounded down) can be used, if it will start the motor.

Figure 18-24(b). Determining the minimum and next size time-delay fuses to start and run a motor. Note that a smaller time-delay fuse than the minimum size (rounded down) may be used, if it will start the motor.

TIME-DELAY FUSES USING MAXIMUM SIZE
430.52(C)(1), Ex. 2(b)

To allow a motor to start and run, the rating of a time-delay fuse shall be permitted to be increased but shall in no case exceed 225 percent of the full-load current, in amps, of the motor. **(See Figure 18-25(b))**

Motors

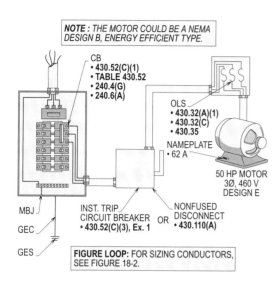

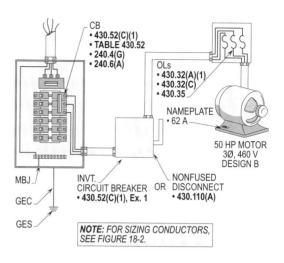

Figure 18-24(c). Determining the minimum and maximum setting for an instantaneous trip circuit breaker to start and run a motor. Note that a smaller than minimum setting may be used, if it will start the motor.

Figure 18-24(d). Determining the minimum and next size inverse time circuit breaker to start and run a motor.

INVERSE TIME CIRCUIT BREAKERS
430.52(C)(1), Ex. 2(c)

The rating for inverse-time circuit breakers shall be permitted to be increased but shall in no case exceed 400 percent for a full-load current of 100 amps or less. Full-load current greater than 100 amps shall be permitted to increase 300 percent. **(See Figure 18-25(c))**

SIZING OCPD FOR TWO OR MORE MOTORS
430.62(A)

To determine the size overcurrent protection device to be installed for a feeder supplying two or more motors the following procedures must be applied:

- Apply **Table 430.52** to select largest motor
- Size largest OCPD for any one motor of group
- Add FLA of remaining motors
- Do not exceed this value with OCPD rating

18-19

See **Figure 18-26** for a detailed procedure of sizing OCPD for a feeder motor circuit.

The largest overcurrent protection device shall be selected based on the motor's full-load current rating times the percentages selected from **Table 430.52**. The next higher standard size rating can be applied per **430.52(C)(1), Ex. 1**. The next higher standard size should not be used if the motor will start. However, it usually can be per NEC. The full-load current ratings of the remaining motors are added to the rating of the largest overcurrent protection device. The next lower standard size overcurrent protection device shall be selected from this total per **240.6(A)**, if this value does not correspond to a standard OCPD.

A larger overcurrent protection device is not allowed to be installed because there is no exception to **430.62(A)** to allow the next size above this rating to be selected. Note that **430.62(A)** only allows a smaller overcurrent protection device to be installed ahead of such feeder-circuit conductors, which is the next size below the computation. **(See Figure 18-27)**

SIZING BRANCH-CIRCUIT TO SUPPLY TWO OR MORE MOTORS
430.52 AND 430.53

Section **430.53** outlines the rules where two or more motors, or one or more motors, and other loads are installed and connected to one branch-circuit protected by an individual overcurrent protection device.

MOTOR NOT OVER 1 HP
430.53(A)

Two or more motors may be installed without individual overcurrent protection devices if rated less than 1 horsepower each and the full-load current rating of each motor does not exceed 6 amps. Motors not rated over 1 horsepower must be within sight of the motor, manually started, and portable. Sections **430.32** and **430.42** must be applied for running overload protection for each motor if these conditions are not met.

The overcurrent protection device rated at 20 amps or less can protect a 120 volt or less branch-circuit supplying these motors. Branch-circuits of 600 volts or less can be protected by a 15 amp or less OCPD. **(See Figure 18-28)**

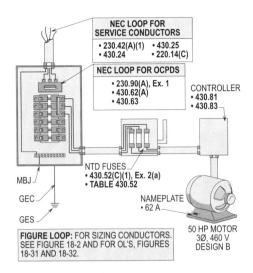

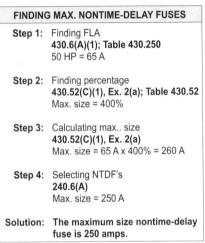

NONTIME-DELAY FUSES USING THE MAXIMUM SIZE
NEC 430.52(C)(1), Ex. 2(a)

Figure 18-25(a). Nontime-delay fuses can be increased to a maximum size of 400 percent of the motor's full-load current rating in amps. Note that for maximum setting of instantaneous trip circuit breaker, see Figure 18-24(c) on page 18-19.

SMALLEST RATED MOTOR PROTECTED
430.53(B)

The branch-circuit overcurrent protection device will be allowed to protect the smallest rated motor of the group for two or more motors of different ratings if the largest motor is allowed to start. The smallest rated motor of the group shall have its overcurrent protection device set at no higher value than allowed per **Table 430.52**. The smallest rated motor and other motors of the group shall be provided with overload protection if necessary per **430.32**. **(See Figure 18-29)**

Motors

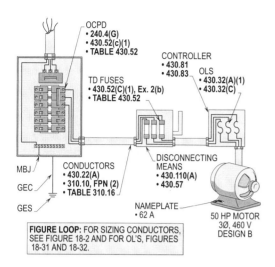

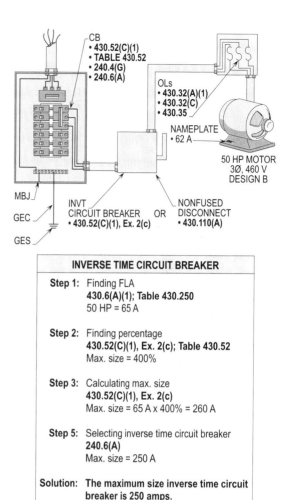

Figure 18-25(b). Determining the maximum size time-delay fuse to start and run a motor.

Figure 18-25(c). An inverse time circuit breaker can be to a maximum size of 400 percent of the motor's full-load current rating in amps because the motor's FLA rating is 100 amps or less.

OTHER GROUP INSTALLATIONS 430.53(C)

Two or more motors of any size are permitted to be installed and connected to an individual branch-circuit. However, the largest motor of the group must be protected by the percentages listed in **Table 430.52** for sizing and selecting fuses and circuit breakers. Each motor controller and component installed in the group shall be approved for such use. The following are elements that must be sized and selected properly:

- Overcurrent protection devices
- Controllers
- Running overload protection devices

The elements may be installed as a listed factory assembly or field installed as separate assemblies listed for such conditions of use.

Design Tip: Any number of motor taps may be installed where a fuse or circuit breaker is installed at the point where each motor is tapped to the line. This type of installation made with a branch-circuit from a feeder per **430.28** and **430.53(D)** is often utilized. **(See Figure 18-30)**

18-21

Stallcup's Electrical Design Book

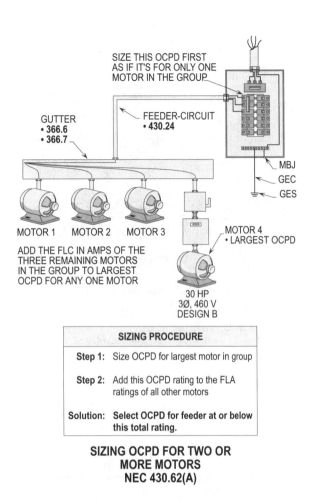

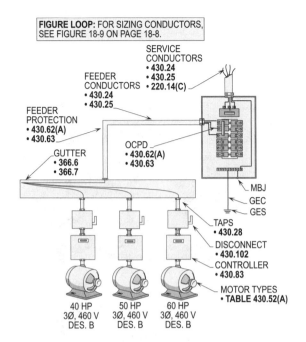

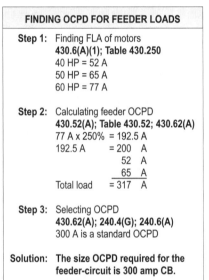

Figure 18-26. Determining OCPD for a feeder-circuit with several motors being protected from short-circuit and ground-fault conditions.

SIZING THE RUNNING OVERLOAD PROTECTION FOR MOTORS
430.32(A)(1)

Devices such as thermal protectors, thermal relays, or fusetrons may be installed to provide running overload protection for motors rated more than 1 horsepower. The service factor or temperature rise of the motor shall be used when sizing and installing the running overload protection for motors. The running overload protection is set to open at 115 percent or 125 percent of the motor's full-load current. Under certain conditions of use, the running overload protection shall be set at 115 percent when the motor is not marked with a service factor or temperature rise. Time-delay fuses selected and sized at these percentages provides overload or back-up overload protection.

Figure 18-27. Sizing an OCPD for a feeder-circuit.

MINIMUM SIZE OVERLOAD PROTECTION
430.32(A)(1)

The amperage for full-load current ratings listed in **Tables 430.247 through 430.250** shall not be used when sizing the running overload protection. The full-load current listed with the motor's nameplate shall be used to size the setting of the running overload protection.

The running overload protection shall be selected and rated no larger than the following minimum percentages based upon the full-load current rating, in amps, listed on the motor's nameplate:
- Motors, with a marked service factor not less than 1.15, use 125 percent
- Motors, with a marked temperature rise not over 40°C, use 125 percent
- All other motors - 115% x FLA

(See Figure 18-31)

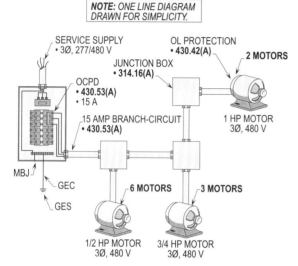

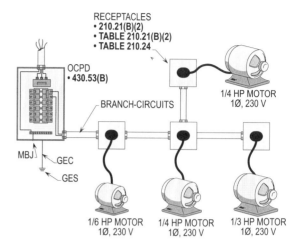

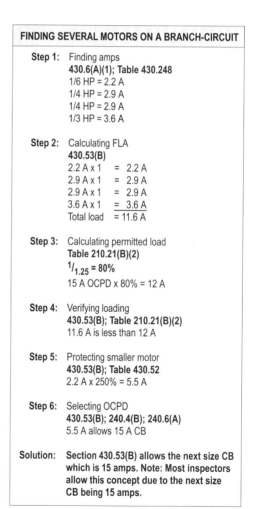

Figure 18-28. Determining the number of motors allowed on a 15 amp branch-circuit.

Figure 18-29. Determining the number of motors allowed on a 15 amp branch-circuit.

Stallcup's Electrical Design Book

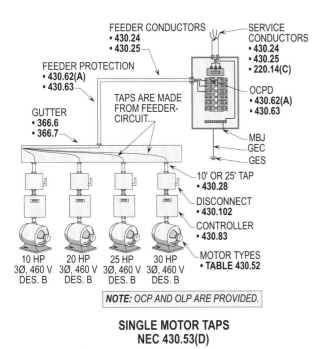

SINGLE MOTOR TAPS
NEC 430.53(D)

Figure 18-30. Taps can be made from a feeder-circuit with the proper size conductors and OCPD with each tap.

MAXIMUM SIZE OVERLOAD PROTECTION
430.32(C)

The running overload protection (overload relay) shall be allowed to be selected at higher percentages if the percentages of **430.32(A)(1)** are not sufficient. The running overload protection shall be selected to trip or shall be rated no larger than the following percentages of the motor's (nameplate) full-load current rating:

- Motors, with marked service factor not less than 1.15, use 140% x FLA
- Motors, with a marked temperature rise not over 40°C, use 140% x FLA
- All other motors - 130% x FLA

(See Figure 18-32)

SIZING THE CONTROLLER TO START AND STOP THE MOTOR
430.81 AND 430.83

The sizes and types of motor controllers are required to be installed with a horsepower rating at least equal to the motor to be controlled. However, there is an Exception to this rule for motors rated at and below a certain horsepower rating.

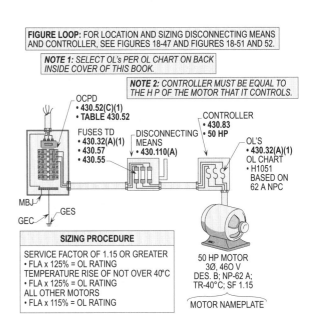

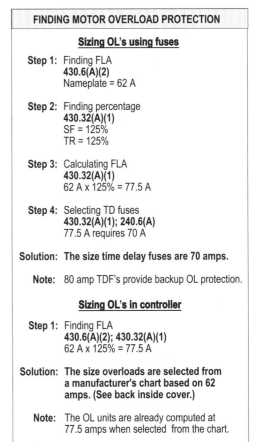

MINIMUM SIZE OVERLOAD PROTECTION
NEC 430.32(A)(1))

Figure 18-31. Determining the minimum size overloads based upon service factor and temperature rise.

18-24

Motors

STATIONARY MOTOR 1/8 HORSEPOWER OR LESS 430.81(B)

The branch-circuit protective device shall be permitted to serve as the controller where the motor is rated 1/8 horsepower or less. **(See Figure 18-33)**

For example, motors less than 1/8 horsepower, where they are mounted stationary or permanent and the construction is such that if one or more should fail during operation, the branch-circuit elements plus the motor(s) won't be damaged. In other words, the components of the circuit won't be burned out, etc.

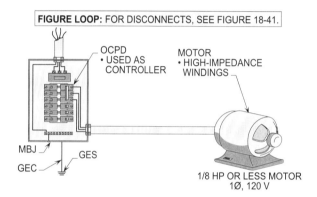

STATIONARY MOTOR 1/8 HORSEPOWER OR LESS NEC 430.81(B)

Figure 18-33. The branch-circuit OCPD can serve as a controller for 1/8 HP or less motor.

PORTABLE MOTOR OF 1/3 HORSEPOWER OR LESS 430.81(C)

The controller shall be permitted to be an attachment plug and receptacle which is acceptable for use with portable motors rated 1/3 horsepower or less. **(See Figure 18-34)**

OTHER THAN HP RATED 430.83(A)

The following five exceptions, other than horsepower rated controllers, are permitted to be used for energizing and deenergizing circuits that supply motors:

- Design E motors, rated more than 2 horsepower
- Inverse time circuit breakers.
- Stationary motors, rated 1/8 horsepower or less or portable motors, rated 1/3 horsepower or less.
- Stationary motors, rated 2 horsepower or less (300

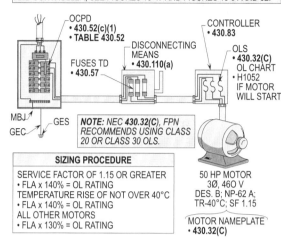

MAXIMUM SIZE OVERLOAD PROTECTION NEC 430.32(C)

Figure 18-32. Determining the maximum size overloads based upon service factor and temperature rise.

18-25

volts or less)
- Torque motors.

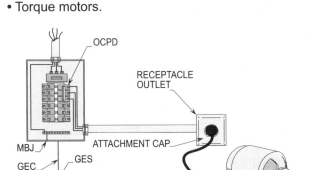

PORTABLE MOTOR OF 1/3
HORSEPOWER OR LESS
NEC 430.81(C)

Figure 18-34. The controller for 1/3 HP motor or less can be an attachment cap and receptacle.

DESIGN E MOTORS
430.83(A)(1)

The controller for Design E motors rated more than 2 horsepower shall be marked for use with Design E motors. The horsepower rating of Design E controllers shall not be less than 1.4 times the motor horsepower for motors rated 3 through 100 horsepower. For motor rated over 100 horsepower, use 1.3 x HP. **(See Figure 18-35)**

INVERSE-TIME CIRCUIT BREAKERS
430.83(A)(2)

Inverse-time circuit breakers shall only be permitted to be installed as a controller where rated in amps. If such circuit breaker is also used for motor overload protection, it shall be sized at 125 percent or less of the motor's nameplate current rating per **430.6(A)(2)** and **430.32(A)(1)**. **(See Figure 18-36)**

When used as a disconnecting means for a motor, it must be sized with a interrupting rating of at least 115 percent of the motor's FLC rating per **430.110(A)**.

STATIONARY AND PORTABLE MOTORS
430.83(A)(3)

Stationary motors rated 1/8 horsepower or less and portable motors rated 1/3 horsepower of less are permitted to serve as controllers and are not required to be horsepower rated. These horsepower rated motors due to smaller locked-rotor currents can be disconnected by cord-and-plug connections.

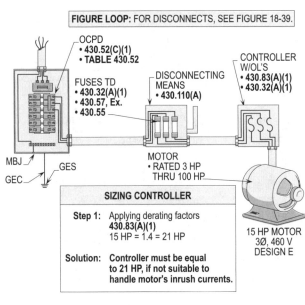

DESIGN E MOTORS
NEC 430.83(A)(1))

Figure 18-35. It may be necessary to increase the HP rating of a controller by 1.4 if it's not marked for Design E use. Note that these multipliers may be applied where the controller is not specifically designed for Design E motors.

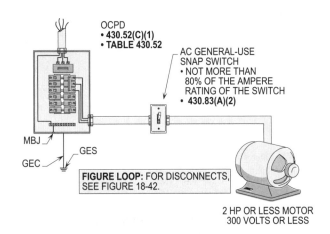

INVERSE-TIME CIRCUIT BREAKERS
NEC 430.83(A)(2)

Figure 18-36. A circuit breaker rated at 125 percent of the motor's FLA can be used as a controller for the motor and also provide overload protection.

STATIONARY MOTORS
430.83(A)(4)

For a stationary motor rated 2 horsepower or less, the controller shall be permitted to be a general-use switch rated for at least twice the motor's full-load current. An AC general-use snap switch may be installed as the controller where the full-load current rating of the switch does not exceed 80 percent (1÷1.25 = 80%) of the branch-circuit rating. (**See Figure 18-37**)

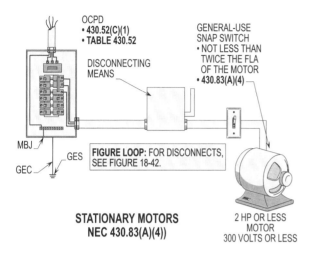

Figure 18-37. For stationary motors rated 2 HP or less, a general-use snap switch may be used if sized not less than twice the motor's full-load current in amps.

TORQUE MOTOR
430.83(A)(5)

The motor controller for a torque motor shall have a continuous duty, full-load current rating not less than the nameplate current rating of the motor. (**See Figure 18-38**)

> **Design Tip:** If the motor controller is rated in horsepower and not marked or rated as above, to determine the amperage or horsepower rating use **Tables 430.247 through 430.250**.

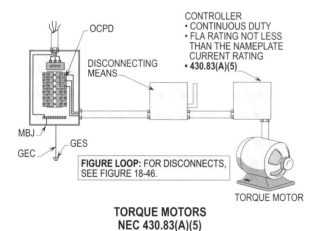

Figure 18-38. The controller for a torque motor must be capable of holding the amps indefinitely.

SIZING THE DISCONNECTING MEANS TO DISCONNECT BOTH THE CONTROLLER AND MOTOR
430.109(A); 430.110(A)

The disconnecting means for motor circuits shall have an ampere rating of at least 115 percent of the full-load current rating of the motor per **430.110(A)**. The disconnecting means shall be horsepower rated and capable of deenergizing locked-rotor currents per **Tables 430.251(A)** and **(B)** of the NEC.

Figure Loop: For sizing controller, see Note 2 to **Figure 18-31**. For sizing control circuits, see **Figures 18-51** and **52**)

OTHER THAN HP RATED
430.109

The following exceptions permit other than a horsepower rated disconnecting means to be used to deenergize the power circuit to certain types of motors:

- Design E motors rated 2 horsepower or less
- Instantaneous trip circuit breakers
- Stationary motors rated 1/8 horsepower or less
- Stationary motors rated 2 horsepower or less (300 volts or less)
- Stationary motors rated 2 horsepower to 100 horsepower
- Stationary motors rated 40 horsepower or more (direct-current) or 100 horsepower or more (alternating-current)
- Cord-and-plug connected motors
- Torque motors

DESIGN E MOTORS
430.109(A)(1)

The disconnecting means for Design E motors rated more than 2 horsepower shall be marked for use with Design E motors. The horsepower rating of Design E disconnecting means shall not be less than 1.4 times the motor's horsepower for motors rated 3 through 100 horsepower. For motors rated over 100 horsepower, use 1.3 x HP. (**See Figure 18-39**)

INSTANTANEOUS TRIP CIRCUIT BREAKER
430.109(A)(4)

An approved instantaneous trip circuit breaker that is part of a listed combination motor controller shall be permitted to serve as a disconnecting means. However, it shall be sized at least 125 percent of the motor's full-load current. (**See Figure 18-40**)

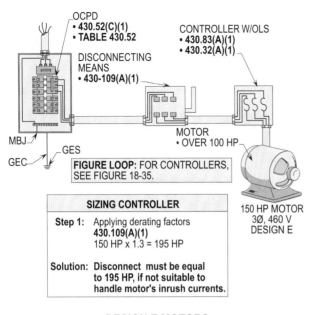

DESIGN E MOTORS
NEC 430.109(A)(1)

Figure 18-39. It may be necessary to increase the HP rating of a controller by 1.3, if it's not specifically marked for Design E use

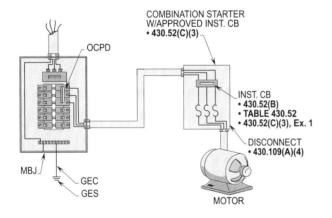

INSTANTANEOUS TRIP CIRCUIT BREAKER
NEC 430.109(A)(4)

Figure 18-40. The disconnecting means for a motor can be an approved instantaneous trip circuit breaker.

STATIONARY MOTORS 430.109(B)

For a stationary motor rated 1/8 horsepower or less, the branch-circuit overcurrent protective device shall be permitted to serve as the disconnecting means. This rule is allowed because the windings of such motors do not produce locked-rotor currents that are high enough to create damage to such motors, circuit conductors, or elements. **(See Figure 18-41)**

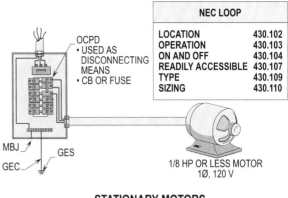

STATIONARY MOTORS
NEC 430.109(B)

Figure 18-41. Motors rated 1/8 HP or less can be disconnected by the OCPD located in the panelboard which is used to supply the circuit.

STATIONARY MOTORS 430.109(C)

For a stationary motor rated 2 horsepower or less, the controller shall be permitted to be a general-use switch rated for at least twice the motor's full-load current. An AC general-use snap switch may be installed as the controller, where the full-load current rating, in amps, of the switch, does not exceed 80 percent (1÷1.25 = 80%) of the branch-circuit rating. **(See Figure 18-42)**

MOTORS OVER 2 HP THRU 100 HP 430.109(D)

Motors rated over 2 horsepower through 100 horsepower are permitted to be installed with a separate disconnecting means (general-use switch) if the motor is equipped with an autotransformer-type controller and all the following conditions are complied with:

- The motor drives a generator that is provided with overload protection.
- The controller is capable of interrupting the locked-rotor current of the motor.
- The controller is provided with a no-voltage release.
- The controller is provided with running overload protection not exceeding 125 percent of the motor's full-load current rating, in amps.
- Separate fuses or an inverse-time circuit breaker is rated at 150 percent or more of the motor's full-load current, in amps. **(See Figure 18-43)**

Motors

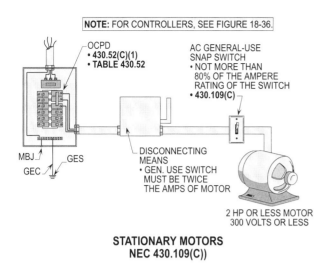

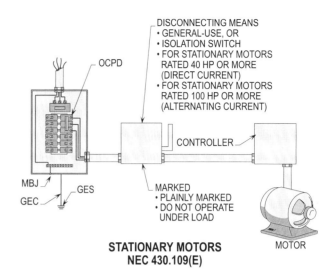

Figure 18-42. The above lists the rules pertaining to the disconnecting means for motors rated 2 HP or less.

Figure 18-44. The above lists the rules for disconnecting means used to disconnect motors rated at 40 HP or more.

CORD-AND-PLUG CONNECTED MOTORS
430.109(F)

Cord-and-plug connected motors are permitted to serve as the disconnecting means for other than Design E motors. Design E motors rated 2 horsepower or less are allowed to be installed with a cord-and-plug disconnecting means. For larger Design E motors, a horsepower rating of 1.4 times the motor rating shall be applied for a cord-and-plug used as the disconnecting means. A cord-and-plug connected appliance is not required to have a horsepower-rated attachment plug and receptacle. The OCPD ahead of the supplying branch-circuit may be utilized for such purposes. For further information see **422.32, 422.31(B), and 440.63**. **(See Figure 18-45)**

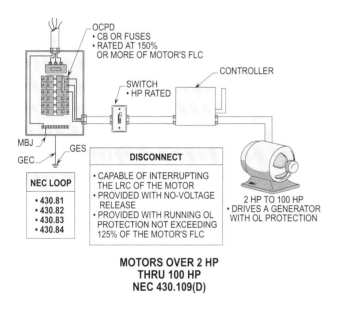

TORQUE MOTOR
430.109(G)

The disconnecting means for a torque motor shall be permitted to be installed as a general-use switch. Such switch must be capable of handling the locked-rotor current, in amps, of the motor indefinitely. **(See Figure 18-46)**

Figure 18-43. The above lists the rules for a disconnecting means and controller used to disconnect and control motors rated 2 HP to 100 HP respectfully.

STATIONARY MOTORS
430.109(E)

The disconnecting means is permitted to be a general-use or isolating switch for DC stationary motors rated at 40 horsepower or greater and AC motors rated 100 horsepower or greater. However, such disconnects must be plainly marked "Do not operate under load." **(See Figure 18-44)**

LOCATION OF THE DISCONNECTING MEANS FOR THE CONTROLLER AND MOTOR
430.102 AND 430.107

A motor and its driven machinery or load shall be installed within sight of the controller for the motor. This rule provides

18-29

safety for electricians and maintenance personnel while servicing such machinery and circuit elements.

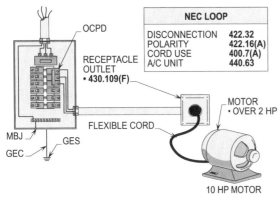

CORD-AND-PLUG CONNECTED MOTORS
NEC 430.109(F)

Figure 18-45. It may be necessary to increase the HP rating of a receptacle and attachment cap used as a disconnecting means for motors rated over 2 HP.

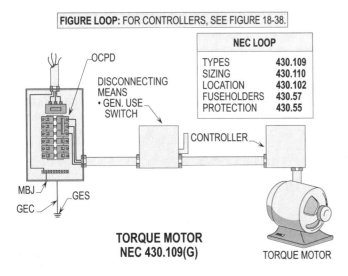

TORQUE MOTOR
NEC 430.109(G)

Figure 18-46. The disconnecting means for a torque motor may be a general-use switch.

WITHIN SIGHT
ARTICLE 100; 430.102

The disconnecting means shall be installed within sight of the motor controller. All of the ungrounded conductors shall be disconnected from both the motor and controller supplying the motor circuit. The disconnecting means shall be installed within sight of the motor and not more than 50 ft. from the motor. If such disconnecting means is not installed within 50 ft. of the motor, other provisions for disconnecting the motor must be made. The controller has a direct relationship to the disconnecting means and must be installed within sight and within 50 ft. of the disconnecting means. The motor does not have a direct relationship with the controller. **(See Figure 18-47)**

Note that a disconnecting means is always required to be located in sight from the controller location. A single disconnecting means is permitted to be located adjacent to a group of coordinated controllers mounted adjacent one to another such as on a multi-motor continuous process machine. For further information pertaining to sizing, selecting, and locating such controllers and disconnecting means, review **430.102, 430.103, 430.107,** and **430.109** very carefully.

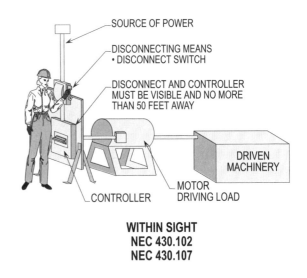

WITHIN SIGHT
NEC 430.102
NEC 430.107

Figure 18-47. The disconnecting means must be within sight and within 50 ft. of the motor and driven machinery.

LOCKED IN THE OPEN POSITION
430.102(A) AND (B)

Section **430.102(A) and (B)** permit the disconnect on the line side of the controller, if within sight and within 50 ft., and capable of being individually locked open, to serve as the disconnecting means for both the controller and motor. In this case, note that the motor shall be installed within sight and within 50 ft. of the disconnecting means of the controller. **(See Figures 18-48(a) and (b))**

Motors

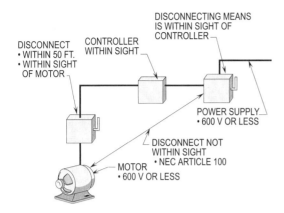

(a) DISCONNECTING MEANS NOT WITHIN SIGHT OF MOTOR

LOCKED IN THE OPEN POSITION
NEC 430.102(A); (B)

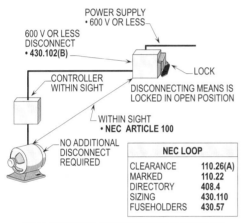

(b) DISCONNECTING MEANS WITHIN SIGHT OF MOTOR
AND LOCKED IN THE OPEN POSITION.

LOCKED IN THE OPEN POSITION
430.102(B)

Figure 18-48(a). Locating the disconnecting means to disconnect power conductors to motors rated 600 volts or less or for motors rated over 600 volts respectfully.

CANNOT BE LOCKED IN THE OPEN POSITION
430.102(A), Ex. 1; 430.102(B)

For motors rated 600 volts or less, an additional disconnecting means must be mounted by the motor and within sight where the disconnecting means installed by the controller cannot be locked in the open position. **(See Figure 18-48(a))** The controller disconnecting means for a motor branch-circuit over 600 volts is allowed to be located out of sight of the motor branch-circuit controller and motor. However, the controller must have a warning label that marks and lists the location and identification of the disconnecting means. To completely satisfy this rule, such disconnecting means must be capable of being locked in the open position. **(See Figure 18-49(a))**

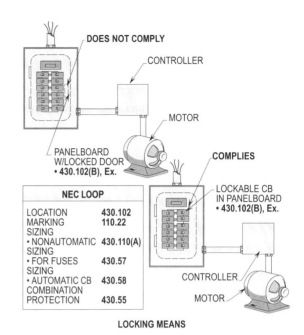

LOCKED IN THE OPEN POSTION
NEC 430.102(B), Ex.

Figure 18-48(b). The locked door of a panelboard cannot serve as the required disconnecting means for a motor, However, an individual locked circuit breaker can serve as such.

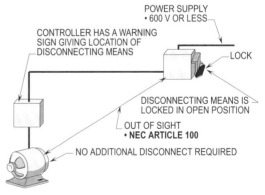

(b) DISCONNECTING MEANS OUT OF SIGHT OF MOTOR
LOCKED IN THE OPEN POSITION.

CANNOT BE LOCKED IN THE
OPEN POSTION
NEC 430.102(A), Ex. 1

Figure 18-49(a). If the disconnecting means is within sight and 50 ft. of controller and can't be locked in the open position, an additional disconnecting means must be installed by the motor.

18-31

APPLYING EXCEPTION
430.102(B)

The **Ex. to 430.102(B)** does not require an additional disconnect to be installed within sight of the motor, where the disconnecting means would be impractical or increase hazards. An additional disconnecting means is not required where it is located in an industrial installation that has written safety procedures and only qualified employees are allowed to work on the equipment involved. **(See Figure 18-49(b))**

Design Tip: Remote-control circuits shall have their disconnecting means located immediately adjacent to the disconnecting means used to disconnect the branch-circuit conductors supplying the controller and motor. Sometimes an interlock in the disconnect for the motor controller is used for this purpose which allows the controller, motor, and remote control circuit to be disconnected simultaneously.

MOTORS, MOTOR CIRCUITS AND CONTROLLERS LOCATION
NEC 430.102(B), Ex.

Figure 18-49(b). Under certain conditions of use, an additional disconnecting means is not required to be installed within sight of the motor.

SIZING CONDUCTORS FOR CONTROL CIRCUIT
430.72 AND 725.23

A motor control circuit tapped on the load side of fuses and circuit breakers utilized for motor branch-circuits shall protect such conductors or supplementary protection devices must be provided.

The size of the control circuit conductors and the rating of the motor's branch-circuit device will be determined by this method of protection. Motor control circuits are classified as remote-control circuits where such circuits derive their power from other than the motor's branch-circuit conductors. Various situations permit fuses or circuit breakers to be utilized to protect remote motor control circuits. For further information see **725.23 and 725.41**.

PROTECTION
430.72(B)

Conductors larger than 14 AWG are selected from **Tables 310.16** through **310.19** for motor control-circuit conductors that are tapped from a motor power circuit. Overcurrent protection for conductors smaller than 14 AWG shall not exceed the values listed in **Table 430.72(B)**, Column A. Conductors 18 AWG and 16 AWG shall be protected at the following amperage ratings:

- 18 AWG shall be protected at 7 amps when used for remote-control circuits
- 16 AWG shall be protected at 10 amps when used for remote-control circuits

Fuses selected at either 1 amp, 3 amp, 6 amp, or 10 amp are normally used to protect these conductors from short-circuits, ground-faults, and overloads. See **240.6(A)** for selection of such fuse sizes.

PROTECTION OF CONDUCTORS
430.72(B), Ex. 2

The secondary conductors of the control transformer circuit may be protected by the primary side of the transformer. The transformer shall be protected per **450.3(B)** and **Table 450.3(B)**. A two-wire secondary for a transformer installed outside or within the control starter enclosure can be protected per **240.4(F)**.

The secondary conductor ampacity shall be multiplied by the secondary-to-primary voltage ratio to provide protection in accordance with **450.3(B)** and **Table 450.3(B)**. Where the rated primary current is 9 amps or greater and 125 percent of this current does not correspond to a standard rating of a fuse or circuit breaker, the next higher standard size can be selected. Where the rated primary current is less than 9 amps, but is 2 amps or greater, an overcurrent protection device rated or set at not more than 167 percent of the primary current can be used. Where the rated primary current is less than 2 amps, an overcurrent protection device rated or set not greater than 300 percent to 500 percent can be used. **(See Figure 18-50)**

For example, if the primary full-load current of a motor control transformer is less than 2 amps, the OCPD can be calculated and sized at 500 percent times such full-load current in amps.

than 14 AWG shall not exceed the values listed in **Table 430.72(B)**, Column B. Conductors rated 18 AWG through 10 AWG can be protected with the following sized OCPD's:

- 18 AWG = 25 amps (7 A x 400% = 28 A and requires 25 A OCPD)
- 16 AWG = 40 amps (10 A x 400% = 40 A and requires 40 A OCPD)
- 14 AWG = 100 amps (25 A x 400% = 100 A and requires 100 A OCPD)
- 12 AWG = 120 amps (30 A x 400% = 120 A and requires 110 A OCPD)
- 10 AWG = 160 amps (40 A x 400% = 160 A and requires 150 A OCPD)
- 8 AWG and larger = 400 percent

Design Tip: The free air ampacities of **Table 310.17** for 60°C wire are used to determine the ampacity ratings for the control circuit conductors. This type of installation has more free space to dissipate the heat where control conductors are installed in the open air space of enclosures instead of enclosed raceways.

See **Figure 18-51** for selecting such conductors based upon the OCPD rating.

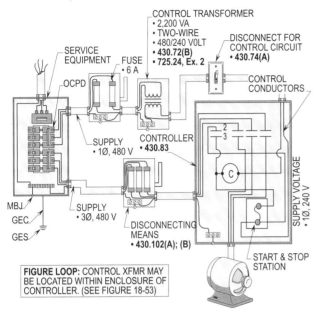

Figure 18-50. Control circuit conductors are supplied by a control transformer and protected by fuses in primary side.

PROTECTION OF CONDUCTORS 430.72(B)(2)

Motor control-circuit conductors that do not extend beyond the control equipment enclosure are permitted to be protected by the motor branch-circuit fuses or circuit breakers. **Table 430.72(B)**, Column B allows this type of installation where the devices do not exceed 400 percent of the ampacity rating of sizes 14 AWG and larger conductors. Overcurrent protection for conductors smaller

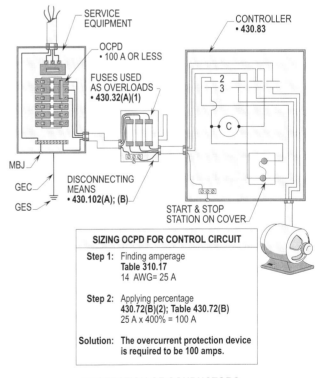

Figure 18-51. Control circuit conductors located in controller and protected by the branch-circuit OCPD.

PROTECTION OF CONDUCTORS 430.72(B)(2)

Motor control-circuit conductors that extend beyond the control equipment enclosure can be protected by the motor's branch-circuit fuse or circuit breaker. **Table 430.72(B), Column C** allows this type of installation where the devices do not exceed 300 percent of the ampacity rating of sizes 14 AWG and larger conductors. Overcurrent protection for conductors smaller than 14 AWG shall not exceed the values listed in **Table 430.72(B), Column C**. Conductors rated 18 AWG through 10 AWG can be protected with the following sized OCPD's:

- 18 AWG = 7 amps and requires 7 A OCPD
- 16 AWG = 10 amps and requires 10 A OCPD
- 14 AWG = 45 amps (15 A OCPD x 300% = 45 A and requires 45 A OCPD)
- 12 AWG = 60 amps (20 A OCPD x 300% = 60 A and requires 60 A OCPD)
- 10 AWG = 90 amps (30 A OCPD x 300% = 90 A and requires 90 A OCPD)
- 8 AWG and larger = 300 percent

Note that the above protection is required anytime the control circuit is used for the remote-control of a coil in the motor controller enclosure.

See Figure 18-52 for selecting such conductors based upon the OCPD rating.

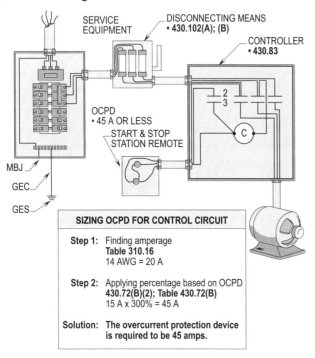

PROTECTION OF CONDUCTORS NEC 430.72(B)(2)

Figure 18-52. Control circuit conductors are run remote and protected by the branch-circuit's OCPD in the panelboard.

TRANSFORMER 430.72(C)(1) THRU (5)

When a motor control-circuit transformer is provided, the transformer shall be protected by the rules and regulations of **Article 450**. A fuse or circuit breaker is permitted to be installed in the secondary circuit of the transformer per **430.72(C)1 through (5)**. The primary OCPD may be used to provide the protection for the conductors tapped from the secondary side of a control transformer, if its rating does not exceed the secondary-to-primary ratio of the transformer. **(See Figure 18-53)**

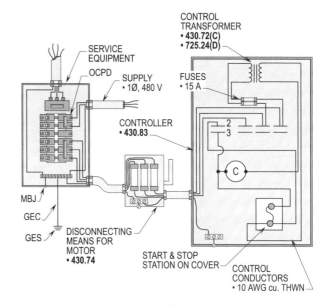

TRANSFORMER NEC 430.72(D) AND (E)

Figure 18-53. The primary OCPD may be used to provide the protection for the conductors tapped from the secondary side of a control transformer, if its rating does not exceed the secondary to primary voltage ratio of the transformer.

MOTOR CONTROL AND MOTOR POWER CIRCUIT CONDUCTORS
430.74

Motor control circuits shall be disconnected from all sources of supply when the disconnecting means is in the open position. The disconnecting means for the starter may be installed to serve as the disconnecting means for the motor circuit conductors if the control-circuit conductors are tapped from the line terminal of the magnetic starter. An auxiliary contact shall be installed in the disconnecting means of the controller or an additional disconnecting means shall be mounted adjacent to the controller to disconnect the motor control-circuit conductors if they are fed from another source and not tapped from the starter conductors. **(See Figure 18-54)**

ROUTING CONTROL CIRCUIT CONDUCTORS
300.3(C)(1) AND 725.26

Conductors of different systems are allowed to occupy the same raceway without regard to AC or DC current per **300.3(C)(1)**. Conductors shall be insulated for the maximum voltage of any one conductor when occupying the same raceway. These conductors shall not exceed 600 volts. Class 1 conductors are allowed to occupy the same raceway when installed with the power conductors supplying the magnetic starter and motor per **725-26**. For control circuit conductors to occupy the same raceway as the motor circuit conductors they must be functionally associated with the motor system. **(See Figure 18-56)**

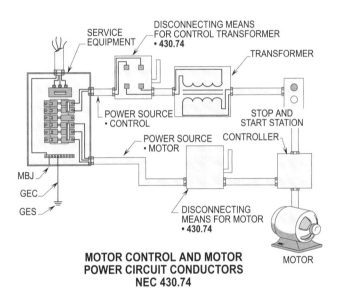

OCCUPYING THE SAME ENCLOSURE
300-3(c)(2)(d) AND 300-32

Motor excitation, magnetic starter, control relay, or ammeter conductors may occupy the same enclosure if rated at 600 volts and less or over 600 volts. These conductors are not allowed to occupy the same raceway where installed as a combination of conductors of 600 volts or less with conductors of over 600 volts. The motor enclosure and the starter enclosure can contain motor excitation, control, relay, and ammeter conductors of 600 volts or less together with power conductors of over 600 volts. **(See Figure 18-57)**

MAGNETIC STARTER CONTACTOR AND ENCLOSURE
725.28(A) AND (B)

When installing four or more current-carrying conductors that are continuously operated in a raceway and used for Class 1 remote-control, signal, and power-limited circuit conductors, they may have to be derated by the derating factors of **310.15(B)(2)(a)** per **725.28(B)(1) or (B)(2)**. Conductors that are noncontinuous operated are not to be derated by the derating factors of **310.15(B)(2)(a)**. When installing power and control conductors in the same raceway, **310.15(B)(2)(a)** shall be applied to all the current-carrying conductors operating for three hours or more. **(See Figure 18-58)**

Figure 18-54. One disconnecting means or a number of disconnects may be required to disconnect a control circuit and power supply to a motor.

> **Design Tip:** Control circuit conductors operating for three hours or more are not required to be derated per **310.15(B)(2)(a)**, if their ampacity does not exceed 10 percent of the control conductors ampacity rating.

CAPACITOR
460.8(A)

The ampacity of capacitor circuit conductors shall not be less than 135 percent of the rated current of the capacitor. The leads for a capacitor that supplies a motor shall not be less than one-third the ampacity of the motor circuit conductors. The larger of the above conductors must be used for the capacitor supply conductors. **(See Figure 18-55)**

Stallcup's Electrical Design Book

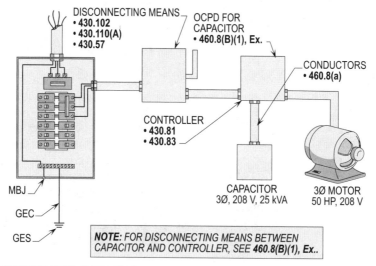

Figure 18-55. There are two calculations to be performed and one of them must be selected to size the capacitor circuit conductors. Note that the greater of the 1/3 calculation or 135 per cent calculation shall be used.

TABLE CURRENT IN AMPS
TABLES 430.247 THRU 430.250

The following methods can be used to determine the full-load current rating in amps for motors that are not listed in **Tables 430.247 through 430.250**:

- The horsepower rating of a listed motor shall be selected which is below the unlisted motor.

- The motor's full-load current rating shall be divided by its horsepower rating to obtain the multiplier.

- The multiplier times the HP of the unlisted motor derives FLC in amps for the unlisted motor.

Design Tip: The full-load current rating of the motor is determined by multiplying these values by the horsepower rating of the unlisted motor. **(See Figure 18-59)**

Motors

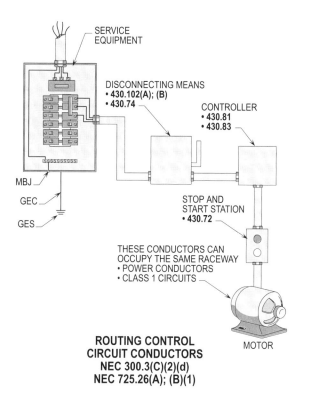

Figure 18-56. When functionally associated with motor operation, a Class 1 control circuit can occupy the same raceway as the motor's power circuit conductors.

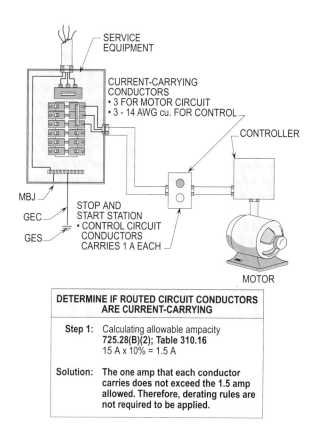

Figure 18-58. Circuit conductors that do not carry more than 10 percent of their ampacity are not considered current-carrying conductors.

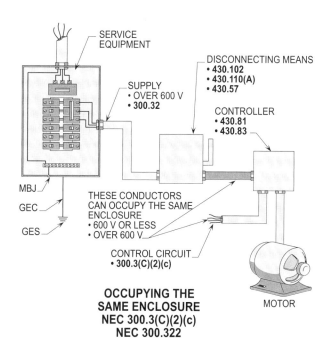

Figure 18-57. Over 600 volts and 600 volts or less conductors can occupy the same enclosure under certain conditions of use.

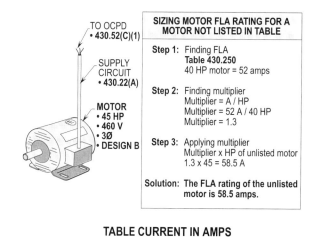

Figure 18-59. The above illustrates the procedure for calculating the FLA of a motor not listed in **Tables 430.248 through 430.250**.

RULE OF THUMB - AMPS

The full-load current of a motor may be found by using the rule of thumb method in **Table 430.248** for single-phase and **Table 430.250** for three-phase. The Table current will not always be exactly the same as the rule-of-thumb amps.

Overcurrent protection devices, conductors, and other elements can be sized with the full-load current ratings obtained by the rule of thumb method. The full-load current ratings are within a usable range when applying the rule of thumb method to determine the full-load current rating in amps. These amperage ratings will provide values to compute elements for a complete and safe electrical motor system.

The following percentages can be applied when using a rule-of-thumb method to derive full-load amps for a particular size motor:

- When installing 550, 575, 600 volt, three-phase motors, the horsepower rating of the motor shall be multiplied by 1.00 to obtain full-load current in amps.

- When installing 440, 460, 480 volt, three-phase motors, the horsepower rating of the motor shall be multiplied by 1.25 to obtain full-load current in amps.

- When installing 220, 230, 240 volt, three-phase motors, the horsepower rating of the motor shall be multiplied by 2.50 to obtain full-load current in amps.

- When installing 220, 230, 240 volt, single-phase motors, the horsepower rating of the motor shall be multiplied by 5.00 to obtain full-load current in amps.

- When installing 110, 115, 120 volt, single-phase motors, the horsepower rating of the motor shall be multiplied by 10.00 to obtain full-load current in amps.

See **Figure 18-60** for a detailed illustration of computing full-load currents in amps for motors using the rule of thumb method.

NUMBER OF MOTORS SERVED BY EACH DISCONNECT AND CONTROLLER
430.87; 430.112

Each controller shall be installed with a disconnecting means located within sight and within 50 ft. Unauthorized energizing of the electrical power circuit must be prevented during maintenance procedures. A single disconnect and controller is permitted to be installed to control and disconnect any number of motors of 600 volts or less per **430.87, Ex.** and **430.112, Ex.** Motors used to drive different parts of the same machine is an example of a number of motors that can be protected by a single disconnect and controller. **(See Figure 18-61)**

A single OCPD is permitted to be installed to protect a group of motors, only if they are rated in the fractional-horsepower range and installed on a general-purpose circuit of 20 amps or less per **430.53(A)**.

Motors installed to be served by the same disconnect and controller shall be within sight and located within one room. Unauthorized energizing of the electrical power circuit can be prevented during maintenance procedures if the above rules are complied with. For lockout and tagout procedures, See OSHA 1910, Subpart S and NFPA 70E, Chapter 1.

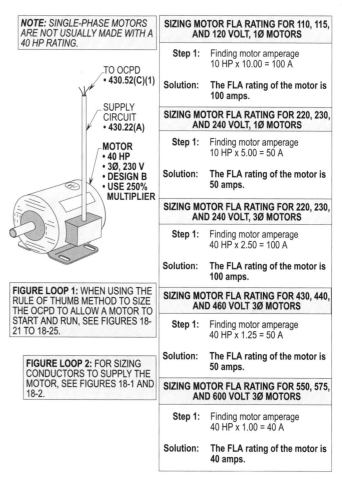

RULE OF THUMB - AMPS
NEC TABLES 430.248 THRU 430.250

Figure 18-60. The above is a rule of thumb method used by the electrical industry to determine the FLA of motors whether found or not found in **Tables 430.248 through 430.250**. Note that this method is used in the field or office when a code book is not available.

Motors

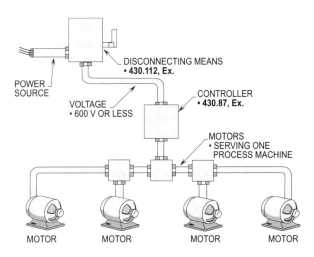

Figure 18-61. Under certain conditions of use, one disconnect and controller can be used for several motors serving a machine.

REDUCED STARTING METHODS

As to methods of starting a motor, there are seven methods used in the electrical industry and they are as follows:

- Full voltage (not reduced at starting)
- Resistor starting
- Reactor starting
- Autotransformer starting
- Solid state starting
- Part winding
- Start wye and run delta

AC motors up to about 10 or 15 HP are usually started on full voltage. For larger motors, a resistance or reactance is inserted in the motor circuit at starting to reduce the starting current. The voltage at starting can be reduced by the use of an autotransformer. A solid-state starter can also be used to reduce starting currents.

ADJUSTABLE-SPEED DRIVE SYSTEMS
PART X TO ARTICLE 430

When designing and installing electrical systems for adjustable-speed drives, the installation provisions of Part I through Part IX are applicable unless modified or supplemented by Part X.

BRANCH/FEEDER CIRCUIT CONDUCTORS
430.122(A)

Circuit conductors supplying power conversion equipment included as part of an adjustable-speed drive system shall have an ampacity not less that 125 percent of the rated input to the power conversion equipment. **(See Figure 18-62)**

Note: Electrical resonance can result from the interaction of the noninusoidal currents from this type of load with power factor correction capacitors.

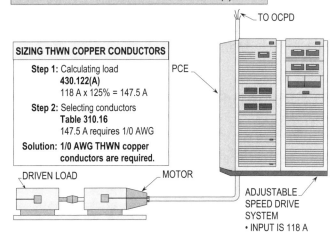

Figure 18-62. The above illustrates the procedure for calculating the load in amps to size the conductors supplying the power conversion equipment.

BYPASS DEVICE
430.122(B)

For an adjustable speed drive system that utilizes a bypass device, the conductor ampacity shall not be less than required by **430.6**. The ampacity of circuit conductors supplying power conversion equipment included as part of an adjustable speed drive system that utilizes a bypass device shall be the larger of either of the following:

(1) 125 percent of the rated input to the power conversion equipment
(2) 125 percent of the motor full-load current rating as determined by **430.6**

For an illustrated description, see **Figure 18-63**.

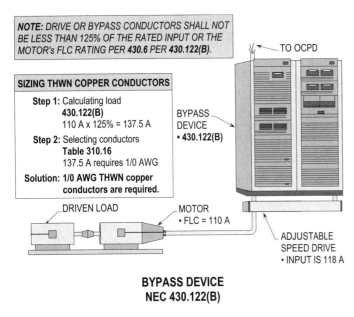

Figure 18-63. The above illustrates the procedures for calculating the load in amps to size the conductors for a bypass ddrive system

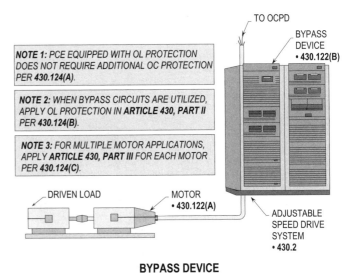

Figure 18-64. The above illustrates the procedure for protecting the motor from overload when using a bypass device.

OVERLOAD PROTECTION
430.124

Overload protection of the motor shall be provided.

INCLUDED IN POWER CONVERSION EQUIPMENT
430.124(A)

Where the power conversion equipment is marked to indicate that motor overload protection is included, additional overload protection shall not be required.

BYPASS CIRCUITS
430.124(B)

For adjustable speed drive systems that utilize a bypass device to allow motor operation at rated full load speed, motor overload protection as described in **Article 430, Part III**, shall be provided in the bypass circuit.

MULTIPLE MOTOR APPLICATIONS
430124(C)

For multiple motor application, individual motor overload protection shall be provided in accordance with **Article 430, Part III**. (See Figure 18-64)

MOTOR OVERTEMPERATURE PROTECTION – GENERAL
430.126(A)

Adjustable speed drive systems shall protect against motor overtemperature conditions. Overtemperature protection is in addition to the conductor protection required in **430.32**. Protection shall be provided by one of the following means.

(1) Motor thermal protector in accordance with **430.32**
(2) Adjustable speed drive controller with load and speed-sensitive overload protection and thermal memory retention upon shutdown or power loss
(3) Overtemperature protection relay utilizing thermal sensors embedded in the motor and meeting the requirements of **430.32(A)(2)** or **(B)(2)**
(4) Thermal sensor embedded in the motor that is received and acted upon by an adjustable speed drive

For a detailed description of these requirements, see **Figure 18-65**.

MOTORS WITH COOLING SYSTEMS
430.126(B)

Motors that utilize extreme forced air or liquid cooling systems shall be provided with protection that shall be continuously enabled or enabled with automatically if the cooling system fails.

Note: Protection against cooling system failure can take many forms. Some examples of protection against inoperative or failed cooling systems are direct sensing of the motor temperature as described in **430.32(A)(1), (A)(3)** and **(A)(4)** or sensing of the presence or absence of the cooling media (flow or pressure sensing).

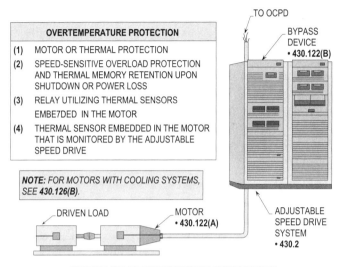

MOTOR OVERTEMPERATURE PROTECTION
NEC 430.126(A)(1) THRU (A)(4)

Figure 18-65. The above illustrates the procedure for determining protection for motor overtemperature problems.

MULTIPLE MOTOR APPLICATIONS
430.126(C)

For multiple motor application, individual motor overtemperature protection shall be provided.

Note: The relationship between motor current and motor temperature changes when the motor is operated by an adjustable speed drive. When operated at reduced speed, overheating of motors may occur at current levels less than or equal to a motor's rated full load current. This is the result of reduced motor cooling when its shaft-mounted fan is operating less than rated nameplate RPM.

AUTOMATIC RESTARTING AND ORDERLY SHUTDOWN
430.126(D)

The provisions of **430.43** and **430.44** shall apply to the motor overtemperature protection means.

DISCONNECTING MEANS
430.128

The disconnecting means shall be permitted to be in the incoming line to the conversion equipment and shall have a rating not less than 115 percent of the rated input current of the conversion unit.

For example, if the rated input current of the conversion unit is 173.9 amps, the disconnecting means shall be rated at least 200 amps (173.9 A x 115% = 199.9 A) per **430.128**.

INDUSTRIAL CONTROL PANELS
ARTICLE 409

Article 409 covers the installation of industrial control panels intended for general use and operating procedures at 600 volts or less. UL 508 A governs the procedures for safety when installing components in industrial control panels.

CONDUCTOR – MINIMUM SIZE AND AMPACITY
409.20

The size of the industrial control panel supply conductor shall have an ampacity not less than 125 percent of the full-load current rating of all resistance heating loads plus 125 percent of the full-load current rating of the highest rated motor plus the sum of the full-load current ratings of all other connected motors and apparatus based on their duty cycle that may be in operation at the same time. **(See Figure 18-66)**

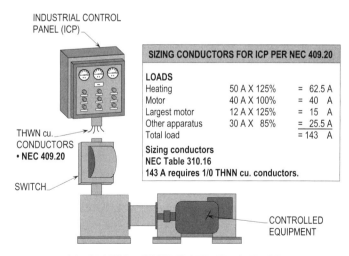

CONDUCTOR – MINIMUM SIZE AND AMPACITY
NEC 409.20

Figure 18-66. The above illustrates the procedure for calculating the load in amps to size the conductors supplying the industrial control panel.

OVERCURRENT PROTECTION – GENERAL
409.21(A)

Industrial control panels shall be provided with overcurrent protection in accordance with **Parts I, II** and **IX** of **Article 240**.

LOCATION
409.21(B)(1); (B)(2)

Overcurrent protection for industrial control panels shall be provided by either of the following:
(1) An overcurrent protective device located ahead of the industrial control panel
(2) A single main overcurrent protective device located within the industrial control

Where overcurrent protection is provided as part of the industrial control panel, the supply conductors shall be considered as either feeders or taps as covered by **240.21**.

RATING
409.21(C); Ex.

The rating or setting of the overcurrent protective device for the circuit supplying the industrial control panel shall not be greater than the sum of the largest rating or setting of the branch-circuit short-circuit and ground-fault protective device provided with the industrial control panel, plus 125 percent of the full-load current rating of all resistance heating loads, plus the sum of the full-load currents of all other motors and apparatus that could be in operation at the same time. **(See Figure 18-67)**

Applying exception to **409.21(C)** – Where one or more instantaneous trip circuit breakers or motor short-circuit protectors are used for motor branch-circuit, short-circuit and ground-fault protection as permitted by **430.52(C)**, the procedure specified above for determining the maximum rating of the protective device for the circuit supplying the industrial control panel shall apply with the following provision:

For the purpose of the calculation, each instantaneous trip circuit breaker or motor short-circuit shall be assumed to have a rating not exceeding the maximum percentage of motor full-load current permitted by **Table 430.52** for the type of control panel supply circuit protective device employed.

Where no branch-circuit short-circuit and ground-fault protective device is provided with the industrial control panel for motor or combination of motor and non-motor loads, the rating or setting of the overcurrent protective device shall be based on **430.52** and **430.53**, as applicable.

DISCONNECTING MEANS
409.30

Disconnecting means that supply motor loads shall comply with **Part IX** of **Article 430**. For example, if the motor loads on an industrial process add up to 173 amps, the size disconnecting would be 200 amps (173 A x 115% = 198.95 A) per **430.110(A)** of the NEC. **(See Figure 18-68)**

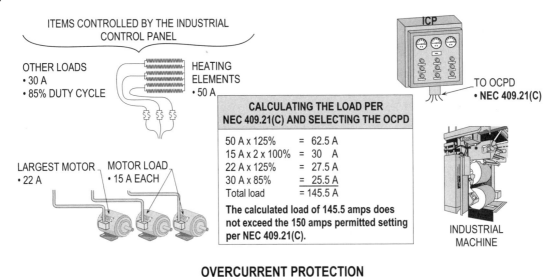

OVERCURRENT PROTECTION
NEC 409.21(C)

Figure 18-67. The above illustrates the procedure for calculating the load in amps to size the overcurrent protection device protecting the industrial control panel.

Motors

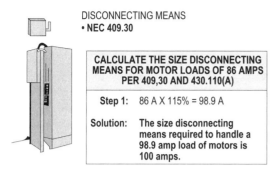

Figure 18-68. The above illustrates the procedure for calculating motor loads in amps to size the disconnecting means.

WIRING SPACE IN INDUSTRIAL CONTROL PANELS – GENERAL 409.104(A)

Industrial control panel enclosures shall not be used as junction boxes, auxiliary gutters, or raceways for conductors feeding through or tapping off to other switches or overcurrent devices, unless adequate space for this purpose is provided. The conductors shall not fill the wiring space at any cross section to more than 40 percent of the cross-sectional area of the space, and the conductors, splices and taps shall not fill the wiring space at any cross section to more than 75 percent of the cross-sectional area of that space. **(See Figure 18-69)**

PHASE CONVERTERS 455.6(A) AND 455.7(A)

Phase converters are used to convert single-phase power to three-phase power. The disconnecting means shall be located within 50 ft. and within sight per **455.8(A)**. Where the voltage is not the same, the output to input ratio shall be applied per **455.6(A)**.

Branch-circuit conductors are sized at 125 percent times the phase converter's nameplate single-phase input full-load current rating, in amps. The overcurrent protection device is sized at 125 percent times the phase converter's nameplate single-phase input full-load amps. The OCPD shall not exceed the 125 percent but shall be equal to or lower than 125 percent. **(See Figure 18-70)** Branch-circuit elements such as OCPD's and conductors supplying specific loads are calculated at 250 percent of the equipment's full-load amp rating. **(See Figure 18-71)**

Feeder-circuit conductors that convert single-phase power to three-phase power to supply power to two or more phase converters are sized at 250 percent times the three-phase amperage of all motors and other loads served. The overcurrent protection device is sized at 250 percent times the full-load three-phase amps of all motors and other loads. If the percentage does not correspond to a standard size, the next size overcurrent protection device above this percentage can be selected per 455.7. **(See Figure 18-72)**

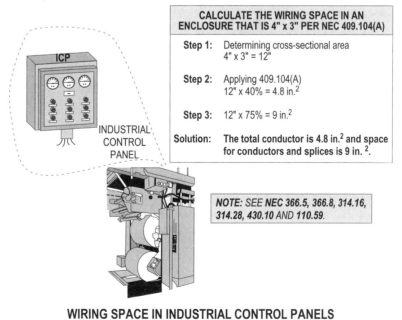

Figure 18-69. The above illustrates the procedure for calculating the wiring space necessary for accomodating conditions and splices in an industrial control panel.

18-43

Stallcup's Electrical Design Book

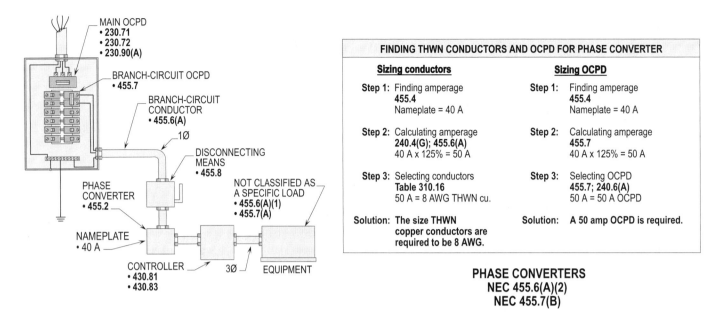

Figure 18-70. Branch-circuit conductors are sized at 125 percent times the phase converter's nameplate single-phase input full-load amperage. The overcurrent protection device is sized at 125 percent times the phase converter's nameplate single-phase input full-load current, in amps.

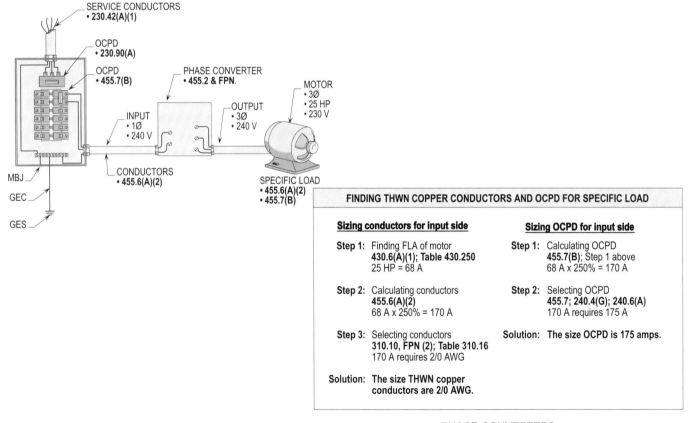

Figure 18-71. Branch-circuit elements such as OCPD's and conductors supplying specific loads are calculated at 250 percent of the equipment's full-load current rating, in amps.

18-44

Motors

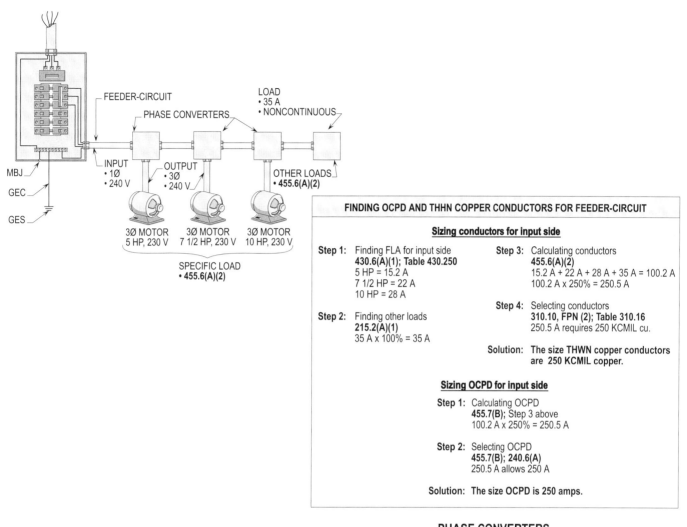

Figure 18-72. Feeder-circuit conductors that convert single-phase power to three-phase power for supplying power to two or more phase converters are sized at 250 percent times the three-phase amperage of all motors and other loads served. The overcurrent protection device is sized at 250 times the full-load three-phase amps of all motors and other loads.

FIRE PUMPS
695

Article 695 covers the installation of electric power sources, interconnecting circuits, and switching and control equipment dedicated to fire pumps.

POWER SOURCES
695.3(A)

Section **695.3(A)** covers power sources which are allowed to supply power to fire pump installations.

Power sources such as a reliable service, an on-site generator, a separately derived system, or a tap ahead of the service disconnecting means are considered dependable power supply systems when serving fire pumps and other related equipment. **(See Figure 18-73)**

Stallcup's Electrical Design Book

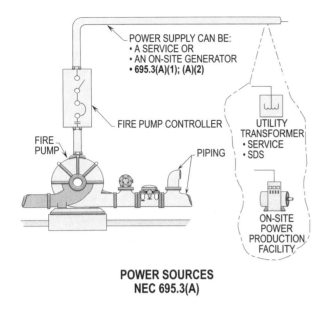

POWER SOURCES
NEC 695.3(A)

Figure 18-73. The above shows power sources that are permitted to supply fire pump installations.

SIZING CONDUCTORS
695.6(C)(1)

Conductors must be sized with enough capacity so that they are protected against short-circuit currents. By sizing the conductors to the fire pump motors at 125 percent of the motor's FLA, this should be accomplished. For sizing the conductors to one motor, see **430.22(A)** and for more than one motor, plus other loads, see **430.24** of the NEC. **(See Figure 18-74)**

SIZING OCPD
695.5(B); (C)(2); 230.90(A), Ex. 4

The OCPD must protect the conductors and fire pump motor and accessories from short-circuits. **(See Figure 18-75)**

SIZING XFMR USED AS A SDS
695.5(A)

Section **695.5(A)** allows a transformer dedicated to supplying a fire pump to be rated at a minimum of 125 percent of the sum of the rated full-load of the fire pump motor(s), the rated full-loads of pressure maintenance pump motor(s), and the full-load amps of any associated fire pump accessory equipment, connected to the transformer.

Secondary overcurrent protection for the transformer is not permitted and the primary OCPD must not be set above 600 percent of the transformer's full-load current rating, in amps.

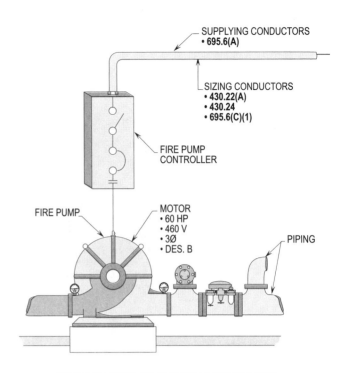

SIZING CONDUCTORS
NEC 695.6(C)(1)
NEC 430.22(A)

Figure 18-74. The above shows the procedure for sizing the conductors to supply a fire pump.

Motors

When separately derived systems are used to supply power to fire pumps and accessories, they are usually installed in the fire pump room with the fire pump controller. Note that the transformer must supply power until pump motor failure. This requirement allows the motor to pump water to fight the fire for as long as possible.

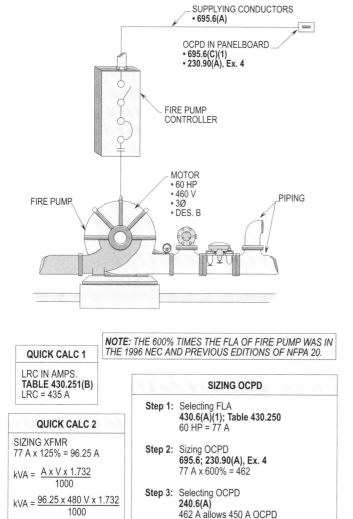

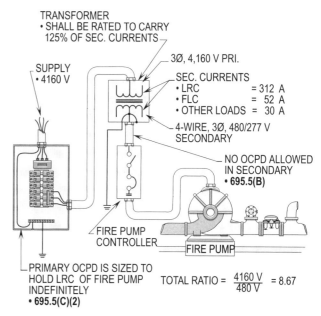

Figure 18-75. The above shows the procedure for sizing the OCPD to protect a fire pump.

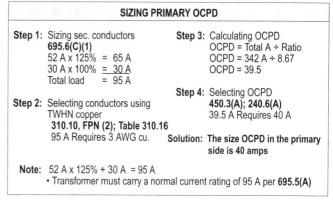

Figure 18-76. The above shows the procedure for sizing the transformer and primary OCPD for a fire pump installation.

SIZING OCPD FOR A SEPARATELY DERIVED SYSTEM
695.5(B)

Section **696.5(B)** requires the OCPD on the primary side of a separately derived system, supplying power to a fire pump installation, to carry the secondary circuit indefinitely.

Note that such secondary currents includes both normal full-load operating current, in amps, as well as the locked rotor current, in amps, of the motor. **(See Figure 18-76)**

18-47

Name   Date

Chapter 18. Motors

Section   Answer

_____   T   F   1. The full-load current in amps for three-phase motors shall be selected from Table **430.148**.

_____   T   F   2. Conductors for a motor used for periodic duty shall be sized with a current-carrying capacity of 125 percent of the motors full-load amps.

_____   T   F   3. When power conversion equipment provides overcurrent protection for the motor, no additional overload protection is required.

_____   T   F   4. Wound-rotor motors are three-phase motors that are installed with two sets of leads.

_____   T   F   5. The motor branch-circuit overcurrent device shall not be required to be capable of carrying the starting current, in amps, of the motor.

_____   T   F   6. Squirrel-cage motors are known in the electrical industry as induction motors.

_____   T   F   7. Code letters, when used, shall be marked on the nameplate of the motor and are tp be used for designing locked-rotor amps.

_____   T   F   8. Time-delay fuses which are sized at 125 percent or less of the motor's FLA rating can provide overload protection for the motor.

_____   T   F   9. The maximum percentage that shall be permitted to be applied for a time-delay fuse is 400 percent of the motor's FLA.

_____   T   F   10. The FLA ratings of the remaining motors are added to the rating of the largest motor's OCPD, when selecting the feeder's OCPD.

_____ _____   11. Two or motors shall be permitted to be installed without individual overcurrent protection devices if rated less than 5 horsepower each and the full-load current rating, in amps, of each motor does not exceed 6 amps.

_____ _____   12. Thermal protectors and thermal relays may be installed to provide running overload protection for motors rated more than 1 horsepower.

_____ _____   13. The controller shall be permitted to be an attachment plug and receptacle which is acceptable for use with portable motors rated 1/8 horsepower or less.

_____ _____   14. A molded case switch rated in amperes shall not be permitted as a controller for all motors.

_____ _____   15. The controller shall be permitted to be a general-use switch rated for at least twice the motor's full-load current, in amps, for stationary motors rated 2 horsepower or less.

18-49

Section	Answer

16. Inverse-time circuit breakers shall only be permitted to be installed as controllers for motors, where they are rated in horsepower.

17. For a stationary motor rated 1/8 horsepower or less, the branch-circuit overcurrent protection device shall be permitted to serve as the disconnecting means.

18. An approved instantaneous trip circuit breaker that is part of a listed combination motor controller shall be permitted to serve as a disconnecting means.

19. The general rule requires the disconnecting means for a motor to be installed within sight of the motor and not more than 50 ft. from the motor.

20. For motors rated 600 volts or less, an additional disconnecting means shall be mounted by the motor and within sight where the disconnecting means installed by controller cannot be locked in the open position.

21. Motor control circuits shall not be required to be disconnected from all sources of supply when the disconnecting means is in the open position.

22. Control circuit conductors and motor circuit conductors can occupy the same raceway if they are functionally associated with the motor system.

23. Phase converters are used to convert single-phase power to three-phase power.

24. Branch-circuit conductors shall be sized at 100 percent times the phase converter's (variable loads) nameplate single-phase input FLA rating.

25. Feeder-circuit conductors supplying power to two or more phase converters are sized at 125 percent times the three-phase amperage of all motors and other loads served.

26. Branch-circuit conductors supply a single motor shall have an ampacity not less than _____ percent of the motor's FLC, in amps.

27. The selection of conductors for wye start and delta run motors between the controller and motor shall be based on _____ percent of the motor's full load current, in amps, times 125 percent for continuous use.

28. The disconnecting means for power conversion equipment shall not be less than _____ percent of the input FLA rating of the conversion unit.

29. The full-load current rating, in amps, of the largest motor shall be multiplied by _____ percent to select the size of conductors for a feeder supplying a group of two or more motors.

30. The motor branch-circuit overcurrent device shall be capable of carrying the _____ current, in amps, of the motor.

31. When a motor won't start and run, a nontime-delay fuse not exceeding 600 amperes in rating shall be permitted to be increased up to _____ percent of the full-load current, in amps, of the motor.

32. When a motor won't start and run, a inverse time circuit breaker greater than 100 amps shall be permitted to be increased up to _____ percent of the motor's full-load current, in amps.

18-50

	Section Answer

33. Three-phase squirrel-cage induction motors have three separate windings per pole on the stator that generate magnetic fields that are _____ degrees out-of-phase with each other.

34. Code letters are installed on motors by manufacturers for calculating the _____, in amps, based upon the kVA per horsepower which is selected from the motors code letter.

35. The locked-rotor current for _____ and _____ motors are selected from **Tables 430.251(A) and (B)** based upon phases, voltage, and horsepower rating of the motor.

36. The OCPD for motors shall protect the branch-circuit conductors from _____-circuits and _____-fault-current.

37. Time-delay fuses will hold _____ times their rating, which will permit most induction motors to start and accelerate their driven loads.

38. Inverse-time circuit breakers will hold about _____ times their rating for different periods of time based upon their frame size.

39. An overcurrent protection device rated at _____ amps or less can protect a 120 volt or less branch-circuit supplying motors rated less than 1 horsepower.

40. The _____ factor or _____ rise of the motor shall be used when sizing and installing the running overload protection for motors.

41. The branch-circuit protective device shall be permitted to serve as the controller for stationary motors where the motor is rated _____ horsepower or less.

42. A branch circuit inverse time circuit breaker rated in amperes shall be permitted as a _____ disconnecting means

43. The motor controller for a torque motor shall have a continuous duty, FLC rating, in amps, of not less than the _____ current rating of the motor.

44. A listed motor circuit switch rated in _____ shall be permitted as a disconnecting means.

45. The disconnecting means shall be permitted to be a general-use or isolating switch for DC stationary motors rated at _____ horsepower or greater and AC motors rated _____ horsepower or greater.

46. A motor and its driven machinery or load shall be installed in _____ of the controller for the motor.

47. The ampacity of capacitor circuit conductors shall not be less than _____ percent of the rated current, in amps, of the capacitor.

48. Conductors shall be _____ for the maximum voltage of any one conductor when occupying the same enclosure.

18-51

Section	Answer

49. When applying the rule of thumb method, the horsepower rating of the motor shall be multiplied by _____ to obtain full-load current in amps for 440, 460, and 480 volt, three-phase motors.

50. When applying the rule of thumb method, the horsepower rating of the motor shall be multiplied by _____ to obtain full-load current in amps for 220, 230, and 240 volt, single-phase motors.

51. Power conversion equipment requires the conductors to be sized at _____ percent of the rated input, in amps, of such equipment.
    (a)   100
    (b)   115
    (c)   125
    (d)   150

52. For Design B motors, the setting of an instantaneous circuit breaker can be adjusted up to _____ percent to allow the motor to start and run.
    (a)   1000
    (b)   1200
    (c)   1300
    (d)   1700

53. A nontime-delay fuse will hold _____ times its rating for approximately 1/4 to 2 seconds based upon type used.
    (a)   2
    (b)   3
    (c)   5
    (d)   10

54. Motors with a marked service factor not less than 1.15 shall have the minimum running overload protection sized at _____ percent.
    (a)   115
    (b)   125
    (c)   130
    (d)   140

55. Motors with a marked temperature rise not over 40°C shall have the minimum running overload protection sized at _____ percent.
    (a)   115
    (b)   125
    (c)   130
    (d)   140

56. Motors with a marked service factor not less than 1.15 shall have the maximum running overload protection sized at _____ percent.
    (a)   115
    (b)   125
    (c)   130
    (d)   140

57. Motors with a marked temperature rise not over 40°C shall have the maximum running overload protection sized at _____ percent.
    (a)   115
    (b)   125
    (c)   130
    (d)   140

	Section	Answer

58. An AC general-use snap switch may be installed as the controller for a stationary motor where the full-load current of the switch does not exceed _____ percent of the branch-circuit rating.
- (a) 50
- (b) 75
- (c) 80
- (d) 90

59. When applying the rule of thumb method, the horsepower rating of the motor shall be multiplied by _____ to obtain full-load current in amps for 220, 230, and 240 volt, three-phase motors.
- (a) 100
- (b) 125
- (c) 250
- (d) 500

60. The OCPD (variable loads) shall be sized at _____ percent times the phase converters nameplate single-phase input FLA rating.
- (a) 100
- (b) 125
- (c) 135
- (d) 150

61. What size THWN branch-circuit copper conductors are required for a 3 HP, 208 volt, single-phase, Design B motor?

62. What size THWN branch-circuit copper conductors are required for a 20 HP, 230 volt, three-phase, Design B motor?

63. What size THWN branch-circuit copper conductors are required for a 75 HP, 460 volt, three-phase, 15 minute rated intermittent duty cycle motor?

64. What size THWN copper conductors are required to supply power conversion equipment with a rated input of 112 amps?

65. What size THWN branch-circuit copper conductors are required to supply a 50 HP, 208 volt, three-phase, Design B part-winding motor?

66. What size THWN branch-circuit copper conductors are required to supply a 30 HP, 40 HP, and 50 HP, 460 volt, three-phase, Design B motors?

67. What size THWN branch-circuit copper conductors are required to supply an 10 HP, 208 volt, three-phase, 5 minute rated intermittent duty cycle, an 15 HP, 208 volt, three-phase, 15 minute rated intermittent duty cycle motor, and a 20 HP, 208 volt, three-phase motor?

68. What is the LRC (maximum) for a 40 HP, 230 volt, three-phase, code letter G motor?

69. What is the LRC for a 40 HP, 230 volt, three-phase, Design letter B motor?

70. What is the the rounded down and rounded up size nontime-delay fuse for a 50 HP, 230 volt, three-phase, Design letter B motor?

71. What is the rounded down and rounded up size time-delay fuse for a 50 HP, 230 volt, three-phase, Design letter B motor?

18-53

Section	Answer

72. What is the minimum and maximum setting for an instantaneous trip circuit breaker for a 50 HP, 230 volt, three-phase, Design letter B motor?

73. What is the rounded down and rounded up size inverse time circuit breaker for a 50 HP, 230 volt, three-phase, Design letter B motor?

74. What is the maximum size nontime-delay fuse for a 50 HP, 230 volt, three-phase, Design letter B motor?

75. What is the maximum size time-delay fuse for a 50 HP, 230 volt, three-phase, Design letter B motor?

76. What is the maximum size inverse time circuit breaker for a 50 HP, 230 volt, three-phase, Design letter B motor?

77. What size OCPD (CB) is required for a feeder-circuit that supplies a 10 HP, 15 HP, 20 HP, and 25 HP, 460 volt, three-phase, Design letter B group of motors?

78. What size overload protection (minimum) is required for a 20 HP, 460 volt, three-phase, Design letter B motor with a nameplate rating of 48 amps, temperature rise of 40°C, and a service factor of 1.15?

79. What size overload protection (maximum) is required for a 20 HP, 460 volt, three-phase, Design letter B motor with a nameplate rating of 48 amps, temperature rise of 40°C, and a service factor of 1.15?

80. What size nonfused disconnect is required for a 50 HP, 460 volt, three-phase, Design letter B motor?

81. What size OCPD (maximum) is required for motor control circuit conductors located in the controller and supplied by a 12 AWG conductor?

82. What size OCPD (maximum) is required for motor control circuit conductors that are run remote and supplied by a 12 AWG conductor?

83. What size OCPD is required for motor control circuit conductors that are supplied by a 2400 VA, 480 volt, two-wire control transformer?

84. What size THWN copper conductors are required to supply a 20 kVA, 208 volt, three-phase capacitor with a 40 HP, 208 volt, three-phase, Design letter B motor?

85. What size FLA rating is required for a 30 HP, 115 volt, three-phase, Design letter B motor using the rule of thumb method?

86. What size FLA rating is required for a 30 HP, 220 volt, three-phase, Design letter B motor using the rule of thumb method?

87. What size FLA rating is required for a 30 HP, 440 volt, three-phase, Design letter B motor using the rule of thumb method?

88. What size FLA rating is required for a 30 HP, 575 volt, three-phase, Design letter B motor using the rule of thumb method?

	Section	Answer
89. What size OCPD and THWN copper conductors are required for a phase converter with a nameplate rating of 35 amps supplying a piece of equipment classified as a variable load?	_____	_____ _____
90. What size OCPD and THWN copper conductors are required for a phase converter supplying a 20 HP, 230 volt, three-phase, Design letter B motor classified as a specific load?	_____	_____ _____

19

Compressor Motors

Article 440 deals with individual or group installations having hermetically sealed motor compressors. The techniques for designing the proper size conductors, disconnecting means, and controllers are discussed.

The conductors supplying power to HACR equipment are sized from the full-load amp (FLA) ratings of the compressor and condenser motor. These FLA ratings are increased by 125 percent per **440.32** to compensate for the starting periods and overload conditions.

The overcurrent protection device (OCPD's) protecting the branch-circuits from short-circuit and ground-fault currents are sized from the provisions listed in **440.22(A)**, which requires the FLA ratings to be increased from 175 up to 225 percent to allow the HACR equipment to start and run without tripping the OCPD ahead of the circuit.

Note that the elements used to supply the branch-circuits to HACR equipment may be required to be selected by the branch-circuit selection currents listed on the nameplate of such equipment per **440.4(C)** and **110.3(B)**.

NAMEPLATE LISTING
440.1

The overcurrent protection devices, running overload protection devices, conductors, disconnecting means, and controllers are sized and selected by the information provided on the nameplate listing for air-conditioning and refrigeration equipment. The information on the nameplate is very important to installers and service personnel, therefore, the nameplate should never be removed from the air-conditioner or refrigeration equipment.

MARKING ON HERMETIC REFRIGERANT MOTOR-COMPRESSORS AND EQUIPMENT
440.4

Hermetic refrigerant motor-compressors shall be provided with a marking on the nameplate giving the manufacturer's name, trademark, or symbol and designating the identification, number of phases, voltage, and frequency. The information provided on the nameplate of the hermetic refrigerant motor-compressor is used to determine the ratings of branch-circuit conductors, ground-fault protection, short-circuits, disconnecting means, controllers, and other elements of the electrical system.

MARKING ON CONTROLLERS
440.5

Controllers shall be marked with information that lists the manufacturer's name, trademark or symbol, identifying voltage, phases, full-load current, locked rotor current rating, or horsepower.

AMPACITY AND RATING
440.6

The full-load current rating listed on the nameplate of the motor-compressor shall be used to determine the branch-circuit conductor rating, short-circuit protection rating, motor overload protection rating, controller rating, or disconnecting means rating. The branch-circuit selection current (if greater) shall be applied if shown instead of the full-load current rating. The full-load current rating, in amps, shall be used to determine the motor's overload protection rating. The full-load current rating listed on the compressor nameplate must be used when the nameplate for the equipment does not list a full-load current rating based upon the BSCS. **(See Figure 19-1)**

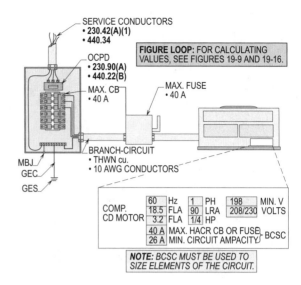

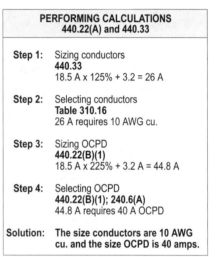

**AMPACITY RATING
NEC 440.6**

Figure 19-1. If the branch-circuit selection current (BCSC) on the nameplate calls for a certain circuit size and OCPD size, this rating shall be used instead of actually calculating such values per **440.22(A)** and **440.32**.

HIGHEST RATED (LARGEST) MOTOR
440.7

When sizing the conductors for a feeder supplying A/C units and motors per **430.24**, the full-load current in amps of the largest motor is multiplied by 125 percent. The full-load current rating of the remaining motors are added to this total to derive the FLA. See Figures 18-9 and 18-27 for an illustration pertaining to this rule.

When sizing the OCPD for two or motors per **430.62(A)**, the full-load current in amps of the largest motor is multiplied

Compressor Motors

by the percentages listed in **Table 430.52**. The full-load current rating of the remaining motors are added to this total to derive the FLA. (Also, see **430.63**)

The full-load current ratings listed on the nameplate of the compressor-motor are used to determine the size conductors and overcurrent protection device using the same procedure. The larger of the two are used.

See Figures 19-2(a) and (b) for a feeder-circuit supplying motors and A/C units. Note that the A/C unit is the largest motor and not one of the motors in the group.

SINGLE MACHINE
440.8

Each motor controller must be provided with an disconnecting means. Air-conditioning and refrigeration systems are considered a single machine even through they consist of any number of motors. The number of disconnecting means to be provided are determined by applying **430.87, Ex.** and **430.112, Ex**.

DISCONNECTING MEANS
440.11

The full-load current rating of the nameplate or the nameplate branch-circuit selection current of the compressor, whichever is greater, shall be used to size the branch-circuit conductors and the disconnecting means to disconnect air-conditioners and refrigeration equipment.

RATING AND INTERRUPTING CAPACITY
440.12

The full-load current rating of the nameplate or the nameplate branch-circuit selection current of the compressor, whichever is greater, shall be sized at 115 percent to size the disconnecting means. A horsepower rated switch, circuit breaker, or other switches may be used as the disconnecting means per **430.109**. **(See Figure 19-3)**

> **Design Tip:** A minimum load is derived when applying 115 percent for sizing the disconnecting means. Therefore, on larger units the 115 percent may not be of sufficient ampacity for opening the circuit under load.

The horsepower amperage rating shall be selected from **Tables 430.247** through **Table 430.250** when corresponding to the nameplate rating or branch-circuit selection current of the motor-compressor or equipment when listed in amperage and not horsepower. The horsepower amperage rating for locked-rotor current must be selected from **Tables 430.251(A)** and **(B)** when the nameplate fails to list the locked-rotor current. Note that the disconnecting means shall be sized with enough capacity in horsepower that is capable of disconnecting the total locked-rotor current. **(See Figure 19-4)**

The full-load current rating of the nameplate may be used to size a circuit breaker at 115 percent or more to disconnect a hermetically sealed motor from the power circuit. **(See Figure 19-5)**

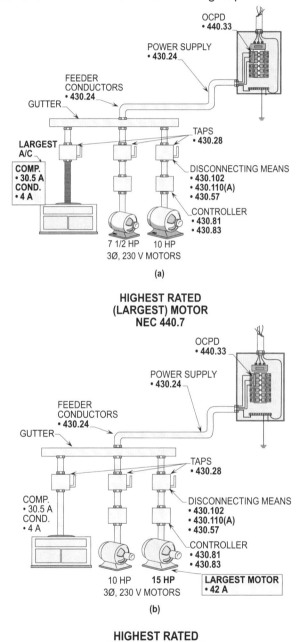

Figure 19-2. The above shows the calculation procedure for sizing conductors and OCPD where the A/C unit or motor is the largest in the group of A/C units and motors.

Design Tip: The circuit breaker shall be sized at 115 percent or more of the branch-circuit selection current if it were greater in rating so as to be capable of disconnecting the circuit safely.

CORD-CONNECTED EQUIPMENT
440.13

Cord-and-plug connected equipment such as room air-conditioners, home refrigerators and freezers, drinking water coolers, and beverage dispensers, cord, plug, or receptacle, which may be of the separable type, can be used to serve as a disconnecting means. **(See Figure 19-7)**

Two or more hermetic motors or combination loads such as hermetic motor loads, standard motor loads, and other loads shall have their separate values totaled to determine the rating of a single disconnecting means. This total rating shall be sized at 115 percent to determine the size disconnecting means required to disconnect the circuits and elements in a safe and reliable manner. **(See Figure 19-6)**

Design Tip: In some cases, room air-conditioners are not permitted to have a cord-and-plug connection to serve as their disconnecting means. Such cases is where their unit switches for manual control are installed in air-conditioners mounted over 6 ft. above finished grade.

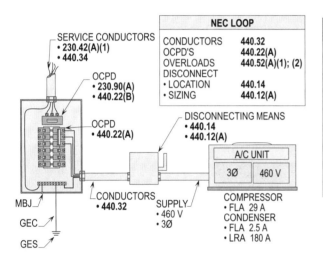
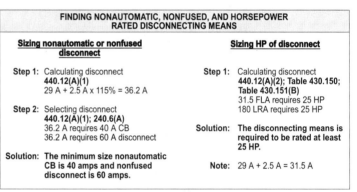

Figure 19-3. The full-load current rating of the nameplate or the nameplate's branch-circuit selection current of the compressor, whichever is greater, shall be sized at 115 percent to size the disconnecting means.

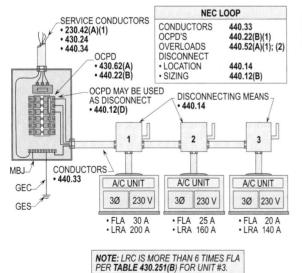
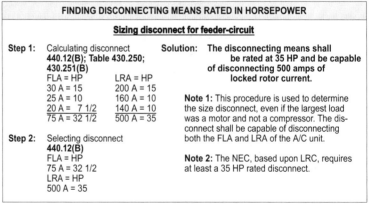

Figure 19-4. Sizing horsepower rating to select disconnecting means based upon locked-rotor current.

Compressor Motors

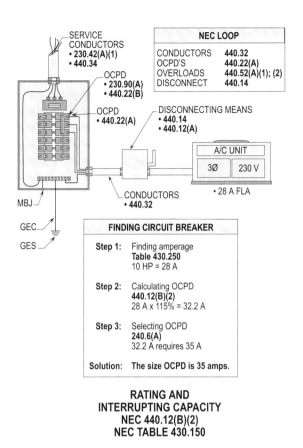

**RATING AND
INTERRUPTING CAPACITY
NEC 440.12(B)(2)
NEC TABLE 430.150**

Figure 19-5. The full-load current rating of the nameplate may be used to size a circuit breaker at 115 percent or more to disconnect a hermetically sealed motor from the power circuit.

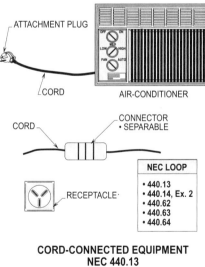

**CORD-CONNECTED EQUIPMENT
NEC 440.13
NEC 440.14, Ex. 2**

Figure 19-7. Cord-and-plug connected equipment such as room air-conditioners, home refrigerators and freezers, drinking water coolers, and beverage dispensers are permitted to be disconnected by a cord and receptacle. A cord, plug, or receptacle, which may be of the separable type, can also be used to serve as such disconnecting means.

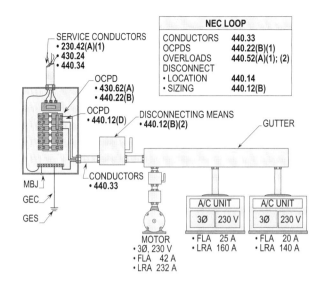

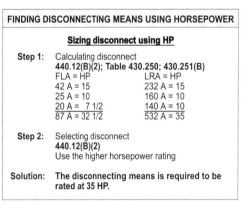

**RATING AND
INTERRUPTING CAPACITY
NEC 440.12(B)(2)**

Figure 19-6. Two or more hermetic motors or combination loads such as hermetic motor loads, standard motor loads, and other loads shall have their separate values totaled to determine the rating of a single disconnecting means.

LOCATION
440.14

The disconnecting means for air-conditioning or refrigeration equipment shall be located within sight and within 50 ft. (15 m) and shall be readily accessible to the user. An additional circuit breaker or disconnecting switch shall be provided at the equipment if the air-conditioning or refrigeration equipment is not within sight or within 50 ft. (15 m). The disconnecting means is permitted to be installed within or on the air-conditioning or refrigeration equipment. For the use of unit switches located in air-conditioning units review **422.34** per the AHJ. **(See Figure 19-8)**

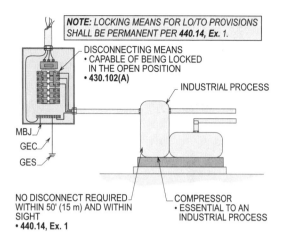

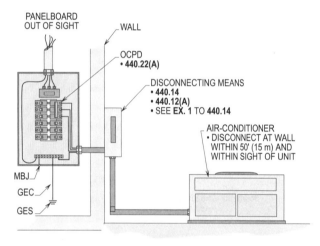

LOCATION
NEC 440.14
NEC 440.14, Ex. 1
NEC 422.34

Figure 19-8. The disconnecting means for air-conditioning or refrigeration equipment shall be located within sight and within 50 ft. (15 m) and shall be readily accessible to the user. An additional circuit breaker or disconnecting switch shall be provided at the equipment if the air-conditioning or refrigeration equipment is not within sight or within 50 ft. (15 m)

APPLICATION AND SELECTION 440.22

The branch-circuit fuse or circuit breaker ratings for hermetically sealed motors shall be size with enough capacity to allow the motor to start and develop speed without tripping open the OCPD due to the momentary inrush current of the compressor and other elements. Maximum protection is always provided by the ratings and settings of the OCPD being sized with values as low as possible. Hermetic refrigerant motor-compressors shall be protected by properly sizing and selecting the ratings and settings of the overcurrent protection devices to protect the branch-circuit conductors and other elements in the circuit from short-circuit and ground-fault conditions.

RATING AND SETTING FOR INDIVIDUAL MOTOR-COMPRESSOR 440.22(A)

The OCPD for hermetic seal compressors shall be selected at 175 percent (for minimum) or 225 percent (for maximum) of the compressor FLA rating or the branch-circuit selection circuit current, whichever is greater. **(See Figure 19-9)**

OCPD's for hermetically sealed compressors may be selected up to 225 percent to allow the motor to start if the compressor will not start and develop speed when applying a lower rating at 175 percent or less.

> **Design Tip:** A normal circuit breaker shall not be installed when the equipment is marked for a particular fuse size or HACR circuit breaker rating. The branch-circuit conductors shall be protected only by that specified fuse size or HACR circuit breaker rating.

RATING OR SETTING FOR EQUIPMENT 440.22(B)

When sizing the overcurrent protection device the rating or setting shall be selected and comply with the number of hermetic motors, or combination of hermetic motors and standard motors installed on a circuit.

SIZING OCPD FOR TWO OR MORE HERMETIC MOTORS 440.22(B)(1)

The OCPD for a feeder-circuit supplying two or more air-conditioning or refrigerating units shall be sized to allow the largest unit to start and allow the other units to start at different intervals of time. The full-load current rating, in amps, of the nameplate or the branch-circuit selection current rating of the largest motor, whichever is greater, shall be size at 175 percent if there are two or more hermetically sealed motors installed on the same feeder-circuit. **(See Figure 19-10)**

OCPD's for hermetically sealed motors may be selected up to 225 percent to allow the motor to start if the motor will not start and develop speed when applying a lower rating which is normally selected at 175 percent or less. **(See Figure 19-11)**

Compressor Motors

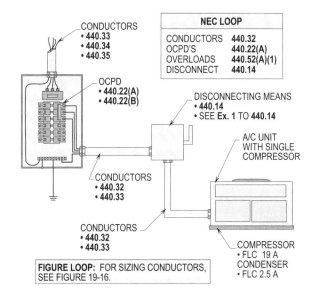

SIZING OCPD FOR HERMETIC MOTORS AND OTHER LOADS WHEN A MOTOR IS THE LARGEST 440.22(B)(2)

When installing hermetically sealed motors and other loads such as motors on the same circuit, and the largest in the group is a motor, the overcurrent protection device is sized and selected based on the percentages from **Table 430.52**. The maximum branch-circuit overcurrent protection device must be used when the standard motor is the largest of the group and the sum of the full-load current ratings of the remaining hermetically sealed motor and other motors of the group added to the largest motor. The next lower standard size overcurrent protection device below this total sum shall be installed per **240.6(A)**. (**See Figure 19-12**)

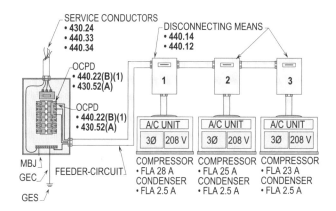

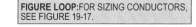

Figure 19-9. The OCPD for hermetically sealed compressors shall be selected at 175 percent (for min.) or 225 percent (for max.) of the compressor's FLA rating or the branch-circuit selection circuit current, whichever is greater.

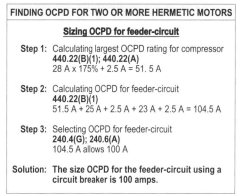

SIZING OCPD FOR HERMETIC MOTOR AND OTHER LOADS WHEN A HERMETICALLY SEALED MOTOR IS THE LARGEST 440.22(B)(1)

When installing hermetically sealed motors and other loads such as motors on the same circuit, and the largest motor of the group is hermetic, the same procedure used for two or more hermetic motors on a feeder-circuit is used to size the overcurrent protection device. The full-load current rating of the nameplate or the branch-circuit selection current rating of the largest hermetic motor, whichever is greater, shall be sized at 175 percent and the sum of the full-load current ratings, in amps, of the other motors added to this largest hermetic motor load.

Figure 19-10. The full-load current rating of the nameplate or the branch-circuit selection current rating of the largest motor, whichever is greater, shall be size at 175 percent if there are two or more hermetically sealed motors installed on the same feeder-circuit.

Stallcup's Electrical Design Book

Design Tip: The next larger standard size is not permitted to be installed for there is not an Exception to allow the next higher size per **440.22(B)(2)** or **430.62(A)**.

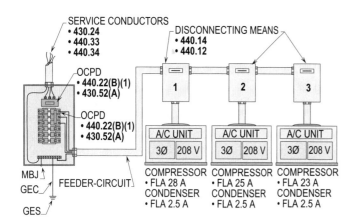

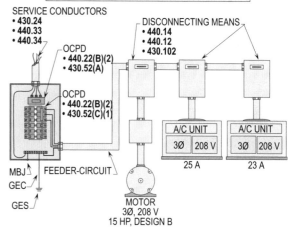

FINDING OCPD FOR TWO OR MORE HERMETIC MOTORS

Sizing OCPD for feeder-circuit

Step 1: Calculating largest OCPD rating for compressor
440.22(B)(1); 440.22(A)
28 A x 225% + 2.5 A = 65.5 A

Step 2: Calculating OCPD for feeder-circuit
440.22(B)(1)
65.5 A + 25 A + 2.5 A + 23 A + 2.5 A = 118.5 A

Step 3: Selecting OCPD for feeder-circuit
240.4(G); 240.6(A)
118.5 A allows 110 A

Solution: The size OCPD for the feeder-circuit using a circuit breaker is 110 amps.

SIZING OCPD FOR TWO OR MORE HERMETIC MOTORS
NEC 440.22(B)(1)

Figure 19-11. OCPD's for hermetically sealed motors may be selected up to 225 percent to allow the motor to start if the motor will not start and develop speed.

USING 15 OR 20 AMP OCPD
440.22(B)(2), Ex. 1

Where the equipment will start, run, and operate on 15 or 20 amp, 120 volt, single-phase branch-circuit, or a 15 amp, 208 volt or 240 volt, a single-phase branch-circuit, with a 15 or 20 amp overcurrent protection device, such device may be used to protect the branch-circuit. However, the values of the overcurrent protection device in the branch-circuit must not exceed the values marked on the nameplate of the equipment. **(See Figure 19-13)**

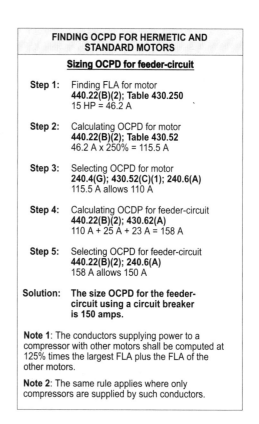

FINDING OCPD FOR HERMETIC AND STANDARD MOTORS

Sizing OCPD for feeder-circuit

Step 1: Finding FLA for motor
440.22(B)(2); Table 430.250
15 HP = 46.2 A

Step 2: Calculating OCPD for motor
440.22(B)(2); Table 430.52
46.2 A x 250% = 115.5 A

Step 3: Selecting OCPD for motor
240.4(G); 430.52(C)(1); 240.6(A)
115.5 A allows 110 A

Step 4: Calculating OCDP for feeder-circuit
440.22(B)(2); 430.62(A)
110 A + 25 A + 23 A = 158 A

Step 5: Selecting OCPD for feeder-circuit
440.22(B)(2); 240.6(A)
158 A allows 150 A

Solution: The size OCPD for the feeder-circuit using a circuit breaker is 150 amps.

Note 1: The conductors supplying power to a compressor with other motors shall be computed at 125% times the largest FLA plus the FLA of the other motors.

Note 2: The same rule applies where only compressors are supplied by such conductors.

SIZING OCPD FOR HERMETIC MOTORS AND OTHER LOADS WHEN A MOTOR IS THE LARGEST
NEC 440.22(B)(2)

Figure 19-12. When installing hermetically sealed motors and other loads such as motors on the same circuit, and the largest in the group is a motor, the overcurrent protection device shall be sized and selected based on the percentages from **Table 430.52**.

Compressor Motors

USING A CORD-AND-PLUG CONNECTION NOT OVER 250 VOLTS
440.22(B)(2), Ex. 2

The rating of the overcurrent protection device must be determined by using the rating of the nameplate of the cord-and-plug connected equipment serving single-phase, 250 volts or less hermetically sealed motors. **(See Figure 19-14)**

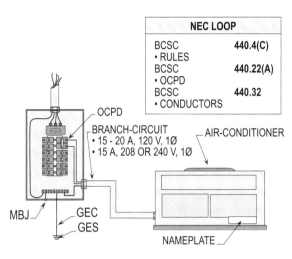

USING 15 OR 20 AMP OCPD
NEC 440.22(B)(2), Ex. 1

Figure 19-13. Where the equipment will start, run, and operate on 15 or 20 amp, 120 volt, single-phase branch-circuit, or a 15 amp, 208 volt or 240 volt, a single-phase branch-circuit, a 15 or 20 amp overcurrent protection device may be used to protect the branch-circuit.

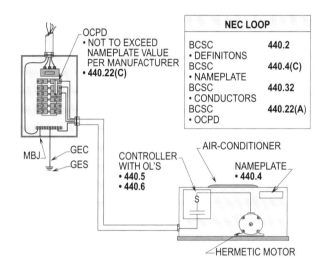

PROTECTIVE DEVICE RATING NOT TO EXCEED THE MANUFACTURER'S VALUES
NEC 440.22(C)

Figure 19-15. The manufacturer's values marked on the equipment shall not be exceeded by the overcurrent protection device rating, where the maximum overcurrent protective device ratings on the manufacturer's heater table for use with a motor controller are less than the rating or setting per **440.22(A) and (B)**.

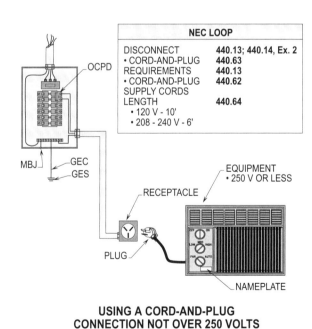

USING A CORD-AND-PLUG CONNECTION NOT OVER 250 VOLTS
NEC 440.22(B)(2), Ex. 2

PROTECTIVE DEVICE RATING NOT TO EXCEED THE MANUFACTURER'S VALUES
440.22(C)

The manufacturer's values marked on the equipment shall not be exceeded by the overcurrent protection device rating, where the maximum overcurrent protective device ratings on the manufacturer's heater table for use with a motor controller are less than the rating or setting per **440.22(A) and (B)**. **(See Figure 19-15)**

Figure 19-14. The rating of the overcurrent protection device shall be determined by using the rating of the nameplate of the cord-and-plug connected equipment having single-phase, 250 volt or less hermetically sealed motors.

BRANCH-CIRCUIT CONDUCTORS 440.31

In general, to prevent conductors and motor elements of the branch-circuit from over heating, the conductors shall be sized with enough capacity to allow a hermetic motor to start and run. To ensure adequate sizing, a derating factor of 80 percent shall be applied to the branch-circuit conductors or such conductors shall be sized at 125 percent of the load.

SINGLE MOTOR-COMPRESSORS 440.32

The conductors supplying power to an air-conditioning or refrigerating unit shall be sized to carry the load of the unit plus an overload for a period of time that won't damage the elements. The full-load current rating of the nameplate or branch-circuit selection current, whichever is greater, shall be sized at 125 percent to size and select the conductors supplying hermetically sealed motors. **(See Figure 19-16)**

Finding Single Motor For Branch-Circuit

Sizing conductor based on controller

Step 1: Calculating FLA
440.32; 440.33
19 A x 125% + 2.5 A = 26.25 A

Step 2: Selecting conductor
240.4(D)
26.25 A requires 10 AWG

Solution: **The size conductors are 10 AWG THHN cooper.**

Sizing conductors based on branch-circuit selection current

Step 1: Selecting conductors
440.4(C)
30 A requires 10 AWG

Solution: **The branch-circuit selection current requires 10 AWG THHN copper conductors.**

SINGLE MOTOR-COMPRESSORS
NEC 440.4(C)
NEC 440.32

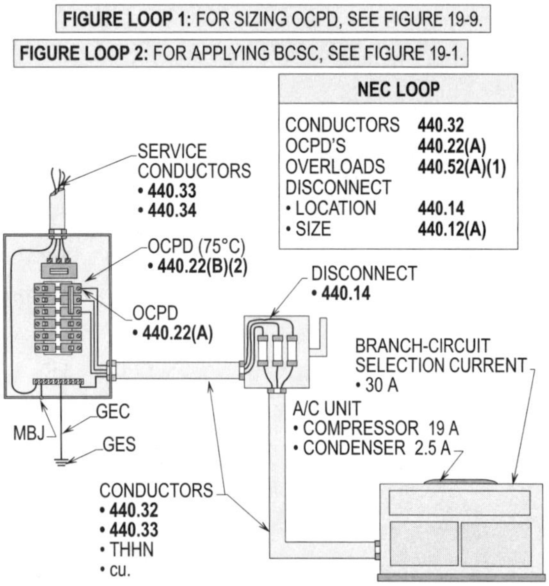

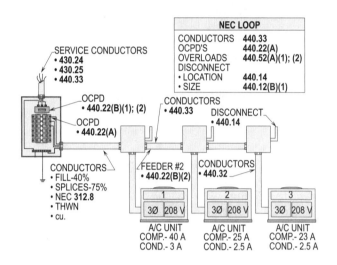

Figure 19-16. The full-load current rating of the nameplate or branch-circuit selection current, whichever is greater, shall be sized at 125 percent to size and select the conductors supplying hermetically sealed motors.

FIGURE LOOP: FOR SIZING OCPD, SEE FIGURES 19-10 AND 11.

Finding Two Or More A/C Units On A Feeder-Circuit

Step 1: Finding FLA
A/C unit #1 = 40 A + 3 A
A/C unit #2 = 25 A + 2.5 A
A/C unit #3 = 23 A + 2.5 A

Step 2: Calculating FLA
440.33

40 A x 125% + 3 A	= 53 A
25 A x 100% + 2.5 A	= 27.5 A
23 A x 100% + 2.5 A	= 25.5 A
Total load	= 106 A

Step 3: Selecting conductors
310.10, FPN (2); Table 310.16
106 A requires 2 AWG cu.

Solution: **The size THWN conductors are 2 AWG copper.**

TWO OR MORE MOTOR COMPRESSORS
NEC 440.33

Figure 19-17. Two or more compressors plus other motor loads can be connected to a feeder-circuit. The largest compressor shall be computed at 125 percent of its FLA and the remaining compressor loads are added to this total at 100 percent of their FLA ratings.

Compressor Motors

TWO OR MORE MOTOR COMPRESSORS
440.33

Two or more compressors plus other motor loads can be connected to a feeder-circuit. The largest compressor is computed at 125 percent of its FLA and the remaining compressor loads are added to this total at 100 percent of their FLA ratings. For units with a branch-circuit selection current, the circuit conductors shall be selected and based upon the nameplate values. **(See Figure 19-17)**

COMBINATION LOAD
440.34

Two or more motor compressors with motor loads plus other loads may be connected to a feeder-circuit or service conductors. The largest compressor or motor load shall be calculated at 125 percent plus 100 percent of the remaining compressors and motors. The other loads shall be computed at 125 percent for continuous and 100 percent for noncontinuous operation per **215.2(A)(1)** and these total values used to select conductors. **(See Figure 19-18)**

MULTIMOTOR AND COMBINATION LOAD EQUIPMENT
440.35

The marking on the nameplate shall be used when sizing the branch-circuit conductors for multimotor and combination load equipment. The conductors shall be installed to have a rating equal to the nameplate rating. Each individual motor or load contained in the unit is not required to be calculated individually to size and select the conductors.

CONTROLLERS FOR MOTOR COMPRESSORS
440.41

When installing the wiring for a motor-controller the circuit supply conductors are run from a motor-controller and connected to the terminals of the compressor. The full-load current rating and the locked rotor current rating of the compressor motor shall be sized at continuous operation.

MOTOR COMPRESSOR CONTROLLER RATING
440.41(A)

The full-load current rating of the nameplate or the branch-circuit selection current ratings, whichever is greater, shall be used to size and select the motor controller. If necessary, the locked rotor current rating of the motor may be used to size and select the motor controller. **(See Figure 19-19)**

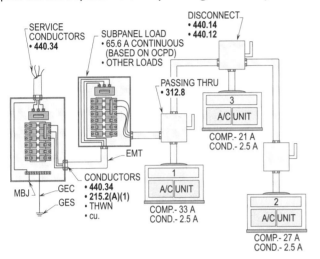

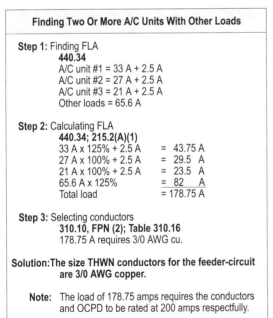

COMBINATION LOAD
NEC 440.34

Figure 19-18. Two or more motor compressors with motor loads plus other loads may be connected to a feeder-circuit or service conductors. The largest compressor or motor load is calculated at 125 percent plus 100 percent of the remaining compressors and motors plus the other loads.

Design Tip: The motor controller shall be sized and selected using the same procedure as used for the sizing of the disconnecting means.

Stallcup's Electrical Design Book

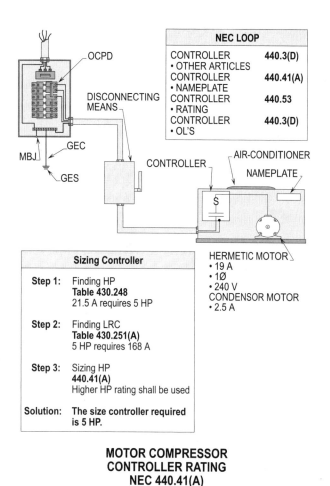

MOTOR COMPRESSOR CONTROLLER RATING NEC 440.41(A)

Figure 19-19. The full-load current rating of the nameplate or the branch-circuit selection current ratings, whichever is greater, shall be used to size and select the motor controller.

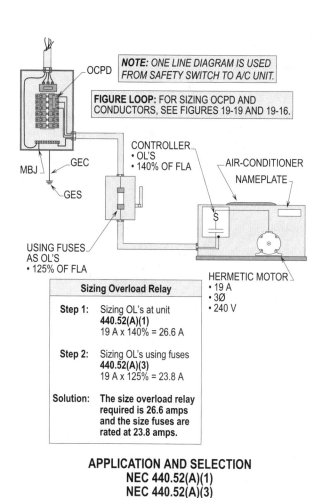

APPLICATION AND SELECTION NEC 440.52(A)(1) NEC 440.52(A)(3)

Figure 19-20. The overload relay for the motor compressor must trip at not more than 140 percent of the full-load current rating. If a fuse or circuit breaker is used for the protection of the motor compressor, it shall trip at not more than 125 percent of the full-load current rating.

MOTOR COMPRESSOR AND BRANCH-CIRCUIT OVERLOAD PROTECTION 440.51

The overload (OL) protection for compressors may be accomplished by using OCPD's in separate enclosures, separate overload relays or thermal protectors which are an integral part of the compressor.

APPLICATION AND SELECTION 440.52

The overload relay for the motor compressor must trip at not more than 140 percent of the full-load current rating. If a fuse or circuit breaker is used for the protection of the motor compressor it must trip at not more than 125 percent of the full-load current rating. **(See Figure 19-20)**

OVERLOAD RELAYS 440.53

Short-circuit and ground-fault protection is not provided by overload relays and thermal protectors. Overload relays and thermal protectors respond to any type of heat build up and open with a delay action which will not operate instantly, even on short-circuits or ground-faults. The branch-circuit overcurrent protection device for the circuit must operate and clear the circuit under short-circuit and ground-fault conditions.

Compressor Motors

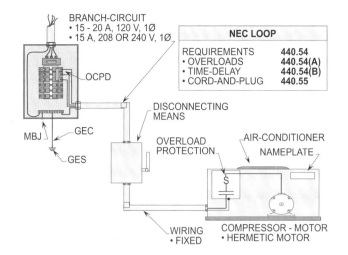

MOTOR COMPRESSORS AND EQUIPMENT ON 15 OR 20 AMP BRANCH-CIRCUIT NOT CORD-AND-PLUG CONNECTED
NEC 440.54

Figure 19-21. Overload protection must be provided for direct or fixed-wired motor compressors and equipment which is connected to 15 or 20 amp, 120 volt, single-phase branch-circuits.

MOTOR COMPRESSORS AND EQUIPMENT ON 15 OR 20 AMP BRANCH-CIRCUIT NOT CORD-AND-PLUG CONNECTED 440.54

Overload protection shall be provided for direct or fixed-wired motor compressors and equipment which is connected to 15 or 20 amp, 120 volt, single-phase branch-circuits. Note that its 15 amp for 240 volt, single-phase branch-circuits.

The full-load current rating of the hermetically sealed motor shall be selected at 140 percent when sizing separate overload relays. Hermetic motors shall be provided with fuses or circuit breaker that provide sufficient time delay to allow the motor to come up to running speed without tripping open the circuit due to the high inrush current. (**See Figure 19-21**)

CORD AND ATTACHMENT PLUG CONNECTED MOTOR COMPRESSORS AND EQUIPMENT ON 15 OR 20 AMP BRANCH-CIRCUITS 440.55

When attachment plugs and receptacles are used for circuit connection they shall be rated no higher than 15 or 20 amp, for 120 volt, single-phase circuit, or 15 amp, for 208 or 240 volt, single-phase branch-circuits. (**See Figure 19-22**)

ROOM AIR-CONDITIONERS 440.60

Room air-conditioners are usually cord-and-plug connected when installed on 120/240 volt, single-phase systems. However, they may be hard-wired. Air-conditioners are always hard-wired when installed on three-phase, or electrical supply systems over 250 volts.

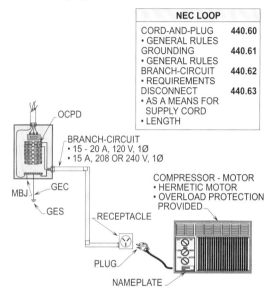

CORD AND ATTACHMENT PLUG CONNECTED MOTOR COMPRESSORS AND EQUIPMENT ON 15 OR 20 AMP BRANCH-CIRCUITS
NEC 440.55

Figure 19-22. When attachment plugs and receptacles are used for circuit connection they shall be rated no higher than 15 or 20 amp, for 120 volt, single-phase circuit, or 15 amp, for 208 or 240 volt, single-phase branch-circuits.

GROUNDING 440.61

The following wiring methods when utilized to wire-in room air-conditioners are required to be grounded:
- Cord-and-plug connected
- Hard-wired (if within reach of the ground or grounded object)
- If in contact with metal
- Operating over 150 volts-to-ground
- Wired with metal-clad wiring
- Located in a hazardous location
- Installed in damp location (within reach of the user)

See Figure 19-23 for wiring methods that can be used to ground room air-conditioners.

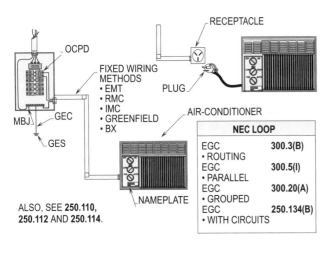

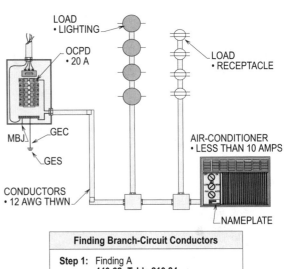

GROUNDING
NEC 440.61

Figure 19-23. The above are wiring methods may be used to ground room air-conditioners.

BRANCH-CIRCUIT REQUIREMENTS
NEC 440.62(C)

Figure 19-24(b). If other loads are served by the branch-circuit, the cord-and-plug connected air-conditioner unit shall not exceed 50 percent of the branch-circuit.

BRANCH-CIRCUIT REQUIREMENTS 440.62

The full-load current rating of the room air-conditioner must be marked on the nameplate and must not operate at more than 40 amps on 250 volts. The branch-circuit overcurrent protection device must be installed with a rating no greater than the circuit conductor's ampacity or the rating of the receptacle serving the unit, whichever is less. The ampacity of a cord-and-plug connected air-conditioning window unit must not exceed 80 percent of the branch-circuit where no other loads are served. If other loads are served by the branch-circuit, the cord-and-plug connected air-conditioner unit must not exceed 50 percent of the branch-circuit. **(See Figures 19-24(a) and (b))**

DISCONNECTING MEANS 440.63

A cord-and-plug may serve as the disconnecting means for the room air-conditioner if all the following conditions are complied with:

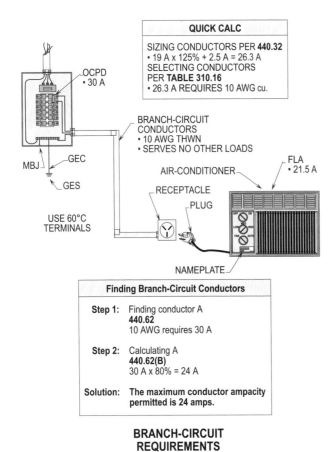

BRANCH-CIRCUIT REQUIREMENTS
NEC 440.62(B)

Figure 19-24(a). The ampacity of a cord-and-plug connected air-conditioning window unit shall not exceed 80 percent of the branch-circuit where no other loads are served.

Compressor Motors

- Operates at 250 volts or less
- Controls are manually operated
- Controls are within 6 ft. of the floor
- Controls are readily accessible to user

Room air-conditioners may be hard-wired and located within sight of the service equipment or it may be wired so that it is readily accessible to an disconnecting switch for the user. However, such switch shall be located within sight and within the unit. **(See Figures 19-25(a) and (b))**

> **Design Tip:** The rules for three-phase room air-conditioners shall not be used for these types of units. Three-phase room air-conditioners are required to be hard-wired and are required to be installed with a disconnecting means which is readily accessible to the user.

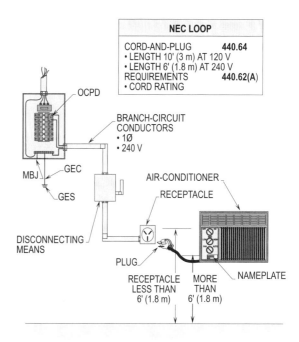

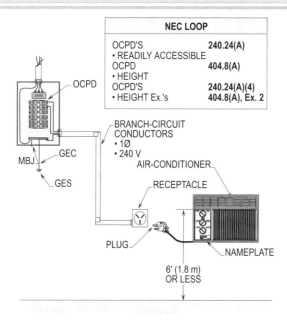

Figure 19-25(a). A cord-and-plug may serve as the disconnecting means for room air-conditioners if it operates at 250 volts or less, controls are manually operated, controls are within 6 ft. (1.8 m) of the floor, and controls are readily accessible to the user.

SUPPLY CORDS
440.64

Room air-conditioners installed with flexible cords shall be a length that is limited to 10 ft. (3 m) for 120 volt circuits and 6 ft. (1.8 m) for 208 or 240 volt circuits. Long cords shall not be used for they are dangerous. Long cords can also be a shock or fire hazard. **(See Figure 19-26)**

Figure 19-25(b). A cord-and-plug may serve as the disconnecting means even if the room air-conditioner's manual controls are located above 6 ft. (1.8 m) from the finished grade.

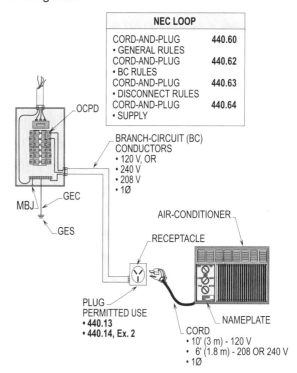

Figure 19-26. Room air-conditioners installed with flexible cords shall have a length that is limited to 10 ft. (3 m) for 120 volt circuits and 6 ft. (1.8 m) for 208 or 240 volt circuits.

19-15

Name Date

Chapter 19. Compressor Motors

Section	Answer		
_____	T F	1.	The overcurrent protection device shall be sized and selected by the information provided on the nameplate listing for air-conditioning and refrigeration equipment.
_____	T F	2.	Each motor controller for an air-conditioning and refrigeration systems shall not be required to provided with an disconnecting means.
_____	T F	3.	A horsepower rated switch shall be permitted to be used as a disconnecting means.
_____	T F	4.	Room air-conditioners shall not be permitted to have a cord-and-plug serve as the disconnecting means for hermetic motors.
_____	T F	5.	An additional CB or disconnecting means shall be provided at the equipment if an A/C unit is not within sight or within 50 ft of the disconnect.
_____	T F	6.	An OCPD for hermetic seal compressors may be selected up to 250 percent of its FLA to allow the motor to start and run.
_____	T F	7.	If a fuse or circuit breaker is used for the protection of the motor compressor, it shall trip at not more than 125 percent of the full-load current rating.
_____	T F	8.	Short-circuit and ground-fault protection for compressor motors can be provided by overload relays and thermal protectors.
_____	T F	9.	The ampacity of a cord-and-plug connected air-conditioning (window) unit shall not exceed 80 percent of the branch-circuit where no other loads are served.
_____	T F	10.	Room air-conditioners installed with flexible cords shall be a length that is limited to 12 ft. for 120 volt circuits.
_____	_____	11.	When sizing the conductors for a feeder supplying A/C units and motors, the full-load current in amps of the largest motor shall be multiplied by _____ percent.
_____	_____	12.	The full-load current rating of the nameplate or the nameplate branch-circuit selection current of the compressor, whichever is greater, shall be sized at _____ percent of the disconnecting means.
_____	_____	13.	The disconnecting means for air-conditioning or refrigeration equipment shall be located within sight and within _____ ft. and shall be readily acces-sible to the user.
_____	_____	14.	The full-load current rating on the nameplate or the branch-circuit selection current rating of the largest motor, whichever is greater, shall be sized at _____ percent, if there are two or more hermetic sealed motors installed on the same feeder-circuit.

19-17

Section	Answer
_____	_____

15. Two or more motor compressors plus other motor loads can be connected to a feeder-circuit with the largest (motor) compressor computed at _____ percent of its FLA and the remaining compressor loads added to this total at _____ percent of their FLA ratings.

16. The overload relay for the motor compressor shall trip at not more than _____ percent of the full-load current rating.

17. When attachment plugs and receptacles are used for circuit connection, they shall be rated no higher than _____ amps, for 208 or 240 volt, single-phase branch-circuits.

18. The full-load current rating of the room air-conditioner shall be marked on the nameplate and must not operate at more than _____ amps on 250 volts single-phase.

19. A cord-and-plug shall be permitted to serve as the disconnecting means, even if the room air-conditioner manual controller is above _____ ft.

20. Room air-conditioners, installed with flexible cord, are to be a length that is limited to _____ ft. for 208 or 240 volt circuits.

21. The OCPD for hermetically sealed compressors shall be selected at _____ pecent (for minimum) of the compressor's FLA rating or the branch-circuit selection current, whichever is greater.
    (a) 125
    (b) 150
    (c) 175
    (d) 225

22. The rating of the overcurrent protection device shall be determined by using the rating on the nameplate of the cord-and-plug connected equipment having single-phase, _____ volt or less hermetically sealed motors.
    (a) 120
    (b) 250
    (c) 480
    (d) 600

23. Overload protection shall be provided for direct or fixed-wired motor compressors and equipment which is connected to _____ or _____ amp, 120 volt, single-phase branch-circuits.
    (a) 15 or 20
    (b) 20 or 30
    (c) 30 or 40
    (d) 40 or 50

24. Room air-conditioners shall be grounded for the protection of _____ .
    (a) equipment
    (b) personnel
    (c) all of the above
    (d) none of the above

	Section	Answer

25. A cord-and-plug shall be permitted to serve as the disconnecting means for a room air-conditioner, if it operates at _____ volts or less.
    (a)   120
    (b)   150
    (c)   250
    (d)   300

26. What size nonautomatic CB is required to disconnect an A/C unit with a compressor rated at 29 amps and the condenser rated at 2.5 amps?

27. What size nonfused disconnect is required for an A/C unit with a compressor rated at 29 amps and the condenser rated at 2.5 amps?

28. What size horsepower rated disconnect is required for an A/C with a compressor rated at 29 amps and the condenser rated at 2.5 amps with a 180 amp locked-rotor current rating? (Note that the supply is 480 V, 3Ø)

29. What size horsepower rated disconnect is required for the following loads on a three-phase, 230 volt system:

• A/C unit with a compressor rated at 29 amps with a 200 amp locked-rotor current rating
• A/C unit with a compressor rating 24 amps with a locked-rotor current of 150 amps
• A/C unit with a compressor rating of 20 amps with a locked-rotor current of 140 amps

30. What size circuit breaker is required to disconnect a hermetic sealed motor for an A/C unit rated at 29 amps? (Note that the supply is 208 V, 3Ø)

31. What size disconnecting means using horsepower is required to disconnect the following loads on a three-phase, 230 volt system:

• Motor rated at 38 amps with a 212 amp locked-rotor current rating
• A/C unit with a compressor rated at 28 amps with a 160 amp locked-rotor current rating
• A/C unit with a compressor rated at 24 amps with a 160 amp locked-rotor current rating

32. What is the minimum size OCPD required for an A/C unit with a compressor rated at 20 amps and the condenser rated at 2.5 amps?

33. What is the maximum size OCPD required for an A/C unit with a compressor rated at 20 amps and the condenser rated at 2.5 amps?

34. What is the minimum size OCPD required for a feeder-circuit with the following loads on a 230 volt, three-phase system:

• A/C unit with a compressor rated at 29 amps and the condenser rated at 2.5 amps
• A/C unit with a compressor rated at 26 amps and the condenser rated at 2.5 amps
• A/C unit with a compressor rated at 22 amps and the condenser rated at 2.5 amps

Section	Answer

35. What is the maximum size OCPD required for a feeder-circuit with the following loads on a 230 volt, three-phase system:

  • A/C unit with a compressor rated at 29 amps and the condenser rated at 2.5 amps
  • A/C unit with a compressor rated at 26 amps and the condenser rated at 2.5 amps
  • A/C unit with a compressor rated at 22 amps and the condenser rated at 2.5 amps

36. What size OCPD (time delay fuse) is required for a feeder-circuit with the following loads on a 230 volt, three-phase system:

  • A/C unit with a compressor rated at 26 amps and the condenser rated at 2.5 amps
  • A/C unit with a compressor rated at 24 amps and the condenser rated at 2.5 amps
  • 10 HP, 230 volt, three-phase, Design letter B motor

37. What size THHN copper conductors are required to supply an A/C unit with a compressor rated at 20 amps and the condenser rated at 2.5 amps?

38. What size THHN copper conductors are required to supply an A/C unit with a branch-circuit selection current of 30 amps?

39. What size THWN copper conductors are required for a feeder-circuit with the following loads on a 208 volt, three-phase system:

  • A/C unit with a compressor rated at 30 amps and the condenser rated at 3 amps
  • A/C unit with a compressor rated at 28 amps and the condenser rated at 2.5 amps
  • A/C unit with a compressor rated at 24 amps and the condenser rated at 2.5 amps

40. What size THWN copper conductors are required for a feeder-circuit with the following loads on a 208 volt, three-phase system:

  • A/C unit with a compressor rated at 30 amps and the condenser rated at 3 amps
  • A/C unit with a compressor rated at 28 amps and the condenser rated at 2.5 amps
  • A/C unit with a compressor rated at 24 amps and the condenser rated at 2.5 amps
  • Other loads of 80 amps (continuous operation)

41. What size controller is required for an A/C unit with a compressor rated at 20 amps and the condenser rated at 2.5 amps on a 230 volt, single-phase system?

42. What size overload relay and fuses is required for an A/C unit with a compressor rated at 20 amps on a 240 volt, three-phase system?

43. What is the allowable ampacity for a 10 AWG THWN copper conductor supplying an air-conditioning window unit?

	Section	Answer
44. What is the allowable ampacity available for an air-conditioning window unit with other loads such as lighting and receptacle loads that are supplied by size 10 AWG THWN copper conductor?	_____	_____

20

Transformers

Transformers must be sized with enough capacity to supply power to loads and allow loads with high inrush currents to start-and-run. In addition, they must be protected by properly sized OCPD's and be equipped with conductors having allowable ampacity ratings to supply the loads. OCPD's and conductors must be designed and installed in such a manner to safely protect the windings of such power sources from dangerous short-circuits, ground-faults, and overloads.

The OCPD's and conductors are sometimes required to be adjusted in size in order to protect the transformer windings or the conductors from overload conditions.

Transformers may be utility or customer owned and fall under the rules of the *National Electrical Safety Code*. If customer owned, they are still required to follow the *National Electrical Safety Code* as well as the *National Electrical Code*. And these requirements must be complied with.

Note: When selecting the actual size CB or fuse for the protection of electrical systems rated over 600 volts, see one of the ANSI C Standards. For example, fuses rated at 100 amps or less, see ANSI C 37.46 and for over 100 amps see ANSI 37.46 and ANSI 37.40. When circuit breakers are used to protect high-voltage systems, see ANSI C 37.06. However, there may be protection designs that require the reference to other ANSI C Standards and the designer must be prepared to refer to such standards.

SIZING TRANSFORMERS

The total volt-amps of all loads in a building shall be used to size a transformer. Depending on the load requirements of a building, single-phase and three-phase currents may be used to supply the building. The windings are connected in configurations necessary to supply voltage and load requirements of the facility.

WYE-CONNECTED SECONDARIES

The size transformers required to supply a wye-connected secondary system can be found by applying the following:

- Adding the total single-phase and three-phase loads together for an individual transformer.
- Dividing the load in VA by 1/3 (.33) to derive three transformers.

The kVA rating of three transformers, if they are separately connected together will add up to one individual transformer. Using this method, a single transformer rating can be sized and selected from the total volt-amps. One transformer with three windings is sized by adding the total VA of all the loads together and selecting the transformers kVA rating based upon this value. **(See Figure 20-1)**

CLOSED DELTA-CONNECTED SECONDARIES

The size transformer required to supply a closed delta-connected secondary system can be found by multiplying the following:

- Multiply the single-phase load in VA by 67 percent.
- Multiply the three-phase load in VA by 33 percent.
- Add the kVA load of (1) and (2) together to derive the two lighting and power transformers.
- Multiply the single-phase and three-phase loads in VA by 33 percent.
- Use this total load in kVA to derive the two power transformer.

See Figure 20-2(a) for the rules and regulations for sizing the transformer used in a closed delta-connected system.

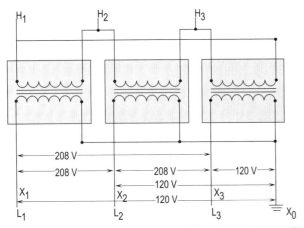

Figure 20-1. Sizing wye-connected transformers with single-phase and three-phase loads.

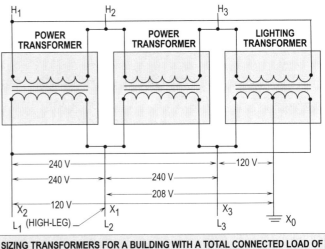

Figure 20-2(a). Sizing closed delta-connected transformers with single-phase and three-phase loads.

OPEN DELTA-CONNECTED SECONDARIES

Open delta-connected secondary systems can be determined by calculating the single-phase load at 100 percent and the three-phase load at 58 percent and use this total value to size the transformer. By adding these two loads together, the size of a mid-tap transformer can be determined. A power transformer can be sized by calculating the three-phase load at 58 percent which is the reciprocal of the square root of 3 (1 ÷ 1.732 = 58%). This reduced total is then used to size the power transformer which will be smaller in rating than the lighting and power transformer. **(See Figure 20-2(b))**

circuit to supply the primary of a transformer to step up or step down the voltage. To determine the FLA of a transformer, the kVA of the transformer must be divided by the voltage x 1.732 if the supply is three-phase. **(See Figures 20-3(a) and (b))**

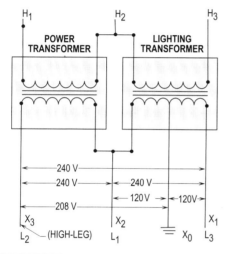

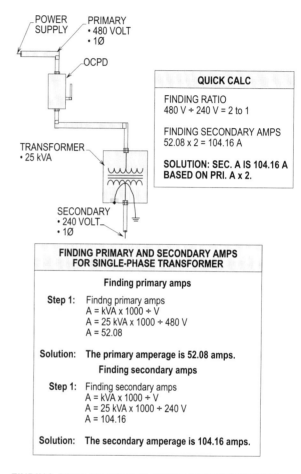

FINDING AMPS OF SINGLE-PHASE TRANSFORMERS

Figure 20-3(a). Finding amps of a single-phase transformer.

FINDING AMPERAGE

The kVA or amp rating for the primary or secondary of a transformer can be determined for a single-phase system by applying the following formula:

kVA = volts x amps ÷ 1000
amps = kVA x 1000 ÷ volts

The following formula shall be applied to determine the ratio of a transformer having a 480 volt primary and 240 volt secondary:

primary ÷ secondary
480 V ÷ 240 V
2:1 ratio

OPEN DELTA-CONNECTED SECONDARIES

Figure 20-2(b). Sizing open delta-connected transformers with single-phase and three-phase loads.

CALCULATING PRIMARY AND SECONDARY CURRENTS

The transformer's primary amp rating shall be equivalent to the amps of the connected load when installing a feeder-

20-3

The amp rating for the primary or secondary can be determined for a three-phase system by applying the following formula:

$$kVA = volts \times 1.732 \times amps \div 1000$$
$$amps = kVA \times 1000 \div volts \times 1.732$$

FINDING AMPS OF THREE-PHASE TRANSFORMERS

Figure 20-3(b). Sizing amps of a three-phase transformer.

INSTALLING TRANSFORMERS
ARTICLE 450

Transformers shall be installed and protected per Art. 450. Transformers shall be permitted to be installed inside or outside buildings based upon their design and type.

LOCATION
450.13

Transformers shall be located where readily accessible to qualified personnel for inspection and maintenance. Where it is necessary to use a ladder, lift, or bucket truck to get to a transformer, it is not considered readily accessible. See definition of readily accessible, in **Article 100** of the NEC. **(See Figure 20-4)**

There are two exceptions to the general rule to the accessibility rules and they are explained in the next two headings.

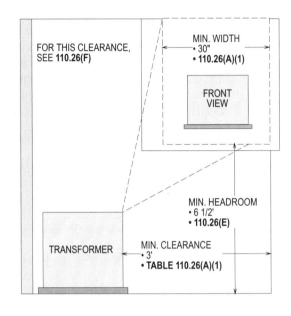

TRANSFORMERS MUST BE READILY ACCESSIBLE

LOCATION
NEC 450.13

Figure 20-4. The general rule of **450.13** requires transformers to be readily accessible for maintenance, repair, and service. See AHJ for this requirement.

HUNG FROM WALL OR CEILING
430.13(A)

- Dry-type transformers, not over 600 volts and located on open walls or steel columns do not have to be readily accessible. It is permissible to gain access to this type of installation using a portable ladder or bucket lift. **(See Figure 20-5(a))**

MOUNTED IN CEILING
450.13(B)

- Dry-type transformers not over 600 volts and 50 kVA may be installed in hollow spaces of buildings. The transformers cannot be permanently closed in and there must be some access to the transformers, but they do not have to be readily accessible per **Article**

100 of the NEC. It was not clear in the 1993 or in previous editions of the NEC if dry-type transformers not exceeding 600 volts, nominal, and rated 50 kVA or less, were permitted to be installed in the space above suspended ceilings with removable panels, even if the transformer is accessible and provided with proper working clearances. Note that the space, where the transformer is installed, shall comply with the ventilation requirements of **450.9** and be designed by the rules of **450.21(A) and (B)**. If such ceiling space is used as a return air space for air-conditioning, **300.22(C)** shall be reviewed and the provisions of this section must be complied with. **(See Figure 20-5(b))**

Design Tip: The two exceptions to the general rule are for dry-type transformers. The exceptions do not apply for oil or askarel-filled transformers due to the damage of a possible oil spillage or the threat of fire because of a rupture occuring in the case.

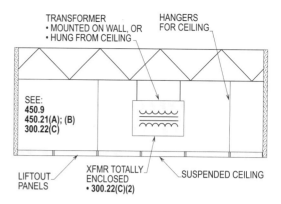

MOUNTED IN CEILING
NEC 450.13(B)

Figure 20-5(b). Transformers mounted in ceiling do not have to be readily accessible.

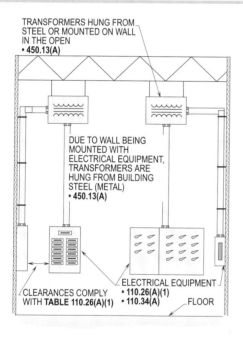

HUNG FROM WALL OR CEILING
NEC 450.13(A)

Figure 20-5(a). Transformers hung from the wall or ceiling, do not have to be readily accessible.

OVERCURRENT PROTECTION 450.3(A); (B)

There are two sets of rules when providing overcurrent protection of transformers. Rules for transformers rated over 600 volts and transformers of 600 volts or less. The OCPD may be placed in the primary only or in the primary and secondary side of the transformer.

PRIMARY ONLY - OVER 600 VOLTS 450.3(A); TABLE 450.3(A)

The term primary is often inferred in the field as being the high side and the term secondary as the low side of the transformer. This is really not the proper terminology. The primary is the input side of the transformer and the secondary is the output side. Thus voltage has nothing to do with the terms whether its high or low.

Each transformer has to be protected by an overcurrent device in the primary side. If the overcurrent protection is fuses, they shall be rated not greater than 250 percent (2.5 times) of the rated primary current of the transformer. When circuit breakers are used they must be set not greater than 300 percent (3 times) of the rated primary current. **(See Figure 20-6)**

This overcurrent protection device may be mounted in the vault or at the transformer, if approved for such purpose. It may also be mounted in the panelboard and be designed to protect the windings and circuit conductors supplying the transformer.

If not installed in a vault, the OCPD can be installed outdoors on a pole, with a disconnecting means installed in the vault to disconnect supply conductors.

APPLYING Note 1
TABLE 450.3(A), Note 1

Where 250 percent (2.5 times) of the rated primary current of the transformer does not correspond to a standard rating of a fuse, the next higher standard rating is permitted, per **240.6(A)**.

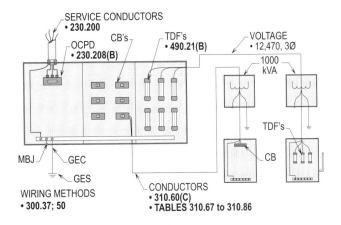

Figure 20-6. If the overcurrent protection is fuses, they shall be rated not greater than 250 percent (2.5 times) of the rated primary current of the transformer. When circuit breakers are used they shall be set at not greater than 300 percent (3 times) of the rated primary current.

PRIMARY AND SECONDARY - OVER 600 VOLTS
450.3(A); TABLE 4503(A)

A transformer over 600 volts, nominal, having an overcurrent protection device on the secondary side rated to open not greater than the values listed in **Table 450.3(A)** or a transformer equipped with a coordinated thermal overload protection by the manufacturers shall not be required to have an individual protection in the primary. However, a feeder overcurrent protection device rated or set to open at not greater than the values listed in **Table 450.3(A)** must be provided.

NONSUPERVISED LOCATIONS
450.3(A); TABLE 450.3(A)

Overcurrent protection for a nonsupervised location may be placed in the primary and secondary side of high-voltage transformers, if the OCPD's are designed and installed according to the provisions listed in **Table 450.3(A)**.

If the secondary voltage is 600 volts or less, the OCPD and conductors on the secondary side shall be sized at 125 percent of the FLC rating. OCPD's sized at 125 percent of the FLC in amps protects the conductors and windings of the transformer from dangerous overload conditions. With higher voltage on the secondary side of the transformer, the percentages for sizing the OCPD's are selected from **Table 450.3(A)** (any location) based upon the particular voltage level. **(See Figure 20-7(a))**

SUPERVISED LOCATIONS
450.3(A); TABLE 450.3(A)

Overcurrent protection may be placed in the primary and secondary side of high-voltage transformers if the OCPD's are designed and installed according to the provisions listed in **Table 450.3(A)**.

Where the facility has trained engineers and maintenance personnel, the OCPD for the secondary can be sized at not more than 250 percent of the FLC for voltage 600 volts or less. With higher voltage on the secondary side of the transformer, the percentages for sizing the OCPD's are selected from **Table 450.3(A)** based upon the particular voltage level. **(See Figures 20-7(b) and (c))**

Note: See Figure 20-8 for certain design conditions which allows the primary OCPD to be used to protect the primary and secondary sides of two-wire to two-wire connected transformers and three-wire to three-wire delta connected transformers per **240.4(F) and 240.21(C)(1)**.

PRIMARY ONLY - 600 VOLTS OR LESS
450.3(B); TABLE 450.3(B)

A transformer 600 volts or less, nominal, having an individual overcurrent protection device on the primary side shall be sized at no more than 125 percent of the transformer's full-load current rating. Note, with the OCPD and conductors sized at 125 percent or less of the transformer's FLC, the

Transformers

supply conductors and transformer windings are considered protected from overload conditions. It appears that individual protection in the primary is not recognized per **450.3(B)** and **Table 450.3(B)**. **(See Figure 20-9)**

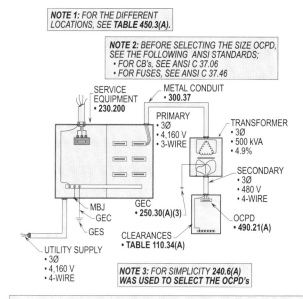

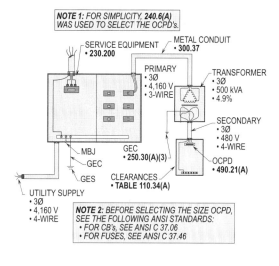

Figure 20-7(a). Sizing the primary and secondary side of a transformer in a nonsupervised (any) location.

Figure 20-7(b). Sizing the primary and secondary side of a transformer in a supervised location.

20-7

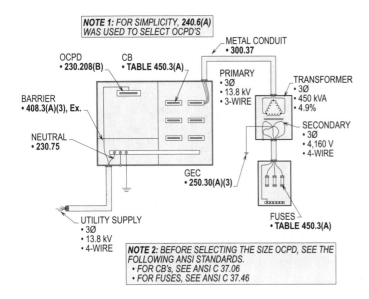

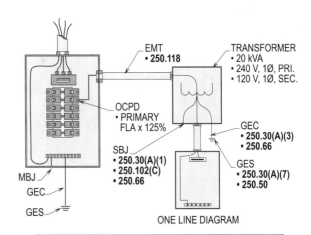

FINDING OCPD FOR PRIMARY SIDE OF TRANSFORMER

Step 1: Finding FLA of primary
FLA = kVA × 1000 ÷ V
FLA = 20 kVA × 1000 ÷ 240 V
FLA = 83.3 A

Step 2: Calculating OCPD
450.3(B); Table 450.3(B)
83.3 A × 125% = 104 A

Step 3: Selecting OCPD
Table 450.3(B), Note 1; 240.6(A)
104 A allows 100 A

Solution: The size OCPD in the primary side is 100 amps.

NEC 450.3(B); TABLE 450.3(B)

Figure 20-9. A transformer 600 volts or less, nominal, having an individual overcurrent protection device on the primary side shall be sized at no more than 125 percent of the transformer's full-load current rating, in amps.

FINDING OCPD FOR THE PRIMARY AND SECONDARY SIDE OF THE TRANSFORMER

Sizing OCPD for primary side

Step 1: Finding FLA of transformer
FLA = kVA × 1000 ÷ V × √3
FLA = 450 × 1000 ÷ 13,800 V × 1.732
FLA = 18.83 A

Step 2: Calculating FLA for OCPD
450.3(A); Table 450.3(A)
FLA = 18.83 A × 600%
FLA = 112.9 A

Step 3: Selecting OCPD
Table 450.3(A), Note 3; 240.6(A)
112.9 A allows 110 A

Solution: The size OCPD for the primary side is a 110 amp circuit breaker.

Sizing OCPD for secondary side

Step 1: Finding FLA of transformer
FLA = 450 × 1000 ÷ 4160 × 1.732
FLA = 62.5 A

Step 2: Calculating FLA for OCPD
450.3(A); Table 450.3(A)
FLA = 62.5 A × 250%
FLA = 156.3 A

Step 3: Selecting OCPD
Table 450.3(A), Note 3; 240.6(A)
156.3 A allows 150 A

Solution: The size OCPD for secondary side is 150 amp fuses.

**SUPERVISED LOCATION
NEC 450.3(A); TABLE 450.3(A)**

Figure 20-7(c). Sizing the primary and secondary side of a transformer in a supervised location.

9 AMPS OR MORE
450.3(B); TABLE 450.3(B)

Where the rated primary current of a transformer is 9 amps or more and 125 percent of this current does not correspond to a standard rating of a fuse or circuit breaker, the next size may be used per **240.6(A)**. Where the rated primary current of a transformer is less than 9 amps but more than 2 amps, an overcurrent device rated or set at no more than 167 percent of primary current may be used. When the rated primary current of a transformer is less than 2 amps, an overcurrent protection device rated or set at not more than 300 percent shall be used. **(See Figures 20-10(a), (b) and (c))**

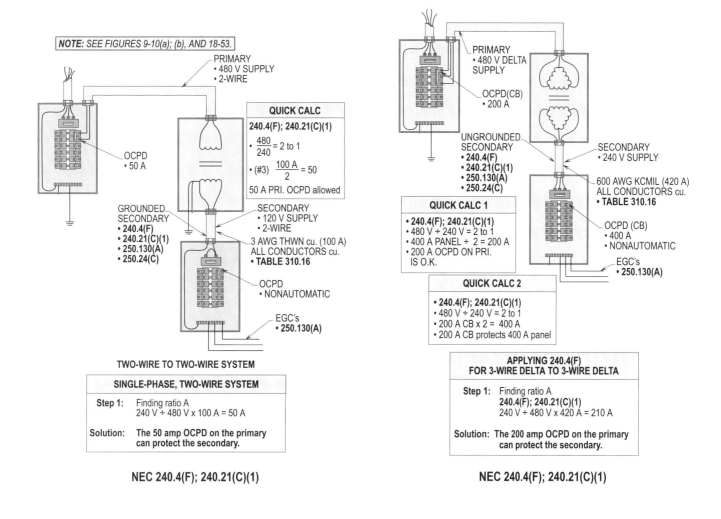

Figure 20-8. Sizing the OCPD for a single-phase, two-wire system and three-phase, three-wire system.

PRIMARY AND SECONDARY - 600 VOLTS OR LESS
450.3(B); TABLE 450.3(B)

Combination protection can be provided for both the primary and secondary sides of a transformer. A current value of 250 percent of the rated primary current of the transformer may be used if 125 percent of the rated primary current of the transformer is not sufficient to allow loads with high inrush currents to start and operate. However, the secondary overcurrent protection device is sized at 125 percent of the rated secondary full-load current of the transformer. Where the rated secondary current of a transformer is less than 9 amps, an overcurrent device rated or set at no more than 167 percent of secondary current may be used. **(See Figure 20-11)**

9 AMPS OR MORE
TABLE 430.3(B), NOTE 1

Where the rated secondary current of a transformer is 9 amps or more and 125 percent of this current does not correspond to a standard rating of a fuse or circuit breaker, the next size may be used per 240.6(A).

GROUNDING AUTOTRANSFORMERS 450.5

Autotransformers are connected to three-phase, three-wire ungrounded systems to derive a three-phase, four-wire grounded system. Three autotransformers connected in a star (wye) configuration to the three-phase ungrounded system converts to a three-phase, four-wire grounded system.

Autotransformers are installed today because many electrical systems are not grounded. Existing ungrounded delta systems are grounded with autotransformers to derive a neutral. Three-phase zigzag transformers are generally installed for this purpose.

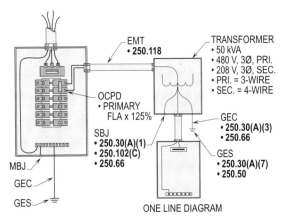

PRIMARY 9 AMPS OR MORE
NEC 450.3(B); TABLE 450.3(B)

Figure 20-10(a). Where the rated primary current of a transformer is 9 amps or more and 125 percent of this current does not correspond to a standard rating of a fuse or circuit breaker, the next size may be used per **240.6(A)** and **Note 1 to Table 450.3(B)**.

THREE-WIRE CIRCUIT TO THREE-PHASE, FOUR-WIRE CIRCUIT
450.5(A)

Grounding autotransformers are connected to derive a neutral from a three-phase, three-wire ungrounded system to a three-phase, four-wire grounded system, the following conditions shall apply:

- Proper connections must be made
- Overcurrent protection must be provided
- Transformer fault sensing installed
- Rating be adequately sized

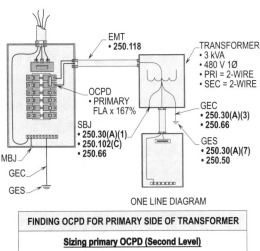

PRIMARY 2 AMPS OR MORE BUT LESS THAN 9 AMPS
NEC 450.3(B); TABLE 450.3(B)

Figure 20-10(b). Where the rated primary current of a transformer is less than 9 amps but 2 amps or more, an overcurrent device rated or set at no more than 167 percent of primary current may be used.

CONNECTIONS
450.5(A)(1)

Transformers shall be directly connected to the ungrounded phase conductors with no switches or overcurrent protection devices installed between the connection and the autotransformer.

OVERCURRENT PROTECTION
450.5(A)(2)

An overcurrent protection sensing device shall be designed to trip at 125 percent of its continuous current per phase or neutral rating. The next higher standard rating may be installed where the input current is 9 amps or more and computed at 125 percent. Input current of 2 amps or less shall not exceed 167 percent. **(See Figure 20-12)**

Transformers

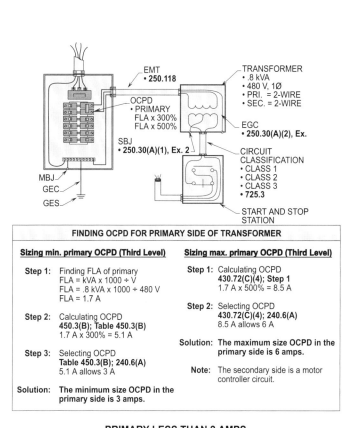

Figure 20-10(c). When the rated primary current of a transformer is less than 2 amps, an overcurrent protection device rated or set at not more than 300 percent shall be used, unless **430.72(C)(4)** is applied.

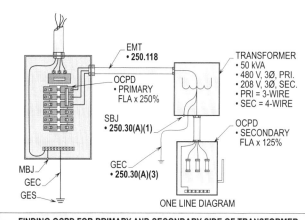

Figure 20-11. Sizing OCPD for the primary and secondary side of a transformer rated 600 volts or less.

TRANSFORMER FAULT SENSING
450.5(A)(3)

A main switch or common-trip overcurrent protection device for the three-phase, four-wire system can be provided with fault sensing systems to guard against single-phasing or internal faults.

RATING
450.5(A)(4)

Autotransformers shall be designed with a continuous neutral current rating sufficient to handle the maximum possible unbalanced neutral load current that could flow in the four-wire system.

DETECTING GROUNDS ON THREE-PHASE, THREE-WIRE SYSTEMS
450.5(B)

The following conditions shall apply when autotransformers are used to detect grounds on three-phase, three-wire systems:
- Proper rating
- Overcurrent protection sized adequately
- Ground reference for damping transitory over voltages

RATING
450.5(B)(1)

Autotransformers shall have a continuous neutral current rating sufficient for the specified ground-fault current that could develop in the system.

20-11

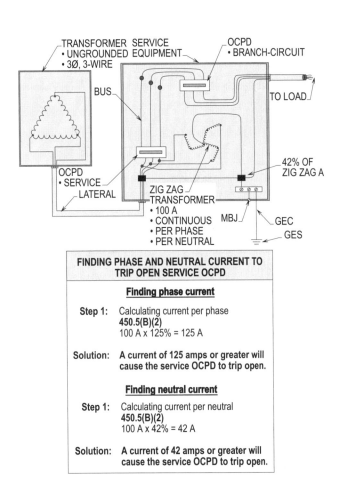

FINDING PHASE AND NEUTRAL CURRENT TO TRIP OPEN SERVICE OCPD

Finding phase current

Step 1: Calculating current per phase
450.5(B)(2)
100 A x 125% = 125 A

Solution: A current of 125 amps or greater will cause the service OCPD to trip open.

Finding neutral current

Step 1: Calculating current per neutral
450.5(B)(2)
100 A x 42% = 42 A

Solution: A current of 42 amps or greater will cause the service OCPD to trip open.

NEC 450.5(B)(2)

Figure 20-13. The overcurrent protection device shall open simultaneously with a common trip all ungrounded conductors and be set to trip at not more than 125 percent of the rated phase current of the transformer. Note that 42 percent of the overcurrent protection device's rating may be used if connected in the autotransformer's neutral connection.

OVERCURRENT PROTECTION
450.5(B)(2)

The overcurrent protection device shall open simultaneously with a common trip all ungrounded conductors and be set to trip at not more than 125 percent of the rated phase current of the transformer. Note that 42 percent of the overcurrent protection device's rating may be used if connected in the autotransformer's neutral connection. **(See Figure 20-13)**

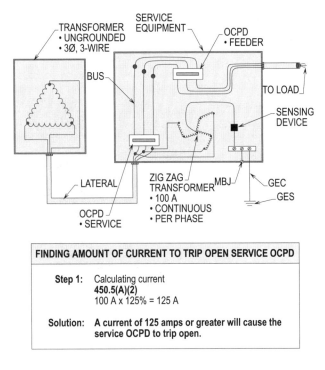

FINDING AMOUNT OF CURRENT TO TRIP OPEN SERVICE OCPD

Step 1: Calculating current
450.5(A)(2)
100 A x 125% = 125 A

Solution: A current of 125 amps or greater will cause the service OCPD to trip open.

NEC 450.5(A)(2)

Figure 20-12. An overcurrent protection sensing device shall be designed to trip at 125 percent of its continuous current per phase or neutral rating.

SECONDARY TIES
450.6

In large industrial plants and facilities a "network" distribution system is usually utilized for supplying power loads. Three-phase banks of transformers are located at various points throughout the plant or facility. There are normally two high-tension primary circuits feeding such transformers. A double throw switch which is located at each transformer bank allows either primary circuit to serve any bank of transformers. The primary circuit conductors are sized with enough capacity so that either circuit is capable of carrying the entire load if a fault develops in the other circuit. Secondary voltage is usually three-phase systems rated 600 volts or less. The transformer secondaries are connected together in a network system, and all transformers are used to feed all the loads involved which can be all at once or as necessary.

Note that the conductors connecting the transformers secondaries together are defined as the secondary ties of the system.

Secondary ties shall be protected at both ends and such protection may be by fuses based upon the current-carrying capacity of the conductors per **450.6(A)** or the ties may be protected by a limiter installed at each end per **450.6(A)(3)**. A limiter protects the elements against short circuit, however it does not provide overload protection. Usually, limiters, rather than fuses, are used for protection of the ties due to the fact they are very current limiting and will protect the circuit elements from damage during short-circuit conditions.

There is normally a load center connected to the tie at the points where a transformer bank connects to the tie. The transformers is protected by a circuit breaker in the secondary leads between the transformer and the load center. Circuit-breaker setting may be up to 250 percent (2.5 times) of transformer's secondary current rating per **450.6(B)**. A reverse power relay must also be provided per **450.6(B)** which opens the circuit in case the transformer should fail for any reason. A reverse power relay is provided to prevent current from being fed to an out of service transformer from the other transformers of the network. Where the secondary voltage is greater than 150 volts to ground, to ensure adequate protection, ties shall be provided with a switch at each end per **450.6(A)(5)**.

The rules for designing and installing secondary ties can be summed up as follows:

- Where transformers are tied together in parallel and connected by tie conductors that do not have overcurrent protection as per **Article 240** of the NEC, the ampacity of the ties connecting conductors shall not be less than 67 percent of the rated secondary current of the largest transformer in the tie circuit. **(See Figure 20-14)**

Design Tip: This applies where the loads are at the transformer supply points per **450.6(A)(1)**.

The paralleling of transformers is common, but great care should be exercised to assure that the transformers are similar in all conditions of use. If they are not, one transformer will try to carry more of the load than the other load. If the transformers are of the same capacity and similar characteristics, they each, in theory, will carry 50 percent of the total load. The 67 percent permits for differences in transformer sizes and this allows for adjusting solutions.

- Where the load is connected to the tie at any point between the transformer supply points and overcurrent protection is not provided by the provisions listed in **Article 240** of the NEC, the rated ampacity of the tie is required to be not less than 100 percent of the rated secondary current of the largest transformer connected to the secondary tie system except as provided in **450.6(A)(5)**. **(See Figure 20-15)**

Design Tip: This rule applies mainly where the loads connected between transformer supply points per **450.6(A)(2)** of the NEC.

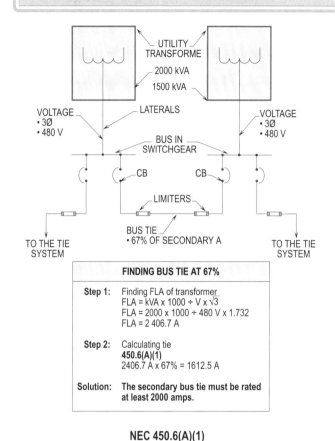

Figure 20-14. The ampacity of the ties connecting conductors must not be less than 67 percent of the rated secondary current of the largest transformer in the tie circuit.

- Sections **450.6(A)(1) and (A)(2)** state that both ends of each tie connection shall be provided with a protective device which opens at a certain temperature of the tie conductor. This prevents damage to the tie conductor and its insulation, and such installations may consist of:

(a) A limiter is a fusible link cable connector. The limiter is selected and designed for the insulation, conductor material, etc., on the tie conductors.

(b) A circuit breaker, actuated by devices having characteristics which are comparable to the above, can be used, if designed and sized properly.

Design Tip: The above applies where the tie circuit protection is provided per **450.6(A)(3)** and the tie conductor must fully comply with all rules and regulations in such sections.

Stallcup's Electrical Design Book

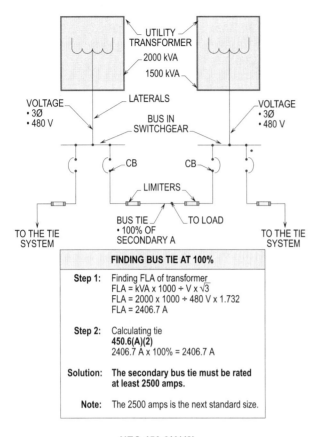

Figure 20-15. The rated ampacity of the tie is required to be not less than 100 percent of the rated secondary current of the largest transformer connected to the secondary tie system except as provided in **450.6(A)(5)**.

- Where the tie consists of more than one conductor per phase, the conductors of each phase shall be interconnected in order to create a load supply point. The protection required in **450.6(A)(3)** is to be provided in each tie conductor at this point, except as follows:

Section **450.6(A)(4)(b)** allows the loads to be connected to the individual conductor(s) of each phase and without the protection listed in **450.6(A)(3)** if, at load connection points, the tie conductors of each phase have a combined capacity of not less than 133 percent of the rated secondary current of the largest transformer connected to the secondary tie system. The total load of such taps shall not exceed the rated secondary current of the largest transformer and the loads are required to be equally divided on each phase and on the individual conductors of each phase as close as possible. **(See Figure 20-16)**

The use of multiple conductors on each phase and the requirement that loads do not have to tap the multiple conductors of the same phase might possibly set up unbalanced current flow in the multiple conductors on the same phase.

In the requirement that the combined capacity of the multiple conductors on the same phase be rated at 133 percent of the secondary current of the largest transformer is satisfied. Limiters are necessary at the tap or connections to the transformers that are tied together to properly protect the elements of the circuit.

Design Tip: The above applies where the interconnection of phase conductors between transformers supply points occur per **450.6(A)(4)(b)**.

- If the operating voltage of secondary ties exceeds 150 volts to ground, there must be a switch ahead of the limiters and tie conductors which is capable of deenergizing the tie conductors and the limiters. This switch shall comply with the following:

(a) The current rating of the switch shall not be less than the current rating of the conductors connected to such switch.

(b) The switch shall be capable of opening its rated current.

(c) The switch shall not open under the magnetic forces caused from short-circuit currents.

Design Tip: The above applies where the tie circuit control is located as mentioned in **450.6(A)(5)**.

- When secondary ties from transformers are used, an overcurrent device in the secondary of each transformer that is rated or set at not greater than 250 percent (2.5 times) of the rated secondary current of the transformer shall be provided. In addition, there must be a circuit breaker actuated by a reverse-current relay, the breaker is to be set at not greater than the rated secondary current of the transformer. Such overcurrent protection protects against overloads and short-circuit conditions, and the reverse-current relay and circuit breaker shall be designed to handle any reversal of current flow into the transformer. **(See Figure 20-17)**

Design Tip: The above applies where overcurrent protection for secondary connections are installed to protect the system as required by **450.6(B)**.

Transformers

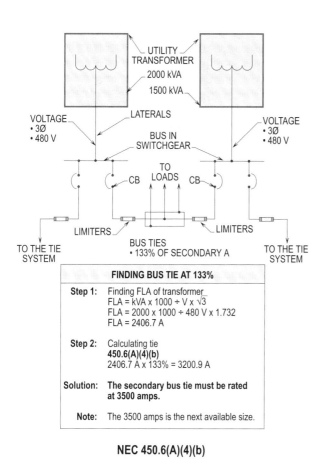

NEC 450.6(A)(4)(b)

Figure 20-16. Section **450.6(A)(4)(b)** allows the loads to be connected to the individual conductor(s) of each phase and without the protection listed in **450.6(A)(3)** if, at load connection points, the tie conductors of each phase have a combined capacity of not less than 133 percent of the rated secondary current of the largest transformer connected to the secondary tie system.

PARALLEL OPERATION
450.7

Transformers are permitted to be connected in parallel and switched as a unit provided each transformer has overcurrent protection that is properly sized and complies with **450.3(A) and 450.3(B). (See Figure 20-18)**

Persons working with paralleled transformers or transformer tie circuits should be extremely careful that there are no feedbacks or other conditions which would affect safety. In order to secure a balance of current between paralleled transformers, all transformers should have characteristics that are very much alike, such as voltage, impedance, and other such pertinent elements.

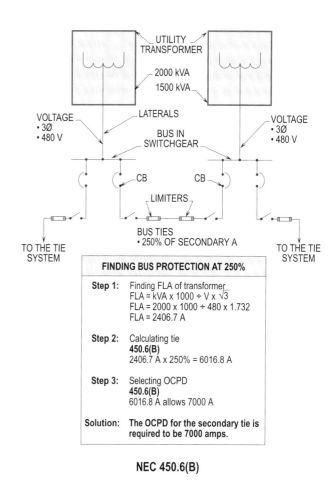

NEC 450.6(B)

Figure 20-17. When secondary ties from transformers are used, an overcurrent device in the secondary of each transformer that is rated or set at not greater than 250 percent (2.5 times) of the rated secondary current of the transformer shall be provided.

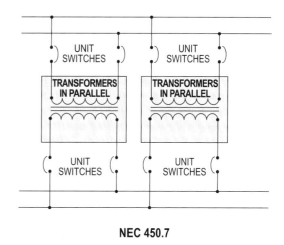

NEC 450.7

Figure 20-18. Transformers are permitted to be connected in parallel and switched as a unit provided each transformer has overcurrent protection that is properly sized.

20-15

GUARDING
450.8

Transformers should be isolated in a room or accessible only to qualified personnel to prevent accidental contact with live parts. To safeguard live parts from possible damage, the transformer may be elevated. The following are acceptable means of safeguarding live parts as required in **110.27(A) and 110.34(E)**.

- Transformers should be isolated in a room or accessible only to qualified personnel.
- Permanent partitions or screens may be installed.
- Transformers shall be elevated at least 8 ft. above the floor to prevent unauthorized personnel from contact.

Design Tip: Signs indicating the voltage of live exposed parts of transformers, or other suitable markings, shall be used in areas where transformers are located.

VENTILATION
450.9

Transformers shall be located and installed in rooms or areas that are not subjected to exceedingly high temperatures to prevent overheating and possible damage to windings. Transformers with ventilating openings shall be installed so that the ventilating openings are not blocked by walls or other obstructions that could block air flow.

GROUNDING
450.10

Transformer cases shall be grounded per **250.110**. Transformers enclosed by fences or guards shall be grounded. Live parts require a guard such as a fence to guard against unauthorized entry and to protect the general public and unqualified persons from dangerous electrical parts. Such fences shall be grounded to prevent metal elements from being accidentally energized with voltage. **(See Figure 20-19)**

TYPES OF TRANSFORMERS
PART II TO ARTICLE 450

Transformers are either used indoors or outdoors based upon their type and condition of use. The type of transformer determines where it can be installed inside or outside of the building. Sometimes it becomes necessary to build rooms of certain fire-rated material or confinement areas to house transformers due to their design and installation.

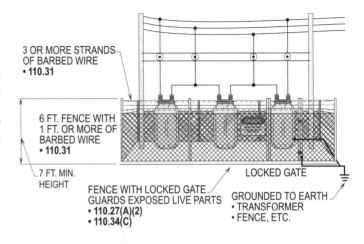

NEC 450.10

Figure 20-19. Transformers enclosed by fences or guards shall be grounded. Live parts require a guard such as a fence to guard against unauthorized entry and to protect the general public and unqualified persons from dangerous electrical parts.

DRY-TYPE TRANSFORMERS INDOORS
450.21

The rules for installing dry-type transformers indoors can be summed up as follows:

- Dry-type transformers greater than 112 1/2 kVA, having Class 155 or higher insulation systems, must have a fire-resistant heat-insulating barrier placed between transformers and combustible material, or, if no barrier, must be separated at least 6 ft. horizontally and 12 ft. vertically from the combustible material per **450.21(B), Ex. 1. (See Figure 20-20)**

- Dry-type transformers, greater than 112 1/2 kVA and having Class 155 or higher of insulation systems, shall be installed in a fire-resistant transformer room per **450.21(B), Ex. 2. (See Figure 20-21)**

- Dry-type transformers, rated at 112 1/2 kVA or less and 600 volts or less, shall have a fire-resistant heat-insulating barrier between transformers and combustible material, or, without a barrier, shall be separated at least 12 in. from the combustible material where the voltage is 600 volts or less per **450.21(A). (See Figure 20-22)**

- Dry-type transformers, rated 112 1/2 kVA or less, and 600 volts or less, shall not be required to have a 12 in. separation or barrier if they are completely enclosed, except for vent openings. **(See Figure 20-23)**

- All indoor dry-type transformers of over 35,000 volts shall be installed in a vault. Vault requirements shall fully comply with Part C of **Article 450** of the NEC. **(See Figure 20-24)**

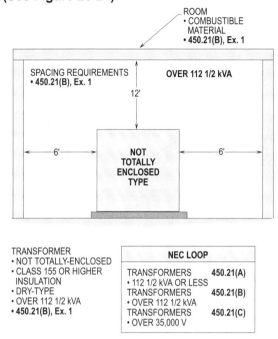

Figure 20-20. Dry-type transformers, greater than 112 1/2 kVA and having Class 155 or higher insulation systems, shall have a fire-resistant heat-insulating barrier placed between transformers and combustible material, or, if no barrier, must be separated at least 6 ft. horizontally and 12 ft. vertically from the combustible material.

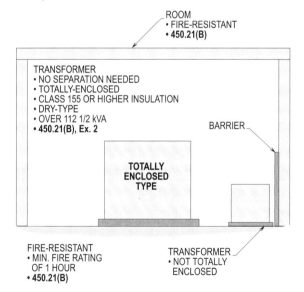

Figure 20-21. Dry-type transformers, greater than 112 1/2 kVA and having Class 155 or higher insulation systems, shall be installed in a fire-resistant transformer room.

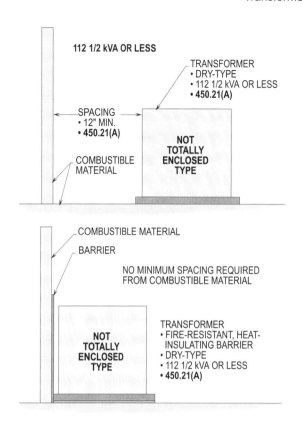

Figure 20-22. Dry-type transformers, rated 112 1/2 kVA or less and 600 volts or less, shall have a fire-resistant heat-insulating barrier between transformers and combustible material, or, without a barrier, shall be separated at least 12 in. from the combustible material where the voltage is over 600 volts.

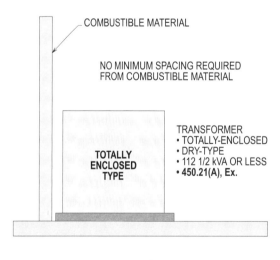

Figure 20-23. Dry-type transformers, rated 112 1/2 kVA or less, and 600 volts or less, shall not be required to have a 12 in. separation or barrier if they are completely enclosed except for vent openings.

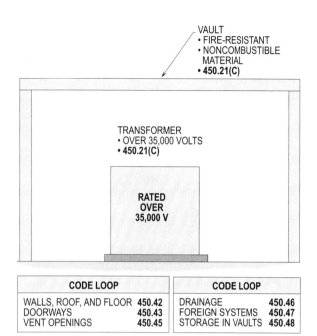

Figure 20-24. All indoor dry-type transformers of over 35,000 volts shall be installed in a vault.

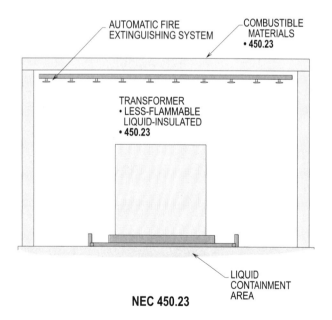

Figure 20-25. Transformers using a "listed" high fire point liquid may be installed indoors, but only in "noncombustible" areas of "noncombustible" buildings. The NEC sets the minimum fire-point at 300°C (572°F).

DRY-TYPE TRANSFORMERS INSTALLED OUTDOORS
450.22

Dry-type transformers installed outdoors must have weatherproof enclosures. See the definition in Article 100 of the NEC for the difference between "weatherproof" and "watertight."

LESS-FLAMMABLE LIQUID-INSULATED TRANSFORMERS
450.23

Transformers using a "listed" high fire point liquid may be installed indoors, but only in "noncombustible" areas of "noncombustible" buildings. The NEC sets the minimum fire-point at 300°C (572°F). This is the minimum temperature at which the liquid ignites. Such transformers may be installed indoors, for voltages up to 35,000. Higher voltages require a vault, if they are installed indoors. This is due to safety that is required for the higher voltage and associated equipment. **(See Figure 20-25)**

NONFLAMMABLE FLUID-INSULATED TRANSFORMERS
450.24

Transformers using a "dielectric" nonflammable liquid may be installed indoors in any location, for voltages up to 35,000. Higher voltages require a vault, when installed indoors due to safety required for the higher voltage and associated equipment.

For the purpose of this Section, a nonflammable dielectric fluid is one which does not have a flash point or fire point, and is not flammable in air.

ASKAREL-INSULATED TRANSFORMERS INSTALLED INDOORS
450.25

Askarel is a liquid that does not burn; therefore, it is safer than oil to be used as a transformer liquid. However, arcing in askarel produces greater gases which are nonexplosive gases.

Askarel-insulated transformers of over 25 kVA shall be furnished with a relief vent such as a chimney to relieve the pressure built up by gases that may be generated within the transformer.

In rooms that are well ventilated the vent may be discharged directly to the room. In rooms that are poorly ventilated the vent shall be piped to a flue or chimney that is capable of carrying the gases out of the room. Or, as an alternative for such ventilating, the transformer can be fitted with a gas absorber placed inside the case. When there is a gas absorber, the vent may also be discharged to the room.

Askarel transformers of more than 35,000 volts shall be installed in a vault due to the oil and higher voltage being a hazard to unqualified personnel. **(See Figure 20-26)**

transformers of 600 volts or less may be installed without a vault. When installed without a vault, the total kVA ratings of all transformers allowed in a room or section of a building is limited to 10 kVA for non-fire-resistant buildings, and 75 kVA for fire-resistant buildings.

See Figure 20-27 for installing rules when applying the **Ex. 1 to 450.26**.

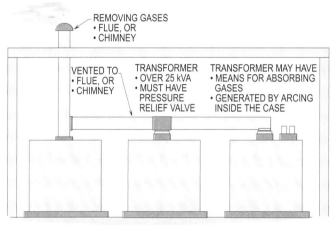

Figure 20-26. Askarel-insulated transformers of over 25 kVA shall be furnished with a relief vent such as a chimney to relieve the pressure built up by gases that may be generated within the transformer.

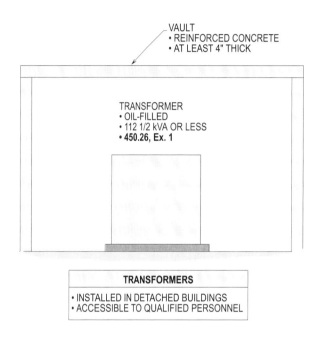

Figure 20-27. Oil-filled transformers rated 112 1/2 kVA or less that are installed in detached buildings and accessible to only qualified personnel shall be installed in a vault with reinforced concrete at least 4 in. thick.

OIL-INSULATED TRANSFORMERS INSTALLED INDOORS
450.26

The rules for installing oil-insulated transformers installed indoors can be summed up as follows:

- Indoor oil-filled transformers greater than 600 volts shall be installed in a vault, with the following exceptions, where, regardless of voltage, a vault is not required:

   (a) Electric furnace transformers with a total rating of 75 kVA or less may be located in a fire-resistant room.
   (b) Oil-filled transformers may be installed in a building without a vault, provided the building is accessible to qualified personnel only, and used solely for providing electric service to other buildings.

- If suitable provisions are provided to prevent a possible oil fire from igniting other materials, oil-filled

OIL-INSULATED TRANSFORMERS INSTALLED OUTDOORS
450.27

When oil-filled transformers are installed on or adjacent to combustible buildings or material, the building or material shall be safeguarded from possible fire originating in a transformer. Fire-resistant barriers, water-spray systems, and enclosures for the transformers are approved safeguards, if, where used, they are installed by the rules of the NEC. **(See Figures 20-28(a) and (b))**

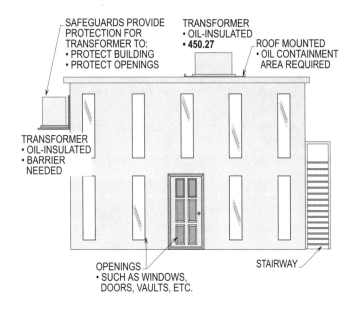

NEC 450.27

Figure 20-28(a). When oil-filled transformers are installed on or adjacent to combustible buildings or material, the building or material shall be safeguarded from possible fire originating in a transformer.

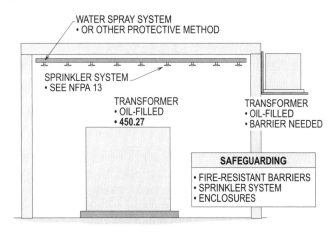

NEC 450.27

Figure 20-28(b). Fire-resistant barriers, water-spray systems, and enclosures for the transformers are approved safeguards, if, where used they are installed by the rules of the NEC.

LOCATION OF TRANSFORMER VAULTS
450.41

Vaults are used to house dry-type transformers that are rated over 35 kV or transformers filled with combustible material used as an aid in cooling their windings.

Vaults are required to be designed and built with specific rules and regulations.

Wherever possible, transformer vaults shall be located at an outside wall of the building. This rule is intended to allow ventilation direct to the outside without using ducts, flues, etc. per **450.45**.

WALLS, ROOF, AND FLOOR
450.42

The rules for construction of vaults are set forth in this section. Floor, walls, and roof shall be of fire-resistant material, such as concrete, and capable of withstanding heat from a fire within for at least three hours. A 6 in. thickness is specified for the walls and roof. The floor, when laid and in contact with the earth, shall be at least 4 in. thick. Walls, roof, and floor shall have at least have a 3 hour fire rating. **(See Figure 20-29)**

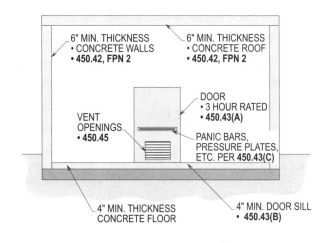

NEC 450.42

Figure 20-29. Floor, walls, and roof shall be of fire-resistant material, such as concrete, and capable of withstanding heat from a fire within for at least three hours. A 6 in. thickness is specified for the walls and roof. The floor, when laid and in contact with the earth, shall be at least 4 in. thick.

DOORWAYS
450.43

The door to a transformer vault shall be built according to the standards of the National Fire Protection Association which requires a 3 hour fire rating. The door sill shall be at least 4 in. high. This is to prevent any oil that may accumulate on the floor from running out of the transformer room and moving to other areas. Doors shall be kept locked at all times to prevent access of unqualified persons to the vault.

> **Design Tip:** Personnel doors shall swing out and be equipped with panic bars, pressure plates or other devices that open under simple pressure.

VENTILATION OPENINGS
450.45

Where ventilation is direct to the outside, without the use of ducts or flues, the vent opening shall have an area of at least 3 square inches for each kVA of transformer capacity, but never less than 1 square foot in area. The vent opening shall be fitted with a screen or grating and an automatic closing damper. If ducts are used in the vent system, the ducts shall have sufficient capacity to maintain a suitable vault temperature. **(See Figure 20-30)**

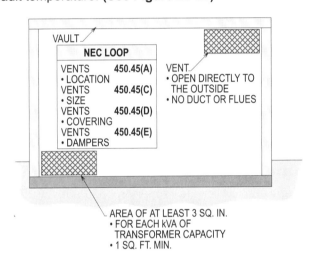

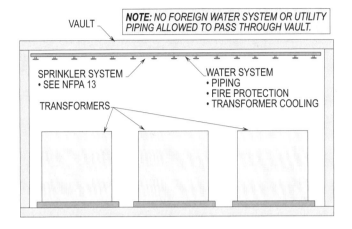

Figure 20-30. Where ventilation is direct to the outside, without the use of ducts or flues, the vent opening shall have an area of at least 3 sq. in. for each kVA of transformer capacity, but never less than 1 square foot in area.

DRAINAGE
450.46

Drains should be provided to drain off oil that might accumulate on the floor due to a leak in a transformer caused by an accident. This rule is designed to prevent a fire hazard from occurring.

WATER PIPES AND ACCESSORIES
450.47

Piping for fire protection within the vault or piping to water-cooled transformers may be present in a vault. No other piping or duct system shall enter or pass through. Valves or other fittings of a foreign piping or duct system are never permitted in a vault containing transformers. **(See Figure 20-31)**

Figure 20-31. Piping for fire protection within the vault or piping to water-cooled transformers may be present in a vault. No other piping or duct system shall enter or pass through.

STORAGE IN VAULTS
450.48

No storage of any kind may be put in a vault other than the transformers and equipment necessary for their operation. This typically means that transformer vaults are not to be used for warehouses or storage areas but to contain transformers and accessories only. The reason the vault is to be kept clear is due to high-voltage and safety measures needed for personnel serving such equipment. Also, consideration must be given to foreign material beings a threat of fire under certain conditions. **(See Figure 20-32)**

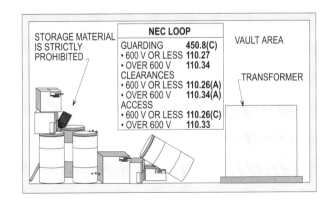

Figure 20-32. No storage of any kind may be put in a vault other than the transformers and equipment necessary for their operation.

CONNECTIONS FROM THE SECONDARY OF TRANSFORMERS 240.21(B); (C)

Overcurrent protection devices of circuits shall be located at the point where the service to those circuits originates. However, it is permitted to make connections from the secondary side of transformers. Such conductors shall be designed and installed by the rules and regulations of **240.21(B) and (C)** of the NEC. Sizing connections, not over 25 ft. long, shall be designed and installed per **240.21(B)(3) and (C)(5)**. Transformer secondary conductors of separately derived systems for industrial locations are sized per **240.21(C)(2), (C)(3) and (C)(6)**. Outside transformer connections are sized per **240.21(C)(4)**. Overcurrent protection shall be provided by **450.3(B)** and **Table 450.3(B)**. (See **249.92(B)** and **(D)** and **Figure 20-33** for illustrated diagrams)

NOT OVER 10 FT. LONG 240.21(C)(2)

Conductors shall be permitted to be connected, without overcurrent protection at the connection, to a feeder or transformer secondary where all the following conditions are met:

- Connecting conductors do not exceed 10 ft. in length.

- Connecting conductors shall have a current rating not less than the combined computed loads of the circuits supplied by connecting conductors. Their ampacity shall not be less than the rating of the overcurrent protection device at the termination of the connecting conductors.

- The connecting conductors shall not extend beyond the switchboard, panelboard, disconnecting means, or control devices they supply.

- Connecting conductors shall be enclosed in a raceway which will extend from the connection to the enclosure of an enclosed switchboard, panelboard, or control devices, or to the back of an open switchboard.

- The rating of the overcurrent device on the secondary side of the connecting conductors shall protect the conductors and the output of the transformer from overload.

If more than 10 percent of the single-pole slots of a panelboard rated at 30 amps or less is used for branch-circuits to lighting and appliance loads with neutral connections, the panelboard is classified as a lighting and appliance panelboard per **408.14 and 408.15**. When the panelboard is classified as a lighting and appliance panelboard, OCPD is restricted to two means or less.

See Figure 20-34 for the proper procedure for making a connection using the 10 ft. rule.

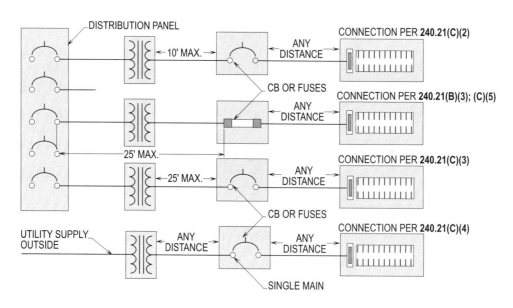

NEC 240.21(B); (C)

Figure 20-33. The illustration above shows the four most used transformer secondary connections that are utilized to supply electrical systems with transformer secondary conductors.

Transformers

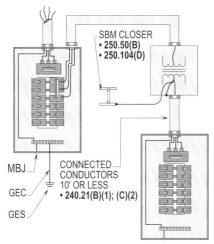

10' CONNECTING RULE
NEC 240.21(C)(2)

Figure 20-34. The above illustration shows the procedure for sizing a 10 ft. connection from the secondary of a transformer.

NOT OVER 25 FT. LONG
240.21(B)(3); (C)(5)

Conductors supplying a transformer shall be permitted to be tapped, without overcurrent protection at the tap, from a feeder where all the following conditions are met:
- Tapped conductors supplying the primary shall have an ampacity at least 1/3 of the rating of the feeder being tapped.
- Connecting conductors supplying the secondary shall have an ampacity at least 1/3 of the rating of the feeder being connected, based on the primary-to-secondary voltage ratio.
- The total length of one primary plus one secondary conductor shall not be over 25 ft.
- The primary and secondary conductors shall be protected from physical damage.
- Secondary conductors shall terminate in a single circuit breaker or set of fuses, sized to protect the secondary.

See Figure 20-35 for the proper procedure for making a tap and connection using the 25 ft. rule.

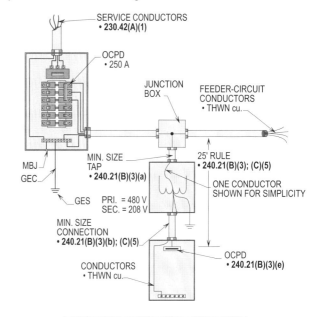

**25' TAP AND CONNECTION RULE
PRIMARY AND SECONDARY**
NEC 240.21(B)(3); (C)(5)

Figure 20-35. The primary tap for this connection rule shall be at least 1/3 of the OCPD device protecting the larger feeder-circuit conductors. The secondary connecting conductors shall be at least 1/3 of the OCPD protecting the feeder-circuit conductors based on the primary-secondary transformer ratio.

20-23

TRANSFORMER SECONDARY CONDUCTORS OF SEPARATELY DERIVED SYSTEMS FOR INDUSTRIAL LOCATIONS
240.21(C)(3)

Conductors shall be permitted to be connected to a transformer secondary of a separately derived system for industrial locations, without overcurrent protection at the connection, where all the following conditions are met:

- Secondary conductors shall not exceed 25 ft. in length.
- Ampacity of connected conductors shall be equivalent to current rating of the transformer and the overcurrent protection devices shall not exceed the ampacity of the connected conductors.
- All overcurrent devices are grouped.
- Connected conductors shall be protected from physical damage.

See Figure 20-36 for the procedure to be applied when a 25 ft. connection rule is installed from the secondary side of a transformer.

OUTSIDE TRANSFORMER CONNECTION
240.21(C)(4)

Outside conductors can be connected to a feeder or be connected at the transformer secondary without overcurrent protection at the connection. However, all of the following conditions shall be complied with:

- The connected conductors are suitably protected from physical damage.

- The conductors terminate at a single circuit breaker or a single set of fuses that will limit the load to the ampacity of the conductors. This single overcurrent protection device can supply any number of additional overcurrent devices of its load side.

- The connecting conductors are installed outdoors, except at the point of termination.

- The overcurrent protection device for the conductors is an integral part of a disconnecting means or shall be located immediately adjacent thereto.
- The disconnecting means for the conductors are installed at a readily accessible location either outside of a building or structure or inside, nearest the point of entrance of the conductors.

See Figure 20-37 for the rule pertaining to outside transformer connections from the secondary side of transformers.

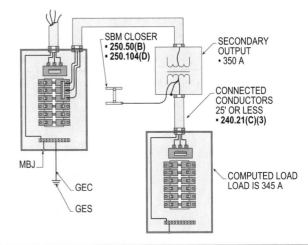

25' CONNECTION RULE
NEC 240.21(C)(3)

Figure 20-36. The above illustration shows the procedure for sizing a 25 ft. connection from the secondary of a transformer.

TRANSFORMER SECONDARY CONDUCTORS IN LENGTHS OF 10 FT. TO 25 FT.
240.21(C)(6)

Conductors over 10 ft. and up to 25 ft. in length can be connected to the secondary size of a transformer. When applying this section of the code, the 25 ft. (7.5 m) secondary connection shall be terminated in a single OCPD (CB or fuses) to limit the load and to also comply with the 1/3 rule when multiplied by the secondary-to-primary voltage ratio. The secondary conductors are required to be protected from physical damage and abuse. **(See Figure 20-38)**

Transformers

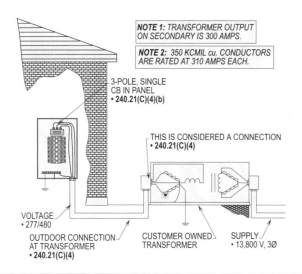

OUTSIDE CONNECTING RULES
NEC 240.21(C)(4)

Figure 20-37. The above illustration shows the rules for sizing the conductors and OCPD for a feeder connection from a transformer located outside.

SUPERVISED INDUSTRIAL INSTALLATIONS FEEDER AND BRANCH-CIRCUIT CONDUCTORS 240.92(A)

Feeder and branch-circuit conductors shall be protected at the point where the conductors receive their supply. However, this permits a variation of requirements for transformer secondary conductors taken from separately derived systems and outside feeder taps. **(See Figure 20-39)**

SUPERVISED INDUSTRIAL INSTALLATIONS AND CONNECTIONS UP TO 100 FT. 240.92(B)(1)(1); (2)

Unprotected lengths of secondary conductors are allowed at up to 100 ft., if the transformer primary overcurrent is sized at a value (reflected to the secondary by the transformer phas voltage ratio) of not more than 150 percent of the secondary conductor ampacity.

Additionally, the conductors shall be protected by a differential relay with a trip setting equal to or less than the conductor ampacity. Note that a differential relay provides superior short-circuit protection at a trip open value that is almost always well below the conductor ampacity. **(See Figure 20-40)**

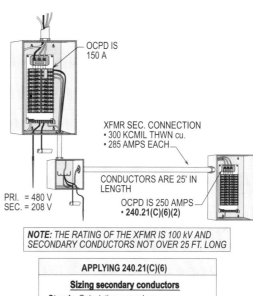

SECONDARY CONDUCTORS
NOT OVER 25' LONG
NEC 240.21(C)(6)

Figure 20-38. The above shows the rules for making a 25 ft. secondary conductor connection in other than industrial locations.

SHORT-CIRCUIT AND GROUND-FAULT PROTECTION 240.92(B)(1)(3)

Conductors up to 100 ft. in length are permitted, if calculations are made under engineering supervision and it is determined that the secondary conductors will be protected within recognized times versus current limits for all short-circuits and ground-fault conditions that could occur. **(See Figure 20-40)**

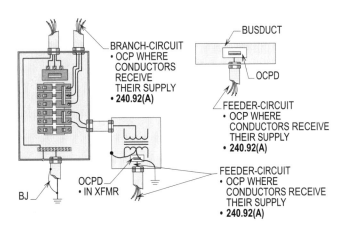

NEC 240.92(A)

Figure 20-39. The above illustration shows feeder and branch-circuit conductors protected at the point where the conductors receive their supply. (Also, See **240.21(C)(4)** for a similar rule).

OVERLOAD PROTECTION 240.92(B)(2)

To provide overload protection, the secondary conductors are permitted to be terminated in a single OCPD or in lugs of the bus, if not more than six OCPD's with a combined rating are installed which do not exceed the ampacity of the conductors. Another method of protection is to provide overload current relaying with the ability (design into) to trip either the primary OCPD's or the downstream OCPD's so that the load current doesn't exceed the conductor's ampacity. **(See Figure 20-41)**

Transformer Tip: In some cases, the short-circuit and ground-fault protective arrangements may provide overload protection. Note that if engineering calculations prove this to be the case, separate overload protection isn't really needed.

SUPERVISED INDUSTRIAL INSTALLATIONS OUTSIDE FEEDER TAPS 240.92(C)

Section **240.92(C)** allows alternate means of protecting transformer secondary conductors in supervised industrial installations where the transformer is located outside. The secondary conductors shall be protected against (1) overloads, with the additional stipulation that (2) they are suitably protected against physical damage. **(See Figures 20-42(a) and (b))**

Transformer Tip: Such protection can be provided by six or less OCPD's, where the total rating does not exceed the ampacity of the conductors routed per **240.92(C)**. Note that up to six OCPD's can be used instead of one OCPD at the feeder termination.

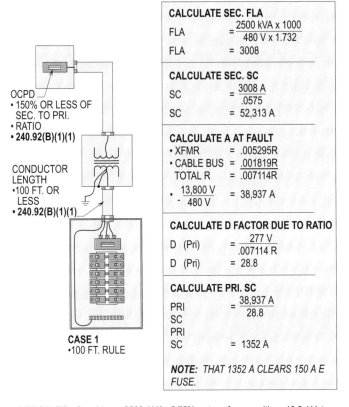

DESIGN TIP: Consider a 2500 kVA, 5.75% z transformer with a 13.8 kV to 480/277 V ratio and 500 mVA available short-circuit current on the primary for 100 circuit feet of 4000 amp cable bus, and a three-phase bolted fault at the end of the bus (worst case), about 38,937 amps will flow from the system, or 1352 amps on the primary, which will clear a typical 150E fuse (which meets the maximum 150 percent requirement) within .42 seconds. This time vs. current value is well within the rating of the secondary conductors. (Cable bus has a resistance of .001819 and the XFMR has a resistance of .005295.)

NEC 240.92(B)(1)(1); (2); (3)

Figure 20-40. The above illustration shows methods of providing short-circuit and ground-fault protection for transformers and conductors.

PROTECTION BY PRIMARY OVERCURRENT DEVICE 2240.92(D)

Conductors supplied by the secondary side of a transformer may be protected by overcurrent protection, provided on the primary (supply) side of the transformer, provided the primary device time-current protection characteristic, multiplied by the maximum effective primary-to-secondary transformer voltage ratio, effectively protects the secondary conductors. **(See Figure 20-43)**

> **Design Tip:** This section recognizes overcurrent protection installations where a device in series, with a transformer primary, is used to protect secondary conductors. the ratios given in **240.92(D)** are recognized in numerous industry standards and references, including IEEE Standard C37.91.

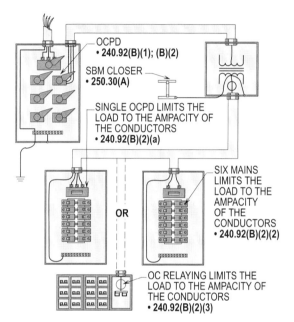

NEC 240.92(B)(2)

Figure 20-41. The above illustration shows methods of providing overload protection.

WIRING AND EQUIPMENT IN SPACES USED FOR ENVIRONMENTAL AIR 300.22(C)(1)

Only wiring methods of a specific type are permitted to be installed in ducts used to transport dust, loose stock, or flammable vapors. Wiring methods of any type are not permitted to be installed in any duct, or shaft containing only such ducts, used for vapor removal or ventilation of commercial type cooking equipment.

The following wiring methods are only permitted to be installed in ducts or plenums used for environmental air:

- MI cable
- MC cable
- AC cable
- Factory assembled multiconductor control or power cable listed for the use
- Listed prefabricated cable assemblies of metallic manufactured wiring systems without nonmetallic sheath
- Electrical metallic tubing
- Flexible metallic tubing
- Intermediate metal conduit
- Rigid metal conduit

Flexible metal conduit and liquidtight flexible metal conduit may be used in lengths not to exceed 4 ft. to connect equipment and devices permitted to be in these ducts and plenum chambers.

Flexible metal conduit and liquidtight flexible metal conduit in single lengths are permitted to be used in air-handling ceiling spaces where not exceeding 6 ft. Cables that have a fire-resistant and low-smoke characteristics may be used in air-handling ceiling spaces without conduit per **725.3(C), 760.3(B)** and **800.3(D)**.

See Figure 20-44 for other types of wiring methods permitted in other types of spaces used for environmental air.

> **Design Tip:** Electrical equipment with metal enclosures or nonmetallic enclosures that are listed are permitted to be installed in other spaces used for environmental air per **300.22(C)(2)**. A transformer that is totally enclosed may be installed to meet this requirement per **450.13(B)** if approved for such use.

Stallcup's Electrical Design Book

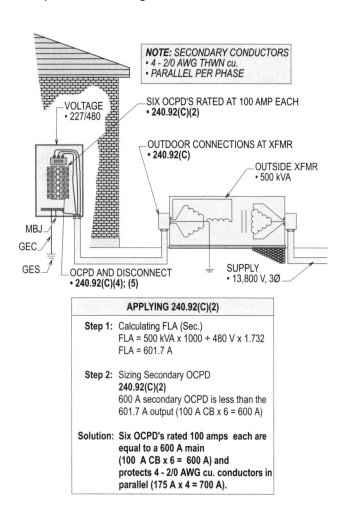

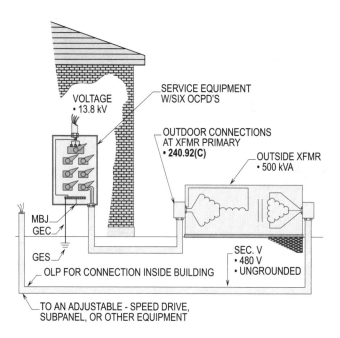

Figure 20-42(a). The above illustration shows alternate means allowed for protecting conductors tapped to a transformer located outside.

Figure 20-42(b). The above illustration shows alternate means allowed for protecting conductors connected to a transformer located outside.

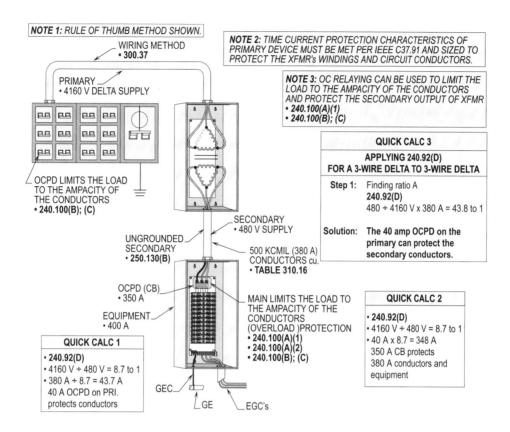

Figure 20-43. The primary protection of a transformer can be used to protect the secondary conductors, provided the primary device time-current protection characteristic, multiplied by the maximum effective primary to secondary voltage ratio will effectively protect the secondary conductors.

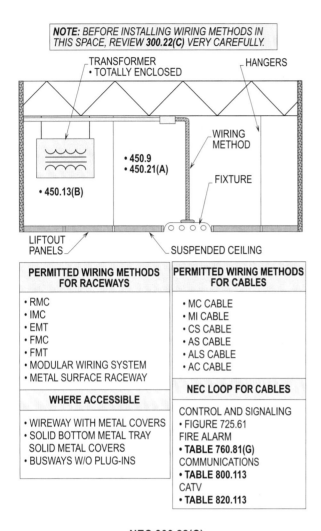

Figure 20-44. The above illustration shows the permitted wiring methods in spaces used for environmental air.

Name

Date

Chapter 20. TRANSFORMERS

Section   Answer

_____   T   F   1. Lighting transformers for a closed delta-connected secondary system can be found by multiplying the three-phase load by 67 percent.

_____   T   F   2. The lighting transformer for an open delta-connected secondary system can be found by multiplying the three-phase load by 58 percent and adding this value to the single-phase load.

_____   T   F   3. Transformers are not required to be readily accessible. (General Rule)

_____   T   F   4. A transformer 600 volts or less, nominal, having an individual OCPD on the primary side shall be sized at not more than 125 percent of the transformer's full-load current rating, in amps. (General Rule)

_____   T   F   5. Combination protection can be provided for both the primary and secondary sides of a transformer rated 600 volts or less.

_____   T   F   6. An overcurrent protection sensing device shall be designed to trip at 250 percent of its continuous current per phase or neutral rating.

_____   T   F   7. Transformers are permitted to be connected in parallel and switched as a unit provided each transformer has overcurrent protection.

_____   T   F   8. Transformers enclosed by fences or guards are not required to be grounded.

_____   T   F   9. Dry-type transformers greater than 112 1/2 kVA rating, are not required to be installed in a fire-resistant transformer room. (General Rule)

_____   T   F   10. Askarel transformers of more than 25,000 volts must be installed in a vault.

_____   T   F   11. Walls, roof, and floor for a transformer vault must at least have a 3 hour fire resistance rating.

_____   T   F   12. Doors to transformer vaults must be kept locked at all times to prevent access of unqualified person to the vault.

_____   T   F   13. Foreign piping or duct systems are permitted to be installed in a transformer vault.

_____   T   F   14. Storage of material is allowed in a transformer vault.

_____   T   F   15. Flexible metal conduit in single lengths are permitted to be used in air-handling spaces where not exceeding 6 ft.

_____ _____ 16. A closed delta-connected secondary system can be found by multiplying the single-phase load by _____ percent.

_____ _____ 17. Dry-type transformers not over 600 volts, which are located on open walls or steel columns, do not have to be _____ accessible.

20-31

Section	Answer

18. If installing fuses, the individual OCPD for the primary side of a transformer rated over 600 volts shall be rated not greater than _____ percent of the rated primary current of the transformer. (Supervised Location)

19. A transformer in a nonsupervised location with a secondary voltage rated 600 volts or less, the OCPD and conductors on the secondary side must be sized at _____ percent of the FLC rating. (Any Location)

20. A transformer 600 volts or less, nominal, where the rated secondary current of a transformer is less than 9 amps, an overcurrent protection device rated or set at no more than _____ percent of secondary current may be used.

21. Autotransformers shall have a _____ neutral current rating sufficient for the specified ground-fault current that could develop in the system.

22. Where transformers are tied together in parallel and connected by tie conductors that do not have overcurrent protection, the ampacity of the ties connecting conductors must not be less than _____ percent of the rated secondary current of the largest transformer in the tie circuit.

23. When secondary ties from transformers are used, an overcurrent protection device in the secondary of each transformer that is rated or set at not greater than _____ percent of the rated secondary current of the transformer must be provided.

24. Transformers shall be elevated at least _____ ft. above the floor or working space to prevent unauthorized personnel from contact.

25. Dry-type transformers greater than 112 1/2 kVA, having Class 155 or higher insulation, must be separated at least _____ ft. horizontally and _____ ft. vertically from the combustible material, if no fire-resistant heat-insulating barrier is provided.

26. All indoor dry-type transformers of over _____ volts are required to be installed in a vault.

27. Dry-type transformers installed outdoors must have _____ enclosures.

28. The walls and roof for a transformer vault must have a _____ in. thickness.

29. The door sills for a transformer vault must be at least _____ in. high.

30. Connecting primary and secondary conductors not over 25 ft. long (taps supplying XFMR) shall have an ampacity of at least _____ of the rating of the feeder's OCPD, based on the primary-to-secondary voltage ratio.

31. The lighting transformer for an open delta-connected secondary system can be determined by multiplying the single-phase load by _____ plus 58 percent of the three-phase load.
    (a)   58 percent
    (b)   67 percent
    (c)   100 percent
    (d)   125 percent

	Section	Answer

32. When you have a transformer with 6 percent impedance, in a supervised location with a secondary voltage rated 600 volts or less, the OCPD on the secondary side must be sized at not more than _____ percent of the FLC rating.
    (a) 100
    (b) 125
    (c) 225
    (d) 250

33. Dry-type transformers rated 112 1/2 kVA or less, must be separated at least _____ in. from the combustible material where the voltage is 600 volts or less.
    (a) 6
    (b) 12
    (c) 18
    (d) 24

34. Askarel-insulated transformers of over _____ kVA must be furnished with a relief vent such as chimney.
    (a) 25
    (b) 35
    (c) 50
    (d) 75

35. The floor for a transformer vault must be at least _____ in. thick.
    (a) 2
    (b) 4
    (c) 6
    (d) 12

36. What is the primary and secondary amperage for a 20 kVA, 480/240 volt, single-phase transformer?

37. What is the primary and secondary amperage for a 20 kVA, 480/240 volt, three-phase transformer?

38. What is the individual overcurrent protection device rating (using CB's and TDF's) for the primary side of a transformer with the following:
    • Supervised location
    • 1500 kVA transformer
    • 12,470 volts
    • Three-phase

39. What is the overcurrent protection device rating for the primary and secondary side of a transformer with the following:
    • 400 kVA transformer
    • 4160/480 volts
    • Three-wire to four-wire
    • Three-phase
    • Nonsupervised location (any location)

20-33

Section	Answer

_____ _____

40. What is the overcurrent protection device rating for the primary and secondary side of transformer with the following:
- 400 kVA transformer
- 4160/480 volts
- Three-wire to four-wire
- Three-phase
- Supervised location

_____ _____

41. What is the overcurrent protection device rating for the primary and secondary side of transformer with the following:
- 400 kVA transformer
- 13,800/4160 volts
- Three-wire to four-wire
- Three-phase
- Supervised location

_____ _____

42. What is the individual overcurrent protection device rating for the primary side of a transformer with the following:
- 25 kVA transformer
- 240/120 volts
- Single-phase

_____ _____

43. What is the individual overcurrent protection device rating for the primary side of a transformer with the following:
- 2 kVA transformer
- 480 volts
- two-wire to two-wire
- Single-phase

_____ _____

44. What is the minimum and maximum individual overcurrent protection device rating for the primary side of a transformer with the following:
- .7 kVA transformer
- 480 volts
- two-wire to two-wire
- Single-phase

_____ _____

45. What is the overcurrent protection device rating for the primary and secondary side of a transformer with the following:
- 40 kVA transformer
- 480/208 volts
- three-wire to four-wire
- Three-phase

_____ _____

46. What size overcurrent protection device and THWN copper conductors are required for a 10 ft. tap with a computed load of 142 amps?

_____ _____

47. What size overcurrent protection device (secondary) and THWN copper conductors (primary and secondary) are required for a 25 ft. tap with the following:
- 200 amp OCPD (primary)
- 480 volt primary
- 240 volt secondary

20-34

21

Hazardous (Classified) Locations

The intent of this chapter is to assist in the classification of areas or locations with respect to hazardous conditions. Such areas or locations are hazardous due to atmospheric concentrations of hazardous gases, vapors, deposits, or accumulations of materials which may be readily ignited.

The requirements are covered for the installation of electrical equipment and wiring in locations that are classified depending on the properties of the flammable vapors, liquids or gases, or combustible dusts which may be present and the likelihood that a flammable or combustible concentration of quantity is present. The hazardous (classified) locations to be covered are assigned the designations as follows:

(1) Class I, Division 1
(2) Class I, Division 2
(3) Class II, Division 1
(4) Class II, Division 2
(5) Class III, Division 1
(6) Class III, Division 2

(For the metric system dimensions, see those listed in the NEC)

GENERAL OVERVIEW

Hazardous areas and locations are classified by group, class, and division. They are determined by the atmospheric mixtures of various gases, vapors, dust, and other materials. The intensity of the explosion that can occur depends upon the concentrations, temperatures, and many other factors which are listed in NFPA codes.

Note that all areas designated as hazardous (classified) locations must be properly documented. Such documentation must be available to those authorized to design, install, inspect, maintain, or operate electrical equipment in the location involved per **500.4(A)**.

DIVISIONS (GASES)
500.5(B)

Class I locations are identified in the NEC as those in which flammable gases or vapors are or may be present in amounts sufficient to create explosive or ignitable mixtures. The amount of vapor varies all the way from being continuously present or never present at all. Naturally, if vapors are not present, the area is not a classified hazardous location. However, where vapors are present, it is very likely that a flammable mixture may or may not be present.

From a designing standpoint, greater care must be exercised if a particular condition is likely to occur, such as the presence of a flammable mixture of vapor and air within the explosive range, than are needed if a flammable mixture within the explosive range is unlikely to occur. Therefore, this is the reason why it is necessary to divide hazardous locations into two divisions.

DIVISION 1 (GASES)
500.5(B)(1)

Division 1 hazardous locations are defined in the NEC as those locations (a) in which ignitable concentrations of flammable gases or vapors can exist under normal operating conditions, or (b) in which ignitable concentrations of such gases or vapors may exist frequently because of repair or maintenance operations or because of leakage, or (c) breakdown or faulty operation of equipment or processes might release ignitable concentrations of flammable gases or vapors and might also cause simultaneous failure of electric equipment. Note that in each case, ignitable concentrations are mentioned. This means concentrations between the lower and upper flammable or explosive limits.

There are minimum and maximum concentration of flammable gases and vapors which are above and below the mixture and will not burn or explode. The lower limit is where the substance is too lean to burn or explode. The upper limit is where the mixture has become too rich to burn or explode.

FPN 1 to **500.5(B)(1)** of the NEC describes a number of areas and occupancies that are normally classified as Division 1 locations. Such locations are as follows:

(1) Petroleum refining facilities
(2) Dip tanks containing flammable or combustible liquids
(3) Dry cleaning plants
(4) Plants manufacturing organic coatings
(5) Spray finishing areas (residue must be considered)
(6) Petroleum dispensing areas
(7) Solvent extraction plants
(8) Plants manufacturing or using pyroxylin (nitrocellulose) type and other plastics
(9) Locations where inhalation anesthetics are utilized
(10) Utility gas plants and operations involving storage and handling of liquefied petroleum and natural gas plants, and
(11) Aircraft hangars and fuel servicing areas

See Figure 21-1 for a detailed illustration of Class I, Division 1 locations.

Design Tip: NEC Article 100 defines a flammable liquid as having a flash point below 38°C (100°F) or a flammable liquid whose temperature is raised above its flash point. Flash point is the minimum temperature to which a combustible or flammable liquid is heated until sufficient vapors are driven off and will flash when brought into contact with a flame, arc, spark, etc. (See Sec. 1-3 of NFPA 497 for more details.)

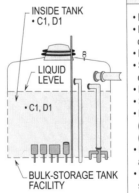

NEC 500.5(B)(1), FPN

Figure 21-1. The facilities that are listed in the above illustration are considered, under certain conditions, to be Class I, Division 1.

DIVISION 2 (GASES)
500.5(B)(2)

The NEC defines Division 2 locations as those locations (a) in which volatile flammable liquids or flammable gases are handled, processed, or used, but will normally be confined within closed containers or closed systems from which they can escape only in case of accidental rupture or breakdown of such containers or systems or in case of abnormal operation of equipment, or (b) in which ignitable concentrations of gases or vapors are normally prevented by positive mechanical ventilation and which might become hazardous through failure or abnormal operation of the ventilating equipment or (c) that are adjacent to a Class I, Division 1 location, and to which ignitable concentrations of gases or vapors might occasionally be communicated unless such communication is prevented by adequate positive-pressure ventilation from a source of clean air and effective safeguards against ventilation failure are provided. FPN's 1 and 2 to **500.7(B)(2)** of the NEC describes a number of areas and occupancies that are normally classified as Division 2 locations. For example, piping without valves, checks, meters, and devices do not usually introduce a hazardous condition even though they are utilized for flammable liquids. They are considered a contained system. Therefore, the area around may be classified as a Division 2 location by the AHJ. The following areas are considered to be Division 2 locations:

(1) Locations where volatile flammable liquids or flammable gases or vapors are used.

(2) Locations where, in the judgment of the AHJ, would become hazardous only in case of an accident or of some unusual operating condition.

(3) Locations where the quantity of flammable material that might escape in case of accident, the adequacy of ventilating equipment, the total area involved, and the record of the industry or business with respect to explosions or fires are all factors that merit consideration in determining the classification and extent of each location.

(4) Locations where piping without valves, checks, meter, and similar devices would not ordinarily introduce a hazardous condition even though used for flammable liquids or gases.

(5) Locations used for the storage of flammable liquids or of liquefied or compressed gases in sealed containers would not normally be considered hazardous unless subject to other hazardous conditions also.

(6) Electrical conduits and their associated enclosures separated from process fluids by a single seal or barrier shall be classified as a Division 2 location if the outside of the conduit and enclosure is an unclassified location.

See **Figure 21-2** for an illustration of Class I, Division 2 locations.

NEC 500.5(B)(2), FPN's 1; 2

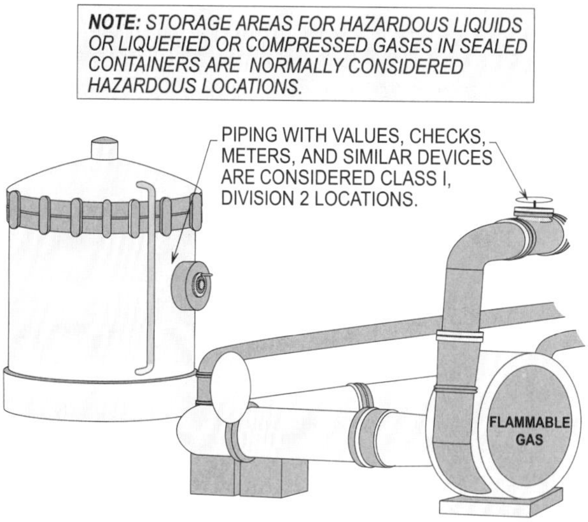

Figure 21-2. Areas in refineries where vessels and pipelines are equipped with valves, meters, and similar devices are usually considered Class I, Division 2 locations.

DIVISION 1 (DUST)
500.5(C)(1)

A Class II, Division 1 location is a location (1) in which combustible dust is in the air under normal operating conditions in quantities sufficient to produce explosive or ignitable mixtures, or (2) where mechanical failure or abnormal operation of machinery or equipment might cause such explosive or ignitable mixtures to be produced, and might also provide a source of ignition through simultaneous failure of electric equipment, operation of protection devices, or from other causes, or (3) in which combustible dusts of an electrically conductive nature may be present in hazardous quantities.

Areas that are considered Class II, Division 1 locations are based upon certain designing techniques and conditions as follows:

(1) Combustible dusts which are electrically nonconductive include dusts produced in the handling and processing of:
  (a) Grain and grain products,
  (b) Pulverized sugar and cocoa,

(c) Dried egg and milk powders,
(d) Pulverized spices, starch and pastes,
(e) Potato and wood flour, oil meal from beans,
(f) Seed, dried hay, and other organic materials that may produce combustible dusts when processed or handled.

(2) Only Group E dusts are considered to be electrically conductive for classification purposes. Dusts containing magnesium or aluminum are particularly hazardous and the use of extreme precaution will be necessary to avoid ignition and explosion.

See Figure 21-3 for a detailed illustration of Class II, Division 1 locations.

AREAS CONSIDERED TO BE CLASS II, DIVISION 1 LOCATIONS

- Grain elevators and bulk handling facilities,
- Manufacture and storage of magnesium,
- Manufacture and storage of starch,
- Fireworks manufacture and storage,
- Flour and feed mills,
- Areas for packaging and handling of pulverized sugar and cocoa,
- Facilities for the manufacture of magnesium and aluminum powder,
- Some coal preparation plants and coal handling facilities,
- Spice grinding plants, and
- Confectionery manufacturing plants.

NEC 500.5(C)(1)

Figure 21-3. The above are facilities which are considered under certain conditions to be Class II, Division 1 locations.

DIVISION 2 (DUST) 500.5(C)(2)

A Class II, Division 2 location is a location where combustible dust is not normally in the air in quantities sufficient to produce explosive or ignitable mixtures and dust accumulations are normally insufficient to interfere with the normal operation of electrical equipment or other apparatus, but combustible dust may be in suspension in the air as a result of infrequent malfunctioning of handling or processing equipment and where combustible dust accumulations on, in, or in the vicinity of the electrical equipment may be sufficient to interfere with the safe dissipation of heat from electrical equipment or may be ignitable by abnormal operation or failure of electrical equipment.

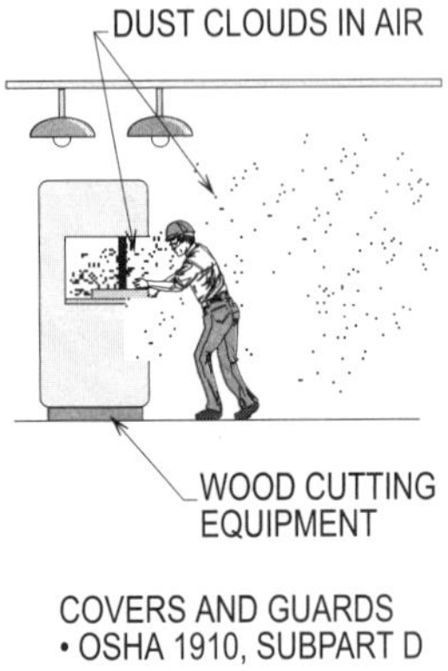

Areas that are considered Class II, Division 2 locations are based upon certain designing techniques and conditions as follows:

(1) The quantity of combustible dust that may be present and the adequacy of dust removal systems are factors that merit consideration in determining the classification and may result in an unclassified area.

(2) Where products such as seed are handled in a manner that produces low quantities of dust, the amount of dust deposited may not warrant classification.

See Figure 21-4 for a detailed illustration of Class II, Division 2 locations.

AREAS CONSIDERED TO BE CLASS II, DIVISION 2 LOCATIONS

- Rooms containing closed spouts and conveyors, closed bins, or hoppers.
- Machines and equipment from which quantities of dust could escape under abnormal operating conditions,
- Rooms adjacent to Class II areas,
- Rooms in which control over suspension of dust in explosive or ignitible quantities is present,
- Warehouses and shipping rooms where dust-producing materials are stored, and
- Any other similar locations.

NEC 500.5(C)(2)

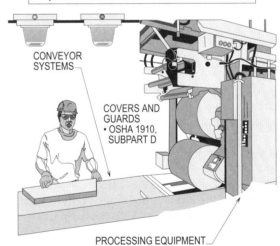

Figure 21-4. The above are facilities which are considered under certain conditions to be Class II, Division 2 locations.

DIVISION 1 (FIBERS) 500.5(D)(1)

A Class III, Division 1 location is a location in which easily ignitable fibers or materials producing combustible flyings are handled, manufactured, or used.

(1) Such locations usually include some parts of:
   (a) rayon, cotton, and other textile mills;
   (b) combustible fiber manufacturing and processing plants;
   (c) cotton gins and cottonseed mills;
   (d) flax-processing plants;
   (e) clothing manufacturing plants;
   (f) woodworking plants, and
   (g) establishments and industries involving similar hazardous processes or conditions.

(2) Easily ignitable fibers and flyings include:
   (a) rayon,
   (b) cotton (including cotton liners and cotton waste,)
   (c) sisal or henequen,
   (d) istle,
   (e) jute, hemp, tow, cocoa fiber, oakum,
   (f) baled waste kapok,
   (g) spanish moss, excelsior, and
   (h) other materials of similar nature.

See **Figure 21-5(a)** for a detailed illustration of Class III, Division 1 locations.

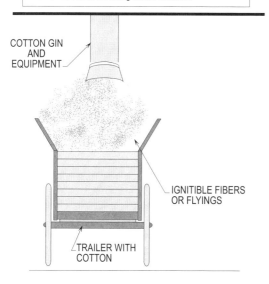

Figure 21-5(a). Listed above are facilities which are considered, under certain conditions, to be Class III, Division 1 locations.

DIVISION 2 (FIBERS)
500.5(D)(2)

A Class III, Division 2 location is a location in which easily ignitable fibers are stored or handled except in process of manufacturing. **(See Figure 21-5(b))**

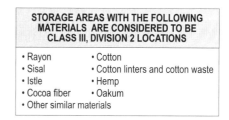

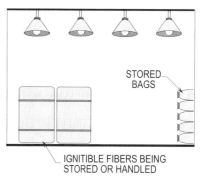

NEC 500.5(D)(2)

Figure 21-5(b). Listed above are facilities which are considered under certain conditions to be Class III, Division 2 locations.

GROUPS (GASES)
500.6(A)

Until publication of the 1937 edition of the NEC, Class I hazardous locations were not divided into groups. All flammable gases and vapors were classified as a single degree of hazard. It was recognized, however, that the degrees of hazard varied and that equipment suitable only for use where gasoline was handled was not necessarily suitable for use where hydrogen or acetylene was handled. It was also recognized that manufacturing equipment and enclosures for use in hydrogen atmospheres was very difficult and that the equipment, even if built, was expensive. It was not logical, from a designing standpoint, to require explosionproof equipment in gasoline filling stations that was also suitable for use in hydrogen atmospheres. Not only would this unnecessarily increase the cost of the electrical installation in one of the most common types of hazardous locations, but it would make some types of equipment unavailable.

By placing flammable materials into groups and classifying

them based upon the explosive characteristics of such gases and vapors, electrical equipment and wiring methods could be selected.

GROUP A (GASES)
500.6(A)(1)

As mentioned above, combustible and flammable gases and vapors are divided into four groups; the classification involving determinations of maximum explosion pressures, and maximum safe clearance between parts of a clamped or threaded joint in an enclosure.

Really, there is no consistent relationship between the Groups A, B, C or D classification and flash point/ignition temperature/explosive limits. Instead, the Groups are classified by chemical families. Certain chemicals produce higher explosive pressures and heat when ignited. Group A gas produces the greater pressures during an explosion—and therefore is the most difficult to handle and control. Group B is next highest in pressure; then Group C; and lastly Group D.

> **Design Tip:** This is the reason why a Group A or B listing is more difficult to get than a Group C or D listing for electrical equipment. **(See Figure 21-6)**

For example, Group A is an atmosphere containing acetylene, which is capable of producing a very dangerous explosive mixture. Note, equipment falling in this group is not available. Therefore, special equipment such as purged and pressurized must be designed and used.

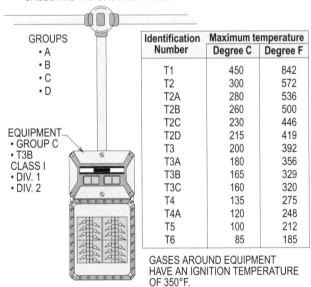

Figure 21-6. The above equipment is required to have a T3B (T-Code) identification number when installed in a Class I, Division 1 location.

GROUP B (GASES)
500.6(A)(2)

Group B is an atmosphere containing hydrogen, fuel and combustible process gases containing more than 30 percent hydrogen by volume, or it contains gases or vapors of equivalent hazard such as butadiene, ethylene oxide, propylene oxide, and acrolein.

GROUP C (GASES)
500.6(A)(3)

Group C is an atmosphere such as ethyl ether, ethylene, or gases or vapors of equivalent hazard. Manufacturers make all types of equipment that fall in this group of explosive equipment. (See **500.7(A)**)

GROUP D (GASES)
500.6(A)(4)

Group D is an atmosphere such as acetone, ammonia, benzene, butane, cyclopropane, ethanol, gasoline, hexane, methanol, methane, natural gas, naphtha, propane, or gases or vapors of equivalent hazard. There are many types of equipment available in this group.

GROUPS (DUST)
500.6(B)

Combustible dusts are divided into three Groups, the classification involving the tightness of the joints of assembly and shaft openings to prevent entrance of dust in the dust-ignition proof enclosure, the blanketing effect of layers of dust on the equipment that may cause overheating and the ignition temperature of the dust are the main concerns.

GROUP E (DUST)
500.6(B)(1)

Group E atmospheres contain combustible metal dusts, including aluminum, magnesium, and their commercial alloys, or other combustible dusts of similar hazard.

> **Design Tip 1:** "Only Group E dusts are considered to be electrically conductive." These dusts are metal dusts, such as aluminum, magnesium and their commercial alloys or other dusts of small particle size, abrasiveness and/or electrical conductivity as to present a similar hazard.

Design Tip 2: When the dust is electrically conductive, care must be taken for these dusts may ignite from bridging the gap between energized terminals, from arcs or from failure of equipment. Where Group E dusts are encountered in hazardous quantities, only Class II, Division 1 electrical equipment can be utilized. There is no such classification as Class II, Group E, Division 2. Either the location has enough electrically conductive dusts to make it a Division 1 location, or there is not enough dust present to even classify it as a hazardous area.

GROUP F (DUST)
500.6(B)(2)

Group F atmospheres contain combustible carbonaceous dusts, including carbon black, charcoal, coal or coke dusts that has more than 8 percent total entrapped volatiles or dusts that have been sensitized by other materials.

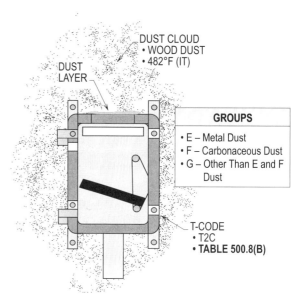

Figure 21-7. The above equipment is required to have a T2C identification number per Figure 21-6.

GROUP G
500.6(B)(3)

Group G atmospheres contain combustible dusts not included in Group E or F, including flour, grain, wood, plastic and chemicals.

Equipment to be used in these atmospheres must not only be approved for Class I, II, or III, but also for the specific group (Class I (A, B, C, or D) and II (E, F, or G)).

Design Tip: The explosion characteristics of air mixtures of dust vary with the materials involved. For Class II locations, Groups E, F, and G, the classification involves the tightness of the joints of assembly and shaft openings to prevent the entrance of dust in the dust-ignition proof enclosure. The blanketing effect of layers of dust on the equipment that may cause overheating and the ignition temperature of the dust is also considered. It is necessary, therefore, that equipment be approved not only for the class, but also for the specific group of dust that will be present.

See Figure 21-7 for a detailed illustration of Groups E, F, and G equipment.

CLASSES AND DIVISIONS
500.5

Classes are utilized to identify hazardous locations which are subjected to materials that can be explosive if mixed with air to create an ignitable mixture.

Hazardous locations must be well understood by anyone designing, installing, working on, or inspecting electrical equipment and wiring methods located in such areas. These locations are dangerous due to the threat of flammable or combustible gases, vapors, or dusts being present some of the time or all of the time.

CLASS I, DIVISION 1
500.5(B)(1)

A location where (a) ignitable concentrations of flammable gases or vapors exist under normal operating conditions, or (b) ignitable concentrations of such gases or vapors may exist frequently because of repair or maintenance operations or because of leakage, or (c) breakdown or faulty operation of equipment or processes might release ignitable concentrations of flammable gases or vapors and might also cause simultaneous failure of electrical equipment. **(See Figure 21-8(a))**

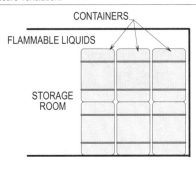

NEC 500.5(B)(2)

Figure 21-8(b). The above is a Class I, Division 2 location due to being confined in containers or closed systems.

NEC 500.5(B)(1)

Figure 21-8(a). The above illustration is a Class I, Division 1 locations due to the presence of gases or vapors during operation, maintenance, or repair.

CLASS I, DIVISION 2
500.5(B)(2)

A location where (a) volatile flammable liquids or flammable gases are handled, processed, or used but will normally be confined within closed containers or closed systems from which they can escape only in case of accidental rupture or breakdown of such containers or systems or in case of abnormal operation of equipment, or (b) where ignitable concentrations of gases or vapors are normally prevented by positive mechanical ventilation and might become hazardous through failure or abnormal operation of the ventilating equipment, or (c) is adjacent to a Class I, Division 1 locations and to which ignitable concentrations of gases or vapors might occasionally be communicated unless such communication is prevented by adequate positive-pressure ventilation from a source of clean air and effective safeguards against ventilation failure are provided. **(See Figure 21-8(b))**

CLASS II, DIVISION 1
500.5(C)(1)

A location where (a) combustible dust is in the air under normal operating conditions in quantities sufficient to produce explosive or ignitable mixtures, or (b) where mechanical failure or abnormal operation of machinery or equipment might cause such explosive or ignitable mixtures to be produced and might also provide a source of ignition through simultaneous failure of electrical equipment, operation of protective devices, or other causes, or (c) where combustible dusts of an electrically conductive nature may be present. **(See Figure 21-9(a))**

CLASS II, DIVISION 2
500.5(C)(2)

A location where combustible dust is not normally in the air in quantities sufficient to produce explosive or ignitable mixtures, and dust accumulations are normally insufficient to interfere with the normal operation of electrical equipment or other apparatus. However, combustible dust that may be in suspension in the air as a result of infrequent malfunctioning of handling or processing equipment and combustible dust accumulations on, in, or in the vicinity of

Hazardous (Classified) Locations

the electrical equipment may be sufficient to interfere with the safe dissipation of heat from electrical equipment or may be ignitable by abnormal operation or failure of electrical equipment. **(See Figure 21-9(b))**

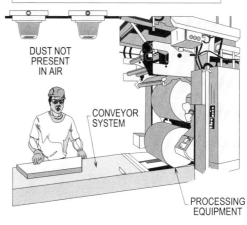

NEC 500.5(C)(2)

Figure 21-9(b). The above is a Class II, Division 2 location due to dust being limited during operation of equipment.

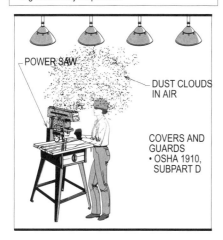

NEC 500.5(C)(1)

Figure 21-9(a). The above is a Class II, Division 1 location due to the presence of dust during operation or mechanical failure of equipment.

CLASS III, DIVISION 1
500.5(D)(1)

A Class III, Division 1 location is a location in which easily ignitable fibers or materials producing combustible flyings are handled, manufactured, or used. **(See Figure 21-10(a))**

CLASS III, DIVISION 2
500.5(D)(2)

A Class III, Division 2 location is a location in which easily ignitable fibers are stored or handled except in process of manufacturer facilities. **(See Figure 21-10(b)**

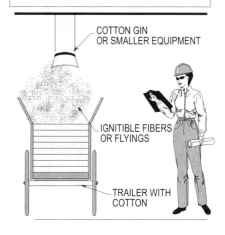

NEC 500.5(D)(1)

Figure 21-10(a). The above illustration is a Class III, Division 1 location due to the presence of ignitible fibers or combustible flyings.

21-9

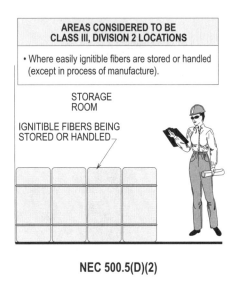

NEC 500.5(D)(2)

Figure 21-10(b). The above is a Class III, Division 2 location due to easily ignitible fibers being stored or handled.

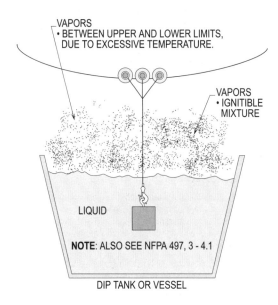

NFPA 497, CH. 1 - 3

Figure 21-11. The above shows a situation where sufficient vapors are driven off by heat due to the surrounding temperature creating an explosive mixture of vapors and air.

EXPLOSIVE PROPERTIES OF GASES AND VAPORS
NFPA 497, CH. 1-3

To select the type of equipment allowed to be installed in Class I, Division 1 and 2 locations, it is necessary to understand the meaning of certain terms and how they are to be applied.

FLASH POINT
NFPA 497, CH. 1

Flash point of a liquid is the minimum temperature at which the liquid gives off sufficient vapor to form an ignitable mixture with the air near the surface of the liquid or within the vessel used. An ignitable mixture is within the flammable range (between upper and lower limits) that is capable of propagating flame away from the source of ignition when ignited. Some evaporation takes place below the flash point but not in sufficient quantities to form an ignitable mixture.

Design Tip: The flash point applies mostly to flammable and combustible liquids, although there are certain solids such as camphor and naphthalene that slowly evaporate or volatilize at ordinary room temperature, and liquids such as benzene that freeze at relatively high temperatures and therefore have flash points while in the solid state. **(See Figure 21-11)**

IGNITION TEMPERATURE
NFPA 497, CH. 1-3; CH. 2-2

Ignition temperature of a substance, whether solid, liquid, or gaseous, is the minimum temperature required to initiate or cause self-sustained combustion independently of the heating or heated element.

Ignition temperatures observed under one set of conditions may be changed substantially by a change of conditions. For this reason, ignition temperatures should be looked upon only as approximations. Some of the variables known to affect ignition temperatures are percentage composition of the vapor or gas-air mixture, shape and size of the space where the ignition occurs, rate and duration of heating, kind and temperature of the ignition source, catalytic or other effect of materials that may be present, and oxygen concentration. As there are many differences in ignition temperature test methods, such as size and shape of containers, method of heating and ignition source, ignition temperatures are affected by the test method. **(See Figure 21-12)** Also, see NFPA 497, 3-4.1 thru 3-4.6.

Design Tip: Ignition temperature of a substance, whether solid, liquid, or gaseous, is the minimum temperature required to initiate or cause self-sustained combustion in the absence of any source of ignition.

Hazardous (Classified) Locations

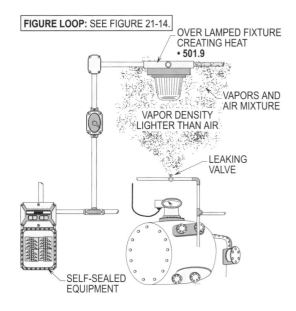

Figure 21-12. The above illustration shows a mixture of vapors and air which could explode due to the excessive heat from the surface temperature of the lighting fixture.

FLAMMABLE (EXPLOSIVE) LIMITS
API 500, DEFINITIONS

In the case of gases or vapors that form flammable mixtures with air or oxygen, there is a minimum concentration of vapor in air or oxygen below which propagation of flame does not occur on contact with a source of ignition. There is also a maximum proportion of vapor or gas in the air above in which propagation of flame does not occur. These boundary-line mixtures of vapor or gas with air, which if ignited will just propagate flame, are know as the "lower and upper flammable or explosive limits" and are usually expressed in terms of percentage by volume of gas or vapor in air. **(See Figure 21-13)**

Design Tip: In popular terms, a mixture below the lower flammable limit is too lean to burn or explode and a mixture above the upper flammable limit is too rich to burn or explode.

VAPOR DENSITY
API 500, DEFINITIONS

Vapor densities are used mainly to determine the settling or rising tendency of a mixture. A figure of less than 1.0 indicates that the vapor is lighter than air and will tend to rise. A figure greater than 1.0 indicates the vapor will settle

and move along the grade. (See NFPA 497, 2-2.1 and 2-2.2)

Design Tip: An example of a lighter-than-air gas is methane natural gas, which has a density of 0.6 and 0.1, respectively. This type of gas in indoor locations will tend to concentrate near the ceiling where lighting fixtures are usually mounted. On the other hand, heavier-than-air mixtures, like ethyl ether vapors, have the ability to travel at low levels for a considerable distance to locations where sources of ignition might be available.

See Figure 21-14 for an illustration showing the different vapor densities of flammable materials.

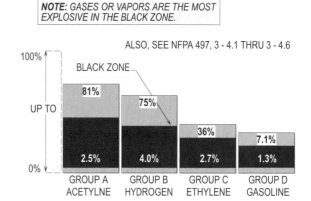

Figure 21-13. The above shows gases or vapors in their lower and upper limits.

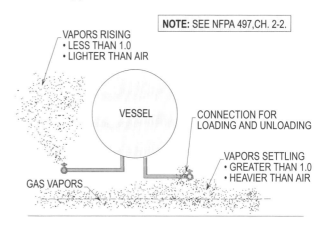

Figure 21-14. The above shows vapors lighter than air rise and vapors heavier than air will settle near the grade.

21-11

COMBUSTION PRINCIPLES
NFPA 497 - CH. 2-2

The following three basic conditions must be satisfied for a fire or explosion to occur:

(1) A flammable liquid, vapor, or combustible must be present in sufficient quantity.

(2) The flammable liquid, vapor, or combustible dust must be mixed with air or oxygen in the proportions required to produce an explosive mixture.

(3) A source of energy must be applied to the explosive mixture.

In applying these principles, the quantity of the flammable liquid or vapor that may be liberated and its physical characteristics shall be recognized. Vapors from flammable liquids have a natural tendency to disperse into the atmosphere and rapidly become diluted to concentrations below the lower explosion limit, particularly when there is natural or mechanical ventilation. Finally, the possibility that the gas concentration may be above the upper explosion limit does not ensure any degree of safety, as the concentration shall first pass through the explosive range to reach the upper explosion limit. (See NFPA 497, 2-2.6 and 2-2.7)

SOURCES OF IGNITION
NFPA 497 - CH. 2-3

A source of energy is all that is needed to touch off an explosion where flammable gases or combustible dust are mixed in the proper proportion with air. One prime source of energy is electricity. Equipment such as switches, circuit breakers, motor starters, push-button stations, or plugs and receptacles can produce arcs or sparks in normal operation when contacts are opened and closed, which could easily cause ignition.

Other hazards are devices that produce heat, such as lighting fixtures and motors. Here, surface temperatures may exceed the safe limits of many flammable atmospheres. Finally, many parts of the electrical system can become potential sources of ignition in the event of insulation failure. This group would include wiring (particularly splices in the wiring), transformers, impedance coils, solenoids, and other low-temperature devices without make-or-break contacts. Nonelectrical hazards such as sparking metal can also easily cause ignition. Thus a hammer, a file or another tool that is dropped on masonry or on a ferrous surface is a hazard unless the tool is made of non-sparking material. For this reason, portable electrical equipment is usually made from aluminum or other material that will not produce sparks if the equipment is dropped. Therefore, electrical safety is of crucial importance.

The electrical installation must prevent accidental ignition of flammable liquids, vapors, and dusts released into the atmosphere. In addition, since much of this equipment is used outdoors or in corrosive atmospheres, the material and finish must be such that maintenance costs and shutdowns are minimized. **(See Figure 21-15)**

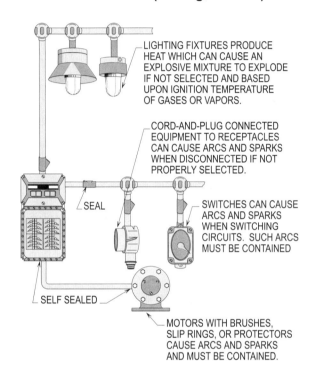

SOURCES OF IGNITION
NFPA 497, CH. 2-3

Figure 21-15. Ground joint construction type enclosures shown above will withstand an internal explosion and cool the hot escaping gases through the ground surface of the joint.

ENCLOSURES
NEMA 250

Each area that contains gases or dusts that are considered hazardous must be carefully evaluated to make certain the correct electrical equipment is selected. To conform with the NEC, the use of fittings and enclosures that are identified or listed for the specific hazardous gas or dust involved is required.

In Class I, Division 1 and 2 locations, conventional relays, contactors, and switches that have arcing contacts must be enclosed in explosionproof housings, except for those few cases where general-purpose enclosures are permitted by the NEC. By definition, enclosures for these locations must prevent the ignition of an explosive gas or vapor that may surround it. In other words, an explosion inside the enclosure must be prevented from starting a larger explosion on the outside. Adequate strength is one requirement for such an enclosure.

For explosionproof equipment, a safety factor of 4 is used. That is, the enclosure must withstand a hydrostatic pressure test of four times the maximum pressure from an explosion within the enclosure. In addition to being strong, the enclosure must be flame-tight. This term does not imply that the enclosure is hermetically sealed but rather that the joints cool the hot gases resulting from an internal explosion so that by the time they reach the outside hazardous atmosphere, they are too cool to effect ignition.

TYPES OF ENCLOSURES
NEMA 250

Type 7 and 10 enclosures, when properly installed and maintained, are designed to contain an internal explosion without causing an external hazard. Type 8 enclosures are designed to prevent combustion through the use of oil-immersed equipment. Type 9 enclosures are designed to prevent the ignition of combustible dust. Descriptions and tests in this standard publication cover equipment that is suitable for installation in locations classified as Division 1 or Division 2. In Division 2 locations, other types of protection and enclosures for nonhazardous locations may be installed if the equipment does not constitute a source of ignition under normal operating conditions.

Intrinsically safe and nonincendive equipment is not capable of releasing sufficient electrical or thermal energy under normal or abnormal conditions to cause ignition of specific hazardous atmospheres and may be installed in any type of enclosure, otherwise suitable for the environmental conditions expected.

Equipment installed in enclosures which are suitable for nonhazardous locations, including the hazardous area, may be reduced or eliminated by adequate positive pressure ventilation from a source of clean air in conjunction with effective safeguards against ventilation failure.

TYPE 7 ENCLOSURES
NEMA 250

Type 7 enclosures are for indoor use in locations classified as Class I, Groups A, B, C, or D, as defined in the NEC.

Type 7 enclosures must be capable of withstanding the pressures resulting from an internal explosion of specified gases and containing such an explosion sufficiently that an explosive gas-air mixture existing in the atmosphere surrounding the enclosure will not be ignited. Enclosed heat generating devices must not cause external surfaces to reach temperatures capable of igniting explosive gas-air mixtures in the surrounding atmosphere. Enclosures must meet explosion, hydrostatic, and temperature design tests.

When Type 7 enclosures are completely and properly installed, they will:

(1) Provide a degree of protection to a hazardous gas environment from an internal explosion or from operation of internal equipment,

(2) Not develop surface temperatures which exceed prescribed limits for the specific gas corresponding to the atmospheres for which the enclosure is intended, when internal equipment is operated at rated load,

(3) Withstand a series of internal explosion design tests which determine:
   (a) The maximum pressure effects of the gas mixture
   (b) Propagation effects of the gas mixtures, and

(4) Withstand, without rupture or permanent distortion, an internal hydrostatic design test based on the maximum internal pressure obtained during explosion tests and on a specified safety factor be marked with the appropriate class and group(s) for which they have been qualified.

TYPE 8 ENCLOSURES
NEMA 250

Type 8 enclosures are for indoor or outdoor use in locations classified as Class I, Groups A, B, C, and D, as defined in the NEC.

Type 8 enclosures and enclosed devices are arranged so that that all arcing contacts, connections, and any other parts that could cause arcing, are immersed in oil. Arcing is confined under the oil so that it will not ignite an explosive mixture of the specified gases in internal spaces above the oil or in the atmosphere surrounding the enclosure. Enclosed heat generating devices must not cause external surfaces to reach temperatures capable of igniting explosive gas-air mixtures in the surrounding atmosphere. Enclosures shall meet operation and temperature design tests. Enclosures intended for outdoor use must also meet the rain test.

When Type 8 enclosures are completely and properly installed, they will:

(1) Provide, by oil immersion, a degree of protection to a hazardous gas environment from operation of internal equipment,

(2) Not develop surface temperatures that exceed prescribed limits for the specific gas corresponding to the atmospheres for which the enclosure is intended when internal equipment is at rated load,

(3) Withstand a series of operation design tests with oil levels arbitrarily reduced and with flammable gas-air mixtures introduced above the oil,

(4) Exclude water when subjected to a water spray design test simulating a beating rain, if installed outdoors, and

(5) Be marked with the appropriate Class and Group(s) for which they have been qualified.

TYPE 9 ENCLOSURES
NEMA 250

Type 9 enclosures are intended for indoor use in locations classified as Class II, Group E or G, as defined in the NEC.

Type 9 enclosures must be capable of preventing the entrance of dust. Enclosed heat generating devices must not cause external surfaces to reach temperatures capable of igniting or discoloring dust on the enclosure or igniting dust-air mixtures in the surrounding atmosphere. Enclosures must meet dust penetration and temperature design tests and prevent aging of gaskets (if used).

When Type 9 enclosures are completely and properly installed, they will:

(1) Provide a degree of protection to a hazardous dust environment from operation of internal equipment,

(2) Not develop surface temperatures that exceed prescribed limits for the Group corresponding to the atmospheres for which the enclosure is intended when internal equipment is operated at rated load,

(3) Withstand a series of operation design tests while exposed to a circulating dust-air mixture to determine that dust does not enter the enclosure and that operation of devices does not cause ignition of surrounding atmosphere, and

(4) Be marked with the appropriate Class and Group(s) for which they have been qualified.

Note: For more information on enclosure types, see www.neca.org.

PREVENTION OF EXTERNAL IGNITION AND EXPLOSION

Explosives are hazardous enough by themselves but around electricity there are even more hazards present. For this reason, special enclosures are necessary to house electrical elements that are capable of producing arcs and sparks. An arc, spark, or hot surface can easily initiate an explosion. Therefore, these ignition sources must be contained or the equipment housing such apparatus must be installed out of the area.

Each area that contains gases or dusts that are considered hazardous must be carefully evaluated to make certain the correct electrical equipment is selected. Many hazardous atmospheres are Class I, Group D, or Class II, Group G. However, certain areas may involve other groups, particularly Class I, Groups B and C. Conformity with the NEC requires the use of fittings and enclosures that are identified or listed for the specific hazardous gas or dust involved. (See NEC **Articles 500 and 501** for more information.)

CLASS I, DIVISION 1 EQUIPMENT
500.2; 500.7; 500.8(A)(1)

Class I, Division 1 equipment must provide a form of construction that will ensure safe performance under conditions of proper use and maintenance and will operate in such a manner to prevent igniting an explosive mixture of gases and air. Note that Class I, division 1 equipment can be used in Class II, Division 2 locations. When approving equipment installed in classified areas, see **500.8(A)(1)** and Figure 21-60 in this chapter.

EXPLOSIONPROOF EQUIPMENT
(CLASS I, DIVISION 1)
500.2; 500.7(A)

The enclosure will withstand an internal explosion of gases or vapors and prevent those gases or vapors from igniting gases and vapors in the surrounding atmosphere outside of the enclosure.

These enclosures are not intended to exclude flammable or combustible gases or vapors, but rather to withstand an internal explosion and prevent the ignition of external gases and vapors. The internal gases are dissipated or cooled before they are released from the enclosure, thereby preventing ignition or an explosion of gases or vapors in the surrounding atmosphere. **(See Figure 21-16)**

Hazardous (Classified) Locations

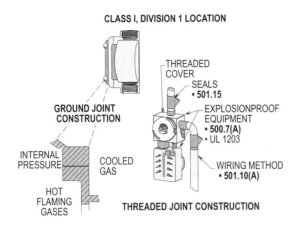

EXPLOSIONPROOF EQUIPMENT
NEC 500.2; 500.7(A)
NEC 500.5(B)(1), FPN 2
UL 1203

Figure 21-16. The above illustration shows an enclosure that will withstand an internal explosion of gases or vapors and prevent those gases or vapors from igniting gases and vapors in the surrounding atmosphere outside of the enclosure.

PURGED AND PRESSURIZED SYSTEMS
(CLASS I, DIVISION 1)
500.2; 500.7(D)

Purged and pressurized systems permit the safe operation of electrical equipment under conditions of hazard for which approved equipment may not be commercially available. For instance, most switchgear units and many large-size motors do not come in designs listed for Class I, Groups A and B. Whether cast metal enclosures for hazardous locations or sheet metal enclosures with pressurization should be used is mainly a question of economics, if both types are available.

As a typical example, if an installation had many electronic instruments that could be enclosed in a single sheet metal enclosure, the installation lends itself to the purging/pressurization system. However, if the instruments due to their nature had to be installed in separate enclosures, the cast metal in hazardous location housing would almost invariably prove more economical. Pressurized enclosures require:

(1) A source of clean air or inert gas,

(2) A compressor to maintain the required pressure on the system, and

(3) Pressure control valves, to prevent the power from being applied before the enclosure have been purged and to deenergize the system should pressure fall below a safe value.

In addition, door-interlock switches must prevent access to the equipment while the circuits are energized. It can readily be seen that all of these accessories can add up to a considerable expenditure.

> **Design Tip:** For a detailed description of purged and pressurized systems see NFPA 496-1896 "Purged and Pressurized Enclosures for Electrical Equipment in Hazardous Locations". **(See Figure 21-17)**

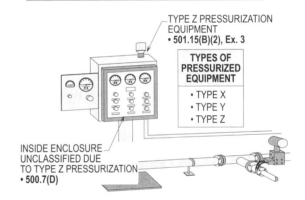

PURGED AND PRESSURIZED SYSTEMS
NEC 500.2; NEC 500.7(D) NFPA 496

Figure 21-17. The above illustration shows the inside of an enclosure being an unclassified area, based upon a Type Z pressurized piece of equipment. Purged or pressurized equipment can be used in Class I, Division 1 or 2 locations.

INTRINSICALLY SAFE EQUIPMENT
(CLASS I, DIVISION 1)
500.2; 500.7(E)

The use of intrinsically safe equipment is primarily limited to process control instrumentation, since these electrical systems lend themselves to the low energy requirements. ANSI/UL 913-1997 and ANSI 12.6 RP provides information on the design test and evaluation. The installation rules are covered in **Article 504** of the NEC. The definition of intrinsically safe equipment and wiring is: "equipment and wiring that are incapable of releasing sufficient electrical energy under normal or abnormal conditions to cause ignition of a specific hazardous atmospheric mixture in its most easily ignited concentration." UL and Factory Mutual list several devices in this category.

> **Design Tip:** The equipment and its associated wiring shall be installed so they are positively separated from the nonintrinsically safe circuits. Induced voltages could defeat the concept of intrinsically safe circuits. **(See Figure 21-18)**

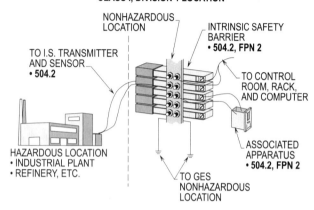

Figure 21-18. Intrinsically safe equipment can be utilized to supply low-voltage instruments, etc. in Class I, Division 1 and 2 locations. See **501.105(B)** and **501.150(B)** of the NEC and page 4-13 of Chapter 4.

NONINCENDIVE EQUIPMENT (CLASS I, DIVISION 2) 500.2; 500.7(F); (G); (H)

Nonincendive equipment is a form of intrinsic safety designed for use in Class I, Division 2 locations. Such equipment is defined as components having contacts that make or break an incendive circuit and elements. The enclosures in which the contacts are enclosed are so constructed that the components are not capable of igniting the surrounding flammable gases and air mixtures.

General-purpose enclosures can be used to house such circuits and components that under normal operating conditions will not release enough energy to ignite a specific explosive mixture.

> **Design Tip:** The housing of nonincendive components is not intended to exclude a flammable atmosphere or contain an explosion. **(See Figure 21-19)**

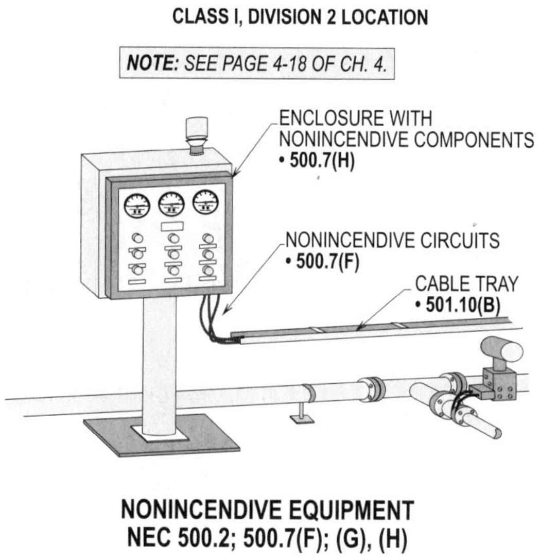

Figure 21-19. The above shows nonincendive components and circuits used in Class I, Division 2 locations.

OIL IMMERSION AND HERMETICALLY SEALED CONTACTS 500.2; 500.7(I); (J)

The general rule requires switches, circuit breakers, and make-and-break contacts of push buttons, relays, alarm bells, and horns to be installed in enclosures that are identified for Class I, Division 1 locations. However, in Class I, Division 2 locations, general-purpose enclosures can be used if the current-interrupting contacts are:

(1) Immersed in oil, or
(2) Enclosed within a chamber hermetically sealed against the entrance of gases or vapors.

See Figure 21-20 for contacts immersed in oil and sealed within an enclosed hermetically sealed chamber.

CLASS I, DIVISION 1 WIRING METHODS 501.10(A)(1)

In Class I, Division 1 locations, threaded rigid metal conduit (RMC), threaded steel intermediate metal conduit (IMC), or Type MI cable with termination fittings that are identified for the location must be the wiring method employed. All boxes, fittings, and joints must be threaded for connection to conduit or cable terminations, and must be of explosionproof type. Threaded joints must be made up with at least five threads fully engaged. Type MI cable must be installed and supported in such a manner to avoid tensile stress at the termination fittings. Where it is necessary to utilize flexible connections, as at motor terminals, flexible fittings that are identified for Class I locations are required to be used.

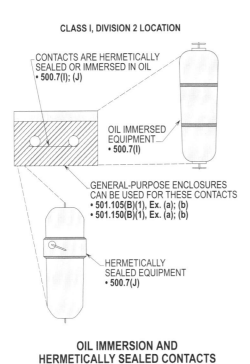

Figure 21-20. The above shows contacts installed in general-purpose enclosures which are immersed in oil and hermetically sealed chambers. Such enclosures and controls may be installed in Class I, Division 2 locations.

RIGID NONMETALLIC CONDUIT (CLASS I, DIVISION 1) 501.10(A)(1)(a), Ex.

the Ex. to **501.10(A)(1)(a)** allows rigid nonmetallic conduit to be installed in Class I, Division 1 areas, if encased in a concrete envelope of at least 2 in. and buried below the surface in not less than 2 ft. of earth.

Design Tip: If rigid nonmetallic conduit is used, threaded rigid metal conduit or IMC must be installed for the last 2 ft. of the underground run to emerge above ground or to connect to the above ground run of conduit. **(See Figure 21-21)**

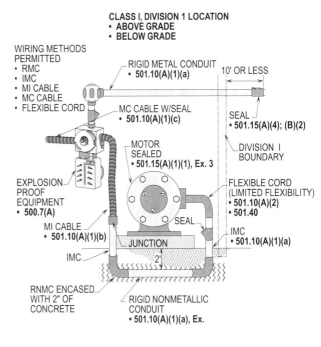

Figure 21-21. As shown above, rigid metal conduit, IMC, rigid nonmetallic conduit, MI and MC cable, and flexible cord under certain conditions of use can be utilized in Class I, Division 1 locations.

MC CABLE AND ITC CABLE (CLASS I, DIVISION 1) 501.10(A)(1); (c); (d)

Section **501.10(A)(1); (C) and (D)** permits Type MC and ITC cable which is listed for such use and equipped with a gas/vapor-tight continuous corrugated aluminum sheath with an overall jacket of suitable polymeric material, to be installed in Class I, Division 1 locations. **(See Figure 21-21)**

WIRING METHODS USED FOR LIMITED FLEXIBILITY (CLASS I, DIVISION 1) 501.10(A)(2)

Flexible fittings which are listed for use in Class I locations are required where it is necessary to employ flexible connections such as at motor terminals, etc.

Section **501.10(A)(2)** refers to **501.140,** which allows flexible cord to be used for that portion of the circuit where the fixed wiring method does not provide the necessary movement for fixed and mobile electrical utilization equipment in an industrial establishment. However, for this rule to be applied, proper maintenance and supervision must be available. (See Figure 21-21)

CLASS I, DIVISION 2 EQUIPMENT
500.2; 500.7

Class I, Division 2 equipment must provide a form of construction that will operate in such a manner to prevent igniting flammable mixtures of gases and air.

TYPES OF EQUIPMENT
(CLASS I, DIVISION 2)
500.7(A); (D); (E)

Equipment installed in Class I, Division 1 areas can be installed in Class I, Division 2 locations. Such equipment is the following:

(1) Explosionproof
(2) Purged and pressurized
(3) Intrinsically safe

Design Tip: See and review the equipment that is either identified or listed and illustrated in this chapter under Class I, Division 1 Equipment.

COMBUSTIBLE GAS DETECTION SYSTEM
500.2; 500.7(K)

Under certain conditions of use, a combustible gas detection system can be used. For example, Class I, Division 2 equipment can be installed in Class I, Division 1 locations and electrical equipment for unclassified locations can be installed in Class I, Division 2 Locations. However, note that the following requirements must be complied with when using a combustible gas detection system:

• In a Class I, Division 1 location that is so classified due to inadequate ventilation, electrical equipment suitable for Class I, Division 2 locations is permitted.

• In a building located in, or with an opening into a Class I, Division 2 location where the interior does not contain a source of flammable gas or vapor, electrical equipment for unclassified locations is permitted.

• In the interior of a control panel containing instrumentation utilizing or measuring flammable liquids, gases or vapors, electrical equipment suitable for Class I, Division 2 locations is permitted.

• Gas detection equipment must be listed for detection of the specific gas or vapor that is to be encountered.

• System must be installed in industrial establishments, with restricted public access and where the conditions of supervision and maintenance ensure that only qualified persons service the installation.

• The types of detection equipment is listing, installation locations(s), alarm and shutdown criteria, and calibration frequency shall be documented when combustible gas detectors are used as a protection technique.

See Figure 21-22 for an illustration of a typical combustible gas detection system.

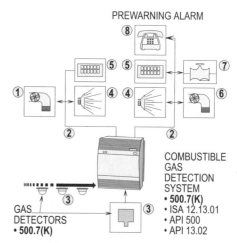

**COMBUSTIBLE GAS
DETECTION SYSTEM
NEC 500.2; 500.7(K)**

Figure 21-22. The above shows a typical combustible gas detection system.

WIRING METHODS CLASS I, DIVISION 2
501.10(B)(I)(1) THRU (6)

In Class I, Division 2 locations, threaded rigid metal conduit, threaded IMC, enclosed gasketed busways, enclosed gasketed wireways, or Type PLTC cable in accordance with the provisions of Article 725, Type MI, MC, MV, TC, or SNM cable with listed termination fittings must be the wiring method utilized. Type ITC, PLTC, MI, MC, MV or TC cable is permitted to be installed in cable tray systems where installed in a manner to avoid tensile stress at the termination fittings.

Design Tip: Boxes, fittings, and joints are not required to be explosion proof except when they are used in areas requiring Class I, Division 1 enclosures.

WIRING METHODS USED FOR LIMITED FLEXIBILITY (CLASS I, DIVISION 2)
501.4(B)(2)(1) THRU (5)

If provision must be made for limited flexibility, as at motor terminals, flexible metal fittings, flexible metal conduit, liquidtight flexible metal conduit, or liquidtight flexible nonmetallic conduit can be used.

Design Tip: The above wiring methods must be provided with listed fittings that provide a proper connection for bonding.

Flexible cord that is listed for extra-hard usage and provided with approved bushed fittings may be used. However, an additional conductor for equipment grounding is required to be included in the flexible cord. **(See Figure 21-23)**

REQUIREMENTS FOR SWITCHES, CIRCUIT BREAKERS, MOTOR CONTROLLERS, AND FUSES
501.115

Enclosures used to house switches, CB's, motor controllers, and fuses must be rated for either Class I, Division 1 or Division 2. The Division determines the type of enclosure, based upon the amount and the time the gas or vapor is present.

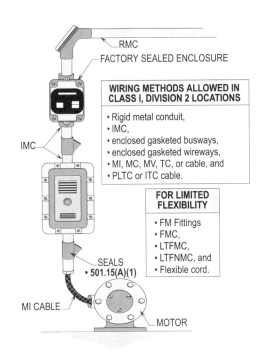

Figure 21-23. As shown above, there are more wiring methods that are permitted in Class I, Division 2 than Class I, Division 1 locations. **See Figure 21-21** for wiring methods permitted in Class I, Division 1 locations.

CLASS I, DIVISION 1
501.115(A)

Identified Class I, Division 1 enclosures must be used for switches, circuit breakers, motor controllers and fuses, including push buttons, relays and similar devices in order to be installed in Class I, Division 1 locations. **(See Figure 21-24)**

CLASS I, DIVISION 2
501.115(B)

Identified Class I, Division 1 enclosures must be used for switches, circuit breakers, motor controllers and fuses, including push buttons, relays and similar devices in order to be installed in Class I, Division 2 locations.

However, general purpose enclosures may be used if the interruption of current occurs in **(1)** a hermetically sealed chamber, or **(2)** the current make-and-break contacts are oil-immersed.

The use of general purpose enclosures are also permitted, if the interruption of current occurs within a factory-sealed explosionproof chamber that is identifed for the location. **(See Figure 21-24)**

Design Tip: The use of disconnects and isolating switches in general purpose enclosures in Class I, Division 2 locations are permitted if neither the switch nor fuse operates as a normal current interrupting device. In such cases, fuses can only be used for short-circuit protection.

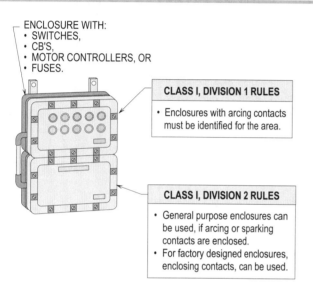

Figure 21-24. Enclosures containing arcing contacts must be identified for Class I, Division 1 locations. In Class I, Division 2 locations, general-purpose and factory enclosures can be used if the contacts are enclosed in such a manner to handle arcing and sparking conditions.

REQUIREMENTS FOR MOTORS AND GENERATORS
501.125

Electric motors are needed to drive pumps, compressors, fans, blowers, and conveyors, so their presence in hazardous atmospheres is sometimes unavoidable.

The types of hazardous atmospheres and corrosive conditions are major factors in motor selection. The hazardous area dictates the type of motor needed to avoid excessive maintenance and expensive shutdowns. The types available vary all the way from "drip-proof" to "totally enclosed" and "fan cooled" motors.

CLASS I, DIVISION 1
501.125(A)

In Class I, Division 1 locations, only the explosionproof, totally enclosed and pressurized with clean air, totally enclosed inert gas filled and special submerged type motors may be used.

The NEC makes it clear that Class I, Division 1 totally enclosed motors must have no external surface operating temperature in excess of 80 percent of the ignition temperature of the gas or vapor involved.

Also required are devices to detect and automatically deenergize the motor (or provide an effective alarm) in case of overheating.

In addition, auxiliary equipment must be of a type that is identified for the location in which it is installed. **(See Figure 21-25)**

CLASS I, DIVISION 2
501.125(B)

Motors for use in Class I, Division 2 locations in which sliding contacts, switching mechanisms, or integral resistance devices are employed, must also be explosionproof (Class I, Division 1) or purged and pressurized.

However, open type motors such as squirrel-cage induction motors without any arcing devices may be used in Class I, Division 2.

Design Tip: UL provides a procedure in which listed explosionproof motors can be repaired. Motor personnel must consult as to which repair shops have been authorized to make such repairs. Unauthorized maintenance of an explosionproof motor can result in voiding the manufacturer's listing.

To enable the AHJ to determine whether or not the maximum temperature exceeds 80 percent of the ignition temperature of the gas or vapor involved, space heaters used in a motor must have a permanent, visible nameplate marking on the motor, indicating the maximum surface temperature (based on 40°C ambient).

If the temperature of the heater does not exceed 80 percent of the ignition temperature, the motor is considered suitable for use in that atmosphere. **(See Figure 21-25)**

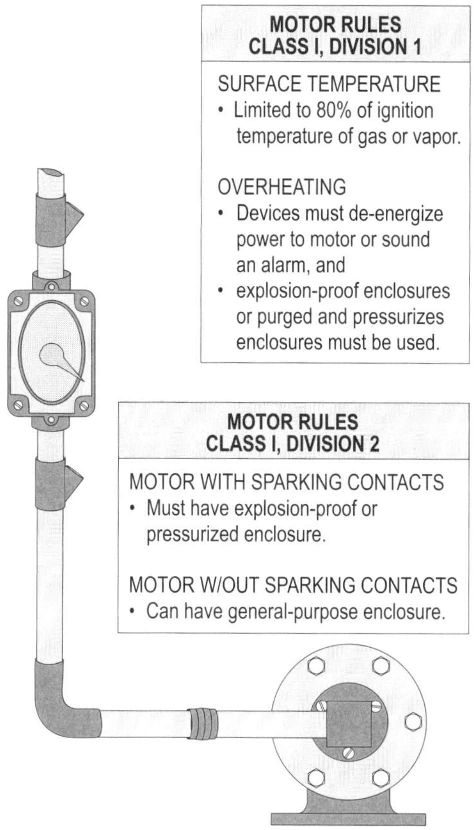

Hazardous (Classified) Locations

**MOTOR RULES
CLASS I, DIVISION 1**

SURFACE TEMPERATURE
- Limited to 80% of ignition temperature of gas or vapor.

OVERHEATING
- Devices must de-energize power to motor or sound an alarm, and
- explosion-proof enclosures or purged and pressurizes enclosures must be used.

**MOTOR RULES
CLASS I, DIVISION 2**

MOTOR WITH SPARKING CONTACTS
- Must have explosion-proof or pressurized enclosure.

MOTOR W/OUT SPARKING CONTACTS
- Can have general-purpose enclosure.

NEC 501.115(A);(B)

Figure 21-25. The above rules apply when installing motors in Class I, Division 1 and 2 locations.

REQUIREMENTS FOR LUMINAIRES (LIGHTING FIXTURES) 501.130

In locations where explosive gases or vapors exist, bare lamps or nonexplosion proof enclosed luminaires can create extreme hazards. Bare lamps may be broken, and the arc or spark can cause explosions. For this reason, lamps must be enclosed in luminaires, which are specifically designed for Class I, Divisions 1 and 2 locations, whichever applies.

CLASS I, DIVISION 1 501.130(A)(1)

Surface temperature of Class I, Division 1 luminaires must not exceed the ignition temperature of the gases or vapors surrounding such luminaires.

Each luminaire must be identified as a complete assembly for the Class I, Division 1 location. In addition, they must be clearly marked to indicate the maximum wattage of lamps for which it is identified.

PENDANT LUMINAIRES (FIXTURES) 501.130(A)(3)

Pendant luminaires for Class I, Division 1 must be suspended and supplied by threaded metal or IMC conduit stems. Threaded joints of Class I, Division 1 luminaires must have setscrews to prevent loosening on rigid stems. Where rigid stems (conduit portion from mounting box to lighting fixture) are longer than 12 in., they must be braced against movement or provided with flexible fittings. **(See Figure 21-26)**

PORTABLE LIGHTING EQUIPMENT 501.130(A)(1)

Fixtures intended for portable use must be specifically identified as a complete assembly for that use. Personnel must not take portable lighting equipment into such classified areas, if they are not identified for such area and use. **(See Figure 21-26)**

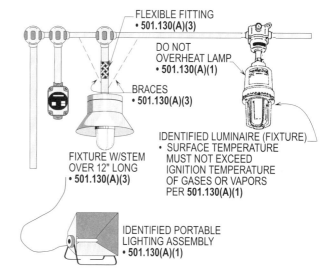

NEC 501.130(A)(1); (A)(2); (A)(3)

Figure 21-26. The above rules and regulations apply when installing luminaires (lighting fixtures) in Class I, Division 1 locations.

CLASS I, DIVISION 2
501.130(B)(1)

For Class I, Division 2 luminaires, it is the lamp surface temperature, and not the luminaire surface temperature, that must not exceed the ignition temperature of the gases or vapors present.

Basically, in locations which are classified Class I, Division 2, fixed luminaires, which have been tested, may operate up to the ignition temperature of the gas or vapor present. However, if the luminaires have not been tested for such use, the maximum temperature must not exceed 80 percent of the ignition temperature of the gas or vapor present. **(See Figure 21-27)**

PENDANT LUMINAIRES (FIXTURES)
501.130(B)(3)

In Class I, Division 2 locations, pendant luminaires must be suspended by flexible hangers unless rigid stems not over 12 in. long are used, or longer stems are permanently braced within 12 in. of the luminaire.

Pendant luminaires are required to be suspended and supplied by threaded metal or IMC conduit stems which are screwed into the threaded joints of the luminaires. Setscrews, to prevent loosening on rigid stems, must also be provided. **See Figure 21-26** for rules pertaining to pendant hung lighting units.

PORTABLE LIGHTING
501.130(B)(4)

Portable lighting used in Class I, Division 2 locations does not have to be identified for Class I, Division 1 if it is mounted on a movable stand and connected by an approved flexible cord. The fixture only needs to be identified for Class I, Division 2, and specify that:

(1) The fixture be protected by a suitable guard or location,

(2) The fixture has a suitable enclosure to prevent sparks or hot metal from a lamp causing ignition of the surrounding atmosphere, and

(3) The fixture does not exceed temperature limitations.

See Figure 21-27 for rules concerning Iluminaires (lighting fixtures) installed and used in Class I, Division 2 locations.

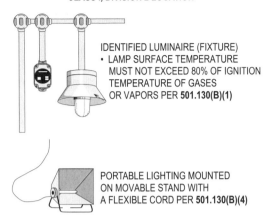

PORTABLE LIGHTING EQUIPMENT
NEC 501.130(B)(1); (B)(4)

Figure 21-27. The above rules and regulations apply when installing luminaires (lighting fixtures) in Class I, Division 2 locations.

REQUIREMENTS FOR UTILIZATION EQUIPMENT
501.135

Utilization equipment installed in hazardous (classified) areas is considered to be appliances specifically designed for such areas.

CLASS I, DIVISION 1
501.135(A)

Utilization equipment, including electrically heated and motor-driven equipment, that is used in Class I, Division 1 locations must be identified for such location. **(See Figure 21-28)**

CLASS I, DIVISION 2
501.135(B)

Utilization equipment installed in Class I, Division 2 locations must comply with certain design criteria.

HEATERS
501.135(B)(1)

Heating appliances must be identified for Class I, Division 1 locations. Heating appliances would include water heaters and room heaters. There are exceptions that permit heaters to be the general purpose type and they are as follows:

(1) If the maximum operating temperature of any exposed surface will not exceed 80 percent of the ignition temperature of the gas or vapor surrounding the heater, when operating continuously at 120 percent rated voltage, the heater may be the general purpose type.

(2) If the maximum operating temperature of any exposed surface will not exceed 80 percent of the ignition temperature of the gas or vapor surrounding the heater, when operating continuously at 100 percent rated voltage, the heater may be the general purpose type, provided it is supplied with a temperature controller.

To meet the requirements of Class I, Division 1 equipment, operating temperature at rated voltage would have to be no greater than about 55 percent of the ignition temperature of the gas or vapor and all heaters must be completely enclosed.

See Figure 21-29 for rules for installing heaters in Class I, Division 2 locations.

APPLIANCES WITH SWITCHES
501.135(B)(3)

If an appliance has a switch, the switch must be identified for Class I locations. The rest of the appliance may in some cases have a general purpose enclosure, but the switch must always be identified for Class I locations.

Design Tip: Except as required by **501.135(B)(1), (2) and (3)** above, appliances in Class I, Division 2 locations may be the general-purpose type. (**See Figure 21-29**)

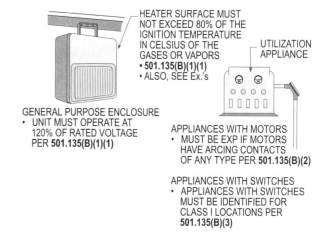

HEATERS
NEC 501.135(B)(1); (B)(2); (B)(3)

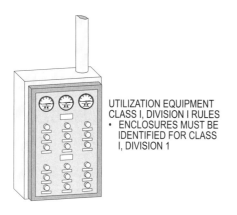

UTILIZATION EQUIPMENT
NEC 501.135(A)

Figure 21-28. Enclosures, used for the components of utilization equipment, must be identified for Class I, Division 1 locations.

Figure 21-29. The above rules and regulations apply when installing utilization equipment (appliances) in Class I, Division 2 locations.

USING FLEXIBLE CORD IN CLASS I, DIVISION 1 AND 2 AREAS
501.140

A flexible cord can only be used for connection between a portable lamp or other portable utilization equipment. The fixed portion of its supply circuit and where such connections are used shall:

(1) Be of a type identified for extra hard-usage,

(2) Contain, in addition to the conductors of the circuit, a grounding conductor conforming to **250.118**, that is, of green color and used for no other purpose than for grounding,

APPLIANCES WITH MOTORS
501.135(B)(2)

Appliances with motors need not be identified for the location unless the motor has brushes, or a centrifugal switch, or a built-in thermal protector, in which case the appliances must be identified for Class I, Division 1 locations. (See **501.135(B)(2)** for further information concerning motors.) (**See Figure 21-29**)

(3) Be connected to terminals or to supply conductors in an approved manner,

(4) Be supported by clamps or other suitable means in such a manner that there will be no tension on the terminal connections, and

(5) Have suitable seals provided where the flexible cord enters boxes, fittings, or enclosures of the explosion-proof type.

In Class I, Division 1 locations, flexible cord can be used for that portion of the circuit where the fixed wiring method of **501.10(A)(1)** cannot provide the necessary degree of movement for fixed and mobile electrical utilization equipment. However, proper maintenance and supervision must be provided and such equipment must be located in an industrial establishment.

CORD-AND-PLUG CONNECTED SUBMERSIBLE PUMPS
501.140(A)(3)

Electric submersible pumps with means for removal without entering the wet pit must be considered portable utilization equipment. The extension of the flexible cord between the wet pit and power source can be enclosed in a suitable raceway.

CORD-AND-PLUG CONNECTED ELECTRIC MIXERS
501.140

Electric mixers traveling in and out of open-type mixing tanks or vats may be classified as portable utilization equipment so that wiring procedures won't have to be too restrictive.

See Figure 21-30 for a detailed illustration of flexible cord being used for such installations.

USING FLEXIBLE CORD TO FACILITATE REPLACEMENTS
501.140, Ex. AND 501.105(B)(6)

To facilitate replacements, process control instruments can be connected through flexible cord, attachment plug, and receptacle. However, the following must be provided:

(1) A switch complying with the rules of 501.4(B) above is provided so that the attachment plug is not depended on to interrupt current,

(2) The current does not exceed 3 amps at 120 volts,

(3) The power supply cord does not exceed 3 ft. and is listed for extra-hard usage or for hard usage if protected by location and is supplied through an attachment plug or receptacle of the locking and grounding type,

(4) Only necessary if receptacles are provided, and

(5) The receptacle carries a label warning against unplugging under load. **(See Figure 21-30)**

USING FLEXIBLE CORD FOR FLEXIBILITY CONNECTIONS
501.140, Ex. AND 501.10(B)

Where provisions must be made for a flexible connection, as at motors, flexible cord may be used but additional grounding must be provided around these flexible connections to ensure an effective ground path for clearing short-circuits and ground-faults.

Extra-hard usage flexible cord with listed bushed fittings and an extra grounding conductor can be used for this purpose and fully complies with such requirements as above. **(See Figure 21-30)**

REQUIREMENTS FOR RECEPTACLES AND ATTACHMENT CAPS
501.145

Arcing at exposed contacts must be prevented in Class I, Division 1 or 2 locations. To accomplish this, receptacles are designed so that plug contacts are safely within an explosionproof enclosure when they are electrically engaged, confining arcing, if any, to the receptacle interior.

RECEPTACLES WITH SWITCHES
501.145; UL 498

With this configuration, the plug cannot be inserted unless the switch is in the OFF position and cannot be withdrawn with the receptacle in the ON position. The reason for this is the receptacle contacts are interlocked with a switch located in an explosionproof enclosure. Therefore, arcing does not occur outside the enclosure due to mated parts being deenergized during plug insertion and removal. **(See Figure 21-31)**

Hazardous (Classified) Locations

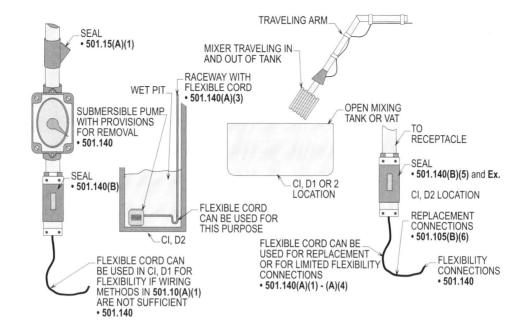

Figure 21-30. The above rules and regulations apply when using flexible cord in Class I, Division 1 and 2 locations.

RECEPTACLES WITHOUT SWITCHES 501.145; UL 498

Receptacles without switches rely on a mechanical means which provides a delayed action to confine arcing to the receptacle interior during plug insertion and withdrawal. The design used in these receptacles prevents removal of the plug until any flame, spark or hot metal from an arc has cooled sufficiently to prevent ignition of the surrounding explosive atmosphere. (**See Figure 21-31**)

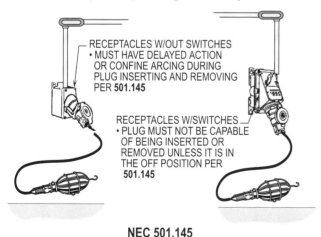

Figure 21-31. The above rules and regulations apply when using cord-and-plug connected equipment in Class I, Division 1 and 2 locations.

WIRING METHODS IN CLASS II, DIVISION 1 AND 2 LOCATIONS 502.10

In Class II, Division 1 and 2 locations, certain types of wiring methods and enclosures must be utilized to properly contain and handle dust related areas.

WIRING METHODS (CLASS II, DIVISION 1) 502.10(A)(1)(1); (2); (3)

In Class II, Division 1 locations, the wiring method must be either metal conduit or Type MI cable (mineral-insulated cable).

Section **502.10(A)(1)(3)** allows Type MC cable, listed for such use and equipped with a gas/vapor-tight continuous corrugated aluminum sheath with an overall jacket of suitable polymeric material, to be installed in Class II, Division 1 locations.

Design Tip: To utilize this wiring method, separate grounding conductors, in compliance with **250.122** and with termination fittings listed for the application, must be used. (**See Figure 21-32**)

FITTINGS AND BOXES (CLASS II, DIVISION 1)
502.10(A)(1)(4)

There are two main parts to this rule that must be applied and they are as follows:

(1) For locations where the dusts is hazardous but not of a combustible electrically conductive nature. This includes grain, cocoa, dried egg and milk dust, starch dust, and hay dust locations.

Boxes and fittings that contain splices or taps must be identified for Class II locations when used in such locations.

Boxes and fittings without splices or taps are not required to be identified, but must have tight-fitting enclosures with no openings such as screw holes.

(2) For locations where dusts of a combustible electrically conductive nature are present. This includes coal dust, coke, carbon black, charcoal dust, magnesium, and aluminum dust locations.

All fittings and boxes, with or without splices or taps, must be identified for Class II locations when installed in these locations. (**See Figure 21-32**)

FLEXIBLE CONNECTIONS
502.10(A)(2)

In Class II, Division 1 locations, liquidtight flexible metal conduit and liquidtight flexible nonmetallic conduit with listed fittings or flexible cord, listed for extra-hard usage and equipped with bushed fittings may be used. Unless other means are provided for grounding, cords shall have a grounding conductor. Where dusts of an electrically conducting nature are present, cords must have dust-tight seals at both ends. (**See Figure 21-32**)

WIRING METHODS (CLASS II, DIVISION 2)
502.10(B)(1) THRU (6)

In Class II, Division 2 locations, the following wiring methods are permitted:

(1) Rigid metal conduit
(2) Intermediate metal conduit (IMC)
(3) Electrical metallic tubing (EMT)
(4) Dust-tight wireways
(5) Type MI cable (mineral-insulated cable)
(6) Type MC cable (metal-clad power cable)
(7) Type SNM cable (shielded nonmetallic-sheathed cable)
(8) Type ITC in cable trays
(9) Type PLTC in cable trays
(10) Type MC or TC cable in cable trays
(11) Nonincendive circuits in any suitable wiring method

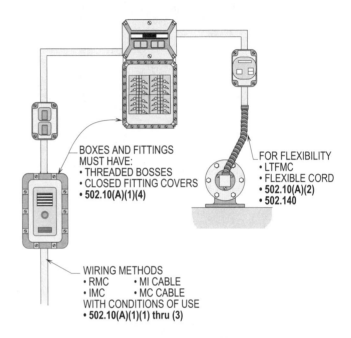

NEC 502.10(A)(1)(1) thru (4)

Figure 21-32. The above rules and regulations apply for wiring methods and equipment installed in Class II, Division 1 locations.

FLEXIBLE CONNECTIONS
502.10(B)(2)

In Class II, Division 2 locations, liquidtight flexible metal conduit and liquidtight flexible nonmetallic conduit with approved fittings or flexible cord, approved for extra-hard usage and equipped with bushed fittings may be used. Unless other means are provided for grounding, cords shall have a grounding conductor. Where dusts of an electrically conducting nature are present, cords must have dust-tight seals at both ends. (**See Figure 21-33**)

WIREWAYS, FITTINGS, AND BOXES
502.10(B)(4)

Wireways, fittings, and boxes in Class II, Division 2 locations are not necessarily required to be approved for Class II locations. However, they must have close-fitting enclosures with no openings such as screw holes that might transmit sparks to the outside of the wireway, fitting, or box. Sparking or burning material must not escape because it could ignite adjacent combustible material. (**See Figure 21-33**)

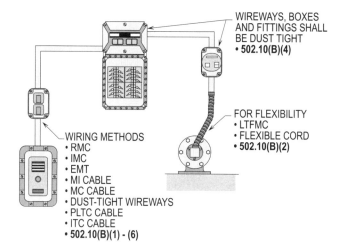

NEC 502.10(B)(1) THRU (6)

Figure 21-33. The above rules and regulations apply for wiring methods and equipment installed in Class II, Division 2 locations.

REQUIREMENTS FOR SEALS AND DRAINS (CLASS I, DIVISIONS 1 AND 2) 501.15

Seals must be provided in conduit and cable systems to minimize the passage of gases or vapors from one portion of the system to another portion. Another purpose of the seals is to keep from transmitting an explosion or to keep ignition from traveling between sections of the system.

NEED FOR SEALS
501.15(A)(1) THRU (A)(4)

Seals and drains are recommended to be installed as follows:

(1) Restrict the passage of gases, vapors, or flames from one portion of the electrical installation to another at atmospheric pressure and normal ambient temperatures,

(2) Limit explosions to the sealed-off enclosure and prevent the compression or "pressure piling" in conduit systems,

(3) While not a code requirement, many engineers consider it good practice to sectionalize long conduit runs by inserting seals not more than 50 to 100 ft. apart, depending on the conduit size, to minimize the effects of "pressure piling." Sealing fittings are required, and

(4) At each entrance to an enclosure housing having an arcing or sparking device and used in Class I, Division 1 and 2 hazardous locations. To be located as close as practical and, in no case, more than 18 in. from such enclosures. (**See Figure 21-34**)

(5) At each entrance of 2 in. size or larger to an enclosure or fitting, housing terminals, splices or taps when used in Class II, Division 1 location. To be located as close as practical and, in no case, more than 18 in. from such enclosures. (**See Figure 21-35**)

(6) In conduit systems when leaving the Class I, Division 1 or Division 2 location. (**See Figure 21-36**), and

(7) In cable systems when the cables either do not have a gas/vapor tight continuous sheath or are capable of transmitting gases or vapors through the cable core where these cables leave the Class I, Division 1 or Division 2 location. (**See Figure 21-37**)

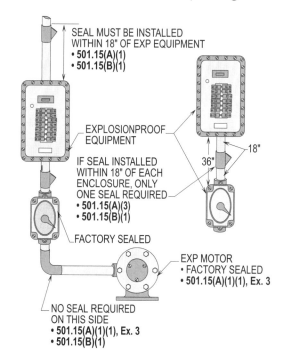

NEC 501.15(A)(1); (A)(3)
NEC 501.15(B)(1)

Figure 21-34. Explosionproof equipment or equipment with arcing or sparking devices must have a seal placed within 18 in. of such equipment.

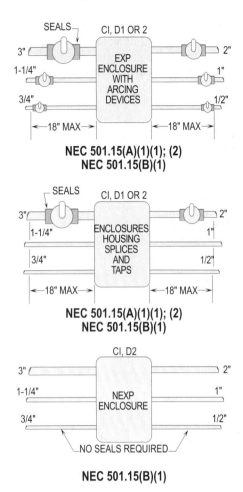

Figure 21-35. The above rules apply for the enclosures of splices, taps, and arcing or sparking devices if they are installed in Class I, Divisions 1 or 2 locations.

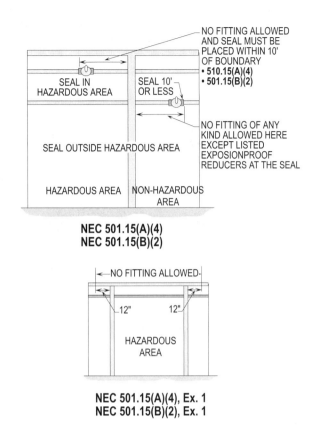

Figure 21-36. The above rules apply when installing seals in conduit runs leaving or passing through a Class I, Divisions 1 and 2 location.

NEED FOR DRAINS 501.15(F)

In humid atmospheres or in wet locations, where it is likely that water can gain entrance to the interiors of enclosures or raceways, the raceways should be inclined so that water will not collect in enclosures or in seals but will be led to low points where it may pass out through ECD drains. Frequently the arrangement of raceway runs makes this method impractical if not impossible. In such instances, Type EZD drain seal fittings must be used. These fittings prevent accumulations of water above the seal.

In locations that usually are considered dry, surprising amounts of water frequently collect in conduit systems. No conduit system is airtight; therefore, it may breathe. Alternate increases and decreases in temperature and/or barometric pressure due to weather changes or due to the nature of the process carried on in the location where the conduit is installed will cause breathing. Outside air is drawn into the conduit system when it breathes in. If this air carries sufficient moisture it will be condensed within the system when the temperature decreases and chills this air. Due to the internal conditions being unfavorable to evaporation, the resultant water accumulation will remain and be added to by repetitions of the breathing cycle. In view of this likelihood, it is good practice to ensure against such water accumulations and probable subsequent insulation failures by installing EZD with drain cover or inspection cover even though conditions prevailing at the time of planning or installing may not indicate their need. **(See Figure 21-38)**

SELECTION OF SEALS AND DRAINS

There are different types of seals and drains which are made to be utilized for vertical or horizontal installations and are to be used only for the purpose for which they are designed. Care shall be taken when selecting and installing such fittings.

The following primary considerations must be used when selecting seals and drains:

Hazardous (Classified) Locations

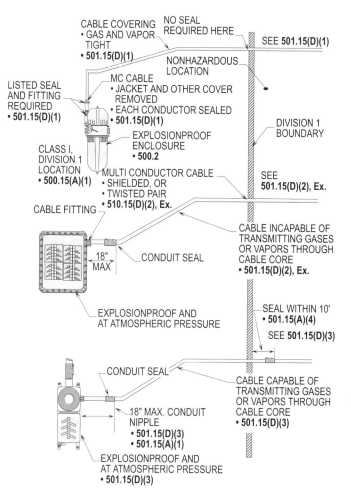

NEC 501.15(D)(1); (D)(2), Ex.; (D)(3)

Figure 21-37. The above rules apply when sealing multiconductor cables leaving a Class I, Division 1 location.

(1) Select the proper sealing fitting for the hazardous vapor involved such as Class I, Groups A, B, C, or D.

(2) Select a sealing fitting for the proper use in respect to mounting position. This is particularly critical when the conduit runs between hazardous and nonhazardous areas. Improper positioning of a seal may permit hazardous gases or vapors to enter the system beyond the seal and permit them to escape into another portion of the hazardous area or to enter a nonhazardous area. Some seals are designed to be mounted in any position; others are restricted to horizontal or vertical mounting.

(3) Install the seals on the proper side of the partition or wall as recommended by the manufacturer.

(4) Only trained personnel can install seals and they must be installed in compliance with the instruction sheets furnished with the seals and sealing compound. Precautionary notes must be included on installation diagrams to stress the importance of following manufacturer's instructions.

(5) Splices or taps in sealing fittings are strictly prohibited.

(6) Sealing fittings are listed by UL for use in Class I hazardous locations with CHICO A compound only. This compound, when properly mixed and poured, hardens into a dense, strong mass, which is insoluble in water, is not attacked by chemicals, and is not softened by heat. It will withstand ample safety factor pressure of the exploding trapped gases or vapor.

Design Tip: Seals must be poured with the compound recommended by its manufacturer.

(7) Conductors sealed in the compound may be approved thermoplastic or rubber insulated types. Both may or may not be lead covered (the lead need not be removed).

See Figure 21-39 for installing seals and drains.

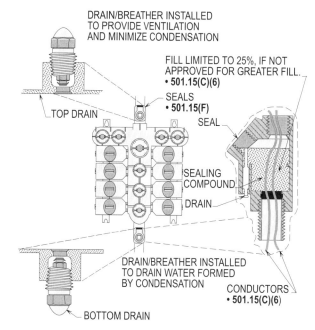

NEC 501.15(F)

Figure 21-38. Drains and seals installed as shown above will minimize water conditions formed by condensation and help prevent mildew from forming.

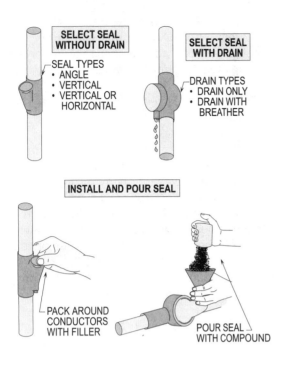

NEC 501.15(C)(1); (C)(2); (C)(3)

Figure 21-39. The above rules must be applied when selecting and installing seals/drains in Class I, Division 1 locations.

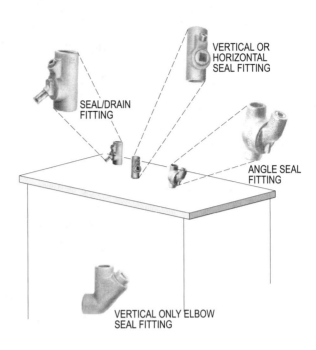

NEC 501.15(C)(1)

Figure 21-40. As shown above, it is most important to select and install the proper seal or seal with drain based upon the type of installation.

TYPES OF SEALING FITTINGS 501.15(C)(1)

The following sealing fittings meet the requirements of the NEC when they are properly installed:

(1) **EYS Sealing Fittings**—A certain style of EYS sealing fittings are for use with vertical or nearly vertical conduit in sizes from 1/2 through 1 in. Other styles are available in sizes 1/2 through 6 in. for use in vertical or horizontal conduits. In horizontal runs, these are limited to face-up openings. Sizes from 1 1/4 through 6 in. have extra large work openings, and separate filling holes so that CHICO X fiber dams are easy to make. Overall diameter of sizes 1 1/4 through 6 in. is scarcely greater than that of unions of corresponding sizes, permitting close conduit spacings.

Note: Check with the manufacturer for the type of compound to be used with a particular seal.

(2) **EZS Sealing Fittings**—EZS seals are for use with conduit running at any angle, from vertical through horizontal.

See Figure 21-40 for the different types of sealing fittings.

REQUIREMENT FOR SEALS (CLASS II, DIVISIONS 1 AND 2) 502.15(1) THRU (4)

In the installation of a dust-ignitionproof enclosure and one that is not, there can be communication of dust between the two enclosures under certain conditions of use. The entrance of dust into the dust-ignitionproof enclosure must be prevented to prevent explosions. This can be accomplished by applying one of the following sealing methods:

(1) By use of a permanent and effective seal, as used in Class I locations,
(2) By connection to a horizontal raceway that is not less than 10 ft. in length, and
(3) By means of a vertical conduit not less than 5 ft. in length, which is installed downward from the dust-ignitionproof enclosure.

Design Tip: Where a conduit provides communication between an enclosure that is required to be dust-ignitionproof and an enclosure in an unclassified location, seals are not required.

COMMERCIAL GARAGES, REPAIR, AND STORAGE
ARTICLE 511

Article 511 of the NEC applies for commercial garages in which repair and service work is done on automobiles, trucks, buses, tractors, etc.

Garages used only for parking or storage are not hazardous areas. The requirements of **Article 511** do not apply to garages in which no maintenance repair or service work is being performed. However, the NEC does require that indoor storage and parking spaces be adequately ventilated to remove exhaust fumes from engines.

EIGHTEEN INCHES ABOVE THE FLOOR
511.3(B)(1); (B)(2)

For each floor, the entire area up to a level of 18 in. above the floor or 18 in from the ceiling is considered to be a Class I, Division 2 location, except where the AHJ determines that there is sufficient mechanical ventilation to provide a minimum of four air changes per hour of the above.

ANY PIT OR DEPRESSION BELOW FLOOR LEVEL
511.3(B)(1) THRU (4)

Pits or depressions below floor level, which extend up to the floor level, must be considered as Class I, Division 1 locations. However, if the pit or depression is ventilated, the AHJ may classify and judge the area to be a Class I, Division 2 or unclassified location.

> **Design Tip:** Any pit or depression in which six air changes per hour are exhausted at the floor level of the pit can be judged by the AHJ to be a Class I, Division 2 location per **511.3(B)(3)(3)**.

In cases where there is adequate and positive ventilation of the area below grade level, the AHJ can classify such an area as unclassified per **511.3(A)(4)** and **Table 514.3(B)(1)**.

ADJACENT AREAS
511.3(A)(3)

Areas adjacent to Class I, Division 2 locations in which there is no likelihood of hazardous vapors being released such as stock rooms, offices, etc. are still classified as Class I, Division 2 locations. If the floor of the adjacent area is elevated 18 in. above the floor of the hazardous area or a separation between the two areas is provided by either an 18 in. tight curb or partition, the area can be classified as nonhazardous.

> **Design Tip:** Gasoline vapors are heavier than air and settle to and move along the floor area. Since these vapors settle to the floor, they may be transmitted into other rooms which may seem otherwise to be nonhazardous areas. Transmittal of such hazardous vapors can be stopped by an elevated floor, tight curb, or a properly built partition of which each must be at least 18 in. high, between the two rooms.

SPECIAL PERMISSION
100; 110.4; 511.3(A); (B)

Adjacent areas, which by reason of adequate ventilation, air pressure differentials or physical spacing, are such that, in the opinion of the AHJ no ignition hazardous condition exists they can be classified as nonhazardous.
See **Figure 21-41** for rules and regulations pertaining to garages.

FUEL DISPENSING UNITS
511.4(B)(1)

When fuel dispensing units other than liquid petroleum gas which is prohibited, are located within buildings, the requirements of Article 514 must be applied.

If adequate mechanical ventilation is provided in the dispensing area, the controls must be interlocked so that the dispenser cannot operate without ventilation as required by **500.5(B)(2)**.

PORTABLE LIGHTING EQUIPMENT
511.4(B)(2)

If fuel pumps are located inside, the requirements of **Article 514** pertaining to Service Stations are required to be applied to the pumps and area around the pumps.

Portable lamps used in garages must:

(1) Be of hard rubber or other nonmetallic material,
(2) Be equipped with a handle,
(3) Have a guard and hook,
(4) Be the unswitched type, and
(5) Have no plug-in.

Design Tip: If portable lamps are to be used in a hazardous area near the floor for instance, they must be a type which are identified for hazardous areas.

If the cord is connected in such a manner that the lamp could not reach a hazardous level when in use, a portable lamp can be the general-purpose type.

WIRING AND EQUIPMENT IN CLASS I LOCATIONS
511.4(A); 511.3

Hazardous vapors in a garage are concentrated in low areas within a short distance above the floor and within the ceiling area. The Code sets the distance as 18 in., and the area below an imaginary plane of 18 in. above the floor is classified as a hazardous area. Raceways beneath the floor or embedded in a masonry are considered to be within the hazardous area if any connections or extensions lead into the hazardous area per **511.3**.

Hazardous vapors seek either a low or high level. Therefore, any ceiling or pit in a garage floor is considered to be hazardous down to the bottom of the pit or up to the ceiling area.

All wiring and equipment located in the space below the hazardous level must be suited for a Division or Zone location, if adequate ventilation is not utilized.

Design Tip: If all of the wiring and equipment is routed and installed above the 18 in. height from the floor or below the 18 in. ceiling area, EMT and general-purpose equipment and devices can be used. If the wiring is installed in conduit below the 18 in. boundary in the floor slab, this part of the wiring system would be a Class location and the wiring method would have to comply with the rules of this classification. To pass from the hazardous area and routed into the nonhazardous area, seals would have to be installed at the 18 in. boundary.

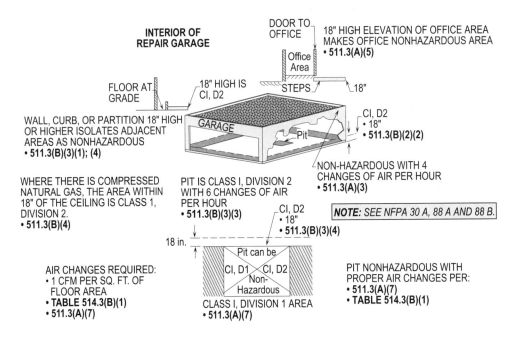

Figure 21-41. In classification of a floor in a garage where vehicles are repaired and volatile flammable fluids are present, the floor area, up to 18 in., is Class I, Division 2. However, if adequate ventilation is present, the 18 in. space above the floor can be considered to be nonhazardous. Likewise, if the pit is adequately ventilated, it can be judged to be a Class I, Division 2 location or a nonhazardous location

SEALING REQUIREMENTS
511.9

Seals must be provided as follows for wiring methods entering or leaving the Class I, Division 2 area:

(1) If conduit is used, a vertical run, unless unbroken, must be sealed at the hazardous boundary level which is 18 in. above the floor.

(2) For a horizontal run, if the run leaves the area at a height below the hazardous boundary level which is less than 18 in. from the floor, the wiring method must be sealed at the point where it leaves the hazardous area.

(3) Seals must be placed in each conduit run entering the enclosures for switches, circuit breakers, fuses, relays resistors, or any device that produces arcing, sparking, or high temperature, and only if the enclosure is located beneath the hazardous boundary.

(4) In each conduit run of 2 in. or larger, seals must be installed at junction boxes and fittings if the box or fitting contains splices, taps, or connections and is located beneath the hazardous boundary.

> **Design Tip:** The seal must be installed within 18 in. of the box or fitting to comply with NEC rules and regulations. (See **501.15** and **501.5(B)(1)**)

WIRING ABOVE CLASS I LOCATIONS
511.7(A)(1)

Above the 18 in. level, the following wiring methods are permitted:

  (1) Metal raceways
  (2) Rigid nonmetallic conduit
  (3) Electrical nonmetallic tubing
  (4) EMT
  (5) Intermediate metal conduit
  (6) Type MC cable (metal clad)
  (7) Type MI cable (mineral-insulated)
  (8) Type TC cable (power and control tray)
  (9) Type PLTC
  (10) ITC
  (11) Flexible metal conduit
  (12) Liquidtight flexible metal conduit
  (13) Liquidtight nonmetallic conduit

> **Design Tip:** Cellular metal floor raceways may be used only to supply ceiling outlets or extensions to areas located below the floor. However, outlets must not be connected above the floor. No electrical conductor can be installed in any cell, header, or duct which contains pipe for any service except electrical or compressed air.

For example, cellular metal floor raceways are a part of the building structure and if the floor in the garage has cellular metal floor raceways installed, the raceway may be used to supply equipment in the room below the garage. However, it may not be used to supply equipment in the garage. Where the cellular raceway is in the floor above, it may be used for ceiling outlets in the garage.

GROUNDED AND GROUNDING CONDUCTORS
511.16

When a circuit which supplies portables or pendants includes an identified grounded conductor (white or gray) as provided in **Article 200**, receptacles, attachment plugs, connectors, and similar devices must be of the polarized type and the grounded conductor of the flexible cord must be connected to the screw shell of any lampholder or to the grounded terminal of any utilization equipment supplied.

> **Design Tip:** Grounding continuity of the grounding conductor to noncurrent-carrying parts must be maintained.

EQUIPMENT ABOVE CLASS I LOCATIONS
511.7(A); (B)

Any equipment that might produce an arc or spark, and which is less than 12 ft. above floor level, must have a tight enclosure that will prevent the escape of any arc or spark. Charging panels, motors, generators, switches, if installed less than 12 ft. above the floor, should be the totally enclosed type, per **511.7(B)(1)(a)**.

Luminaires (lighting fixtures) must be the totally enclosed type or of a construction that does not allow the escape of arcs or sparks, if located less than 12 ft. above lanes where vehicles are commonly driven or if located where exposed to physical damage, per **511.7(B)(1)(b)**. (**See Figure 21-42**)

> **Design Tip:** Fluorescent and incandescent luminaires (fixtures) in Class I locations must be fitted with a lens or globe. Exposed tubes and light bulbs are not permitted.

AIRCRAFT HANGARS
ARTICLE 513

Article 513 applies to hangars that are used for servicing or storing aircraft containing gasoline or other hazardous liquids or gases. Certain areas in aircraft hangars are a Class I location, except as follows:

(1) Locations used exclusively for aircraft which has never contained such volatile flammable gases or liquids.

(2) Aircraft which has been drained or properly purged.

Adjacent areas into which hazardous vapors are likely to be communicated or released, such as stockrooms and electrical control rooms, are considered a Class I, Division 2 location up to a height of 18 in. These adjacent areas need not be classified as hazardous if they are effectively cut off from the hangar by walls or partitions and are adequately ventilated per **513.3(D)**. (See **513.3(B)** above) **(See Figure 21-43)**

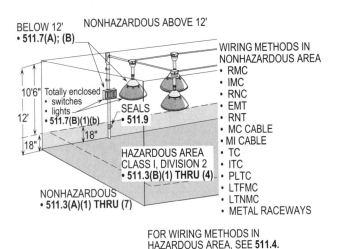

Figure 21-42. The above are rules and regulations pertaining to wiring methods and equipment installed in and above the hazardous area in garages.

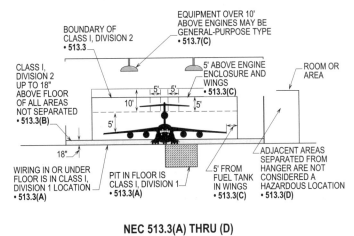

Figure 21-43. The above are the rules and regulations for equipment and wiring methods installed in aircraft hangars.

CLASSIFICATION OF LOCATIONS 513.3(A) THRU (D)

Pits or depressions located below the hangar floor level are considered as Class I, Division 1 locations, and this classification extends to the floor level per **513.3(A)**.

The entire area of the hangar is considered a Class I, Division 2 location to a height of 18 in. above the floor and includes adjacent areas into which the hazards may be communicated into unless suitably cut off from the hangar per **513.3(B) and (D)**.

The area immediately adjacent to the aircraft is also a Class I, Division 2 location. Such an area is defined as being within 5 ft. horizontally from aircraft power plants, aircraft fuel tanks, or aircraft structures containing fuel. These locations must extend from the floor level vertically to a level of 5 ft. above the wings and above the engine enclosures per **513.3(C)**.

WIRING AND EQUIPMENT IN CLASS I LOCATIONS 513.4

Section **513-3** gives the classification of areas within a hangar as either a Class I, Division 1 or Class I, Division 2 location. All wiring and equipment within a hazardous area in a hangar must comply with the requirements for either a Class I, Division 1, or Class I, Division 2 location, based upon the classification of the area in which the wiring and equipment is installed per **513.4**. Note that wiring in or under the hangar floor must be classified as Class I, Division 1 per **513.8**.

Attachment plugs and receptacles in hazardous areas are subject to the approval of the AHJ. Such devices may be either identified for Class I locations, or, as an alternative, may be of a type so designed that they cannot be plugged or unplugged unless the circuit is deenergized.

WIRING NOT WITHIN CLASS I LOCATIONS 513.7(A)

Wiring that is outside the hazardous areas in a hangar may be as follows:

Hazardous (Classified) Locations

(1) Metal raceways,
(2) Type MI cable (mineral-insulated),
(3) Type TC cable (powe rand control tray cable), and
(4) Type MC cable (metal-clad).

This requirement is for nonhazardous areas in the room where the aircraft is stored, such as areas above the 18 in. level. In adjoining nonhazardous rooms, the wiring method may be any approved type.

GROUNDED AND GROUNDING CONDUCTORS
513.16

When circuits supplying portable equipment and pendants include an identified grounded conductor (white or gray) per **Article 200**, they must have the following devices of the polarized type and the grounded conductor of the flexible cord must be connected to the screw shell of any lampholder or to the identified terminal of utilization of equipment such as:

(1) Receptacles,
(2) Attachment plugs,
(3) Connectors, and
(4) Similar devices.

Grounding continuity must be provided between fixed raceways and the noncurrent-carrying metal parts of:

(1) Pendant fixtures,
(2) Portable lamps, and
(3) Portable utilization equipment.

This may be accomplished by properly bonding the equipment grounding conductor to the metal raceway system. The insulation of EGC's must be green or green with one or more yellow stripes.

EQUIPMENT NOT WITHIN CLASS I LOCATIONS
513.7(C); (D)

In locations other than those described in **513.3**, equipment capable of producing arcs, sparks, or particles of hot metals, such as lamps and lampholders for fixed lighting, cutouts, switches, receptacles, charging panels, generators, motors, or other equipment having make-and-break or sliding contacts, must be of such a type that the escape of sparks or hot metal particles will be prevented. Equipment of the totally enclosed type, or so constructed as to prevent the escape of sparks or hot metal particles, must comply with this rule. However, in areas described in **513.3(D)**, equipment may be of the general-purpose type.

Design Tip: Equipment that might produce an arc or spark and which is less than 10 ft. above wings and engine of aircraft, must have a tight enclosure that will prevent the escape of hot particles from an arc or spark.

STANCHIONS, ROSTRUMS, AND DOCKS
513.7(E)

Where these items are located, or likely to be located within 5 ft. of aircraft fuel tanks or engines outlined as in **513.3(C)**, the entire stanchion, rostrum, or dock is considered to be a hazardous location. All wiring and equipment must comply with the rules and regulations for Class I, Division 2 locations.

Where they are not located, or not likely to be located, as described in **513.7**, they are considered hazardous only up to a height of 18 in. above the floor. Above the 18 in. level, wiring and equipment may be installed by the provision of **513.4(B)**. Wiring and equipment below the 18 in. level must comply with the requirements for Class I, Division 2 locations. **(See Figure 21-44)**

Design Tip: Mobile stanchions, containing electrical equipment which conforms to **513.3(C)**, **513.7**, and **513.4(B)**, must carry at least one sign, permanently affixed to the stanchion, which must read as follows:

**WARNING
KEEP 5 FT. CLEAR
OF AIRCRAFT ENGINES
AND FUEL TANK AREAS**

SEALING
513.9

A seal is required at every point where a conduit is routed from a hazardous area to a nonhazardous area or from a Division 1 to a Division 2 location. Note that conduits beneath the hangar floor are considered to be in a Class I, Division 1 location per **501.15** and **505.16**.

Conduits that are embedded in or under concrete will be classified as being in the same hazardous area as that installed above the floor, when any connections lead into or through such areas.

Conduits emerging from such areas are consider as passing from a Class I, Division 1 area to a Class I, Division 2 area.

Design Tip: Seals are required when connecting conduits to an enclosure where there is a possibility of arcs or sparks or when passing from a hazardous to a nonhazardous area.

WARNING
KEEP 5 FT. CLEAR
OF AIRCRAFT ENGINES
AND FUEL TANK AREAS

MOBILE SERVICING EQUIPMENT WITH ELECTRICAL COMPONENTS
513.10(D)

Mobile servicing equipment which is not suitable for Class I, Division 2 locations must be so designed and mounted that all fixed wiring and equipment will be at least 18 in above the floor. This equipment includes vacuum cleaners, air compressors, air movers, and similar type equipment.

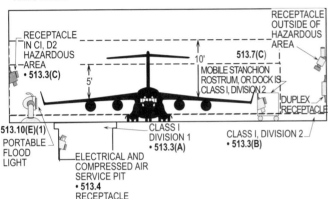

Figure 21-44. The above shows wiring methods, equipment, and portable equipment installed in hazardous and nonhazardous areas in aircraft hangers.

Design Tip: Unless such equipment is approved by the AHJ for Class I, Division 2 locations, the electrical equipment must be kept at least 5 ft. clear of aircraft engines and fuel tank areas.

Flexible cords for mobile equipment must be suitable for the type of service, be listed for extra-hard usage and include an equipment grounding conductor. Receptacles and attachment plugs must be suitable for the type of service and approved by the AHJ for the location in which they are installed.

Design Tip: Equipment which is not of a type suitable for Class I, Division 2 locations should not be operated in areas where maintenance operations are likely to release hazardous vapors in progress of repair. Such equipment must carry a warning sign to read as follows:

POWER SYSTEMS
513.10(A) THRU (E)

When an aircraft is stored in a hangar, the battery must be disconnected. It should also be disconnected, if possible, when the aircraft is brought into the hangar for repairs or maintenance.

Aircraft batteries must not be charged when the aircraft is in a hangar. Such charging must be done when the aircraft is entirely outside the hangar. Battery chargers should be kept in a separate building when they are not in use.

Electrical equipment on energizers must be kept at least 5 ft. clear of aircraft engines and fuel tank areas, and in addition must be at least 18 in. above the floor.

Design Tip: Power systems must have a permanently affixed sign which reads as follows:

WARNING
KEEP 5 FT. CLEAR
OF AIRCRAFT ENGINES
AND FUEL TANK AREAS

GROUNDING
513.16

All metallic raceways, and all noncurrent-carrying metallic portions of fixed or portable equipment, regardless of voltage, must be grounded according to the provision of **Article 250**. The insulation of EGC's must be green or green with one or more yellow stripes.

GASOLINE DISPENSING AND SERVICE STATIONS
514

One of the most common applications of the requirements in **Article 514** would be gasoline service stations. Basically, hazardous locations are areas where fuel (gasoline) is transferred to the fuel tanks of self-propelled vehicles or to auxiliary tanks. **Articles 510 and 511** applies to locations in filling stations such as lubritoriums, service rooms, repair rooms, offices, salesrooms, compressor rooms, storage rooms, rest rooms, as well as other associated areas.

Gasoline service stations are not the only areas subjected to the requirements of **Article 514**. Any location having a gasoline dispensing pump, whether in a service station or elsewhere, would come under the rules of this Article.

CLASS I LOCATIONS
TABLE 514.3(B)(1)

Table 514.3(B)(1) must be applied where Class I liquids are stored, handled, or dispensed and must be used to delineate and classify service stations. A Class I location does not extend beyond an unpierced wall, roof, or other solid partition. Locations, where the AHJ can satisfactorily determine that flammable liquids such as gasoline having a flash point below 38°C. (100°F.), will not be present, can be classified as nonhazardous. (See **Table 514.3(B)(1)** for the dimensions that are used in the following text - up to **514.4**.)

THE SPACE WITHIN A GASOLINE PUMP

The space within a gasoline pump is generally classified as a Class I, Division 1 location up to a height of 4 ft. above the base and immediately underneath the base. Normally, the space surrounding a gasoline pump, 18 in. out from the pump and 4 ft. up from its base, is Class I, Division 2. (See ANSI 87 and **Table 514.3(B)(1)** for these dimensions)

OUTSIDE AREA OF PUMP

In addition to the above, the circular outside area of 20 ft. horizontally surrounding the gasoline pump and out from the 18 in. circle is classified as a Class I, Division 2 location up to a height of 18 in. above ground. A building, or any part of a building, falling within this circle is a Class I, Division 2 location, if not suitably cut off by a barrier. In most cases, a solid wall or an 18 in. elevation above grade provides an adequate barrier.

OUTSIDE AREA OF FILL PIPES

The circular outside area surrounding a fill pipe, 10 ft. out from the pipe, is considered a Class I, Division 2 location up to a height of 18 in. above ground. The space within a building that is within the 10 ft. circle is Class I, Division 2, if not suitably cut off.

> **Design Tip:** A wall or an 18 in. elevation above grade would in most cases provide a suitable barrier.

SUSPENDED PUMP

When a gasoline pump is suspended from a canopy, the entire area within the enclosure and 18 in. out from the enclosure in all directions is considered Class I, Division 1. The space 2 ft. beyond the 18-in. limit and down to grade is Class I, Division 2.

SURROUNDING SPACE FROM DISPENSER

The surrounding space out to a distance of 20 ft., measured from a point vertically below the edge of the dispenser enclosure, is Class I, Division 2 up to a height of 18 in. above grade. If a building or part of a building is constructed within this space, it is considered nonhazardous, if suitably cut off by a ceiling or wall which acts as a barrier.

OUTSIDE AREA OF VENT PIPES

The area 3 ft. in all directions from the discharge ends of a vent pipe is considered Class I, Division 1. The area 2 ft. beyond the 3 ft. range is considered Class I, Division 2.

AREA AROUND A LUBRICATION ROOM

A lubrication room is considered Class I, Division 2 up to a height of 18 in. above the floor. A pit in a lubrication room is considered Class I, Division 1 from the floor down to the bottom of the pit. See **Figure 21-45** for a detailed illustration of classifying area around dispensers and gas stations.

WIRING AND EQUIPMENT WITHIN CLASS I LOCATIONS
514.4

Electrical equipment and wiring installed in the Class I, Division 1 and Class I, Division 2 locations, must be approved and suitable for such locations by the AHJ. These wiring methods are acceptable in these areas:

(1) Rigid metal conduit, and
(2) Type MI cable.

Note: See **Article 514.3** and **Article 501**.

> **Design Tip:** Underwriter's Laboratories has a directory with listings on types of insulations on conductors that are approved for these locations.

WIRING AND EQUIPMENT ABOVE CLASS I LOCATIONS
514.7

The rules for wiring and equipment above hazardous areas (above the 18 in. level) are the same as for commercial garages per **514.3** and **511.7**.

These wiring methods are permitted in these locations:

(1) Metal raceways, and
(2) Type MI cable.

CIRCUIT DISCONNECTS
514.11; 514.13

Each circuit leading to or through a dispensing pump must be provided with a switch or other acceptable means to disconnect simultaneously from the source of supply all conductors of the circuit including the grounded neutral, if used per **514.11** and **514.13**.

The intent of this requirement is to ensure that the supply is disconnected from the source and there are no conductors connected that lead to or through a dispensing pump.

> **Design Tip:** There are special breakers available that have a pigtail which ties to the neutral bus and ensures that the ungrounded and grounded (neutral) conductor(s) are disconnected when the breaker is in the OFF position.

Therefore, the threat of an arc or spark is limited, should the grounded neutral be disconnected with another circuit still energized, which is sharing the neutral with other circuits. **(See removing voltage in Figure 21-46)**

> **Design Tip:** Each dispensing device must be provided with a means to remove all external voltage sources, including feedback, during periods of maintenance and service of the dispensing equipment per **514.13**.

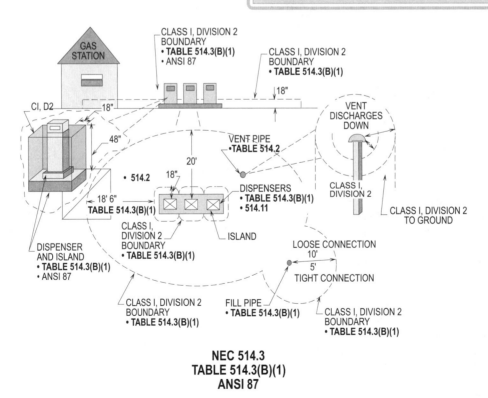

Figure 21-45. The above illustrates the classified area around dispensers and gas stations.

Hazardous (Classified) Locations

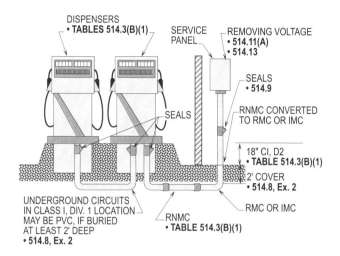

Figure 21-46. The above illustrates the rules and regulations for wiring methods and seals used for dispensers and boundaries to gas stations.

Design Tip: Emergency controls must shut off all power to all dispensing equipment at the station. Controls must be manually reset only in a manner approved by the AHJ. (Review **514.11** and **514.13** of the NEC.)

SEALING AT DISPENSER
514.9(A)

An approved seal must be provided in each conduit run entering or leaving a dispenser or any cavities or enclosures in direct communications therewith. The sealing fitting must be the first fitting after the conduit emerges from the earth or concrete.

Design Tip: All seals are to be readily accessible and no seal is to be buried or installed in an inaccessible wall. This rule makes it very clear that no fitting of any kind can be installed in the conduit between the floor and seal. For example, if a conduit run is below a panelboard and there is a hazardous location below, a seal must be installed at 18 in. or more above the floor without a fitting between the seal and floor. (**See location of seals in Figure 21-46**)

ATTENDED SELF-SERVICE STATIONS
514.11(B)

For attended self-service stations, the emergency controls specified in **514.11(A) and (B)** and **514.13** must be installed at a location acceptable to the Authority Having Jurisdiction (AHJ). Such emergency controls must not be located more than 100 ft. (30 m) from the dispensers.

UNATTENDED SELF-SERVICE STATIONS
514.11(C)

For unattended self-service stations, the emergency controls specified in **514.11(A) and (C)** must be installed at a location acceptable to the AHJ. Such emergency controls must be more than 20 ft. (7 m) but less that 100 ft. (30 m) from the dispensers.

Additional emergency controls must be installed on each group of dispensers or the indoor equipment must be used to control the dispenser.

The main emergency disconnect usually is an across-the-line device. However, a remote control device is allowed for the additional controls required but they must be capable of being manually reset.

SEALING AT BOUNDARY
514.9(B)

Additional seals shall be provided as required in **501.15**, **501.15(A)(4)** and **501.15(B)(2)**. Such sealing rules must apply to both horizontal and vertical boundaries between the hazardous and nonhazardous areas. (**See Figure 21-47**)

GROUNDING
514.16

Metallic portions of dispensing pumps, metallic raceways, and all noncurrent-carrying metal parts of electric equipment, regardless of voltage, must be grounded as required in **Article 250**. Such grounding must extend back to the service equipment and service ground.

In case of an accidental fault, a poor grounding connection can create an arc and cause an explosion to occur.

UNDERGROUND WIRING
514.8 and Ex. 2

Underground wiring must be installed in rigid metal conduit, threaded steel intermediate metal conduit, or, where buried under not less than 2 ft. of earth, it can be installed in rigid nonmetallic conduit if it complies with **Article 352**. Where rigid nonmetallic conduit is used, threaded rigid metal conduit, or threaded steel intermediate metal conduit must be used for the last 2 ft. of the underground run to emergence or to the connection to the aboveground raceway, an equipment grounding conductor must be included to provide electrical continuity of the raceway system and for grounding noncurrent-carrying metal parts. (**See Figures 21-46 and 47**)

Design Tip: Nonmetallic conduit was added to the NEC to take the place of metallic conduit and ease the problem of corrosion which might let fumes or gasoline into the conduit system. The use of nonmetallic conduit does not change the rules for sealing. To protect from physical damage, metallic conduit must be brought up into the dispenser or the nonhazardous location.

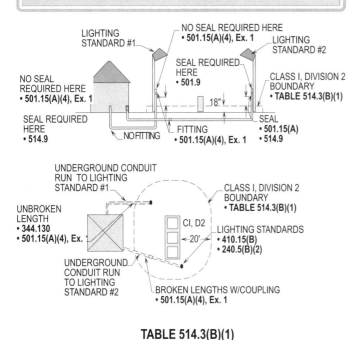

Figure 21-47. The above illustrates additional seals that must be provided.

BULK-STORAGE PLANTS
ARTICLE 515

A bulk storage plant is that portion of a property where flammable liquids are received by tank vessel, pipelines, tank car, or tank vehicle and are stored or blended in bulk for the purpose of distributing such liquids by tank vessel, pipeline, tank car, tank vehicle, portable tank, or container. For example, such designation includes locations where gasoline or other volatile flammable liquids are stored in tanks having an aggregate capacity of one carload or more and from which these products are distributed normally by tank truck, rail tank, etc.

CLASS I LOCATIONS
515.3; TABLE 515.3

Table 515.3 must be applied where Class I liquids are stored, handled, or dispensed and must be used to delineate and classify bulk storage plants. The Class I location is not required to extend beyond an unpierced wall, roof, or other solid partition.

PUMPS, BLEEDERS, WITHDRAWAL FITTINGS, METERS, AND SIMILAR DEVICES LOCATED INDOORS
TABLE 515.3

Such devices when installed in a gasoline pipeline can be subject to leakage, and therefore present a hazardous condition. Where such devices are located indoors, the hazardous area around the devices extends as follows:

(1) 5 ft. in all directions, to which this boundary includes an imaginary sphere around such device, and

(2) 25 ft. out horizontally and 3 ft. up from the floor grade, to which this boundary includes an imaginary circular area around such device.

The above areas are considered Class I, Division 2 areas and all wiring and equipment installed in these areas must comply with the requirements for a Class I, Division 2 location.

PUMPS, BLEEDERS, WITHDRAWAL FITTINGS, METERS, AND SIMILAR DEVICES LOCATED OUTDOORS
TABLE 515.3

Such devices, when located outdoors, the Class I, Division 2 hazardous area around these devices extends as follows:

(1) 3 ft. in all directions, forming an imaginary sphere, and
(2) 10 ft. out along the ground and 18 in. up from the ground forming an imaginary circular area.

See Figure 21-48 for a detailed illustration of such devices installed outdoors.

Hazardous (Classified) Locations

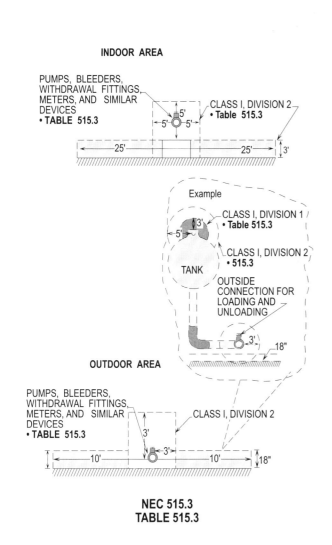

Figure 21-48. Classifying the hazardous area around a device or fitting located indoors or outdoors.

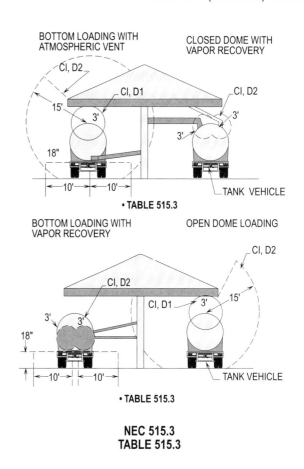

Figure 21-49. The above illustrates the hazardous areas, under certain conditions, around tank and car vehicles, based on top or bottom loading or unloading procedures.

TANK VEHICLES AND TANK CARS IN OUTSIDE LOCATIONS
TABLE 515.3

The hazardous area extends 15 ft. in all directions around an open dome or vent forming a 15 ft. sphere. The first 3 ft. is considered Class I, Division 1 with the remaining 12 ft. considered Class I, Division 2.

For bottom loading or unloading, the hazardous area extends 3 ft. in all directions from the fixed connection and within a 10 ft. radius and 18 in. up from the ground. These hazardous areas are considered Class I, Division 2, but only during loading and unloading operations. **(See Figure 21-49)**

ABOVEGROUND TANKS
TABLE 515.3

The space above the roof and within the shell of a floating roof type tank is classified as a Class I, Division 1 location.

For other tanks of this type, the area within 10 ft. in all directions is Class I, Division 2, except the area within 5 ft. in all directions is considered Class I, Division 1. The area between 5 and 10 ft. out is classified Class I, Division 2. **(See Figure 21-50)**

PITS
TABLE 515.3

A pit that is located in the 10 or 25 ft. sphere around pumps, bleeders, meters, etc. is considered as a Class I, Division 1 location.

A pit located in the 10 ft. sphere around a fill pipe opening is a Class I, Division 1 location.

A pit located in a nonhazardous area can be considered as nonhazardous where there are no piping, valves, or fittings in the pit. If the pit has piping, valves or fittings, it must be classified as a Class I, Division 2 area.

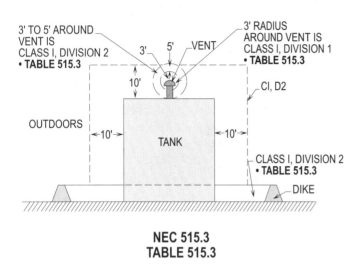

Figure 21-50. The above illustrates the hazardous area around an outdoor bulk storage tank.

GARAGES FOR TANK VEHICLES
TABLE 515.3

Storage and repair garages for tank vehicles are considered as Class I, Division 2 up to a height of 18 in. above the floor and must be wired with wiring methods and equipment rated for such locations.

ADJACENT LOCATIONS
TABLE 515.3

Office buildings, boiler rooms, etc. are not considered hazardous locations, if they are not used for handling and storage of gasoline or other flammable material.

WIRING AND EQUIPMENT WITHIN IN CLASS I LOCATIONS
515.3

All wiring and equipment in a hazardous area must comply with the requirements of a Class I, Division 1 or 2 location, or Article 505 for the Zones, whichever applies, based upon the flammable material.

WIRING AND EQUIPMENT ABOVE HAZARDOUS LOCATIONS
515.4

The requirements of this Section are basically the same as those for commercial garages. The main concern is sparks, arcs, or hot metal particles that might drop into the hazardous area and create an explosion, if there was a proper mixture of gases and air.

Electrical equipment which might produce arcs, sparks, or particles of hot metal, such as lamps and lampholders for fixed lighting, cutouts, switches, receptacles, motors, or other such equipment having make-and-break or sliding contacts, must be totally enclosed or constructed in such a manner as to prevent the escape of sparks or hot metal particles.

Design Tip: Portable lamps or utilization equipment must be suitable and must comply to the provisions of Article 501 for the hazardous location in which they are to be used.

The wiring methods used above the 3 ft. and 18 in. levels to connect the electrical equipment listed above can be one of the following or any combination of such:
  (1) Metal raceways,
  (2) Type MI cable,
  (3) Type TC cable,
  (4) Type MC cable,
  (5) PVC Sch. 80, and
  (6) Rigid nonmetallic conduit.

UNDERGROUND WIRING
515.8(A) THRU (C)

Underground wiring routed to and around above ground storage tanks:

(1) Where buried less than 2 ft. in the earth, rigid or IMC conduit must be used.

(2) Where buried 2 ft. or more, nonmetallic conduit may be used. However, a change must be made to metal conduit for the last 2 ft. before emergence from the ground. Cable requires 2 ft. of cover.

(3) Where nonmetallic conduit is used, a grounding conductor must be routed with the circuit conductors to ground all noncurrent-carrying parts.

(4) Where underground cable, without a metal sheath

is used, the cable must have a grounding conductor to ensure proper grounding of equipment with metal parts.

(5) Where cable is used, it must be protected by rigid metal conduit or IMC at points where it emerges from the earth.

(6) Conductor insulation must be a type approved for the location, to protect from gasoline in case of leakage or an accidental rupture of equipment.

> **Design Tip:** Nonmetallic conduit and cables must be buried at least 2 ft. in the earth or have a 2 ft. cover.

SEALING REQUIREMENTS
515.9

Seals must be provided in accordance with **501.15**. Sealing requirements in **501.15(A)(4)** and **501.15(B)(2)** apply to horizontal as well as to vertical boundaries of the defined Class I locations. Buried raceways under defined hazardous locations are considered to be within such locations.

Seals when required must be installed as follows:
(1) Seals must be placed in each conduit run entering the enclosure for switches, circuit breakers, fuses, relays, resistors, or any other device that is capable of producing arcing, sparking, or high temperature and which is located in a hazardous area. These seals must be placed no further than 18 in. from such enclosures.

(2) A seal is required at the point where the conduit leaves a hazardous area and passes into a nonhazardous area. The seal is not required to be installed at the exact boundary of the hazardous area. It can be installed at a convenient point in the run, on either side of the boundary. However, it is recommended to place the seal as near to the boundary as possible.

A vertical run must be sealed where it enters or leaves the hazardous level either at 18 in. or 36 in. above the floor or ground. For a horizontal run, if the run leaves the area at a height below the hazardous level, it is mandatory that the conduit be sealed where it leaves the hazardous area.

An exception allows a conduit which passes through a hazardous area, with no fitting within or 12 in. beyond the hazardous boundary, to be installed without seals.

(3) For 2 in. and larger conduit runs, seals are required at the entry to an enclosure, if the box or fitting contains splice, tap, or connection and is located within a hazardous area. Note, such seal must be installed within 18 in. of the enclosure.

> **Design Tip:** Conduits installed underneath a hazardous area is considered to be in the hazardous area beneath which it is buried and must be treated as such.

GASOLINE DISPENSING PUMPS
515.10

When gasoline is dispensed from a bulk storage plant, the rules of **Article 514**, which covers the dispensing of gasoline, must be applied.

In other words, if there are gasoline pumps on the premises, the pump area must comply with the rules and regulations for service stations.

GROUNDING REQUIREMENTS
515.16

All metal raceways, metal-jacketed cables, and all noncurrent-carrying metal parts of fixed or portable electrical equipment, regardless of voltage, must be grounded by the provisions in **Article 250**. Grounding in Class I locations must comply with the rules of **501.30**.

> **Design Tip:** This grounding technique is required for all wiring in order to provide proper grounding for raceways and equipment where there is no raceway continuity.

The grounding continuity must be complete back to the service equipment and the service ground for nonhazardous as well as hazardous areas.

FINISHING PROCESSES
ARTICLE 516

Article 516 applies for locations where paint, lacquer, or powdered finishing materials are frequently applied by brushing, spraying, or dipping. Conditions of process must always be evaluated based upon use and the effects of the material.

For example, the vapors from finishing processes and the storage of paints, lacquers, or other flammable finishes must be considered.

Another problem is the residue which is the result of the

finishing process. Vapors can be easier to control than the residue because residue can be ignited by heat and cause fires to occur. The right mixture of vapors and air can create a dangerous explosion when ignited by an arc or spark.

> **Design Tip:** NFPA 33 states that water-base paints, when mixed create no problem. However, when sprayed, a residue condition exists and this could cause a problem.

CLASS I OR II LOCATIONS
516.3

Classification is in respect to the effects of and exposure to flammable gases and vapors, and in some cases deposits of paint spray.

> **Design Tip:** For deposits and residues requirements, review **516.3(B) and (C)**.

CLASS I OR II, DIVISION 1 LOCATIONS
516.3(B)(1) THRU (6)

Where brushing, spraying, or dipping is not performed in a spray booth, the hazardous areas are to be as follows:

(1) The space around a spraying operation which contains a dangerous quantity of spray vapor is considered a Class I, Division 1 location. The NEC does not define a definite boundary for such space. The extent of the hazardous boundary is always subject to the judgment of the AHJ.

(2) The space beyond the Class I, Division 1 space is Class I, Division 2 up to 20 ft. beyond, and up from the floor to a height of 10 ft. above the spraying area.

(3) For dipping operations, all space within 5 ft. in any direction from the liquid, whether the liquid is in the tank or as a wet coating on the drain board or on the dipped object is classified as a Class I, Division 1 location.

> **Design Tip:** Where the brushing, spraying, or dipping is performed in a spray booth, the entire spray booth is classified as Class I, Division 1. Exhaust ducts for the spray booth are also considered as Class I, Division 1. **(See Figure 21-51)**

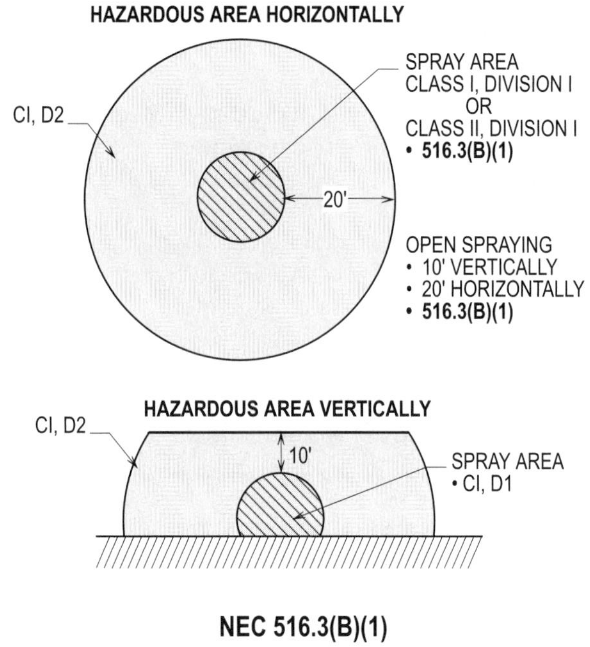

Figure 21-51. The above illustrates the hazardous area for an unenclosed area used for spraying.

CLASS I OR II, DIVISION 2 LOCATIONS
516.3(B)(2)(a); (b)

Where the spray booth is equipped with an open front, the space outside the booth is considered Class I, Division 2 and rules are applied as follows:

(1) For spray booths with an interlocked vent system, so designed and arranged that spraying cannot be performed except when the vent system is operating, under these conditions, the space outside the booth must be classified as a Class I, Division 2.

(2) For spray booths without an interlocked vent system, the space outside the spray booth is considered a Class I, Division 2, due to the possibility of spraying without proper ventilation.

(3) For open-top spray booths, the Class I, Division 2 hazardous area is considered to extend to 3 ft. above the booth and 3 ft. from all openings.

See Figures 21-52(a); (b); and (c) for a detailed illustration of hazardous area around spray booths.

Hazardous (Classified) Locations

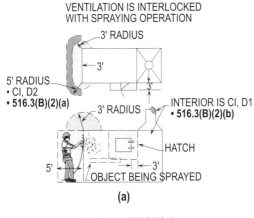

NEC 516.3(B)(2)(a)

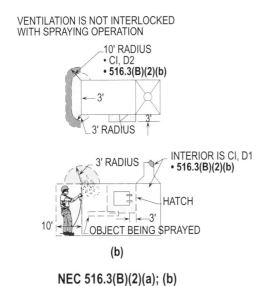

NEC 516.3(B)(2)(a); (b)

Figure 21-52(a). The above illustrates the hazardous areas around spray booths, based upon the ventilation being interlocked and not interlocked with spray operation.

NEC 516.3(B)(5); (B)(6)

Figure 21-52(c). The above illustrates the hazardous area in and around dip tanks and drain boards.

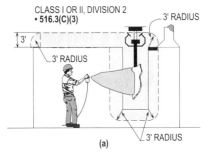

NEC 516.3(C)(3)

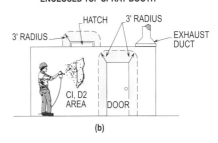

NEC 516.3(C)(4)

Figure 21-52(b). The above illustrates the hazardous areas in and around open and enclosed spray booths.

WIRING AND EQUIPMENT IN CLASS I LOCATIONS
516.4

All electric wiring and equipment installed within the Class I location containing vapor only and not residue must comply with the following rules:

(1) All wiring and equipment in the hazardous areas must comply with the NEC requirements for a Class I, Division 1 or Class I, Division 2 location, whichever applies.

(2) Any equipment, installed in a location where paint or lacquer could accumulate on the equipment during operation to such an extent that a hazard might exist, must be approved for such condition of use.

(3) Lighting fixtures within a spray booth are required to be approved for Class I, Division 1 locations, and are usually of an explosionproof type. To avoid use of such expensive lighting fixtures, lighting can be accomplished through glass panels in the wall or ceiling of the spray booth. If the lighting fixture is installed outside of the Class I, Division 1 location, a less expensive lighting fixture can be used. (**See Figure 21-53**)

21-45

(4) Portable lamps must be of a type which are approved for Class I, Division 1 locations. However, portable appliances can be used in a spray booth only, when no spraying, brushing, or dipping operation is being performed.

Where deposits of residue present a heating problem, illumination can be provided through panels of glass or other approved transparent or translucent material, if the following rules are followed:

(a) Fixed lighting units are utilized.

(b) The panel is of a material that breakage will be unlikely during condition of use.

(c) Panels properly isolate the hazardous location from the area in which the lighting units are installed.

(d) The accumulation of hazardous residues on the surface of the panel will not rise to a dangerous temperature during spray operations.

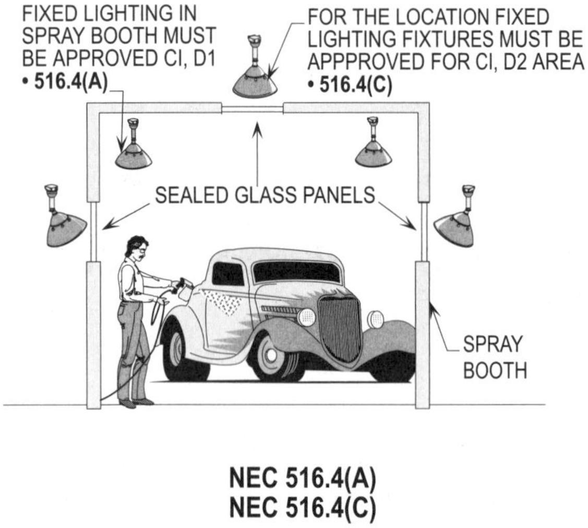

NEC 516.4(A)
NEC 516.4(C)

Figure 21-53. The above illustrates the rules for installing lighting fixtures inside or outside of spray booths.

WIRING ABOVE CLASS I LOCATIONS
516.7(A)

Above hazardous areas, the following wiring methods are allowed as follows:
(1) Metal raceways,
(2) Type MI cable,
(3) Type TC cable,
(4) Type MC cable,
(5) Electrical nonmetallic tubing, and
(6) Rigid nonmetallic conduit.

Design Tip: Leads without seals may be brought out of a cellular metal raceway in the ceiling. However, such leads can only be brought out of a cellular metal raceway in the floor, if seals are properly installed.

CLASSIFICATION OF AREA

There are basically seven steps to be applied for classifying an area and they are as follows:

(1) Determining the governing codes, standards, and authorities,
(2) Determining flammable materials,
(3) Determining group,
(4) Determining class,
(5) Determining division,
(6) Determining extent (boundaries) of classified areas, and
(7) Preparing plans and specifications.

DETERMINING THE GOVERNING CODES, STANDARDS AND AUTHORITIES

The first step in area classification consists of establishing the relevant codes, standards, and the authority having jurisdiction (AHJ), and obtaining all permits and other licenses. There are many US companies that have their own operating standards and these should also be consulted.

The AHJ is the organization, office, or individual responsible for approving equipment, installation, etc. The AHJ can be federal, state, or local, a fire chief or marshal, fire preventive bureau, labor department, a rating bureau, etc. The following are some of the relevant codes, standards, and testing laboratories:

(1) Underwriters Laboratories (UL)
(2) Factory Mutual Research Corporation (FM)
(3) Electrical Testing Laboratories (ETL)
(4) American Gas Association (AGA)
(5) National Electrical Manufacturers Association (NEMA)
(6) American Petroleum Institute (API)
(7) Occupational Safety and Health Administration (OSHA)
(8) Mine Safety and Health Administration (MSHA)
(9) United States Coast Guard (USCG)
(10) Instrument Society of America (ISOA)
(11) Institute of Electrical and Electronic Engineers (IEEE)
(12) International Electrotechnical Commission (IEC)

(13) Local Building Codes
(14) Manufacturer's data
(15) National Fire Protection Association (NFPA)
- National Electrical Code (NEC®)

DETERMINING FLAMMABLE MATERIALS

Flammable liquids are defined as being any liquid having a closed cup flash point below 100°F and a vapor pressure not exceeding 40 lb/in.^{2} absolute at 100°F. Flammable and combustible liquids are subdivided into Classes I, II, and III. Classes are used to identify flammable and combustible liquids and should not be confused with class as defined in the National Electrical Code (NEC®).

Flammable liquids (Class I) are those having flash points below 73°F and having boiling points below 100°F. Class I(B) liquids are those having flash points below 73°F and having boiling points at or above 100°F. Class I(C) liquids are those having flash point at or above 73°F and below 100°F.

Combustible liquids (Class II and III) are those having flash points at or above 100°F and below 140°F. Class III(A) liquids are those having flash points at or above 140°F and below 200°F. Class III(B) liquids are those having flash points at or above 200°F.

DETERMINING GROUP

Equipment must be selected, tested, and approved for the type of flammable material involved. Maximum explosive pressures, MESG, safe operating temperatures (ignition temperature), and ignition energy must be considered in the design and testing of safe electrical equipment for installation in classified areas. The NEC has a partial list of the most commonly encountered materials which have been tested. They are grouped on the basis of their flammability characteristics:

Group A	acetylene
Group B	hydrogen
Group C	carbon monoxide, ethylene
Group D	gasoline, benzine, propane, ethyl alcohol, methane
Group E-G	dusts

DETERMINING CLASS

The *National Electrical Code* (NEC) refers to three classes:

Class I	flammable gases and vapors
Class II	combustible dusts
Class III	easily ignitable fibers or flyings

DETERMINING DIVISION

The *National Electrical Code* (NEC) recognizes two divisions and they are as follows:

Division 1: A location that is likely to have flammable gases or vapors present under normal conditions.

Division 2: A location that is likely to have flammable gases or vapors present only under abnormal conditions.

DETERMINING EXTENT (BOUNDARIES) OF CLASSIFIED AREAS

It is important to exercise sound engineering experience and judgment in determining the extent (boundaries) of a classified area. Having established the presence of flammable substances and determined the perimeter of the classified area, the class, group, and division is selected. The next step is to select the equipment and installation techniques. The following steps will aid in determining the extent of the boundaries of the classified area:

(1) The flammable materials involved,
(2) The type of installation,
(3) The Class (I, II, or III),
(4) The Group,
(5) The Division (1 or 2),
(6) Flash point of material,
(7) Ignition of sources,
(8) Volume and pressure release,
(9) Ventilation,
(10) Fire walls and barriers, and
(11) Purging and pressurization.

PREPARING PLANS AND SPECIFICATIONS

Plans and specifications should be prepared so as to outline the hazardous area and the extent of the classified boundaries. In addition, plans that clearly illustrate the divisions and the extent of the areas should be prepared. Tables should also be developed with properties of the flammable materials and a written description based on the classification which explains the methods and procedures of each area being classified.

QUESTIONS AND ANSWERS FOR CLASSIFYING DIVISIONS

The following questions can be used to determine the Division of a Classified Location.

ASSIGNMENT OF DIVISION 1 LOCATIONS

(1) Is an ignitable atmospheric mixture likely to exist under "normal" operating conditions?

(2) Is an ignitable atmospheric mixture likely to occur frequently due to repair, maintenance, or leakage?

(3) Would failure of process equipment, piping or vessels be likely to cause a failure of the electrical system simultaneous with the release of the combustible material?

(4) Is a piping system containing combustible material in an inadequately ventilated space and is the system likely to leak?

(5) Is the space or area in question below grade level such that heavier-than-air vapors may accumulate there?

(6) Is the space or area in question above grade level such that lighter-than-air vapors may accumulate there?

An answer of yes to any of these questions would require a Division 1 classification.

ASSIGNMENT OF DIVISION 2 LOCATIONS

(1) Is a piping system containing combustible material in an inadequately ventilated space and is the system **not** likely to leak?

(2) Is a process equipment system containing a combustible material in an inadequately ventilated area and can the material escape only during abnormal situations such as failure of gaskets or packing?

(3) In the location adjacent and open to a Division 1 location by trenches, pipes or ducts?

(4) If mechanical ventilation is used, can failure or abnormal operation of the ventilation equipment permit an ignitible atmospheric mixture?

An answer of yes to any of these questions would require a Division 2 classification.

GENERAL REQUIREMENTS 501.1

Article 501 in Chapter 5 of the NEC covers the general requirements and how such rules apply to the electrical wiring and equipment installed in locations classified as Class I, Zone 0, Zone 1, or Zone 2. The concept of using Zones, per the IEC, instead of Divisions, per the NEC, has been used internationally for many years. In the past, the International Electrotechnical Commission (IEC) has regulated the requirements for such a system.

The IEC system uses two groups to identify the hazards involved. Group I is used for mining while group II is used for surface industries and offshore installations.

Group II consists of three subgroups which are A, B, and C respectfully. Subgroups A, B, and C are designed to represent categories of flammable gases or vapors that are based upon the minimum ignition energy of the hazard. Subgroup A represents the most difficult flammable gas or vapor to ignite and Subgroup C is the easiest to ignite. **(See Figure 21-54)**

Design Tip: Check with local NEC and OSHA inspectors to obtain their permission to use the zone concept, when designing and installing electrical equipment and wiring methods in hazardous locations.

	NEC GROUP	REPRESENTATIVE GAS	IEC GROUP	
GROUP A AND B IS MORE IGNITIBLE • 500.6(A)(1); (A)(2)	A	ACETYLENE	IIC	GROUP IIC IS MORE IGNITIBLE • 505.6(A), FPN
	B	HYDROGEN		
GROUP C • 500.6(A)(3)	C	ETHYLENE	IIB	GROUP IIB • 505.6(B), FPN
GROUP D • 500.6(A)(4)	D	PROPANE	IIA	GROUP IIA • 505.6(C), FPN

TABLE SHOWING A COMPARISON OF NEC AND IEC GROUPING AND CLASSIFICATION SCHEME NEC 505.6(A) thru (C)

Figure 21-54. The Table above shows a comparison of the NEC and IEC grouping and classification scheme.

ZONE CLASSIFICATION 505.5(A); (B)

The IEC system uses three Zones to represent the different levels of risk. Zone 0 represents areas in which an explosive gas/air mixture is continuously present or present for long periods of time. Zone 1 represents areas in which an explosive gas/air mixture is likely to occur in normal operation. Zone 2 represents areas in which gas/air mixtures are not likely to occur and should they occur, they exist only for a short period of time. See **500.5(B)(1)** and **(B)(2)** to define Divisions 1 and 2 for similar situations, where gas/air mixtures present risk levels based upon conditions. **(See Figure 21-55)**

TABLE 1	
IEC ZONE CHART	
ZONES	HAZARDOUS RISK LEVELS
ZONE 0	IN WHICH AN EXPLOSIVE GAS-AIR MIXTURE IS CONTINUOUSLY PRESENT. • 505.5(B)(1)
ZONE 1	IN WHICH AN EXPLOSIVE GAS-AIR MIXTURE IS LIKELY TO OCCUR IN NORMAL OPERATION. • 505.5(B)(2)
ZONE 2	IN WHICH AN EXPLOSIVE GAS-AIR MIXTURE IS NOT LIKELY TO OCCUR IN NORMAL OPERATION AND IF IT OCCURS WILL ONLY LAST FOR A SHORT PERIOD. • 505.5(B)(3)

TABLE 2			
COMPARISON OF IEC ZONES AND NEC DIVISIONS			
IEC ZONES	0	1	2
NEC DIVISIONS		1	2

NEC 505.5(B)(1) THRU (B)(3)

Figure 21-55. The Table above can be used to define the different Zones, based upon how often a hazardous gas or vapor is present, using the IEC system concept.

CLASS I, ZONE 0 LOCATIONS
505.5(B)(1)

A Class I, Zone 0 location is a location in which an explosive gas atmosphere is present continuously or for long periods of time.

Note: This usually applies to the inside of containers or apparatus such as vaporizers, reactors, storage tanks, etc. Long periods of time could be considered as the time needed for gases and vapors to mix with air and create an explosive mixture. **(See Figure 21-56)**

CLASS I, ZONE 1 LOCATIONS
NEC 505.5(B)(2)

A Class I, Zone 1 location can be defined as follows:

(1) Ignitable concentrations of flammable gases or vapors are likely to exist under normal operating conditions.

(2) Ignitable concentrations of flammable gases or vapors may exist frequently because of repair or maintenance operations because of leakage.

(3) Equipment is operated or processes are carried on, of such a nature that equipment breakdown or faulty operations could result in the release of ignitable concentrations of flammable gases or vapors and also cause simultaneous failure of electrical equipment in a mode to cause the electrical equipment to become a source of ignition.

(4) Adjacent to a Class I, Zone 0 locations, which ignitable concentrations of vapor could be communicated, unless communication is prevented by adequate positive-pressure ventilation from a source of clean air and effective safeguards against ventilation failure are provided. **(See Figure 21-57)**

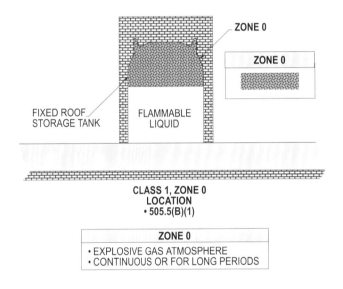

Figure 21-56. The above illustration shows a Class I, Zone 0 location.

CLASS I, ZONE 2 LOCATIONS
505.5(B)(3)

A Class I, Zone 2 location can be defined as follows:

(1) Ignitable concentrations of flammable gases or vapors are not likely to occur in normal operation and if they do occur, will exist only for a short period.

(2) Volatile flammable liquids, flammable gases, or flammable vapors are handled, processed, or used, but in which the liquids, gases, or vapors normally are confined within closed containers or closed systems from which they can escape only as a result of accidental rupture or breakdown of the containers or system, or as the result of the abnormal operation of the equipment with which the liquids or gases are handled, processed, or used.

(3) Ignitable concentrations of flammable gases or vapors normally are prevented by positive mechanical ventilation, but which may become hazardous as the result of failure or abnormal operation of the ventilation equipment.

(4) Is adjacent to a Class I, Zone 1 location, from which ignitable concentrations of flammable gases or vapors could be communicated, unless such communication is prevented by adequate positive-pressure ventilation from a source of clean air, and effective safeguards against ventilation failure are provided. **(See Figure 21-58)**

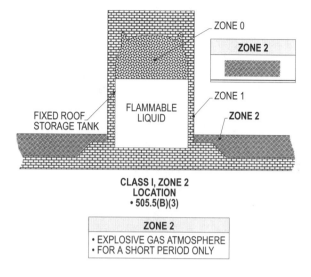

Figure 21-58. The above illustration sows a Class I, Zone 2 location.

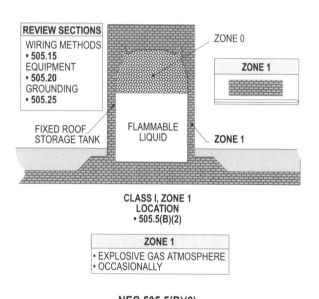

Figure 21-57. The above illustration shows a Class I, Zone 1 location.

CLASS I, ZONE 0, 1, and 2 EQUIPMENT 505.8(A) THRU (I)

Equipment installed in the zone locations must be provided a form of constuction that will ensure safe performance under conditions of use. Proper maintenance of the equipment is also necessary to prevent the possible igniting of an explosive mixture of gases and air. **(See Figure 21-59)**

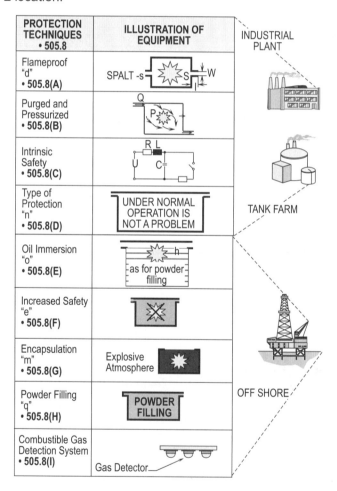

Figure 21-59. The illustration above shows equipment that is permitted in the zone locations.

Hazardous (Classified) Locations

SUITABILITY OF EQUIPMENT
505.9(A); (B)

Suitability of identified equipment shall be determined by one of the following:

(1) Equipment listing or labeling
(2) Evidence of equipment evaluation from a qualified testing laboratory or inspection agency concerned with product evaluation
(3) Evidence acceptable to the AHJ, such as a manufacturer's self-evaluation or an owner's engineering judgment. **(See Figure 21-60)**

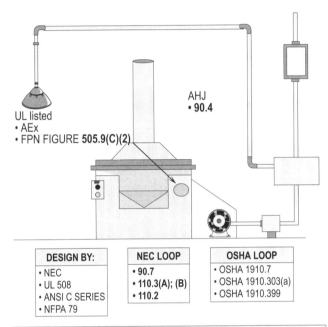

Figure 21-60. The above illustration shows how equipment must be accepted in Division and Zone locations.

LISTING
505.9(B)(1); (B)(2)

Section **505.9(B)(1)** allows equipment that is listed for a Zone 0 location to be used in Zone 1 or Zone 2 locations of the same gas group. Likewise, equipment listed or otherwise acceptable for a Zone 1 location is allowed to be used in a Zone 2 location of the same gas group. **(See Figure 21-61)**

Equipment must be listed for a specific gas or vapor, specific mixtures of gases or vapors, or any combination of gases or vapors per **505.9(B)(2)**. FPN: One common example is equipment marked for "IIB + H2".

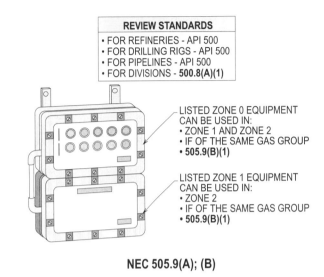

Figure 21-61. Equipment, under certain conditions, can be used from one zone to another.

MARKING
505.9(C)

Sections **505.9(C)(1) or (C)(2)** requires electrical equipment, approved for operation at ambient temperatures exceeding 45°C, to be marked with the maximum ambient temperature for which the equipment is approved and the operating temperature range at that specific ambient temperature.

Design Tip: Electrical equipment must be selected by a T-code number so that the enclosure won't overheat and cause an explosion to occur. The outer surface will not exceed the ignition temperature of the specific gas or vapor if the proper T-code number is selected and used. **(See Figure 21-62)**

WIRING METHODS FOR CLASS I, ZONE 0, 1, AND 2
505.15(A) THRU (C)

Wiring methods that are permitted to be used in a Class I, Zones 0, 1, or 2 locations are listed in **Figure 21-63** and are based upon area classification.

IEC SYSTEM		NEC SYSTEM	
TEMPERATURE CLASS	MAX. SURFACE TEMPERATURE °C	TEMPERATURE IDENTIFICATION NUMBER	IGNITION TEMPERATURE OF GASES OR VAPORS (°C)
T1	450	T1	450
T2	300	T2	300
		> T2A	> 280
		> T2B	> 260
		> T2C	> 230
		> T2D	> 215
T3	200	T3	200
		> T3A	> 180
		> T3B	> 165
		> T3C	> 160
T4	135	T4	135
		> T4A	> 120
T5	100	T5	100
T6	85	T6	85

>: NEC T CODE NUMBERS FOR TEMPERATURE IDENTIFICATION WHICH ARE NOT LISTED BY THE IEC SYSTEM, BUT THEY ARE LISTED IN THE NEC PER TABLE 500.8(B).

WHAT T NUMBER IS REQUIRED FOR IGNITION TEMPERATURE OF A GAS RATED AT 300°C?

STEP 1: FINDING T NUMBER
TABLE 505.9(D)(1)
300°C REQUIRES T2

SOLUTION: THE T CODE NUMBER IS REQUIRED TO BE T2.

NEC 505.9(D)(1)

Figure 21-62. The above illustration shows the procedure for selecting the temperature identification number for electrical equipment.

WIRING METHODS ALLOWED
ZONE 0
• INTRINSICALLY SAFE WIRING PER 505.15(A) • RMC • IMC • MI CABLE PER 505.15(B)(1)(C) • NONINCENDIVE • OPTICAL FIBER CABLE PER
ZONE 1
WIRING METHODS ALLOWED ARE THOSE IN: • CLASS I, DIVISION 1 • CLASS I, ZONE 0 • PLUS SEALING PER 505.16(A)
ZONE 2
WIRING METHODS ALLOWED ARE THOSE IN: • CLASS I, DIVISION 2 • CLASS I, DIVISION 1 • CLASS I, ZONE 0 • CLASS I, ZONE 1 • PLUS SEALING PER 505.16

NEC.'S 505.15(A) THRU (C)

Figure 21-63. The above chart lists the equipment allowed in Class I, Zones 0, 1, and 2.

EQUIPMENT 505.20(A) THRU (D)

Equipment must be listed and marked specifically for the Zone in which it is going to be installed. **Ex.'s to 505.20(A), (B) and (C)** allow equipment, under certain conditions, to be used from one Zone to another Zone.

Design Tip: Electrical equipment installed using the zone concept must comply with **Article 505** and be listed for use by a testing laboratory such as UL utilizing the symbol AEx. **(See Figure 21-64)**

CLASSIFICATION OF AREA

When applying the zone concept, the protection methods required to ensure safety are related to the likelihood of the presence of an explosive gas/air mixture. The higher the probability, greater is the care to ensure safety.

Installations where flammable gases, vapors, or liquids are present can be categorized into areas where the probability of the presence of explosive gas/air mixtures is high or low, and the electrical apparatus must be selected and installed accordingly. It is clear that the overall risk of exposure to an explosive atmosphere is assessed to both the anticipated frequency of release of flammable material and the probable duration of its presence.

See Figure 21-65 for an illustrated comparison between the IEC and NEC classifications, based upon the presence of gas/air mixtures and the conditions of exposure to personnel and equipment.

Listing, Marking, and Documentation 505.9(A); (B); (C); 505.4(A)

Zone equipment, if used instead of Division equipment, must be marked with Class, Zone, and the Symbol AEx to verify to the user that such equipment complies with American Standards as well as IEC standards. **(See Figures 21-65 and 66 for a detailed illustration of such markings)**

Hazardous (Classified) Locations

COMPARISON BETWEEN EQUIPMENT USING THE IEC ZONE CONCEPT OR 1996 NEC DIVISION CONCEPT			
EQUIPMENT ALLOWED BY IEC		**EQUIPMENT ALLOWED BY NEC**	
ZONE 0	ONLY INTRINSICALLY SAFE EQUIPMENT IS ALLOWED	CLASS I, DIVISION 1	THE FOLLOWING EQUIPMENT CAN BE USED: • EXPLOSIONPROOF ENCLOSURES • PURGING • INTRINSIC SAFETY
ZONE 1	EQUIPMENT ALLOWED TO BE USED: "d" FLAMEPROOF ENCLOSURE "p" PRESSURIZED APPARATUS "i" INTRINSIC SAFETY "o" OIL IMMERSION "e" INCREASED SAFETY "q" POWDER FILLING		
ZONE 2	ALL EQUIPMENT CERTIFIED FOR ZONE 0 OR 1 REQUIREMENTS CAN BE USED IN ZONE 2	CLASS I, DIVISION 2	ALL EQUIPMENT CERTIFIED FOR DIVISON 1 CAN BE USED PLUS: • HERMETICALLY SEALED CONTACTS • OIL IMMERSION OF CONTACTS • NONINCENDIVE CIRCUITS

NEC.'S 505.20(A) THRU (C)
NEC.'S 505.8(A) THRU (I)
TABLE 505.9(C)(2)(4)
FPN FIGURE 505.9(C)(2)

Figure 21-64. the above chart lists the equipment allowed in Class I, Zones 0, 1, and 2.

	IEC (Publication 79-10)		NEC (Section **500.5(B)(1); (B)(2)**)
Zone 0	Area in which an explosive gas-air mixture is continuously present or present for long periods. (Example: vapor space of a process vessel or storage tank.)	Class I, Division 1	A Class I, Division 1 location • in which ignitible concentrations of flammable gases of vapors exist under normal operating conditions; or • in which ignitible concentrations of such gases or vapors may exist frequently because of repair or maintenance operations or because of leakage; or • in which breakdown or faulty operation of equipment or processes might release ignitible concentrations of flammable gases or vapors, and might also cause simultaneous failure of electric equipment.
Zone 1	Area in which an explosive gas-air mixture is likely to occur in normal operation.		
Zone 2	Area in which an explosive gas-air mixture is not likely to occur, and if it occurs it will only exist for a short time. NOTE: ALSO, SEE **505.5(B)(1) THRU (B)(3)**	Class I, Division 2	A Class I, Division 2 location • in which volatile flammable liquids or flammable gases are handled, processed or used but in which the liquids, vapors or gases will normally be confined within closed containers or closed systems from which they can escape only in case of accidental rupture or breakdown of such containers or systems, or in case of abnormal operation of equipment; or • in which ignitible concentrations of gases or vapors are normally prevented by positive mechanical ventilation, and which might become hazardous through failure or abnormal operation of the ventilating equipment; or • that is adjacent to a Class I, Division 1 location, and to which ignitible concentrations of gases or vapors might occasionally be communicated unless such communication is prevented by adequate positive-pressure ventilation from a source of clean air, and effective safeguards against ventilation failure are provided.

Figure 21-65. The above Table illustrates a comparison between IEC and NEC area classifications.

Explanation of Markings
• FPN Figure 505.9(C)(2)

Example: Class I, Zone O   AEx   ia   IIC   T6

- Area classification
- Symbol for equipment built to American Standards
- Type(s) of protection designation
- Gas classification group (not required for protection techniques indicated in **505.6, FPN 2**)
- Temperature classification

Figure 21-66. The above Table illustrates an explanation of markings that are used on Zone equipment to identify certain characteristics.

Name   Date

Chapter 21. HAZARDOUS (CLASSIFIED) LOCATIONS

Section	Answer		
_____	T	F	1. Class I, Division 1 locations are those locations in which volatile flammable liquids or flammable gases are handled, processed, or used.
_____	T	F	2. Class II, Division 1 locations are those locations in which combustible dust is in the air under normal operation conditions in quantities sufficient to produce explosive or ignitable mixtures.
_____	T	F	3. Class III, Division 2 locations are those locations in which easily ignitable fibers are stored or handled except in process of manufacturing.
_____	T	F	4. Equipment used in the Divisions or Zones shall be identified and suitable for the location.
_____	T	F	5. Group E atmospheres contain combustible metal dusts.
_____	T	F	6. Vapor densities are used mainly to determine the settling or rising tendency of a mixture.
_____	T	F	7. Type 8 enclosures are used only for indoor locations classified as Class I, Groups A, B, C, and D.
_____	T	F	8. Type 9 enclosures are intended for indoor use in locations classified as Class II, Groups E or G.
_____	T	F	9. Nonincendive equipment is a form of intrinsic safety designed for use in Class I, Division 1 locations.
_____	T	F	10. Type MI cable shall be installed and supported in such a manner to avoid tensile stress at the termination fittings in Class I, Division 1 locations.
_____	T	F	11. Flexible cord that is listed for extra-hard usage and provided with listed bushed fittings may shall be permitted to be used in Class I, Division 1 locations.
_____	T	F	12. Surface temperature of Class I, Division 1 luminaires (fixtures) shall not exceed the ignition temperature of the gases or vapors surrounding such units.
_____	T	F	13. Heating appliances shall not be required to be identified for Class I, Division 1 locations.
_____	T	F	14. Where provisions must be made for a flexible connection, such as at a motor, flexible cord may be used, additional grounding is not required around these flexible connections.
_____	T	F	15. In Class II, Division 1 locations, the wiring method shall be either metal conduit or MI cable.

Section	Answer		
_____	T F	16.	Seals shall be provided in conduit and cable systems to minimize the passage of gases or vapors from one portion of the system to another portion.
_____	T F	17.	EYS sealing fittings are for use with vertical or nearly vertical conduit in sizes from 3/4 to 1 in.
_____	T F	18.	All garages used only for parking or storage are hazardous areas.
_____	T F	19.	Pits or depressions below floor level, which extend up to the floor level, shall be considered as Class I, Division 1 locations. (General Rule)
_____	T F	20.	In commercial garages, EMT shall be permitted to be installed above the 18 in. level.
_____	T F	21.	A seal shall be required at every point in an aircraft hanger where a conduit is routed from a hazardous area to a nonhazardous area.
_____	T F	22.	A Class I location for a service station does extend beyond an unpierced wall, roof, or other solid partition.
_____	T F	23.	A lubrication room shall be considered Class I, Division 2 up to a height of 24 in. above the floor.
_____	T F	24.	Metallic portions of dispensing pumps and all noncurrent-carrying metal parts of electric equipment shall be grounded.
_____	T F	25.	For open-top spray booths, the Class I, Division 2 hazardous area shall be considered to extend to 5 ft. above the booth and 5 ft. from all openings.
_____	T F	26.	Luminaires (lighting fixtures), within a spray booth, shall be identified for Class I, Division 1 locations.
_____	T F	27.	Equipment shall be selected, tested, and identified for the type of flammable material involved.
_____	_____	28.	A Class II, Division 2 location is a location where combustible _____ is not normally in the air in quantities sufficient to produce explosive or ignitable mixtures.
_____	_____	29.	A Class III, Division 1 location is a location in which easily ignitable _____ or materials producing combustible flyings are handled, manufactured, or used.
_____	_____	30.	Group B is an atmosphere containing flammable gases or _____ liquids.
_____	_____	31.	Group F atmospheres contain combustible _____ dusts.
_____	_____	32.	Flash point of a liquid is the minimum _____ at which the liquid gives off sufficient vapor to form an ignitable mixture with the air near the surface of the liquid or within the vessel.
_____	_____	33.	Vapor _____ are used mainly to determine the settling or rising tendency of a mixture.

	Section	Answer

34. Explosionproof enclosures will withstand an _____ explosion of gases or vapor and prevent those gases or vapors from igniting gases and vapors in the surrounding atmosphere outside of the enclosure.

35. Threaded joints of rigid metal conduit shall be made up with at least _____ threads fully engaged.

36. Rigid nonmetallic conduit may be installed in Class I, Division 1 locations if encased in a concrete envelope of at least _____ in. and buried below the surface in not less than _____ ft. of earth.

37. Equipment installed in Class I, Division 1 areas shall be permitted to be installed in Class _____, Division _____ locations.

38. In Class I, Division 2 locations, pendant luminaires (fixtures) shall be suspended by a flexible hanger unless rigid stems not over _____ in. long are used.

39. Boxes and fittings that contain _____ or _____ shall be identified for Class II locations when used in such locations.

40. Explosionproof equipment or equipment with arcing or sparking devices shall have a seal placed within _____ in. of such equipment.

41. In classification of a floor of a garage where vehicles are repaired and volatile flammable fluids are present, the floor area up to _____ in. is Class I, Division 2.

42. Any equipment in commercial garages that might produce an arc or spark, and which is less than _____ ft. above the floor level, shall have a tight enclosure that will prevent the escape of any arc or spark.

43. Any equipment installed in aircraft hangars that is over _____ ft. above engines shall be permitted to be the general-purpose type.

44. The surrounding space out to a distance of _____ ft., measured from a point vertically below the edge of the dispenser enclosure, is Class I, Division 2 up to a height of _____ in. above grade.

45. Emergency controls for unattended self-service stations shall be more than _____ ft. but less than _____ ft. from the dispensers.

46. A pit for a bulk-storage tank that is located in the _____ or _____ ft. sphere (horizontally) around pumps, bleeders, meters, etc. shall be considered as a Class I, Division 1 location.

47. Storage and repair garages for tank vehicles shall be considered as Class I, Division 2 up to a height of _____ in. above the floor.

48. Underground wiring routed to and around aboveground storage tanks may be installed with rigid metal conduit where buried less than _____ ft. in the earth.

49. Portable lamps within a spray booth shall be of a type which are _____ for Class I, Division 1 locations.

Section **Answer**

_____ _____ **50.** Receptacles used to cord-and-plug connect equipment in Class I, Division 1 or 2 locations shall be _____ for such use.

22

Residential Calculations

Residential calculations are the most difficult to perform because the rules and regulations of the NEC are more restrictive than those for commercial and industrial facilities. The standard or optional calculation shall be permitted to be used to calculate the loads to size and select the elements for the feeder or service. The optional calculation seems to be the favorite method used by designers and electricians. This is true, because once loads are calculated, it produces smaller VA or amps than the more complicated standard calculation. Therefore, smaller components are required in the electrical system and greater savings in wiring methods are achieved. The procedure for laying out residential calculations will be different in some ways than those used for commercial and industrial.

APPLYING THE STANDARD CALCULATION PARTS II AND III TO ARTICLE 220

Residential occupancies are known in the industry as dwelling units or single family dwellings. This chapter mainly deals with one and two family dwelling units.

When using the standard calculation for calculating loads for a residential occupancy, all loads are divided into three groups and four columns. The groups are as follows:

Group 1
   General lighting and receptacle loads
Group 2
   Small appliance loads
Group 3
   Special appliance loads

See Figure 22-1 for a detailed illustration of the three groups of loads.

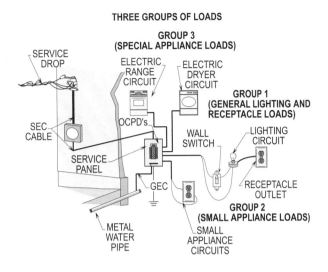

APPLYING THE STANDARD CALCULATION
ARTICLE 220, PARTS I & II

For example: Calculate the general lighting and receptacle load for a 60 ft. x 55 ft. dwelling unit.

Step 1: Calculating the load in VA
220.12
60' x 55' = 3300 sq. ft.

Step 2: Calculating VA
Table 220.12
3300 sq. ft. x 3 VA = 9900 VA

Solution: **Table 220.12 requires 9900 VA for the general lighting and receptacle loads.**

Figure 22-1. The loads in a dwelling unit are divided into three groups and four columns in which demand factors shall be permitted to be applied. The demand factors reduce the loads by a percentage based on the NEC.

GENERAL LIGHTING LOAD
220.12; TABLE 220.12

The general lighting load for a dwelling unit shall be determined by multiplying the square footage by 3 VA per sq. ft. per **Table 220.12**. The required square footage per unit load VA (volt-amps) is found in **Table 220.12**. All lighting loads and general-purpose receptacle loads in a dwelling shall be determined by the general lighting load. All lighting loads per **210.70(A)**, and general-purpose receptacle loads per **Article 100** and **210.52(A) through (H)** shall be calculated per **Table 220.12**. The locations in which general-purpose receptacle outlets are installed are bedrooms, bathrooms, dens, living rooms, halls, garages and outside areas.

See **Figure 22-2** for calculating the general-purpose lighting and receptacle loads in a dwelling unit.

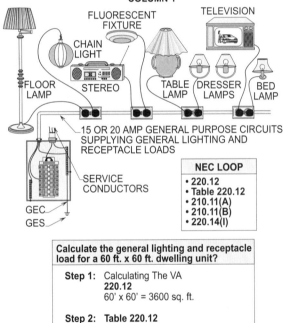

GENERAL LIGHTING LOAD
NEC 220.12
TABLE 220.12

For example: Calculate the general lighting and receptacle load for an 2400 sq. ft. dwelling unit.

Step 1: Calculating the load in VA
Table 220.12
2400 sq. ft. x 3 VA = 7200 VA

Solution: **Table 220.12 requires 7200 VA for the general lighting and receptacle loads.**

Figure 22-2. The general lighting load for a dwelling unit shall be calculated by multiplying the square footage by 3 VA per sq. ft. per **Table 220.12**. All lighting loads and general-purpose receptacle loads in a dwelling shall be calculated by the general lighting load calculation. See the asterisk by the dwelling units in **Table 310.16** and the Note below the Table.

Residential Calculations

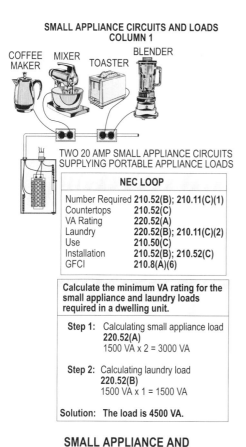

Figure 22-3. All small appliance and laundry loads shall be calculated at 1500 VA to determine the feeder conductors and elements to size the service. At least three small appliance circuits shall be required. However, more circuits are permitted.

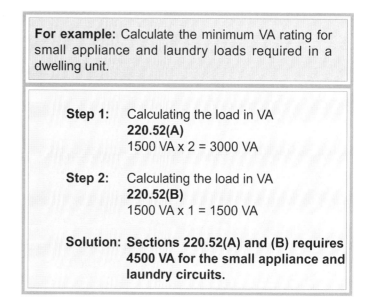

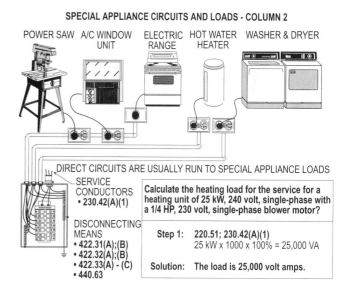

Figure 22-4. All direct circuit loads shall not be permitted to be connected to the general lighting load. The kilowatt rating listed on the nameplate for special appliance loads shall be calculated at 125 percent for continuous operation and 100 percent noncontinuous operation.

SMALL APPLIANCE AND LAUNDRY LOADS
220.52(A); (B)

All small appliance circuits are located in a dwelling unit per **210.52(B)(1)** and **210.11(C)(1)**. At least two small appliance circuits shall be required to supply receptacle outlets located in the kitchen, breakfast room, pantry and dining room per **220.52(A)**. A laundry room receptacle outlet shall be required per **220.52(B)**. All small appliance and laundry loads shall be calculated at 1500 VA to determine the size feeder conductors and elements to size the service.

See Figure 22-3 for calculating the small appliance and laundry loads in a dwelling unit.

SPECIAL APPLIANCE LOADS
220.14(A) THRU (14)(C); 230.42(A); (B)

Special appliance loads are usually supplied by direct circuits. These loads are water heaters, heating units, ranges, air-conditioners, dishwashers, motors, etc. Direct circuit loads shall not be connected to the general lighting

22-3

load. The VA rating for the OCPD for special appliance loads shall be calculated at 125 percent for continuous operation and 100 percent for noncontinuous operation, with demand loads being determined by the kilowatt rating listed on the nameplate times a percentage.

See Figure 22-4 for calculating the special appliance loads.

For example: Calculating the special appliance load for a 8 kW water heater.

Step 1: Calculating the VA
210.19(A)(1); 230.42(A)(1); 422.13
8 kW water heater = 8000 watts (volt-amps)

Solution: Sections 230.42(A)(2), 210.19(A)(1) and 220.14(A) requires 8000 VA for the water heater load.

For example: Calculate the heating load for the service and branch-circuit load for a heating unit of 20 kW, 240 volt, single-phase with a 1/4 HP, 230 volt, single-phase blower motor?

Step 1: Calculating the VA
220.51
20 kW x 1000 x 100% = 20,000 VA

Solution: Section 220.51 requires 20,000 VA for the service load.

Step 1: Calculating the branch-circuit load
424.3(B); Table 430.248
Heating load = 20,000 VA
Motor load (2.9 A x 240 V) = 696 VA
Total load = 20,696 VA

Step 2: Calculating the total VA
424.3(B)
20,696 VA x 125% = 25,870 VA

Solution: Section 424.3(B) requires 25,870 VA for the branch-circuit load.

Design Tip: Some designers and inspectors include the blower motor load with the kW rating of heating unit. Verify with local codes for interpretation.

DEMAND FACTORS
ARTICLE 220, PART III

The following four loads are separated into two columns of loads and demand factors are applied:

Column 1:
General lighting and receptacle loads
and small appliance loads   **Table 220.42**

Column 2:
Cooking equipment loads   **220.55; Table 220.55**
Fixed appliance loads   **220.53**
Dryer loads   **220.54; Table 220.54**

General lighting and receptacle loads and the small appliance circuits plus laundry circuit shall be permitted to have demand factors applied.

The total number of fixed appliance loads shall be permitted to have demand factors applied. The total number of ranges and dryers in a dwelling unit shall be permitted to be reduced by a percentage.

See Figure 22-5 for the four loads that shall be permitted to have demand factors applied.

FOUR DEMAND LOADS - COLUMNS 1 AND 2

SERVICE CONDUCTORS
• TABLE 220.42

GENERAL LIGHTING AND SMALL APPLIANCE LOAD COLUMN 1, DEMAND LOAD 1
3 VA per sq. ft.
• **Table 220.12**
1500 VA for each app. cir.
• **220.52(A);(B)**

COOKING EQUIPMENT LOADS (Range, Cooktop, Oven, etc.)
COLUMN 2 DEMAND LOAD 2
Units rated 3 1/2 - 8 3/4 kW
• use Col. A, **Table 220.55**
Units rated less than 3 1/2 kW
• use Col. B, **Table 220.55**
Units rated 9-12 kW
• use Col. C, **Table 220.55**
Units with greater ratings
• use Notes 1, 2, 3, or 4

GEC
GES

FIXED APPLIANCE LOADS COLUMN 2, DEMAND LOAD 4
Four or more may have 75% Demand factor applied to total VA rating of units
• **220.17**

DRYER LOAD COLUMN 2, DEMAND LOAD 3
Three or less calculated at 100%, four or more units shall be permitted to have demand factors applied, based on the number • **Table 220.54**

DEMAND FACTORS
TABLE 220.12
TABLE 220.55
TABLE 220.54
220.53

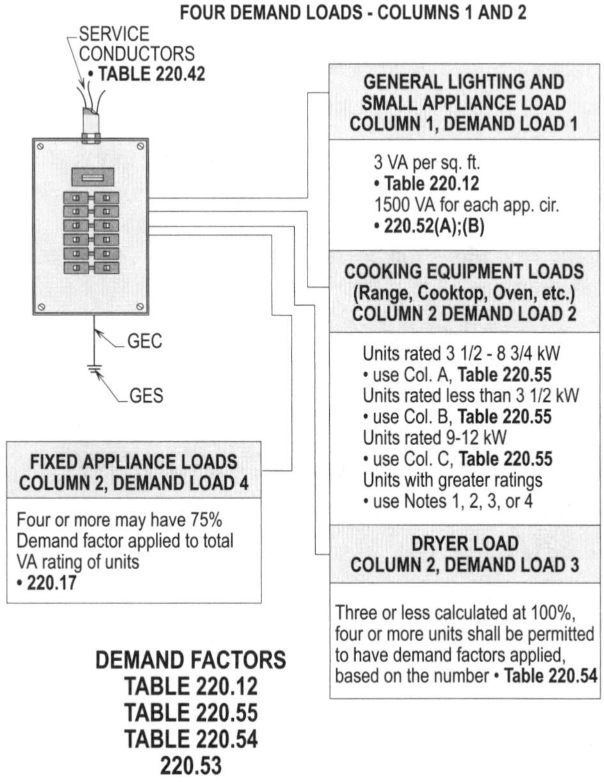

Figure 22-5. General lighting and receptacle loads and small appliance loads, cooking equipment loads, fixed appliance loads and dryer loads are four loads in Columns 1 and 2 to which demand factors shall be permitted to be applied.

GENERAL LIGHTING AND RECEPTACLE LOADS AND SMALL APPLIANCE AND LAUNDRY LOADS
COLUMN 1 - TABLE 220.42; 220.52(A);(B)

The general lighting load for a dwelling unit shall be calculated by multiplying the square footage by 3 VA per sq. ft. per **Table 220.12**. The required square footage per unit load (volt-amps) is found in **Table 220.12**. All small appliance and laundry loads shall be calculated at 1500 VA per **220.52(A)** and **(B)**. A demand factor shall be permitted per **Table 220.42**. The demand factors for the general lighting and receptacle loads per **Table 220.42** for dwelling units are as follows:

```
           Volt-amps
       0 -   3,000 at 100%
   3,001 - 120,000 at  35%
 120,001 -         at  25%
```

See **Figure 22-6** for calculating the general-purpose lighting and receptacle loads including the small appliance and laundry loads.

> **For example:** Calculate the demand load for the general lighting load for a 2800 sq. ft. dwelling unit.

Column 1 and Demand load 1

Step 1: General lighting load
Table 220.12
2800 sq. ft. x 3 VA = 8400 VA

Step 2: Small appliance load
220.52(A)
1500 VA x 2 = 3000 VA

Step 3: Laundry load
220.52(B)
1500 VA x 1 = 1500 VA

Step 4: Total load
General lighting load = 8,400 VA
Small appliance load = 3,000 VA
Laundry load = 1,500 VA
Total load = 12,900 VA

Step 5: Applying demand factors
Table 220.42
First 3000 VA x 100% = 3000 VA
Next 9900 VA x 35% = 3465 VA
Total load = 6465 VA

Solution: Demand load 1 requires 6465 VA for the general lighting and receptacle loads including the small appliance and laundry loads.

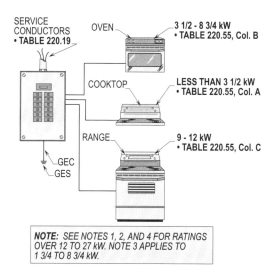

COOKING EQUIPMENT LOADS
NEC 220.55
TABLE 220.55

Figure 22-7. The demand factors listed in **Table 220.55** shall be used to apply the demand loads for cooking equipment. The demand factors of **Table 220.55, Columns A, B or C** or based on the kW rating of the cooking equipment. If the kW rating is greater than the kW rating of the equipment listed in **Columns A, B or C**, **Notes 1, 2, 3, or 4 to Table 220.55** shall be applied.

COOKING EQUIPMENT LOADS
COLUMN 2 - 220.55; TABLE 220.55

The cooking equipment loads and Demand load 2 are separated from Group 3 and placed in Column 2 and demand factors applied accordingly.

The demand factors listed in **Table 220.55** apply to the demand loads for cooking equipment. The **Footnotes** are based on the kW rating and number of units. Ranges, wall-mounted ovens and counter-mounted cooktops are units of cooking equipment per **Table 220.55**.

See **Figure 22-7** for the demand factors listed in **Table 220.55, Columns A, B** or **C**.

Stallcup's Electrical Design Book

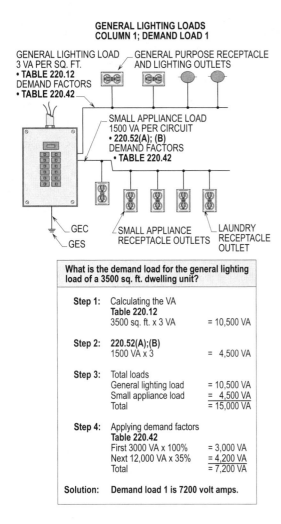

GENERAL LIGHTING AND RECEPTACLE LOADS AND SMALL APPLIANCE AND LAUNDRY LOADS
TABLE 220.42
NEC 220.52(A); (B)

Figure 22-6. The general lighting and receptacle loads including the small appliance loads are calculated in VA when determining the size service-entrance equipment and conductors. This is Demand load 1 in the standard calculation.

DEMAND LOAD 2
TABLE 220.55, COLUMN C

The demand load of cooking equipment is calculated in kW. The maximum demand for the size and number of ranges is already calculated for selecting the elements of the feeder or service. Therefore, cooking equipment does not need to be multiplied by the kW rating of the unit by a percentage until **Columns A** or **B** are utilized.

See Figure 22-8 for demand loads that shall be permitted to be applied in **Table 220.55, Column C**.

For example: Calculating the demand load for a 11.5 kW range.

Step 1: Calculating the load in VA
Table 220.55, Column C
11.5 kW = 8 kVA

Solution: Column C to Table 220.55 allows 8 kVA for the range.

For example: Calculating the demand load for a 12 kW and 10 kW cooking unit.

Step 1: Calculating VA
Table 220.55, Column C
12 kW and 10 kW = 11 kVA

Solution: Column C to Table 220.55 allows 11 kVA for the two cooking units.

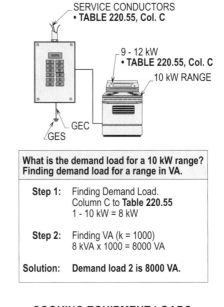

COOKING EQUIPMENT LOADS
TABLE 220.55, Col. C

Figure 22-8. The demand load in kW for cooking equipment in **Column C** is already calculated per **Table 220.55**.

DEMAND LOAD 2
TABLE 220.55, COLUMN A

The nameplate (kW) wattage rating of a range shall be calculated by the number of cooking units times the percentage factor found in **Table 220.55**. This calculation is used to obtain the maximum demand load.

Residential Calculations

See **Figure 22-9** for demand loads to be applied in **Table 220.55, Column A**.

For example: What is the percentage allowed for three cooking units rated at 3 kW, 2 1/2 kW and 2 kW respectively? The first step is to select the percentage to be applied, based upon the number of cooking units. Three cooking units permit a demand factor of 70 percent per **Table 220.55, Column A**.

For example: Calculating the demand load for a piece of cooking equipment with a 2.5 kW rating.

Step 1: Calculating the VA
Table 220.55, Column A
2.5 kW x 80% = 2 kVA

Solution: Column A to Table 220.55 allows 2 kVA for the cooking equipment.

DEMAND LOAD 2
TABLE 220.55, COLUMN B

The nameplate (kW) wattage rating of a range shall be calculated by the number of cooking units times the percentage factor applied in **Table 220.55**. This computation derives the maximum demand load.

See **Figure 22-10** for demand loads to be applied from **Table 220.55, Column B**.

For example: Calculating the demand load for a piece of cooking equipment with a 8.5 kW rating.

Step 1: Calculating the VA
Table 220.55, Column B
8.5 kW x 80% = 6.8 kVA

Solution: Column B to Table 220.55 allows 6.8 kVA for the cooking equipment.

DEMAND LOAD 2
TABLE 220.55, NOTE 1

For cooking equipment rated over 12 kW to 27 kW in **Column C**, each kW over 12 kW shall be increased 5 percent to calculate the demand load per **Note 1**.

See **Figure 22-11** for demand factors to be applied from **Table 220.55, Column C, Note 1**.

For example: Calculating the demand load for a range with a 25 kW rating.

Step 1: Calculating the percentage
Table 220.55, Note 1
25 kW - 12 kW = 13 kW
13 kW x 5% = 65%

Step 2: Calculating the VA
Table 220.55, Column C
8 kW x 165% = 13.2 kVA

Solution: Note 1 to Table 220.55 allows 13.2 kVA for the range.

DEMAND LOAD 2
TABLE 220.55, NOTE 2

Cooking equipment of unequal values rated over 12 kW to 27 kW in **Column C, Note 2** shall be calculated by adding the kW ratings of all units and dividing by the number of units, all ranges below 12 kW shall be calculated at 12 kW. When an average rating is found, the number of units shall be increased by 5 percent for each kW exceeding 12 kW to derive the allowable kW.

See **Figure 22-12** for demand factors that are applied per **Table 220.55, Column C, Note 2**. **Note:** The demand factor selected from **Table 220.55, Column C** shall be based upon the number of units.

For example: Determine the demand load for three pieces of cooking equipment with a 12 kW, 14 kW, and 20 kW rating respectfully.

Step 1: Calculating the percentage
Table 220.55, Note 2
Total kW rating
12 kW + 14 kW + 20 kW = 46 kW
Average rating
46 kW ÷ 3 = 15.3 (round up)
16 - 12 = (4 x 5%) = 20%

Step 2: Calculating the VA
Table 220.55, Column C
14 kW x 120% = 16.8 kVA

Solution: Note 2 to Table 220.55 allows 16.8 kVA demand load for three pieces of cooking equipment.

Design Tip: It is permissible per **Note 3 to Table 220.55** to add all pieces of cooking equipment with ratings over 1 3/4 kW through 8 3/4 kW together and multiply by the percentage of **Columns A or B**, whichever produces the smaller kW rating. This calculation shall be permitted to be applied if it provides the smaller kW rating of all the methods available in **Table 220.55** and **Notes**.

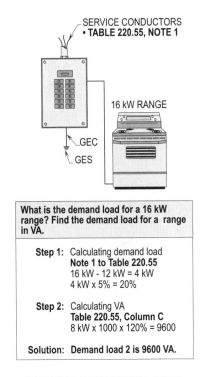

COOKING EQUIPMENT LOADS
TABLE 220.55, NOTE 1

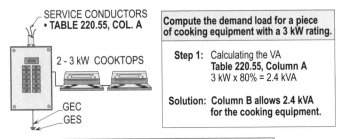

COOKING EQUIPMENT LOADS
TABLE 220.55, Col. A

Figure 22-9. The maximum demand load is calculated by the number of cooking units times the percentage factor from **Table 220.55, Column A**.

Figure 22-11. Cooking equipment rated over 12 kW to 27 kW in **Column C**, each kW over 12 kW shall be increased 5 percent to calculate the demand load per **Note 1**. This percentage times 8 kW will calculate the demand load to be used to size the elements.

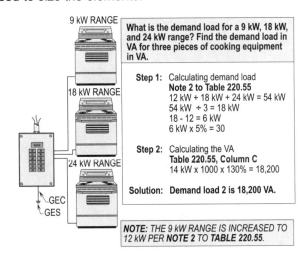

COOKING EQUIPMENT LOADS
TABLE 220.55, NOTE 2

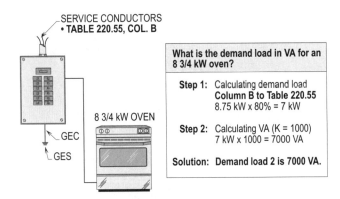

COOKING EQUIPMENT LOADS
TABLE 220.55, Col. C

Figure 22-10. The maximum demand load shall be calculated by the number of cooking units times the percentage factors from **Table 220.55, Column B**. This calculation produces smaller volt-amp ratings and allows smaller elements.

Figure 22-12. For cooking equipment of unequal values rated over 12 kW to 27 kW in **Column C, Note 2** shall be calculated by adding the kW ratings of all units and dividing by the number of units, all ranges below 12 kW shall be calculated at 12 kW. When an average rating is determined, the number of units shall be increased 5 percent for each kW exceeding 12 kW.

Residential Calculations

DEMAND LOAD 2
TABLE 220.55, NOTE 4

The demand load for a cooktop and two or less wall-mounted ovens shall be calculated by finding the average rating of each unit in kW. Each kilowatt that exceeds 12 kW shall be be increased by 5 percent. The nameplate rating of the appliance shall be calculated per **Table 220.55, Column C**. **Sections 240.21(A)** and **210.19(A)(3), Ex. 1** list requirements for each piece of cooking equipment to be supplied by a tap. Each piece of cooking equipment shall be installed in the same room in order to comply with **Note 4 to Table 220.55**. The demand load for each kW exceeding 12 kW is found by multiplying the kW rating by 5 percent to determine the average rating per **Table 220.55, Column C**.

See **Figure 22-13** for demand factors to be applied from **Table 220.55, Column C, Note 4**.

> **For example:** Calculating the demand load for a 12 kW cooktop and 2 - 6 kW ovens for a dwelling unit.
>
> **Step 1:** Calculating the percentage
> **Table 220.55, Note 4**
> Total kW rating
> 12 kW + 6 kW + 6 kW = 24 kW
> 24 kW - 12 kW = 12 kW
> 12 kW x 5% = 60%
>
> **Step 2:** Calculating the VA
> **Table 220.55, Column C**
> 8 kW x 160% = 12.8 kVA
>
> **Solution:** Note 4 to Table 220.55 allows 12.8 kVA demand load for one cooktop and two ovens.

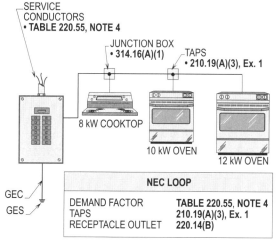

**COOKING EQUIPMENT LOADS
TABLE 220.55, NOTE 4**

See **Figure 22-13**. The demand load for a cooktop and two or less wall-mounted ovens shall be calculated by finding the amperage rating of each unit in kW. Each kilowatt that exceeds 12 kW shall be increased by 5 percent to obtain the multiplier. The multiplier times the demand for one unit in **Column C to Table 220.55** derives the demand load for the elements of the circuit.

FIXED APPLIANCE LOAD
COLUMN 2 - 220.53

The fixed appliance load for cooking equipment loads, dryer equipment loads, air-conditioning loads and heating equipment loads for three or less fixed appliances is determined by adding wattage (volt-amps) values by the nameplate ratings for each appliance. A 75 percent demand factor shall not be applied for these loads per **220.53**. However, these fixed appliance loads shall be permitted to have demand factors applied if there are four or more. All other fixed appliances of four or more grouped into a special appliance load shall be found by adding wattage ratings from appliance nameplates and multiplying the total wattage (volt-amps) by 75 percent to obtain demand load.

See **Figure 22-14** for demand factors to be applied per **220.53**. Note that this is demand load 4.

Three or less fixed appliance loads shall be calculated at 100 percent. The 75 percent per **220.55** shall not be applied unless there are four or more present.

Fixed appliance loads include dishwashers, compactors, disposals, water heaters, attic fans, water circulating pumps, etc. It shall be permitted to apply a demand factor to these loads to calculate the elements of the service or feeder equipment. However, the branch-circuit elements shall be calculated and sized according to other Sections in the NEC.

For example: The elements of a water heater shall be calculated at 125 percent times the amperage of the total elements load per **422.13**. A dishwasher shall be calculated at 100 percent of its nameplate rating per **422.10(A)** and **422.62**.

Stallcup's Electrical Design Book

An attic fan shall be calculated per **430.22(A)** based upon the FLA selected from **Table 430.248** for single-phase motors used for such purposes.

Design Tip: The elements for branch-circuits shall be calculated at 125 percent for continuous operation and 100 percent for noncontinuous operation. This includes the conductors and OCPD's feeding and protecting branch-circuits.

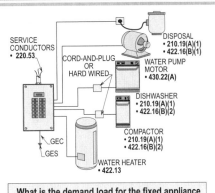

What is the demand load for the fixed appliance load in VA?

5000 VA A/C	240 V, single-phase
15,000 VA heating unit	240 V, single-phase
6000 VA water heater	240 V, single-phase
9000 VA oven	240 V, single-phase
7500 VA cooktop	240 V, single-phase
1600 VA water pump	240 V, single-phase
900 VA disposal	120 V, single-phase
1200 VA compactor	120 V, single-phase
1000 VA microwave	120 V, single-phase
5000 VA dryer	240 V, single-phase
700 VA blower motor	240 V, single-phase

Step 1: Finding the demand load
220.53
Remove the following fixed appliances

Heating load — 15,000 VA heating unit
Air-conditioning load — 5000 VA A/C unit
Dryer load — 5000 VA dryer
Cooking equipment load — 9000 VA oven and 7500 VA cooktop

Step 2: Compute the VA
220.53

water heater load	6,000 VA
water pump load	1,600 VA
disposal load	900 VA
compactor load	1,200 VA
microwave load	1,000 VA
blower motor load	700 VA
Total load	11,400 VA

Step 3: Applying demand factors
220.53
11,400 VA x 75% = 8550 VA

Solution: Demand load 4 is 8550 VA

**FIXED APPLIANCE LOAD
NEC 220.53**

Figure 22-14. All fixed appliances of four or more grouped in a special appliance load shall be found by adding wattage ratings from appliance nameplates and multiplying the total wattage (volt-amps) by 75 percent. Calculation derives demand load.

For example: Calculating the demand load for the following fixed appliances.

6000 VA A/C	240 V, single-phase
10,000 VA heating unit	240 V, single-phase
5000 VA water heater	240 V, single-phase
8000 VA oven	240 V, single-phase
8500 VA cooktop	240 V, single-phase
2600 VA water pump	240 V, single-phase
1000 VA disposal	120 V, single-phase
1200 VA compactor	120 V, single-phase
1600 VA dishwasher	120 V, single-phase
1000 VA microwave	120 V, single-phase
5000 VA dryer	240 V, single-phase
800 VA blower motor	240 V, single-phase

Step 1: Special appliance loads, removing the following loads:
220.53
Heating load
• 10,000 VA heating unit
Air-conditioning load
• 6000 VA A/C unit
Dryer load
• 5000 VA dryer
Cooking equipment load
• 8000 VA oven
• 8500 VA cooktop

Step 2: Totaling the VA
220.53

Water heater load	=	5,000 VA
Water pump load	=	2,600 VA
Disposal load	=	1,000 VA
Compactor load	=	1,200 VA
Dishwasher load	=	1,600 VA
Microwave load	=	1,000 VA
Blower motor load	=	800 VA
Total load	=	13,200 VA

Step 3: Applying demand factors
220.53
13,200 VA x 75% = 9900 VA

Solution: Section 220.53 allows 9900 VA demand load for the fixed appliance.

Design Tip: After the heating load, A/C load, cooking equipment load and dryer load has been removed from the fixed appliance load, all other appliances shall be considered fixed appliances per **220.53**.

Residential Calculations

DRYER LOAD
COLUMN 2 - 220.54, TABLE 220.54

The demand load for household dryers shall be calculated at 5 kVA or the nameplate rating, whichever is greater. Dryer equipment of four or fewer dryers shall be calculated at 100 percent of the nameplate rating. Dryer equipment of five or more dryers shall be permitted to have a percentage applied based on the number of units per **Table 220.54**.

See **Figure 22-15** for demand factors to be applied in **220.54** and **Table 220.54**.

For example: Calculate the demand load for a 4 kW dryer.

Step 1: Calculating the kW
220.54
4 kW = 5 kW

Step 2: Applying demand factors
Table 220.54
Four or fewer dryers = 100%
5 kW x 100% = 5 kVA

Solution: Table 220.54 allows 5 kVA demand load for the dryer.

For example: Calculate the demand load for 5 dryers rated at 8,000 VA each.

Step 1: Selecting percentage
220.54; Table 220.54
5 dryers = 85%

Step 2: Applying demand factors
Table 220.54
8000 VA x 5 x 85% = 34,000 VA

Solution: Table 220.54 allows 34,000 VA demand load for 5 dryers.

Note: A dryer rated at 4500 VA shall be calculated at 5000 VA for determining the components for sizing a branch-circuit, feeder or service. Where applying the optional calculation, the volt-amp rating shall be calculated at the nameplate values only.

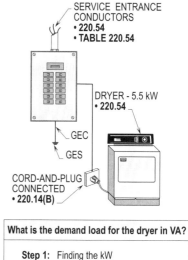

DRYER EQUIPMENT – COL. 2, DEMAND LOAD 3

What is the demand load for the dryer in VA?

Step 1: Finding the kW
220.54
5.5 kW = 5.5 kW

Step 2: Calculating the VA
Four or fewer dryers = 100%
5.5 kW x 1000 x 100% = 5500 VA

Solution: Demand load 3 is 5500 VA.

DRYER LOAD
NEC 220.54
TABLE 220.54

Figure 22-15. The demand load for household dryers shall be calculated at 5 kVA or the nameplate rating, whichever produces the greater rating. Four or fewer dryers shall be calculated at 100 percent of the nameplate rating. Five or more dryers shall be permitted to have a percentage applied based on the number of units per **Table 220.54**. The Table is based on five dryers being used at different times and the load is limited to about 85 percent of total. **Table 220.54** is used for dwelling units in apartment complexes.

LARGEST LOAD BETWEEN HEATING AND A/C
COLUMN 3 - 220.60

Section 220.60 shall be applied when determining the largest load between heat and A/C. The heating and A/C loads shall be calculated at 100 percent and the smaller of the two loads is dropped.

See **Figure 22-16** for demand factors to be applied per **220.60**.

Design Tip: Some designers recognize the A/C unit as still being eligible for the largest motor per **220.50**, even if its been dropped per **220.60**.

22-11

For example: Calculating the load for a 10 kW heating unit and a 5.5 kW A/C unit load in a dwelling unit.

Step 1: Calculating the VA
220.60
Heating load
10 kW x 100% = 10 kVA
A/C load
5.5 kW x 100% = 5.5 kVA

Solution: Section 220.60 requires 10 kVA load for the largest load between the heating and A/C load.

See **Figure 22-17** for demand factors to be applied in 220.50.

For example: Calculating the VA load for a 3 HP, 230 volt, single-phase motor which is to be used for the largest load.

Step 1: Finding FLA
Table 430.248
3 HP = 17 A

Step 2: Calculating A
220.50; 430.22(A), 430.24
17 A x 25% = 4.25 A

Step 3: Calculating VA
Text
4.25 A x 240 V = 1020 VA

Solution: Section 220.50 requires 1020 VA load to supply the largest motor load.

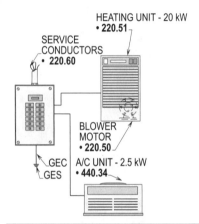

LARGEST BETWEEN HEATING AND A/C
NEC 220.60

Figure 22-16. The heating and A/C loads shall be calculated at 100 percent and the smaller of the two dropped and not used again until the branch-circuit for each is calculated. **Note:** Heat pumps are usually calculated with the heating load. (Their total VA ratings are combined)

LARGEST MOTOR LOAD
COLUMN 4 - 220.50

The motor's total full-load current rating, in amps, shall be calculated at 25 percent per **220.50** which is required per **430.24** and **430.25**, for calculating the load of one or more motors with other loads.

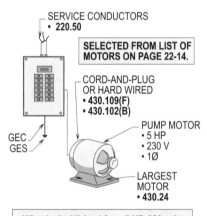

LARGEST MOTOR LOAD
COLUMN 4
NEC 220.50

Figure 22-17. The motors total full-load current rating, in amps, shall be calculated at 25 percent per **220.50** which refers to **430.24**, for calculating the load of one or more motors with other loads.

> **For example:** Calculating the load for a 7.5 HP, 230 volt, single-phase motor with 1400 VA of other loads.

> **Step 1:** Finding FLA
> **430.6(A)(1); Table 430.248**
> Motor load
> 7.5 HP = 40 A
> 40 A x 240 V   = 9,600 VA
> Other loads
> 1400 VA   = 1,400 VA
> Total loads   = 11,000 VA
>
> **Step 2:** Calculating the VA
> **220.50; 430.24; 430.22(A)**
> Total loads   = 11,000 VA
> Largest motor load
> 9600 VA x 25%   = 2,400 VA
> Total load   = 13,400 VA
>
> **Solution:** Section 220.50 requires 13,400 VA demand load to supply the largest motor load.

Select largest motor from fixed appliance load

20 kW heating unit	240 V, single-phase
5.5 kW A/C unit	240 V, single-phase
6 kW water heater	240 V, single-phase
1.2 kW compactor	120 V, single-phase
1.8 kW dishwasher	120 V, single-phase
1.7 kW disposal	120 V, single-phase
1.6 kW attic fan	120 V, single-phase
1.3 kW microwave	120 V, single-phase
5 HP pump motor	240 V, single-phase
12 kW range	240 V, single-phase

The first step is to select the largest motor from the list of loads that are eligible to be considered the largest motor. The 20 kW heating unit is larger than the 5.5 kW A/C unit which eliminates the A/C unit per **220.60**. After scanning the other motors to find the largest, the **5 HP motor** becomes the largest per **220.50**.

The largest motor load shall always be added into the calculation and it really doesn't matter what size the largest motor is. The NEC doesn't distinguish between the size of the motor, but requires one to be selected and calculated based upon HP and voltage. **(See Design Tip on Page 22-10)**

ONE-FAMILY DWELLING STANDARD CALCULATION

For example: Calculating the following loads of a 2800 sq. ft. dwelling unit and size the service-entrance conductors required for Phases A and B and the Neutral.

Loads grouped

Group 1:
General lighting and receptacle load
Group 2:
Small appliance and laundry load
Group 3:
Special appliance load

6000 VA A/C	240 V, single-phase
10,000 VA heating unit	240 V, single-phase
5000 VA water heater	240 V, single-phase
8000 VA oven	240 V, single-phase
8500 VA cooktop	240 V, single-phase
2600 VA water pump	120 V, single-phase
1000 VA disposal	120 V, single-phase
1200 VA compactor	120 V, single-phase
1600 VA dishwasher	120 V, single-phase
1000 VA microwave	120 V, single-phase
5000 VA dryer	240 V, single-phase
800 VA blower motor	240 V, single-phase

Column 1
General lighting and receptacle load

Table 220.12
2800 sq. ft. x 3 VA = 8,400 VA

Small appliance and laundry loads
220.52(A); (B)
1500 VA x 2 = 3,000 VA
1500 VA x 1 = 1,500 VA
Total load = 4,500 VA

Applying demand factors for demand load 1
Table 220.42
General lighting load = 8,400 VA
Small appliance and
laundry load = 4,500 VA
Total load = 12,900 VA

First 300 VA x 100% = 3,000 VA
Next 9900 VA x 35% = 3,465 VA
Total load = **6,465 VA** • √

Column 2
Special appliance loads

Cooking equipment loads and demand load 2
Table 220.55, Col. B
Total kW rating
8 kW + 8.5 kW = 16.5 kW
16.5 kW x 65% = 10.725 kVA •
10.725 kVA x 100% x 1000 = **10,725 kVA** •

Fixed appliance load and demand load 3
220.53
Water heater = 5,000 VA
Water pump = 2,600 VA √
Disposal = 1,000 VA √
Compactor = 1,200 VA √
Dishwasher = 1,600 VA √
Microwave = 1,000 VA √
Blower motor = 800 VA
Total load = 13,200 VA
13,200 VA x 75% = **9,900 VA** •

Dryer load and demand load 4
220.54; Table 220.54
5000 VA x 100% = **5,000 VA** •

Column 3
Largest between heating or A/C load

220.60
Heating load
10,000 VA x 100% = **10,000 VA** •

Column 4
Largest motor load

220.50
Water pump
2600 VA x 25% = **650 VA** •

Finding total VA for Phases A and B

General lighting load = 6,465 VA •
Cooking equipment = 10,725 VA •
Fixed appliance load = 9,900 VA •
Dryer load = 5,000 VA •
Heating load = 10,000 VA •
Largest motor load = 650 VA •
Total load = 42,740 VA

Finding amps for Phases A and B
220.5(B)

I = VA ÷ V
I = 42,740 VA ÷ 240 V
I = 178 A

Total VA loads (Neutral)

Column 1
General lighting load and demand load 1

220.61(A)
6465 VA   = **6,465 VA** √

Column 2
Cooking equipment load and demand load 2

(Use 70% of cooking load)
220.61(B)(1)
10,725 VA x 70%   = **7,508 VA** √

Column 2 - Use 120 V loads (√)
Fixed appliance load and demand load 3

(Use 75% of 120 V load)
7400 VA x 75% = 5550 VA
220.53
5550 VA   = **5,550 VA** √

Column 2
Dryer load and demand load 4

(Use 70% of dryer load)
5000 VA x 70%   = **3,500 VA** √

Column 4
Largest motor load

(Use 25% of largest motor load)
2600 VA x 25%   =   **650 VA** √

Finding total VA for neutral

General lighting load	= 6,465 VA √
Cooking equipment	= 7,508 VA √
Dryer load	= 3,500 VA √
Fixed appliance load	= 5,550 VA √
Largest motor load	= 650 VA √
Total load	= 23,673 VA

Finding amps for neutral
I = VA ÷ V
I = 23,673 ÷ 240
I = 99 A

Table 310.16 and Table 310.15(B)(6)
Phases A and B

Table 310.16 allows
3/0 AWG THWN cu. conductors

Table 310.15(B)(6) allows
2/0 AWG THWN cu. conductors

Neutral is 3 AWG THWN cu. conductor

See Design Problem 22-1 for a detailed illustration of calculating the loads for a dwelling unit using the standard calculation.

APPLYING THE OPTIONAL CALCULATION
220.82

The optional calculation for dwelling units provides a easier method for calculating the load per **220.82(B)** and **(C)**. The loads are separated into two columns. The first column of loads consist of all loads except heating and A/C loads which are described as general loads. Heating and A/C loads are the second column of loads. The elements of the service shall be determined by using the total loads of these two columns.

The optional calculation has percentages applied that are derived from the demand factors. The total kVA of a dwelling unit shall be used in determining the demand factors to be applied. The percentages in **220.82(B)** and **(C)** shall be used to calculate the load in dwelling units. Separate units located in a multifamily dwelling shall be permitted to have the optional calculation applied per **220.82(B)** and **(C)** . The optional calculation method shall only be applied when an ampacity of at least 100 amps is applied to the service conductors. This method shall be permitted to be applied to newly constructed or older existing dwelling units. **(See Design Problem 22-2)**

GENERAL LOADS
220.82(B)

Other loads are the first columns of loads to be calculated. General lighting and general-purpose receptacle loads shall be calculated at 3 VA per sq. ft. per **220.82(B)(2)**. Small appliance loads shall be calculated at 1500 VA per **220.82(B)(1)** and **220.52(A)** and **(B)**. These loads are then applied to the other loads in the column. Heating and A/C loads shall not be applied to the other loads in the column. Heating and A/C loads shall not be applied to other loads per **220.82(B)(3)**. Special appliance loads shall be added to general loads per **220.82(B)(3)**. **Section 220.82(B)(4)** includes motor loads in the general loads column. **Section 220.50** shall not require the largest motor load at 25 percent to be added to general loads. The nameplate ratings in the general columns of loads shall be calculated at 100 percent. The first 10,000 VA of the general loads shall be calculated at 100 percent of the demand factor and the remaining VA shall be calculated at 40 percent of the demand factor allowed per **220.82(B)**.

HEAT OR A/C LOADS
220.82(C)(1) THRU (C)(5)

The second column of loads consist of heating, A/C or heating or heat pumps. The loads are determined by the following steps:

- Three or fewer units shall be calculated at 65 percent of the total kW rating of the heating load.

- Four or more units shall be calculated at 40 percent of the total kW rating of the heating load.

- The A/C load shall be calculated at 100 percent of its kVA rating.

- The total load VA rating of the heating unit shall be compared to the A/C load and the smaller load is dropped per **220.82(C)**.

- Heat pumps that operate with the heating unit shall be figured at 100 percent and added to the heating load. **(See Design Problem 22-3)**

APPLYING THE OPTIONAL CALCULATION FOR EXISTING UNITS
220.83

Existing dwelling units per **220.83** shall be permitted to use the optional calculation to size the components for the service equipment to determine if additional loads can be added to the service. A 120/240 or 120/208 volt, three-wire, single-phase system shall be required to supply the service. Two columns are used in separating the loads in the dwelling unit. General lighting and general-purpose receptacle loads shall be calculated at 3 VA per sq. ft. per **220.82(A)(1)**. All other loads shall be calculated at 100 percent of their nameplate ratings. Heating and A/C loads shall be calculated at 100 percent and the smaller load is dropper per **220.83(B)**.

OTHER LOADS
220.83

Existing loads are the first column to be selected. This column shall consist of adding 3 VA per sq. ft. to the small appliance loads plus the nameplate ratings of each special appliance load to determine the existing load in a dwelling unit. The demand factors in **220.83** shall be used in reducing the VA rating to the existing load and this value is then added to the second column which contains the appliance load to be added.

ADDED APPLIANCE LOAD
220.83

Added appliance loads are the second column to be applied. This column shall be determined by adding the total VA rating of the load at 100 percent. The existing loads shall be added to the added loads to determine the total VA rating of the dwelling unit. The added load usually consist of an heating unit, air-conditioner, dryer or some other type of special appliance load.

From the following loads in an existing dwelling unit. It will be determined if a 5040 VA A/C unit load can be added to the service conductors. (Service is rated 100 amps)

- 1800 sq. ft. dwelling unit
- 2 small appliance circuits
- 1 laundry circuit

- 12 kW range   240 V, single-phase
- 5 kW dryer   240 V, single-phase
- 1000 VA compactor   120 V, single-phase
- 900 VA disposal   120 V, single-phase
- 1200 VA dishwasher 120 V, single-phase

See Design Problem 22-4 to verify if the A/C load can be added without upgrading the size of the service conductors. It is a different story, if a heat pump is involved with the heating unit load. There are two methods in which a heat pump shall be calculated in relation to the heating unit. The first method is where the heat pump operates independently and does not operate with the heating unit. In this case, it is not added to the heating unit load at 100 percent to derive the total load. If it does operate with the heating load, it is added to the heating load at 100 percent to derive the total load.

Heat pumps are usually not effective when the temperature is below freezing. Heating elements are normally installed to melt the frozen ice from the coil. Section **220.82(C), item 4** addresses the supplemental heating required to provide this service. However, the supplementary heating elements performing this duty shall be permitted to be disconnected when the temperature reaches a certain temperature and the heating unit can operate independently.

ONE-FAMILY DWELLING OPTIONAL CALCULATION

For example: Calculating the following loads of a 2800 sq. ft. dwelling unit and size the service-entrance conductors required for Phases A and B and the Neutral.

Loads grouped

Group 1:
- General lighting and receptacle loads
- Small appliance and laundry loads
- Special appliance loads

5000 VA water heater	240 V, single-phase
8000 VA oven	240 V, single-phase
8500 VA cooktop	240 V, single-phase
2600 VA water pump	120 V, single-phase
1000 VA disposal	120 V, single-phase
1200 VA compactor	120 V, single-phase
1600 VA dishwasher	120 V, single-phase
1000 VA microwave	120 V, single-phase
5000 VA dryer	240 V, single-phase
800 VA blower motor	240 V, single-phase

Group 2:
6000 VA A/C	240 V, single-phase
10,000 VA heating unit	240 V, single-phase

Column 1
General (other) loads

General lighting and receptacle load
220.82(B)(1); Table 220.12
2800 sq. ft. x 3 VA = 8,400 VA •

Small appliance and laundry loads
220.82(B)(2); 220.52(A); (B)
1500 VA x 2 = 3,000 VA •
1500 VA x 1 = 1,500 VA •

Special appliance loads
Table 220.82(B)(3); (B)(4)
Oven = 8,000 VA •
Cooktop = 8,500 VA •
Water heater = 5,000 VA •
Water pump = 2,600 VA •
Disposal = 1,000 VA •
Compactor = 1,200 VA •
Dishwasher = 1,600 VA •
Microwave = 1,000 VA •
Dryer = 5,000 VA •
Blower Motor = 800 VA •
Total load = 47,600 VA

Column 2
Largest load between heating and A/C load

220.82(C)(1); (C)(2); (C)(4)
A/C unit   (6000 VA x 100%) = 6,000 VA
Heating unit (10,000 VA x 65%) = 6,500 VA
Total load   (heating selected) = 6,500 VA

Calculating general loads and largest load per 220.82(B)

First 10,000 VA x 100% = 10,000 VA
Next 37,600 VA x 40% = 15,040 VA

Largest load = 6,500 VA
(Between heat and A/C)

Calculating the VA

General loads = 10,000 VA
 = 15,040 VA
Largest load = 6,500 VA
(Between heating and A/C)
Total load = 31,540 VA

Calculating the amps

$I = VA \div V$
$I = 31{,}540 \text{ VA} \div 240 \text{ V}$
$I = 131 \text{ A}$

Table 310.16 and Table 310.15(B)(6)
Phases A and B

Table 310.16 allows
1/0 AWG THWN cu. conductors

310.15(B)(6) allows
1 AWG THWN cu. conductors

Neutral is 3 AWG THWN cu. conductor

Note: See the calculation for the neutral in the previous explanation of the standard calculation on **pages 22-14 and 15**.

MULTIFAMILY STANDARD CALCULATION PARTS III AND IV TO ARTICLE 220

When applying the standard calculation for multifamily dwelling units, the loads shall be calculated the same as using the standard calculation for one-family dwellings. The only difference is to calculate the loads of each unit and multiply the total number of dwelling units and pieces of electrical equipment together to derive the total VA or amps to size the elements of the service or feeder-circuit. **(See Design Problem 22-5)**

MULTIFAMILY OPTIONAL CALCULATION PARTS III AND IV TO ARTICLE 220

When applying the optional calculation for multifamily dwelling units, the loads shall be calculated the same as using the optional calculation for one-family dwellings. The only difference is to calculate the total number of dwelling units and pieces of electrical equipment together and apply a percentage based upon the number to derive the total VA or amps to size the elements of the service or feeder-circuit. **(See Design Problem 22-6)**

FEEDER TO MOBILE HOME STANDARD CALCULATION 550.18

Service calculation for mobile homes are performed at the factory. However, it is the responsibility of the designer or electrician to size and select the proper size feeder-circuit to supply power to the mobile home using the standard calculation. Such feeder-circuit shall be a four-wire circuit with all conductors insulated. The size of the conductors shall be permitted to be selected from **Table 310.16** or **310.15(B)(6)**. The service equipment on a pole or pedestal shall be rated at least 100 amps. **(See Design Problem 22-7)**

MOBILE HOME PARK SERVICE AND FEEDERS - OPTIONAL CALCULATION 550.31

The elements of the service shall be permitted to be calculated using the optional calculation. All loads are added together based upon the total number of mobile homes and multiplied by a percentage to derive the total load. Use 16,000 VA for each mobile home if the calculated load per **550.31** produces a lower VA rating. **(See Design Problem 22-8)**

CALCULATION PROBLEMS PARTS III AND IV TO ARTICLE 220

The elements of electrical systems shall be permitted to be calculated by using the standard or optional calculation. The size of these elements are determined by whichever method the designer chooses to calculate these loads.

The following calculations are typical examples of how these loads are calculated, sized and selected. The step-by-step procedures are easy to follow and have condensed the more complicated rules pertaining to calculating loads into a compact listing, which provides easier understanding of how to perform calculations according to the provisions of the NEC.

A broad assortment of basic NEC calculations have been selected to represent the main principles of designing and installing electrical systems in residential occupancies.

MINIMUM RATING OF SERVICE EQUIPMENT FOR A MOBILE HOME 550.32(A)

Generally, the mobile home service equipment shall be located adjacent to the mobile home and not mounted in or on the mobile home. The service equipment shall be located in sight from and not more than 30 ft. (9 m) from the exterior wall of the mobile home it serves. Mobile home service equipment shall be rated at not less than 100 amps at 120/240 volts, and provisions shall be made for connecting a mobile home feeder assembly by a permanent wiring method. Note that 50 amp power outlets shall also be permitted to be used as service equipment.

Residential Calculations

DESIGN PROBLEM 22-1. What is the load in VA and amps for a residential dwelling unit with the following loads?

General lighting and receptacle load
- 2500 sq. ft. dwelling unit
- 2 small appliance circuits
- 1 laundry circuit

120 V, single-phase loads
- 2600 VA water pump
- 1000 VA disposal
- 1200 VA compactor
- 1600 VA dishwasher

240 V, single-phase loads
- 5000 VA dryer
- 6000 VA A/C unit
- 20,000 VA heating unit
- 6000 VA water heater
- 10,000 VA oven
- 9000 VA cooktop
- 800 VA blower motor
- 1000 VA pool pump

Sizing phases = •
Sizing neutral = √

Column 1
Calculating general lighting and receptacle load

Step 1: General lighting and receptacle load
Table 220.12
2500 sq. ft. x 3 VA = 7,500 VA

Step 2: Small appliance and laundry load
220.52(A); (B)
1500 VA x 2 = 3,000 VA
1500 VA x 1 = 1,500 VA
Total load = 12,000 VA

Step 3: Applying demand factors
Demand load 1; **Table 220.42**
First 3000 VA x 100% = 3,000 VA
Next 9000 VA x 35% = 3,150 VA
Total load = 6,150 VA • √

Column 2
Calculating cooking equipment load

Step 1: Applying demand factors for phases A and B
Demand load 2; **Table 220.55, Col. C**
9 kW and 10 kW = 11,000 VA •

Step 2: Applying demand factors for neutral
220.61(B)(1)
11,000 VA x 70% = 7,700 VA √

Column 2
Calculating dryer load

Step 1: Applying demand factors for phases A and B
Demand load 3; **Table 220.54**
5000 VA x 100% = 5,000 VA •

Step 2: Applying demand factors for neutral
220.61(B)(1)
5000 VA x 70% = 3,500 VA • √

Column 2
Calculating fixed appliance load

Step 1: Applying demand factors for phases A and B
Demand load 4; **220.53**
2600 VA x 75% = 1,950 VA √
1000 VA x 75% = 750 VA √
1200 VA x 75% = 900 VA √
1600 VA x 75% = 1,200 VA √
 800 VA x 75% = 600 VA
1000 VA x 75% = 750 VA
6000 VA x 75% = 4,500 VA
Total load = 10,650 VA •

Step 2: Applying demand factors for neutral
Demand load 4; **220.53; 220.61(A)**
2600 VA x 75% = 1,950 VA
1000 VA x 75% = 750 VA
1200 VA x 75% = 900 VA
1600 VA x 75% = 1,200 VA
Total load = 4,800 VA √

Column 3
Largest load between heating and A/C load

Step 1: Selecting largest load
Demand load 5; **220.51; 220.60**
Heating unit
20,000 VA x 100% = 20,000 VA •

Column 4
Calculating largest motor load

Step 1: Selecting largest motor load for phases A and B
220.50; 430.24
2600 VA x 25% = 650 VA •

Step 2: Selecting largest motor load for neutral
220.50; 430.24
2600 VA x 25% = 650 VA √

Calculating phases (add all •)

- General lighting load = 6,150 VA •
- Cooking load = 11,000 VA •
- Dryer load = 5,000 VA •
- Appliance load = 10,650 VA •
- Heating load = 20,000 VA •
- Largest motor load = 650 VA •
Total load = 53,450 VA

Calculating neutral (add all √)

- General lighting load = 6,150 VA √
- Cooking load = 7,700 VA √
- Dryer load = 3,500 VA √
- Appliance load = 4,800 VA √
- Largest motor load = 650 VA √
Total load = 22,800 VA

Finding amps for phases A and B

I = VA ÷ V
I = 53,450 VA ÷ 240 V
I = 223 A

Finding amps for neutral

I = VA ÷ V
I = 22,800 VA ÷ 240 V
I = 95 A

DESIGN PROBLEM 22-2. What is the load in VA and amps for a residential dwelling unit with the following loads? (See Standard Calculation in Design Problem 22-1 for sizing the neutral)

General lighting and receptacle load
- 2500 sq. ft. dwelling unit
- 2 small appliance circuits
- 1 laundry circuit

120 V, single-phase loads
- 2600 VA water pump
- 1000 VA disposal
- 1200 VA compactor
- 1600 VA dishwasher

240 V, single-phase loads
- 5000 VA dryer
- 6000 VA A/C unit
- 20,000 VA heating unit
- 6000 VA water heater
- 10,000 VA oven
- 9000 VA cooktop
- 800 VA blower motor
- 1000 VA pool pump

Sizing phases = •

COLUMN 1
GENERAL LOAD

Step 1: General lighting load
220.82(B)(1); Table 220.12
2500 sq. ft. x 3 VA = 7,500 VA

Step 2: Small appliance and laundry load
220.82(B)(2); 220.52(A); (B)
1500 VA x 2 = 3,000 VA
1500 VA x 1 = 1,500 VA

Step 3: Appliance load
220.82(B)(3); (B)(4)
Cooktop load = 9,000 VA
Oven load = 10,000 VA
Dryer load = 5,000 VA
Water heater load = 6,000 VA
Disposal load = 1,000 VA
Compactor load = 1,200 VA
Dishwasher load = 1,600 VA
Pool pump load = 1,000 VA
Blower motor load = 800 VA
Water pump load = 2,600 VA
Total load = 50,200 VA

Step 4: Applying demand load
220.82(B)
First 10,000 VA x 100% = 10,000 VA
Next 40,200 VA x 40% = 16,080 VA
Total load = 26,080 VA •

COLUMN 2
LARGEST LOAD BETWEEN HEATING AND A/C LOAD

Step 5: Selecting largest load
220.82(C)(1); (C)(2); (C)(4)
Heating load
20,000 VA x 1 x 65% = 13,000 VA •
A/C load
6000 VA x 1 x 100% = 6,000 VA
Total load = 13,000 VA •

TOTALING COLUMNS 1 AND 2
220.82(B); 220.82(B); (C)

Col. 1 ld. = 26,080 VA •
Col. 2 ld. = 13,000 VA •
Total load = 39,080 VA

FINDING AMPS FOR PHASES A AND B

I = VA ÷ V
I = 39,080 VA ÷ 240 V
I = 163 A

FINDING AMPS FOR NEUTRAL

I = VA ÷ V
I = 22,800 VA ÷ 240 V
I = 95 A

Note: See "Calcualting neutral" and "Finding amps for neutral" in Design Problem 22-1 on page 22-19.

DESIGN PROBLEM 22-3. What is the load in VA and amps for a residential dwelling unit with the following loads? (See Standard Calculation in Design Problem 22-1 for sizing the neutral)

General lighting and receptacle load
- 2500 sq. ft. dwelling unit
- 2 small appliance circuits
- 1 laundry circuit

120 V, single-phase loads
- 2600 VA water pump
- 1000 VA disposal
- 1200 VA compactor
- 1600 VA dishwasher

240 V, single-phase loads
- 5000 VA dryer
- 6000 VA heat pump
- 20,000 VA heating unit
- 6000 VA water heater
- 10,000 VA oven
- 9000 VA cooktop
- 800 VA blower motor
- 1000 VA pool pump

Sizing phases = •

Note: Heat pump may be used with heating unit.

COLUMN 1
GENERAL LOADS

Step 1: General lighting load
220.82(B)(1); Table 220.12
2500 sq. ft. x 3 VA = 7,500 VA

Step 2: Small appliance and laundry load
220.82(B)(2); 220.52(A); (B)
1500 VA x 2 = 3,000 VA
1500 VA x 1 = 1,500 VA

Step 3: Appliance load
220.82(B)(3); (B)(4)
Cooktop load = 9,000 VA
Oven load = 10,000 VA
Dryer load = 5,000 VA
Water heater load = 6,000 VA
Disposal load = 1,000 VA
Compactor load = 1,200 VA
Dishwasher load = 1,600 VA
Pool pump load = 1,000 VA
Blower motor load = 800 VA
Water pump load = 2,600 VA
Total load = 50,200 VA

Step 4: Applying demand load
220.82(B)
First 10,000 VA x 100% = 10,000 VA
Next 40,200 VA x 40% = 16,080 VA
Total load = 26,080 VA •

COLUMN 2
LARGEST LOAD BETWEEN HEATING AND A/C LOAD

Step 5: Selecting largest load
220.82(C)(1); (C)(2); (C)(4)
Heating load
20,000 VA x 1 x 65% = 13,000 VA
Heat pump
6000 VA x 1 x 100% = 6,000 VA
Total load = 19,000 VA •

TOTALING COLUMNS 1 AND 2
220.82(B); (C)

Col. 1 ld. = 26,080 VA •
Col. 2 ld. = 19,000 VA •
Total load = 45,080 VA

FINDING AMPS FOR PHASES FOR PHASES A AND B

I = VA ÷ V
I = 45,080 VA ÷ 240 V
I = 188 A

FINDING AMPS FOR PHASES FOR NEUTRAL

I = VA ÷ V
I = 22,800 VA ÷ 240 V
I = 95 A

Note: The neutral load in VA and amps is calculated as shown in Design Problem 22-1 on page 22-19. See "Calculating neutral" and "Finding amps for neutral."

DESIGN PROBLEM 22-4: Can a 5040 VA combination A/C unit heat pump be added to the existing dwelling unit without upgrading the service elements? (Service is rated 100 amps)

General lighting and receptacle load
- 1800 sq. ft. dwelling unit
- 2 small appliance circuits
- 1 laundry circuit

120 V, single-phase loads
- 900 VA disposal
- 1000 VA compactor
- 1200 VA dishwasher

240 V, single-phase loads
- 5000 VA dryer
- 12,000 VA range
- 5040 VA A/C unit and heat pump to be added

COLUMN 1
CALCULATING EXISTING LOADS

Step 1: General lighting load
220.83(B)(1)
1800 sq. ft. x 3 VA = 5,400 VA

Step 2: Small appliance and laundry load
220.83(B)(2)
1500 VA 2 = 3,000 VA
1500 VA x 1 = 1,500 VA

Step 3: Existing load
220.83(B)(3); (B)(4)
Range load = 12,000 VA
Dryer load = 5,000 VA
Disposal load = 900 VA
Compactor load = 1,000 VA
Dishwasher load = 1,200 VA
Total load = 30,000 VA

Step 4: Applying demand load
220.83(B)
First   8000 VA x 100% = 8,000 VA
Next 22,000 VA x 40% = 8,800 VA
Total load = 16,800 VA •

COLUMN 2
ADDED LOAD

Step 5: Calculating added load
220.83(B)
Fixed appliance load
5040 VA x 100% = 5,040 VA •

TOTALING COLUMNS 1 AND 2
220.83(B)(1) THRU (B)(4)

Col. 1 ld. = 16,800 VA •
Col 2 ld. = 5,040 VA •
Total load = 21,840 VA

FINDING AMPS FOR PHASES A AND B

I = VA ÷ V
I = 21,840 VA ÷ 240 V
I = 91 A
Existing service = 100 A
New calculated = 91 A

Note: If calculated load is less than service load, the new load can be added. Since 91 amps is less than the 100 amp service listed in Design Problem 22-4, the A/C unit heat pump load of 5040 VA may be added to the existing service.

Residential Calculations

DESIGN PROBLEM 22-5. What is the load in VA and amps for 25 multifamily dwelling units with the following loads? **Note:** Parallel service conductors, 6 times per phase.

General lighting and receptacle load
- 25 - 1000 sq. ft. dwelling unit
- 2 small appliance circuits per unit
- 1 laundry circuit per unit

120 V, single-phase loads
- 25 - 1000 VA dishwashers
- 25 - 1200 VA disposals

240 V, single-phase loads
- 25 - 12,000 VA ranges
- 25 - 6000 VA water heaters
- 25 - 20,000 VA heating units

Sizing phases = •
Sizing neutral = √

COLUMN 1
CALCULATING GENERAL LIGHTING AND RECEPTACLE LOAD

Step 1: General lighting and receptacle load
Table 220.12
1000 sq. ft. x 3 VA x 25 = 75,000 VA

Step 2: Small appliance and laundry load
220.52(A); (B)
1500 VA x 2 x 25 = 75,000 VA
1500 VA x 1 x 25 = 37,500 VA
Total load = 187,500 VA

Step 3: Applying demand factors
Demand load 1; **Table 220.42**
First 3000 VA x 100% = 3,000 VA
Next 117,000 VA x 35% = 40,950 VA
Remaining 67,500 VA x 25% = 16,875 VA
Total load = **60,825 VA** • √

COLUMN 2
CALCULATING COOKING EQUIPMENT LOAD

Step 1: Applying demand factors for phases A and B
Demand load 2; **Table 220.55, Col. C**
25 - 12,000 VA ranges = 40,000 VA •

Step 2: Applying demand factors for neutral
220.61(B)(1)
40,000 VA x 70% = 28,000 VA √

COLUMN 2
CALCULATING FIXED APPLIANCE LOAD

Step 1: Applying demand factors for phases A and B
Demand load 4; **220.53**
1000 VA x 25 x 75% = 18,750 VA
1200 VA x 25 x 75% = 22,500 VA
6000 VA x 25 x 75% = 112,500 VA
Total load = 153,750 VA •

Step 2: Applying demand factors for neutral
Demand load 4; **220.61(A)**
1000 VA x 25 x 75% = 18,750 VA
1200 VA x 25 x 75% = 22,500 VA
Total load = 41,250 VA √

COLUMN 3
LARGEST LOAD BETWEEN HEATING AND A/C LOAD

Step 1: Selecting largest load
Demand load 5; **220.60**
Heating unit
20,000 VA x 25 x 100% = 500,000 VA •

COLUMN 4
CALCULATING LARGEST MOTOR LOAD

Step 1: Selecting largest motor load for phases A and B
220.50; 430.24
1200 VA x 25% = 300 VA •

Step 2: Selecting largest motor load for neutral
220.50; 430.24
1200 VA x 25% = 300 VA √

CALCULATING PHASES A AND B (add all •)

- General lighting load = 60,825 VA •
- Cooking load = 40,000 VA •
- Appliance load = 153,750 VA •
- Heating load = 500,000 VA •
- Largest motor load = 300 VA •
- Total load = 754,875 VA

CALCULATING NEUTRAL (add all √)

- General lighting load = 60,825 VA √
- Cooking load = 28,000 VA √
- Appliance load = 41,250 VA √
- Largest motor load = 300 VA √
- Total load = 130,375 VA

FINDING AMPS FOR PHASES A AND B

I = VA ÷ V
I = 754,875 VA ÷ 240 V
I = 3145 A

FINDING AMPS FOR NEUTRAL

I = VA ÷ V
I = 130,375 VA ÷ 240 V
I = 543 A

FINDING SIZE CONDUCTORS IN PARALLEL FOR PHASES A AND B

Phases A and B
310.4
I = 3145 ÷ 6 (No. runs per phase)
I = 524 A

Neutral (applying demand factors)
220.61(B)(2)

543 A
First 200 A x 100% = 200 A
Next 343 A x 70% = 240 A
Total load = **440 A**
310.4; 250.24(C)(2)

I = 440 A ÷ 6 (No. runs per phase)
I = 73 A

Table 310.16

Phases A and B
6 - 1000 KCMIL THWN copper conductors per phase
Neutral
6 - 1/0 AWG THWN copper conductors per phase

Note: The neutral conductor shall be 2/0 AWG per **250.24(C)(2)**. Column 1.

DESIGN PROBLEM 22-6. What is the load in VA and amps for 25 multifamily dwelling units with the following loads? (See Standard Calculation in Design Problem 22-5 for sizing the neutral) **Note:** Parallel service conductors, 6 times per phase.

General lighting and receptacle load
- 25 - 1000 sq. ft. dwelling unit
- 2 small appliance circuits per unit
- 1 laundry circuit per unit

120 V, single-phase loads
- 25 - 1000 VA dishwashers
- 25 - 1200 VA disposals

240 V, single-phase loads
- 25 - 12,000 VA ranges
- 25 - 6000 VA water heaters
- 25 - 20,000 VA heating units

Sizing phases = •

COLUMN 1
CALCULATING GENERAL LIGHTING AND RECEPTACLE LOAD

Step 1: General lighting and receptacle load
220.84(C)(1); Table 220.12
1000 sq. ft. x 3 VA x 25 = 75,000 VA

Step 2: Small appliance and laundry load
220.84(C)(2); 220.52(A); (B)
1500 VA x 2 x 25 = 75,000 VA
1500 VA x 1 x 25 = 37,500 VA
Total load = 187,500 VA •

COLUMN 2
CALCULATING COOKING EQUIPMENT LOAD

Step 1: Applying demand factors for phases A and B
Demand load 2; **220.82(C)(3); Table 220.55, Col. C**
12,000 VA x 25 = 300,000 VA •

COLUMN 2
CALCULATING FIXED APPLIANCE LOAD

Step 1: Applying demand factors for phases A and B
Demand load 4; **220.82(C)(3); 220.53**
1000 VA x 25 = 25,000 VA
1200 VA x 25 = 30,000 VA
6000 VA x 25 = 150,000 VA
Total load = 205,000 VA •

COLUMN 3
LARGEST LOAD BETWEEN HEATING AND A/C LOAD

Step 1: Selecting largest load
Demand load 5; **220.82(C)(5); 220.60**
Heating unit
20,000 VA x 25 x 100% = 500,000 VA •

CALCULATING PHASES

- General lighting load = 187,500 VA •
- Cooking load = 300,000 VA •
- Appliance load = 205,000 VA •
- Heating load = 500,000 VA •
Total load = 1,192,500 VA

APPLYING DEMAND FACTORS
TABLE 220.84

I = VA ÷ V
I = 1,192,500 VA ÷ 240
I = 4969

FINDING AMPS FOR PHASES A AND B
TABLE 220.84

I = A x %
I = 4969 x 35%
I = 1739

FINDING SIZE CONDUCTORS FOR PHASES A AND B

Phases A and B
310.4
I = 1739 ÷ 6 (No. runs per phase)
I = 290 A

Table 310.16
Phases A and B
6 - 350 KCMIL THWN copper conductors

Note: The neutral in VA and amps is calculated as shown in Design Problem 22-5 on page 22-23. See "Calculating neutral," "Finding amps for neutral" and "Sizing conductors for neutral."

Residential Calculations

DESIGN PROBLEM 22-7: What is the load in VA and amps for a mobile home with the following loads?

General lighting and receptacle load
- 800 sq. ft. dwelling unit
- 2 small appliance circuits
- 1 laundry circuit
- 8500 VA range
- 6000 VA water heater
- 540 disposal
- 800 VA dishwasher
- 5500 VA heating

Sizing phases = •
Sizing neutral = √

CALCULATING GENERAL LIGHTING AND RECEPTACLE LOAD

Step 1: General lighting and receptacle load
550.18(A)(1)
800 sq. ft. x 3 VA = 2,400 VA

Step 2: Small appliance and laundry load
550.18(A)(2); (A)(3)
1500 VA x 2 = 3,000 VA
1500 VA x 1 = 1,500 VA
Total load = 6,900 VA

Step 3: Applying demand factor
550.18(A)(5)
First 3000 VA x 100% = 3,000 VA
Next 3900 VA x 35% = 1,365 VA
Total load = **4,365 VA** • √

CALCULATING SPECIAL APPLIANCE LOADS

Step 1: Applying demand factors for phases A and B
550.18(B)(2); (B)(3); (B)(4)
Water heater = 6,000 VA
Dishwasher = 800 VA √
Disposal = 540 VA √
Heating = 5,500 VA
Total load = **12,840 VA** •

CALCULATING RANGE LOAD

Step 1: 550.18(B)(5)
8500 VA x 80% = **6,800 VA** •
6800 VA x 70% = **4,760 VA** √

CALCULATING LOAD FOR PHASES A AND B (add all •)

- General lighting load = 4,365 VA •
- Special appliance load = 12,975 VA •
- Range load = 6,800 VA •
- Largest motor load = 135 VA •
Total load = **24,275 VA**

FINDING AMPS FOR PHASES A AND B

I = VA ÷ V
I = 24,140 VA ÷ 240 V
I = 101 A

CALCULATING NEUTRAL LOAD (add all √)

- General lighting load = 4,365 VA √
- Dishwasher load = 800 VA √
- Disposal load = 540 VA √
- Range = 4,760 VA √
- Largest motor = 135 VA √
Total load = **10,600 VA**

FINDING AMPS FOR NEUTRAL

I = VA ÷ V
I = 10,600 VA ÷ 240 V
I = 44 A

Finding size conductors for phases A and B
Table 310.16
Phases A and B
101 A requires 2 AWG THWN cu.
Neutral
44 A requires 8 AWG THWN cu.

310.15(B)(6)
Phases A and B
101 A requires 3 AWG THWN cu.
Neutral
44 A requires 8 AWG THWN cu.

DESIGN PROBLEM 22-8: What is the load in VA and amps for 28 mobile homes with the following loads?

Mobile homes
- 28 units
- 17,000 VA each based upon calculated load

FINDING TOTAL VA LOAD

Step 1: Finding VA load
550.31
Mobile home = 17,000 VA

Step 2: Calculating VA load
550.31(2); Table 550.31
17,000 VA x 28 x 24% = **114,240 VA**

SIZING CONDUCTORS

Step 1: Finding VA for phases A and B
550.31
Total VA = 114,240 VA

Step 2: Calculating A for phases A and B
550.31
A = 114,240 VA ÷ 240 V
A = 476

Step 3: Selecting size conductors for phases A and B
Table 310.16; 310.4
Phases A and B (Parallel)
A = 476 ÷ 3 (No. runs per phase)
A = 158.7 (round up)
159 A requires 2/0 AWG THWN cu.

SIZING NEUTRAL BASED UPON CALCULATED LOAD

Step 1: Finding VA for neutral
550.31
Total VA = 114,240 VA

Step 2: Calculating A for neutral
550.31
A = 114,240 VA ÷ 240 V
A = 476

Step 3: Applying demand factors
220.61(B)
First 200 A x 100% = 200 A
Next 276 A x 70% = 193 A
Total load = 393 A

Step 4: Selecting conductors for phases A and B and neutral
Table 310.16; 310.4
Neutral (parallel)
A = 393 A ÷ 3 (No. runs per phase)
A = 131
131 A requires 1/0 AWG THWN cu.

Residential Calculations

SIZING ELEMENTS
DESIGN PROBLEM 22-1

Using the service amps of Design Problem 22-1 on page 22-19 of this Chapter, size the elements of the following:

- Size THWN copper conductors per **Table 310.16**
- Size overcurrent protection device
- Size panelboard
- Size conduit using RMC
- Size grounding electrode conductor (cu.)
- Size supplement ground
- Size grounded conductor

SIZING THE CONDUCTORS FOR THE SERVICE PER TABLE 310.16 AND TABLE 310.15(B)(6)

Step 1: Calculated loads
Phases = 223 A
Neutral = 95 A

Step 2: Selecting conductors
Table 310.16
Phases = 223 A requires 4/0 AWG THWN cu.
Neutral = 95 A requires 3 AWG THWN cu.

Step 3: Applying **Table 310.15(B)(6)**
Phases = 223 A requires 3/0 AWG THWN cu.
Neutral = 95 A requires 3 AWG THWN cu.

Solution: The phase conductors are **4/0 AWG THWN copper conductors and the neutral is a 2 AWG copper conductor, per 250.24(C)(1).**

Note that 4/0 AWG THWN copper conductors were selected based upon Step 2 above and not Step 3.

SIZING THE OCPD FOR THE SERVICE CONDUCTORS PER TABLE 310.16 AND 110.14(C)(1) AND (C)(2)

Step 1: Amperage of load or conductor
240.4(A) - (G); 240.6(A)
223 A requires 225 A OCPD

Solution: The size OCPD required is 225 amps.

SIZING THE PANELBOARD FOR SERVICE PER 240.4(A) - (G); 408.14(A); 408.16(A)

Step 1: Amperage of load or conductor
240.4(A) - (G); 240.6(A)
223 A requires 225 A panelboard

Solution: The size panelboard required is 225 amps.

SIZING THE CONDUIT FOR THE SERVICE CONDUCTORS USING RIGID METAL CONDUIT PER TABLES 5 AND 4 TO CHAPTER 9

Step 1: Different size conductors (max. size)
Table 5 and Table 4 to Chapter 9
4/0 AWG THWN
.3237 sq. in. x 2   = .6474
2 AWG THWN
.1158 sq. in. x 1   = .1158
Total sq. in. area   = .7632

Step 2: Selecting size
Table 4 to Chapter 9
.7632 sq. in requires 1 1/2" (41) conduit

Solution: The size rigid metal conduit required is 1 1/2 in. (41).

SIZING THE GEC TO GROUND THE SERVICE TO A METAL WATER PIPE PER 250.66 AND TABLE 250.66

Step 1: Size of phase conductors (max. size)
250.66; Table 250.66
4/0 AWG THWN requires 2 AWG cu.

Solution: The size GEC required is 2 AWG cu.

SIZING THE GEC REQUIRED TO GROUND THE SERVICE TO A DRIVEN ROD TO SUPPLEMENT THE METAL WATER PIPE PER 250.104(A) AND 250.50(A)

Step 1: Size of phase conductors (max. size)
250.66(A)
4/0 AWG THWN requires 6 AWG cu.

Solution: The size of the grounding electrode conductor is 6 AWG cu.

> **SIZING THE GROUNDED CONDUCTOR REQUIRED TO CLEAR A GROUND-FAULT PER 220.22 AND 250.24(B)(1)**
>
> **Step 1:** Size of phase conductors
> **250.24(C)(1); 220.61(B)(2);**
> **Table 250.66**
> 4/0 AWG requires 2 AWG cu.
>
> **Solution: The size grounded service conductor is required to be 2 AWG copper.**

Note 1: The calculated neutral per **220.61(B)(2)** produces the smaller grounded (neutral) conductor. However, such grounded (neutral) conductors shall be sized per Table **250.66**, because it is used as a neutral and EGC. See the neutral calculation of Design Problem 22-1 on Page 22-19.

Note 2: For selecting the proper size neutral to carry the amount of fault current that it might be called on to carry, see pages 11-15 thru 11-18 of Chapter 11.

Name   Date

Chapter 22. Residential Calculations

Section   Answer

_____   T   F   1. The general lighting load for a dwelling unit shall be determined by multiplying the square footage by 2 VA per **Table 220.3(A)**.

_____   T   F   2. All small appliance and laundry loads shall be computed at 1500 VA to determine the size feeder conductors and elements to size the service.

_____   T   F   3. The VA rating for the OCPD for special appliance loads shall be calculated at 100 percent for continuous duty and 125 percent for noncontinuous duty, with demand loads determined by the kilowatt rating listed on the nameplate times a percentage.

_____   T   F   4. The total VA for a number of ranges and dryers in a dwelling unit shall be permitted to be reduced by a percentage.

_____   T   F   5. Ranges, wall-mounted ovens and counter-mounted cooktops shall be considered as units of cooking equipment per **Table 220.18**.

_____   T   F   6. For four or more fixed appliances which are grouped into a special appliance load can be found by adding wattage ratings from appliance nameplates and multiplying the total wattage (volt-amps) by 65 percent to obtain the demand load.

_____   T   F   7. The demand load for household dryers shall be calculated at 3 kVA or the nameplate rating, whichever is greater.

_____   T   F   8. The heating and A/C loads shall be calculated at 125 percent and the smaller of the two is dropped.

_____   T   F   9. The optional calculation method shall only be permitted to be applied when an ampacity of at least 50 amps is applied to the service conductors.

_____   T   F   10. When adding appliance loads for the optional calculation method, the total VA rating of the load shall be determined at 100 percent.

_____ _____   11. The VA rating for the OCPD for special appliance loads shall be calculated at _____ percent for continuous duty and _____ percent for noncontinuous duty, if a demand factor shall not be applied based on a number.

_____ _____   12. The demand factors listed in _____ apply to the demand loads for cooking equipment installed in dwelling units.

_____ _____   13. The fixed appliance load for three or less fixed appliances shall be determined by adding wattage (volt-amps) values by _____ percent.

_____ _____   14. Dryer equipment of four or fewer units shall be calculated at _____ percent of the nameplate rating or 5000 VA, whichever is greater.

22-29

Section	Answer

15. The largest motor load shall be calculated at _____ percent, for computing the load of one or more motors with other loads.

16. The general lighting load for a dwelling unit shall be determined by multiplying the square footage by _____ VA.
    - (a) 1
    - (b) 2
    - (c) 3
    - (d) 4

17. At least two small appliance circuits are required to supply receptacle outlets located in the:
    - (a) Kitchen
    - (b) Pantry
    - (c) Dining room
    - (d) All of the above

18. The demand load for four or more fixed appliances grouped into a special appliance load is found by adding wattage ratings from the appliance nameplates and multiplying the total wattage (volt-amps) by:
    - (a) 65 percent
    - (b) 75 percent
    - (c) 80 percent
    - (d) 100 percent

19. The demand load for a household dryer shall be calculated by the nameplate rating or _____ kVA, whichever is greater.
    - (a) 2
    - (b) 3
    - (c) 4
    - (d) 5

20. The heating and A/C load shall be calculated at _____ percent and the smaller of the two dropped.
    - (a) 80
    - (b) 100
    - (c) 125
    - (d) 150

21. What is the general lighting and receptacle load in VA for a 50 ft. x 50 ft. dwelling unit?

22. What is the small appliance load in VA for two small appliance circuits and one laundry circuit?

23. What is the largest load between the heating and A/C load in VA of a 20 kW heating unit and a 6000 VA A/C unit?

24. What is the demand load in VA for a 9 kW range?

25. What is the demand load in VA for a 3 1/2 kW cooktop?

26. What is the demand load in VA for two 3 1/2 kW cooktops?

	Section	Answer

27. What is the demand load in VA for a 8 3/4 kW oven? _____ _____

28. What is the demand load in VA for a 18 kW range? (Use **Note 1 to Table 220.19**) _____ _____

29. What is the demand load in VA for a 8 kW, 16 kW, and 24 kW range? (Use **Note 2 to Table 220.19**) _____ _____

30. What is the demand load in VA for a 8 kW and 10 kW oven and a 10 kW cooktop? (Use **Note 4 to Table 220.19**) _____ _____

31. What is the demand load for the fixed (special) appliance load in VA for a 6000 VA A/C unit, 20,000 VA heating unit, 5000 VA water heater, 10,000 VA oven, 8000 VA range, 1400 VA water pump, 1000 VA compactor, 800 VA blower motor, 1200 VA microwave, 900 VA disposal and a 4500 VA dryer? _____ _____

32. What is the demand load in VA for a 4500 VA dryer? _____ _____

33. What is the demand load in VA for a 5500 VA dryer? _____ _____

34. What is the demand load in VA for six dryer rated at 6000 VA? _____ _____

35. What is the largest motor load in VA for a 5 HP, 230 volt, single-phase motor? _____ _____

36. What is the amp load for a 3000 sq. ft. dwelling unit with a 240 volt, single-phase service with the following loads? (use the standard calculation)

    • Two small appliance circuits
    • One laundry circuit
    • 6,000 VA A/C unit   240 volt, single-phase
    • 35,000 VA heating unit   240 volt, single-phase
    • 5000 VA water heater   240 volt, single-phase
    • 12,000 VA oven   240 volt, single-phase
    • 9000 VA cooktop   240 volt, single-phase
    • 5000 VA dryer   240 volt, single-phase
    • 2400 VA sump pump   240 volt, single-phase
    • 1000 VA pool pump   240 volt, single-phase
    • 900 VA microwave   120 volt, single-phase
    • 1600 VA dishwasher   120 volt, single-phase
    • 1050 VA disposal   120 volt, single-phase
    • 900 VA compactor   120 volt, single-phase

    (a) What size (minimum and maximum) THWN copper conductors are required. _____ _____

    (b) What size OCPD based on computed load is required. _____ _____

    (c) What size panelboard is required. _____ _____

    (d) What size RMC conduit is required. _____ _____

    (e) What size cu. grounding electrode conductor is required. _____ _____

Section	Answer

_____   _____

37. What is the amp load for a 3000 sq. ft. dwelling unit with a 240 volt, single-phase service with the following loads? (use the optional calculation)

- Two small appliance circuits
- One laundry circuit
- 6 000 VA A/C unit   240 volt, single-phase
- 35,000 VA heating unit   240 volt, single-phase
- 5000 VA water heater   240 volt, single-phase
- 12,000 VA oven   240 volt, single-phase
- 9000 VA cooktop   240 volt, single-phase
- 5000 VA dryer   240 volt, single-phase
- 2400 VA sump pump   240 volt, single-phase
- 1000 VA pool pump   240 volt, single-phase
- 900 VA microwave   120 volt, single-phase
- 1600 VA dishwasher   120 volt, single-phase
- 1050 VA disposal   120 volt, single-phase
- 900 VA compactor   120 volt, single-phase

_____   _____ (a) What size (max.) THWN copper conductors are required?

_____   _____ (b) What size OCPD, based on computed load, is required?

_____   _____ (c) What size panelboard is required?

_____   _____ (d) What size RMC conduit is required?

_____   _____ (e) What size cu. grounding electrode conductor is required?

Note: Use the standard calculation to size the neutral. (See problem 36)

_____   _____

38. From the following loads in an existing dwelling unit, determine if a 25 amp, A/C unit load (computed load) can be added to the service conductors. The existing service conductors are 3 AWG THWN copper conductors. (use the existing optional calculation)

- 2000 sq. ft. dwelling unit
- Two small appliance circuits
- One laundry circuit
- 11,000 VA range   240 volt, single-phase
- 5000 VA dryer   240 volt, single-phase
- 1200 VA compactor   120 volt, single-phase
- 800 VA disposal   120 volt, single-phase
- 1400 VA dishwasher   120 volt, single-phase

22-32

	Section	Answer

39. What is the amp load for twenty 1000 sq. ft. multifamily dwelling units with a 240 volt, single-phase service with the following loads? (Use the standard calculation - parallel service conductors, six times per phase)

- Two small appliance circuits
- One laundry circuit
- 20 - 11,000 VA ranges   240 volt, single-phase
- 20 -   5000 VA water heaters 240 volt, single-phase
- 20 - 25,000 VA heating units 240 volt, single-phase
- 20 -   1000 VA dishwashers   120 volt, single-phase
- 20 -   1200 VA disposals   120 volt, single-phase

40. What is the amp load for twenty 1000 sq. ft. multifamily dwelling units with a 240 volt, single-phase service with the following loads? (Use the optional calculation - parallel service conductors, six times per phase)

- Two small appliance circuits
- One laundry circuit
- 20 - 11,000 VA ranges   240 volt, single-phase
- 20 -   5000 VA water heaters 240 volt, single-phase
- 20 - 25,000 VA heating units 240 volt, single-phase
- 20 -   1000 VA dishwashers   120 volt, single-phase
- 20 -   1200 VA disposals   120 volt, single-phase

41. What is the amp load for a 1000 sq. ft. mobile home with a 240 volt, single-phase service with the following loads?

- Two small appliance circuits
- One laundry circuit
- 8500 VA range   240 volt, single-phase
- 5500 VA water heater   240 volt, single-phase
- 600 VA disposal   120 volt, single-phase
- 800 VA dishwasher   120 volt, single-phase
- 6000 VA heating unit   240 volt, single-phase

42. What is the amp load for twenty-six mobile homes with following loads?
- 26 units
- 15,000 VA each based on computed load
- Per **550.18** in the NEC

22-33

23

Commercial Calculations

Commercial facilities such as offices, banks, stores, and restaurants have diverse loads. These loads are classified continuous or noncontinuous, or such loads may cycle on and off, allowing demand factors to be applied.

The procedure and manner in which the lighting, receptacle, and equipment loads are used in the electrical system determines how they are classified.

Loads shall be calculated based upon the type of occupancy and the requirements of the equipment supplied. Either the standard or optional calculation is utilized to calculate the loads in VA or amps to size the service equipment and associated elements.

APPLYING THE STANDARD CALCULATION PART III TO ARTICLE 220

The standard calculation can be used to calculate the VA or amp rating to size and select the elements of the service equipment and associated components. The selection of loads for applying the standard calculation are arranged in a different manner for commercial loads than for the loads used in residential occupancies. There are seven loads utilized to determine the service load. Based upon conditions of use, demand factors can be applied to certain loads which reduces the VA or amps.

The loads are grouped into seven individual loads, and the proper NEC rule is applied to each of these loads based upon use. The loads are grouped and classified as follows:

(1) Lighting loads
- General lighting load per **220.12**
- Show window load per **220.43(A)**
- Lighting track load per **220.43(B)**
- Low-voltage lighting load per **Article 411**
- Outside lighting load per **230.42(A)(1); (A)(2)**
- Outside sign lighting load per **220.14(F)**

(2) Receptacle loads
- General-purpose receptacle load
  (Noncontinuous per **220.14(I); Table 220.44**
  and **230.42(A)(1); (A)(2)**)
- General-purpose receptacle load
  (Calculated per **220.14(I), 230.42(A)(1);**
  (A)(2) and **Table 220.44**)
- Multioutlet assembly load
  (Used simultaneously per **220.14(H)(2)**)
  (Not used simultaneously per **220.14(H)(1)**)

(3) Special appliance loads
- Noncontinuous load per **230.42(A)(1);**
  220.14(A)
- Continuous load per **230.42(A)(1)**
- Demand factor per various Sections of the NEC

(4) Compressor loads
- Refrigeration per **440.34** and **230.42(A)(1)**
- Cooling per **440.34** and **230.42(A)(1)**

(5) Motor loads
- Single-phase per **430.25**
- Three-phase per **430.25**

(6) Heat or A/C loads
- Heating per **220.51**
- A/C per **440.34**
- Heat pump per **440.34**

(7) Largest motor loads
- Taken from loads (4), (5), or (6) per **220.14**

The seven loads shall be calculated at continuous operation or noncontinuous operation. Demand factors shall be applied to the noncontinuous operated loads by specific Sections of the NEC.

Generally, no demand factors can be applied to the loads for commercial occupancies, as the loads are usually used at continuous operation. However, **Table 220.42** permits the general lighting loads in hospitals, hotels, motels, and warehouses to have demand factors applied because of load diversity.

LIGHTING LOADS
ARTICLES 220; 410; AND 411

Lighting loads are the first of the loads to be calculated. Six lighting loads shall be calculated to derive the total lighting load in commercial facilities. These six loads are as follows:

- General lighting loads per **Table 220.12**
- Show window loads per **220.43**
- Lighting track loads per **220.43**
- Low-voltage lighting per **Article 411**
- Outside lighting load per **230.42(A)(1); (A)(2)**
- Sign lighting load per **220.14(F)**

Each lighting load shall be calculated by the operation in which it is used. Loads shall be calculated at continuous or noncontinuous operation and other lighting loads can have demand factors applied, where permitted by the NEC.

GENERAL LIGHTING LOADS
220.12; TABLE 220.12

General lighting loads consist of lighting units installed inside the facility. This load shall be calculated by the VA rating times sq. ft. from **Table 220.12** based on the type of commercial occupancy.

The general lighting load found in listed occupancies is calculated according to the number of VA per sq. ft. and not based upon the number of outlets served. When fluorescent lighting is used, either the area load of VA per sq. ft. of the facility or the total connected load of each ballast is used, whichever is greater in rating.

Lighting loads either the VA per sq. ft. or individual units shall be calculated at 100 percent for noncontinuous operation or 125 percent for continuous operation.

For example: A noncontinuous lighting load of 60 amps and a continuous load of 100 amps is calculated as follows:

Step 1: Calculating load
230.42(A)(1)
60 A x 100% = 60 A
100 A x 125% = 125 A
Total load   = 185 A

Solution: The calculated load for the lighting is 185 amps.

NONCONTINUOUS AND CONTINUOUS OPERATION
230.208(B); 230.42(A)(1); (A)(2)

The requirements for derating the OCPD by 80 percent for continuous operation is no longer found in Sec. 384-16(d), as it was in the 1999 NEC. For OCPD's rated over 600 volts, the rules are listed in **230.208(B)**, and under certain conditions, the 80 percent derating rule does not apply. For feeder-circuits rated over 600 volts, see **215.2(B)**.

To apply the 80 percent derating rule for loads rated at 600 volts or less, **230.42(A)(1)** requires such loads to be calculated at 125 percent (1 ÷ 80% = 1.25) if they operate for three hours or more. Loads operating for a period of less than three hours shall be calculated at 100 percent per **230.42(A)(2)**. The ampacity of the conductors as well as the OCPD shall be calculated and sized at 125 percent and 100 percent or a combination of both based on the operation of such loads. (For feeders, see **215.2(A)(1)**.)

Section 230.90(A) with Ex.'s refers to various Sections of the NEC to calculate the size of OCPD's. Loads that are not motor related are calculated based upon their operation procedures which are listed in **230.42(A)(1)** and **(A)(2)**. **(See Figure 23-1)**

LISTED OCCUPANCIES
220.12; TABLE 220.12

Table 220.12 is used to select the VA rating for listed occupancies. For example, a store building shall be calculated at 3 VA per sq. ft. and an office building shall be calculated at 3.5 VA per sq. ft. Other types of occupancies have different VA ratings per sq. ft. based on the type of occupancy involved.

> **For example:** What is the load in VA for the general-purpose lighting load in an 8000 sq. ft. office per **Table 220.12**?
>
> **Step 1:** Calculating load
> **Table 220.12**
> 8000 sq. ft. x 3.5 = 28,000 VA
>
> **Solution:** The general-purpose lighting load is 28,000 VA. Note: If the number of receptacle outlets are not known, an extra 1 VA per sq. ft. shall be added to the above.

See Figure 23-2 for a detailed illustration of calculating the lighting load for a listed occupancy.

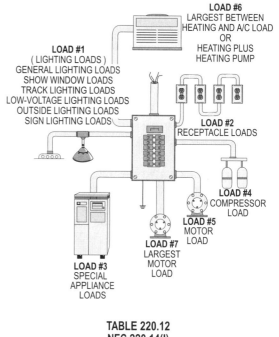

Figure 23-1. As outlined above, there are seven loads to be calculated when applying the standard calculation to determine the service load in VA or amps.

UNLISTED OCCUPANCY
220.18; 220.14(L)

If an occupancy is not listed in **Table 220.12**, the general-purpose lighting load shall be calculated in the following manner.

- Lamps for incandescent lighting per **220.14(L)**
- Lamps for recessed fixtures per **220.14(D)**
- Ballasts for electric discharge lighting per **220.18(B)**
- Track lighting unit per **220.43(B)**
- Low-voltage systems per **Article 411**

An unlisted VA rating shall be calculated at 125 percent for sizing the OCPD and conductors for continuous operation and 100 percent for noncontinuous operated loads per **230.42(A)(1)**.

For example: What is the load in VA for 60, 120 volt, lighting ballast's rated at 1.5 amps each and used for 12 hours a day?

Step 1: Calculating load in A
220.18(B)
60 x 1.5 A = 90 A

Step 2: Calculating continuous load
230.42(A)(1)
90 A x 125% = 112.5 A

Step 3: Calculated VA
112.5 A x 120 V = 13,500 VA

Solution: The lighting loads for the unlisted occupancy is 13,500 VA.

See **Figure 23-3** for a detailed illustration of calculating the lighting load for an unlisted occupancy.

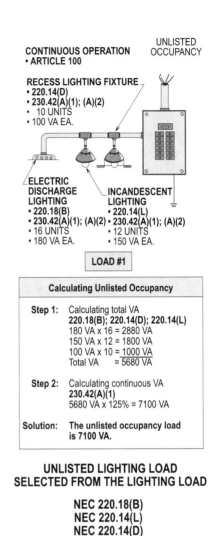

Figure 23-3. The unlisted lighting load shall be permitted to be calculated by multiplying the VA of each lighting load by the number and increasing this value by 125 percent if used for three hours or more.

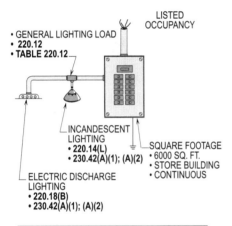

Figure 23-2. The general lighting load (listed occupancy) shall be calculated by multiplying the sq. ft. of the facility by the VA per sq. ft. per **Table 220.12**.

SHOW WINDOW LIGHTING LOAD
220.43(A)

The lighting load in VA for the show window is calculated by multiplying the linear feet of the show window by 200 VA per foot. Such lighting load shall be calculated at 100 percent for noncontinuous operation and 125 percent for continuous operation per **230.42(A)(1)**. Conductors and OCPD's shall be increased to comply with **240.3** and **4** in the NEC.

Commercial Calculations

For example: What is the lighting load in VA for a 80 ft. show window used at noncontinuous or continuous operation?

> **Step 1:** Calculating noncontinuous load
> 220.43(A); 230.42(A)(1)
> 80' x 200 VA x 100% = 16,000 VA
>
> **Step 2:** Calculating continuous load
> 220.43(A); 230.42(A)(1)
> 80' x 200 VA x 125% = 20,000 VA
>
> **Solution:** The noncontinuous load is 16,000 VA and the continuous load is 20,000 VA.

If the number of lighting outlets are known, the VA rating of each luminaire (fixture) shall be multiplied by 125 percent for sizing the show window load. If the VA is not known, each outlet shall be calculated at 180 VA times 125 percent to obtain the lighting load in VA for the show window. Note that the greater between the 200 VA per linear foot or each individual calculation shall be used.

See **Figure 23-4** for a detailed illustration of calculating the show window lighting load.

TRACK LIGHTING LOAD
220.43(B)

The lighting load in VA for lighting track shall be calculated by multiplying the lighting track by 150 VA and dividing by 2. Such VA rating shall be multiplied by 100 percent or 125 percent based upon noncontinuous or continuous operation.

> **For example:** What is the load in VA for 80 ft. of lighting track used at noncontinuous or continuous operation?
>
> **Step 1:** Calculating noncontinuous load
> 220.43(B); 230.42(A)(1)
> 80' ÷ 2' x 150 VA x 100% = 6000 VA
>
> **Step 2:** Calculating continuous load
> 220.43(B); 230.42(A)(1)
> 80' ÷ 2' x 150 VA x 125% = 7500 VA
>
> **Solution:** The noncontinuous load is 6000 VA and the continuous load is 7500 VA.

See **Figure 23-5** for a detailed illustration of calculating the track lighting load.

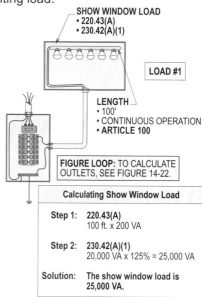

**SHOW WINDOW LIGHTING
SELECTED FROM THE LIGHTING LOAD
NEC 220.43(A)
NEC 230.42(A)(1)**

Figure 23-4. The show window load shall be calculated by multiplying the linear foot by 200 VA. This value shall be increased by 125 percent as such load is continuous.

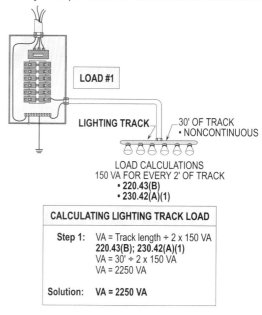

**LIGHTING TRACK LOAD
SELECTED FROM THE LIGHTING LOADS
NEC 220.43(B)
NEC 230.42(A)(1)**

Figure 23-5. The lighting track load shall be calculated by dividing the length of the track by 2 and multiplying by 150 VA. This value shall be increased by 125 percent if it is used for three hours or more.

LOW-VOLTAGE LIGHTING LOAD
ARTICLE 411; 230.42(A)(1); (A)(2)

The lighting load in VA for low-voltage lighting systems shall be calculated by multiplying the FLA of the isolation transformer by 100 percent for noncontinuous operation and 125 percent for continuous operation.

> **For example:** What is the load in VA for a low-voltage lighting system supplied by an isolation transformer with a FLA of 50 amps used at noncontinuous or continuous operation?
>
> **Step 1:** Calculating noncontinuous load
> **Article 411; 230.42(A)(1)**
> 50 A x 100% = 50 A
>
> **Step 2:** Calculating continuous load
> **Article 411; 230.42(A)(1)**
> 50 A x 125% = 62.5 A
>
> **Solution:** The noncontinuous load is 50 amps and the continuous load is 62.5 amps.

See Figure 23-6 for a detailed illustration of calculating the low-voltage lighting load.

OUTSIDE LIGHTING LOAD
220.18(B); 230.42(A)(1); (A)(2)

The lighting load in VA for outside lighting loads shall be calculated by multiplying the VA rating of each lighting unit by 100 percent for noncontinuous operation and 125 percent for continuous operation.

> **For example:** What is the lighting load in VA for 30 continuous operated luminaires (lighting fixtures) with a 75 VA ballast in each unit and 10 noncontinuous operated luminaires (lighting fixtures) with each ballast having a rating of 75 VA?
>
> **Step 1:** Calculating load
> **220.18(B); 230.42(A)(1)**
> 75 x 30 x 125%   = 2812.5 VA
> 75 x 10 x 100%   = 750 VA
> Total load   = 3562.5 VA
>
> **Solution:** The total outside lighting load is 3562.5 VA.

See Figure 23-7 for a detailed illustration for calculating the outside lighting load at continuous operation.

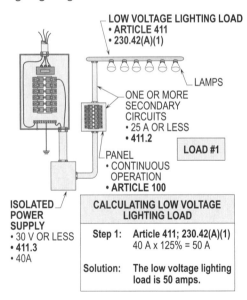

Figure 23-6. The low-voltage lighting load shall be calculated by multiplying the FLA of the isolation transformer by 125 percent if used for three hours or more.

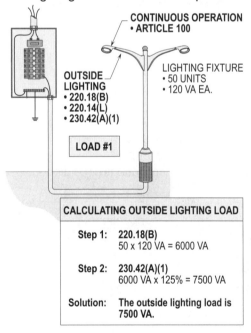

Figure 23-7. The outside lighting load shall calculated by multiplying the number of ballast by 125 percent if the load can operate for three hours or more.

OUTSIDE SIGN LIGHTING LOAD
220.14(F); 230.42(A)(1); (A)(2)

The lighting loads for signs shall be calculated by the commercial occupancy or facility having ground floor footage accessible to pedestrians. Occupancies with grade level access for pedestrians shall have a minimum of 1200 VA provided for a sign lighting load. This VA rating shall be multiplied by 125 percent for signs operating for three hours or more and 100 percent for those operating less than three hours.

See **Figure 23-8** for a detailed illustration for calculating the outside sign lighting load.

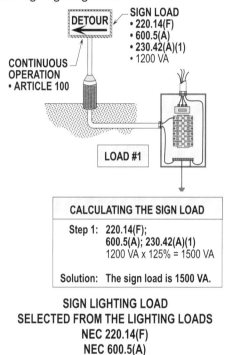

Figure 23-8. The outside sign lighting load shall be calculated by multiplying the VA of the sum by 125 percent if it operates for three hours or more.

RECEPTACLE LOADS
220.14(I); TABLE 220.44

Receptacle loads are the second group of loads to be calculated. Such loads are divided into two subgroups as follows:
- General-purpose receptacle outlets
- Multioutlet assemblies

Each load in the subgroup shall be calculated differently to derive total VA. The load in VA for the general-purpose receptacle load shall be calculated by multiplying the number of outlets times 180 VA each times 100 percent for noncontinuous operation and 125 percent for continuous operation.

Commercial Calculations

For example: What is the load in VA for 48 general-purpose receptacles used to serve noncontinuous and continuous related loads?

Step 1: Calculating noncontinuous load
220.14(I); 230.42(A)(1)
180 VA x 48 x 100% = 8640 VA

Step 2: Calculating continuous load
220.14(I); 230.42(A)(1)
180 VA x 48 x 125% = 10,800 VA

Solution: The noncontinuous load is 8640 VA and the continuous load is 10,800 VA.

See **Figure 23-9** for a detailed illustration for calculating the general-purpose receptacle load.

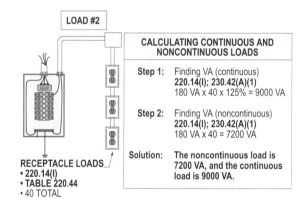

Figure 23-9. The general-purpose receptacle load shall be calculated by multiplying the VA rating of each receptacle by the number and increasing this value by 125 percent if used for three hours or more.

APPLYING DEMAND FACTORS
220.14(I); TABLE 220.44

General-purpose receptacle outlets for cord-and-plug connected loads used at noncontinuous operation are calculated per **220.14(I)** and **Table 220.44**. Noncontinuous operated receptacles with a VA rating of 10,000 VA or less shall be calculated at 100 percent. If the VA rating of the receptacle load exceeds 10,000 VA, a demand factor of 50 percent shall be permitted to be applied to all VA exceeding 10,000 VA per **Table 220.44**.

For example: What is the VA rating for 125 general-purpose receptacle outlets to cord-and-plug connect loads used at noncontinuous operation?

Step 1: Calculating load
220.14(I); 230.42(A)(1)
125 x 180 VA = 22,500 VA

Step 2: Applying demand factors
Table 220.44
First 10,000 VA x 100% = 10,000 VA
Next 12,500 VA x 50% = 6,250 VA
Total load = 16,250 VA

Solution: The demand load is 16,250 VA.

See **Figure 23-10(a)** for a detailed illustration for calculating the (noncontinuous) receptacle demand load.

For example: What is the load in VA for 100 ft. of multioutlet assembly used to cord-and-plug connect loads that are not used simultaneously and used simultaneously? (**220.14(H)(1)**)

Step 1: Connecting load for non-simultaneously use
VA = length ÷ 5' x 180 VA
VA = 100' ÷ 5' x 180 VA
VA = 3600

Step 2: Calculating load for simultaneously use
VA = length x 1' x 180 VA
VA = 100' x 1' x 180 VA
VA = 18,000

Solution: The load in VA for the non-simultaneously load is 3600 VA and for simultaneously load is 18,000 VA.

See **Figures 23-10(b) and (c)** for a detailed illustration for calculating the multioutlet assembly load.

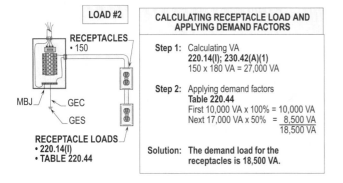

THE DEMAND LOAD FOR THE GENERAL PURPOSE RECEPTACLE SELECTED FROM THE RECEPTACLE LOADS
NEC 220.14(I)
NEC TABLE 220.44

Figure 23-10(a). The demand load for general-purpose receptacles shall be calculated by taking the first 10,000 VA at 100 percent and all remaining VA above 10,000 VA at 50 percent.

MULTIOUTLET ASSEMBLIES
220.14(A)(1); (2)

For connected loads not operating simultaneously, the VA rating shall be calculated by dividing the length of the assembly by 5 ft. and multiplying by 180 VA. For connected loads operating simultaneously, each foot of multioutlet assembly is multiplied by 180 VA. The fixed multioutlet assembly load shall be permitted to be added to the noncontinuous receptacle load and demand factors applied per **Table 220.44**.

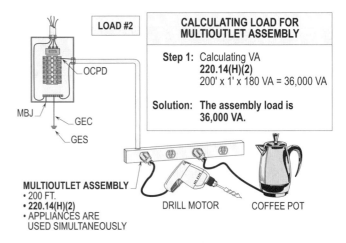

MULTIOUTLET ASSEMBLY LOAD SELECTED FROM RECEPTACLE LOADS
NEC 220.14(H)(2)

Figure 23-10(b). The multioutlet assembly load shall be calculated by multiplying the total length of the assembly (each 1 ft.) by 180 VA where appliances are likely to be used simultaneously.

SPECIAL APPLIANCE LOADS
230.42(A)

Special appliance loads are the third group of loads to be calculated. These loads, which include calculators, processing machines, etc. are usually served by individual circuits.

Commercial Calculations

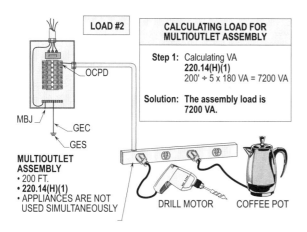

Figure 23-10(c). The multioutlet assembly load shall be calculated by multiplying the total length of the assembly by 180 VA and dividing by 5 where appliances are not likely to be used simultaneously.

CONTINUOUS AND NONCONTINUOUS OPERATION
230.42(A)(1); (A)(2)

The load in VA for special appliance loads shall be calculated by multiplying the VA rating of each load by 100 percent for noncontinuous operation and 125 percent for continuous operation. To determine classification, special appliance loads operating for less than three hours shall be classified as a noncontinuous operated load. However, a special appliance load operating for three hours or more shall be classified as continuous operated load.

For example: What is the VA rating for a 208 volt, three-phase, 65 amp special appliance load operating for ten hours and supplied by an individual branch-circuit?

Step 1: Calculating VA
 220.5(A)
 VA = V × 1.732 (360 V) × I
 VA = 208 V × 1.732 × 65
 VA = 23,400

Step 2: Calculating continuous load
 230.42(A)(1)
 23,400 VA × 125% = 29,250 VA

Solution: The load at continuous operation is 29,250 VA.

For example: Consider and calculate the VA rating for a special appliance load of 52 amps operating at 480 volts, three-phase, for a period of 2 1/2 hours every four hours?

Step 1: Calculating VA
 220.5(A)
 VA = V × 1.732 (831 V) × I
 VA = 480 V × 1.732 × 52 A
 VA = 43,212

Step 2: Calculating noncontinuous load
 230.42(A)(1)
 43,212 VA × 100% = 43,212 VA

Solution: The load in VA for the noncontinuous load is 43,212 VA.

See **Figure 23-11** for a detailed illustration for calculating the VA or a continuous special appliance load.

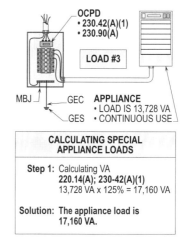

Figure 23-11. The appliance load shall be calculated by multiplying the VA of the appliance by 125 percent if used for three hours or more.

APPLYING DEMAND FACTORS
TABLE 220.56

Demand factors shall be permitted to be applied to cooking equipment in restaurants with three or more cooking units. The load in VA for cooking equipment shall be permitted to be calculated by applying the optional calculation listed in **Table 220.56**.

For example: What is the demand load in amps for 10 - 208 volt, three-phase cooking units rated 8 kW each?

Step 1: Calculating VA
220.56
8 kW x 10 = 80 kW

Step 2: Calculating amps
220.5(A)
I = kW x 1000 ÷ V x 1.732
I = 80 kW x 1000 ÷ 208 V x 1.732
I = 222 A

Step 3: Applying demand factors
Table 220.56
222 A x 65% = 144.3 A

Solution: The demand load is 144.3 A.

See **Figure 23-12** for a detailed illustration for calculating demand loads for certain types of equipment.

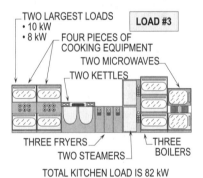

DEMAND LOAD FOR COOKING EQUIPMENT
SELECTED FROM SPECIAL APPLIANCE LOADS
NEC 220.56
NEC TABLE 220.56

Figure 23-12. The demand load shall be calculated based upon the number and the percentage per **Table 220.56**.

COMPRESSOR LOADS
440.34

Compressor loads are the fourth group of loads to be calculated. Special considerations shall be applied when calculating loads for hermetic sealed compressors supplying refrigerant and cooling related equipment.

CONTINUOUS OR NONCONTINUOUS OPERATION
440.34; 230.42(A)(1); (A)(2)

Compressor related equipment shall be calculated at 100 percent of their VA or amps. If one of such is the largest motor per load seven, it is added to the total calculation of all loads at 125 percent of its FLA rating.

For example: What is the load in VA for six compressors rated at 26.5 amps each and supplied by a 480 volt, three-phase supply?

Step 1: Calculating VA
220.5(A); 440.34
26.5 A x 480 V x 1.732 x 125% = 27,539 VA

Step 2: Calculating continuous load
230.42(A)(1); 440.34
132,129 VA x 125% = 165,161 VA
27,539 VA + (26.5 x 5 x 5 x 480 V x 1.732 x 100) = 49,570 VA

Solution: The continuous load rating is 49,750 VA.

See **Figure 23-13** for a detailed illustration for calculating compressor related loads.

MOTOR LOADS
220.50; 430.24

Motor loads are the fifth group of loads to be calculated. The VA rating of motors is converted from FLA to VA by multiplying the FLA from **Table 430.248** for single-phase or **Table 430.250** for three-phase by the supply voltage.

Design Tip: Motors can be used as a single unit to drive a piece of equipment. Motors used in an approved assembly such as a processing machine are not usually considered individual motor loads.

Commercial Calculations

For example: What is the VA rating for a group of 480 volt, three-phase motors rated at 30 HP, 20 HP and 15 HP respectfully?

Step 1: Finding FLA
Table 430.250
30 HP = 40 A
20 HP = 27 A
15 HP = 21 A

Step 2: Calculating total VA
220.5(A)
VA = V x 1.732 x I
30 HP
480 V x 1.732 x 40 A = 33,240 VA
20 HP
480 V x 1.732 x 27 A = 22,437 VA
15 HP
480 V x 1.732 x 21 A = 17,451 VA
Total VA   = 73,128 VA

Solution: **The total load for the motors is 73,128 VA.**

See Figure 23-14 for a detailed illustration for calculating individual motor loads.

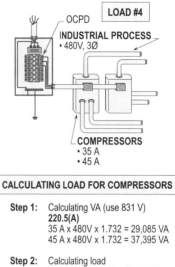

CALCULATING LOAD FOR COMPRESSORS

Step 1: Calculating VA (use 831 V)
220.5(A)
35 A x 480V x 1.732 = 29,085 VA
45 A x 480V x 1.732 = 37,395 VA

Step 2: Calculating load
29,085 VA x 100% = 29,085 VA
37,395 VA x 100% = 37,395 VA
Total Load   = 66,480 VA

Solution: The compressor load is 66,480 VA, however, if one of the units is the largest motor, it shall be increased by 25% or 125% per load seven.

**COMPRESSOR LOAD CONSISTS
OF ALL COMPRESSORS
NEC 440.34
NEC 220.50**

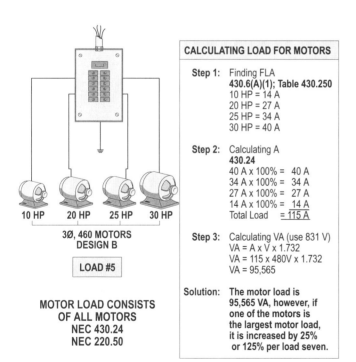

Figure 23-13. The compressor load shall be calculated at 100 percent for each compressor and if one of such is the largest motor, it shall be increased by 25 percent (or 125 percent) and added to the other loads.

HEAT OR A/C LOADS
220.60

Heating or A/C loads is the sixth group of loads to be calculated. The largest VA rating between the heating or A/C load shall be selected and the smaller of the two loads is dropped. To determine the largest of the two loads, the VA rating of each load shall be calculated at 100 percent and the largest load of the two is selected. The load dropped is not used again in the calculation.**(See Figure 23-15)**

Design Tip: There is no need to calculate both loads to select the elements of the service equipment for the loads are never used simultaneously in the electrical system.

Figure 23-14. The motor load shall be calculated at 100 percent for each motor and if one of such is the largest motor load, it shall be increased by 25 percent (or 125 percent) and added to the other loads.

23-11

Stallcup's Electrical Design Book

For example: What is the largest load between a 30 kW heating unit and a 32.5 amp A/C unit? The voltage is supplied by 208 volts, three-phase system.

Step 1: Selecting largest load
220.60
Heating load
30 kW x 1000 x 100% = **30,000 VA**
A/C load
208 V x 1.732 x 32.5 A = **11,700 VA**

Solution: The 30,000 VA heating unit is the largest load.

See **Figure 23-15** for a detailed illustration for calculating the largest VA rating between the heat and A/C load.

LARGEST MOTOR LOAD
220.50; 430.24

The largest motor load in VA is the seventh of the loads to be calculated. The largest motor load shall be selected from one of the motor related loads listed in the fourth, fifth, or sixth loads. The VA rating of the largest motor shall be calculated by multiplying the amperage of the unit by the voltage times 25 percent.

For example: What is the largest motor from the following loads?
  Fourth load = compressor of 35 A
  Fifth load = motor of **40 A**
  Sixth load = A/C unit = 30 A
  (larger than the gas heating)

Step 1: Selecting largest load
220.50; 440.34; 430.24
The motor load of 40 A is the largest load

Solution: The largest motor load is 40 amps.

See **Figure 23-16** for a detailed illustration of calculating the largest motor load in amps.

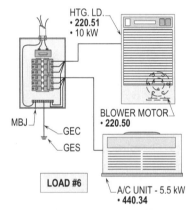

HEATING OR A/C UNIT LOAD IS DETERMINED FROM LOAD SIX
NEC 220.60

Figure 23-15. The largest between the heating and A/C unit in VA shall be selected and the smaller load dropped.

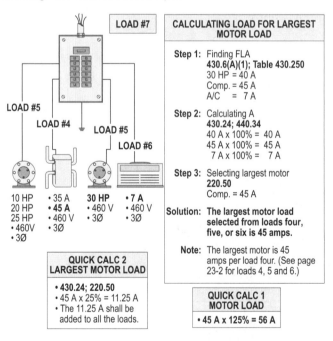

THE LARGEST MOTOR LOAD SELECTED FROM LOADS FOUR, FIVE, OR SIX
• NEC 220.60; NEC 430.24;
NEC 440.34; NEC 220.50

Figure 23-16. The largest motor load is the largest between the compressor loads, A/C unit loads, where eligible, or the motor loads, whichever is greater.

Commercial Calculations

APPLYING THE OPTIONAL CALCULATION PART IV TO ARTICLE 220

The load in VA and amps can be calculated by the optional calculation instead of the standard calculation. The optional calculation is based upon specific use of the electrical system or type of occupancy and its use and operation.

KITCHEN EQUIPMENT
220.56; TABLE 220.56

Table 220.56 in the NEC shall be permitted to be used for load calculation for commercial electrical cooking equipment, such as dishwashers, booster heaters, water heaters and other kitchen equipment. The demand factors shown in that Table are applicable to all equipment that is thermostatically controlled or is only intermittently used as part of the kitchen equipment. In no way do the demand factors apply to the electric heating, ventilating or air-conditioning equipment. In calculating the demand, the demand load shall not be permitted to be less than the sum of the two largest kitchen equipment loads. **(See Figure 23-12)**

SCHOOLS
220.86; TABLE 220.86

Table 220.86 shall be permitted to be used to calculate the service or feeder loads for schools if they are equipped with electric space heating or air-conditioning or both. The demand factors in **Table 220.86** apply to both interior and exterior lighting, power, water heating, cooking or other loads, and the larger of the space heating load or the air-conditioning load.

When using this optional calculation, the neutral of the service or feeder loads shall be permitted to be calculated as required in **220.61**. Feeders within the building or structure where the load is calculated by this optional method may use the reduced ampacity as connected, but the ampacity of any feeder need not be larger than the individual ampacity for the entire building. Portable classrooms or buildings are not included in this Section. **(See Figure 23-17)**

RESTAURANTS
220.88; TABLE 220.88

When calculating the service or feeder load for a new restaurant, and the feeder carries the entire load, **Table 220.88** shall be permitted to be used to size the elements necessary to supply the load. Overload protection shall be in accordance with **230.90, 215.3,** and **240.4**. Also, feeder or subfeeder conductors do not have to be larger than service conductors, regardless of calculations. **(See Figure 23-18)**

TOTAL VA OF SCHOOL
• 801,693 VA

CALCULATING LOAD IN VA

Step 1: Calculating sq. ft.
  Classroom area = 30,000 sq. ft.
  Auditorium area = 5,000 sq. ft.
  Cafeteria area = 3,000 sq. ft.
  Hall area = 1,000 sq. ft.
  Total area = 39,000 sq. ft.

Step 2: Calculating VA per sq. ft.
  Table 220.86
  VA per sq. ft. = Total VA ÷ Total sq. ft.
  VA per sq. ft. = 801,693 VA ÷ 39,000 sq. ft.
  VA per sq. ft. = 20.55623

Step 3: Applying demand factors
  Table 220.86
  First 3 VA ÷ sq. ft. @ 100%
  Next 17.55623 @ 75%
  3 VA x 39,000 sq. ft. x 100% = 117,000 VA
  17.55623 VA x 39,000 sq. ft. x 75% = 513,519.72 VA
  Total VA = 630,519.72 VA

Solution: The demand load in VA for the school is 630,519.72 VA.

TOTAL SQUARE FOOTAGE OF SCHOOL

Classroom area	= 30,000 sq. ft.
Auditorium area	= 5,000 sq. ft.
Cafeteria area	= 3,000 sq. ft.
Hall area	= 1,000 sq. ft.

NEC 220.86
NEC TABLE 220.86

Figure 23-17. The above calculation shows the optional calculation being applied for a school.

OPTIONAL CALCULATIONS FOR ADDITIONAL LOADS TO EXISTING INSTALLATIONS
220.87

When additional loads are added to existing facilities having feeders and service as originally calculated, the maximum kVA calculations in determining the load on the existing feeders and service shall be permitted to be used if the following conditions are complied with:

• If the maximum data of the demand in kVA is available for a minimum of one year, such as demand meter ratings.

• If the demand ratings for that period of one year at

23-13

125 percent and the addition of the new load does not exceed the rating of the service. Where demand meters are used, in most cases the load as calculated will probably be less than the demand meter indications.

- If the overcurrent protection meets **230.90, 215.3** and **240.4** for the feeder or service.

Design Tip: By measurement or calculation, the larger of the heating or cooling equipment load shall be included in the load to be added. **(See Figure 23-20)**

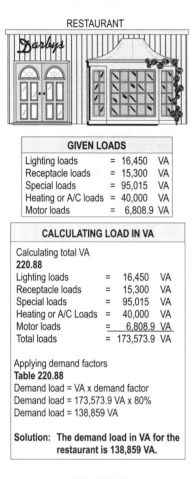

Figure 23-18. The above calculation shows the optional calculation being applied for a restaurant.

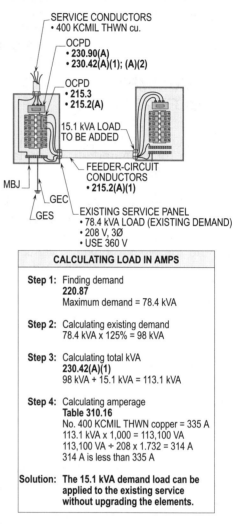

Figure 23-19. The above illustration is the calculation for a service or feeder-circuit.

APPLYING Ex. TO 220.87
220.87, Ex.

If the maximum demand data for a one year period is not available, the calculated load shall be permitted to be based on the maximum demand (measure of average power demand over a 15-minute period) continuously recorded over a minimum 30 day period using a recording ammeter or power meter connected to the highest loaded ungrounded (phase) of the feeder or service, based on the initial loading at the start of the recording. **(See Figure 23-19)**

CALCULATING THE NEUTRAL
220.61; 310.15(B)(4)(c)

For a service or feeder, the maximum unbalanced load controls the ampacity selected for the grounded (neutral) conductor. Grounded (neutral) feeder load shall be considered wherever a grounded (neutral) conductor is used in conjunction with one or more ungrounded (phase) conductors. On a single-phase feeder using one ungrounded (phase) conductor and a grounded (neutral) conductor, the grounded (neutral) conductor will carry the same amount of current as the ungrounded (phase) conductor. A two-wire feeder is seldom used, so in considering the grounded (neutral) feeder current, always assume that there is a grounded (neutral) conductor and two or more ungrounded (phase) conductors. If there are two ungrounded (phase) conductors that are connected to the same phase, and a grouned (neutral) conductor, the grounded (neutral) conductor would be required to carry the total current from both ungrounded (phase) conductors, which would not be an accepted practice.

For three-wire DC or single-phase AC, four-wire three-phase, three-wire two-phase, and five-wire two-phase systems, a further demand factor of 70 percent shall be permitted to be applied to that portion of the unbalanced load in excess of 200 amperes. There shall be no reduction of the grounded (neutral) conductor capacity for that portion of the load which consists of electric-discharge lighting, electronic calculator/data processing or similar equipment, when supplied by four-wire, wye-connected, three-phase systems. **(See Figure 23-21)**

For example, on a four-wire, three-phase wye circuit where the major portion (over 50 percent) of the load consists of nonlinear loads, there are harmonic currents present in the grounded (neutral) conductor, and the grounded (neutral) conductor shall be considered to be a current-carrying conductor. In other words, the ampacity of the conductor shall be derated per **310.15(B)(2)(a)**.

Review the rules and examples of **Chapters 14** and **15** for the sizing and use of the grounded (neutral) conductor in service, feeder-circuits and branch-circuit installations.

As an example, the grounded (neutral) conductor served by a 277/480 volt supply would be calculated per **Figure 23-21,** where there is electrical discharge lighting (inductive), incandescent lighting (resistive) and other resistive related loads.

CALCULATION PROBLEMS
PARTS III AND IV TO ARTICLE 220

The elements of electrical systems shall be permitted to be calculated by using the standard or optional calculation. The size of these elements are determined by which method the designer chooses to calculate these loads. The following calculations are typical examples of how these loads are calculated, sized and selected. The step-by-step procedures are easy to follow and have condensed the more complicated rules pertaining to calculating loads into a compact listing, which provides easier understanding of how to perform calculations according to the provisions of the NEC. A broad assortment of basic code calculations have been selected to represent the main principles of designing and installing electrical systems per NEC rules.

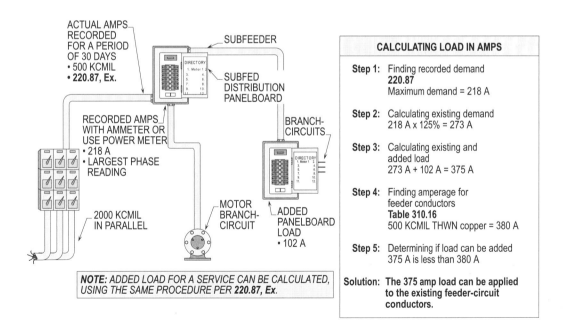

Figure 23-20. The above calculation shows the optional calculation being applied for adding a load to an existing feeder-circuit.

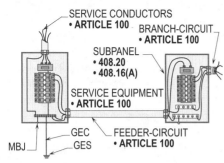

Figure 23-21. The above calculation shows the procedure for calculating the neutral for a service or feeder and applying demand factors, where permitted.

PROCEDURE FOR CALCULATING THE LOADS IN VA AND AMPS
PARTS III AND IV TO ARTICLE 220

Commercial loads shall be calculated by applying the following steps:

- The first step is to calculate the general-purpose lighting load by the VA per sq. ft. according to the provisions of **Table 220.12** and **230.42(A)(1)** and **(A)(2)** if the occupancy is listed.

- The second step is to calculate the general-purpose receptacle load by multiplying the total number by 180 VA by the provisions of **220.14(I)** and **230.42(A)(1)** and **(A)(2)**. If the total exceeds 10 kVA, a demand factor of 50 percent shall be permitted to be applied to all VA ratings exceeding 10 kVA. **Tables 220.42** and **220.44** permit the reduction of such loads.

- The third step is to calculate the total load for all special appliances available in the facility. These appliances shall be calculated at nameplate rating based upon noncontinuous and continuous or a combination of both.

- The fourth step is to calculate the compressor load at noncontinuous use and set aside as the largest in case it is the largest motor load.

- The fifth step is to calculate the motor loads. Each motor in VA or amps shall be added together at 100 percent value to obtain the total and set aside as the largest in case it is the largest motor load.

- The sixth step is to calculate the largest of the air-conditioning (A/C) or electric heating load, and drop the smallest of the two.

- The seventh step is to calculate the largest motor load by 25 percent (or 125 percent) and add to the other loads in the steps above. The largest motor load is selected from one of the loads in steps four, five and six respectfully. (Apply 25 percent rule)

The final step is to total the calculated loads in all the steps and divide by the voltage if in VA and size the electrical elements for a service or feeder-circuit.

STORE BUILDING SUPPLIED BY 120/240 VOLT POWER SOURCE TABLE 220.12; (LISTED OCCUPANCY)

It is the loads in a facility that determines the voltage for designing and sizing the service or feeder-circuit elements.

For example, 120/240 volt supply is usually used to supply a facility having an equal number of three-phase and single-phase related loads.

The disadvantage of using a 120/240 volt supply is it produces larger ratings for selecting the elements for service equipment and feeder-circuits. **(See Design Problem 23-1)**

STORE BUILDING SUPPLIED BY 120/208 VOLT POWER SOURCE TABLE 220.12; (LISTED OCCUPANCY)

A facility equipped with a large number of 120 volt and 208 volt three-phase loads are normally supplied by a four-wire, three-phase wye-connected system.

On a three-phase wye-connected system, three 120 volt ungrounded (phase) conductors are obtained from ground; therefore, three ungrounded (phase) conductors may be routed with each grounded (neutral) conductor. This type of installation requires less grounded (neutral) conductors and therefore saves on the amount of copper or aluminum that would otherwise be needed if one grounded (neutral) conductor was pulled with each ungrounded (phase) conductor.

See Design Problem 23-2 for calculating the total load for sizing the elements to be used with 120/208 volt, three-phase, four-wire systems.

STORE BUILDING SUPPLIED BY 277/480 VOLT POWER SOURCE TABLE 220.12; (LISTED OCCUPANCY)

Smaller store buildings are usually supplied by 120/208 volt power systems. Loads such as lighting, receptacles and other related loads are served by the lower voltage of 120 volts. Equipment is normally served by the higher voltage rated at 208 volts phase-to-phase.

Larger office buildings utilize 277/480 volt, three-phase, four-wire systems. The higher 480 volt, three-phase voltage supplies the heavier equipment and the 277 volts, single-phase voltages supplies lighting. Transformers are used to step down the 480 volts to 120/208 volts or 120/240 volts to supply the lower voltage loads and equipment. **(See Design Problem 23-3)**

STORE BUILDING SUPPLIED BY 120/240 VOLT, THREE-PHASE POWER SOURCE TABLE 220.12; (LISTED OCCUPANCY)

Store buildings with a greater number of three-phase loads than single-phase loads may be supplied by a three-phase, four-wire, 120/240 volt service. The 120 volts may be obtained from two of the ungrounded (phase) conductors to ground. This phase-to-ground voltage is taken from the lighting and power transformer.

Transformers on a delta-connected system are connected in an open or closed delta configuration.

Closed delta systems require three transformers, while open delta systems require only two transformers. One of the advantages of closed delta systems is that if one transformer fails, the remaining two may be connected in an open delta configuration, and continue to supply the load until a replacement transformer is purchased and installed. **(See Design Problem 23-4)**

OFFICE BUILDING SUPPLIED BY 277/480 VOLT POWER SOURCE TABLE 220.12; (LISTED OCCUPANCY)

Office buildings consist mostly of lighting loads and receptacle loads supplying small sensitive electronic machines which are used for office related work. Heavier loads such as equipment, heating, A/C and other three-phase loads are supplied by the higher voltage.

The advantage of 277/480 volt systems is that they allow OCPD's, conductors and other electrical elements to be smaller in size. Naturally, these smaller elements reduce the cost of installation. **(See Design Problem 23-5)**

SCHOOL BUILDING SUPPLIED BY 277/480 VOLT POWER SOURCE TABLE 220.12; (LISTED OCCUPANCY)

School buildings supplied by 277/480 volt services are calculated with the basic steps used to calculate the load for any other commercial occupancy. **(See Design Problem 23-6)**

RESTAURANT SUPPLIED BY 120/208 VOLT POWER SOURCE TABLE 220.12; (LISTED OCCUPANCY)

Small restaurants consist of lighting and receptacle loads, with the larger loads being the cooking equipment and other such pertinent apparatus.

Based upon the number of cooking units, demand factors shall be permitted to be applied to the total load per **220.56** and **Table 220.56**. Larger restaurants are supplied with 277/480 volt services with step down transformers being utilized to serve smaller 120 volts loads. **(See Design Problem 23-7)**

HOSPITAL BUILDING SUPPLIED BY 277/480 VOLT POWER SOURCE TABLE 220.12; (LISTED OCCUPANCY)

The loads in hospitals are calculated by their conditions of use. They are either calculated at continuous operation, noncontinuous operation or demand factors shall be permitted to be applied for certain loads. The service voltage is determined by the size of the facility and related equipment. The procedure for calculating the load is to use the seven steps listed in this chapter for calculating the total load for a premises. **(See Design Problem 23-8)**

WELDING SHOPS SUPPLIED BY 120/208 VOLT POWER SOURCE
PARTS II AND III TO ARTICLE 220;
(UNLISTED OCCUPANCY)

Unlisted occupancies are those not appearing in **Table 220.12**, therefore, their VA ratings shall be calculated from other Sections than those listed in **Table 220.12**. The load for unlisted occupancies shall be calculated by applying the same step procedure listed in this chapter. The only difference is the lighting is not obtained per **Table 220.12**. **(See Design Problem 23-9)**

SIZING ELEMENTS

Design problems 23-1 through 23-9 cover specific questions on problems pertaining to sizing elements that are found in commercial facilities. Questions are based on the calculations of design problems 23-1 through 23-9. Elements such as OCPD's, conductors, conduits, etc. are determined using step by step procedures.

SIZING ELEMENTS
DESIGN PROBLEM 23-1

(1) What size THWN copper conductors are required for the service when they are paralleled 5 times per phase? (See page 23-25)

Step 1: Paralleling conductors
310.4; page 23-25
Amps of conductors = service A ÷ # in parallel
A = 1274 A ÷ 5
A = 254.8

Step 2: Selecting conductors size
Table 310.16; 220.5(B)
254.8 A requires 250 KCMIL
250 KCMIL = 255 A

Solution: It takes 5 - 250 KCMIL THWN copper conductors to supply a load of 1274 amps (255 A x 5 = 1275 A).

---

(2) What size OCPD is required for the service based upon the ampacity of the conductors?

Step 1: Sizing OCPD
230.90(A); 240.4(C); 240.6(A);
page 23-25
1274 A requires 1250 A

Solution: The size OCPD based upon ampacity of conductors is 1250 amps.

---

(3) What size neutral conductor is required for the service when they are paralleled 5 times per phase?

Step 1: Selecting conductor size
310.4; page 23-25
Amps of conductors = service A ÷ # in parallel
A = 759 A ÷ 5
A = 151.8

Step 2: Selecting conductor size
Table 310.16; 220.5(B)
152 A requires 2/0 AWG cu.
2/0 AWG cu. = 175 A

Solution: It takes 5 - 2/0 AWG THWN copper conductors to supply a load of 759 amps (175 A x 5 = 875 A).

---

(4) What size rigid metal conduit is required to enclose the conductors for each run?

Step 1: Sizing sq. in. area
Table 5, Ch. 9
250 KCMIL cu.   = 0.397
2/0 AWG THWN cu. = 0.2223

Step 2: Calculating total sq. in. area
Table 5, Ch. 9
0.3970 x 2 = .794 sq. in.
0.2223 x 1 = .2223 sq. in.
Total sq. in. = 1.0163 sq. in.

Step 3: Selecting RMC
Table 4, Ch. 9
1.0163 sq. in. = 2" (53)

Solution: The size rigid metal conduit for each is 2 in. (53).

Commercial Calculations

SIZING ELEMENTS
DESIGN PROBLEM 23-2

(1) What size THWN copper conductors are required for the service when they are paralleled 5 times per phase? (See page 23-26)

Step 1: Paralleling conductors
310.4; page 23-26
Amps of conductors = service A ÷ # in parallel
A = 849 A ÷ 5
A = 169.8

Step 2: Selecting conductors size
Table 310.16
169 A requires 2/0 AWG cu.
2/0 AWG cu. = 175 A

Solution: It takes 5 - 2/0 AWG THWN copper conductors to supply a load of 849 amps (175 A x 5 = 875 A).

(2) What size OCPD is required for the service based upon the calculated load?

Step 1: Sizing OCPD
230.90(A); 240.4(C); 240.6(A);
page 23-26
849 A requires 800 A

Solution: The size OCPD based upon calculated load is 800 amps.

(3) What size neutral conductor is required for the service when they are paralleled 5 times per phase?

Step 1: Selecting conductor size
310.4; page 23-26
Amps of conductors = service A ÷ # in parallel
A = 506 A ÷ 5
A = 101

Step 2: Selecting conductor size
310.4; Table 310.16; 220.5(B); 250.24(C)(2)
101 A requires 1/0 AWG cu.
1/0 AWG cu. = 150 A

Solution: It takes 5 - 1/0 AWG THWN copper conductors to supply a load of 506 amps (150 A x 5 = 750 A).

(4) What size rigid metal conduit is required to enclose the conductors for each run?

Step 1: Sizing sq. in. area
Table 5, Ch. 9
2/0 AWG THWN = 0.2223
1/0 AWG THWN = 0.1855

Step 2: Calculating total sq. in. area
Table 5, Ch. 9
0.2223 x 3 = .6669 sq. in.
0.1855 x 1 = .1885 sq. in.
Total sq. in. = .8554 sq. in.

Step 3: Selecting RMC
Table 4, Ch. 9
.8554 sq. in. = 2" (53)

Solution: The size rigid metal conduit for each run is 2 in. (53).

(5) What size copper grounding electrode conductor is required to ground the service to building structural steel (BSS)?

Step 1: Sizing GEC
Table 8, Ch. 9
2/0 AWG cu. = 133,100 CM
CM = CM x # of conductors
CM = 133,100 CM x 5
CM = 665,500 CM
KCMIL = CM ÷ 1,000
KCMIL = 665,500 CM ÷ 1000
KCMIL = 665.5

Step 2: Selecting GEC
Table 250.66
665.5 KCMIL = 2/0 AWG cu.

Solution: The size of the GEC required to ground to BSS is 2/0 AWG copper.

23-19

SIZING ELEMENTS
DESIGN PROBLEM 23-3

(1) What size THWN copper conductors are required for the service when they are paralleled 2 times per phase? (See page 23-27)

Step 1: Paralleling conductors
310.4; page 23-27
Amps of conductors = service A ÷ # in parallel
A = 369 A ÷ 2
A = 184.5

Step 2: Selecting conductors size
Table 310.16; 220.5(B)
185 A requires 3/0 AWG THWN cu.
3/0 AWG THWN cu. = 200 A

Solution: It takes 2 - 3/0 AWG THWN copper conductors to supply a load of 369 amps (200 A x 2 = 400 A).

(2) What size OCPD is required for the service based upon the ampacity of conductors?

Step 1: Sizing OCPD
230.90(A); 240.4(B); 240.6(A);
page 23-27
369 A requires 400 A

Solution: The size OCPD based upon the ampacity of the conductors is 400 amps.

(3) What size neutral conductor is required for the service when they are paralleled 5 times per phase?

Step 1: Selecting conductor size
310.4; page 23-27
Amps of conductors = service A ÷ # in parallel
A = 150 A ÷ 2
A = 75

Step 2: Selecting conductor size
310.4; Table 310.16; 250.24(C)(2)
75 A requires 1/0 AWG cu.
1/0 AWG cu. = 150 A

Solution: It takes 2 - 1/0 AWG THWN copper conductors to supply a load of 150 amps (150 A x 2 = 300 A).

(4) What size rigid metal conduit is required for each run?

Step 1: Sizing sq. in. area
Table 5, Ch. 9
3/0 AWG THWN = 0.2679
1/0 AWG THWN = 0.1855

Step 2: Calculating total sq. in. area
Table 5, Ch. 9
0.2679 x 3 = .8037 sq. in.
0.1855 x 1 = .1855 sq. in.
Total sq. in. = .9892 sq. in.

Step 3: Selecting RMC
Table 4, Ch. 9
.9892 sq. in. = 2" (53)

Solution: The size rigid metal conduit for each run is 2 in. (53).

(5) Will the 3/0 AWG THHN copper conductors supply the service load of 356 amps if four conductors in each RMC run are current-carrying?

Step 1: Applying derating factors
310.15(B)(2)(a); 110.14(C)
3/0 AWG THWN cu. = 225 A
225 A x 80% = 180 A

Step 2: Checking amps
310.4
180 A x 2 = 360 A

Solution: Yes, the deration of 360 amps will supply a load of 356 amps.

Commercial Calculations

SIZING ELEMENTS
DESIGN PROBLEM 23-4

(1) What size THWN copper conductors are required to supply Phases A and C when they are paralleled four times per phase? (See page 23-28)

Step 1: Paralleling conductors
310.4; page 23-28
Amps of conductors = service A ÷ # in parallel
A = 1121 A ÷ 4
A = 280.25

Step 2: Selecting conductors
Table 310.16; 220.5(B)
280 A requires 300 KCMIL
300 KCMIL THWN cu. = 285 A

Solution: It takes 4 - 300 KCMIL THWN copper conductors to supply a load of 1121 amps (285 A x 4 = 1140 A).

(2) What size THWN copper conductors are required to supply the high-leg which is Phase B? (Phase B is also paralleled four times per phase)

Step 1: Paralleling conductors
310.4; 220.5(B); page 23-28
# of conductors = 217 A ÷ 4 = 54 A

Solution: Section 310.4 requires at least 1/0 AWG conductors to be connected in parallel. Therefore, Phase B requires 4 THWN copper conductors per phase.

(3) What size THWN copper conductors are required to supply the neutral load?

Step 1: Paralleling conductors
310.4; 220.5(B); page 23-28
Amps of conductors = service A ÷ # in parallel
A = 759 A ÷ 4
A = 189.7

Step 2: Selecting conductors
Table 310.16; 250.24(C)(2)
190 A requires 3/0 AWG cu.
3/0 AWG cu. THWN = 200 A

Solution: It takes 4 - 3/0 AWG THWN copper conductors to supply a load of 759 amps (200 A x 4 = 800 A).

(4) What color is the high-leg?

Step 1: Determining color code
110.15; 230.56; 408.3(E)
Orange shall be used

Solution: Orange is the color required for the high-leg when used in an open or closed delta-connected system.

Stallcup's Electrical Design Book

SIZING ELEMENTS
DESIGN PROBLEM 23-5

(1) What size GEC is required per **Table 250.66** when the service conductors are paralleled 6 times per phase? (See page 23-30)

Step 1: Sizing conductors in parallel
310.4; Table 310.16; 220.5(B);
page 23-30
Amps of = A of conductors ÷ # in parallel
A = 1042 A ÷ 6
A = 174 A
2/0 AWG THWN cu. = 175 A
175 A supplies 174 A

Step 2: Calculating KCMIL
Table 8, Ch. 9
2/0 AWG cu. = 133,100 CM
CM = 133,100 CM x 6
CM = 798,600
KCMIL = 798,600 ÷ 1000
KCMIL = 798.6

Step 3: Selecting GEC
Table 250.66; 220.5(B)
798.6 KCMIL requires 2/0 AWG cu.

Solution: The size GEC is 2/0 AWG copper.

(2) What size copper equipment bonding jumper is required to bond all of the rigid metal conduits to the grounded busbar?

Step 1: Selecting EBJ
250.102(C); Table 250.66
798.6 KCMIL requires 2/0 AWG cu.

Solution: The size equipment bonding jumper is 2/0 AWG copper.

(3) What size THWN copper grounded (neutral) conductors are required per phase?

Step 1: Selecting grounded conductor
250.24(C)(2); Table 250.66
798.6 KCMIL requires 2/0 AWG cu.

Step 2: Calculating A
310.4; 220.5(B); page 23-30
Amps of conductors ÷ # in parallel
A = 634 A ÷ 6
A = 106 A
106 A requires 1/0 AWG cu. per
250.24(B)(2)

Solution: The size grounded (neutral) conductor per phase is 1/0 AWG THWN copper which is the correct size per 250.24(C)(2).

SIZING ELEMENTS
DESIGN PROBLEM 23-6

(1) What is the allowable ampacity for 4 THHN cu. current-carrying conductors routed through an ambient temperature of 120°F? (See page 23-31)

Step 1: Selecting phase conductors
310.10, FPN(2); Table 310.16;
page 23-31
335 A requires 400 KCMIL cu.

Step 2: Applying adjustment factors
Table 310.15(B)(2)(a)
380 A x 80% = 304 A

Step 3: Applying correction factors
Table 310.16; 220.5(B) (too small)
304 A x 82% = 249 A

Step 4: Applying adjustment and correction factors (large enough)
Table 310.16; Table 310.15(B)(2)(a)
700 KCMIL cu. = 520 A
520 A x 80% x 82% = 341 A

Solution: The size THHN copper conductors are 700 KCMIL. Note: It would be more logical to parallel the number of conductors necessary to supply the load.

Commercial Calculations

SIZING ELEMENTS DESIGN PROBLEM 23-7

(1) What size kVA transformer is required to supply the calculated load? (Based on calculated load)

Step 1: Calculating the load in VA
Load = 137,185 VA

Step 2: Sizing transformer
page 23-32
kVA = VA ÷ 1000
kVA = 137,185 VA ÷ 1000
kVA = 137.185

Step 3: Selecting transformer
Chart (inside back cover of book)
137.185 kVA requires 150 kVA

Solution: The size transformer to supply the load is 150 kVA.

(2) What size panelboard is required to supply the calculated load?

Step 1: Calculating the load in amps
220.5(A); page 23-32
I = VA ÷ V x √3
I = 137,185 VA ÷ 208 V x 1.732
I = 380.7 A

Step 2: Selecting panelboard
408.16; 240.4(B); 220.5(B)
Chart (inside back cover of book)
381 A requires 400 A

Solution: The size panelboard to handle the calculated load is 400 amps.

SIZING ELEMENTS DESIGN PROBLEM 23-8

(1) What size busway is required to supply the calculated load? (See page 23-33)

Step 1: Calculating the load in amps
Chart
I = VA ÷ V x √3
I = 1,142,563 VA ÷ 480 V x 1.732
I = 1375 VA

Step 2: Selecting busway
Chart (inside back cover of book)
1375 A requires 1600 A

Solution: The size busway required to supply the calculated load is 1600 amps.
Note: If available by the manufacturer, a 1500 amp busway may be used.

(2) What size disconnect switch is required ahead of the distribution panelboard?

Step 1: Sizing fuses
240.6(A)
Chart (inside back cover of book)
1375 A requires 1400 A

Solution: The size disconnect switch required to hold the fuses is 1600 amps.
Note: 240.6(A) allows a listed manufacturer fuse of 1400 amps to be used.

SIZING ELEMENTS
DESIGN PROBLEM 23-9

(1) What is the load in amps for Phases A, B, and C when adding the amps of each individual load?

Step 1: Calculating amps
220.5(A); page 23-34
General lighting load (•)

I = VA ÷ V x √3
I = 13,800 VA ÷ 208 V x 1.732 (360)
I = **38.33 A**
Receptacle loads (•)

I = 13,500 VA ÷ 360 V
I = **37.5 A**

Special loads (•)

I = 56,468 VA ÷ 360 V
I = **156.86 A**

Compressor and motor loads (•)

I = 11,088 VA ÷ 360 V
I = **30.8 A**

Heating load (•)

I = 20,000 VA ÷ 360 V
I = **55.55 A**

Largest motor load (•)

I = 2178 VA ÷ 360 V
I = **6.05 A**
Total load = **325.1 A**

Solution: Round down to 325 amps per Tables, Ch. 9 and 220.5(B). Note that .1 is dropped.

(2) Does the added amps per phase for each load equal the calculated amps in Design Problem 23-9 on page 23-34?

Step 1: Calculating amps
Added amps per phase equals calculated amps per phase

Solution: Yes, the amps per phase are equal.

Commercial Calculations

DESIGN PROBLEM 23-1: What is the load in VA and amps to calculate and size the elements for 120/240 volt, single-phase service supplying a 40,000 sq. ft. store with 20,000 sq. ft. of warehouse space?

120 V, single-phase loads

- 80 linear feet of show window (noncontinuous operation)
- 120' of lighting track
- 30 - 180 VA ballasts outside lighting (continuous operation)
- 3600 VA sign lighting (continuous operation)
- 65 receptacles (noncontinuous operation)
- 28 receptacles (continuous operation)
- 80' multioutlet assembly (heavy-duty)

240 V, single-phase loads

- 7380 VA freezer
- 5580 VA ice cream boxes
- 1 - 1/2 HP exhaust fan
- 10,000 VA water heater
- 9540 VA walk-in cooler
- 50,000 VA heating unit
- 22,320 VA A/C unit
- 1 - 2 HP water pump

Sizing phases = •
Sizing neutral = √
Sizing total load = *

Note 1: For sizing elements, see pages 23-18 of this chapter.

Note 2: Add Asterisks and and checks to obtain total load.

CALCULATING LIGHTING LOAD

Step 1: General lighting load
Table 220.12; 230.42(A)(1)
40,000 sq. ft. x 3 VA = 120,000 VA √
120,000 VA x 125% = 150,000 VA *
20,000 sq. ft. x 1/4 VA = 5,000 VA √
5,000 VA x 125% = 6,250 VA *

Step 2: Show window load
220.43(A)
80' x 200 = 16,000 VA * √

Step 3: Track lighting load
220.43(B)
120' ÷ 2 x 150 VA = 9,000 VA * √

Step 4: Outside lighting load
230.42(A)(1)
30 x 180 VA = 5,400 VA √
5400 VA x 125% = 6,750 VA *

Step 5: Sign lighting load
220.14(F); 230.42(A)(1)
3600 VA x 100% = 3,600 VA √
3600 VA x 125% = 4,500 VA *

Total load = 192,500 VA •

CALCULATING RECEPTACLE LOAD

Step 1: Noncontinuous operation
220.14(H)(2); 220.14(I); 230.42(A)(1)
65 x 180 VA = 11,700 VA
80' x 180 VA = 14,400 VA
Total Load = 26,100 VA
Table 220.44
First 10,000 VA x 100% = 10,000 VA
Next 16,100 VA x 50% = 8,050 VA
Total load = 18,050 VA * √

Step 2: Continuous operation
220.14(I); 230.42(A)(1)
28 x 180 VA = 5,040 VA √
5040 VA x 125% = 6,300 VA *
Total load = 24,350 VA •

CALCULATING SPECIAL LOAD

Step 1: Water heater load
10,000 VA x 100% = 10,000 VA * •

CALCULATING COMPRESSOR LOADS

Step 1: Freezer load
230.42(A)(1); 440.34
7380 VA x 100% = 7,380 VA *

Step 2: Ice cream boxes
5580 VA x 100% = 5,580 VA *

Step 3: Walk-in cooler
9540 VA x 100% = 9,540 VA *
Total load = 22,500 VA •

Calculating motor loads

Step 1: Water pump load
430.24; Table 430.248
12 A x 240 V x 100% = 2,880 VA *

Step 2: Exhaust fan load
4.9 A x 240 V x 100% = 1,176 VA *
Total load = 4,056 VA •

CALCULATING HEATING OR A/C LOAD

Step 1: Heating load selected
220.60
50,000 VA x 100% = 50,000 VA * •
23,320 VA x 100% = 22,320 VA

CALCULATING LARGEST MOTOR LOAD

Step 1: Walk-in cooler
220.50; 430.24; 440.34
9540 VA x 25% = 2,385 VA * •

CALCULATING FOR PHASES A AND B (ADD ALL •)

Lighting loads = 192,500 VA •
Receptacle loads = 24,350 VA •
Special loads = 10,000 VA •
Compressor loads = 22,500 VA •
Motor loads = 4,056 VA •
Heating load = 50,000 VA •
Largest motor load = 2,385 VA •
Total load for facility = 305,791 VA

FINDING AMPS FOR PHASES A AND B

I = VA ÷ V
I = 305,791 VA ÷ 240 V
I = 1274 A

CALCULATING NEUTRAL (ADD ALL √)

Lighting load = 159,000 VA √
Receptacle load = 23,090 VA √
Total load = 182,090 VA

FINDING AMPS FOR NEUTRAL (CALCULATED AT 100% OF VA)

I = VA ÷ V
I = 182,090 VA ÷ 240 V
I = 759 A

See **220.5(B), 220.61** and Design Tip No. 5 on page 23-29.

DESIGN PROBLEM 23-2: What is the load in VA and amps to calculate and size the elements for 120/208 volt, three-phase, four-wire service

120 V, single-phase loads

- 80 linear feet of show window (noncontinuous operation)
- 120' of lighting track
- 30 - 180 VA ballasts outside lighting (continuous operation)
- 3,600 VA sign lighting (continuous operation)
- 65 receptacles (noncontinuous operation)
- 28 receptacles (continuous operation)
- 80' multioutlet assembly (heavy-duty)

208 V, three-phase loads

- 7,380 VA freezer
- 5,580 VA ice cream boxes
- 1 - 3/4 HP exhaust fan
- 10,000 VA water heater
- 9,540 VA walk-in cooler
- 50,000 VA heating unit
- 22,320 VA A/C unit
- 1 - 2 HP water pump

Sizing phases = •
Sizing neutral = √
Sizing total load = *

Note 1: For sizing elements, see page 23-19 of this chapter.

Note 2: Add Asterisks and checks to obtain total load.

CALCULATING LIGHTING LOAD

Step 1: General lighting load
Table 220.12; 230.42(A)(1)
40,000 sq. ft. x 3 VA = 120,000 VA √
120,000 VA x 125% = 150,000 VA *
20,000 sq. ft. x 1/4 VA = 5,000 VA √
5000 VA x 125% = 6,250 VA *

Step 2: Show window load
220.43(A)
80' x 200 = 16,000 VA * √

Step 3: Track lighting load
220.43(B)
120 ÷ 2 x 150 VA = 9,000 VA * √

Step 4: Outside lighting load
230.42(A)(1)
30 x 180 VA = 5,400 VA √
5400 VA x 125% = 6,750 VA *

Step 5: Sign lighting load
220.14(F); 230.42(A)(1)
3600 VA x 100% = 3,600 VA √
3600 VA x 125% = 4,500 VA *
Total load = 192,500 VA •

CALCULATING RECEPTACLE LOAD

Step 1: Noncontinuous operation
220.14(H)(2); 220.14(I); 230.42(A)(1)
65 x 180 VA = 11,700 VA
80' x 180 VA = 14,400 VA
Total Load = 26,100 VA
Table 220.44
First 10,000 VA x 100% = 10,000 VA
Next 16,100 VA x 50% = 8,050 VA
Total load = 18,050 VA * √

Step 2: Continuous operation
220.14(H); 230.42(A)(1)
28 x 180 VA = 5,040 VA √
5040 VA x 125% = 6,300 VA *
Total load = 24,350 VA •

CALCULATING SPECIAL LOADS

Step 1: Water heater load
10,000 VA x 100% = 10,000 VA * •

CALCULATING COMPRESSOR LOADS

Step 1: Freezer load
230.42(A)(1); 440.34
7380 VA x 100% = 7380 VA *

Step 2: Ice cream boxes
5580 VA x 100% = 5,580 VA *

Step 3: Walk-in cooler
9540 VA x 100% = 9,540 VA *
Total load = 22,500 VA •

CALCULATING MOTOR LOADS

Step 1: Water pump load
430.24; Table 430.250
7.5 A x 360 V x 100% = 2,700 VA *

Step 2: Exhaust fan load
3.5 A x 360 V x 100% = 1,260 VA *
Total load = 3,960 VA •

CALCULATING HEATING OR A/C LOAD

Step 1: Heating load selected
220.60
50,000 VA x 100% = 50,000 VA * •
23,320 VA x 100% = 22,320 VA

CALCULATING LARGEST MOTOR LOAD

Step 1: Walk-in cooler
220.50; 430.24; 440.34
9540 VA x 25% = 2,385 VA * •

CALCULATING FOR PHASES A, B AND C (ADD ALL •)

Lighting loads = 192,500 VA •
Receptacle loads = 24,350 VA •
Special loads = 10,000 VA •
Compressor loads = 22,500 VA •
Motor loads = 3,960 VA •
Heating load = 50,000 VA •
Largest motor load = 2,385 VA •
Total load for facility = 305,695 VA

FINDING AMPS FOR PHASES A, B AND C

I = VA ÷ V
I = 305,695 VA ÷ 208 V x 1.732 (360 V)
I = 849 A

CALCULATING NEUTRAL (ADD ALL √)

Lighting load = 159,000 VA √
Receptacle load = 23,090 VA √
Total load = 182,090 VA

FINDING AMPS FOR NEUTRAL

I = VA ÷ V
I = 182,090 VA ÷ 208 V x 1.732 (360 V)
I = 506 A

See Design Tip 5 on page 23-29.

Commercial Calculations

DESIGN PROBLEM 23-3: What is the load in VA and amps to calculate and size the elements for 277/480 volt, three-phase, four-wire service supplying a 40,000 sq. ft. store with 20,000 sq. ft. of warehouse space with 277 volt lighting?

120 V, single-phase loads

- 80 linear feet of show window (noncontinuous operation)
- 120' of lighting track
- 30 - 180 VA ballasts outside lighting (continuous operation)
- 3,600 VA sign lighting (continuous operation)
- 65 receptacles (noncontinuous operation)
- 28 receptacles (continuous operation)
- 80' multioutlet assembly (heavy-duty)

480 V, three-phase loads

- 7,380 VA freezer
- 5,580 VA ice cream boxes
- 1 - 1 HP exhaust fan
- 10,000 VA water heater
- 9,540 VA walk-in cooler
- 50,000 VA heating unit
- 22,320 VA A/C unit
- 1 - 2 HP water pump

Sizing phases = •
Sizing neutral = √
Sizing total load = *

Note 1: For sizing elements, see page 23-20 of this chapter.

Note 2: Add Asterisks and checks to obtain total load.

CALCULATING LIGHTING LOAD

Step 1: General lighting load
Table 220.12; 230.42(A)(1)
40,000 sq. ft. x 3 VA = 120,000 VA √
120,000 VA x 125% = 150,000 VA *
20,000 sq. ft. x 1/4 VA = 5,000 VA √
5000 VA x 125% = 6,250 VA *

Step 2: Show window load
220.43(A)
80' x 200 = 16,000 VA *

Step 3: Track lighting load
220.43(B)
120 ÷ 2 x 150 VA = 9,000 VA *

Step 4: Outside lighting load
230.42(A)(1)
30 x 180 VA = 5,400 VA
5400 VA x 125% = 6,750 VA *

Step 5: Sign lighting load
220.14(F); 230.42(A)(1)
3600 VA x 100% = 3,600 VA
3600 VA x 125% = 4,500 VA *
Total load = 192,500 VA •

CALCULATING RECEPTACLE LOAD

Step 1: Noncontinuous operation
220.14(H)(2); 220.14(I); 230.42(A)(1)
65 x 180 VA = 11,700 VA
80' x 180 VA = 14,400 VA
Total Load = 26,100 VA
Table 220.13
First 10,000 VA x 100% = 10,000 VA
Next 16,100 VA x 50% = 8,050 VA
Total load = 18,050 VA *

Step 2: Continuous operation
220.14(I); 230.42(A)(1)
28 x 180 VA = 5,040 VA
5040 VA x 125% = 6,300 VA *
Total load = 24,350 VA •

CALCULATING SPECIAL LOADS

Step 1: Water heater load
10,000 VA x 100% = 10,000 VA * •

CALCULATING COMPRESSOR LOADS

Step 1: Freezer load
230.42(A)(1); 440.34
7380 VA x 100% = 7,380 VA *

Step 2: Ice cream boxes
5580 VA x 100% = 5,580 VA *

Step 3: Walk-in cooler
9540 VA x 100% = 9,540 VA *
Total load = 22,500 VA •

CALCULATING MOTOR LOADS

Step 1: Water pump load
430.24; Table 430.250
3.4 A x 480 V x 1.732 (831 V) = 2,825 VA *

Step 2: Exhaust fan load
2.1 A x 480 V x 1.732 (831 V) = 1,745 VA *
Total load = 4,570 VA •

CALCULATING HEATING OR A/C LOAD

Step 1: Heating load selected
220.60
50,000 VA x 100% = 50,000 VA * •
23,320 VA x 100% = 22,320 VA

CALCULATING LARGEST MOTOR LOAD

Step 1: Walk-in cooler
220.50; 430.24
9540 VA x 25% = 2,385 VA * •

CALCULATING FOR PHASES A, B AND C (ADD ALL •)

Lighting loads = 192,500 VA•
Receptacle loads = 24,350 VA•
Special loads = 10,000 VA•
Compressor loads = 22,500 VA•
Motor loads = 4,570 VA•
Heating load = 50,000 VA•
Largest motor load = 2,385 VA•
Total load for facility = 306,305 VA

FINDING AMPS FOR PHASES A, B AND C

I = VA / V
I = 306,305 VA ÷ 480 V x 1.732 (831 V)
I = 369 A

CALCULATING NEUTRAL (ADD ALL √)

Lighting load = 125,000 VA √
Total load = 125,000 VA

FINDING AMPS FOR NEUTRAL

I = VA / V
I = 125,000 VA ÷ 480 V x 1.732 (831 V)
I = 150 A

DESIGN PROBLEM 23-4: What is the load in VA and amps to calculate and size the elements for 120/240 volt, three-phase, four-wire service supplying a 40,000 sq. ft. store with 20,000 sq. ft. of warehouse space?

120 V, single-phase loads

- 80 linear feet of show window (noncontinuous operation)
- 120' of lighting track
- 30 - 180 VA ballasts outside lighting (continuous operation)
- 3,600 VA sign lighting (continuous operation)
- 65 receptacles (noncontinuous operation)
- 28 receptacles (continuous operation)
- 80' multioutlet assembly (heavy-duty)

240 V, three-phase loads

- 7,380 VA freezer
- 5,580 VA ice cream boxes
- 1 - 3/4 HP exhaust fan
- 10,000 VA water heater
- 9,540 VA walk-in cooler
- 50,000 VA heating unit
- 22,320 VA A/C unit
- 1 - 3 HP water pump

Sizing phases = •
Sizing neutral = √
Sizing total load = *

Note 1: For sizing elements, see pages 23-21 of this chapter.

Note 2: Add Asterisks and checks to obtain total load.

CALCULATING LIGHTING LOAD

Step 1: General lighting load
Table 220.12; 230.42(A)(1)
40,000 sq. ft. x 3 VA = 120,000 VA √
120,000 VA x 125% = 150,000 VA *
20,000 sq. ft. x 1/4 VA = 5,000 VA √
5,000 VA x 125% = 6,250 VA *

Step 2: Show window load
220.43(A)
80' x 200 = 16,000 VA * √

Step 3: Track lighting load
220.43(B)
120 ÷ 2 x 150 VA = 9,000 VA * √

Step 4: Outside lighting load
230.42(A)(1)
30 x 180 VA = 5,400 VA √
5400 VA x 125% = 6,750 VA *

Step 5: Sign lighting load
220.14(F); 230.42(A)(1)
3600 VA x 100% = 3,600 VA √
3600 VA x 125% = 4,500 VA *
Total load = **192,500 VA •**

CALCULATING RECEPTACLE LOAD

Step 1: Noncontinuous operation
220.14(H)(2); 220.14(I); 230.42(A)(1)
65 x 180 VA = 11,700 VA
80' x 180 VA = 14,400 VA
Total load = 26,100 VA
Table 220.13
First 10,000 VA x 100% = 10,000 VA
Next 16,100 VA x 50% = 8,050 VA
Total load = 18,050 VA * √

Step 2: Continuous operation
220.14(I); 230.42(A)(1)
28 x 180 VA = 5,040 VA √
5040 VA x 125% = 6,300 VA *
Total load = **24,350 VA •**

CALCULATING SPECIAL LOADS

Step 1: Water heater load
10,000 VA x 100% = **10,000 VA * •**

CALCULATING COMPRESSOR LOADS

Step 1: Freezer load
230.42(A)(1); 440.34
7380 VA x 100% = 7,380 VA *

Step 2: Ice cream boxes
5580 VA x 100% = 5,580 VA

Step 3: Walk-in cooler
9540 VA x 100% = 9,540 VA *
Total load = **22,500 VA •**

CALCULATING MOTOR LOADS

Step 1: Water pump load
430.24; Table 430.250
9.6 A x 240 V x 1.732 x 100% = 3,994 VA *

Step 2: Exhaust fan load
3.2 A x 240 V x 1.732 x 100% = 1,331 VA *
Total load = **5,325 VA •**

CALCULATING HEATING OR A/C LOAD

Step 1: Heating load selected
220.60
50,000 VA x 100% = 50,000 VA * •
23,320 VA x 100% = 22,320 VA

CALCULATING LARGEST MOTOR

Step 1: Walk-in cooler
220.50; 430.24
9540 VA x 25% = **2,385 VA * •**

SINGLE-PHASE LOADS

Lighting loads = 192,500 VA •
Receptacle loads = 24,350 VA •
Total load = **216,850 VA**

THREE-PHASE LOADS (SPECIAL LOADS)

Water heater load = 10,000 VA •
Compressor load = 22,500 VA •
Motor load = 5,325 VA •
Heating load = 50,000 VA •
Largest motor load = 2,385 VA •
Total load = **90,210 VA**

SINGLE-PHASE NEUTRAL LOADS

Lighting loads = 159,000 VA √
Receptacle loads = 23,090 VA √
Total load = **182,090 VA**

CALCULATING SINGLE-PHASE LOAD

I = 216,850 VA ÷ 240 V
I = 904 A

CALCULATING THREE-PHASE LOAD (HIGH-LEG)

I = 90,210 VA ÷ 240 V x 1.732 (416 V)
I = 217 A

CALCULATING NEUTRAL LOAD

I = 182,090 VA ÷ 240 V
I = 759 A

CALCULATING PHASES A AND C

Single-phase load	= 904 A
Three-phase load	= 217 A
Total load	**= 1,121 A**

CALCULATING PHASE B (HIGH-LEG)

Three-phase load   **= 217 A**

---

(1)
Design Tip (OPEN DELTA SYSTEM): The power and lighting transformer will consist of 120/240 volt single-phase loads plus the three-phase loads at 240 volts, respectively. The power transformer will consist of the three-phase loads only. Note that there will be one larger transformer (power plus lighting loads) and one smaller transformer (power 3Ø loads only)

(2)
Design Tip (ADDING LOADS TOGETHER ON AN OPEN DELTA SYSTEM): The larger transformer is determined by adding the single-phase and three-phase loads together. The smaller transformer is determined by adding the three-phase loads together which will not include the single-phase loads.

(3)
Design Tip (CLOSED DELTA SYSTEM): There will be three transformers on a closed delta system that are connected at each corner of the windings to form a closed delta system. A closed delta connected system should be used when the greater of the loads are three-phase motors, compressors, etc.

(4)
Design Tip (ADDING LOADS TOGETHER ON A CLOSED DELTA SYSTEM): The two larger transformers supplying power to the single-phase and three-phase loads is determined by adding the single-phase and three-phase loads together.

The smaller transformer(s) is determined by adding the three-phase loads together. However, in some installations, it is possible to add the 240 volt single-phase loads of phases A and C to phase B which is the high-leg. When this is done, the high-leg load will be greater in size. Note that phases A and C in a delta system will usually always have a greater calculated load than phase B.

(5)
Design Tip (CALCULATING LOAD FOR THE NEUTRAL): There is really no reason for calculating the load in VA or amps at 125 percent to size the grounded (neutral) conductor. Remember that the grounded (neutral) conductor connects to the lugs of the busbar and not to the terminals of an OCPD. However, there are designers who will calculate such load at 125 percent. It is your choice, whether to calculate the grounded (neutral) conductor at 125 percent or not. The grounded (neutral) conductor in this book is calculated at 100 percent of the VA or amps. **Section 366.23(A)** in the NEC does not require a busbar (bare copper) to be derated 80 percent of its rating. Therefore, the 125 percent rule in **210.19(A)(1)** and **210.20** or **215.2(A)(1)** and **215.3** does not necessarily have to be applied.

DESIGN PROBLEM 23-5: What is the load in VA and amps to calculate and size the elements for 277/480 volt, three-phase, four-wire service supplying a 150,000 sq. ft. office facility with 3000 sq. ft. hall area equipped with 277 volt lighting units?

120 V, single-phase loads

- 60' of lighting track
- 20 - 180 VA ballasts outside lighting (continuous operation)
- 4800 VA sign lighting (continuous operation)
- 182 receptacles (noncontinuous operation)
- 121 receptacles (continuous operation)
- 6000 VA isolation transformer for LVLS continuous operation

Sizing phases = •
Sizing neutral = √
Sizing total load = *

208 V, three-phase loads

- 200' multioutlet assembly (heavy-duty)
- 5 - 1450 VA copying machine
- 8500 VA water heater
- 25 - 225 VA data processors
- 10 - 175 VA word processor
- 4 - 1200 VA printers

480 V, three-phase loads

- 40 HP elevator (15 minute intermittent duty)
- 40 kW heating units
- 12,000 VA A/C unit

Note 1: For sizing elements, see page 23-22 of this chapter

Note 2: Add Asterisks and checks to obtain total load.

CALCULATING LIGHTING LOAD

Step 1: General lighting load
Table 220.12; 230.42(A)(1)
150,000 sq. ft. x 3.5 VA = 525,000 VA √
525,000 VA x 125% = 656,250 VA *
3,000 sq. ft. x 1/2 VA = 1,500 VA √
1,500 VA x 125% = 1,875 VA *

Step 2: Track lighting load
220.43(B)
60 ÷ 2 x 150 VA = 4,500 VA *

Step 3: Low-voltage lighting load
Art. 411; 230.42(A)(1)
6000 VA x 100% = 6,000 VA
6000 VA x 125% = 7,500 VA *

Step 4: Outside lighting load
230.42(A)(1)
20 x 180 VA = 3,600 VA
3600 VA x 125% = 4,500 VA *

Step 5: Sign lighting load
220.14(F); 230.42(A)(1)
4800 VA x 100% = 4,800 VA
4800 VA x 125% = 6,000 VA *
Total load = 680,625 VA •

CALCULATING RECEPTACLE LOAD

Step 1: Noncontinuous operation
220.14(I); 230.42(A)(1)
182 x 180 VA = 32,760 VA
200' x 180 VA = 36,000 VA
Total load = 68,760 VA
Table 220.13
First 10,000 VA x 100% = 10,000 VA
Next 58,760 VA x 50% = 29,380 VA
Total load = 39,380 VA *

Step 2: Continuous operation
220.14(I); 230.42(A)(1)
121 x 180 VA = 21,780 VA
21,780 VA x 125% = 27,225 VA *
Total load = 66,605 VA •

CALCULATING SPECIAL LOADS

Step 1: Copying machine load
230.42(A)(1)
1450 VA x 5 = 7,250 VA
7250 VA x 125% = 9,063 VA *

Step 2: Water heater load
422.13; 230.42(A)(1)
8500 VA x 100% = 8,500 VA *

Step 3: Data processor load
225 VA x 25 = 5,625 VA
5625 VA x 125% = 7,031 VA *

Step 4: Word processor load
230.42(A)(1)
175 VA x 10 = 1,750 VA
1750 VA x 125% = 2,188 VA *

Step 5: Printer load
230.42(A)(1)
1200 VA x 4 = 4,800 VA
4800 VA x 125% = 6,000 VA *
Total load = 32,782 VA •

CALCULATING MOTOR LOADS

Step 1: 40 HP elevator
430.24; 430.22(B); Table 430.22(B)
52 A x 480 V x 1.732 x 85% = 36,730 VA * •

CALCULATING HEATING OR A/C LOAD

Step 1: Heating load selected
220.60; 220.51
40,000 VA x 100% = 40,000 VA * •

CALCULATING LARGEST MOTOR

Step 1: 40 HP elevator
220.50; 430.24
36,730 VA x 25% = 9,183 VA * •
Total load for facility = 865,925 VA

FINDING AMPS FOR PHASES A, B AND C

$I = VA \div V \times \sqrt{3}$
$I = 865,925 \text{ VA} \div 480 \text{ V} \times 1.732 \text{ (831 V)}$
I = 1,042 A

CALCULATING NEUTRAL

General lighting load
(office building) = 525,000 VA √
(halls) = 1,500 VA √
Total load = 526,500 VA

FINDING AMPS FOR NEUTRAL

$I = VA / V \times \sqrt{3}$
$I = 526,500 \text{ VA} \div 480 \text{ V} \times 1.732 \text{ (831 V)}$
I = 634 A

Commercial Calculations

DESIGN PROBLEM 23-6: What is the load in VA and amps to calculate and size the elements for 277/480 volt, three-phase, four-wire service supplying a 30,000 sq. ft. classroom area, 5,000 sq. ft. auditorium area, and 1,000 sq. ft. assembly hall area? (School building general lighting load is supplied by 277 volt fixtures)

120 V, single-phase loads
- 200 receptacles (noncontinuous duty)
- 50 receptacles (continuous duty)
- 200' multioutlet assembly (heavy-duty)

Single-phase and three-phase motor loads
- 4 - 1 HP hood fans
  208 V, single-phase
- 3 - 3/4 HP grill vent fans
  208 V, single-phase
- 20 - 3/4 HP exhaust fans
  480 V, three-phase

Cooking equipment
- 2 - 1 kW toaster
  120 V, single-phase
- 4 - 1.5 kW refrigerators
  120 V, single-phase
- 3 - 1.5 kW freezers
  120 V, single-phase
- 4 - 12 kW ranges
  208 V, single-phase
- 3 - 9 kW ovens
  208 V, single-phase
- 4 - 4 kW fryers
  208 V, single-phase

Note: All loads are continuous

Sizing phases = •
Sizing neutral = √
Sizing total load = *

Note 1: For sizing elements, see page 23-22 of this chapter.
Note 2: Add Asterisks and checks to obtain total load.

CALCULATING LIGHTING LOADS

Step 1: General lighting load
Table 220.12; 230.42(A)(1)
30,000 sq. ft. x 3 VA = 90,000 VA √
90,000 VA x 125% = 112,500 VA *
5,000 sq. ft. x 1 = 5,000 VA √
5,000 VA x 125% = 6,250 VA *
1,000 sq. ft. x 1 = 1,000 VA √
1,000 VA x 125% = 1,250 VA *
Total load = **120,000 VA** •

CALCULATING RECEPTACLE LOADS

Step 1: Noncontinuous operation
220.14(I); 230.42(A)(1)
200 x 180 VA = 36,000 VA
200' x 180 VA = 36,000 VA
Total load = 72,000 VA
Table 220.13
First 10,000 VA x 100% = 10,000 VA
Next 62,000 VA x 50% = 31,000 VA
Total load = **41,000 VA** *

Step 2: Continuous operation
220.14(I); 230.42(A)(1)
50 x 180 VA = 9,000 VA
9000 VA x 125% = 11,250 VA *
Total load = **52,250 VA** •

CALCULATING SPECIAL LOADS

Step 1: Kitchen equipment
220.56
Toasters
   2 x 1 kW x 1000 = 2,000 VA
Refrigerators
   4 x 1.5 kW x 1000 = 6,000 VA
Freezers
   3 x 1.5 kW x 1000 = 4,500 VA
Ranges
   4 x 12 kW x 1000 = 48,000 VA
Ovens
   3 x 9 kW x 1000 = 27,000 VA
Fryers
   4 x 4 kW x 1000 = 16,000 VA
Total load = 103,500 VA

Step 2: Applying demand factors
103,500 VA x 65% = 67,275 VA * •

CALCULATING MOTOR LOADS
TABLES 430.248 AND 430.250

Step 1: Exhaust fans
32 A x 100% x 480 V = 15,360 VA
15,360 VA x 1.732 = 26,604 VA *
(1.6 A x 20 = 32 A)

Step 2: Hood fans
35.2 A x 100% x 208 V = 7,322 VA *
(8.8 A x 4 = 35.2 A)

Step 3: Grill vent fans
22.8 A x 100% x 208 V = 4,742 VA *
(7.6 A x 3 = 22.8)
Total load = **38,668 VA** •

CALCULATING LARGEST MOTOR LOAD

Step 1: Hood fan
8.8 A x 100% x 208 V = 1,830 VA
1830 VA x 25% = 458 VA * •
Total load for facility = **278,651 VA**

FINDING AMPS FOR PHASES A, B AND C

I = VA ÷ V x √3
I = 278,651 VA ÷ 480 V x 1.732 (831 V)
I = 335 A

CALCULATING NEUTRAL

General lighting load = 96,000 VA √
Largest motor load = 375 VA √
Total load = **96,375 VA**

FINDING AMPS FOR NEUTRAL

I = VA ÷ V x √3
I = 96,375 VA ÷ 480 V x 1.732 (831 V)
I = 116 A

Note: Largest 120 volt motor is calculated by taking 1.5 kW x 1000 x .25% = 375 VA.

DESIGN PROBLEM 23-7: What is the load in VA and amps to calculate and size the elements for 120/208 volt, three-phase, four-wire service supplying a restaurant with an area of 5600 sq. ft.?

120 V, single-phase loads

Lighting load

30' lighting track (continuous)
10 - 180 VA outside lighting (continuous)
1200 VA sign lighting (continuous)

Receptacle load

35 receptacles (noncontinuous)
25 receptacles (continuous)
20' multioutlet assembly (heavy-duty)

208 V, three-phase loads

Special loads

2 - 20 kW heating units
208 V, three-phase
2 - 8650 VA A/C units
208 V, three-phase

Motor loads

7322 VA hood fans
208 V, single-phase
4742 VA grill vent fans
208 V, single-phase

208 V, three-phase loads

Kitchen equipment

3800 VA boiler
2 - 2700 deep fat fryers
20 A walk-in cooler
6000 VA water heater

208 V, single-phase loads

13 A freezer
11,000 VA cooktop
2 - 9000 VA ovens
12,000 VA range
14 A refrigerator
3650 VA ice cream box

Note 1: For sizing elements, see page 23-23 of this Chapter.

Note 2: Add Asterisks and checks to obtain total load.

CALCULATING LIGHTING LOADS

Step 1: General lighting load
Table 220.12; 230.42(A)(1)
5600 sq. ft. x 2 VA = 11,200 VA √
11,200 VA x 125% = 14,000 VA *

Step 2: Track lighting load
220.43(B); 230.42(A)(1)
30 ÷ 2 x 150 VA = 2,250 VA √
2250 VA x 125% = 2,813 VA *

Step 3: Outside lighting load
220.14(L); 230.42(A)(1)
180 VA x 10 = 1,800 VA √
1800 VA x 125% = 2,250 VA *

Step 4: Sign lighting load
220.14(F); 230.42(A)(1); (A)(1)
1200 VA x 100% = 1,200 VA √
1200 VA x 125% = 1,500 VA *
Total load = **20,563 VA •**

CALCULATING RECEPTACLE LOADS

Step 1: Receptacle load (noncontinuous)
220.14(I); 230.42(A)(2)
35 x 180 VA = 6,300 VA * √

Step 2: Receptacle load (continuous)
220.14(I); 230.42(A)(1)
25 x 180 VA = 4,500 VA √
4500 VA x 125% = 5,625 VA *

Step 3: Multioutlet assembly
220.14(H)(2); 230.42(A)(1)
20' x 180 VA = 3,600 VA * √
Total load = **15,525 VA •**

CALCULATING SPECIAL LOAD

Step 1: Kitchen equipment
220.56
Boiler = 3,800 VA
Deep fat fryer = 5,400 VA
Walk-in cooler = 7,200 VA
Water heater = 6,000 VA
Ice cream box = 3,650 VA
Freezer = 2,704 VA
Cooktop = 11,000 VA
Ovens = 18,000 VA
Range = 12,000 VA
Refrigerator = 2,912 VA
Total load = 72,666 VA

Applying demand factor
Table 220.56
72,666 VA x 65% = **47,233 VA * •**

CALCULATING MOTOR LOAD

Step 1: Hood fans
430.22(A); 430.24; 430.25
7322 VA x 100% = 7,322 VA *

Step 2: Grill vent fans
4742 VA x 100% = 4,742 VA *
Total load = **12,064 VA •**

CALCULATING HEATING OR A/C LOAD

Step 1: Heating load
220.60
20 kW x 2 x 1000 = **40,000 VA * •**

CALCULATING LARGEST MOTOR LOAD

Step 1: Walk-in cooler
220.50; 430.22(A); 440.34
20 A x 100% x 208 V x 1.732 = 7,200 VA
7200 VA x 25% = 1,800 VA * •
Total load = **137,185 VA**

CALCULATING VA LOAD (NEUTRAL)
220.61; 230.42(A)(1); (A)(2)

Lighting load = 16,450 VA √
Receptacle load = 14,400 VA √
Total load = 30,850 VA

FINDING AMPS FOR PHASES A, B AND C

$I = VA \div V \times \sqrt{3}$
I = 137,185 VA ÷ 208 V x 1.732 (360 V)
I = 381 A

FINDING AMPS FOR NEUTRAL

$I = VA \div V \times \sqrt{3}$
I = 30,850 VA ÷ 208 V x 1.732 (360 V)
I = 86 A

Commercial Calculations

DESIGN PROBLEM 23-8: What is the load in VA and amps to calculate and size the elements for 277/480 volt, three-phase service supplying a hospital with an office area of 150,000 sq. ft. illuminated by 277 volt lighting units?

277 V, single-phase loads
- 150,000 sq. ft. office
- 3000 sq. ft. halls
- 800 sq. ft. of closets
- 1000 sq. ft. of hallways
- 4000 sq. ft. of storage space

Note 1: For sizing elements, see page 23-23 of this chapter.

Note 2: Add Asterisks and checks to obtain total load.

480 V, three-phase motor loads
- 6 - 40 HP elevators (15 minute intermittent duty)
- 12,000 VA A/C unit

Sizing phases = •
Sizing neutral = √
Sizing total load = *

CALCULATING LIGHTING LOAD

Step 1: General lighting load
Table 220.12; 230.42(A)(1)
(office)
150,000 sq. ft. x 3.5 VA = 525,000 VA √
525,000 VA x 125% = 656,250 VA *
(halls)
3000 sq. ft. x 1/2 VA = 1,500 VA √
1500 VA x 125% = 1,875 VA *
(closets)
800 sq. ft. x 1/2 VA = 400 VA √
400 VA x 125% = 500 VA *
(stairways)
1000 VA x 1/2 VA = 500 VA √
500 VA x 125% = 625 VA *
(storage space)
4000 VA x 1/4 VA = 1,000 VA √
1000 VA x 125% = 1,250 VA *
Total load = **660,500 VA •**

CALCULATING SPECIAL LOADS

Step 1: Emergency system loads
230.42(A)(1)
(Life safety branch)
60,000 VA x 125% = 75,000 VA *
(Critical branch)
35,000 VA x 125% = 43,750 VA *
(Life support equipment)
30,000 VA x 125% = 37,500 VA *
(Essential system)
45,000 VA x 125% = 56,250 VA *
Total load = **212,500 VA •**

CALCULATING MOTOR LOADS

Step 1: 40 HP elevators
430.24; 430.22(B); Table 430.22(B)
52 A x 480 V x 1.732 x 85% = 36,730 VA
36,730 VA x 6 = **220,380 VA * •**

CALCULATING HEATING OR A/C LOAD

Step 1: Heating load
220.60; 220.51
40,000 VA x 100% = **40,000 VA * •**

CALCULATING LARGEST MOTOR LOAD

Step 1: 40 HP elevator
220.50; 430.24
36,730 VA x 25% = **9,183 VA * •**

CALCULATING FOR PHASES A, B AND C

General lighting load = 660,500 VA •
Special loads = 212,500 VA •
Motor loads = 220,380 VA •
Heating loads = 40,000 VA •
Largest motor load = 9,183 VA •
Total load = **1,142,563 VA**

CALCULATING NEUTRAL

General lighting load
(Office) = 525,000 VA √
(Halls) = 1,500 VA √
(Closets) = 400 VA √
(Stairways) = 500 VA √
(Storage) = 1,000 VA √
Total load = **528,400 VA**

FINDING AMPS FOR PHASES A, B AND C

$I = VA \div V \times \sqrt{3}$
$I = 1{,}142{,}563 \text{ VA} \div 480 \text{ V} \times 1.732 \ (831)$
I = 1375 A

FINDING AMPS FOR NEUTRAL

$I = VA \div V \times \sqrt{3}$
$I = 528{,}400 \text{ VA} \div 480 \text{ V} \times 1.732 \ (831)$
I = 636 A

DESIGN PROBLEM 23-9: What is the load in VA and amps to calculate and size the elements for 120/208 volt, three-phase service supplying a welding shop?

120 V, single-phase loads

- 9000 VA inside lighting loads (continuous operation)
- 6 - 180 VA outside lighting loads (continuous operation)
- 1200 VA sign lighting loads (noncontinuous operation)
- 60 receptacles (continuous operation)

Sizing phase = •
Sizing neutral = √
Sizing total load = 13,800 VA
Sizing total load = *

Note 1: A welding shop is not a listed occupancy per **Table 220.12**.
Note 2: For sizing elements, see page 23-24 of this chapter.
Note 3: Add Asterisks and checks to obtain total load

208 V, three-phase loads

- 2 - 10 kW heating units
- 5400 VA A/C units
- 7.5 HP air-compressor
- 2 - 1 1/2 HP grinders
- Welders - resistance (50% duty cycle)
- 12 kW
- 8 kW
- Welders - motor-generator arc (90% duty cycle)
- 14 kW
- 12 kW
- Welders - AC transformer and DC rectifier (80% duty cycle)
- 13 kW
- 9 kW

CALCULATING LIGHTING LOADS

Step 1: Inside lighting load
230.42(A)(1)
9000 VA x 100%   = 9,000 VA √
9000 VA x 125%   = 11,250 VA *

Step 2: Outside lighting load
230.42(A)(1)
6 x 180 VA   = 1,080 VA √
1080 VA x 125%   = 1,350 VA *

Step 3: Sign lighting load
220.14(F); 230.42(A)(1)
1200 VA x 100%   = 1,200 VA * √
Total load   = **13,800 VA** •

CALCULATING RECEPTACLE LOAD

Step 1: Continuous duty
220.14(I); 230.42(A)(1)
60 x 180 VA   = 10,800 VA √
10,800 VA x 125%   = **13,500 VA** * •

CALCULATING SPECIAL LOADS

Step 1: Welders - resistance
630.31(A); (B)
12,000 VA x 71%   = 8,520 VA *
8000 VA x 71% x 60% = 3,408 VA *

Step 2: Welders - motor-generator arc
630.11(A); (B)
14,000 VA x 96%   = 13,440 VA *
12,000 VA x 96%   = 11,520 VA *

Step 3: Welders - nonmotor-generator arc
630.11(A); (B)
13,000 VA x 89%   = 11,570 VA *
9000 VA x 89%   = 8,010 VA *
Total load   = **56,468 VA** •

CALCULATING MOTOR LOADS

Step 1: Air-compressor
430.24; 430.22(E); Table 430.22(E)
24.2 A x 100% x 208 V x 1.732 = 8,712 VA *

Step 2: Grinders
6.6 A x 100% x 208 V x 1.732 = 2,376 VA *
Total load   = **11,088 VA** •

CALCULATING HEATING OR A/C LOAD

Step 1: Heating load
220.60; 220.51
20,000 VA x 100%   = **20,000 VA** * •

CALCULATING LARGEST MOTOR LOAD

Step 1: Air-compressor
24.2 A x 208 V x 1.732 x 25% = **2,178 VA** * •

CALCULATING TOTAL LOAD

Lighting loads   = 13,800 VA •
Receptacle loads   = 13,500 VA •
Special loads   = 56,468 VA •
Comp. and Motor load   = 11,088 VA •
Heating loads   = 20,000 VA •
Largest motor load   = 2,178 VA •
Total load   = 117,034 VA

FINDING AMPS FOR PHASES A, B AND C

I = VA ÷ V x √3
I = 117,034 VA ÷ 208 V x 1.732 (360 V)
I = 325 A

CALCULATING NEUTRAL

220.61
Lighting loads (9000 + 1080 + 1200 VA) = 11,280 VA √
Receptacle loads   = 10,800 VA √
Total load   = 22,080 VA

FINDING AMPS FOR NEUTRAL

I = VA ÷ V x √3
I = 22,080 VA ÷ 208 V x 1.732 (360 V)
I = 61 A

Name	Date

Chapter 23. Commercial Calculations

Section   Answer

_____ T   F   1. **Table 220.12** shall be used to calculate the general lighting load for an unlisted occupancy.

_____ T   F   2. When using the standard calculation to determine the total lighting load, there is a possibility that six lighting loads will have to be calculated.

_____ _____ 3. There are _____ loads to be calculated when using the standard calculation to determine the load in VA or amps for the receptacle load.
   (a) 2
   (b) 3
   (c) 4
   (d) 5

_____ _____ 4. Multioutlet assemblies with cord-and-plug connected appliances that operate simultaneously shall be calculated at _____ VA per foot.

_____ _____ 5. In a commercial building, receptacles shall be calculated at _____ VA for each outlet. (General Rule)
   (a) 150
   (b) 180
   (c) 200
   (d) 225

_____ _____ 6. The total VA rating for a 8000 sq. ft. office is _____ per **Table 220.12**.

_____ _____ 7. The total rating for 60 ft. of show window area is _____ VA if used noncontinuous.
   (a) 9000
   (b) 10,800
   (c) 12,000
   (d) 15,000

_____ _____ 8. The total rating for 70 ft. of multioutlet assembly having connected loads that are not simultaneously is _____ VA.
   (a) 2520
   (b) 2530
   (c) 2560
   (d) 2620

_____ _____ 9. The total VA rating for 40 ft. of lighting track is _____ VA when used at continuous operation.
   (a) 2000
   (b) 2500
   (c) 3000
   (d) 3750

Section	Answer

10. The demand load for 130 general-purpose receptacles that are used noncontinuous in an office is _____ VA.
    (a) 16,600
    (b) 16,650
    (c) 16,700
    (d) 23,400

11. The total load for 130 receptacles used at continuous operation in a bank is _____ VA.
    (a) 16,000
    (b) 16,700
    (c) 23,400
    (d) 29,250

12. If a 460 volt, three-phase compressor (Largest motor load) is rated at 42 amps, the total rating for the largest motor load is _____ VA.
    (a) 8640
    (b) 8726
    (c) 34,680
    (d) 34,902

13. The total rating for a continuous 120 volt, single-phase low-voltage isolation transformer with a nameplate rating of 20 amps is _____ VA.
    (a) 2200
    (b) 2400
    (c) 2500
    (d) 3000

14. The total rating for a service load having a three-phase 35,000 VA heating unit and a three-phase 8280 VA heat pump is _____ amps.(277 / 480 volts)
    Note: The units can operate togehter.
    (a) 10
    (b) 42
    (c) 52
    (d) 65

15. The total rating for 50 ft. of lighting track operating for 15 hours a day is _____ VA.
    (a) 3550.5
    (b) 3660.8
    (c) 3750
    (d) 4687.5

16. The demand for one of the following receptacle load is _____ VA.
    • 150 ft. of multioutlet assembly appliances used simultaneously
    • 100 ft. of multioutlet assembly appliances not used simultaneously
    • 70 general-purpose receptacles operated at noncontinuous use
    (a) 11,300
    (b) 22,100
    (c) 39,600

17. The total rating for 45 receptacles used at continuous operation in a bank is _____ VA.
    (a) 8000
    (b) 8100
    (c) 10,125

23-36

	Section	Answer

18. The total rating for the largest motor load based on the following motors is _____ amps. (Apply 25 percent rule)
    - 20 HP, three-phase, 460 volt motor
    - 15 HP, three-phase, 208 volt motor
    - 10 HP, three-phase, 230 volt motor
    - (a) 6.75
    - (b) 7
    - (c) 11.5

19. The total rating for the above motor loads is _____ amps for a 480 volt, three-phase supply. (round up amps)
    - (a) 35
    - (b) 61
    - (c) 68

20. The total load for a 40 ft. x 200 ft. bank with 60 general-purpose receptacles is _____ VA. (Calculate VA without applying 125 percent rule)
    - (a) 35,600
    - (b) 38,400
    - (c) 38,800

21. The total load for a 10,000 sq. ft. office with 50 general-purpose receptacles is _____ VA. (Calculate VA without applying 125 percent rule)
    - (a) 44,000
    - (b) 52,750
    - (c) 55,000

22. The total load for a 400 ft. hallway, 600 ft. of stairway areas, and 200 ft. of storage space is _____ VA. (Calculate VA without applying 125 percent rule)
    - (a) 550
    - (b) 600
    - (c) 750

23. The total load for 2000 sq. ft. church with an 8000 sq. ft. auditorium is _____ VA. (Calculate for continuous load)
    - (a) 9500
    - (b) 10,000
    - (c) 12,500

24. The total load for 40 ballasts rated at .86 amp each and used for ten hours a day is _____ amps. (Calculated applying 125 percent rule)
    - (a) 33
    - (b) 34
    - (c) 43

25. The total load for a small 20,000 sq. ft. school with 2400 sq. ft. of assembly hall area is _____ VA. (Calculate VA without applying 125 percent rule)
    - (a) 62,400
    - (b) 75,000
    - (c) 76,200

Section	Answer

26. What is the amp load for a 50,000 sq. ft. store including a 30,000 sq. ft. warehouse space with a 120/240 volt, single-phase service with the following loads:

    120 V, single-phase loads
    - 100 linear feet of show window (noncontinuous operation)
    - 120 ft. of lighting track (noncontinuous operation)
    - 40 - 180 VA ballasts outside lighting (continuous operation)
    - 4200 VA sign lighting (continuous operation)
    - 74 receptacles (noncontinuous duty)
    - 24 receptacles (continuous duty)
    - 100 ft. multioutlet assembly (simultaneously operated)

    240 V, single-phase loads
    - 12,000 VA water heater
    - 60,000 VA heating unit
    - 24,800 VA A/C unit
    - 7240 VA freezer
    - 6480 VA ice-cream box
    - 9560 VA walk-in cooler
    - 1 - 1/2 HP exhaust fan
    - 1 - 2 HP water pump

27. What is the amp load for a 50,000 sq. ft. store including a 30,000 sq. ft. warehouse space with a 120/208 volt, three-phase service with the following loads:

    120 V, single-phase loads
    - 100 linear feet of show window (noncontinuous operation)
    - 120 ft. of lighting track
    - 40 - 180 VA ballasts outside lighting (continuous operation)
    - 4200 VA sign lighting (continuous operation)
    - 74 receptacles (noncontinuous duty)
    - 24 receptacles (continuous duty)
    - 100 ft. multioutlet assembly (simultaneously operated)

    208 V, three-phase loads (Use 208 V x 1.732 = 360 V)
    - 12,000 VA water heater
    - 60,000 VA heating unit
    - 24,800 VA A/C unit
    - 7240 VA freezer
    - 6480 VA ice-cream box
    - 9560 VA walk-in cooler
    - 1 - 1/2 HP exhaust fan
    - 1 - 2 HP water pump

28. What is the amp load for a 50,000 sq. ft. store including a 30,000 sq. ft. warehouse space with a 277 volt lighting system and a 277/480 volt, three-phase service with the following loads:

    120 V, single-phase loads
    - 100 linear feet of show window (noncontinuous operation)
    - 120 ft. of lighting track
    - 40 - 180 VA ballasts outside lighting (continuous operation)
    - 4200 VA sign lighting (continuous operation)
    - 74 receptacles (noncontinuous duty)
    - 24 receptacles (continuous duty)
    - 100 ft. multioutlet assembly (simultaneously operated)

480 V, three-phase loads (Use 480 V x 1.732 = 831 V)
- 12,000 VA water heater
- 60,000 VA heating unit
- 24,800 VA A/C unit
- 7240 VA freezer
- 6480 VA ice-cream box
- 9560 VA walk-in cooler
- 1 - 1/2 HP exhaust fan
- 1 - 2 HP water pump

29. What is the amp load for a 50,000 sq. ft. store including a 30,000 sq. ft. warehouse space with a 120/240 volt, three-phase service with the following loads:

120 V, single-phase loads
- 100 linear feet of show window (noncontinuous operation)
- 120 ft. of lighting track
- 40 - 180 VA ballasts outside lighting (continuous operation)
- 4200 VA sign lighting (continuous operation)
- 74 receptacles (noncontinuous duty)
- 24 receptacles (continuous duty)
- 100 ft. multioutlet assembly (simultaneously operated)

240 V, three-phase loads (Use 240 V x 1.732 = 416 V)
- 12,000 VA water heater
- 60,000 VA heating unit
- 24,800 VA A/C unit
- 7240 VA freezer
- 6480 VA ice-cream box
- 9560 VA walk-in cooler
- 1 - 1/2 HP exhaust fan
- 1 - 2 HP water pump

30. What is the amp load for a 150,000 sq. ft. office facility with a 2500 sq. ft. of halls, equipped with a 277 volt lighting system supplied with a 277/480 volt, three-phase service with the following loads:

120 V, single-phase loads
- 50 ft. of lighting track (noncontinuous operation)
- 20 - 180 VA ballasts outside lighting (continuous operation)
- 4200 VA sign lighting (continuous operation)
- 174 receptacles (noncontinuous duty)
- 114 receptacles (continuous duty)
- 6000 VA isolation transformer for LVLS (continuous operation)
- 160 ft. multioutlet assembly (simultaneously operated)

208 V, three-phase loads
- 4- 1275 VA copying machines (noncontinuous load)
- 1- 8000 VA water heater (noncontinuous load)
- 22 - 225 VA data processors (continuous load)
- 8 - 175 VA work processors (continuous load)
- 2 - 1000 VA printers (continuous load)

480 V, three-phase loads
- 40 HP elevator (15 minute intermittent duty)
- 40 kW heating unit

Section	Answer
_____	_____

31. What is the amp load for a 20,000 sq. ft. classroom area, a 4000 sq. ft. auditorium area, and 1000 sq. assembly hall area with a 277/480 volt, three-phase service with the following loads: (lighting supplied by 277 volts)

    120 V, single-phase loads
    • 170 receptacles (noncontinuous duty)
    • 40 receptacles (continuous duty)
    • 80 ft. multioutlet assembly (simultaneously operated)

    120 V, single-phase cooking equipment
    • 2 - 1 kW toasters
    • 4 - 1.5 kW refrigerators
    • 2 - 1.5 kW freezers

    208 V, single-phase cooking equipment
    • 4 - 9 kW ranges
    • 3 - 10 kW ovens
    • 4 - 3 kW fryers

    208 V, single-phase motor loads
    • 3 - 1 HP vent-hood fans
    • 3 - 3/4 HP grill-vent fans

    480 V, three-phase motor loads
    • 18 - 3/4 HP exhaust fans

Section	Answer
_____	_____

32. What is the amp load for a 6000 sq. ft. restaurant with a 120/208 volt (use 360 volt), three-phase service which is equipped with the following loads:

    120 V, single-phase loads
    • 35 ft. of lighting track (continuous duty)
    • 10 - 180 VA outside lighting (continuous duty)
    • 1200 VA sign lighting (continuous duty)
    • 38 receptacles (noncontinuous duty)
    • 30 receptacles (continuous duty)
    • 20 ft. multioutlet assembly (simultaneously operated)

    208 V, single-phase loads
    • 6,950 VA vent-hood fans
    • 4758 VA grill-vent fans
    • 13 A freezer
    • 8000 VA cooktop
    • 2 - 10,000 VA range
    • 11,000 VA range
    • 14 A refrigerator
    • 3750 VA ice cream box

    208 V, three-phase loads
    • 2 - 25 kW heating units
    • 2 - 7850 VA A/C units
    • 3600 VA boiler
    • 2 - 2600 VA deep fat fryers
    • 20 A walk-in cooler
    • 6500 VA water heater

23-40

	Section	Answer

33. What is the amp load for a hospital having an office area of 150,000 sq. ft. that is illuminated by a 277 volt lighting system with a 277/480 volt (use 831 volt), three-phase service with the following loads:

277 V, single-phase loads
- 150,000 sq. ft. office space
- 2500 sq. ft. hall space
- 600 sq. ft. closet space
- 1200 sq. ft. hallway
- 3800 sq. ft. of storage space

480 V, three-phase motor loads
- 4 - 40 HP elevators (15 minute intermittent duty)
- 11,000 VA A/C unit

480 V, three-phase loads (emergency system loads)
- 50 kW heating unit (continuous duty)
- 50,000 VA life safety branch (continuous duty)
- 35,000 VA critical branch (continuous duty)
- 40,000 VA life support equipment (continuous duty)
- 40,000 VA essential system load (continuous duty)

34. What is the amp load for a welding shop with a 120/208 volt, three-phase service with the following loads:

120 V, single-phase loads
- 8500 VA inside lighting loads (continuous duty)
- 6 - 180 VA outside lighting loads (continuous duty)
- 1200 VA sign lighting load (noncontinuous duty)
- 50 receptacles (continuous duty)

208 V, three-phase loads
- 2 - 12 kW heating units
- 6,000 VA A/C unit
- 7.5 HP air-compressor
- 2 - 1 1/2 HP grinders
- 2 - welders - resistance (30% duty cycle)
    - 11 kW
    - 9 kW
- 2 - welders - motor generator arc (80% duty cycle)
    - 12 kW
    - 10 kW
- 2 - welders - nonmotor-generator arc (90% duty cycle)
    - 12 kW
    - 9 kW

24

Industrial Calculations

The NEC recognizes certain rules for calculating loads for sizing and selecting elements of electrical systems used to supply power to industrial occupancies. According to NEC requirements, each service and feeder shall be calculated and sized with enough capacity to carry a load current that is not less than the sum of all branch-circuits it supplies in the electrical system. These calculations vary, depending upon the type of facility, and the size and nature of the total load served.

In any electrical system, the distribution system consists of the equipment and wiring methods used to carry power from the supply transformer to the service equipment's overcurrent devices.

Distribution systems are used to carry power to lighting panelboards, power panelboards, switchboards and motor control centers which houses feeder-circuits and branch-circuit protective devices for supplying individual and multiple power loads. Adequate calculations will ensure that the elements of the system will provide the right amount of power at the right voltage to each distribution point.

LAYING OUT THE LOADS
PART III TO ARTICLE 220

Due to industrial occupancies not being a listed occupancy, lighting loads shall be calculated without the use of **Table 220.12**. However, the other loads shall be calculated in the same manner as commercial facilities. They are laid out using the standard calculation as follows:
- Lighting loads
- Receptacle loads
- Special loads
- Compressor loads
- Motor loads
- Largest between heat and A/C loads
- The largest motor load

Note: See Chapter 23 for a detailed description of these loads and the number and types of loads that are associated with each specific load.

UNLISTED OCCUPANCY
220.14(A) THRU (L)

If an occupancy is not listed in **Table 220.12**, the general-purpose lighting load shall be calculated in the following manner per **220.14**.
- Lamps for incandescent lighting per **220.14(L)**
- Lamps for recessed lighting per **220.14(D)**
- Ballasts for electric discharge lighting per **220.18(B)**
- Show window per **220.43(A)**
- Track lighting unit per **220.43(B)**
- Low-voltage systems per **Article 411**

An unlisted VA rating shall be calculated at 125 percent for sizing the OCPD and conductors for continuous operation and 100 percent for noncontinuous operated loads per **215.2(A)(1)** and **230.42(A)(1)**. **(See Figure 23-3)**

For example: What is the load in VA for 160, 120 volt, lighting ballasts rated at 1.5 amps each and used for 12 hours a day?

Step 1: Calculating load in A
220.18(B)
160 x 1.5 A = 240 A

Step 2: Calculating continuous load
215.2(A)(1); 230.42(A)(1)
240 A x 125% = 300 A

Step 3: Calculated VA
300 A x 120 V = 36,000 VA

Solution: The lighting loads for the unlisted occupancy is 36,000 VA.

SHOW WINDOW LIGHTING LOAD
220.43(A)

The lighting load in VA for the show window shall be calculated by multiplying the linear feet of the show window by 200 VA per foot. Such lighting load shall calculated at 100 percent for noncontinuous operation and 125 percent for continuous operation per **215.2(A)(1)** and **230.42(A)(1)**. Conductors and OCPD's shall be increased to comply with **240.3** and **240.4**.

For example: What is the lighting load in VA for a 40 ft. show window used at noncontinuous or continuous operation?

Step 1: Calculating noncontinuous load
220.43(A); 215.2(A)(1); 230.42(A)(1)
40' x 200 VA x 100% = 8000 VA

Step 2: Calculating continuous load
220.43(A); 215.2(A)(1); 230.42(A)(1)
40' x 200 VA x 125% = 10,000 VA

Solution: The noncontinuous load is 8000 VA and the continuous load is 10,000 VA.

If the number of lighting outlets is known, the VA rating of each luminaire (fixture) shall be multiplied by 125 percent for sizing the show window load. If the VA is not known, each outlet shall be calculated at 180 VA times 125 percent to obtain the lighting load in VA for the show window. Note that the greater between the 200 VA per linear foot or each individual calculation shall be used.

See Figure 23-4 for a detailed illustration of calculating the show window lighting load.

TRACK LIGHTING LOAD
220.43(B)

The lighting load in VA for lighting track shall be calculated by multiplying the lighting track by 150 VA and dividing by 2. Such VA rating shall be multiplied by 100 percent or 125 percent based upon noncontinuous or continuous operation per **90.7** and **110.3(B)**. **(See Figure 23-5)**

> **For example:** What is the load in VA for 180 ft. of lighting track used at noncontinuous or continuous operation?
>
> **Step 1:** Calculating noncontinuous load
> 220.43(B); 215.2(A)(1); 230.42(A)(1)
> 180' ÷ 2' x 150 VA x 100% = 13,500 VA
>
> **Step 2:** Calculating continuous load
> 220.43(B); 215.2(A)(1); 230.42(A)(1)
> 180' ÷ 2' x 150 VA x 125% = 16,875 VA
>
> **Solution:** The noncontinuous load is 13,500 VA and the continuous load is 16,875 VA.

> **For example:** What is the lighting load in VA for 130 continuous operated luminaires (fixtures) with a 175 VA ballast in each unit and 110 noncontinuous operated units with each ballast having a rating of 175 VA?
>
> **Step 1:** Calculating load
> 220.18(B); 215.2(A)(1); 230.42(A)(1)
> 175 x 130 x 125% = 28,437.5 VA
> 175 x 110 x 100% = 19,250   VA
> Total load   = 47,687.5 VA
>
> **Solution:** The total outside lighting load is 47,687.5 VA.

LOW-VOLTAGE LIGHTING LOAD
ARTICLE 411; 215.2(A)(1); 230.42(A)(1)

The lighting load in VA for low-voltage lighting systems shall be calculated by multiplying the full-load amps of the isolation transformer by 100 percent for noncontinuous operation and 125 percent for continuous operation.

> **For example:** What is the load in VA for a low-voltage lighting system supplied by an isolation transformer with a FLA of 150 amps used at noncontinuous or continuous operation?
>
> **Step 1:** Calculating noncontinuous load
> Article 411; 215.2(A)(1); 230.42(A)(1)
> 150 A x 100% = 150 A
>
> **Step 2:** Calculating continuous load
> Art. 411; 215.2(A)(1); 230.42(A)(1)
> 150 A x 125% = 187.5 A
>
> **Solution:** The noncontinuous load is 150 amps and the continuous load is 187.5 amps.

OUTSIDE LIGHTING LOAD
220.18(B); 215.2(A)(1); 230.42(A)(1)

The lighting load in VA for outside lighting loads shall be calculated by multiplying the VA rating of each lighting unit by 100 percent for noncontinuous operation and 125 percent for continuous operation. **(See Figure 23-7)**

OUTSIDE SIGN LIGHTING LOAD
220.14(F); 215.2(A)(1); 230.42(A)(1)

The lighting loads for signs shall be calculated based on the commercial occupancy or facility having ground floor footage accessible to pedestrians. Occupancies with grade level access for pedestrians shall have a minimum of 1200 VA provided for a sign lighting load. This VA rating is multiplied by 125 percent for signs operating for three hours or more and 100 percent for those operating less than three hours. **(See Figure 23-8)**

RECEPTACLE LOADS
220.14(I); TABLE 220.44

Receptacle loads are the second group of loads to be calculated. Such loads are divided into two subgroups as follows:

- General-purpose receptacle outlets
- Multioutlet assemblies

Each load in the subgroup shall be calculated differently to derive total VA.

GENERAL-PURPOSE RECEPTACLE LOADS
220.14(I); 215.2(A)(1); 230.42(A)(1)

The load in VA for the general-purpose receptacle load shall be calculated by multiplying the number of outlets times 180 VA each times 100 percent for noncontinuous operation and 125 percent for continuous operation. (See **Figure 23-9**)

For example: What is the load in VA for 55 general-purpose receptacles used to serve noncontinuous and continuous related loads?

Step 1: Calculating noncontinuous load
220.14(I); 215.2(A)(1); 230.42(A)(1)
180 VA x 55 x 100% = 9900 VA

Step 2: Calculating continuous load
220.14(I); 215.2(A)(1); 230.42(A)(1)
180 VA x 55 x 125% = 12,375 VA

Solution: The noncontinuous load is 9900 VA and the continuous load is 12,375 VA.

Note that a demand factor as listed in **Table 220.44** shall not be permitted to be applied to the continuous load of 12,375 VA even if this value exceeds 10,000 VA.

APPLYING DEMAND FACTORS
220.14(I); TABLE 220.44

General-purpose receptacle outlets for cord-and-plug connected loads used at noncontinuous operation shall be calculated per **220.14(I)** and **Table 220.44**. Noncontinuous operated receptacles with a VA rating of 10,000 VA or less shall be calculated at 100 percent. If the VA rating of the receptacle load exceeds 10,000 VA, a demand factor of 50 percent shall be permitted to be applied to all VA exceeding 10,000 VA per **Table 220.44**. (See Figure 23-10(a))

For example: What is the VA rating for 225 general-purpose receptacle outlets to cord-and-plug connect loads used at noncontinuous operation?

Step 1: Calculating load
220.14(I)
225 x 180 VA = 40,500 VA

Step 2: Applying demand factors
Table 220.44
First 10,000 VA x 100% = 10,000 VA •
Next 30,500 VA x 50% = 15,250 VA •
Total load   = 25,250 VA

Solution: The demand load is 25,250 VA.

MULTIOUTLET ASSEMBLIES
220.14(H); (2)

For connected loads not operating simultaneously, the VA rating shall be calculated by dividing the length of the assembly by 5 ft. and multiplying by 180 VA. For connected loads operating simultaneously, each foot of multioutlet assembly shall be multiplied by 180 VA. The fixed multioutlet assembly load shall be permitted to be added to the noncontinuous receptacle load and a demand factor applied per **Table 220.44**. (See Figures 23-10(b) and (c))

For example: What is the load in VA for 200 ft. of multioutlet assembly used to cord-and-plug connect loads that are not used simultaneously and used simultaneously? (**220.14(A)(1)**)

Step 1: Calculating load for non-simultaneous use
VA = length ÷ 5' x 180 VA
VA = 200' ÷ 5' x 180 VA
VA = 7200

Step 2: Calculating load for simultaneously use
VA = length x 180 VA
VA = 200' x 180 VA
VA = 36,000

Solution: The load in VA for the non-simultaneously load is 7200 VA and for the simultaneously load is 36,000 VA.

SPECIAL LOADS
215.2(A)(1); 230.42(A)(1)

Special appliance loads are the third group of loads to be calculated. These loads, which include computers, processing machines, etc. are usually served by individual circuits.

Industrial Calculations

CONTINUOUS AND NONCONTINUOUS OPERATION
215.2(A)(1); 230.42(A)(1)

The load in VA for special appliance loads shall be calculated by multiplying the VA rating of each load by 100 percent for noncontinuous operation and 125 percent for continuous operation. To determine the classification, special appliance loads operating for less than three hours shall be classified as a noncontinuous operated load. However, a special appliance load operating for three hours or more shall be classified as continuous operated load. **(See Figure 23-11)**

For example: What is the VA rating for a 208 volt, three-phase, 165 amp special appliance load operating for ten hours and supplied by an individual branch-circuit? (Use 360 V (208 x 1.732))

Step 1: Calculating VA
220.5(A)
VA = V x 1.732 x I
VA = 208 x 1.732 x 165
VA = 59,400

Step 2: Calculating continuous load
215.2(A)(1); 230.42(A)(1)
59,400 VA x 125% = 74,250 VA

Solution: The load at continuous operation is 74,250 VA.

For example: Consider and calculate the VA rating for a special appliance load of 82 amps operating at 480 volts, three-phase, for a period of 2 1/2 hours every four hours? (use 831 V (480 V x 1.732))

Step 1: Calculating VA
220.5(A)
VA = V x 1.732 x I
VA = 480 V x 1.732 x 82 A
VA = 68,142

Step 2: Calculating noncontinuous load
215.2(A)(1); 230.42(A)(1)
68,142 VA x 100% = 68,142 VA

Solution: The load in VA for the noncontinuous load is 68,142 VA.

COMPRESSOR LOADS
440.34

Compressor loads are the fourth group of loads to be calculated. Special considerations shall be applied when calculating loads for hermetic sealed compressors supplying refrigerant and cooling related equipment.

CONTINUOUS OR NONCONTINUOUS OPERATION
440.34; 215.2(A)(1); 230.42(A)(1)

Compressor related equipment shall be calculated at 100 percent of their VA or amps. If one of such is the largest motor per load seven, it is added to the total calculation of all loads at 125 percent of its nameplate rating. **(See Figure 23-13)**

For example: What is the load in VA for 6 compressors rated at 42 amps each and supplied by a 480 volt, three-phase supply? (All used in one process at the same time)

Step 1: Calculating VA
220.5(A)
42 A x 6 x 480 V x 1.732 (831 V) = 209,412 VA

Step 2: Calculating continuous load
440.34; 215.2(A)(1); 230.42(A)(1)
209,412 VA x 125% = 261,765 VA

Solution: The load in VA for the 6 compressors used in an industrial process at the same time is 261,765 VA.

MOTOR LOADS
220.50; 430.24

Motor loads are the fifth group of loads to be calculated. The VA rating of motors is converted from amperage to VA by multiplying the amperage from **Table 430.248** for single-phase or **Table 430.250** for three-phase by the supply voltage. **(See Figure 23-14)**

24-5

> **For example:** What is the VA rating for a group of 480 volt, three-phase motors rated at 125 HP, 40 HP and 30 HP?

> **Step 1:** Finding FLA
> **Table 430.250**
> 125 HP = 156 A
> 40 HP = 52 A
> 30 HP = 40 A
>
> **Step 2:** Calculating total VA
> **220.5(A)**
> VA = V x 1.732 x I
> 125 HP (use 831 V)
> 480 V x 1.732 x 156 A = 129,636 VA •
> 40 HP (use 831 V)
> 480 V x 1.732 x 52 A = 43,212 VA •
> 30 HP (use 831 V)
> 480 V x 1.732 x 40 A = 33,240 VA •
> Total VA = 206,088 VA
>
> **Solution:** The total load for the motors is 206,088 VA.

> **Design Tip:** Motors can be used as a single unit to drive a piece of equipment. Motors used in an approved assembly such as a processing machine are not usually considered individual motor loads.

HEAT OR A/C LOADS
220.60

Heating or A/C loads is the sixth group of loads to be calculated. The largest VA rating between the heating or A/C load is selected and the smaller of the two loads is dropped. To determine the largest of the two loads, the VA rating of each load shall be calculated at 100 percent and the largest load of the two is selected. The load dropped is not used again in the calculation. **(See Figure 23-15)**

> **Design Tip:** There is no need to calculate both loads to select the elements of the service equipment for the loads are never used simultaneously in the electrical system. For heat pump rules, **see Figure 23-16**.

> **For example:** What is the largest load between a 240 kW heating unit and a 97 amp A/C unit? The voltage is supplied by 480 volts, three-phase system.

> **Step 1:** Selecting largest load
> **220.60**
> Heating load
> 240 kW x 1000 x 100% = 240,000 VA
> A/C load (use 831 V)
> 97 A x 480 V x 1.732 = 80,607 VA
>
> **Solution:** The 240,000 VA heating unit is the largest load.

LARGEST MOTOR LOAD
220.50; 430.24

The largest motor load in VA is the seventh of the loads to be calculated. The largest motor load is selected from one of the motor related loads listed in the fourth, fifth, or sixth loads. The VA rating of the largest motor shall be calculated by multiplying the amperage of the unit by the voltage times 25 percent. **(See Figure 23-16)**

> **For example:** What is the largest motor from the following loads?

> Fourth load = compressor of 52 A
> Fifth load = motor of 65 A
> Sixth load = A/C unit = 23 A
>
> **Step 1:** Selecting largest load
> **220.50; 440.34; 430.24**
> The motor load of 65 A is the largest load
>
> **Solution:** The largest motor load is 65 amps.

OPTIONAL CALCULATIONS FOR ADDITIONAL LOADS TO EXISTING INSTALLATIONS
220.87

When additional loads are added to existing facilities having feeders and service as originally calculated, the maximum kVA calculations in determining the load on the existing feeders and service can be used if the following conditions are complied with:
• If the maximum data of the demand in kVA is available for a minimum of one year, such as demand meter ratings.

- If the demand ratings for that period of one year at 125 percent and the addition of the new load does not exceed the rating of the service. Where demand meters are used, in most cases the load as calculated will probably be less than the demand meter indications.

- If the OCPD meets **230.90, 215.3** and **240.4** for feeders or a service. **(See Figure 23-19)**

APPLYING Ex. TO 220.87
220.87, Ex.

If the maximum demand data for a one year period is not available, the calculated load shall be based on the maximum demand (measure of average power demand over a 15 minute period) continuously recorded over a minimum 30-day period using a recording ammeter or power meter connected to the highest loaded phase of the feeder or service, based on the initial loading at the start of the recording. **(See Figure 23-20)**

> **Design Tip:** By measurement or calculation the larger of the heating or cooling equipment load shall be included if it is not in the demand data.

CALCULATING THE NEUTRAL
220.61; 310.15(B)(4)(c) TO TABLE 310.16

For a service or feeder, the maximum unbalanced load controls the ampacity of the neutral. Neutral feeder load shall be considered wherever a neutral is used in conjunction with one or more ungrounded conductors. On a single-phase feeder using one ungrounded conductor and a neutral, the neutral will carry the same amount of current as the ungrounded conductor. A two-wire feeder is seldom used, so in considering the neutral feeder current, always assume that there is a neutral and two or more ungrounded (phase) conductors. If there are two ungrounded conductors that are connected to the same phase, and a neutral, the neutral would be required to carry the total current from both ungrounded (phase) conductors, which would not be an accepted practice.

For three-wire DC or single-phase AC, four-wire three-phase, three-wire two-phase, and five-wire two-phase systems, a further demand factor of 70 percent may be applied to that portion of the unbalanced load in excess of 200 amperes. There shall be no reduction of the neutral capacity for that portion of the load which consists of electric-discharge lighting, electronic computer/data processing, or similar equipment, when supplied by four-wire, wye-connected, three-phase systems.

For example, on a four-wire, three-phase wye circuit where the major portion (over 50 percent) of the load consists of nonlinear loads, there are harmonic currents present in the neutral conductor, and the neutral is considered to be a current-carrying conductor. In other words, the ampacity of the conductor shall be derated per **310.15(B)(2)(a)**.

Review the rules and examples of Chapters 14 and 15 for the sizing and use of the neutral in service, feeder-circuit and branch-circuit installations.

USING ONE-LINE DIAGRAM
90.8

When calculating the load for an existing electrical system or to design an entirely new system, an important tool is a good, up-to-date line diagram of the system. It indicates by single lines and standard symbols the routed course and components parts of an electric circuit or system or circuits.

A schedule of loads and values must be developed and such load values calculated to size and select components and wiring methods. **(See Figure 24-1)**

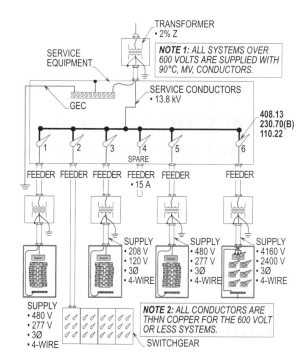

ONE-LINE DIAGRAM

Figure 24-1. The above is an one-line diagram depicting the electrical equipment in an industrial plant.

CALCULATING FEEDER NO. 1

(1) Lighting loads

Number	Type load	Phases	Volts	Amps
• 60	2 ballast	1Ø	277 V	.38 A ea.
• 42	HID 1000 W	1Ø	480 V	.86 A ea.
• 6	Large equip.	3Ø	480 V	21 A ea.
• 10	process units	3Ø	480 V	34 A ea.

Note: All loads are used at continuous operation.

(2) Calculating total amps for feeder No. 1
220.18(B); 215.2(A)(1); 440.34
(2 ballast at .38 A ea.)
• 60 x 2 x .38 x 125%   = 57   A •
(HID 1000 W at .86 A ea.)
• 42 x .86 A x 125%   = 45.15 A •
• 6 x 21 A x 125%   = 157.5 A •
• 10 x 34 A x 125%   = 425   A •
Total load   = 684.65 A •

Rounded up to 685 amps per **220.2(B)**

(3) Sizing conductors for feeder No. 1
Table 310.16

Step 1: Paralleling three per phase
310.4
A of conductors = A ÷ # of conductors
A = 685 A ÷ 3
A = 229

Step 2: Selecting conductors for feeder No. 1
Table 310.16
229 A requires 4/0 AWG cu.
4/0 AWG THHN cu. = 230 A

Step 3: Total amps of conductors
310.4
4/0 AWG cu. = 230 A x 3 = 690 A
690 A supplies 685 A

Solution: The size THHN copper conductors are 4/0 AWG paralleled 3 times per phase.

(4) Sizing OCPD based upon load
240.4(B); 240.6(A)

685 A or 690 A requires 700 A OCPD

Solution: The size OCPD for 3 - 4/0 paralleled THHN copper conductors is 700 amps.

CALCULATING FEEDER NO. 2

(1) What size 90°C, MV, copper conductors and OCPD is required to supply power to the switchgear based on the following loads per Part III to **Article 220**.

Number	Type load	Phases	Volts	Amps
• 2	heavy Equip.	3Ø	13.8 kV	2 A ea.
• 2	large motors	3Ø	13.8 kV	1.5 ea.
• 1	distribution center	3Ø	13.8 kV	15 A

Note: All loads are used at continuous operation.

(2) Calculating total amps for feeder No. 2
230.202; 230.208(B); 215.2(A)(1); 430.24
(2 heavy equipment at 2 A ea.)
• 2 x 2 A x 125%   = 5   A
(2 large motors at 1.5 ea.)
• 2 x 1.5 A x 125%   = 3.75 A
(1 distribution center at 15 A)
• 1 x 15 A x 125%   = 18.75 A
(largest motor load)
• 1.5 A x 25%   = .375 A
Total loads   = 27.875 A

(2) What size 90°C, MV copper conductors are required for an underground run in conduit to supply the distribution equipment?

Number	Type load	Phases	Volts	Loads
• 1	distribution run	3Ø	13.8 kV	27.875 A

(3) Calculating load for underground feeder
Figure 310.60, Detail 3; Table 310.77

27.875 A requires 6 AWG cu.

Solution: The size 90°C, MV conductors for the underground run is 6 AWG copper.

(4) How many mains are allowed for disconnecting the feeder-circuit if it were run to a separate building?

Step 1: Finding number of mains
225.33(A); 225.34(A)
The number of mains is 6

Solution: Six mains are allowed to disconnect the feeder-circuit.

CALCULATING FEEDER-CIRCUIT NO. 3
PART II TO ARTICLE 220

Number	Type load	Phases	Volts	Amps
• 150	4 - 34 W fluorescent	1Ø	120 V	.86 A ea.
• 50	recess fixtures	1Ø	120 V	1.25 A ea.
• 40	150 W bulbs	1Ø	120 V	1.25 A ea.
• 200	receptacles (180 W ea.)	1Ø	120 V	1.5 A ea.
• 10	100 ft. multi-outlet assembly	1Ø	120 V	1.5 per ft.
• 10	isolation receptacles	1Ø	120 V	1.5 A ea.

Note: All the loads except the receptacles and bullet four are used at continuous operation.

(1) Calculating total amps for feeder No. 3
220.14(I); 220.14(H)(2); 215.2(A)(1); Table 220.44
(150 fluorescent at .86 A ea.)
• 150 x .86 A x 125%   = 161.25 A
(50 recess fixtures at 1.25 A ea.)
• 50 x 1.25 A x 125%   = 78.125 A
(40 - 150 W bulbs at 1.25 A ea.)
• 40 x 1.25 A x 125%   = 62.5 A
(200 receptacle at 1.5 A ea.)
• 200 x 180 W   = 36,000 W
First 10,000 W x 100%   = 10,000 VA
Next 26,000 W x 50%   = 13,000 VA
Total load   = 23,000 VA

I = 23,000 VA ÷ 208 x 1.732 = 63.9 A
(10 - 10 ft. multioutlet assembly at 1.5 per foot)
• 10 x 10' x 1.5 A   = 150 A
(10 isolation receptacles at 1.5 A ea.)
• 10 x 1.5 x 125%   = 18.75 A
Total load   = 534.525 A

(2) What is the minimum size OCPD required on the primary side of the transformer for feeder No. 3?

Step 1: Sizing the transformer based on Sec. GE manual (based on 125% of LD.)
kVA = I x V x √3 ÷ 1000
kVA =
534.525 A x 208 V x 1.732 ÷ 1000
kVA = 192.566

Step 2: Selecting size of the transformer ACME chart (based on 125% of LD.)
192.566 kVA requires 225 kVA

Solution: The size transformer to supply the load is 225 kVA.

Step 3: Finding FLA of transformer
I = kVA x 1000 ÷ V x √3
I = 225 kVA x 1000 ÷ 13,800 V x 1.732
I = 9.4 A

Step 4: Finding size of OCPD for primary
Table 450.3(A); 240.6(A)
9.4 A x 600% = 56.4 A
56.4 A requires 50 A

Solution: Under certain conditions, 230.208(B) allows the load for over 600 volt systems to be calculated at 100 percent. Size OCPD is 50 amp.

(3) What size 90°C, MV copper conductors, based on the OCPD, are required to supply the primary of the transformer for Feeder No. 3?

Step 1: Selecting conductors based on OCPD
Table 310.77; 110.40
50 A OCPD requires 6 AWG cu.

Solution: The size of the conductors for feeder No. 3 are 6 AWG copper based on the OCPD.

CALCULATING FEEDER NO. 4
PART II TO ARTICLE 220

Number	Type load	Phases	Volts	Amps
• 1	Spare	3Ø	13.8 kVA	15 A

(1) Providing load capacity for Feeder No. 4

A load capacity of 15 amps is provided for future use

CALCULATING FEEDER NO. 5
PART III TO ARTICLE 220

Number	Type load	Phases	Volts	Amps
• 10	10 HP motors	3Ø	460 V	14 A ea.
• 12	15 HP motors	3Ø	460 V	21 A ea.
• 15	5 HP motors	3Ø	460 V	7.6 A ea.
• 16	3 HP motors	3Ø	460 V	4.8 A ea.

(1) What size conductors using the 10 ft. (3 m) connection rule are required to supply the motors on the secondary side of the transformer?

Step 1: Calculating motor load
430.24
10 x 14 A = 140 A
12 x 21 A = 252 A
15 x 7.6 A = 114 A
16 x 4.8 A = 76.8 A
21 A x 25% = 5.3 A
Total load = 588.1 A

(2) Using the 10 ft. (3 m) connection rule, how many 3/0 AWG, THHN copper conductors are required for a parallel hook-up between the transformer and panel?

Step 1: Calculating the number of conductors
310.4; Table 310.16
(3/0 AWG = 200 A)
# of conductors = total A ÷ A of conductors
# of conductors = 588.1 A ÷ 200 A
# of conductors = 2.94 A

Step 2: Finding the number of conductors
310.4; Table 310.16
2.94 requires 3

Solution: The parallel hook-up requires 3 - 3/0 AWG per phase.

(3) What size OCPD (CB) is required for the connected secondary conductors?

Step 1: Sizing OCPD for largest motor load
430.62(A); 430.52(C)(1), Table 430.52
21 A x 250% = 52.5 A

Step 2: Selecting size OCPD
430.52(C)(1); Table 430.52
52.5 A requires 50 A

Step 3: Sizing OCPD for tapped conductors
430.28(1); 430.62(A)
Largest OCPD = 50 A
Other motors = 140 A
(252 A - 21 A = 231 A) = 231 A
= 114 A
= 76.8 A
Total load = 611.8 A

Step 4: Selecting OCPD for the secondary conductors
430.62(A); 240.6(A)
611.8 A requires 600 A

Solution: The size OCPD for the feeder-circuit using a CB is 600 amps. Note: There is no Ex. to 430.62(A) that will allow the next size OCPD above 611.8 amps to be used.

(4) What size copper grounding electrode conductor, bonding jumper, and grounded conductor are required to ground the secondary to building steel?

Step 1: Sizing the GEC
250.30(A)(4); Table 250.66; Table 8, Ch. 9
3/0 AWG = 167,800 CM x 3 in parallel = 503,400 CM
KCMIL = 503,400 CM ÷ 1000 = 503.4 KCMIL
503.4 KCMIL requires 1/0 AWG cu.

Solution: The size of the GEC is 1/0 AWG copper.

Step 1: Sizing the BJ
250.30(A)(1); 250.102(C); Table 250.66
503.4 KCMIL requires 1/0 AWG cu.

Solution: The size of the bonding jumper must be at least 1/0 AWG copper.

Step 1: Sizing the grounded conductor
250.24(C)(2); Table 250.66; 310.4
503.4 KCMIL requires 1/0 AWG copper

Solution: The size of the grounded conductor must be 1/0 AWG copper in each conduit run.

Industrial Calculations

CALCULATING ELEMENTS FOR FEEDER NO. 6

Number	Type load	Phases	Volts	Amps
2	Heavy load (machine)	3Ø	4160 V	12 A
4	Large motors (in a process)	3Ø	4160 V	
1	motor			20 A
1	motor			26 A
1	motor			15 A
1	motor			25 A
3	motors			50 A
1	distribution panel	3Ø	4160 V	50 A

1 motor control center
- 1 - 200 HP   3Ø   4160 V   27.9 A
- 1 - 150 HP   3Ø   4160 V   20.9 A
- 1 - 125 HP   3Ø   4160 V   18.5 A
- 2 - 50 HP   3Ø   4160 V   7.6 A

(1) What is the total load for Feeder No. 6

Step 1: Calculating total load
215.2(B); 430.24; 430.25
(2 heavy loads)
2 x 12 A = 24 A
(7 large motors)
1 x 20 A = 20 A
1 x 26 A = 26 A
1 x 15 A = 15 A
1 x 25 A = 25 A
3 x 50 A = 150 A
(1 distribution panel at 50 A)
1 x 50 A = 50 A
(motor control center - 5 motors)
1 x 27.9 A = 27.9 A
1 x 20.9 A = 20.9 A
1 x 18.5 A = 18.5 A
2 x 7.6 A = 15.2 A
(largest motor load)
50 A x 25% = 12.5 A
Total load = 405 A

CALCULATING THE SERVICE LOAD IN AMPS FOR EACH VOLTAGE
230.42(A)(1); 430.24; 440.34

Step 1: Calculating amps of each feeder
220.5(A)
Feeder No. 1 = 684.65 A
(277 / 480 V)
Feeder No. 2 = 27.875 A
(13.8 kV)
Feeder No. 3 = 534.525 A
(120 / 208 V)
Feeder No. 4 = 15 A
(277 / 480 V)
Feeder No. 5 = 588.05 A
(13.8 kV)
Feeder No. 6 = 405 A
Total load = 2255.1 A

Step 2: Calculating kVA of feeders 1, 3, 5 and 6 based on supply voltage and load
220.5(A)
Feeder No. 1 on page 24-8:
kVA = 684.65 A x 480 V x 1.732 ÷ 1000
kVA = 569.19
Feeder No. 3 on page 24-9:
kVA = 534.525 A x 208 V x 1.732 ÷ 1000
kVA = 192.57
Feeder No. 5 on page 24-10:
kVA = 588.05 x 480 V x 1.732 ÷ 1000
kVA = 488.88
Feeder No. 6 on page 24-11:
kVA = 405 x 4160 V x 1.732 ÷ 1000
kVA = 2918.07

Step 3: Calculating amps of each feeder based on supply voltage and load
220.5(A)
Feeder No. 1:
I = kVA x 1000 ÷ V x $\sqrt{3}$
I = 569.19 x 1000 ÷ 13,800 x 1.732
I = 23.82 A
Feeder No. 2 on page 24-8:
Amps at 13,800 V = 27.88 A
Feeder No. 3:
I = kVA x 1000 ÷ V x $\sqrt{3}$
I = 192.57 x 1000 ÷ 13,800 x 1.732
I = 8.06 A
Feeder No. 4 on page 24-9:
Amps at 13,800 V = 15 A
Feeder No. 5:
I = kVA x 1000 ÷ V x $\sqrt{3}$
I = 488.88 x 1000 ÷ 13,800 x 1.732
I = 20.46 A
Feeder No. 6:
I = kVA x 1000 ÷ V x $\sqrt{3}$
I = 2918.07 x 1000 ÷ 13,800 x 1.732
I = 122.09 A
Total amps = 217.31 A

Solution: The service load in amps to size the elements of the service is **217.31 amps.**

Step 4: Calculating the 90°C, MV, conductors of the service
Table 310.77, Col. 4
217.31 A at 13.8 kV requires 4 AWG

Solution: The size conductors for the service are **4 AWG copper.**

WHAT SIZE TRANSFORMER IS REQUIRED TO SUPPLY THE SERVICE EQUIPMENT LOADS
220.5(A); PAGE 24-11

Step 1: kVA = I x V x $\sqrt{3}$
kVA = 217.31 A x 13,800 V x 1.732
kVA = 5194.06

Step 2: Selecting transformers
ACME chart
use a transformer bank

Solution: Use a transformer bank that can handle 5194.06 at 13,800 volts.

WHAT IS THE FLC IN AMPS OF THE TRANSFORMER

Step 1: FLC = kVA x 1000 ÷ V x $\sqrt{3}$
FLC = 5194.06 kVA x 1000 ÷ 13,800 V x 1.732
FLC = 217.31 A

Solution: The FLC of the transformer is 217.31 amps.

WHAT IS THE AVAILABLE FAULT-CURRENT AT THE TERMINALS OF THE TRANSFORMER

Step 1: AFC = FLA of transformer ÷ Z
AFC = 217.31 A ÷ .02
AFC = 10,865.5 A

Solution: The AFC at the terminals of the transformer is 10,865.5 amps.

SEE FIGURE 24-2 FOR THE SIZE ELEMENTS OF EACH FEEDER
220.2

Note 1: Figure 24-2 has the size elements of each feeder shown for easy identification after such elements have been calculated and selected.

Note 2: Figure 24-2 has a summary of the elements in Figure 24-1 after they were sized from the load calculations.

Note 3: The NEC defines low-voltage systems as 600 volts or less and high-voltage systems are those operating over 600 volts.

Note 4: IEEE-141 defines low-voltage systems as those which are 600 volts or less and medium-voltage as systems over 600 volts up to 69,000 volts. High-voltage systems are greater than 69,000 volts.

Feeder No. 1 on page 24-8

Size OCPD for the panelboard
- 700 A

Size conductors for panelboard
- 3 - 4/0 AWG THWN cu. conductors per phase

Feeder No. 2 on page 24-8

Size OCPD for the switchgear
- 30 A or 50 A based on conductors

Size conductors for the switchgear
- 3 - 6 MV cu. conductors

Feeder No. 3 on pages 24-9

Size OCPD for the primary of transformer
- 50 A

Size conductors for the primary of transformer
- 6 MV cu. conductors

Feeder No. 4 on page 24-9

Size OCPD for the future load
- 15 A

Size conductors for the future load
- 6 MV cu. based on NEC

Feeder No. 5 on page 24-10

Size OCPD for the connection
- 600 A

Size conductors for the connection
- 3 - 3/0 THWN cu. conductors per phase

Feeder No. 6 on page 24-11

Total load for feeder in amps
- 405 A

Service elements on pages 24-11 and 12

Size conductor for the service
- 4 MV cu. conductors per phase

Size transformer for the service
- transformer bank

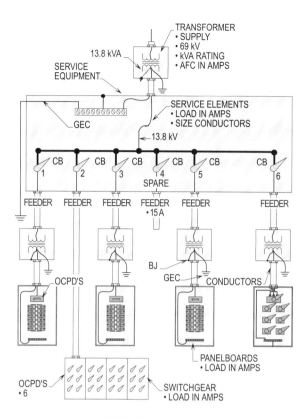

ONE-LINE DIAGRAM

Figure 24-2. One-line diagram showing electrical equipment and wiring methods with the size of individual elements of service and feeder-circuits after they were sized from load calculations. For calculation procedure for each element, refer to the page number listed below the sizes shown.

Name   Date

Chapter 24. Industrial Calculations

Section	Answer		
_____	T	F	**1.** What is the load in VA for an unlisted occupancy feeder-circuit having 140 - 120 volt, lighting ballasts rated at 1.5 amps each and used for 6 hours a day?
_____	T	F	**2.** What is the lighting load in VA for an unlisted occupancy feeder-circuit having 50 ft. show window used at noncontinuous or continuous operation?
_____	T	F	**3.** What is the load in VA for 200 ft. of lighting track on a service used at noncontinuous or continuous operation?
_____	T	F	**4.** What is the load in VA for 50 general-purpose receptacles on a service used to serve noncontinuous and continuous related loads?
_____	T	F	**5.** What is the load in VA for 200 general-purpose receptacle outlets to cord-and-plug connect loads used at noncontinuous operation?
_____	T	F	**6.** What is the load in VA for 150 ft. of multioutlet assembly used to cord-and-plug connect loads that are not used simultaneously and used simultaneously?
_____	T	F	**7.** What is the VA rating for a 208 volt, three-phase, 155 amp special appliance load operating for eight hours and supplied by an individual branch-circuit?
_____	T	F	**8.** What is the VA rating for a 208 volt, three-phase, 155 amp special appliance load operating for two hours and supplied by an individual branch-circuit?
_____	T	F	**9.** What is the load in amps for four compressors rated at 38 amps each and supplied by a 480 volt, three-phase supply? (All used in one process at the same time)
_____	T	F	**10.** What is the VA rating for a group of 480 volt, three-phase motors rated at 100 HP, 50 HP, and 25 HP? (Calculate for a feeder-circuit)

Appendix

Table A

Size AWG/ KCMIL	Ohms to Neutral per 1000 feet					
	AC Resistance for Uncoated Copper Wires			AC Resistance for Aluminum Wires		
	PVC Conduit	Alu. Conduit	Steel Conduit	PVC Conduit	Alu. Conduit	Steel Conduit
14	3.1	3.1	3.1	—	—	—
12	2.0	2.0	2.0	3.2	3.2	3.2
10	1.2	1.2	1.2	2.0	2.0	2.0
8	0.78	0.78	0.78	1.3	1.3	1.3
6	0.49	0.49	0.49	0.81	0.81	0.81
4	0.31	0.31	0.31	0.51	0.51	0.51
3	0.25	0.25	0.25	0.40	0.41	0.40
2	0.19	0.20	0.20	0.32	0.32	0.32
1	0.15	0.16	0.16	0.25	0.26	0.25
1/0	0.12	0.13	0.12	0.20	0.21	0.20
2/0	0.10	0.10	0.10	0.16	0.16	0.16
3/0	0.077	0.082	0.079	0.13	0.13	0.13
4/0	0.062	0.067	0.063	0.10	0.11	0.10
250	0.052	0.057	0.054	0.085	0.090	0.086
300	0.044	0.049	0.045	0.071	0.076	0.072
350	0.038	0.043	0.039	0.061	0.066	0.063
400	0.033	0.038	0.035	0.054	0.059	0.055
500	0.027	0.032	0.029	0.043	0.048	0.045
600	0.023	0.028	0.025	0.036	0.041	0.038
750	0.019	0.024	0.021	0.029	0.034	0.031
1000	0.015	0.019	0.018	0.023	0.027	0.025

Note: See Table 9 to Ch. 9 of the NEC.

Table B

(Three Single Conductors) "C" Value for Conductors and Busways

Copper AWG or kcmil	Three Single Conductor Conduit Steel			Nonmagnetic		
	600V	5KV	15KV	600V	5KV	15KV
14	389	389	389	389	389	389
12	617	617	617	617	617	617
10	981	981	981	981	981	981
8	1557	1551	1557	1558	1555	1558
6	2425	2406	2389	2430	2417	2406
4	3806	3750	3695	3825	3789	3752
3	4760	4760	4760	4802	4802	4802
2	5906	5736	5574	6044	5926	5809
1	7292	7029	6758	7493	7306	7108
1/0	8924	8543	7973	9317	9033	8590
2/0	10755	10061	9389	11423	10877	10318
3/0	12843	11804	11021	13923	13048	12360
4/0	15082	13605	12542	16673	15351	14347
250	16483	14924	13643	18593	17120	15865
300	18176	16292	14768	20867	18975	17408
350	19703	17385	15678	22736	20526	18672
400	20565	18235	16365	24296	21786	19731
500	22185	19172	17492	26706	23277	21329
600	22965	20567	17962	28033	25430	22690
750	24136	21386	18888	28303	25430	22690
1000	25278	22539	19923	31490	28083	24887
Aluminum						
14	236	236	236	236	236	236
12	375	375	375	375	375	375
10	598	598	598	598	598	598
8	950	950	951	951	950	951
6	1480	1476	1472	1481	1478	1476
4	2345	2332	2319	2350	2341	2333
3	2948	2948	2948	2958	2958	2958
2	3713	3669	3626	3729	3701	3672
1	4645	4574	4497	4678	4631	4580
1/0	5777	5669	5493	5838	5766	5645
2/0	7186	6968	6733	7301	7152	6986
3/0	8826	8466	8163	9110	8851	8627
4/0	10740	10167	9700	11174	10749	10386
250	12122	11460	10848	12862	12343	11847
300	13909	13009	12192	14922	14182	13491
350	15484	14280	13288	16812	15857	14954
400	16670	15355	14188	18505	17321	16233
500	18755	16827	15657	21390	19503	18314
600	20093	18427	16484	23451	21718	19635
750	21766	19685	17686	23491	21769	19976
1000	23477	21235	19005	28778	26109	23482

Table B

(Three Conductor Cable) "C" Value for Conductors and Busways

Copper AWG or kcmil	Three Single Conductor Conduit Steel			Nonmagnetic		
	600V	5KV	15KV	600V	5KV	15KV
14	389	389	389	389	389	389
12	617	617	617	617	617	617
10	981	981	981	981	981	981
8	1559	1557	1559	1559	1558	1559
6	2431	2424	2414	2433	2428	2420
4	3830	3811	3778	3837	3823	3798
3	4760	4790	4760	4802	4802	4802
2	5989	5729	5827	6087	6022	5957
1	7454	7364	7188	7579	7507	7364
1/0	9209	9086	8707	9472	9372	9052
2/0	11244	11045	10500	11703	11528	11052
3/0	13656	13333	12613	14410	14118	13461
4/0	16391	15890	14813	17482	17019	16012
250	18310	17850	16465	19779	19352	18001
300	20617	20051	18318	22524	11938	20163
350	22646	21914	19821	24904	24126	21982
400	24253	23371	21042	26915	26044	23517
500	26980	25449	23125	30028	28712	25916
600	28752	27974	24896	32236	31258	27766
750	31050	20024	26932	32404	31338	28303
1000	33864	32688	29320	37197	35748	31959
Aluminum						
14	236	236	236	236	236	236
12	375	375	375	375	375	375
10	598	598	598	598	598	598
8	951	951	951	951	951	951
6	1481	1480	1478	1482	1481	1479
4	2351	2347	2339	2353	2349	2344
3	2948	2956	2948	2958	2958	2958
2	3733	3719	3693	3739	3724	3709
1	4686	4663	4617	4699	4681	4646
1/0	5852	5820	5717	5875	5851	5771
2/0	7327	6968	6733	7301	7152	6986
3/0	9077	8980	8750	9242	9164	8977
4/0	11184	10121	10642	11408	11277	10968
250	12796	12636	12115	13236	13105	12661
300	14916	14698	13973	15494	15299	14658
350	15413	16490	15540	17635	17351	16500
400	18461	18063	16921	19587	17321	16233
500	18755	16827	15657	21390	19243	18154
600	23633	23195	21348	25750	25243	23294
750	26431	25789	23750	25682	25141	23491
1000	29864	29049	26608	32938	31919	29135

Note : These values are equal to one over the impedence per foot for impedences found in IEEE. std. 241-1990. IEEE Recommended Practice for Commercial Building Power System.

Table B

AWG or MCM	Copper Three Single conductors				Copper Three-Conductor Cable	
	Steel conduit		Nonmagnetic Conduit		Steel Conduit	Nonmagnetic Conduit
	600 V and 5 kV Nonshielded	5 kV Shielded and 15 kV	600 V and 5 kV Nonshielded	5 kV Shielded and 15 kV	600 V and 5 kV Nonshielded	600 V and 5 kV Nonshielded
12	588		588			
10	909		1449			
8	1429	1230	1449	1230	1230	1230
6	2222	1940	2273	1940	1950	1950
4	3333	3040	3448	3070	3080	3090
3	4167	3830	4348	3870	3880	3900
2	5000	4670	5263	4780	4830	4850
1	6250	5750	6250	5920	6020	6100
1/0	7692	6990	7692	7250	7410	7580
2/0	9091	8260	9091	8770	9090	9350
3/0	10638	9900	11364	10700	11100	11900
4/0	12500	10800	13514	12600	13400	14000
250	13699	12500	17857	14000	14900	15800
300	15385	13600	16949	15500	16700	17900
350	16667	14700	18868	17000	18600	20300
400	17857	15200	20408	17900	19500	21100
500	20000	16500	23256	19700	21900	24000
600	21277	17200	25000	20900	23300	25700
750	23256	18300	27778	22500	25600	28200
1000	25000		31250			

AWG or MCM	Aluminum-Three Single Conductors Or Three-Conductor Cables	
	Steel Conduit	Nonmagnetic Conduit
	600 V and 5 kV Nonshielded	600 V and 5 kV Nonshielded
12	357	357
10	555	556
8	909	909
6	1388	1408
4	2173	2173
3	2702	2702
2	3333	3333
1	4000	4166
1/0	5000	5000
2/0	6250	6250
3/0	7142	7692
4/0	9090	9090
250	10000	10638
300	11363	12195
350	12500	13698
400	13698	15384
500	15625	17543
600	17241	19607
750	19230	22222
1000	21739	25641

Note: These tables were used for fault calculations before Tables B on page IV and V were available.

Table C

Size AWG/ KCMIL	Area Cir. Mills	Conductors Standing Quantity	Conductors Standing Diam. In.	Conductors Overall Diam. In.	Conductors Overall Area Sq. In.^{2}	DC Resistance at 75°C (167°F) Copper Uncoated ohm/kFT	DC Resistance at 75°C (167°F) Copper Coated ohm/kFT	DC Resistance at 75°C (167°F) Aluminum ohm/kFT
18	1620	1	---	0.040	0.001	7.77	8.08	12.8
18	1620	7	0.015	0.046	0.002	7.95	8.45	13.1
16	2580	1	---	0.051	0.002	4.89	5.08	8.05
16	2580	7	0.019	0.058	0.003	4.99	5.29	8.21
14	4110	1	---	0.064	0.003	3.07	3.19	5.06
14	4110	7	0.024	0.073	0.004	3.14	3.26	5.17
12	6530	1	---	0.081	0.005	1.93	2.01	3.18
12	6530	7	0.030	0.092	0.006	1.98	2.05	3.25
10	10380	1	---	0.102	0.008	1.21	1.26	2.00
10	10380	7	0.038	0.116	0.011	1.24	1.29	2.04
8	16510	1	---	0.128	0.013	0.764	0.786	1.26
8	16510	7	0.049	0.146	0.017	0.778	0.809	1.28
6	26240	7	0.061	0.184	0.027	0.491	0.510	0.808
4	41740	7	0.077	0.232	0.042	0.308	0.321	0.508
3	52620	7	0.087	0.260	0.053	0.245	0.254	0.403
2	66360	7	0.097	0.292	0.067	0.194	0.201	0.319
1	83690	19	0.066	0.332	0.087	0.154	0.160	0.253
1/0	105600	19	0.074	0.373	0.109	0.122	0.127	0.201
2/0	133100	19	0.084	0.419	0.138	0.0967	0.101	0.159
3/0	167800	19	0.094	0.470	0.173	0.0766	0.0797	0.126
4/0	211600	19	0.106	0.528	0.219	0.0608	0.0626	0.100
250	---	37	0.082	0.575	0.260	0.0515	0.0535	0.0847
300	---	37	0.090	0.630	0.312	0.0429	0.0446	0.0707
350	---	37	0.097	0.681	0.364	0.0367	0.0382	0.0605
400	---	37	0.104	0.728	0.416	0.0321	0.0331	0.0529
500	---	37	0.116	0.813	0.519	0.0258	0.0265	0.0424
600	---	61	0.099	0.893	0.626	0.0214	0.0223	0.0353
700	---	61	0.107	0.964	0.730	0.0184	0.0189	0.0303
750	---	61	0.111	0.998	0.782	0.0171	0.0176	0.0282
800	---	61	0.114	1.03	0.834	0.0161	0.0166	0.0265
900	---	61	0.122	1.09	0.940	0.0143	0.0147	0.0235
1000	---	61	0.128	1.15	1.04	0.0129	0.0132	0.0212
1250	---	91	0.117	1.29	1.30	0.0103	0.0106	0.0169
1500	---	91	0.128	1.41	1.57	0.00858	0.00883	0.0141
1750	---	127	0.117	1.52	1.83	0.00735	0.00756	0.0121
2000	---	127	0.126	1.63	2.09	0.00643	0.00662	0.0106

Note: See Table 8, Ch. 9 of the NEC.

Abbreviations

A

A - amps
AC - alternating current
A/C - air-conditioning
AEGCP - assured equipment grounding conductor program
AHJ - authority having jurisdiction
Alu. - aluminum
ASCC - available short-circuit current
AWG - American Wire Gauge

B

BC - branch-circuit
BCSC - branch-circuit selection current
BJ - bonding jumper
BK - black
BL - blue
BR - brown

C

°C - Celsius
CB - circuit breaker
CEE - concrete-encased electrode
CL - code letter
CM - circular mills
CMP - Codemaking Panel
Comp. - compressor
Cond. - condenser
Cont. - continuous
cu. - copper
cu. in. - cubic inches

D

dia. - diameter
DC - direct-current
DPCB - double pole circuit breaker

E

EBJ - equipment bonding jumper
Eff. - efficiency
EGB - equipment grounding bar
EGC - equipment grounding conductor
EMT - electrical metallic tubing
ENT or ENMT - electrical nonmetallic tubing
Ex. - Exception
EExde - increase safety
Eexe - flameproof/increased safety components
Epf - explosionproof

F

°F - Fahrenheit
FC - feeder-circuit
FLA - full-load amperage
FLC - full-load current
FMC - flexible metal conduit
FPN - fine print note
ft. - foot

G

G - ground
GE - grounding electrode
GEC - grounding electrode conductor
GES - grounding electrode system
GFL - ground-fault limiter
GFCI - ground-fault circuit interrupter
GFPE - ground-fault protection of equipment
GR - green
GRY - gray
GSC - Grounded service conductor

H

H - hot conductor
HACR - heating, air-conditioning, cooling, and refrigeration
HP - horsepower
Htg. - heating
Hz - hertz

I

I - amperage or current
IEC - International Electrotechnical Commission
IG - Isolated ground
in. - inches
INVT - inverse-time circuit breaker
INST. CB - instantaneous trip circuit breaker
IRA - inrush amps
ITSC - intrinsically safe circuits

K

kFT - 1,000'
kV - kilo volts
kVA - kilo volt-amps
kW - kilo watts
kWH - kilo watt-hour

L

L - length of conductor
LRA - locked rotor amps
Ld. - load
LPB - lighting panelboard
LRC - locked-rotor current
LFMC - liquidtight flexible metal conduit
LFNC - liquidtight flexible nonmetallic conduit
LTR - long time rated

M

MA - milli-amps
max. - maximum
MEL - maximum energy level
mf - microfarads
MGFA - maximum ground-fault available
min. - minimum
min. - minute
MR - momentary rated
Mt. - motor

N

N - neutral
NACB - nonautomatic circuit breaker
NB - neutral bar
NEC - *National Electrical Code*
NEMA - National Electrical Manufacturers Association
NFD - nonfused disconnect
NFPA - National Fire Protection Association
NLTFMC - nonmetallic liquidtight flexible metal conduit
NTDF - nontime-delay fuse
NPC - nameplate current of motor

O

OCP - overcurrent protection
OCPD - overcurrent protection device
OSHA - Occupational Safety and Health Administration
OL - overload
OLP - overload protection
OL's - overloads
OR - orange

P

Ph. - phases (hots)
pri. - primary
PSA - power supply assembly
PF - power factor
PU - purple

R

R - ohms or resistance
RD - red
RMC - rigid metal conduit

S

SBS - structural building steel
SIA - seal-in amps
SP - single-pole
S/P - single-phase
SCC - short-circuit current
sec. - secondary
SF - service factor
SPCB - single-pole circuit breaker
sq. ft. - square foot (feet)
STR - short-time rated
sq. in. - square inches
SWD - switched disconnect
SWG - switchgear
SDS - separately derived system

T

TDC - time-delay cycle
TDF - time-delay fuse
TDL - time-delay limiter
TP - thermal protector
TR - temperature rise
Tran. - transformer
TS - trip setting
TV - touch voltage

U

UF - underground feeder

V

V - volts
VA - volt-amps
VD - voltage drop

W

W - watts
WT - white
WP - weatherproof

X

XTMR - transformer

Y

YEL - yellow

Glossary

A

Active power is the true electrical power or real power supplying the load.

Across-the-line starter is a device consisting of contactor and overload relay which is used to start an electric motor by connecting it directly to the supply line.

Air gap is the air space between two electrically related parts such as the space between poles of a magnet or poles in an electric motor.

Alternating current (AC) is the current in an electrical circuit that alternates in flowing first with a positive polarity and then with a negative polarity.

Alternator is a rotating machine whose output is AC.

Ampere is an unit of measure for current flow. Note that one ampere equals a flow of one coulomb of charge per second.

Ambient conditions is the condition of the atmosphere adjacent to electrical equipment.

Ambient temperature is the ambient temperature is the temperature of the surrounding atmosphere cooling medium, which comes into contact with the heated parts of equipment.

Ambient temperature compensated is a device, such as an overload relay, which is not affected by the temperature surrounding it.

Ampacity is the current-carrying capacity in amperes.

Ampere is a unit of intensity of electrical current produced in a conductor by an applied voltage.

Armature is a special designed rotor.

Apparent power is the sum of active power and reactive power. It is determined by multiplying voltage times current.

Arc-chute is a cover around contacts designed to protect surrounding parts from arcing effects.

Armature reaction is the reaction of the magnetic field produced by the current on the magnetic lines of force which are produced by the field coil of an electric motor or generator.

Apparatus is a set of control devices used to help perform the intended control functions.

Automatic is a means of self acting that operates by its own mechanism, as such a change in pressure or temperature, etc.

Automatic controller is a motor or other control mechanism which uses automatic pilot devices as activating devices. These devices may be pressure switches, level switches or thermostats.

Autotransformer starter is equipped with an autotransformer which is designed to reduce the voltage to the motor terminals and reduce the starting current and still start the motor, etc.

Auxiliary contacts are contacts in addition to the main-circuit contacts and function with the movement of the latter.

Auxiliary device is any device other than motors and motor starters necessary to fully operate the machine or equipment.

Auxiliary interlock:

- Mechanical - A physical device or arm so arranged that it cannot close both starter circuits at the same time.

- Electrical - An additional contact mounted on the side of a magnetic starter.

B

Bearings is a device used to support the motor shaft, and allows it to rotate smoothly.

Bimetallic disc is a disc made up of two strips of dissimilar metals combined to form a single strip.

Breakdown torque is the maximum torque which a motor develops under increasing load conditions at rated voltage and frequency without an sudden drop in rotating speed.

Brushes are sliding contacts, usually made of carbon which are located between a commutator and the outside circuit in a generator or motor.

Branch-circuit is that portion of a wiring system extending beyond the final overcurrent device protecting the circuit.

C

Commutator is a device which reverses the connections to the revolving loops on the armature.

Capacitance is the ability to store electricity in an electrostatic field.

Capacitor is a device which is designed to introduce capacitance into an electric circuit.

Capacitor-start motor is an AC split-phase induction motor that has a capacitor connected in series with an auxiliary winding which provides a way for it to start. This auxiliary circuit is designed to disconnect to the motor when it reaches its speed.

Circuit is an electrical network of conductors that provides one or more paths for current.

Circuit breaker is a device designed to open and close a circuit either by a nonautomatic means or by an automatic means due to a predetermined overload of current.

Combination starter is a magnetic starter having a manually operated disconnecting means built into the enclosure housing the magnetic contactor or starter.

Compensating windings are the windings embedded in the main pole pieces of a compound DC motor.

Component is the smallest element of a circuit.

Contacts are connecting parts which co-act with other parts to connect or disconnect a circuit.

Contactor is a electro-mechanical device for connecting and disconnecting an electric power circuit.

Control is a device or group of devices which is used in some predetermined manner to govern the electric power delivered to an apparatus.

Control circuit is the control circuit of the control apparatus which carries the signals directing the performance of the controller.

Control, two-wire is a control function which utilizes a maintained contact type pilot device to provide undervoltage release.

Control circuit transformer is a control circuit transformer utilized to supply a reduced voltage suitable for the operation of control devices.

Control circuit voltage is the control circuit voltage providing the operation to the coils of magnetic devices.

Control, three-wire is a control function which utilizes a momentary contact pilot device and a holding circuit contact to provide voltage to the coil of a controller. The holding circuit maintains the control circuit voltage.

Controller is a device, or group of devices, which is used in some predetermined manner to connect and disconnect the electric power delivered to the apparatus.

Copper loss is the electrical power lost through the resistance of the coils due to the current flowing through the wire of the coils.

Core is the magnetic path through the center of coil or transformer.

Core losses is the loss of power in the coil (core) due to eddy currents and hysteresis.

Core transformer is an electrical transformer which has the core inside of the coils.

Counterelectromotive force is the voltage induced in the armature coil of an electric motor.

Counter torque is a repulsion force between two magnetic fields.

D

Delta-delta connected is a coil connection in which the primary and secondary coils are delta connected.

Delta-wye connected is a connection in which the primary is delta connected while the secondary windings are wye connected.

Diagram is a connection diagram showing the electrical connection between the parts of the control, and external connections.

Direct-current (DC) is a current that always flows in only one direction only.

Disconnect means is a motor circuit switch which is intended to connect and disconnect a circuit to a motor. It must be rated in horsepower and capable of interrupting the maximum current.

E

Efficiency is the ratio of output power (watts) to input power (watts).

Electric motor is a machine which converts electrical energy to mechanical energy.

Electromagnet is a magnet comprised of a coil of wire wound around a soft-iron core. When current is passed through the wire, a magnetic field is produced.

Eddy currents are the electrical currents circulating in the core of a transformer as the result of induction.

Electricity is the electrical charges in motion. Such movement is called current and is measured in amps.

Electrolytic capacitor is a capacitor which uses a liquid or past as one of its electrical storage plates.

Electromotive force is a voltage or force which causes free electrons to move in a conductor.

Electromagnet is a temporary magnet created by passing an electric current through a coil wound around a soft iron or other magnetic core.

Electromagnetism is a magnetic field which exists around a wire or other conductor if a current is passing through it.

Electromechanical is a device which uses electrical energy to create mechanical motion of force.

Electron is a negative electric charge.

Electron flow is the flow of electrons from a negative point to a positive point in a conductor.

Electrostatic charge is the electrical charge stored by a capacitor.

Electrostatic field is the stored electrical charges on the surface of an insulator.

Excitation is creating a magnetic field to be used to create electromagnetics when an electric current is passed through a coil.

F

Feeder-circuit is the conductors between the service equipment and the branch-circuit overcurrent device.

Float switch is a switch which is operated by a float and is responsive to the level of liquid.

Foot switch is a switch which is suitable for operation by an operators foot.

Frequency is the rate at which AC changes its direction of flow, it is normally expressed in terms of hertz (cycles) per second.

Fuse is an overcurrent protection device with a circuit opening fuseable member which opens when overheated by current passing through it.

G

Gate is one of the leads on a thyristor. This lead is the one that normally controls output when it is correctly biased.

Generator is a rotating machine which changes mechanical energy into DC.

Generator action is inducing voltage into a wire that is cutting a magnetic field.

Ground is any point on a motor component where the ohmic resistance between the component and the motor frame is one megaohm or less.

Guarded is covered, shielded, fenced, enclosed, or otherwise protected by means of suitable covers or casings, barriers, rails or screens, mats or platforms to remove the likelihood of dangerous contact or approach by persons or objects to a point of danger.

H

Hertz is a measurement of frequency and it actually means "cycles per second" of AC.

Hermetic refrigerant motor compressor is a combination consisting of a compressor and motor, both of which are enclosed in the same housing, with no external shaft or shaft seals, the motor operating in the refrigerant.

High side in a transformer, this marking indicates the high-voltage winding.

Horsepower is a unit of measure for power and it represents the force times distance times time. For example, one horsepower (HP) equals 746 watts, or 33,000 ft. lb. per minute, or 550 ft. lb. per second.

Hysteresis is the property of a magnetic substance that causes the magnetization to lag behind the magnetizing force.

I

Induced current is the current which flows in a conductor because of a changing magnetic field.

Inductance is electromotive force resulting from a change in magnetic flux surrounding a circuit or conductor.

Induction is the generation of electricity by magnetism.

Inductive reactance is the opposition in ohms to an AC as a result of induction. This is voltage resulting from cutting lines of magnetic force.

Interlock is an electrical or mechanical device actuated by the operation of a different device in which it is directly related.

Intermittent duty is a requirement or service that demands operations for alternate intervals of (1) load and no-load; or (2) load and rest; or (3) load, no-load and rest; such alternate intervals being definitely specified.

Isolating transformer is an isolating transformer used to electrically isolate one circuit from another.

Inverter is a circuit capable of receiving a positive signal and sends out a negative one or vice versa. It is a device which changes AC to DC or vice versa.

Impedance is the total opposition to current flow in a circuit and is measured in ohms.

J

Jogging is the rapid and repeated opening and closing of the circuit to start a motor from rest for the purpose of creating small movements of the motor or driven load.

K

kVA is the term used to rate transformers.

KVAR is the reactive power in a circuit.

kW is used to rated the load of certain types of equipment, etc.

L

Lamination consist of sheet material which is sandwiched together to construct a stator or rotor of a rotating machine.

Limit switch is a switch which is operated by a part or motion of a power-driven piece of equipment. Such operation alters the electric or electronic circuits related to the equipment.

Locked-rotor current of a motor is the current taken from the line when the motor starts or the rotor becomes locked in place.

Low side in a transformer, this marking indicates which is the low voltage winding.

M

Magnetic starter is a starter which is actuated by an electromagnetic means.

Manual controller is a device which manually closed or opened.

Manual reset is a device which requires manual action to re-engage the contacts after an overload.

Magnetic drive (magnetic clutch) is electromagnetic device that is connected between a three-phase motor and its load. Its main purpose is regulating the speed at load rotated speed.

Magnetic field is the invisible lines of force found between the north and south poles of a magnet.

Magnetic lines of force in a magnetic field, are imaginary lines which show the direction of the magnetic flux.

Maintained contacts closes the circuit when the push button pressed and will open the circuit when the push button is pressed again.

Megaohm is one million ohms.

Motor action is the mechanical force that exist between magnets. Two magnets approaching each other will either pull toward or push away from the other. In other words, there is a pull and push action between the rotor and field poles of the motor.

Multi-speed motor is a motor which is capable of operating at two or more fixed speeds.

N

NEMA (National Electrical Manufacturers Association) is an organization which establishes certain voluntary standards relating to motors; such as operating characteristics, terminology, basic dimensions, ratings, and testing.

No-load speed is the speed reached by the rotor or armature when it rotates.

Non-automatic requires personal action and operation of devices for its control means.

Nonreversing is a control function which provides for operation in one direction only.

Normally open and normally closed is a term when applied to a magnetically operated switching device, signifies the position that the contacts are in.

Normally closed contacts are motor control contacts (set) that are open when the push button is depressed.

Normally open contacts are contacts (set) which are closed when the push button is depressed.

O

Ohm is a unit of electrical resistance of a conductor.

Out-of-phase is a condition where two or more phases of AC is changing direction at different intervals of time.

Over excited is a condition where a synchronous motor is equipped with a DC field which supplies more magnetization than is needed.

Overload protector is a device affected by an abnormal operating condition which causes the interruption of current flow to the device governed.

Overload relay is a device that provides overload protection for conductors and electrical equipment.

Overcurrent protective device (OCPD) is a device that operates on excessive current which causes the interruption of power to the circuit if necessary.

P

Parallel circuit is a circuit in which all positive terminals are connected at a common point and all negative terminals are connected to another point.

Periodic duty is a type of intermittent duty in which the load conditions are regularly recurrent.

Permanent-capacitor motor is a single-phase electric motor which uses a phase winding and capacitor in conjunction with the main winding. The phase winding is controlled by the capacitor which remains in the circuit at all times.

Permanent magnetism is magnet that will keep its magnetic properties indefinitely.

Permeability is a condition in which domains in a magnetic core can be made to line up to create magnetism.

Phase is the relationship of two wave forms having the same frequency.

Phase angle is the difference in angle between two sine wave vectors.

Phase shift is the creation of a lag or advance in voltage or current in relation to another voltage or current in the same electrical circuit.

Phase voltage is the voltage across a coil.

Polarity is a condition where a magnet has north and south poles which are positive and negative charge.

Polyphase is more than one phase, usually three-phase when related to generators, transformers, and motors.

Pounds force is an English unit of conventional measurement for force.

Power factor is the figure which indicates what portion of the current delivered to the motor is used to do work.

Primary coil is one of two coils in a transformer.

Prime mover is the primary power source which can be used to drive a generator.

Pull-in to torque is the maximum torque at which an induction motor will pull into step.

Pull-up torque is the minimum torque developed by an induction motor during the period of acceleration from rest to full speed.

Pull-out torque is the maximum torque developed by a motor for one minute, before it pulls out of step due to an overload.

Push button control is the control and operation of equipment through push buttons used to activate relays.

Push button switch is a switch utilizing a button for activating a coil and contact to open or close a circuit.

R

Rainproof is an enclosure so constructed, as to prevent rain from interfering with operation of the apparatus.

Raintight is an enclosure so constructed to exclude rain under specified test conditions.

Rated-load current is the load rated-load current for a hermetic refrigerant motor-compressor is the current resulting when the motor-compressor is operated at the rated load, rated voltage, and rated frequency of the equipment it serves.

Rating is a designated limit of operating characteristics based on conditions of use such as load, voltage, frequency, etc.

Rating, continuous is the rating which defines the substantially constant load which can be carried for an indefinitely long time.

Reactive power is the reactive voltage times the current, or voltage times the reactive current, in an AC circuit.

Rectifier is an electrical device which converts AC to DC by allowing the current to move in only one direction.

Relay is a device that operates by a variation of a condition which effects the operation of other devices in an electric circuit.

Relay contacts are the contacts that are closed or opened by movement of a relay armature.

Reluctance is the ratio between the magnetomotive force and the resulting flux.

Reset is to restore a mechanism or device to a prescribed state.

A-15

Reset, automatic is a function which operates automatically to re-establish certain circuit conditions.

Reset, manual is a function which requires a manual operation to re-establish certain circuit conditions.

Residual magnetism is the magnetism remaining in the core of a coil or an electromagnet after the current flow has been removed.

Resistance is a property of conductors which makes them resist the movement of current flow.

Resistance starting is a reduced-voltage starting method employing resistances which are short-circuited in one or more steps to complete the starting cycle of a motor, etc.

Resistors is an electrical-electronic device which is attached to a circuit to produce resistance to current flow.

Rheostat is a variable resistor with a fixed terminal and a movable contact.

Rotor is the rotating section which rotates within the stator of a motor.

Rotor impedance is the phasor sum of resistance and inductive reactance.

RPM is the revolutions per minute.

Running torque is the torque or turning effort determined by the horsepower and speed of a motor at any given point of operation.

S

Saturated is where an electrical or magnetic component can not receive any more electrical current or magnetism.

Saturation is a point at which a magnet will receive any more flux density.

Sealing, voltage or current is the voltage or current required to seat the armature of a magnetic circuit closing device to the make position.

Service factor is the number by which the horsepower rating is multiplied to determine the maximum safe load that a motor can carry continuously at its rated voltage and frequency.

Secondary coil is the coil that is connected to the load in the electrical circuit.

Self excitation is a condition of supplying excitation voltages by a device on the generator rather than from an outside source.

Self induction is a counterelectromotive force produce in a conductor when the magnetic field produced by the conductor collapses or expands after a change in current flow.

Separate excitation is a condition of producing generator field current from an independent source.

Short-time rating is referring to the motor load which can be carried for a short and definitely specified time.

Series circuit is a circuit in which all resistances and other components are connected so that the same current flows from point to point.

Series field is the total magnetic flux caused by the action of the series winding in a rotating piece of machine.

Series motor is a motor in which the field and armature circuits are connected in series.

Shaded-pole motor is a single-phase squirrel cage induction motor with stator poles slotted and used to create two sections in each pole.

Shading coil is a copper ring or coil that is set into a section of the pole piece and its function is to produce the lagging part of a rotating magnetic field for starting torque.

Short is any two points of a motor where there is zero, or extremely low, resistance between them or between two motor components.

Shunt field is a type of field coil designed for a DC motor which is connected in parallel with the armature.

Silicon controlled rectifier (SCR) is a semiconductor device which has the ability to block a voltage that is applied in either direction. On a signal applied to its gate, it is capable of conducting current even when the signal has been removed.

Single-phase is having only one AC or voltage in a circuit.

Slip is the difference between the synchronous speed of a motor and the speed at which it operates.

Slip ring motor has a rotor with the same number of magnetic poles at the stator.

Slip rings are equipped with circular bands on a rotor which are used to transmit current from rotor coils to brushes.

Slip speed is the difference between the rotor speed and synchronous speed in an induction motor.

Soft neutral position is a condition where the brushes of a repulsion electric motor are aligned with the stator field.

Solid state devices contain circuits and components using semiconductors.

Solid state controls are devices that control current to motors through semiconductors.

Solid state relay is a relay that uses semiconductor devices.

Split-phase (resistance-start) motor is a single-phase induction motor equipped with an auxiliary winding which is connected in parallel with the main winding.

Squirrel-cage rotor is designed with a rotor which is made up of metal bars which are short-circuited at each end.

Starter is a controller for accelerating a motor from rest to its running speed.

Starter, automatic is a starter which automatically controls the starting of a motor.

Starter, autotransformer is a starter which is provided with an autotransformer that provides a reduced voltage for starting.

Starter, part-winding applied voltage to partial sections of the primary winding of an AC motor.

Starter, reactor includes a resistor connected in series with the primary winding of an induction motor to provide reduced voltage for starting.

Starter, wye-delta connects the motor leads in a wye configuration for reduced voltage starting and reconnects the leads in a delta configuration for the run position.

Stall torque is the torque which the rotor of an energized motor produces when the rotor is not rotating.

Starting torque is the amount of torque produced by a motor as it breaks the motor shaft from standstill and accelerates to its running speed.

Stator is the portion which contains the stationary parts of the magnetic circuit with their associated windings when installed in a motor.

Static electricity is electricity at rest. It is also known in the industry as a static charge.

Stator field contains a magnetic field which is set up in the electric motor when the motor is energized and electric current is flowing.

Stator poles are the shoes on an electric motor stator which hold the windings and the magnetic poles of the stator.

Switch is a device for making, breaking or changing the connections in an electric circuit.

Switch, float is responsive to the level of a liquid.

Switch, foot is a switch that is operated by an operator's foot.

Switch, general-use is a general-use type non-horse rated switch which is capable of interrupting the rated current at the rated voltage.

Switch, limit is operated by some part or motion which alters the electrical circuit associated with the equipment.

Switch, master controls the operation of contactors, relays, or other similar operated devices.

Switch, motor circuit is rated in horsepower, and is capable of interrupting the maximum operating current of the motor.

Switch, pressure is operated by fluid pressure, etc.

Switch, selector is a manually operated multiposition switch which is used for selecting an alternative control circuit.

Synchronous is a condition where the currents and voltages are in-step or in-phase.

Synchronous motor is an induction motor which runs at synchronous speed.

Synchronous speed is the constant speed to which an AC motor adjusts itself, depending on the frequency of the power source and the number of poles in the motor.

T

Tachometer is a device which is capable of measuring rotational speed of rotating machines.

Tap changer is a mechanical device which has the ability to change the voltage output of a transformer.

Taps are fixed electrical connections which are located at specific positions on a transformers coil.

Temperature, ambient is the temperature of the medium such as air, oil, etc. into which the heat of the equipment is dissipated.

Terminal is a point at which an electrical element may be connected to another electrical element.

Terminal board is an insulating base equipped with one or more terminal connectors used for making electrical connections.

Thermocouple is a device which consists of two unlike metals which are joined together and when heat is applied, a current will flow.

Thermal, cutout is an overcurrent protective device having a heater element which affects a fusible member that open the circuit due to an overload.

Thermal protector is a protective device that is an integral part of the motor which is designed to protect the motor windings from dangerous overloads.

Thermostat is an instrument which responds to changes in temperature to effect control over an operating condition.

Three-phase alternator is a rotating machine which generates three separate phases of AC.

Three-phase electric motor is a motor which operates from a three-phase power supply.

Timer is a device which is designed to delay the closing or opening of a circuit for a specific period of time.

Torque is a force that produces a rotating or twisting action.

Torque, breakdown is the maximum torque which a motor develops with rated voltage when applied at rated frequency.

Torque, locked rotor is the minimum torque which a motor develops at standstill when rated voltage is applied at rated frequency.

Turns ratio is the ratio of the number of turns in the primary winding of a transformer to the number of turns in the secondary winding.

Two-capacitor motor is an induction motor which uses one capacitor for starting and one for running.

U

Under excited is a term used to describe the magnetizing power of synchronous motor.

Unity power factor is a power factor of 1 and this is the best PF that can be obtained in an electrical system.

Undervoltage protection is a device which operates on the reduction or failure of voltage, and has the ability to maintain the interruption of power.

Undervoltage release is a device which operates on the reduction or failure of voltage and has the ability to interrupt the power but not to prevent the re-establishment of the circuit.

Vector is an in-phasor diagram having lines with a specific length and direction.

Voltage is a force which, when applied to a conductor, produces a current in the conductor.

W

Watt is a unit of electrical power and is the product of voltage and amperage.

Wattmeter is an instrument used for measuring electrical power.

Wye or star connection is an electrical connection in which all of the terminals are joined at the neutral junction. After it is connected it resembles a wye connection.

Wye-wye connection is where the coil arrangement in which both the primary and the secondary coils are wye-connected.

Transformer is an device designed to change the voltage in an AC electrical circuit. Step-up transformers increase the voltage and lowers the current. Step-down transformers decrease the voltage and raises the current.

Transformer efficiency is the ratio of input to output power.

Index

A

AC circuits.......11-9, 11-10, 11-14

AC grounded service.......11-68

accessible.......6-28

adjustable trip cb's.......9-10

air conditioning.......9-4, 17-9, 19-6

aircraft hangars.......21-2, 21-34

angle pull.......12-17, 12-18, 12-19

appliance load.......22-4, 22-5, 22-9, 22-10, 22-13, 22-14, 22-15, 22-16, 22-19, 22-22, 22-23, 22-24, 22-25, 23-9

appliances.......6-20, 6-28, 6-29, 7-6, 7-7, 8-13, 9-1, 9-8, 9-22, 12-10, 13-7, 13-20, 14-2, 14-9, 14-10, 14-15, 14-23, 15-12, 15-14, 16-1, 16-2, 16-5, 16-6, 16-8, 16-9, 16-12, 16-13, 16-14, 16-15, 16-16, 18-6, 21-22, 21-23, 21-46, 22-9, 22-10, 23-8, 23-9, 23-16

Askarel-insulated transformers.......20-18, 20-19

attachment plugs.......19-13, 21-33, 21-43

attics.......13-18

autotransformers.......20-9, 20-10, 20-11, 20-12

Auxiliary gutters.......6-7

auxiliary gutters.......12-22

B

bathroom.......6-20, 12-13, 16-14, 17-5, 17-6, 17-10, 22-2

batteries.......4-7, 13-6, 13-10, 21-36

bonding.......6-6, 6-31, 7-3, 9-28, 10-27, 11-1, 11-22, 11-24, 11-42, 11-47, 11-48, 11-52, 11-53, 11-54, 11-55, 11-56, 11-57, 11-66, 11-68, 11-69, 11-73, 11-74, 11-75, 12-2, 13-3, 13-10, 21-19, 21-35, 22-27, 23-22

bonding jumper.......11-22, 11-23

boxes.......5-11, 6-18, 7-1, 7-8, 7-9, 7-10, 11-52, 11-53, 11-54, 11-67, 11-69, 11-70, 11-74, 12-1, 12-2, 12-3, 12-4, 12-5, 12-6, 12-7, 12-8, 12-9, 12-10, 12-11, 12-12, 12-13, 12-14, 12-15, 12-16, 12-17, 12-18, 13-1, 13-2, 13-13, 13-14, 13-15, 13-16, 13-17, 13-19, 13-20, 17-1, 17-2, 17-3, 17-19, 17-21, 17-22, 21-16, 21-24, 21-26, 21-33, 23-25, 23-26, 23-27, 23-28

> boxes and fittings.......13-4
>
> boxes used as supports.......13-2

branch-circuit conductors.......8-12, 9-12, 11-74, 12-8, 14-11, 14-20, 14-26, 14-27, 14-29, 14-31, 14-34, 14-35, 15-1, 16-3, 18-2, 18-3, 18-14, 18-32, 19-2, 19-3, 19-6, 19-10, 19-11

branch-circuit taps.......9-12

branch-circuits.......6-1, 7-2, 7-6, 8-7, 8-13, 9-7, 9-12, 9-30, 11-67, 11-68, 12-6, 12-8, 13-4, 13-23, 14-2, 14-3, 14-8, 14-10, 14-11, 14-14, 14-17, 14-20, 14-21, 14-23, 14-24, 14-26, 14-27, 15-11, 16-2, 16-3, 16-4, 16-5, 16-8, 16-20, 17-3, 18-20, 18-32, 19-1, 19-13, 20-22, 22-10

busbars.......6-29, 7-1, 7-2, 7-5, 9-28, 11-43, 15-9

bushings.......6-17, 6-18, 11-54, 12-16, 12-17

busways.......6-7, 6-14, 6-15, 6-31, 9-18, 9-22, 21-19

C

cabinets.......6-18, 7-1, 7-8, 11-52, 11-54

cable.......6-6, 6-7, 6-12, 6-14, 6-16, 6-17, 6-18, 6-19, 7-2, 7-9, 8-4, 8-6, 8-9, 8-10, 8-13, 8-15, 9-18, 11-43, 11-52, 11-54, 11-59, 11-60, 11-65, 11-66, 11-67, 11-72, 12-3, 12-4, 12-8, 12-10, 12-11, 12-12, 12-23, 12-24, 12-25, 12-26, 12-27, 13-1, 13-3, 13-4, 13-5, 13-6, 13-7, 13-15, 13-17, 13-18, 13-19, 13-20, 14-11, 14-13, 15-4, 15-9, 15-11, 16-3, 16-8, 16-11, 17-2, 17-18, 17-19, 20-13, 20-27, 21-16, 21-17, 21-19, 21-25, 21-26, 21-27, 21-35, 21-38, 21-42, 21-43, 21-46

cable systems.......13-3, 13-15, 13-17

cable trays.......12-23

cartridge fuses.......9-25

circuit and system grounding.......11-2

circuit breaker ratings.......19-6

circuit breakers.......5-2, 5-15, 5-17, 5-18, 6-3, 6-20, 6-22, 6-24, 6-26, 6-27, 6-28, 6-29, 7-2, 7-7, 9-3, 9-9, 9-10, 9-12, 9-18, 9-25, 9-26, 9-27, 9-29, 14-26, 14-29, 14-30, 16-11, 18-6, 18-10, 18-11, 18-15, 18-16, 18-17, 18-19, 18-21, 18-25, 18-26, 18-27, 18-32, 18-33, 20-5, 20-6, 21-12, 21-16, 21-19, 21-33, 21-43

circuits.......3-3, 5-4, 5-18, 6-1, 6-2, 6-4, 6-11, 6-12, 6-19, 6-20, 6-28, 7-1, 7-2, 7-5, 7-6, 8-3, 8-5, 8-7, 8-10, 8-13, 9-1, 9-7, 9-8, 9-9, 9-10, 9-11, 9-12, 9-19, 9-25, 9-26, 9-27, 9-29, 9-30, 11-1, 11-9, 11-10, 11-12, 11-21, 11-34, 11-53, 11-55, 11-61, 11-65, 11-67, 11-68, 11-70, 11-75, 12-6, 12-8, 13-4, 13-10, 13-14, 14-1, 14-2, 14-3, 14-5, 14-6, 14-8, 14-10, 14-11, 14-14, 14-15, 14-16, 14-17, 14-18, 14-20, 14-21, 14-22, 14-23, 14-24, 14-26, 14-27, 14-29, 15-1, 15-6, 15-7, 15-8, 15-10, 15-11, 16-2, 16-3, 16-4, 16-5, 16-8, 16-9, 16-16, 16-19, 16-20, 16-21, 17-3, 17-14, 17-18, 18-1, 18-6, 18-14, 18-15, 18-16, 18-20, 18-27, 18-32, 18-35, 19-1, 19-2, 19-4, 19-12, 19-13, 19-15, 20-1, 20-12, 20-15, 20-22, 21-15, 21-16, 21-24, 21-26, 21-35, 21-38, 22-3, 22-4, 22-10, 22-16, 22-19, 22-20, 22-21, 22-22, 22-23, 22-24, 22-25, 23-8, 23-15, 23-16

circuits not to be grounded.......11-12

class 1 circuits.......4-9

clear spaces.......5-4

clearances.......6-8, 7-6

closed delta systems.......4-3

code letters.......18-2, 18-13, 18-14

combined rating.......6-24

concrete encased electrode.......11-46

condensation.......6-17, 8-16, 21-29

conductor fill........12-2

conductor identification.......8-6

conductor protection.......9-30

conductors.......5-10, 6-1, 6-3, 6-4, 6-5, 6-6, 6-7, 6-8, 6-9, 6-10, 6-11, 6-12, 6-13, 6-14, 6-15, 6-16, 6-17, 6-18, 6-20, 6-22, 6-23, 6-24, 6-25, 6-26, 6-27, 6-28, 6-29, 6-31, 6-33, 7-1, 7-2, 7-3, 7-4, 7-5, 7-9, 7-10, 7-11, 8-1, 8-2, 8-3, 8-4, 8-5, 8-6, 8-7, 8-8, 8-9, 8-10, 8-11, 8-12, 8-13, 8-14, 8-15, 8-16, 8-17, 8-18, 8-19, 8-20, 9-1, 9-2, 9-3, 9-4, 9-6, 9-7, 9-8, 9-9, 9-10, 9-12, 9-13, 9-14, 9-15, 9-16, 9-17, 9-18, 9-19, 9-20, 9-21, 9-23, 9-24, 9-25, 9-26, 9-28, 9-29, 9-30, 10-27, 11-1, 11-10, 11-12, 11-17, 11-22, 11-24, 11-32, 11-34, 11-40, 11-41, 11-43, 11-44, 11-45, 11-46, 11-47, 11-49, 11-52, 11-54, 11-55, 11-57, 11-59, 11-60, 11-61, 11-64, 11-65, 11-66, 11-67, 11-69, 11-70, 11-72, 11-73, 11-74, 12-1, 12-2, 12-3, 12-4, 12-5, 12-6, 12-7, 12-8, 12-9, 12-10, 12-11, 12-12, 12-13, 12-14, 12-15, 12-16, 12-17, 12-18, 12-19, 12-20, 12-21, 12-22, 12-23, 12-24, 12-25, 12-27, 12-28, 13-1, 13-2, 13-3, 13-4, 13-5, 13-7, 13-12, 13-13, 13-14, 13-15, 13-19, 14-1, 14-6, 14-11, 14-12, 14-13, 14-14, 14-15, 14-16, 14-17, 14-19, 14-20, 14-24, 14-26, 14-27, 14-29, 14-31, 14-34, 14-35, 14-36, 14-37, 15-1, 15-2, 15-4, 15-6, 15-7, 15-8, 15-9, 15-10, 15-11, 15-12, 15-

13, 16-2, 16-3, 16-11, 16-20, 17-2, 17-5, 17-15, 17-18, 18-1, 18-2, 18-3, 18-4, 18-5, 18-6, 18-7, 18-8, 18-9, 18-14, 18-20, 18-24, 18-28, 18-30, 18-31, 18-32, 18-33, 18-34, 18-35, 18-36, 18-37, 18-38, 18-43, 18-44, 19-1, 19-2, 19-3, 19-6, 19-10, 19-11, 19-14, 20-1, 20-5, 20-6, 20-7, 20-10, 20-12, 20-13, 20-14, 20-15, 20-22, 20-23, 20-24, 20-25, 21-23, 21-24, 21-25, 21-38, 21-42, 22-3, 22-6, 22-10, 22-14, 22-15, 22-16, 22-17, 22-18, 22-23, 22-24, 22-25, 22-26, 22-27, 22-28, 23-3, 23-13, 23-14, 23-15, 23-17, 23-18, 23-19, 23-20, 23-21, 23-22

conductors in parallel.......8-15

conductors of different circuits.......4-9

conduit bodies......12-16

conduit system.......8-9, 16-11, 21-28, 21-40

connections

    receptacle grounding terminals to box.......11-68

    up to 50 ft........9-20

    up to 75 ft........9-20

continuous loads.......8-4, 8-13, 14-18, 15-4

control and signaling cables.......12-25

control circuits.......8-5, 13-10, 18-1, 18-32, 18-35

controllers.......18-1, 18-24, 18-25, 18-26, 18-30, 19-1, 19-2, 21-19

cord-and-plug connected equipment.......16-6, 19-4, 19-5, 19-9, 21-25

cords.......9-7, 9-8, 11-74, 13-20, 13-21, 13-22, 14-23, 16-1, 16-5, 16-6, 16-7, 16-10, 16-17, 16-20, 19-15, 21-26, 21-36

countertops.......16-9, 16-16

D

definitions.......8-4

delta breakers.......7-8

demand factors.......9-23, 9-25, 14-3, 14-5, 15-2, 15-4, 15-5, 15-12, 22-2, 22-4, 22-5, 22-7, 22-9, 22-10, 22-11, 22-12, 22-14, 22-15, 22-16, 22-19, 22-23, 22-24, 22-25, 22-26, 23-1, 23-2, 23-8, 23-10, 23-13, 23-17, 23-31

derating

    circuit conductors.......8-9

    ambient temperature correction factor.......8-10

design letters.......18-14, 18-15, 18-17

disconnecting means.......1-6, 1-19, 5-4, 5-17, 6-13, 6-14, 6-15, 6-19, 6-20, 6-22, 6-23, 6-24, 6-25, 6-27, 6-28, 6-29, 6-30, 6-31, 6-32, 6-33, 9-17, 9-18, 9-28, 11-22, 11-24, 11-30, 11-68, 12-13, 14-29, 14-30, 14-33, 18-1, 18-4, 18-26, 18-27, 18-28, 18-29, 18-30, 18-31, 18-32, 18-35, 18-38, 18-43, 19-1, 19-2, 19-3, 19-4, 19-5, 19-6, 19-11, 19-14, 19-15, 20-5, 20-22, 20-24

donut method.......6-31, 9-28

dwelling unit.......6-13, 6-24, 6-26, 6-27, 11-74, 12-10, 12-13, 13-1, 14-1, 14-2, 14-3, 14-7, 14-9, 14-10, 14-17, 14-21, 14-23, 15-1, 15-5, 16-1, 16-3, 16-4, 16-5, 16-6, 16-9, 16-11, 16-12, 16-13, 16-14, 16-15, 16-20, 16-21, 17-4, 17-5, 17-6, 17-7, 17-8, 17-9, 17-10, 17-14, 22-1, 22-2, 22-3, 22-4, 22-5, 22-9, 22-11, 22-12, 22-14, 22-15, 22-16, 22-17, 22-18, 22-19, 22-20, 22-21, 22-22, 22-23, 22-24, 22-25

E

ear protection.......3-2

electric-discharge.......14-15, 14-16, 14-17, 14-18, 23-15

electrical equipment.......1-5, 1-10, 1-12, 1-13, 1-15, 1-16, 1-17, 1-18, 3-3, 5-1, 5-2, 5-3, 5-4, 5-5, 5-6, 5-8, 5-9, 5-10, 5-13, 5-15, 9-2, 11-74, 12-1, 12-18, 12-21, 14-17, 15-1, 16-15, 16-16, 16-22, 17-12, 17-13, 17-18, 21-1, 21-4, 21-6, 21-7, 21-8, 21-9, 21-12, 21-14, 21-15, 21-35, 21-36, 21-42, 21-43, 21-47, 21-48, 21-49, 21-51, 22-18

electrical metallic tubing.......13-12

electrical service.......6-30

electrical systems.......1-1, 1-2, 1-10, 1-18, 5-17, 8-17, 9-29, 9-30, 9-31, 11-1, 11-42, 13-3, 13-6, 14-12, 14-17, 15-8, 20-10, 20-22, 21-15, 22-18, 23-15

electromagnetic interference.......11-53

electromagnetic interference.......7-10, 7-11, 11-53, 11-69

EMT.......1-5, 1-8, 6-7, 6-14, 6-17, 8-2, 8-10, 9-18, 17-2, 21-26, 21-32

enclosure.......1-16, 5-8, 5-9, 5-11, 5-12, 5-15, 5-16, 5-17, 6-13, 6-14, 6-18, 6-19, 6-20, 6-22, 6-24, 6-26, 6-28, 6-29, 6-32, 7-1, 7-2, 7-5, 7-6, 7-8, 7-9, 7-10, 7-11, 8-4, 9-7, 9-10, 9-23, 9-25, 9-26, 11-12, 11-31, 11-32, 11-43, 11-44, 11-45, 11-49, 11-52, 11-53, 11-54, 11-56, 11-57, 11-58, 11-65, 11-66, 11-67, 11-68, 11-71, 11-73, 12-5, 12-21, 12-24, 13-1, 13-2, 13-3, 13-7, 13-8, 13-9, 13-13, 13-14, 13-16, 18-32, 18-33, 18-34, 18-35, 18-37, 19-12, 20-18, 20-19, 20-20, 20-22, 20-27, 21-3, 21-5, 21-6, 21-7, 21-12, 21-13, 21-14, 21-15, 21-16, 21-17, 21-19, 21-20, 21-22, 21-23, 21-24, 21-25, 21-26, 21-27, 21-28, 21-30, 21-33, 21-34, 21-35, 21-36, 21-37, 21-39, 21-43, 21-51

engineering supervision......9-20

entrance and access to workspace.......5-10
equipment
  approval.......4-14
  connected to supply side.......6-24
  grounding conductor.......8-4, 11-31
  grounding conductor installation.......11-60
  installation.......4-15
  examination, installation and use.......1-16
eye and face protection.......3-2

F

fans.......13-1, 17-10, 17-11, 17-14, 17-15, 17-17, 21-20, 22-9, 23-31, 23-32
feeder taps
  10 ft. or less in length.......9-13
  over 10 ft. to 25 ft. in length.......9-13
  over 25 ft. up to 100 ft. in length.......9-15
feeder-circuit conductors.......12-13, 15-1, 15-4, 15-9, 18-6, 18-20, 18-43, 20-23
feeder circuits.......13-22
feeders.......6-1, 7-2, 13-4, 23-13, 23-14
fire alarm circuit conductors.......9-7
fire protective signaling circuits.......4-13
fire pumps.......6-2
fittings.......6-11, 6-14, 6-15, 6-17, 7-5, 11-9, 11-47, 11-49, 11-56, 11-59, 11-61, 11-73, 12-1, 12-4, 12-6, 12-10, 12-11, 13-8, 13-14, 13-15, 13-16, 13-17, 13-18, 13-19, 13-20, 20-21, 21-12, 21-14, 21-16, 21-17, 21-19, 21-21, 21-24, 21-25, 21-26, 21-27, 21-28, 21-29, 21-30, 21-33, 21-42
fixed appliance load.......22-4, 22-9, 22-10, 22-13, 22-19, 22-23, 22-24
fixture whip.......9-8
fixture wires.......9-7, 12-3, 12-7, 12-8
flexible cords.......9-7, 11-74, 13-20, 13-21, 21-36, 13-22, 16-5, 19-15
flexible metal conduit (greenfield).......13-9
fluorescent.......5-5, 9-27, 14-15, 14-19, 14-23, 17-9, 17-11, 17-12, 17-13, 17-20, 21-33, 23-2
foor and leg protection.......3-3
full-load current.......8-2, 14-8, 14-29, 14-30, 14-31, 14-34, 14-35, 14-36, 14-37, 18-2, 18-3, 18-5, 18-6, 18-10, 18-11, 18-16, 18-18, 18-19, 18-20, 18-21, 18-22, 18-23, 18-24, 18-26, 18-27, 18-28, 18-33, 18-36, 18-38, 19-2, 19-3, 19-4, 19-5, 19-6, 19-7, 19-10, 19-11, 19-12, 19-13, 19-14, 20-6, 20-8, 20-9, 22-12
fuse markings.......9-25

fuseholders.......7-2
fuses.......6-25, 6-29, 9-23

G

garages.......1-6, 6-25, 13-6, 14-2, 16-11, 16-12, 16-13, 16-14, 17-5, 17-6, 17-8, 21-31, 21-38, 21-42, 22-2
general lighting loads.......23-2, 23-33
general-purpose receptacle loads.......22-2, 22-15, 22-16
general-use switch.......18-26, 18-28, 18-29, 18-30
generators.......9-1, 11-32, 11-33, 21-20, 21-33, 21-35
GFCI.......9-29, 11-64, 11-75, 12-10, 13-2, 13-13, 16-1, 16-3, 16-4, 16-9, 16-10, 16-11, 16-12, 16-13, 16-14, 16-15, 16-16, 16-17, 16-18, 16-19, 16-20, 16-21, 16-22, 17-14, 17-17, 17-18
ground grid.......10-27
ground-fault
  detection.......6-31, 9-28
  equipment.......9-28
  protection.......6-29
grounded conductor.......6-22, 7-2, 8-4, 8-6, 9-20, 11-15, 11-32
grounded systems.......9-12, 11-14
grounding.......1-10, 1-11, 6-6, 6-19, 7-2, 7-3, 7-6, 7-9, 7-10, 7-11, 8-4, 8-6, 8-9, 8-15, 9-21, 10-27, 11-1, 11-12, 11-14, 11-21, 11-22, 11-24, 11-29, 11-31, 11-32, 11-33, 11-40, 11-41, 11-42, 11-43, 11-44, 11-45, 11-46, 11-47, 11-49, 11-52, 11-53, 11-56, 11-57, 11-58, 11-59, 11-60, 11-61, 11-63, 11-64, 11-65, 11-66, 11-67, 11-68, 11-69, 11-70, 11-71, 11-72, 11-73, 11-74, 11-75, 12-5, 13-5, 13-7, 13-10, 13-13, 15-10, 15-11, 16-2, 16-3, 16-20, 17-2, 17-3, 17-17, 17-22, 21-19, 21-23, 21-24, 21-25, 21-26, 21-33, 21-35, 21-36, 21-39, 21-40, 21-42, 21-43, 22-27, 23-19
grounding conductors.......6-6, 7-9, 7-10, 7-11, 8-4, 8-6, 8-15, 11-12, 11-32, 11-40, 11-47, 11-49, 11-52, 11-60, 11-61, 11-64, 11-65, 11-66, 11-70, 11-73, 11-74, 12-5, 15-10, 16-20, 21-25
grounding electrode system.......7-9, 11-12, 11-14, 11-32, 11-43, 11-63, 11-70, 11-75
grounding electrodes.......11-12, 11-13, 11-29, 11-40, 11-42, 11-45, 11-47, 11-57
grounding equipment.......11-67, 13-5
grounding of panelboards.......7-9
grounding path.......6-19
  for grounding electrode.......11-12
guarding of live parts.......5-6
gutters.......6-7, 7-1, 12-21, 12-22

H

HACR.......19-1, 19-6

hand and arm protection.......3-3

hard hats.......3-2

harmonics.......15-8, 15-9

hazardous areas.......21-2, 21-29, 21-31, 21-32, 21-34, 21-35, 21-36, 21-38, 21-39, 21-41, 21-43, 21-44, 21-45, 21-46

hazardous locations.......1-11, 1-15, 21-2, 21-5, 21-7, 21-13, 21-15, 21-27, 21-29, 21-37, 21-42, 21-43, 21-48

head protection.......3-1

headroom.......5-6

heart compression.......2-3

hermetic motors.......19-4, 19-5, 19-6, 19-7

high-frequency reference grid.......11-72

high-voltage.......1-9, 1-10, 5-10, 5-12, 5-13, 5-16, 6-31, 6-33, 9-29, 9-30, 9-31, 11-11, 11-21, 20-6

hot tub.......14-24, 16-18, 17-17, 17-21, 17-22

hotel.......9-22, 17-12, 23-2

hydromassage.......11-74, 11-75, 16-19, 17-10, 17-18

I

identification.......4-13, 4-17

illumination.......5-5, 5-16

IMC.......6-7, 6-11, 6-14, 6-17, 6-31, 13-16, 17-15, 21-16, 21-17, 21-19, 21-21, 21-22, 21-26, 21-42, 21-43

incandescent.......5-5, 14-15, 14-16, 14-20, 14-23, 17-9, 17-11, 17-12, 17-13, 17-20, 17-21, 21-33, 23-3, 23-15

individual circuits.......8-13

induced currents.......17-18

inductive effect.......7-2, 17-18

inhalation treatment.......2-3

inrush current.......6-25, 6-26, 9-24, 9-26, 9-30, 14-29, 14-31, 14-35, 18-4, 18-10, 19-6, 19-13, 20-1, 20-9

instantaneous trip circuit breaker.......18-10, 18-11, 18-16, 18-19, 18-27, 18-28

insulation.......1-12, 1-16, 6-5, 6-8, 6-11, 6-13, 6-16, 7-5, 8-1, 8-2, 8-3, 8-4, 8-6, 8-7, 8-8, 8-10, 8-15, 8-16, 8-17, 9-1, 9-2, 9-3, 9-23, 9-25, 9-29, 9-30, 11-19, 11-20, 11-54, 11-60, 12-1, 12-3, 12-4, 12-18, 12-19, 12-22, 12-23, 13-7, 13-8, 13-9, 13-10, 13-12, 13-13, 14-11, 14-12, 14-29, 16-2, 16-6, 17-13, 18-2, 20-13, 20-16, 20-17, 21-12, 21-28, 21-35, 21-36, 21-38, 21-43

intermediate metal conduit (IMC).......13-7

intrinsically safe circuits.......21-16

intrinsically safe equipment.......21-15

Inverse time circuit breakers.......18-19, 18-25

isolated equipment grounding conductors.......7-10

isolating switches.......6-32, 21-20

J

junction boxes.......1-5, 5-11, 7-9, 7-10, 11-67, 11-74, 12-13, 12-14, 12-15, 12-17, 12-18, 13-1, 21-33

K

L

lampholders.......14-9, 14-10, 14-15, 14-20, 17-11, 21-35, 21-42

laterals.......6-1, 6-2, 6-3, 6-5, 6-11, 13-7

lighting fixtures.......1-5, 5-6, 5-16, 9-22, 11-75, 12-2, 12-6, 12-10, 13-1, 13-10, 13-11, 13-20, 14-8, 14-15, 14-16, 14-17, 14-18, 14-20, 17-1, 17-2, 17-3, 17-10, 17-11, 17-12, 17-13, 17-14, 17-15, 17-18, 17-20, 17-22, 21-11, 21-12, 21-21, 21-22, 21-45, 21-46, 23-6

lighting loads.......14-18, 14-19, 14-21, 17-20, 17-21, 22-2, 23-2, 23-4, 23-6, 23-7, 23-17, 23-31, 23-32, 23-33, 23-34

lighting outlets.......14-18, 16-12, 16-20, 17-1, 17-4, 17-5, 17-6, 17-7, 17-8, 17-9, 17-10, 17-11, 17-12, 17-14, 17-17, 17-18, 17-22, 23-5

locked rotor current.......6-2, 19-2, 19-11

low-voltage.......5-16, 6-31, 6-33, 11-10, 11-11, 17-17, 21-16, 23-6

M

made electrodes.......11-46

magnetism and electromagnetism.......10-1, 11-1, 12-1, 13-1, 14-1, 15-1, 16-1, 17-1, 19-1, 20-1, 21-1, 22-1, 23-1, 24-1

matching temperature markings.......8-3

metal enclosure.......5-12, 11-44, 11-66, 12-5, 21-15

metal water pipe.......11-46

minimum wire bending space.......7-4

mobile homes.......1-11, 22-18, 22-26

motor circuit.......14-33, 18-5, 18-12, 18-20, 18-30, 18-35, 18-39 motor circuit taps.......9-18

motor compressors.......19-1, 19-11, 19-13

motor control circuit.......9-3, 18-32

motor controllers.......18-24, 21-19

motors.......1-18, 1-19, 6-8, 6-12, 9-1, 9-5, 9-9, 9-21, 9-22, 9-24, 9-25, 9-26, 13-8, 13-9, 14-30, 14-31, 15-12, 15-14, 16-10, 16-19, 17-20, 18-1, 18-2, 18-3, 18-4, 18-5, 18-6, 18-7, 18-8, 18-9, 18-10, 18-11, 18-12, 18-13, 18-14, 18-15, 18-17, 18-19, 18-20, 18-21, 18-22, 18-23, 18-24, 18-25, 18-26, 18-27, 18-28, 18-29, 18-30, 18-31, 18-36, 18-37, 18-38, 18-39, 18-43, 19-2, 19-3, 19-4, 19-5, 19-6, 19-7, 19-8, 19-9, 19-10, 19-11, 19-13, 21-12, 21-15, 21-20, 21-21, 21-23, 21-24, 21-33, 21-35, 21-42, 22-3, 22-10, 22-12, 22-13, 23-10, 23-11

multifamily dwelling.......1-12, 13-5, 15-5, 17-7, 22-15, 22-18, 22-23, 22-24

multiple-conductor cables.......12-23

multiwire.......8-7, 9-12, 14-26

N

nameplates.......22-9, 22-10

neutral load.......7-7, 8-5, 9-12, 14-26, 15-5, 15-6, 15-7, 15-9, 20-11, 23-21, 23-28, 23-29

neutral wire.......13-7

nipples.......12-20, 12-21

NM cable.......13-5, 21-19, 21-26

NMC cable.......13-5

nonincendive
    circuits.......4-19
    components.......4-21
    contacts.......4-19
    equipment.......4-21

nonmetallic sheathed cable (romex)......13-5

nonpower-limited signaling circuits.......4-13

nontime delay fuses.......18-16, 18-20, 18-21

O

objectionable current.......11-5

OCPD'S for systems over 600 volts.......9-29

octagon boxes.......12-6, 12-7, 12-8, 12-9

open delta systems.......4-4

outlet boxes.......1-5, 11-67, 12-6, 13-14, 17-19

overcurrent protection.......1-4, 5-2, 5-4, 5-10, 6-1, 6-2, 6-6, 6-20, 6-25, 6-26, 6-27, 6-28, 6-29, 6-31, 6-33, 7-1, 7-7, 7-8, 7-9, 8-1, 8-3, 8-8, 8-12, 8-13, 8-14, 9-1, 9-2, 9-3, 9-5, 9-7, 9-11, 9-12, 9-13, 9-14, 9-16, 9-17, 9-18, 9-19, 9-20, 9-22, 9-23, 9-25, 9-28, 9-29, 11-22, 11-24, 11-33, 11-61, 11-68, 11-71, 11-73, 11-74, 14-1, 14-2, 14-8, 14-9, 14-12, 14-13, 14-14, 14-17, 14-19, 14-20, 14-22, 14-23, 14-24, 14-26, 14-27, 14-28, 14-29, 14-30, 14-31, 14-34, 14-35, 14-36, 14-37, 15-1, 15-8, 15-12, 15-13, 16-8, 18-4, 18-6, 18-10, 18-13, 18-14, 18-15, 18-19, 18-20, 18-32, 18-38, 18-43, 18-44, 19-1, 19-2, 19-3, 19-6, 19-7, 19-8, 19-9, 19-12, 19-14, 20-5, 20-6, 20-8, 20-9, 20-10, 20-11, 20-12, 20-13, 20-14, 20-15, 20-22, 20-23, 20-24, 22-27, 23-14

overcurrent protection devices.......1-4, 5-2, 5-4, 5-10, 6-1, 6-25, 6-26, 6-27, 6-28, 6-29, 6-33, 7-1, 7-7, 7-8, 7-9, 8-1, 8-12, 9-1, 9-2, 9-12, 9-16, 9-17, 9-22, 9-23, 9-25, 9-28, 9-29, 11-33, 14-8, 14-17, 14-19, 14-24, 14-27, 14-29, 14-30, 15-1, 18-15, 18-20, 19-2, 19-6, 20-10, 20-12, 20-24

overhead conductors.......11-10

overhead services.......6-8

overheating and inductive heating.......7-2

overload protection.......6-2, 9-25, 18-1, 18-4, 18-5, 18-15, 18-20, 18-21, 18-22, 18-23, 18-24, 18-26, 18-28, 19-2, 20-6, 20-13

overload relays.......14-19, 19-12, 19-13

P

panelboards.......1-5, 5-4, 5-5, 5-6, 5-17, 6-2, 6-18, 6-31, 8-7, 11-69, 12-23, 15-1, 17-12, 17-19, 17-20, 17-21, 17-22

panels.......1-1, 1-18, 5-2, 5-3, 5-8, 5-13, 5-15, 7-5, 7-7, 13-14, 15-1, 16-7, 20-5, 21-33, 21-35, 21-45, 21-46

parallel operation.......20-15

phase converter.......9-3, 18-43, 18-44

plenums.......20-27

plugs.......13-20, 16-20, 19-13, 21-12, 21-33, 21-34, 21-35, 21-36

portable and vehicle-mounted generators.......11-32

power loss hazard.......9-3

power panel.......7-7

primary current.......14-34, 14-35, 18-32, 20-5, 20-6, 20-8, 20-9, 20-10, 20-11

protection
    by primary overcurrent device.......9-15
    of attachment.......11-9
    of conductors.......9-2
    of equipment.......9-1
    of fixture wires and cords.......9-7

pull boxes.......5-11, 7-9, 7-10

PVC.......6-17, 7-9, 11-67, 13-6, 13-11, 13-16, 17-15, 17-16, 21-42

R

raceway systems.......13-16

raceways.......1-5, 1-8, 1-16, 6-31, 6-32, 7-5, 7-6, 8-3, 8-6, 8-9, 8-10, 8-15, 9-18, 11-52, 11-53, 11-55, 11-59, 11-61, 11-65, 12-17, 12-18, 13-6, 13-8, 13-14, 13-16, 13-19, 13-20, 17-2, 18-33, 21-28, 21-33, 21-35, 21-36, 21-38, 21-39, 21-42, 21-43, 21-46

raintight service head.......6-17

ranges and clothes dryers.......11-67

reactor starting.......18-39

readily accessible.......5-16, 6-2, 6-3, 6-8, 6-20, 6-22, 6-23, 6-25, 6-28, 9-11, 9-17, 9-22, 9-23, 14-27, 14-28, 16-10, 16-12, 16-13, 16-14, 16-15, 19-5, 19-6, 19-15, 20-4, 20-5, 20-24, 21-39

receptacle grounding terminals.......11-68

receptacle loads.......7-7, 7-8, 11-21, 14-9, 15-5, 22-1, 22-2, 22-4, 22-5, 22-6, 22-15, 22-16, 22-17, 23-17, 23-31, 23-32

receptacle outlets.......7-6, 11-75, 13-13, 14-1, 14-2, 14-24, 15-5, 16-1, 16-2, 16-3, 16-4, 16-5, 16-6, 16-7, 16-8, 16-9, 16-10, 16-11, 16-12, 16-13, 16-15, 16-16, 16-17, 16-19, 16-20, 16-21, 17-1, 17-3, 17-4, 17-5, 17-6, 22-2, 22-3, 23-3, 23-7, 23-8

receptacles.......6-20, 6-31, 9-29, 11-33, 11-64, 11-69, 11-75, 12-1, 12-2, 12-4, 12-6, 12-9, 12-10, 13-2, 13-13, 13-14, 13-20, 14-1, 14-26, 15-5, 15-12, 16-1, 16-2, 16-3, 16-5, 16-6, 16-7, 16-10, 16-11, 16-12, 16-13, 16-14, 16-15, 16-16, 16-19, 16-21, 17-1, 17-7, 18-6, 19-13, 21-12, 21-24, 21-25, 21-33, 21-34, 21-35, 21-42, 23-7, 23-8, 23-17, 23-25, 23-26, 23-27, 23-28, 23-30, 23-31, 23-32, 23-34

resistor.......9-11, 18-6, 18-12, 18-39

rigid metal conduit (RMC).......13-8

rigid nonmetalic conduit (PVC).......13-11

rosettes.......17-11

rotor.......6-2, 6-26, 6-27, 18-1, 18-4, 18-5, 18-7, 18-11, 18-12, 18-13, 18-14, 18-15, 18-16, 18-17, 18-26, 18-27, 18-28, 18-29, 19-2, 19-3, 19-4, 19-11

rough-in.......13-13

S

sealing.......6-17, 21-29, 21-30, 21-39, 21-40

secondary current.......9-16, 18-6, 20-9, 20-13, 20-14, 20-15

secondary ties.......20-12, 20-13, 20-14, 20-15

separately derived.......6-24, 7-10, 8-4, 9-16, 11-11, 11-22, 11-24, 11-63, 11-68, 11-69, 15-1, 20-24

service conductors.......6-1, 6-7, 6-25, 6-26, 6-31, 6-33, 11-22, 11-43, 13-13, 19-11, 22-15, 22-16, 22-23, 22-24, 23-13, 23-22

service disconnecting means.......6-13, 6-20, 6-22, 6-23, 6-24, 6-27, 6-29, 6-30, 6-33, 9-28, 11-68, 12-13

service drop conductors.......6-8, 13-13, 13-14

service drops.......6-1, 6-3, 6-5, 6-8

service equipment.......1-4, 1-5, 5-6, 5-10, 6-1, 6-2, 6-3, 6-11, 6-13, 6-18, 6-19, 6-20, 6-21, 6-25, 6-29, 6-31, 6-33, 7-2, 7-3, 7-5, 7-6, 7-7, 7-9, 7-10, 7-11, 8-4, 9-28, 11-12, 11-32, 11-46, 11-49, 11-56, 11-57, 11-63, 11-64, 11-67, 11-68, 11-69, 11-71, 11-73, 13-14, 14-27, 15-1, 15-5, 17-12, 19-15, 21-39, 21-43, 22-16, 22-18, 23-1, 23-11, 23-16

service equipment terminals.......6-19, 6-21

service lateral conductors.......6-12

service loads.......6-5

service mast.......6-10, 6-11, 13-8, 13-9

service point.......6-31, 6-33

service raceway.......6-6, 6-18, 11-67

service-entrance cable.......6-6, 13-3, 13-6, 13-7, 13-15

service-entrance conductors.......6-1, 6-6, 6-9, 6-11, 6-13, 6-14, 6-16, 6-17, 6-18, 6-20, 6-22, 6-25, 6-26, 6-27, 6-28, 6-31, 9-18, 11-46, 13-7, 13-14, 22-14, 22-17

service-entrance disconnecting means.......6-22

service-entrance raceway.......6-17

services.......1-3, 1-4, 2-1, 6-2, 6-4, 6-30, 6-31, 6-32, 9-22, 9-25, 9-28, 11-12, 11-53, 13-4, 23-17

show window lighting load.......23-5

sign.......6-32, 14-22, 14-23, 21-35, 21-36, 23-2, 23-7, 23-25, 23-26, 23-27, 23-28, 23-30, 23-32, 23-34

single family dwelling units.......6-13

single-phase systems.......4-2

single-phase transformer.......9-5, 20-3

single-pole circuit breakers.......9-12

small appliance load.......7-8, 22-4, 22-6, 22-16

snap switches.......6-25, 7-8, 7-9, 17-19

spas.......11-74, 16-18, 17-17, 17-21, 17-22

special appliance load.......7-6, 8-13, 22-3, 22-4, 22-9, 22-10, 22-16, 22-25, 23-9

specific circuits.......6-28

splices.......9-15, 12-16, 21-12, 21-26, 21-27, 21-28, 21-33

split bus panel.......7-8

square boxes.......12-9, 12-10

stator.......18-11, 18-12, 18-13

storage areas.......20-21

storage battery.......13-6

store building.......23-3, 23-17

straight pull.......12-17, 12-18

structural steel.......11-57

subfeeders.......6-1

supplementary overcurrent protection.......9-11

surface mounted boxes.......11-69

swimming pools.......11-74, 16-17, 16-18, 17-14, 17-15, 17-20, 17-21

switchboards.......5-4, 5-5, 5-6, 5-15, 6-2, 6-18, 7-5, 7-6, 17-12

switches.......5-2, 5-10, 5-15, 5-18, 6-3, 6-18, 6-20, 6-22, 6-24, 6-25, 6-32, 7-8, 7-9, 9-17, 9-25, 9-26, 9-27, 11-69, 11-75, 12-1, 12-4, 12-6, 12-9, 12-10, 13-2, 13-14, 15-1, 17-1, 17-3, 17-7, 17-8, 17-18, 17-19, 17-20, 17-21, 17-22, 19-3, 19-4, 19-5, 20-10, 21-12, 21-15, 21-16, 21-19, 21-20, 21-25, 21-33, 21-35, 21-42, 21-43

T

tap conductors.......9-15, 9-17, 9-18, 11-44, 11-45, 14-6, 20-22, 20-23, 20-24

taps.......9-7, 9-12, 9-14, 9-15, 11-44, 11-45, 12-16, 13-21, 14-6, 14-7, 15-1, 15-12, 18-21, 20-14, 20-22, 20-24, 21-26, 21-27, 21-28, 21-29, 21-33

temporary pole.......13-13

terminal ratings.......8-8

terminals.......6-5, 6-19, 6-21, 6-33, 7-1, 8-2, 8-6, 8-7, 8-8, 8-9, 9-2, 11-68, 12-4, 12-6, 12-11, 13-11, 14-11, 14-12, 14-13, 14-29, 16-3, 17-15, 17-18, 19-11, 21-7, 21-17, 21-19, 21-24, 21-27

thermal devices.......18-6

thermal expansion.......13-12

threadless fittings.......11-54

three-phase, three-wire systems.......20-11

three-wire circuit.......14-16, 15-7, 15-8, 20-10

time-delay fuses.......9-25, 18-15, 18-17

transfer equipment.......6-24

transformer case.......20-16

transformer vault.......5-16, 20-20, 20-21

transformers.......6-24, 6-31, 6-32, 7-3, 9-6, 11-21, 14-23, 17-14, 20-2, 20-3, 20-4, 20-5, 20-6, 20-8, 20-9, 20-10, 20-11, 20-12, 20-13, 20-14, 20-15, 20-16, 20-17, 20-18, 20-19, 20-20, 20-21, 20-22, 20-24, 21-12, 23-17

U

underground service.......6-11, 6-13

underground wiring.......1-10, 21-42

underwater fixtures.......17-15

ungrounded conductors.......6-29, 6-33, 8-7, 9-12, 9-20, 9-21, 9-28, 11-10, 15-7, 18-30, 20-12, 23-14, 23-15, 23-17

ungrounded system.......11-10, 11-32, 11-63, 20-9, 20-10

uninsulated grounded conductor.......6-11

USE cable.......13-6

V

ventricular fibrilation.......2-2

voltage drop.......6-2, 9-21, 15-9, 15-10, 15-11, 16-2

W

wall switches.......17-1, 17-21, 17-22

warning signs.......6-32

water heaters.......6-20, 8-13, 14-26, 14-27, 21-22, 22-3, 22-9, 22-23, 22-24, 23-13

welding shop.......14-34, 23-34

wet-niche fixtures.......11-74, 17-15

wireways.......6-7, 21-19, 21-26

wiring.......1-2, 1-3, 1-4, 1-5, 1-6, 1-9, 1-10, 1-11, 1-12, 1-14, 1-15, 1-16, 2-1, 6-1, 6-6, 6-7, 6-14, 6-15, 6-16, 6-31, 7-1, 7-5, 7-6, 7-9, 8-6, 8-7, 8-17, 11-41, 11-52, 11-53, 11-60, 11-64, 11-65, 11-66, 11-75, 12-1, 13-3, 13-4, 13-5, 13-6, 13-7, 13-10, 13-12, 13-13, 13-15, 13-20, 13-21, 14-2, 14-11, 16-3, 16-11, 16-19, 17-2, 17-3, 17-18, 19-13, 19-14, 20-27, 21-1, 21-6, 21-7, 21-12, 21-15, 21-16, 21-18, 21-19, 21-24, 21-25, 21-26, 21-32, 21-33, 21-34, 21-35, 21-36, 21-37, 21-38, 21-

39, 21-40, 21-42, 21-43, 21-45, 21-46, 21-48, 22-1

wiring methods.......1-2, 1-3, 1-4, 1-5, 1-6, 1-9, 1-11, 1-15, 6-1, 6-6, 6-31, 7-9, 8-17, 11-52, 11-53, 11-65, 11-66, 11-75, 12-1, 13-3, 13-13, 13-21, 16-19, 17-2, 17-18, 19-13, 19-14, 20-27, 21-6, 21-7, 21-19, 21-25, 21-26, 21-33, 21-34, 21-36, 21-38, 21-42, 21-46, 21-48, 22-1

working clearances.......5-1

working space.......5-1, 5-2, 5-3, 5-4, 5-5, 5-6, 5-10, 5-13, 5-15, 17-12

wye system.......8-5, 8-7, 9-11, 9-28, 11-10

wye-connected secondary.......20-2